Coulson and Richardson's Chemical Engineering

Coulson and Richardson's Chemical Engineering

Volume 3B: Process Control

Fourth edition

Sohrab Rohani

Butterworth-Heinemann is an imprint of Elsevier
The Boulevard, Langford Lane, Kidlington, Oxford OX5 1GB, United Kingdom
50 Hampshire Street, 5th Floor, Cambridge, MA 02139, United States

Library of Congress Cataloging-in-Publication Data
A catalog record for this book is available from the Library of Congress

British Library Cataloguing-in-Publication Data
A catalogue record for this book is available from the British Library

ISBN: 978-0-08-101095-2

For information on all Butterworth-Heinemann publications
visit our website at https://www.elsevier.com/books-and-journals

Publisher: Joe Hayton
Acquisition Editor: Fiona Geraghty
Editorial Project Manager: Ashlie Jackman
Production Project Manager: Mohana Natarajan
Designer: Vicky Pearson

Typeset by SPi Global, India

Contents

Contributors

Dominique Bonvin Ecole Polytechnique Fédérale de Lausanne (EPFL), Lausanne, Switzerland
Panagiotis D. Christofides University of California, Los Angeles, CA, United States
Jean-Pierre Corriou Lorraine University, Nancy Cedex, France
Victoria M. Ehlinger University of California, Berkeley, CA, United States
Matthew Ellis University of California, Los Angeles, CA, United States
Grégory François The University of Edinburgh, Edinburgh, United Kingdom
Liangfeng Lao University of California, Los Angeles, CA, United States
Ali Mesbah University of California, Berkeley, CA, United States
Sohrab Rohani Western University, London, ON, Canada
Yuanyi Wu Western University, London, ON, Canada

About Prof. Coulson

John Coulson, who died on January 6, 1990 at the age of 79, came from a family with close involvement with education. Both he and his twin brother Charles (renowned physicist and mathematician), who predeceased him, became professors. John did his undergraduate studies at Cambridge and then moved to Imperial College where he took the postgraduate course in chemical engineering—the normal way to qualify at that time—and then carried out research on the flow of fluids through packed beds. He then became an Assistant Lecturer at Imperial College and, after wartime service in the Royal Ordnance Factories, returned as Lecturer and was subsequently promoted to a Readership. At Imperial College, he initially had to run the final year of the undergraduate course almost single-handed, a very demanding assignment. During this period, he collaborated with Sir Frederick (Ned) Warner to write a model design exercise for the I. Chem. E. Home Paper on "The Manufacture of Nitrotoluene." He published research papers on heat transfer and evaporation, on distillation, and on liquid extraction, and coauthored this textbook of Chemical Engineering. He did valiant work for the Institution of Chemical Engineers which awarded him its Davis medal in 1973, and was also a member of the Advisory Board for what was then a new Pergamon journal, *Chemical Engineering Science*.

In 1954, he was appointed to the newly established Chair at Newcastle-upon-Tyne, where Chemical Engineering became a separate Department and independent of Mechanical Engineering of which it was formerly part, and remained there until his retirement in 1975. He took a period of secondment to Heriot Watt University where, following the splitting of the joint Department of Chemical Engineering with Edinburgh, he acted as adviser and *de facto* Head of Department. The Scottish university awarded him an Honorary DSc in 1973.

John's first wife Dora sadly died in 1961; they had two sons, Anthony and Simon. He remarried in 1965 and is survived by Christine.

JFR

About Prof. Richardson

Professor John Francis Richardson, Jack to all who knew him, was born at Palmers Green, North London, on July 29, 1920 and attended the Dame Alice Owens School in Islington. Subsequently, after studying Chemical Engineering at Imperial College, he embarked on research into the suppression of burning liquids and of fires. This early work contributed much to our understanding of the extinguishing properties of foams, carbon dioxide, and halogenated hydrocarbons, and he spent much time during the war years on large-scale fire control experiments in Manchester and at the Llandarcy Refinery in South Wales. At the end of the war, Jack returned to Imperial College as a lecturer where he focused on research in the broad area of multiphase fluid mechanics, especially sedimentation and fluidization, two-phase flow of a gas and a liquid in pipes. This laid the foundation for the design of industrial processes like catalytic crackers and led to a long lasting collaboration with the Nuclear Research Laboratories at Harwell. This work also led to the publication of the famous paper, now common knowledge, the so-called Richardson-Zaki equation which was selected as the Week's citation classic (*Current Contents*, February 12, 1979)!

After a brief spell with Boake Roberts in East London, where he worked on the development of novel processes for flavors and fragrances, he was appointed as Professor of Chemical Engineering at the then University College of Swansea (now University of Swansea), in 1960. He remained there until his retirement in 1987 and thereafter continued as an Emeritus Professor until his death on January 4, 2011.

Throughout his career, his major thrust was on the wellbeing of the discipline of Chemical Engineering. In the early years of his teaching duties at Imperial College, he and his colleague John Coulson recognized the lack of satisfactory textbooks in the field of Chemical Engineering. They set about rectifying the situation and this is how the now well-known Coulson-Richardson series of books on Chemical Engineering was born. The fact that this series of books (six volumes) is as relevant today as it was at the time of their first appearance is a testimony to the foresight of John Coulson and Jack Richardson.

Throughout his entire career spanning almost 40 years, Jack contributed significantly to all facets of professional life, teaching, research in multiphase fluid mechanics and service to the Institution of Chemical Engineers (IChem E, UK). His professional work and long standing public service was well recognized. Jack was the president of IChem E during the period 1975–76 and was named a Fellow of the Royal Academy of Engineering in 1978. He was also awarded OBE in 1981.

In his spare time, Jack and his wife Joan were keen dancers, being the founder members of the Society of International Folk Dancing and they also shared a love for hill walking.

RPC

Preface

The present volume in the series of Coulson and Richardson's Chemical Engineering deals with the fundamentals and practices of process dynamics and control in the process industry including the chemical industry, pharmaceutical industry, biochemical industry, etc. The primary audience of the book is the undergraduate and postgraduate students in chemical engineering discipline who pursue an undergraduate or a postgraduate degree in this discipline. The book is also of value to the practitioners in the process industry.

Chapter 1 provides an introduction to the two main areas of process control, namely, the theoretical development of dynamic models and control system design theory, while the second area deals with the implementation of the control systems. In the context of the latter, the basics to developing piping and instrumentation diagram (P&ID) and block diagrams for feedback and feedforward control systems are discussed.

Chapter 2 discusses the required instrumentation and control systems to monitor the process variables and implement the control systems. Although not exhaustive, it describes the principles of operation of transducers to measure the main process variables, temperature, level, pressure, flow rate, and concentration. A brief discussion of the control system architecture for the single-input single-output (SISO) and multi-input multi-output (MIMO) systems, including the LabView and distributed control system (DCS) environments, is provided. Transducers' accuracy, reproducibility, and the steady-state (instrument gain) and dynamic models (transfer function) of various components in a typical control loop are discussed.

Chapter 3 deals with the dynamic modeling of the chemical and biochemical processes based on the first principles, namely, the conservation of mass, energy, momentum, and particle population—along with the auxiliary equations describing the transfer of mass, heat, and momentum; equations of state; reaction kinetics, etc. The chapter discusses a generic 8-step procedure to model the lumped parameter systems, the stage-wise processes, and the distributed parameter systems. The resulting equations describe the systems in a dynamic fashion and are often nonlinear. Therefore, simple methods to solve the resulting dynamic equations using Matlab and Simulink are discussed. For each category of the systems, many examples are presented to convey the concepts in a clear manner.

In Chapter 4, methods to develop linear models for dynamical processes in the state-space and Laplace transfer domain (transfer function) are discussed. The linear models are either obtained theoretically by linearizing the nonlinear dynamic models discussed in Chapter 3, or experimentally by graphical or numerical analysis of the input-output data sets (process identification). The derivation of the transfer function of simple processes is presented in detail with many examples. The transfer functions of first order, second order, higher order systems with or without delays, and processes with inverse response are derived. The general presentations of the state-space and transfer functions for MIMO processes are provided.

In Chapter 5, various control system design methodologies including the basic proportional-integral-derivative (PID) feedback controllers, cascade controllers, selective controllers, and feedforward controllers in the Laplace domain for the SISO and MIMO systems are discussed. The implementation of the ideal PID control law in analog and digital controllers is discussed.

Chapters 1–5 provide the materials necessary for a first compulsory undergraduate course for the chemical engineering students.

Chapter 6 discusses the fundamentals of the stability and the design of digital controllers in the discrete Laplace domain, z-domain. Digital sampling, filtering, and the design of SISO and MIMO feedback and feedforward controllers are discussed. The design of model-based digital feedback controllers such as the deadbeat controllers, Dahlin controller, the Smith dead-time compensator, the Kalman controller, the internal model controllers (IMC), and the pole-placement controllers is discussed.

In Chapter 7, the stability analysis and the controller design in the state-space and frequency domain are briefly discussed. Concepts such as the controllability, the observability, the design of the state feedback regulators, and the time-optimal controllers of dynamical systems are introduced. The frequency response analysis for the stability of dynamical systems is introduced and the basic controller design methodology in the frequency domain is discussed.

Chapter 8 deals with the stochastic systems involving uncertainties due to the presence of the process and sensor noise. The time series formulation for the stochastic processes is introduced and the numerical analytical methods for the identification of the dynamical models for the process and the noise part of the systems are introduced. Parameter estimation techniques such as the least squares method, the weighted least square method, and the maximum likelihood are discussed. The recursive versions of the parameter estimation algorithms for the online process identification are also introduced. The design of stochastic feedback controllers such as the minimum variance controllers, the generalized predictive controllers, and the pole-placement controllers for the stochastic processes is discussed.

Chapter 9 is dedicated to the design of model predictive controllers (MPC) for the chemical processes. A detailed example for the design of a SISO MPC feedback controller for a batch crystallization process provides the necessary steps for the design and implementation of MPC controllers.

Chapters 6–9 contain materials appropriate for a second undergraduate technical elective course or a first course at the graduate level in process control for the chemical engineering students.

Chapters 10–13 deal with the advanced topics in the nonlinear control, optimal control, optimal control of batch processes, and the control of distributed parameter systems. The contents of these chapters are appropriate for a graduate course or a series of graduate courses, depending on the extent to which the topics in each chapter are covered.

Throughout the book, attempt has been made to present the concepts in a clear manner. Many examples are provided to enable the students to grasp both the fundamentals and the implementation of the control system design. In many examples, simulation case studies are provided in the Matlab and Simulink environments to facilitate the understanding of difficult concepts.

Sohrab Rohani
Western University,
London, ON, Canada

Introduction

Welcome to the next generation of Coulson-Richardson series of books on *Chemical Engineering*. I would like to convey to you all my feelings about this project which have evolved over the past 30 years, and are based on numerous conversations with Jack Richardson himself (1981 onwards until his death in 2011) and with some of the other contributors to previous editions including Tony Wardle, Ray Sinnott, Bill Wilkinson, and John Smith. So what follows here is the essence of these interactions combined with what the independent (solicited and unsolicited) reviewers had to say about this series of books on several occasions.

The Coulson-Richardson series of books has served the academia, students, and working professionals extremely well since their first publication more than 50 years ago. This is a testimony to their robustness and to some extent, their timelessness. I have often heard much praise, from different parts of the world, for these volumes both for their informal and user-friendly yet authoritative style and for their extensive coverage. Therefore, there is a strong case for continuing with its present style and pedagogical approach.

On the other hand, advances in our discipline in terms of new applications (energy, bio, microfluidics, nanoscale engineering, smart materials, new control strategies, and reactor configurations, for instance) are occurring so rapidly as well as in such a significant manner that it will be naive, even detrimental, to ignore them. Therefore, while we have tried to retain the basic structure of this series, the contents have been thoroughly revised. Wherever, the need was felt, the material has been updated, revised, and expanded as deemed appropriate. Therefore the reader, whether a student or a researcher or a working professional should feel confident that what is in the book is the most up-to-date, accurate, and reliable piece of information on the topic he/she is interested in.

Evidently, this is a massive undertaking that cannot be managed by a single individual. Therefore, we now have a team of volume editors responsible for each volume having the individual chapters written by experts in some cases. I am most grateful to all of them for having joined us in the endeavor. Further, based on extensive deliberations and feedback from a large number of individuals, some structural changes were deemed appropriate, as detailed here. Due to their size, each volume has been split into two sub-volumes as follows:

Volume 1A: Fluid Flow
Volume 1B: Heat and Mass Transfer
Volume 2A: Particulate Technology and Processing
Volume 2B: Separation Processes
Volume 3A: Chemical Reactors
Volume 3B: Process Control

Undoubtedly, the success of a project with such a vast scope and magnitude hinges on the cooperation and assistance of many individuals. In this regard, we have been extremely fortunate in working with some outstanding individuals at Butterworth-Heinemann, a few of whom deserve to be singled out: Jonathan Simpson, Fiona Geraghty, Maria Convey, and Ashlie Jackman who have taken personal interest in this project and have come to help us whenever needed, going much beyond the call of duty.

Finally, this series has had a glorious past but I sincerely hope that its future will be even brighter by presenting the best possible books to the global Chemical Engineering community for the next 50 years, if not for longer. I sincerely hope that the new edition of this series will meet (if not exceed) your expectations! Lastly, a request to the readers, please continue to do the good work by letting me know if, no not if, when you spot a mistake so that these can be corrected at the first opportunity.

Raj Chhabra
Editor-in-Chief
Kanpur, July 2017.

CHAPTER 1

Introduction

Sohrab Rohani, Yuanyi Wu
Western University, London, ON, Canada

In this chapter, the basic definitions and concepts discussed in the book are introduced, and the incentives for implementing automation in the process industry are briefly discussed.

1.1 Definition of a Chemical/Biochemical Process

In an industrial setup, the name process applies to a series of events or operations run in a continuous, semicontinuous/semibatch, or a batch mode of operation, to convert a given raw material or a few raw materials to a useful final product and by-products. In a chemical/biochemical process, the raw materials and finished products are various chemical elements or molecules that undergo physical, chemical, and biochemical changes. In the simplest form, a process consists of a single unit such as a chemical reactor, a distillation column, a crystallizer, etc.

1.1.1 A Single Continuous Process

A single continuous process receives inputs from upstream units and continuously processes the materials received and sends out the product to the downstream units. An example of a continuous chemical process is a continuous stirred tank reactor (CSTR) shown in Fig. 1.1 in which reactant A is converted to product B. The reaction is assumed to be exothermic, therefore, cooling water with a volumetric flow rate F_c (m^3/s) and inlet temperature $T_{c,in}$ (°C) is supplied to the cooling jacket of the reactor in order to maintain the reactor temperature, T (°C), at a desired value. The reactant enters the reactor at a flow rate F_i (m^3/s), a reactant concentration $C_{A,i}$ ($kmol/m^3$), and a temperature T_i (°C) and leaves the reactor with an effluent flow rate F (m^3/s), a reactant concentration C_A, ($kmol/m^3$), and a temperature T (°C). In this simple system, there are a number of output variables that need to be controlled in order to maintain a stable and steady-state operation: the reactor volume, V (m^3); the reactant concentration, C_A; and the reactor temperature, T.

Coulson and Richardson's Chemical Engineering. http://dx.doi.org/10.1016/B978-0-08-101095-2.00001-1

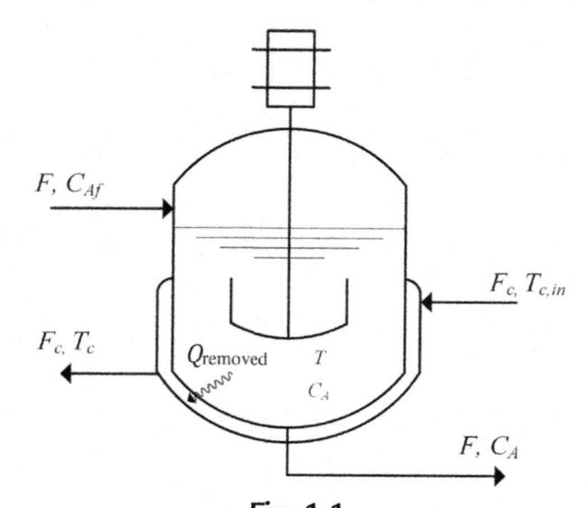

Fig. 1.1
A continuous stirred tank reactor (CSTR) as a typical continuous process.

1.1.1.1 A continuous chemical plant

An example of a chemical plant is a urea or carbamide plant that uses liquid ammonia and carbon dioxide in an exothermic reaction to form ammonium carbamate and its subsequent conversion in an endothermic reaction to form urea and water. Fig. 1.2 shows the simplified process flow diagram (PFD) of a urea plant.

1.1.1.2 A continuous biochemical process

An example of a biochemical process is the large-scale synthesis of insulin. In the human body, insulin is produced in the pancreas and regulates the amount of glucose in blood. For the industrial-scale production of insulin, a multistep biochemical process using recombinant DNA

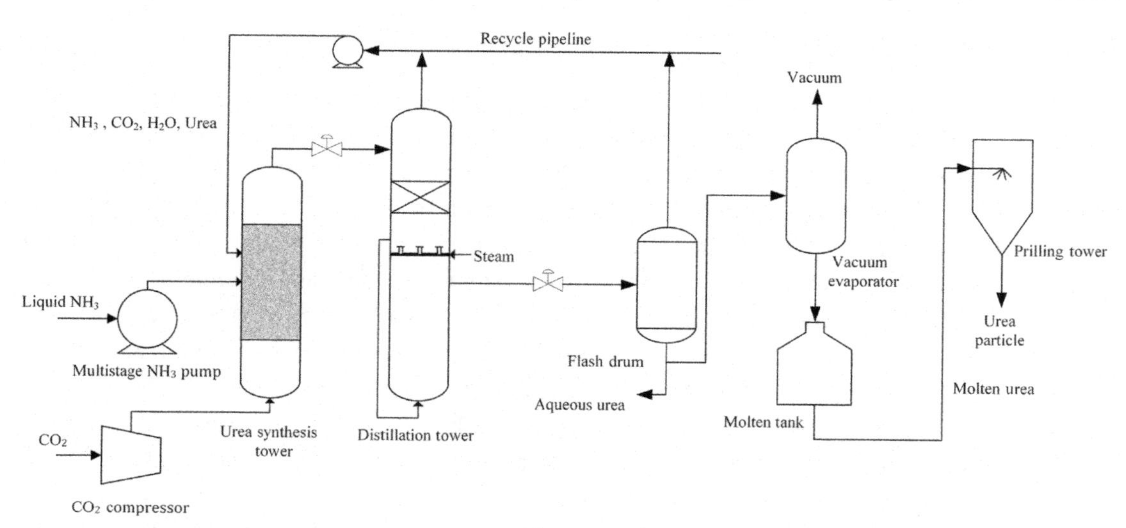

Fig. 1.2
Process flow diagram (PFD) of a continuous urea plant.

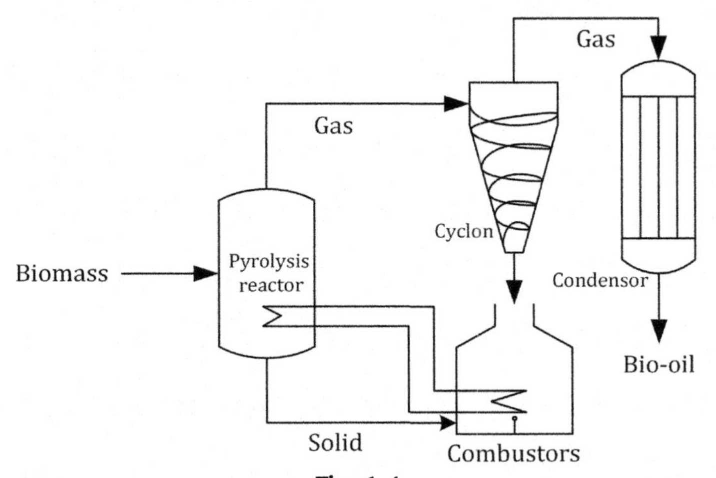

Fig. 1.4
Process flow diagram of a green process for the production of bio-oil from biomass.

is used. The process involves inserting the insulin gene into the *Escherichia coli* bacterial cell in a fermentation tank. Fig. 1.3 shows the simplified PFD of a biochemical process for the production of insulin.

1.1.1.3 A continuous green process

An example of a green process is the conversion of the agricultural wastes (biomass) to bio-oil using an ultra-fast pyrolysis process. Fig. 1.4 shows the simplified PFD of a green process for the production of bio-oil from biomass.

1.1.2 A Batch and a Semibatch or a Fed-Batch Process

In a batch process, a specific recipe is followed to convert an initial charge of reactants to products. In a semibatch or fed-batch operation, one or more reactants are added continuously during the operation.

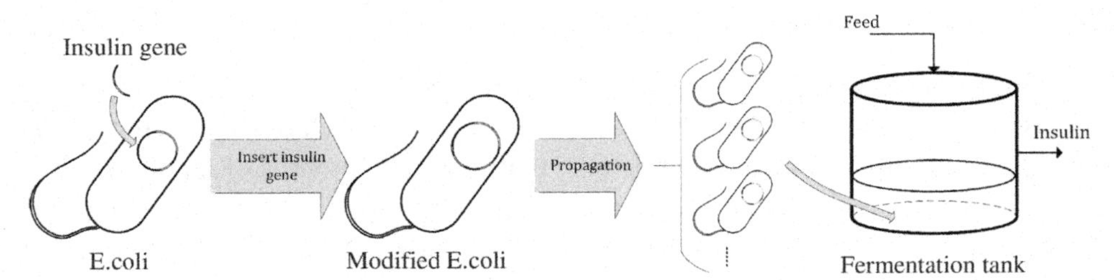

Fig. 1.3
Process flow diagram (PFD) of a biochemical plant for the production of insulin.

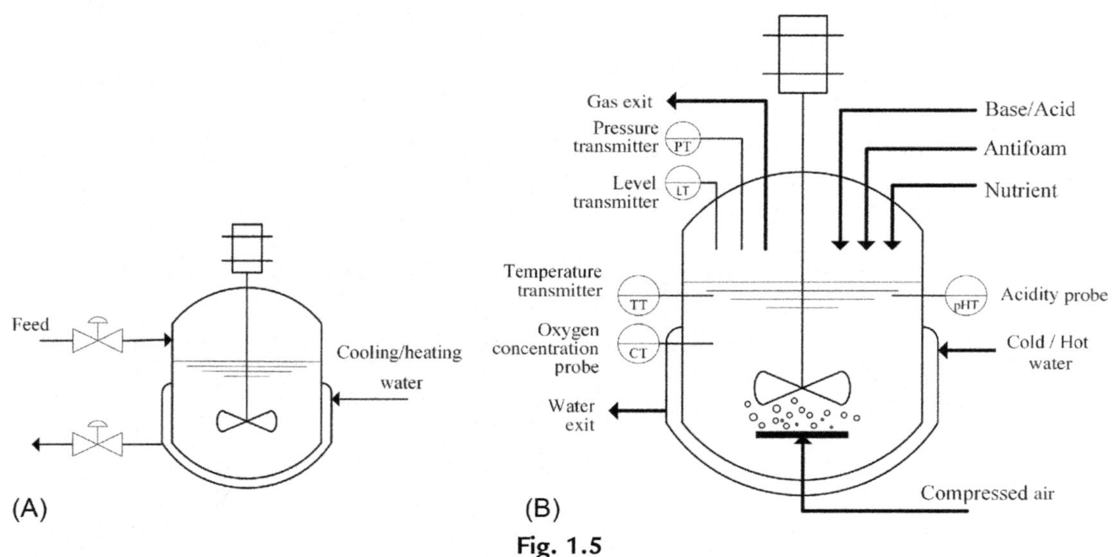

Fig. 1.5
A Batch reactor (A), a Fed-Batch bioreactor (B).

Batch and semibatch or fed-batch processes are used in the fine chemical, pharmaceutical, microelectronic, and specialty chemical industries. An example of a batch reactor and a semibatch fermenter is shown in Fig. 1.5.

1.2 Process Dynamics

The subject of process dynamics deals with the study of the dynamic behavior of various processes in the chemical, biochemical, petrochemical, food, and pharmaceutical industries. The objective of running a process is to convert the given raw materials to useful finished products, safely, economically, and with the minimum impact on the environment. The production rate and the quality of the product are functions of the operating conditions such as the process temperature, pressure, energy input, and the purity of raw materials. Understanding the effects of such variables on the product quality and the production rate is of paramount importance to the successful operation of the process. A mathematical description of the relationships between the process input variables and the output variables, in a dynamic fashion, is the subject of process dynamics. In such mathematical models, time is always an independent and often an implicit variable.

It is clear that for a complex process with hundreds of process variables, developing a complete dynamic mathematical model that expresses the interrelationships among all process inputs and outputs is a formidable task that may take months of a competent engineering team. In such cases, it is advisable to use the experimental input–output data from the process and fit them to simple linear dynamic models (black-box modeling approach). The latter approach is referred to as the 'process identification' technique which will be discussed later in the book.

1.2.1 Classification of Process Variables

Process variables can be divided into input variables and output variables. The input variables are further divided into disturbances or loads, and manipulated variables.

The *loads or disturbances* are input variables that affect the process outputs in an uncontrolled and random fashion. The disturbances are represented by the letter (d or D).

The *manipulated variables* are shown by the letter (u or U) and are manipulated by the controllers to achieve the control objectives.

The *output variables* (y or Y) can be either controlled to achieve the control objectives or not controlled. The output variables are often measured and monitored in an online fashion.

The *block diagram* of a process, whether continuous or batch, along with the *input* and *output* variables is represented in Fig. 1.6.

If there is only one controlled variable and one manipulated variable (u and y will be scalar quantities), the process is referred to as a single-input, single-output (SISO) system. If there are more than one manipulated variables and controlled variables (u and y will be vectors with a given dimension), the process is referred to as a multiple-input, multiple-output (MIMO) system.

1.2.2 Dynamic Modeling

A mathematical relationship between the input and output variables in which the time is an independent and often implicit variable is referred to as the dynamic model of the process. A dynamic model can be developed based on the first principles, i.e., mass, energy, momentum, etc., balances or based on an empirical approach using the input–output data of the process (process identification).

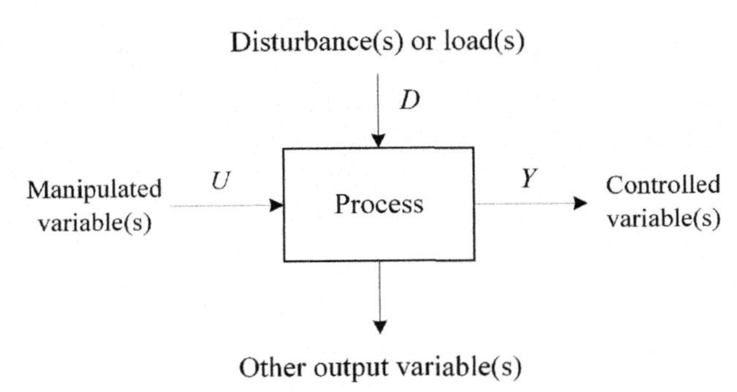

Fig. 1.6
The *block diagram* of a typical process with the designated *input* and *output* variables.

A dynamic model derived from the first principles involves the time derivative of the output variable(s) as a *nonlinear* function of the input and output variables:

$$\frac{dy(t)}{dt} = f\{u(t), d(t), y(t)\}; \text{ I.C.} \tag{1.1}$$

where I.C. represents the initial condition of the equation, i.e., $y(t=0)$. Eq. (1.1) describes the dependence of $y(t)$ on $u(t)$ and $d(t)$ in a convoluted and nonlinear manner. Such a model usually involves nonlinear terms and therefore is not useful for the design of *linear* controllers. The classical process control theory is based on the *linear* input-output models. Using a technique called Taylor series expansion, it is possible to linearize all the nonlinear terms appearing in a *nonlinear* dynamic model and convert the nonlinear model to a *linearized* dynamic model of the form

$$\tau \frac{dy'(t)}{dt} + y'(t) = K_p u'(t) + K_d d'(t); \text{ I.C.} \tag{1.2}$$

where τ, K_p, and K_d are constants, and the superscript (′) on each variable represents the value of the variable in terms of its *deviation* or *perturbation* from its corresponding steady-state value. This approach is used to get rid of the constant terms resulting from the linearization operation. Eq. (1.2) can be Laplace transformed to derive linear input-output models in the s-domain. In the Laplace domain, the linear input-output models are referred to as Transfer Function models. Defining $Y(s) = \mathcal{L}[y'(t)]$, $L(s) = \mathcal{L}[l'(t)]$, and $\mathcal{L}\left[\frac{dy'(t)}{dt}\right] = sY'(s) - y'(0)$, and assuming that $y'(0) = 0$, the transfer functions can be obtained:

$$(\tau s + 1)Y(s) = K_p U(s) + K_d D(s) \tag{1.3}$$

$$Y(s) = \frac{K_p}{\tau s + 1} U(s) + \frac{K_d}{\tau s + 1} D(s) \tag{1.4}$$

Note that in the Laplace domain, for convenience, we drop the superscript (′); however, it is understood that all variables are expressed in deviation or perturbation form. In general, one can write the algebraic equations relating the input and output variables in the Laplace domain by:

$$Y(s) = G_p(s)U(s) + G_d(s)D(s) \tag{1.5}$$

where $G_p(s)$ is the process transfer function and $G_d(s)$ is the disturbance transfer function. The transfer functions can be represented pictorially in the form of a *block diagram* (Fig. 1.7).

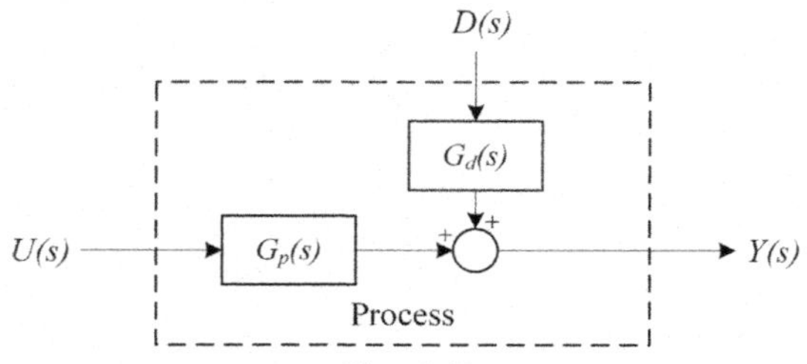

Fig. 1.7
The block diagram of a SISO system and the corresponding *transfer functions*.

Example 1.1

Consider a liquid storage tank in a continuous operation. A liquid stream enters the tank with a volumetric flow rate, F_i, in m^3/s. The effluent stream from the tank is given by F (m^3/s). Both F_i and F are functions of time, i.e., $F_i(t)$ and $F(t)$. Since the liquid in the tank does not undergo any reactions/mixing or temperature changes, we may assume that the density of the liquid, ρ (kg/m^3), remains constant. Under such conditions, if $F(t) = F_i(t)$, even though they may change with time, the mass of the liquid in the tank, $\overline{m}$ (kg), remains constant. However, if $F(t)$ is not equal to $F_i(t)$, then the mass of liquid in the tank changes with time, i.e., the accumulation of mass in the tank will not be zero. The total mass balance for this simple system can be expressed by the following equation:

$$\frac{dm(t)}{dt} = \rho F_i(t) - \rho F(t); \text{ I.C.} \tag{1.6}$$

In order to solve the previous equation, the initial condition, i.e., the mass of liquid in the tank at time 0, $m(t=0)$, is needed. Since $m(t) = \rho V(t)$, where $V(t)$ is the volume of the liquid in the tank, the previous equation can be further simplified to:

$$\frac{dV(t)}{dt} = F_i(t) - F(t); \text{ I.C.} \tag{1.7}$$

If the tank has a constant cross-sectional area, for example, if it has a cylindrical shape, Eq. (1.7) can be expressed in terms of the liquid level, $l(t)$, in the tank.

$$A\frac{dl(t)}{dt} = F_i(t) - F(t) \tag{1.8}$$

where A is the cross-sectional area of the tank which is constant for a cylindrical tank (see Fig. 1.8)

If the storage tank is not cylindrical, then the cross-sectional area will be a function of the liquid level. For example, for a spherical tank (see Fig. 1.9), the liquid volume in the tank is given in terms of the liquid level and the radius of the sphere, R, by:

$$V = \frac{\pi l^2}{3}(3R - l) \tag{1.9}$$

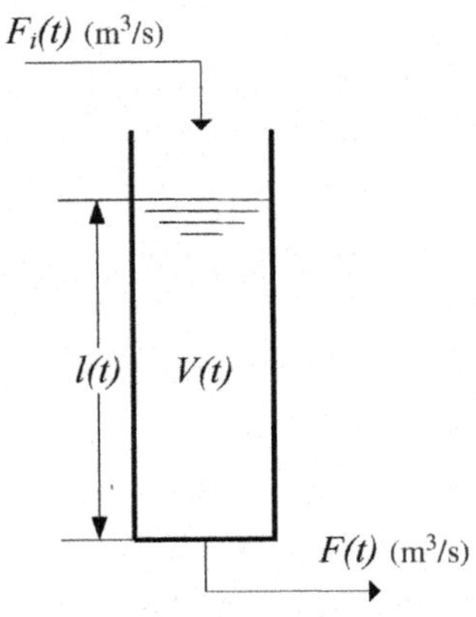

Fig. 1.8
Schematics of a liquid storage tank with a constant cross-sectional area.

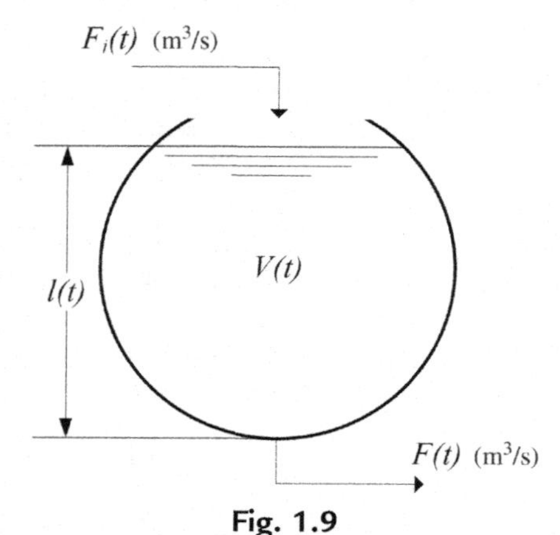

Fig. 1.9
Schematics of a liquid storage tank with a variable cross-sectional area.

And the mass balance is given by:

$$\frac{\pi}{3}\cdot\frac{d[l^2(t)\cdot 3R - l^3(t)]}{dt} = F_i(t) - F(t) \tag{1.10}$$

The previous equations express the relationships between the liquid level in the tank with the inlet and outlet flow rates, $F_i(t)$ and $F(t)$. Usually $F(t)$ is a nonlinear function of the liquid level (hydrostatic head) in the tank, in the form of $F(t) = \alpha\sqrt{l(t)}$, where α is a constant. With the given dependence of $F(t)$ on $l(t)$, Equation (1.10) involves a nonlinear term that has to be linearized before a transfer function can be derived between $L(s)$ and $F_i(s)$.

1.3 Process Control

Process control deals with the science and technology to study and implement automation in the process industry to ensure the product quality, maximize the production rate, meet the environmental regulations and operational constraints, and maximize the profit. In order to meet these challenges, two complementary approaches are used.

The first approach is theoretical rendering dynamic models to design effective controllers to ensure the overall process objectives. The starting point in this approach relies heavily on the process dynamic models briefly discussed earlier. A wealth of controller design techniques in the time domain (state-space controller design techniques), Laplace domain (transfer function controller design techniques), and the frequency domain, both in the continuous and discrete time domains have been developed and are commonly used.

The second approach involves the actual implementation of the control strategies using a host of instrumentations including the sensors/transducers to measure the process variables such as temperature (T), pressure (P), level (L), flow rate (F), and concentration (C); the design of controllers with different architectures; the data acquisition systems (DAQ systems); the transmission lines; and the final control elements such as the control valves (CV) and the variable speed pumps. A controller may be designed and an actual control loop may be implemented with little theoretical consideration given to the dynamics of the process. However, for a sound design and for more complex systems, the design and implementation of controllers must be based on a thorough understanding of the process dynamics, and therefore, using the first approach prior to embarking on the implementation phase is advisable.

In the implementation phase, analog, digital, or a combination of both units are used. For example, one may use an analog pneumatic or electronic transducer to measure a process variable with an output ranging over 3–15 psig, 0–5 V, or 4–20 mA signal. There are also *smart transducers* with digital outputs. The classical controllers were analog pneumatic or electronic units with input-output signals in the range of 3–15 psig or 4–20 mA. However, since 1970s, various digital controller architectures have become the industry norm. The CV are primarily pneumatically actuated. Signal conversion and conditioning to convert the pneumatic and analog signals to and from digital signals, or vice versa, require instrumentation such as pneumatic to electrical converters (P/I), electrical to pneumatic converters (I/P), analog-to-digital converters (A/D), and digital-to-analog converters (D/A).

1.3.1 Types of Control Strategies

The majority of controllers used in an industrial plant are feedback controllers. In such controllers the process variables that are to be controlled are measured directly or inferred from other easily measurable variables, and the information is fed back to the controller. If the process is continuously disturbed by a few disturbances, it is beneficial to complement the feedback controllers with feedforward controllers (FFCs) that measure the disturbances and compensate for their adverse effect.

1.3.1.1 Feedback control

A feedback control strategy is based on the measurement of the controlled variable (process variable, PV) by a proper sensor and transmitting the measured signal to the controller. The desired value of the controlled variable is also made available to the feedback controller as the set point (SP). The controller subtracts the measured controlled variable, $y_m(t)$, from its set point, y_{sp}, and calculates the error signal, $e(t) = y_{sp} - y_m(t)$. The error signal triggers the control law. The controller calculates a corrective action that is implemented by throttling a final control element. Therefore in a feedback control system three tasks are performed, *measurement* of a process variable of interest (directly or indirectly) by a sensor/transducer; compare the measured variable with its set point and calculate the corrective action (*decision*); and *implement* the corrective action, using a final control element such as a CV. The controller output, $P(t)$, which is related to the manipulated variable, $u(t)$, is calculated from the error signal, $e(t)$, based on the employed *control law*. The majority of the feedback controllers in the process industry are proportional-integral-derivative (PID) type controllers. The simplest feedback controller is an *on-off* controller whose output is either at its maximum or minimum, depending on the sign of the error signal. In a proportional controller, the controller output is proportional to the magnitude of the error signal; in an integral controller, the controller output is proportional to the duration of the error signal; and in a derivative controller, the controller output is proportional to the time rate of change of the error signal.

- ***On-off*** controllers are those in which the controller output is either at its minimum or maximum depending on whether the error signal is positive or negative.

$$P(t) = \begin{cases} \min, & e(t) \le 0 \\ \max, & e(t) > 0 \end{cases} \tag{1.11}$$

- ***Proportional (P)*** controllers' output is proportional to the *magnitude* of the error signal. The proportionality constant, K_c, is referred to as the controller proportional gain, $\overline{P}$ is the *controller bias* which is the controller output when the error signal is zero.

$$P(t) = \overline{P} + K_c\, e(t) \tag{1.12}$$

- ***Proportional–Integral (PI)*** controllers' output is proportional to the *magnitude* and *time integral* or *duration* of the error signal. τ_I is referred to as the controller *integral time* or *rest time*.

$$P(t) = \overline{P} + K_c\, e(t) + \frac{K_c}{\tau_I}\int_0^t e(t)dt \tag{1.13}$$

- ***Proportional–Integral–Derivative (PID)*** controllers' output is proportional to the magnitude, duration (time integral), and time rate of change (derivative) of the error signal. τ_D is referred to as the controller *derivative time* or *rate time*.

$$P(t) = \overline{P} + K_c e(t) + \frac{K_c}{\tau_I}\int_0^t e(t)dt + K_c\tau_D\frac{de(t)}{dt} \tag{1.14}$$

The earlier PID controller equations represent the operation of analog controllers. If digital controllers are used, due to their sampled-data nature, the corresponding PID control laws must be discretized. Using a rectangular rule for the integral part and a back difference equation for the derivative part, Eq. (1.14) is discretized as follows:

$$P_n = \overline{P} + K_c\, e_n + \left(\frac{K_c}{\tau_I}\right)\sum_{i=1}^{n} e_i\,\Delta t + K_c\tau_D\left(\frac{e_n - e_{n-1}}{\Delta t}\right) \tag{1.15}$$

where Δt is the sampling/control interval, $\overline{P}$ is the controller bias, when error is zero, $P_n = P(t = n\cdot\Delta t)$, and $e_n = e(t = n\cdot\Delta t)$.

Eq. (1.15) is referred to as the **position** form of the **discrete** PID law. The 'position' form of the discrete PID has two drawbacks, first, it requires the value of $\overline{P}$, which is not known, and second all the past values of the error signal must be stored in the memory. In order to circumvent these limitations, the **velocity** form of the discrete PID law is derived by subtracting the control action at the $(n-1)$st from Eq. (1.15) to obtain ΔP_n, which is the relative change in the controller action in the nth interval compared to its value at the $(n-1)$st interval.

$$P_{n-1} = \overline{P} + K_c e_{n-1} + \left(\frac{K_c}{\tau_I}\right)\sum_{i=1}^{n-1} e_i \Delta t + K_c \tau_D \left(\frac{e_{n-1} - e_{n-2}}{\Delta t}\right) \tag{1.16}$$

$$\Delta P_n = P_n - P_{n-1} = K_c(e_n - e_{n-1}) + \left(\frac{K_c}{\tau_I}\right) e_n \Delta t + K_c \tau_D \left(\frac{e_n - 2e_{n-1} + e_{n-2}}{\Delta t}\right) \tag{1.17}$$

The required controller action at the nth interval is, therefore, the sum of the past controller action at the $(n-1)$st interval plus the relative change of the controller action in the nth interval compared to its value at the $(n-1)$st interval, ΔP_n.

$$P_n = P_{n-1} + \Delta P_n \tag{1.18}$$

1.3.1.2 Feedforward control

In a feedforward control algorithm, instead of the controlled variable, the major disturbances are measured. A FFC is unaware of the whereabouts of the controlled variable. The FFC receives information on the measured disturbances and the set point and calculates the necessary corrective action to maintain the controlled variable at its set point in the presence of disturbances. The FFC is predictive in nature and therefore, perfect control is achievable, theoretically. The performance of the FFC depends on the model accuracy and precision of the measuring devices. The FFC can only correct for the measured disturbances. Therefore the FFC should always be used in conjunction with a feedback controller.

1.4 Incentives for Process Control

There are various incentives to employ an effective control system in an industrial plant. The following is a list of the general incentives that will be elaborated upon throughout the book:

(1) To ensure plant safety
(2) To meet the product specification
(3) To meet the environmental constraints
(4) To meet the operational constraints
(5) To maximize the profit using an optimization algorithm in a "supervisory control" manner

  max profit $= f$ (yield, purity, energy consumption, etc.)
  subject to equality constraints (process model) and inequality constraints

The result of the optimization step is the optimum operating conditions (i.e., T_{opt}, P_{opt}, F_{opt}, etc.) that are used as set points for the low-level T, P, F, etc., feedback controllers.

1.5 Pictorial Representation of the Control Systems

The analysis and design of control systems are facilitated by the use of pictorial representation either in the form of a *block diagram* or a *piping and instrumentation diagram* (P&ID). For the P&ID representation, there are standard symbols that must be used. Tables 1.1 and 1.2 list some of the symbols that are commonly used in constructing a P&ID.

Table 1.1 Examples of the letters used in a P&ID [1,2]

	First Letter	Subsequent Letters	Examples
A	Analyzer/analog		A/D: analog-to-digital signal converter AT or CT: analyzer/concentration transducer
C	Concentration	Controller	
D	Digital		D/A: digital-to-analog signal converter
F	Flow rate		FC: flow controller (feedback) FIC: flow indicator and controller FFC: feedforward controller FT: flow transducer
I	Current	Indicator	
L	Level		
P	Pressure, pneumatic		P/I: pneumatic-to-current convertor PT: pressure transducer
T	Temperature	Transducer	TIC: temperature indicator and controller TT: temperature transducer

Table 1.2 Symbols used in the P&ID

Symbol	Meaning
	Piping/Process connection
	Electrical (analog or digital signal)
	Pneumatic signal
	Control valve
 FC 101	Flow controller 101 mounted in the field
 FC 101	Flow controller 101 mounted in the control room and accessible to the operator
 FC 101	Flow controller 101 mounted in the control room but normally not accessible to the operator (behind the panel)
 FC 101	DCS (Distributed Control System) flow controller 101 mounted in the control room and accessible to the operator

Example 1.2

Sketch the block diagram and the P&ID of a feedback temperature control system for a CSTR having a pneumatic temperature transducer, an analog electrical controller, and a pneumatically actuated CV. Include the necessary signal converters. On the block diagram, mark the nature of the signal at any point around the control loop (Figs. 1.10 and 1.11).

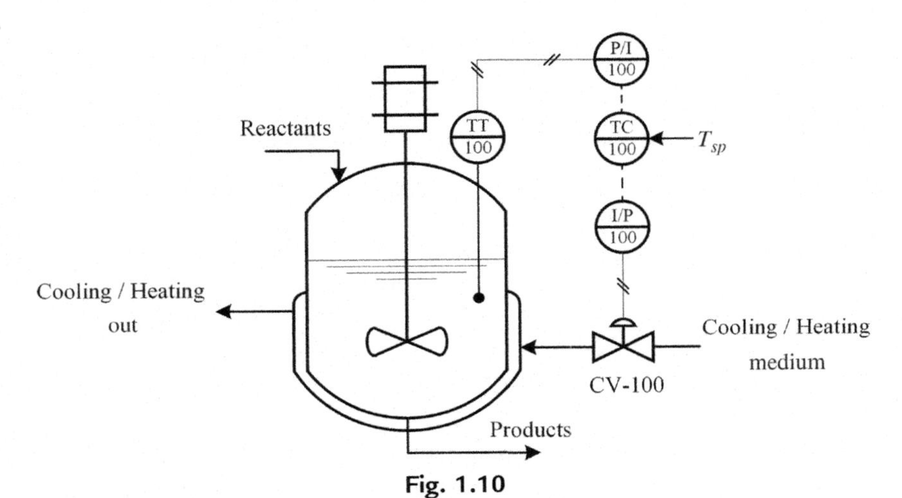

Fig. 1.10
The P&ID of a feedback temperature control system.

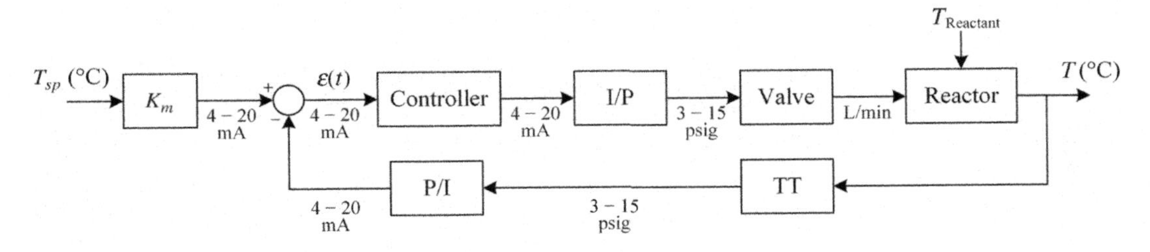

*K_m is the calibration constant converting the process variable in engineering unit (°C) to the required signal by the controller. It is the steady state gain of the transducer

Fig. 1.11
The block diagram of a feedback temperature control system.

Example 1.3

Sketch the block diagram and the P&ID of the temperature control system of the previous example having an analog electrical temperature transducer, a digital controller, and a pneumatically actuated CV. Include the necessary signal converters (Figs. 1.12 and 1.13).

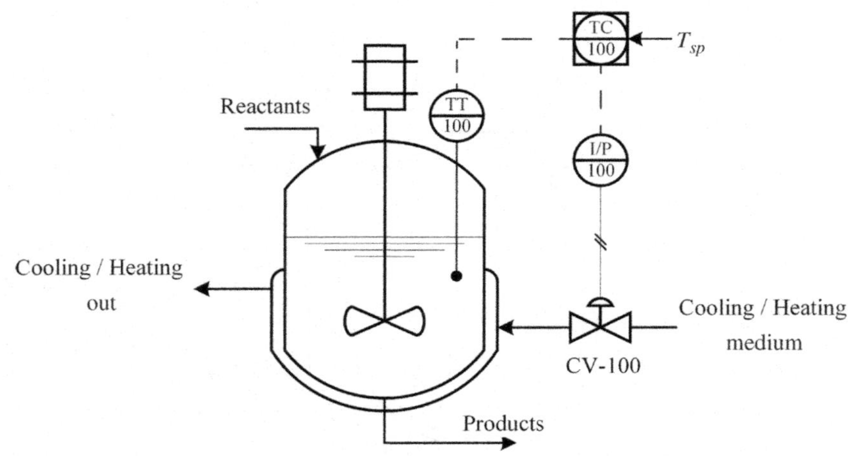

Fig. 1.12
The P&ID of a feedback temperature control system using a digital controller.

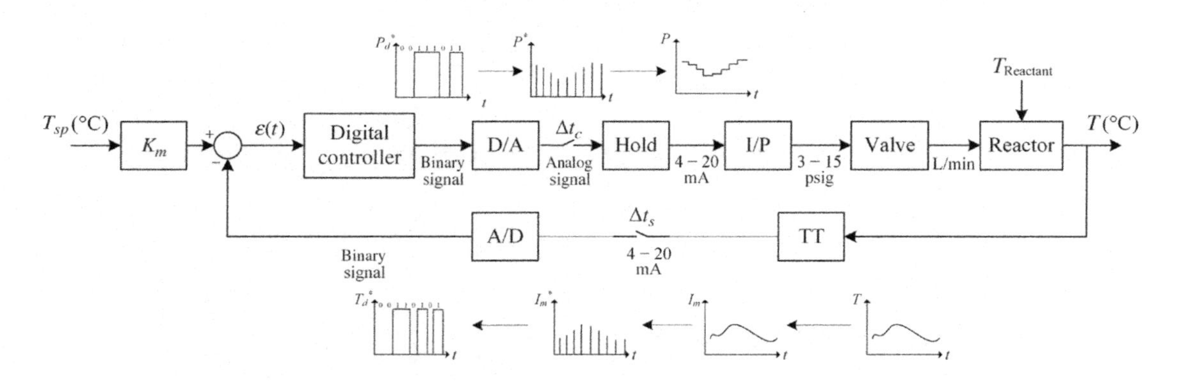

*The hold element is to maintain the controller output signal between the control intervals.
T_{sp}: Temperature set point
Δt_s : sampling interval in second.
Δt_c : control interval in second.
subscript 'd': represents a digital signal
superscript '*' : represents a discrete signal

Fig. 1.13
The block diagram of a feedback temperature control system using a digital controller with the required data acquisition system (A/D, D/A, samplers, and the hold system).

Example 1.4

Sketch the block diagram and the P&ID of a feedback plus feedforward temperature control system of a CSTR having analog electrical temperature transducers, an analog electrical controller, and a pneumatically actuated CV. The FFC corrects for the changes in the reactants' temperature. Include the necessary signal converters. On the block diagram, mark the nature of the signal in the control loop (Figs. 1.14 and 1.15).

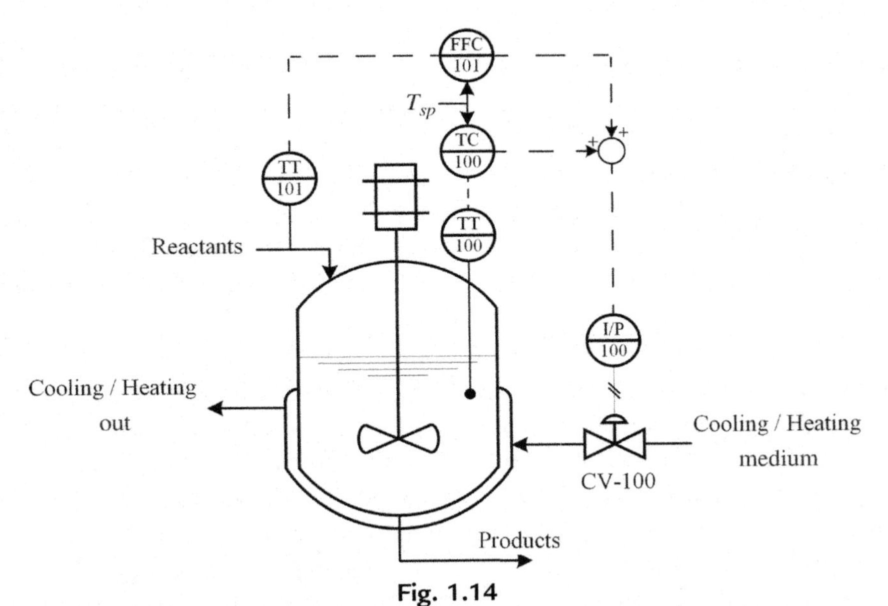

Fig. 1.14
The P&ID of the feedback plus feedforward temperature control system.

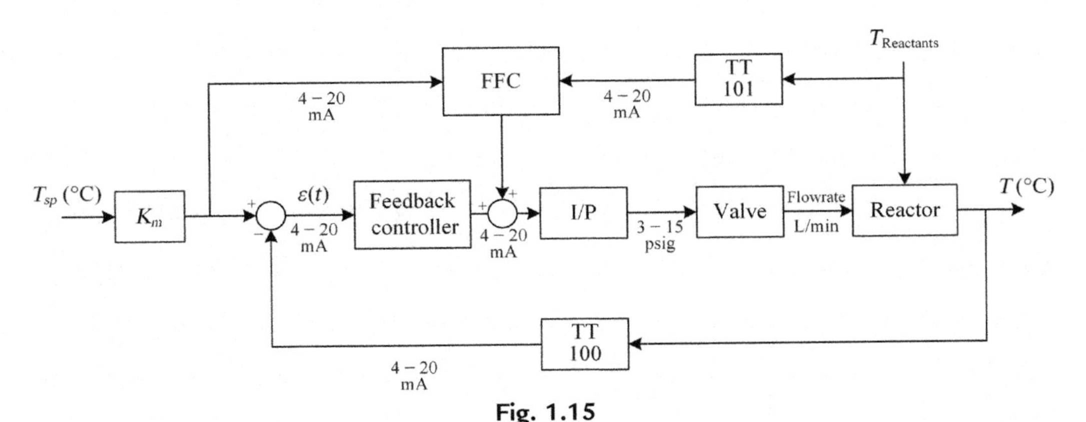

Fig. 1.15
The block diagram of the feedback plus feedforward temperature control system.

1.6 Problems

(1) A digital controller is used to control the temperature of a highly exothermic reaction carried out in a CSTR. There are significant variations in the composition **and** the flow rate of the reactants entering the reactor. Therefore it is recommended that a feedback plus FFC be used for this system. The flow transmitter has an output of 0–5 V, the temperature transmitter has an output in the range of 4–20 mA, and reactant concentration measuring device has an output of 0–5 V. The 'data acquisition board' accepts 0–5 V and 4–20 mA input signals and has an analog output module with 4–20 mA output signal. The CV supplying the cooling water to the reactor jacket is pneumatically actuated accepting a signal in the range of 3–15 psig. Sketch the P&ID and the block diagram of the combined feedback/FFC for this system.

(2) For an automatic washer and dryer system used in a household, identify the control objective, the measuring device, the controller, and the final control element. Also draw the corresponding P&ID and the block diagram for both the washing machine and the dryer.

(3) The outlet temperature of the heating oil from a parabolic solar collector is controlled by adjusting the inlet flow rate of the heating oil. The inlet temperature and the solar radiations are the two main disturbances affecting the outlet oil temperature. Develop the P&ID and the block diagram of a feedback plus feedforward control system that controls the outlet oil temperature by manipulating the inlet oil flow rate. The solar radiation is measured by a photo cell with an output signal range of 0–5 V. The temperatures are measured by temperature transmitters with an inlet range of 30°C–350°C and an outlet range of 0–5 V, the controller is digital, and the final control element is a pneumatic CV with an input signal range of 3–15 psig.

References

International Society of Automation (ISA). *Instrumentation symbols and identification.* Research Triangle Park, NC: Standard ISA-5.1 1984 (R1992); 1992.

Perry RH, Green DW, editors. *Chemical engineers' handbook.* 8th ed. New York: McGraw-Hill; 2008.

CHAPTER 2

Hardware Requirements for the Implementation of Process Control Systems

Sohrab Rohani, Yuanyi Wu
Western University, London, ON, Canada

In this chapter a brief description of the hardware components necessary for the implementation of SISO and MIMO control systems will be presented. For each component, in addition to a short description of the hardware, the transfer function of the component is also derived or presented to analyze the behavior of the entire closed-loop control system.

2.1 Sensor/Transmitter

In the chemical process industry, the most common variables that are measured and controlled are temperature (T), pressure (P), level (L), flow rate (F), and less frequently the concentration of a species (C). The sensors measure the changes in the process variables using a physical (mechanical or electrical) phenomenon. The "transmitters" convert the measured mechanical or electrical phenomenon to a 3–15-psig pneumatic signal, a 4–20-mA electrical current signal, a 0–5-V DC voltage, or a digital signal that can be transmitted to the controller. The combination of the sensor and transmitter is referred to as a *transducer*. The output signal from a transducer is either an analog or a digital signal:

Analog	Pneumatic (3–15-psig signal)
	Electronic (4–20-mA or 0–5-V signal)
Digital	Smart transmitters (digital signal)

In what follows, a brief account of the most common transducers is presented. For the detailed description of the process instrumentation, refer to many books on this subject including Soloman,[1] Liptak,[2] and Shuller and Kargi.[3] In addition, there is ample information on the internet.

2.1.1 Temperature Transducers

Among the popular temperature sensors are liquid-filled thermometers (based on the measurement of the volumetric expansion of a liquid such as mercury or an alcohol), the thermocouples (based on the Seebeck's effect—if two dissimilar metals or metal alloys form a

Coulson and Richardson's Chemical Engineering. http://dx.doi.org/10.1016/B978-0-08-101095-2.00002-3

loop with two junctions, an electromotive force is generated in the loop that is proportional to the difference between the temperatures of the two junctions), the resistance temperature detectors (RTDs) (metal objects such as platinum, copper, and nickel whose electrical resistance increases with temperature), the thermistors (ceramic metal oxides whose electrical resistance decreases with temperature), and pyrometers (optical sensors measuring the wavelength and the radiation intensity of a hot object). The output of the temperature sensors is often in the mV range and has to be amplified and conditioned using a signal conditioning unit that consists of an amplifier. Thermocouples are of different types, T, J, K, each having a different temperature range. Thermocouples must also be equipped with a "cold junction compensator."

2.1.2 Pressure Transducers

The pressure sensors include bellows, bourdon tubes, diaphragms, and strain gages that measure the mechanical deflection of a metallic object with the applied pressure or a pressure difference. A pneumatic or electronic transmitter converts the mechanical deflection of the sensor to a 3–15-psig pneumatic signal or an electrical signal (4–20 mA).

2.1.3 Liquid or Gas Flow Rate Transducers

The flow rate of a liquid stream can be measured indirectly, using the measured pressure drop across an orifice or a venturi (Fig. 2.1) according to Eq. (2.1). In this case the volumetric flow rate of the liquid is proportional to the square root of the measured pressure drop across the orifice. The venturi meter operates similar to an orifice meter with an accuracy to within $\pm 2\%$.

$$F = \alpha \sqrt{\frac{\Delta P}{\text{specific gravity}}} \tag{2.1}$$

The liquid flow rate can also be measured using the pulse outputs generated by the rotation of a turbine or the deflection of a vane inserted in a pipe such as in a vortex shedding meter.

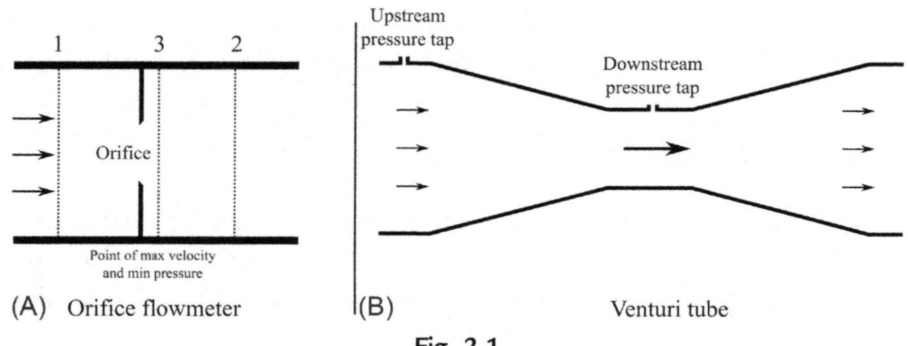

Fig. 2.1
Flow rate transducers.

Mass flow meters are used for the measurement of mass flow rate of a gas and are independent of changes in the pressure, temperature, viscosity, and density of the process fluid. One type of a mass flow meter is based on the rate of heat transfer from a heated element. Another type of a mass flow meter is called a Coriolis meter that consists of a vibrating flow loop whose amplitude of vibration is converted to a voltage.

2.1.4 Liquid Level Transducers

Differential pressure transducers connected to the top and the bottom of a tank and through pressure taps can be used to measure the level. In addition, capacitive and radiation meters can detect the level of a liquid or solid storage tank.

2.1.5 Chemical Composition Transducers

Chemical composition is the most challenging process variable for online measurement. There are numerous online chemical composition analyzers. They are usually quite expensive and require high level of maintenance. In addition, they introduce a time delay in the feedback control loop. For example, a gas chromatograph (GC), a high-performance liquid chromatography (HPLC), and analyzers based on mass spectroscopy (MS), Fourier transform infrared (FTIR) spectroscopy, and Raman spectroscopy have found applications in the chemical, oil, pharmaceutical, and specialty chemical industries.

Transmitters. After the process variable is sensed (measured) by a sensor, a pneumatic/electronic transmitter converts the measured mechanical phenomenon to a pneumatic (3–15 psig) or an electrical signal (4–20 mA or 0–5 V). A pneumatic transmitter consists of a series of pneumatic elements such as bellows, nozzles, flappers, and boosters. An electronic transmitter consists of an electronic circuit such as a Wheatstone bridge that converts the measured low voltage/current signal generated by a sensor to a 4–20-mA signal or a 0–5-V signal.

A smart transducer is one whose output is a digital signal that can directly be sent to a digital controller.

2.1.6 Instrument or Transducer Accuracy

A number of parameters are defined for a transducer:

- the zero of a transducer,
- the input span of a transducer,
- the output span of a transducer,
- the gain of a transducer,
- the precision of a transducer,
- the accuracy (expressed in terms of the absolute or relative error) of a transducer, and
- the reproducibility of a transducer.

Let us consider an example to clarify the previous definitions.

Example 2.1

An electronic transducer (output 4–20 mA) is used to measure the temperature of a process stream over a range of 20°C–380°C. Its output at 108°C (measured by an accurate thermometer) is 8.1 mA. Determine all the previous parameters for this measuring device/transducer.

The zero of the transducer is the minimum value of the process variable, i.e., 20°C. The input span of the transducer is the maximum minus minimum value of the process variable, i.e., $380°C - 20°C = 360°C$. The output span is the maximum minus minimum value of the transducer output which is $20\,mA - 4\,mA = 16\,mA$. The transducer gain is the ratio of the output span to the input span that has the unit of mA/°C.

$$\text{Transducer Gain, } K_m = \text{output span/input span} = 16\,mA/360°C$$

The precision of the transducer output is the smallest measurable change in the transducer output which is 0.1 mA. Note that this is deduced from the stated significant figure after the decimal point in the problem statement, i.e., 8.1 mA.

The accuracy of the transducer is calculated using the given information, signifying that 8.1 mA corresponds to a precise temperature of 108°C. Having calculated the transducer gain, an output reading of 8.1 mA corresponds to a temperature of 112.25°C [$T = 20°C + [(380 - 20)\,°C/(20 - 4)\,mA] \times (8.1 - 4)\ mA = 112.25°C$]. Therefore the absolute error is $112.25°C - 108°C = 4.25°C$ and the relative error with respect to the input span is $4.25/360 = 0.012$ or 1.2%. If the reading of the measuring element is not biased in one direction and it is equally spread over the positive and negative error values, the absolute and relative errors are $\pm 4.25°C$ and ± 0.012 or $\pm 1.2\%$, respectively.

The resolution of the transducer is defined as the smallest increment in the process variable that results in a measurable and detectable change in the transducer output signal. In this case, the precision is 0.1 mA that corresponds to a resolution of 2.25°C, i.e., $\{0.1\,mA \times [(380 - 20)°C/(20 - 4)mA\} = 2.25\,°C$.

The reproducibility of the transducer is defined as the ability of the transducer to generate the same output for the same given input over a number of measurements.

2.1.7 Sources of Instrument Errors

The transducers may suffer from a number of limitations, nonlinearity, hysteresis (backlash), deadband, drift, and dynamic lag or time delay. Nonlinearity refers to a nonlinear relationship between the transducer's output and the measured process variable (transducer's input). Hysteresis occurs when the transducer output depends on the direction (increasing or decreasing) of the measured process variable (the same absolute change). Hysteresis is caused by the electrical or magnetic components of the transducer. Backlash is the mechanical equivalent of the hysteresis due to the friction of the mechanical components in a transducer.

A "deadband" occurs when the output does not change up to a certain threshold in the input change (process variable). A "drift" refers to a slowly changing instrument output when the instrument input (process variable) is constant, usually due to the temperature sensitive electrical components of the transducer.

2.1.8 Static and Dynamic Characteristics of Transducers

The transfer function of a measuring element (transducer) is often approximated by a pure gain or a first-order model whose parameters are to be determined experimentally.

$$G_m(s) = \frac{K_m}{\tau_m s + 1} \tag{2.2}$$

where K_m is the steady-state gain and τ_m is the time constant of the measuring element.

$$K_m = \frac{\text{span or output range}}{\text{span or input range}} \tag{2.3}$$

Certain measuring elements, such as a gas chromatograph, involve a dead time (the analysis time), θ, and therefore the transfer function of such analyzers is expressed in the form of:

$$G_m(s) = K_m e^{-\theta s} \tag{2.4}$$

2.2 Signal Converters

In a feedback control system, it is often necessary to convert a pneumatic signal (3–15 psig) to an electronic signal (4–20 mA), or vice versa. This is achieved by certain hardware units called a pneumatic-to-current (P/I) and a current-to-pneumatic (I/P) converters. In addition, it is necessary to convert an analog to a digital signal by an (A/D) and a digital to an analog signal by a (D/A) converter. The use of an analog and/or a digital filter is also quite common to get rid of the undesirable process and transducer noise. Due to the fast response of a signal converter, no dynamics is associated with these units.

Resolution of an (A/D) and a (D/A) converter. Depending on the type of the (A/D) converter, that is being an 8-bit, a 16-bit, or a 32-bit converter ($n = 8$, 16 or 32), the resolution of a 0–5 V or a 4–20 mA of a digital converter is expressed by:

$$\text{Resolution} = \frac{(5-0)\,V}{2^n - 1} \text{ or } \frac{(20-4)\text{ mA}}{2^n - 1} \tag{2.5}$$

2.3 Transmission Lines

Sensors/transducers with local display units and manual valves require no signal transmission. However, most sensors/transmitters and control valves require signal transmission, so that personnel in a single location (control room) can manage the entire process. Reliable, accurate, and rapid signal transmission is essential for the implementation of a good process control system.

Typically, the measured process variable is converted to a "signal" that can be transmitted over a long distance without being corrupted by other units in the plants. The signal can be an analog electronic (usually 4–20 mA) or a digital entity. The signal is sent from the plant site to the control room, where it can be employed for many purposes (see Fig. 2.2). When the signal is used for control purposes, the measured controlled variable is used by the controller to calculate the error signal and subsequently the controller output. The controller output is transmitted back to the plant and to the final control element, which is often a control valve, but could be a motor or a variable speed pump. For a pneumatically actuated control valve, the valve stem position is moved by the instrument air pressure; therefore the electrical signal from the controller must be converted to an air pressure signal. This conversion is achieved at the site. The valve stem moves the valve plug, changes the resistance to flow, and the flow rate of the manipulated variable changes.

Nowadays, the analog pneumatic transmission lines are quite rare and are made of ½ or ¾-in. nylon tubing that are restricted to distances less than 100 m to maintain the dynamic error reasonably low. The analog electronic transmission lines are made of twisted pair 4–20-mA cables. Voltage signals are prone to more corruption by electrical and magnetic high frequency noise, therefore the industry standard is to transmit the analog electronic signals in the form of an electrical current, namely, 4–20 mA.

Digital signals are transmitted by data highways or fiber optics. The fieldbus technology is a digital, serial, two-wire communication protocol configured like a local area network. Fieldbus technology has become the industry norm for the distributed control systems (DCS). This technology has much greater flexibility for transmitting multiple variable values, in both

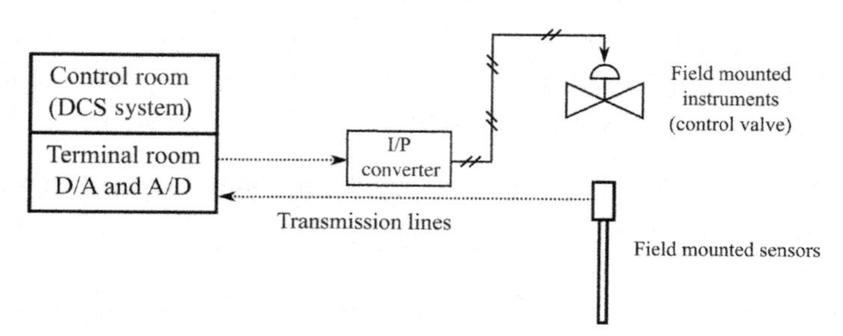

Fig. 2.2
The layout of a typical chemical plant, instrumentation in the site and the control room.

directions. Two-way communication and computation at any control device in the system (not just the controller) provides the opportunity for diagnostic information to be communicated about the performance of the sensor and the final control element.

Analog	Pneumatic (3–15 psig), nylon tubing
	Electronic (4–20 mA), twisted pair cable
Digital	Coaxial cable
	Fiber optics (data highway)

2.4 The Final Control Element

The final control element or the actuator throttles the manipulated variables based on the command signal it receives from the controller. In the chemical industry, the manipulated variable is often the flow rate of a process stream (e.g., the steam or cooling water flow rate for temperature control, a gas or liquid flow rate of reactants, the solid flow rate of a catalyst using a slide valve or a screw conveyor, etc.). The most commonly used final control element for throttling a flow rate is a control valve. Other final control elements include a variable speed pump, a silicon-controlled rectifier (SCR), and an immersion heater.

2.4.1 Control Valves

A control valve consists of *valve body*, *trim/stem*, *plug* and *seat*, and an *actuator* (see Fig. 2.3).

The *valve body* consists of an orifice through which the manipulated variable passes. The *trim* or *stem* is connected to the diaphragm and the valve *plug* that can be in the form of a ball, a disk, or a gate. The valve *seat* provides a tight seal for a valve shutoff position. Control valves are either *reciprocating* (rising stem) such as ball valves, gate valves, butterfly valves,

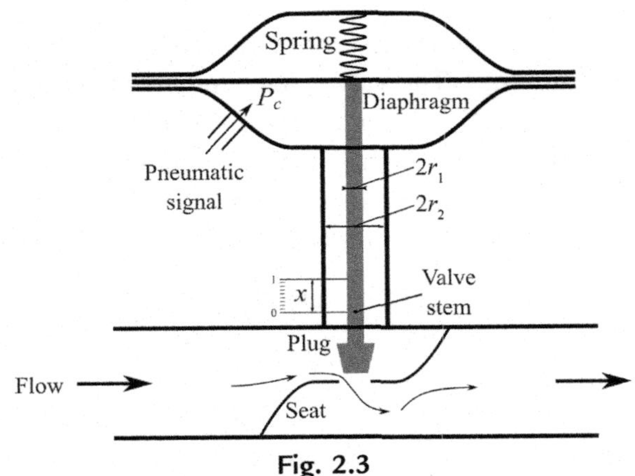

Fig. 2.3
Schematic of a control valve.

or *rotary* such as globe valves. The valve can be either *fail-open* (air-to-close, A–C) or *fail-closed* (air-to-open, A–O). The choice is made based on the process safety. The valve *actuator* can be a pneumatically operated diaphragm, DC motor, or a stepping motor (useful with digital controllers).

A control valve has certain accessories that can be purchased separately. These include a valve *positioner* (consisting of a local feedback controller to ensure the position of the valve stem) and a valve *booster* (to speed up the response of the valve).

2.4.1.1 Selection and design of a control valve

In the design of a control valve, there are a number of issues to be considered:

- The choice of a proper actuator: pneumatic, hydraulic, electromagnetic.
- The correct *action* for the valve based on the process safety, fail-open (air-to-close) or fail-closed (air-to-open).
- The material of construction of the body, seat, and plug of the valve.
- Valve sizing that also includes the selection of the valve characteristics.

The equation that describes the flow of a liquid medium through a control valve is the same as the orifice equation:

$$q = C_v f(l)\sqrt{\frac{\Delta P_v}{s.g.}} \tag{2.6}$$

where q is the flow rate of the liquid through the valve in (gpm), C_v is the valve constant which is the number of gpm of H_2O (with a specific gravity, $s.g. = 1$) in a fully open valve $[f(l) = 1]$ and a pressure drop across the valve of 1 psi, $\Delta P_v = 1$ psi, $f(l)$ is the valve characteristic.

2.4.1.2 Valve characteristic

The valve characteristic is the relation between the flow of the liquid stream, $f(l)$, through a valve and the valve stem position, l. The valve characteristic is chosen such that the overall process steady-state gain $(K_c K_v K_p K_m)$ remains relatively constant over the entire operating range. Often the process or the transducer characteristics are nonlinear. In such cases the valve characteristic is chosen from among the *linear*, *square root*, or *equal percentage* categories (shown in Fig. 2.4) to render the control loop nearly linear over the entire operating range.

The pressure drop across the valve should be selected such that

$$\left(\frac{1}{4} < \frac{\Delta P_v}{P_s} < \frac{1}{3}\right) \tag{2.7}$$

where P_s is the pressure drop in the line (pump supply pressure $= P_s + \Delta P_v$). The pressure drop across the valve $\Delta P_v = P_1 - P_2$, should be selected to avoid flashing and cavitation of the

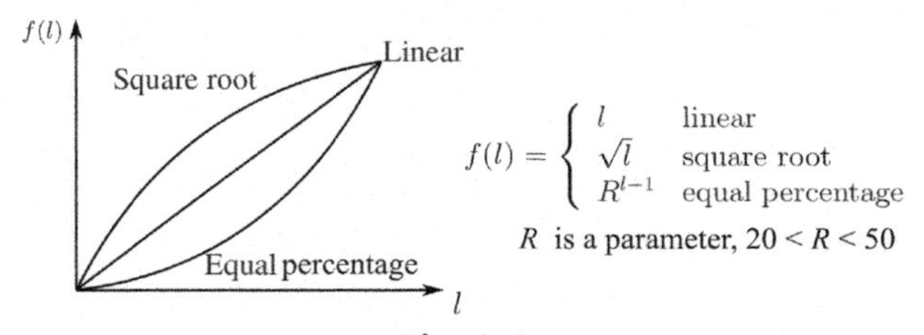

Fig. 2.4
The valve characteristic curve.

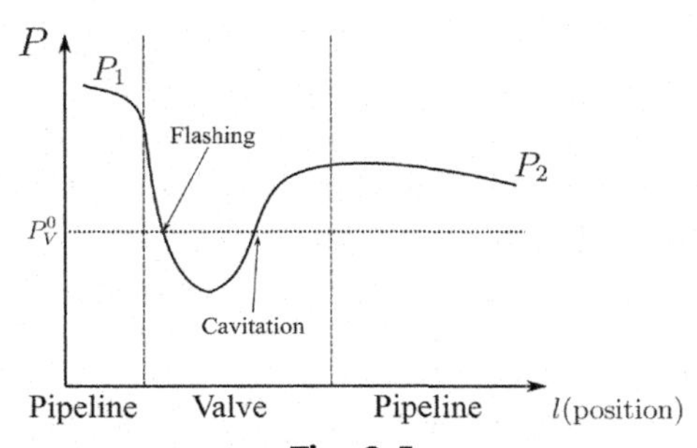

Fig. 2.5
Changes in the liquid pressure as it moves through the control valve.

liquid in the valve. If P_V^0 represents the vapor pressure of the liquid passing through the valve, flashing and cavitation may happen as the liquid passes through the valve and during the pressure recovery stage. Cavitation causes wear and tear on the valve and should be avoided (see Fig. 2.5).

In designing a control valve (selecting the control valve size, C_v), as the valve size, C_v, decreases, the pumping cost increases, and the capital cost (price of the valve) decreases.

2.4.1.3 The transfer function of a control valve

$$G_v(s) = \frac{\mathcal{L}\,(\text{manipulated variable in deviation form})}{\mathcal{L}\,(\text{controller output in deviation form})} = \frac{q(s)}{P(s)} = \frac{q(s)\,X(s)}{X(s)\,P(s)} \tag{2.8}$$

Where $P(s)$ is the Laplace transform of the controller output (corrective action) sent to the valve diaphragm in 3–15 psig, $X(s)$ is the Laplace transform of the valve stem position changing between 0 and 1, and $q(s)$ is the Laplace transform of the manipulated variable in, e.g., L/min.

As shall be discussed in Chapter 3, the derivation of the actual transfer function of a control valve based on the momentum balance leads to a second-order overdamped transfer function; however, the valve's transfer function is usually approximated by a first-order lag term or even a steady-state gain:

$$G_v(s) = \frac{K_v}{\tau_v s + 1} \tag{2.9}$$

If the valve time constant is much smaller than the process time constant, $\tau_v \ll \tau_{\text{process}}$, the control valve can be approximated by a pure gain $G_v(s) \approx K_v$.

2.5 Feedback Controllers

Single loop controllers can be analog (pneumatic or electronic) or digital. The digital controllers are more commonly used either in a dedicated single loop architecture or a DCS platform. The DCS has become the industry norm. It is based on microprocessor technology and supports low-level regulatory PID controllers, historical data collection, database, display units, process flow diagrams, alarms, and logic controllers. Logic controllers are implemented by programmable logic controllers (PLC) that are primarily used for batch processes and continuous processes during the start-ups and shut downs as well as implementation of logics for plant safety.

The most common feedback control law used in the chemical industry is the proportional-integral-derivative (PID) control law.

2.5.1 The PID (Proportional-Integral-Derivative) Controllers

The PID controllers can be a single loop analog or digital controller. The analog PID controllers either operate pneumatically with a range of 3–15 psig or electronically over a range of 4–20 mA.

Analog PID controllers
Pneumatic controllers accept the measured process variables (PV) and the set points (SP) over the range of 3–15 psig and issue the output command signals over the same range (corresponding to 0%–100% input and output range)
Electronic controllers accept the measured process variables and the set points over the range of 4–20 mA or 0–5 V and issue the output command signals over the same range (corresponding to 0%–100% input and output range)

Digital PID controllers
A host of digital computer architectures and platforms may be used to implement the various types of control laws including the PID control law. Personal computers, dedicated process control computers (centralized control), microprocessor-based instrumentation (DCS), and the PLC are among the various digital controllers that can be used in the laboratory, pilot plants, and the full-scale industrial-scale process industries

Fig. 2.6
A single loop (left) controller and a distributed digital control (right) system (DCS).

Most process control systems have a distributed architecture, with sensors and valves located in the plant site and the computer control and monitors in the central control room. Therefore the measured key process variables are transmitted from the process site to the control room and the command signals are sent back from the control room to the control valves (or other final control elements such as the variable speed pumps and the stepping motors) located in the plant site. A schematic of a typical control system with a dedicated single loop controller and a DCS digital controller system is shown in Fig. 2.6.

2.5.2 The PID Controller Law

The equation for an "analog" PID controller is:

$$P(t) = \overline{P} + K_c\varepsilon(t) + \left(\frac{K_c}{\tau_I}\right)\int_0^t \varepsilon(t)dt + K_c\tau_D\frac{d\varepsilon(t)}{dt} \qquad (2.10)$$

where K_c is the controller Proportional gain, τ_I is the controller Integral or reset time, τ_D is the controller Derivative or rate time, and $\overline{P}$ is the controller bias which is the command signal when the error signal $\varepsilon(t)$ is zero.

The transfer function of an "ideal" PID controller can be derived as:

$$p'(t) = P(t) - \overline{P}$$

$$p'(t) = K_c\varepsilon(t) + \left(\frac{K_c}{\tau_I}\right)\int_0^t \varepsilon(t)dt + K_c\tau_D\frac{d\varepsilon(t)}{dt}$$

$$P(s) = K_cE(s) + \left(\frac{K_c}{\tau_I}\right)\frac{E(s)}{s} + K_c\tau_D[s\cdot E(s)] \qquad (2.11)$$

$$G_c(s) = \frac{P(s)}{E(s)} = K_c + \frac{K_c}{\tau_I s} + K_c\tau_D s = K_c\left(1 + \frac{1}{\tau_I s} + \tau_D s\right) \qquad (2.12)$$

A commercial controller uses a lead-lag term instead of the derivative term (which is physically unrealizable).

$$G_c(s) = K_c\left(1 + \frac{1}{\tau_I s}\right)\left(\frac{\tau_D s + 1}{\alpha \tau_D s + 1}\right) \tag{2.13}$$

where α is an adjustable parameter which has a value between 0.01 and 1. As $\alpha \to 0$ the lead-lag term approaches a derivative action.

2.5.3 The Discrete Version of a PID Controller

For a digital PID controller the discrete versions of the PID control law must be implemented. A finite difference approximation of the integral and derivative terms is used. Let P_n be the controller output at the nth sampling/control interval, i.e., $P_n = P(t = n\Delta t)$.

$$P_n = \overline{P} + K_c \varepsilon_n + \left(\frac{K_c}{\tau_I}\right)\sum_{i=1}^{n} \varepsilon_i \Delta t + K_c \tau_D \left(\frac{\varepsilon_n - \varepsilon_{n-1}}{\Delta t}\right) \tag{2.14}$$

We refer to the previous equation (2.14) as the **position form of a discrete PID controller.** The major problems with the position form of a PID controller are as follows:

- The controller bias, $\overline{P}$, must be known.
- All the previous values of the error signal must be stored and the summation $\sum_{i=1}^{n} \varepsilon_i \Delta t$ may lead to *reset or integral windup*.

Because of these problems, it is preferred to use the **velocity form of a PID controller**. In order to derive the velocity form of a PID controller, the controller output at the $(n-1)$th interval, P_{n-1}, is subtracted from its value at the nth interval, P_n, given in Eq. (2.14):

$$P_{n-1} = \overline{P} + K_c \varepsilon_{n-1} + \left(\frac{K_c}{\tau_I}\right)\sum_{i=1}^{n-1} \varepsilon_i \Delta t + K_c \tau_D \left(\frac{\varepsilon_{n-1} - \varepsilon_{n-2}}{\Delta t}\right)$$

$$\Delta P_n = P_n - P_{n-1} = K_c(\varepsilon_n - \varepsilon_{n-1}) + \left(\frac{K_c}{\tau_I}\right)\varepsilon_n \Delta t + K_c \tau_D \left(\frac{\varepsilon_{n-1} - 2\varepsilon_{n-1} + \varepsilon_{n-2}}{\Delta t}\right) \tag{2.15}$$

where ΔP_n represents the increment in the controller output at $t = n\Delta t$ with respect to its value at $t = (n-1)\Delta t$. Therefore the controller output at $t = n\Delta t$ can be obtained:

$$P_n = P_{n-1} + \Delta P_n \tag{2.16}$$

Eqs. (2.15), (2.16) provide the **velocity form of the Discrete PID controller.**

2.5.4 Features of the PID Controllers

The main features of a PID controller are discussed as follows.

2.5.4.1 The reset or integral windup

According to Eq. (3.49), if the error signal persists for a long time, no matter how small it might be, the controller output $P(t)$, eventually hits either its upper or its lower limit and saturates and eventually exceeds the minimum or maximum of the controller output range. This oversaturation is referred to as the *reset windup* and should be avoided. In an analog PID, the **antireset windup** module is often an extra option that must be separately purchased. For a digital PID controller, the **antireset windup** can be programmed by inserting two "IF" statements.

$$\text{IF } (P_n > P_{\max}) \text{ THEN } P_n = P_{\max}$$

$$\text{IF } (P_n < P_{\min}) \text{ THEN } P_n = P_{\min}$$

The integral windup occurs when the I-action accumulates the error and causes the controller output exceed the range of the actuator. The oversaturation should be offset first before the controller output returns to the valid range, leading to undesired dead time and usually higher overshoot.

The antireset windup is used to solve the problem by limiting the controller output. The integral windup and the effect of antireset windup are shown in Fig. 2.7.

In Simulink, the antiwindup of the "PID Controller" block is configured by double clicking the PID block to display the parameters. In the "PID Advanced" tab, check "Limit output" and select a method in "Antiwindup method" dropdown window.

2.5.4.2 The derivative and proportional kicks

If the operator introduces a sudden change in the set point, this causes a large change in the proportional and the derivative terms of the controller and causes the controller to saturate. This undesired situation is referred to as the "proportional and derivative kicks."

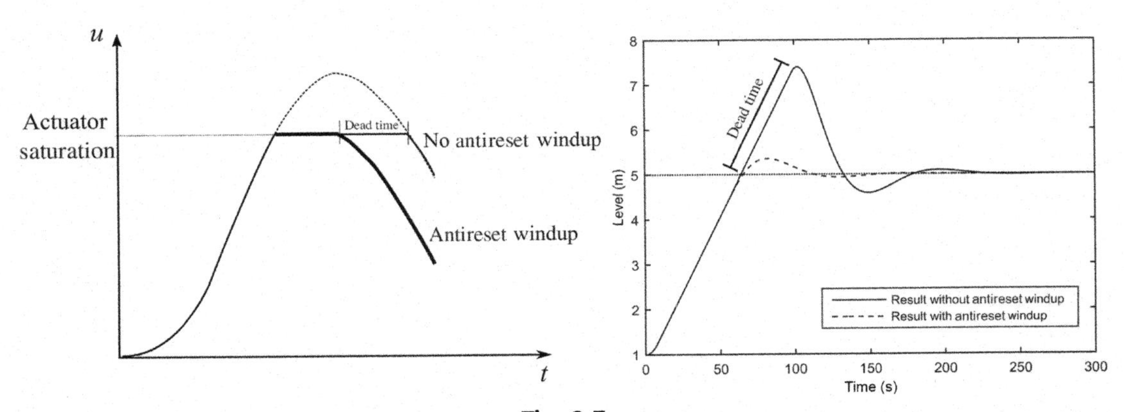

Fig. 2.7
Effect of integral windup and antireset windup.

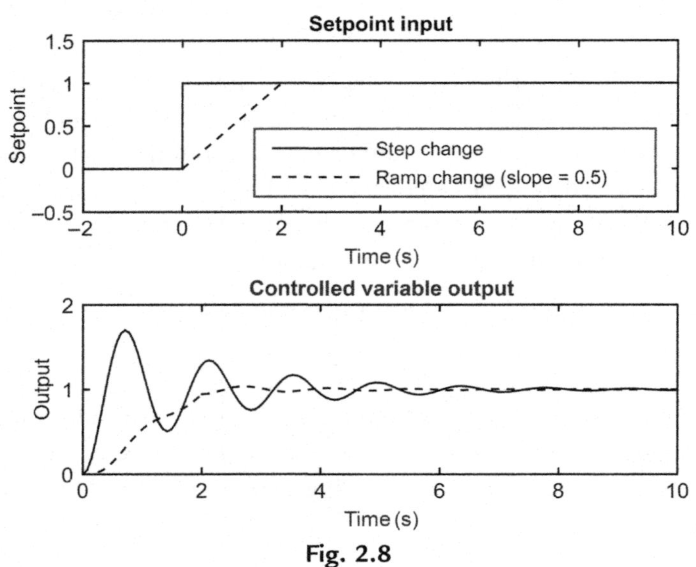

Fig. 2.8
A ramp change in the set point reduces the oscillations in the controlled output variable.

In order to avoid such a sudden jump in the controller output, the following remedial actions may be taken:

(a) Instead of a step change in the set point a ramp change must be used (see Fig. 2.8).
(b) In the case of a digital PID controller, the error signal in the proportional and derivative terms is replaced by the negative of the measured signal in the equation of the PID controller, i.e., replace $\varepsilon_n = y_{sp} - y_n$ by $-y_n$ in the proportional and derivative terms of Eq. (2.15):

$$P_n = P_{n-1} + K_c(-y_n + y_{n-1}) + \left(\frac{K_c}{\tau_I}\right)\varepsilon_n \Delta t + K_c \tau_D \left(\frac{-y_{n-1} + 2y_{n-1} - y_{n-2}}{\Delta t}\right) \tag{2.17}$$

The response of a general PI controller and a modified PI controller by replacing $\varepsilon_n = y_{sp} - y_n$ by $-y_n$ in the proportional term, using the same parameters, is shown in Fig. 2.9. The overshoot is suppressed and the response is less oscillating.

2.5.4.3 Caution in using the derivative action

The D-action should not be used if the measured signal is noisy because the derivative of the error signal becomes large due to the rapid changes of the noisy signal and causes the controller to saturate. Fig. 2.10 shows a controller tracking a unit step change in the set point while a high frequency noise is added to the measured signal. Without a low-pass filter, the output of the controller is quite erratic and the command-following capability of the

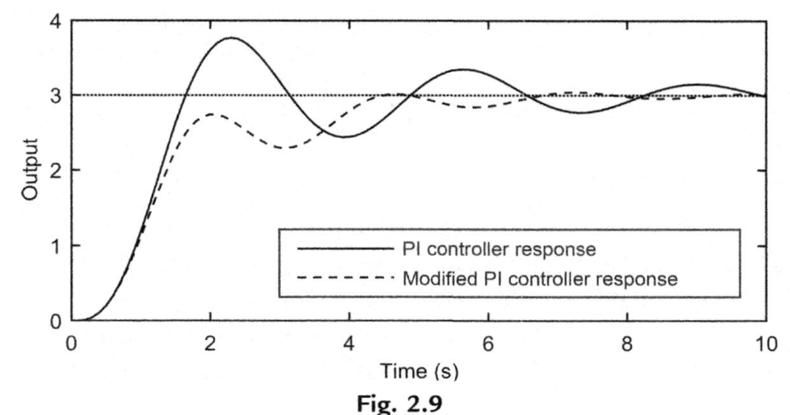

Fig. 2.9
Modified PI controller responses (set point step change of 3 units).

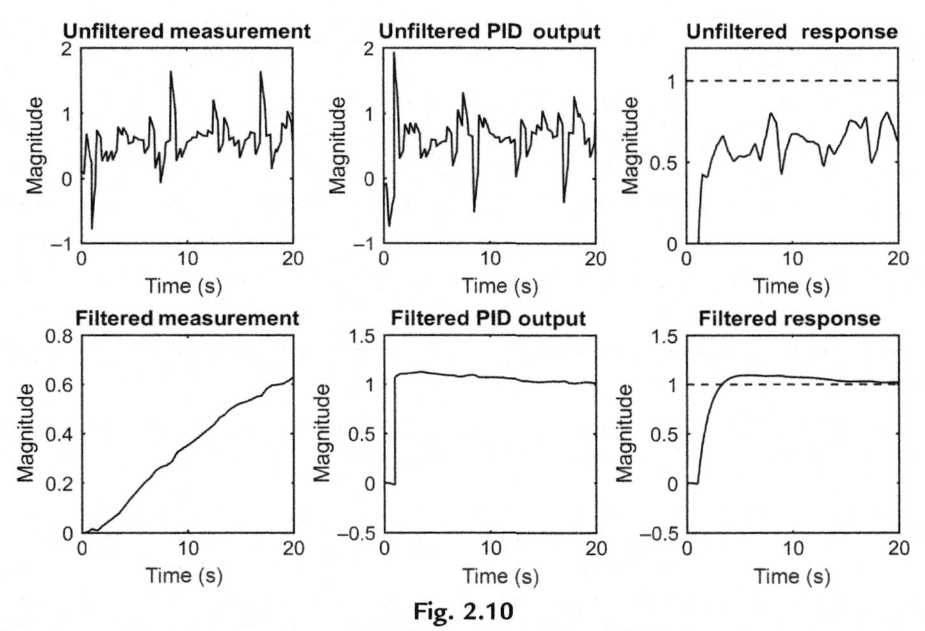

Fig. 2.10
The performance of a PID controller with unfiltered and filtered high frequency measurement noise.

controller is not acceptable. However, when the measured noisy signal is passed through a low-pass filter, the controller performance is improved.

2.5.4.4 Auto and manual modes of the controller

The analog controllers are equipped with a switch that can set the controller in the "auto" or "manual" mode of operation. In the "manual" mode the controller becomes a remote valve positioner that can send a constant signal to the remote control valve. In a digital controller including the DCS systems the switching between the auto and manual modes of the controller operation can be achieved by proper configuration of the PID block. The "manual" mode of the controllers is useful in the start-up and shut down operations of a process.

2.5.4.5 The reverse or direct controller action

The controller action should be chosen after the correct action of the control valve has been selected and by considering the sign of the process gain. The process safety dictates whether a control valve should be "fail-open" or "fail-closed" corresponding to a negative or a positive control valve gain, respectively. The controller action is chosen such that the overall steady-state gain of the feedback control system is positive (to ensure a negative feedback control strategy).

$$K_c K_v K_p K_m > 0 \tag{2.18}$$

where K_m is the gain of the measuring element and is often positive; the control valve action determines the sign of K_v; and the inherent characteristics of the process determines the sign of the process gain, K_p. Once the signs of the control valve and the process gains have been determined, the sign of K_c is chosen such that the overall steady-state gain of the feedback control system is positive. A positive controller gain is referred to as a *reverse* controller action. The following examples demonstrate how the sign or the action of the controller should be selected. The analog controllers are equipped with a switch between the *direct* and *reverse* actions. In a digital controller the sign is implemented in the controller equation.

Example 2.2

Determine the controller action in a temperature control system of a reactor system in which an exothermic reaction takes place using a cooling water jacket.

The process safety requires a fail-open (FO) valve, so that in the case of a system failure, the cold water continues to be supplied at its maximum rate to the cooling jacket of the reactor to avoid a run-away condition. The valve must be fail-open or air-to-close corresponding to a negative value of K_v, which means that as the supplied pressure to the valve decreases, the flow rate of the cooling water increases. The process gain is also negative, since higher cooling water flow rates result in a decrease in the reactor temperature. Assuming that the sensor gain is positive, in order to ensure that the overall gain of the control system remains positive, $K_c K_v K_p K_m > 0$, the controller gain must be positive, that is the controller "*reverse*" acting.

Example 2.3

In a surge tank with a control valve at its effluent stream, in order to avoid the possibility of a liquid spill over, safety considerations require a fail-open (FO) valve (Fig. 2.11). Therefore if the control valve air pressure increases, the valve opening decreases. Such a choice results in a negative $K_v < 0$. The process gain is also negative, as F (the effluent flow rate) increases, the liquid level (l) decreases. Therefore $K_p < 0$. The sensor gain K_m is positive. To meet the requirements of Eq. (2.18), K_c must be positive, indicating a *reverse* acting controller. Recall that for a P controller,

$$P(t) = \overline{P} + K_c\lfloor l_{sp} - l(t)\rfloor \tag{2.19}$$

Therefore if the level in the tank increases the controller output decreases. Because the valve is fail-open, decreasing the controller output increases the effluent flow rate and brings the level closer to its set point.

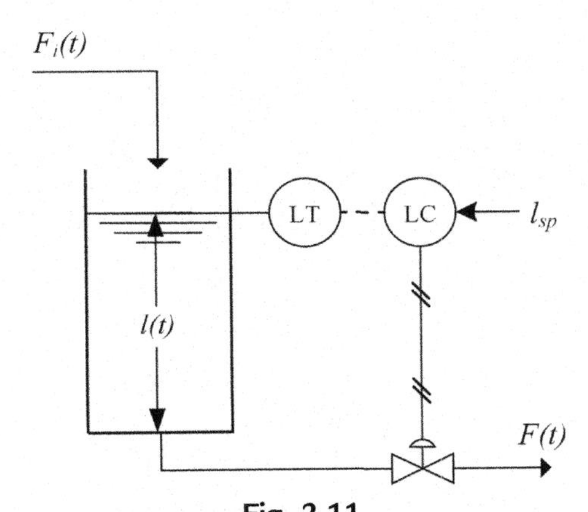

Fig. 2.11
A tank with a fail-open valve.

Example 2.4

Determine the correct action of a temperature controller in a reactor with an endothermic reaction using the flow rate of steam supplied to the heating jacket of the reactor as the manipulated variable.

Due to the safety considerations, the control valve must be fail-closed (air-to-open)—corresponding to a positive K_v. As the supplied pressure to the valve increases, the flow rate of steam increases. In the case of a sensor malfunction leading to the loss of control, the valve will shut down to stop the flow of steam to the process to avoid a possible hike in the temperature and eventual explosion. The process gain is also positive, higher steam flow rates result in an increase in the reactor temperature. Assuming that the sensor gain is positive, in order to ensure that the overall gain of the control system remains positive, $K_cK_vK_pK_m > 0$, the controller gain must be positive, that is the controller will be "*reverse*" acting.

Example 2.5

Determine the correct action of a temperature controller in a polymerization reactor shown in Fig. 2.12.

In a polymerization reactor, if the temperature of the reacting mixture falls below a limit, the viscosity increases and the mixture may solidify. Safety requires a fail-open valve (i.e., a negative K_v). The process gain is positive, $K_p > 0$, as the steam flow rate supplied to the heating coil increases, the temperature increases. In order to have an overall positive gain, $K_cK_vK_pK_m > 0$, K_c must be negative, $K_c < 0$, i.e., a *direct* acting controller.

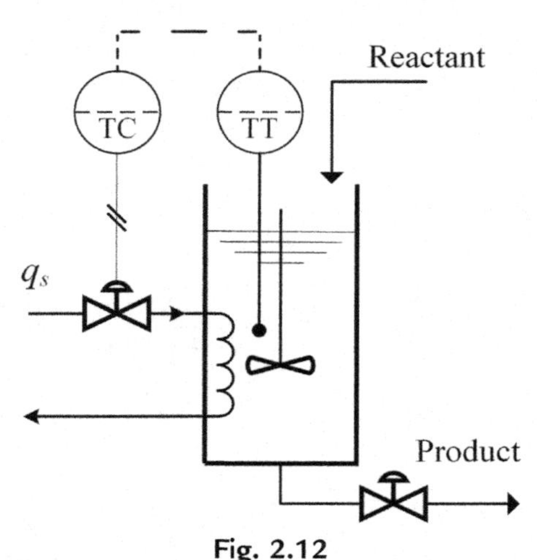

Fig. 2.12
A polymerization reactor with a steam coil or a heating jacket.

2.6 A Demonstration Unit to Implement A Single-Input, Single-Output PID Controller Using the National Instrument^R Data Acquisition (NI-DAQ) System and the LabVIEW

Among the diverse control system architectures available in the market, the National Instrument4 has gained widespread acceptance for the implementation of control systems in the laboratory and pilot plant scales. National Instrument has a wide selection of data acquisition (DAQ) systems with different numbers of analog inputs (AI), analog outputs (AO), digital inputs (DI), and digital outputs (DO). In some units, the communication between the DAQ system and the computer is facilitated by a USB port. The signal processing and controller design are implemented by a powerful software language called LabVIEW which is a symbolic programming language. Once a new LabVIEW program with an extension.vi (standing for a virtual instrument) is prompted, two windows appear, a Front Panel and a Block Diagram. In the Front Panel which works as the operator's interface, various icons such as a Numeric or a Slide control can be utilized to change the desired set point by the operator. In addition, various types of charts and graphs can be incorporated in the Front Panel to monitor the process variables and the set points. The controller output range and the controller parameters can be created in the Block Diagram but can be accessed and changed in the Front Panel.

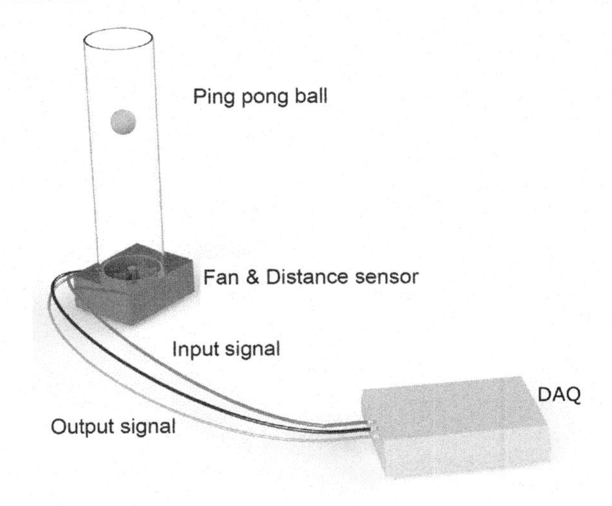

Fig. 2.13
The schematics of the ball levitation demonstration unit.

Fig. 2.13 shows the schematic of a ball levitation demonstration unit. An IR (infrared) emitter/ detector unit measures the position of a ball in a Plexiglass tube. The measured signal is passed on to the NI-DAQ unit (USB-6009 14 bit 48 Ks/S) which is connected to a personal computer via a USB port. The steps to develop a LabVIEW program to read in the position of the ball and to control the speed of the fan used as the manipulated variable to position the ball along a Plexiglass tube are briefly described later.

Fig. 2.14 depicts the P&ID and the block diagram and Fig. 2.15 shows the Front panel and the Block Diagram panel of the LabVIEW program for the implementation of a PI controller on the ball levitation feedback control system.

2.7 Implementation of the Control Laws on the Distributed Control Systems

The implementation of the various control loops/configurations on the DCS systems varies from one manufacturer to the next. Each DCS manufacturer has developed their own software systems. The process variables in most DCS systems are expressed in engineering units or percentage of the whole range. For example, reading the temperature in a process as an analog input (AIN) in the input channel 3 with a low value (zero) of 50°C and an input span of 100°C can be implemented by the following commands

```
T=AIN(input channel number, low value, span of transmitter)
T=AIN(3,50,100)
```

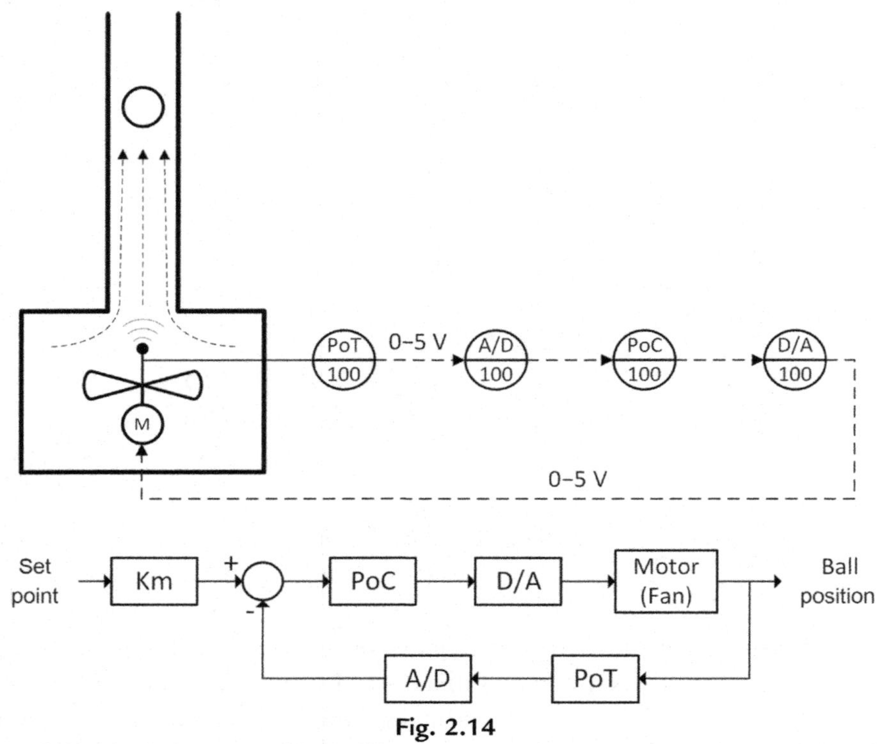

Fig. 2.14
The P&ID and the block diagram of the ball levitation demonstration unit. PoT: position transducer, PoC: position controller.

The controller action is sent out as an analog output (AOUT) to a given channel number, in percentage.

```
AOUT (output channel #, out variable in percentage)
```

The control law can either be programmed or a PID block can be used and properly configured. Other computing blocks such as various mathematical operations or the transfer function of a lead-lag unit can either be implemented using mathematical blocks or programmed using a given syntax. Consider, for example, the implementation of a concentration feedback control system in a mixing tank shown in Fig. 2.16 using a PID control law in a language-oriented DCS system.

```
CA = AIN (4, 0.05, 0.45)              ; reads the input channel 4, the concentration of
                                        component A
CASET = 0.02                          ; concentration set point of component A
m = PID (CA, CASET, 0.05, 0.45)       ; output of the concentration controller
AOUT(1, m)                            ; send out the command signal to channel 1
```

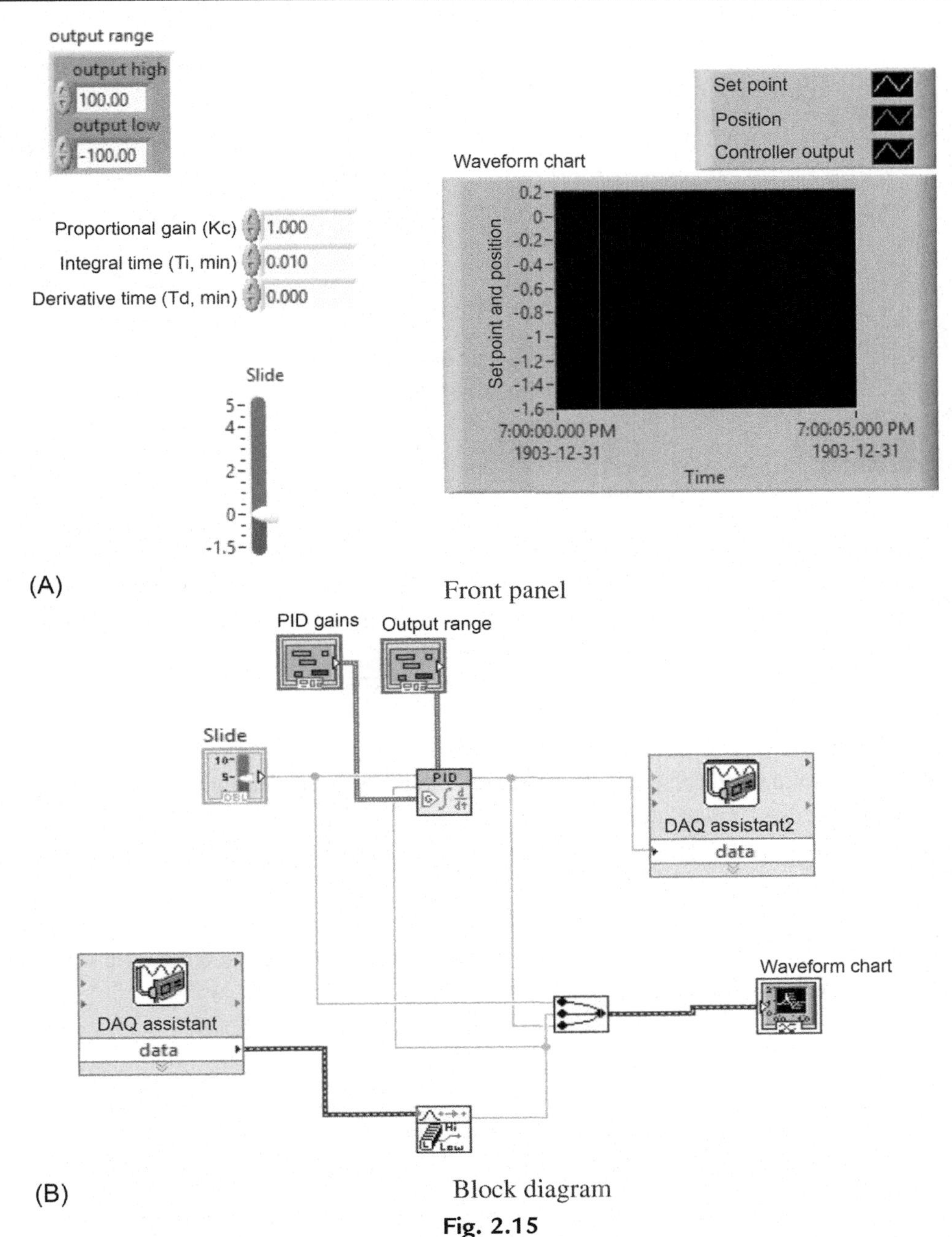

Fig. 2.15
The Front panel (top) and the Block Diagram (bottom) LabVIEW program.

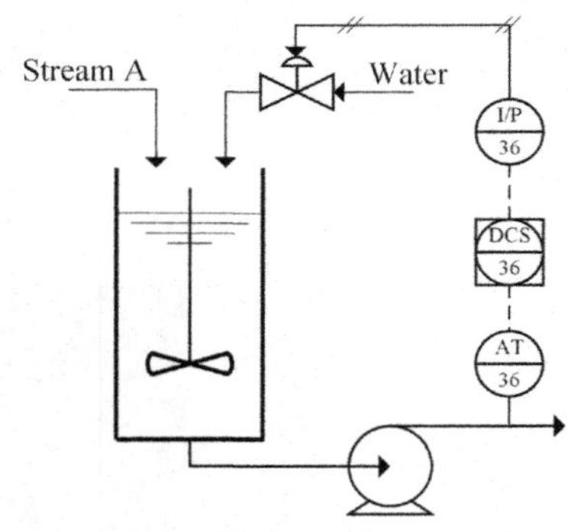

Fig. 2.16
The implementation of a simple feedback control law in a DCS system.

2.8 Problems

(1) A pressure sensor/transmitter has an input range of 200–1000 kPa and an output range of 4–20 mA, respectively. The output reading of the sensor/transmitter at a pressure of 650 kPa (measured with an accurate sensor without any error) is 12.6 mA. Obtain the following quantities for the sensor/transmitter:
  (a) The zero, input span, and output span of the sensor/transmitter.
  (b) The gain of the sensor/transmitter.
  (c) The resolution of the sensor/transmitter.
  (d) The absolute and relative errors (in percentage of the maximum pressure, 1000 kPa) of the sensor/transmitter.

(2) For a 12-bit A/D converter, obtain the resolution of an A/D converter in terms of mA. The input signal has a range of 4-20 mA.

(3) Write a LabVIEW program to acquire an analog signal over the range of 4–20 mA, configure a PID controller to generate an output signal, and send out the output signal as an analog signal over the range of 4–20 mA to the plant.

References

1. Soloman S. *Sensors handbook.* New York: McGraw-Hill; 1999.
2. Liptak BG, editor. *Instrument engineer handbook.* 3rd ed. vol. 1. Philadelphia, PA: Chilton Books; 2000 [Process measurement].
3. Shuller MP, Kargi F. *Bioprocess engineering.* Upper Saddle River, NJ: Prentice-Hall; 2002.
4. National Instrument, http://www.ni.com/en-ca.html.

CHAPTER 3

Theoretical Process Dynamic Modeling

Sohrab Rohani, Yuanyi Wu
Western University, London, ON, Canada

There are two approaches to develop a dynamic model for a given process:

Theoretical approach: This approach involves the use of first principles based on the conservation of mass, energy, and momentum, together with the necessary auxiliary equations representing mass transfer, heat transfer, momentum transfer, phase equilibrium, kinetics, etc. The resulting models are nonlinear, relating the input and output process variables. Such models are useful for process simulation, process optimization, and design but are seldom used in the design of control systems.

The detailed nonlinear dynamic models can be linearized to develop simple input-output or cause-and-effect dynamic models in the time domain (linear state-space models) or in the Laplace domain (transfer function models). These linear models are approximation of the detailed nonlinear models but are useful for the design of controllers.

Empirical approach: The second approach that is often used in the design of control systems for complex processes is based on the use of measured input and output data of an existing process. Statistical methods such as the regression and time series analyses are used to fit the input-output data to linear models between the input and output variables. This approach is referred to as Process Identification technique.

In this chapter, we shall cover the detailed dynamic modeling using a theoretical approach. In this chapter we shall deal with the linearization of the detailed models and the development of linear state-space models and transfer functions. At the end of this chapter, we shall study the process identification technique.

3.1 Detailed Theoretical Dynamic Modeling

A systematic procedure is adopted in this book to develop the dynamic models of chemical processes. This procedure involves eight steps.

(1) Define the system, its surroundings, its inputs and outputs. Divide the system into appropriate subsystems to simplify the modeling effort. For example, a reactor that is equipped with a cooling/heating jacket is often divided into two subsystems: the

Coulson and Richardson's Chemical Engineering. http://dx.doi.org/10.1016/B978-0-08-101095-2.00003-5

reacting mixture inside the reactor and the cooling/heating jacket. In a multiphase system having vapor, liquid, and solid phases the appropriate conservation equations are applied to each phase separately. In a tray distillation column, the conservation principles are applied to each tray and each phase (vapor and liquid) separately.
In this step, the general control objectives for the entire system are reduced to specific controller equations in terms of the specific controlled variables.

(2) Use reasonable assumptions to simplify the model. No model is exact; the best model is the simplest model that works.

(3) Based on the simplifying assumptions, determine whether the system *is* or *can be approximated* by a *lumped parameter system*, a *stage-wise process*, or a *distributed parameter system*. In a lumped parameter system, the process variables do not have any position or spatial dependence, the process is assumed to be homogeneous. For example, in a *well-mixed* reactor, the process variables such as the reactor temperature and species concentrations are only a function of time, $T=T(t)$; $C=C(t)$, and not a function of position in the reactor. A stage-wise process consisting of many well-mixed homogenous stages can be modeled as a series of lumped parameter systems. Examples of stage-wise processes are a tray distillation column, a series of continuous stirred tank reactors (CSTRs), or a cascade of evaporators. In a distributed parameter system, however, the process variables are position dependent, e.g., in a heat exchanger, the temperature varies with the position along the length of the heat exchanger, $T=T(t,x)$, where x is the axial dimension along the length of the heat exchanger. Other examples of distributed parameter systems are a tubular fixed-bed reactor and a packed distillation or absorption column.

(4) Apply the relevant conservation principles, i.e., the total mass balance, component mass balance, energy balance, momentum balance, and population balance, to the system or subsystems under consideration. These principles must be applied to the entire system or subsystems in the case of a lumped parameter system. For a stage-wise process, the conservation principles are applied to a typical stage, and in a distributed parameter system, the conservation principles are applied to an infinitesimal control volume within the system.

Rate of accumulation of a quantity	$=$	Rate of inflow of the quantity into the system by molecular and bulk motions	$-$	Rate of outflow of the quantity from the system by molecular and bulk motions	$\pm$	Rate of generation or consumption of the quantity within the system

Note that the *quantity* in the previous equation may refer to the component mass/mole or the population density of a particulate entity. In the case of the total mass and energy balance the generation/consumption term is not needed.
In the previous equation, if the rate of accumulation is zero, a steady-state model is obtained, which renders the expected steady-state conditions for a given set of parameters.

The resulting conservation principles for a lumped parameter system and a stage-wise process in a dynamic model will be a set of ordinary differential equations (ODSs) and for a distributed parameter system will be a set of partial differential equations (PDEs).

(5) Apart from the conservation principles, other auxiliary equations describing the transport processes due to the convection and diffusion; phase and chemical equilibria and reaction kinetics must be also included to reduce the degrees of the freedom of the system.

(a) Transport phenomena

- Heat transfer
  - Conduction (molecular transfer in terms of thermal conductivity)
  - Convection (bulk transfer in terms of the heat transfer coefficient)
  - Radiation
- Mass transfer
  - Molecular diffusion
  - Bulk motion (in terms of the mass transfer coefficient)
- Momentum transfer
  - Molecular level (viscous force)
  - Bulk level (friction factor)

(b) Equations of state

Thermodynamics provides relationships to calculate the density, enthalpy, and other physical properties of various streams in terms of state variables such as the temperature and pressure.

- Density $= \rho =$ function $(T, P, \text{conc.})$
- Enthalpy $= H =$ function $(T, P, \rho, \text{conc.})$

  Phase and chemical equilibria provide additional equations. For example, the relationship between the concentration of a component in the liquid phase and its partial pressure in the vapor phase (Raoult's Law) can be represented by thermodynamics.
- Thermodynamics
  - Phase equilibrium
  - Chemical equilibrium

(c) Kinetics

The rate of a chemical reaction (chemical kinetics) depends on the speed of the reaction that is determined by the species concentration, reaction temperature, pressure, and catalyst concentration.

(6) Check the model consistency which manifests itself in two requirements. First, the units of each term appearing in an equation must be the same. Second, in order to obtain a unique solution to a model, the degrees of freedom of the model must be zero. The degree of freedom of a system is defined as the number of variables minus the number of equations and the number of disturbances for which specific values are assigned. The equations include the *state equations* which result from the

conservation principles, the *auxiliary equation*, and the *control equations* that are the relationships between the *manipulated variables* and the *controlled variables* through the *control law*.

$$N_f = N_V - N_E - N_d - N_c \tag{3.1}$$

Degrees of freedom = Number of variables
– Number of equations (*state equations* and *auxiliary equations*)
– Number of disturbances
– Number of necessary control equations to make the degrees of freedom zero

The *controller* equations translate the control objectives to individual equations relating the manipulated and controlled variables through the control law. Once the required number of *controller equations* has been determined, a *process block diagram* can be drawn that clearly demonstrates the *manipulated variables*, the *disturbances*, and the *controlled variables*.

(7) The mathematical model developed must be solved to provide the trends and the quantitative relationships between the input and output variables. If the model is nonlinear and complex, only a numerical approach can be used to solve the model. In certain limited cases, if the model turns out to be linear, an analytical method can be used to solve the model. The numerical solution of the process dynamic model requires the solution of ODEs or PDEs. Therefore before considering examples of each category of systems, a brief description of a powerful numerical simulator, MATLAB and Simulink, will be provided.

(8) In order to verify the validity of the model, *theoretical* results or model predictions must be compared with the *experimental* observations. Once this has been verified, the developed model can be used to *predict* the process behavior. No model is validated without demonstrating that it has the required *predictive capability*.

3.2 Solving an ODE or a Set of ODEs

There are different algorithms for the solution of a nonlinear ODE or a set of ODEs and PDEs.[1,2] There exist also a number of powerful software packages that facilitate the solution of differential equations. Among them is MATLAB and Simulink (a graphical equation solver in MATLAB).

3.2.1 Solving a Linear or a Nonlinear Differential Equation in MATLAB

Within MATLAB, there are a number of ODE solvers. The list includes solvers for both nonstiff and stiff differential equations.

Nonstiff: ode45, ode23, ode113s.
Stiff: ode15s, ode23s.

Although there are different ways to solve differential equations in MATLAB, one simple approach is to create a new "function" which is a script file, and define the differential equations in that "function." The "function" is then saved in the "working directory" of the MATLAB. From the "command window" of MATLAB, the required commands are issued to solve the differential equations. In order to establish proper communication between the "command window" and the "function" file, it is necessary to define all the variables and parameters as "global" variables.

Example 3.1

Solve the following set of ODES over the interval 0–20 s.

$$\begin{cases} \dfrac{dy_1}{dt} = 0.2 \cdot \exp(-3 \cdot t) - 0.1 \cdot y_3 & ; y_1(0) = 1 \\ \dfrac{dy_2}{dt} = 0.6 \cdot \exp(-t \cdot y_2) & ; y_2(0) = 0.5 \\ \dfrac{dy_3}{dt} = y_1 \cdot y_3 & ; y_3(0) = 1 \end{cases} \tag{3.2}$$

Create a "function" in the "Work directory" of MATLAB called "threeodes" in define the ODEs:

```
ODE Function File: threeodes.m

function dy = threeodes(t,y)
    dy = zeros(3,1);    % a column vector to store the results
    dy(1) = 0.2*exp(-3*t)-0.1*y(3);
    dy(2) = 0.6*exp(-t*y(2));
    dy(3) = y(1)*y(3);
```

In the MATLAB "command window" issue the following commands:

```
Run commands:

% The option statement sets the relative and absolute error for the ode solver
options = odeset('RelTol',1e-4,'AbsTol',[1e-4 1e-4 1e-5]);

% [0 20] is the integration span time interval, [1 0.5 1] specifies the initial conditions
    of y1, y2 and y3, respectively, ode45 is an ode solver
[t,y] = ode45(@threeodes,[0 20],[1 0.5 1],options);

% plot y1 (solid line), y2 (dash-dot line) and y3 (dash line)
plot(t,y(:,1),'-',t,y(:,2),'-.',t,y(:,3),'--');

% create the legend and axis label.
legend('y_1','y_2','y_3');
xlabel('time (second)');
```

The solution to the set of the three ODEs is given in Fig. 3.1

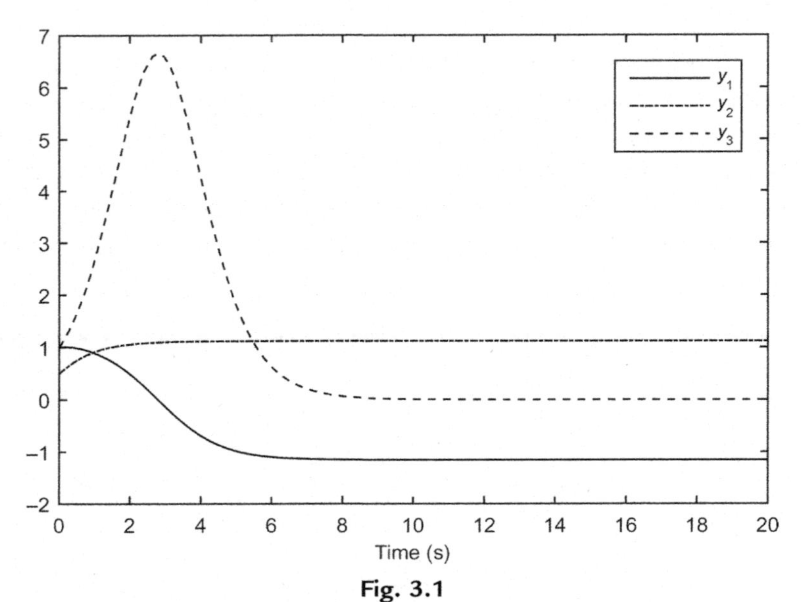

Fig. 3.1
The solution of the set of three ODEs given in Eq. (3.2).

3.2.2 Solving a Linear or a Nonlinear Differential Equation on Simulink

(1) On a sheet of paper, rearrange the given differential equation to have the highest derivative on the left-hand side of the equation.
(2) Open Simulink, either from the menu bar or by typing "Simulink" in the MATLAB "Command Window".
(3) Open a new Simulink model file.
(4) Open the Simulink library.
(5) Using various blocks from the Simulink library such as an Integrator, an Add, a Product/ Divide blocks; construct the differential equation in the Simulink model window. Most of these blocks can be found in the "Most Commonly Used" block, the "Continuous" block, and the "Math Operations" block.
(6) In order to show the results, one can use a "Scope," "Simout," or an "X-Y Graph," all available in the "Sinks" library. "Constants" and "Step Change" are in the "Source"

library. If Simout is used, the simulation results will be stored in working space of MATLAB and a plot can be generated by issuing a `plot(y)` command in the "Command window" of MATLAB.

Example 3.2

Solve the following ODE on Simulink:

$$\frac{d^2y}{dt^2} + 1.4\frac{dy}{dt} + 2y = -3; \; y(0) = 0; \; \left.\frac{dy}{dt}\right|_{t=0} = 2 \qquad (3.3)$$

Rearrange the equation so that the highest order of derivative is on the left-hand side of the equation:

$$\frac{d^2y}{dt^2} = -1.4\frac{dy}{dt} - 2y - 3 \qquad (3.4)$$

Assuming that the second derivative signal is available, integrate that signal as many times as needed (in the case of a second-order ODE, two successive integrations are needed) to obtain the function "*y*." Note that the initial conditions of the function and its first derivative must be specified in the integrator blocks either "internally" or "externally." Fig. 3.2 shows the solution of the previous ODE in the Simulink environment, and Fig. 3.3 depicts the variations of "*y*" with respect to time, the solution of the ODE. The solution of the ODE can be monitored on the "Scope," or on the "To Workspace" by passing the "*y*" and "time" vectors to the "Workspace" of MATLAB and then plot the results in the "Command Window."

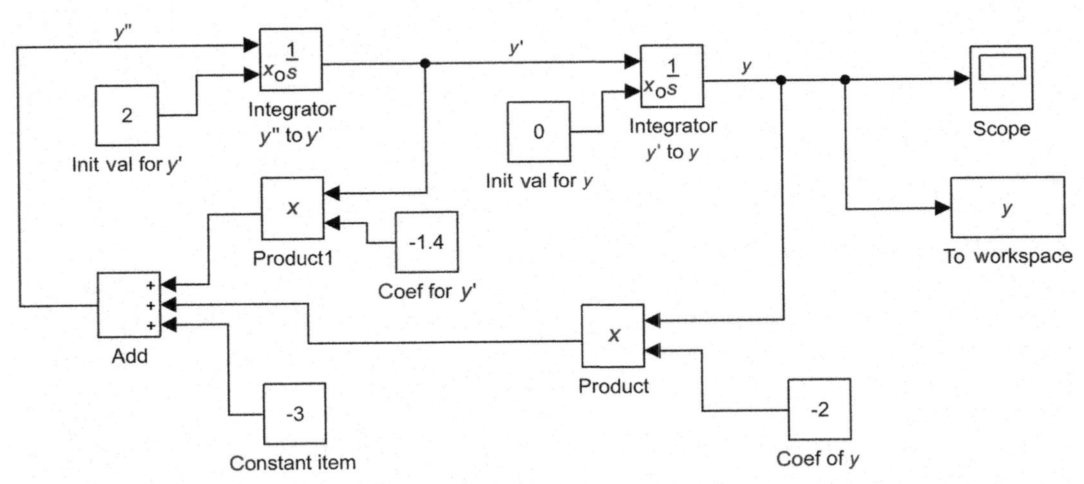

Fig. 3.2
Simulink program for the solution of Eq. (3.3).

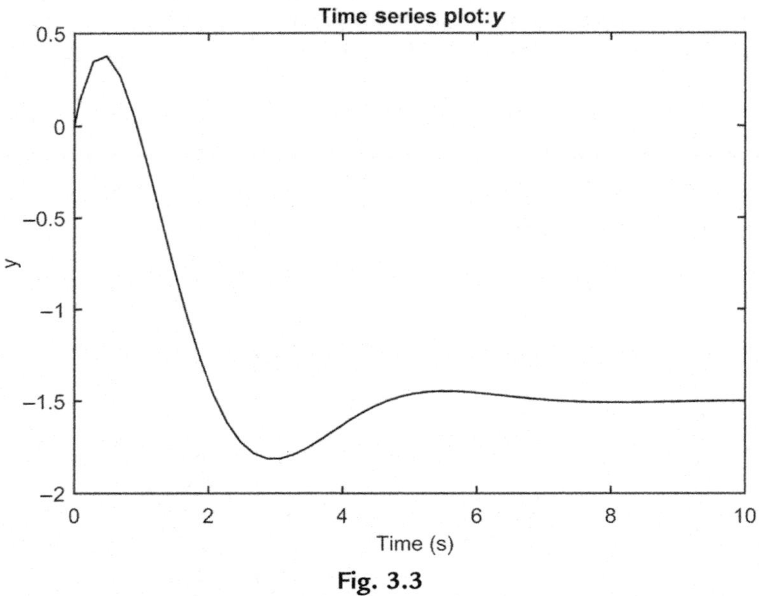

Fig. 3.3
Solution of the ODE given in Eq. (3.3).

In what follows, a series of examples belonging to the lumped parameter systems, stage-wise processes, and distributed parameter systems will be considered and the earlier systematic eight-step procedure will be applied (except the final step which requires experimental data). In a few examples the numerical solution by MATLAB is not provided to save space.

3.3 Examples of Lumped Parameter Systems

3.3.1 A Surge Tank With Level Control

Develop a detailed dynamic mathematical model of a liquid storage tank shown in Fig. 3.4 both for the open-loop case and for the closed-loop case where the liquid level is controlled by a controller. The parameters of the simulation are given in Table 3.1.

The control objective is to maintain the level of the liquid in the tank at l_{sp}. Write a MATLAB program to solve the open-loop and the closed-loop system using a proportional P and a PI controller. In the open-loop case, investigate the effect of various process parameters such as the tank cross-sectional area, the control valve size, and the changes in the inlet flow rate of the liquid on the level in the tank. For the closed-loop simulation, use the velocity form of the discrete PI-control law, discussed in Chapter 1.

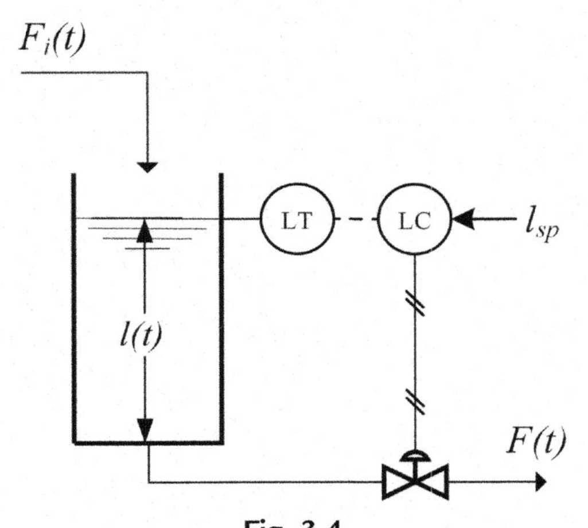

Fig. 3.4
Surge tank with a level controller.

Table 3.1 Parameters for the tank with level control

Symbol	Description	Units	Nominal Data	Data Value or Range
A	Tank cross-sectional area	m^2	3	1.5–6
F_i	Inlet flow rate	m^3/s	0.2	0.1–0.4
C_v	Valve constant	$m^{2.5}/s$	0.4	0.2–0.8
$\bar{x}$	Controller bias, steady-state valve opening		0.5	–
K_c	Controller proportional gain		0.5	0.2–1
τ_I	Controller reset time	s	10	
Δt	Sampling/control interval ($\Delta t_s = \Delta t_c = \Delta t$)	s	1	–
$l_0 = \bar{l}$	Initial liquid level, steady-state level	m	1	–

The eight-step systematic approach is used to develop and solve the dynamic mathematical model.

Step 1: The system is the surge tank. The inlet and outlet flow rates are F_i and F (m^3/s), respectively.
Step 2: Constant density and constant tank cross-sectional area, A, are the main assumptions.
Step 3: The system is a lumped-parameter homogeneous process. There are no variations of temperature, concentration, etc., with position in the tank.
Step 4: The only relevant conservation equation is the total mass of the liquid in the tank:

$$\begin{aligned} \rho A \frac{dl(t)}{dt} &= \rho F_i(t) - \rho F(t) \\ A \frac{dl(t)}{dt} &= F_i(t) - F(t) \end{aligned} \tag{3.5}$$

Step 5: Either the inlet or the outlet flow rate can be used as the manipulated variable to control the liquid level in the tank. If the effluent stream is chosen as the manipulated variable, the flow through the control valve installed on the effluent stream depends on the valve stem position (the valve's opening), $x(t)$; the pressure drop across the valve, ΔP; and the specific gravity of the liquid trough the valve, $s.g.$

$$F(t) = C_1 \cdot x(t)\sqrt{\Delta P/(s.g.)} \quad (3.6)$$

The pressure drop across the valve depends on the hydrostatic head or the liquid level in the tank, $l(t)$.

$$F(t) = C_v \cdot x(t)\sqrt{l(t)} \quad (3.7)$$

where C_v is the valve constant (a parameter representing the size of the valve has to be specified to solve the problem). The control signal from the controller will move the valve stem position (the valve's opening) and causes a change in the outlet flow rate. Therefore the manipulated variable is $x(t)$ or $F(t)$ as shown in Fig. 3.5.
Step 6: The units of each term in Eqs. (3.5), (3.6) must be checked for consistency. Each term in Eq. (3.5) has the units of m^3/s. In Eq. (3.6), $x(t)$ is unit-less, l is in m, therefore C_v must have a unit of $m^{2.5}/s$.
Degrees of freedom analysis:

Number of variables	4	$F_i(t)$, $F(t)$, $x(t)$, $l(t)$
Disturbance	1	$F_i(t)$

There is one state equation (the total mass balance), and one auxiliary equation, Eq. (3.6), therefore there must be one control equation to reduce the degrees of freedom to zero and ensure a unique solution:

Control equation P-control $x_n = \bar{x} - K_c e_n$ (3.8)

PI-control $x_n = x_{n-1} - K_c(e_n - e_{n-1}) - \left(\frac{K_c}{\tau_I}\right) \cdot e_n \cdot \Delta t$ (3.9)

Where $e_n = l_{sp} - l_n$, is the error signal at $t = n \cdot \Delta t$, i.e., at the nth sampling interval. l_n is the measured liquid level in the tank at the nth sampling interval. K_c and τ_I are

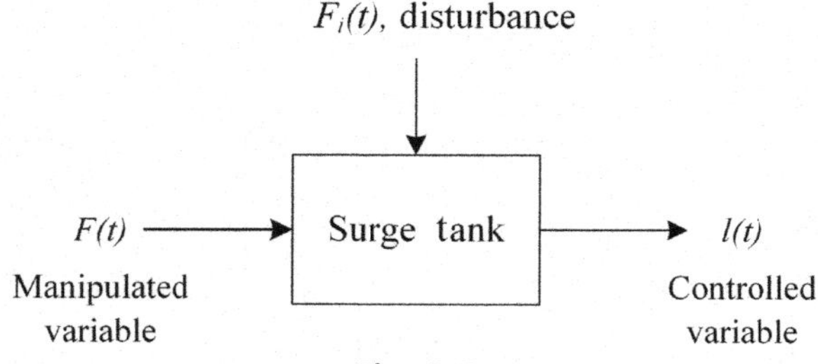

Fig. 3.5
The block diagram of the surge tank level control system.

the controller proportional gain and the integral time. Note that the sign of K_c must be negative for an air-to-open control valve. If the measured level is larger than the set point level, $x(t)$ and hence $F(t)$ must increase to let more liquid leave the tank to bring the level to its set point value.

Step 7: In order to solve the dynamic model in the open- and closed-loop control cases, MATLAB and Simulink will be used. Details of simulation in each case are presented for this example.

(a) The open-loop simulation in MATLAB

The open-loop simulation in MATLAB is conducted numerically. In the example code, the created function OpenLoopODE uses the measured level and parameters C_v, x, Fi, A to solve the dynamic model of the process.

ODE function file: OpenLoopODE.m

```
function [ dlevel ] = OpenLoopODE( time, level, Cv, x, Fi, A)
    % This function is used to simulate an open-loop surge tank.
    % time: simulation time (s)
    % level: liquid level in the tank (m)
    % Cv: valve coefficient (m^2.5/s)
    % x  : valve opening (0 to 1)
    % Fi: inlet flowrate (m^3/s)
    % A  : tank cross sectional area (m^2)

    %1. Update the auxiliary equation Equation3.6.
    F = Cv * x * sqrt(level);

    %2. Update the system state using Equation3.7.
    dlevel = (Fi - F) / A;
end
```

The next step is to call an ODE solver. The example script code demonstrates how to use "ode45" to simulate the system to change a parameter (e.g., the cross-sectional area of the tank) to observe its effect on the dynamics of the system output (liquid level). The effect of the system parameters such as the valve size coefficient, C_v, and the disturbances (the inlet flow rate), F_i, can be tested similarly.

Script file for open-loop simulation:

```
clear;
%Assign parameters
A=3; Cv=0.4; x=0.5; Fi=0.2;
%Initialize the liquid level, starting from an empty tank
l_init=0;
%Specify the simulation time span, from 0 to 200 s
tspan=[0 200];
```

```
%ODE solver options, leave it blank for default values
options=[];

%Call ode45 solver.
[time,level]=ode45(@OpenLoopODE, tspan, l_init, options, Cv, x, Fi, A);

%change the cross-sectional area of tank and recalculate the results
A=1.5;
[t_SmallArea,l_SmallArea]=ode45(@OpenLoopODE,tspan,l_init,options,Cv,x,Fi,A);
A=6;
[t_LargeArea,l_LargeArea]=ode45(@OpenLoopODE,tspan,l_init,options,Cv,x,Fi,A);

%plot the results in one figure. Use 'hold on' to prevent the figure from
%being overwritten.
hold on;
plot(t_SmallArea, l_SmallArea);
plot(time,level,'--');
plot(t_LargeArea, l_LargeArea);
hold off;
%Use legend to label the results
legend('Reference A=1.5', 'A=3', 'A=6');
```

Fig. 3.6 shows that the tank cross-sectional area does not affect the steady-state value of the liquid level but the speed to reach the steady state increases as the area

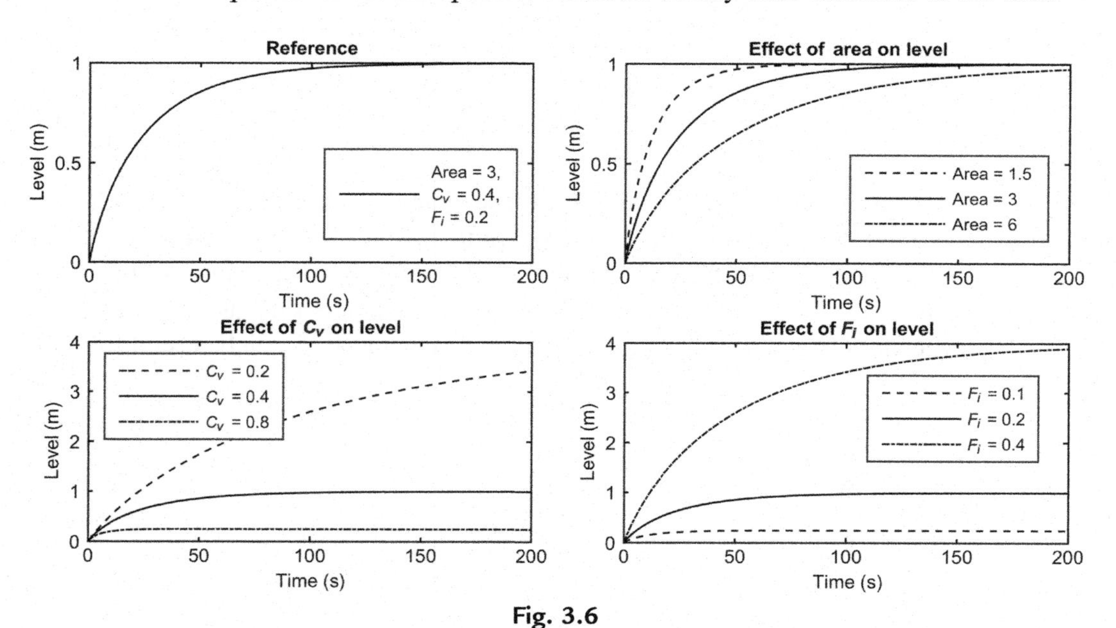

Fig. 3.6
Open-loop simulation results and the effect of various parameters on the liquid level dynamics. Reference curve is plotted with parameter set $C_v = 0.2$, $F_i = 0.1$, area $= 3$.

decreases. The valve coefficient affects both the steady-state value and the speed of response. As C_v increases (a larger valve) the response time decreases (faster acting valve) and the steady-state value of the liquid level decreases. As the inlet flow rate increases the final steady-state value of the liquid level increases but the speed of the response decreases.

(b) The closed-loop simulation with MATLAB

In the closed-loop simulation, a feedback controller (PID controller) is used to control the manipulated variable (x) based on the measured output variable (l). This can be achieved by modifying the open-loop ODE *function*.

The ODE *function* code with a P controller is given later. The argument list includes all the parameters required by the open-loop simulation expect the manipulated variable, x, which is calculated by the control law given in Eq. (3.3). The controller proportional gain K_c, the level set point, l_{sp}, and the controller bias, $\bar{x}$, must be specified. It is assumed that the system is at steady state before simulation starts. Thus the controller bias $\bar{x}$ can be calculated by setting $dl/dt = 0$ in Eq. (3.5). The time 0^- represents the time just before simulation starts.

$$F_i(0^-) = C_v \cdot x(0^-)\sqrt{l(0^-)} \tag{3.10}$$

$$x(0^-) = \frac{F_i(0^-)}{C_v\sqrt{l(0^-)}} \tag{3.11}$$

By substituting the given conditions the controller bias is calculated to be $x(0^-) = 0.5$.

The closed loop simulation with a P-Controller: ClosedLoopODEP.m

```
function [ dlevel ] = ClosedLoopODEP( time, level, Kc, l_sp, x_bar, Cv, Fi, A )
    % Calculate the error signal
    e = l_sp - level;

    % update the manipulated variable using the P-controller law.
    x = x_bar - Kc * e;
    % The valve position is between 0 and 1 (closed and open valve).
    % x = min(1, max(0, x));
    if x > 1
        x = 1;
    end
    if x<0
        x = 0;
    end
    % Call open loop function to calculate the derivative of the level.
    dlevel = OpenLoopODE(time, level, Cv, x, Fi, A);
end
```

Script file for the closed loop simulation with a P-controller:

```
clear;
%Assign parameters
A = 3; Cv = 0.4; x = 0.5; Fi = 0.2 ; l_init = 1;

%Assign the controller parameters, Kc is negative as discussed before. The
%set point of the level is 5 m.
Kc = 0.2; l_sp = 5; x_bar = 0.5;

Options = [];
%Simulation time span
TimeSpan = [0 200];

%simulate and plot the results
[t,y]=ode45(@ClosedLoopODEP,TimeSpan,l_init,Options,Kc,l_sp,x_bar ,Cv,Fi,A);
plot(t,y);
xlabel('Time (second)');
ylabel('Level (m)');
```

The next set of codes simulate a PI controller. As discussed in Chapter 1, the *position form* of the PID controller requires storing the past error signals. However, the *velocity form* of the PID controller only requires the controller output at the last time instant and the two past error signals. At each sampling interval Δt_s, the controller received one sample (the measured liquid level) and calculates an impulse signal which is the manipulated variable every Δt_c. With a *zero order hold* device, the impulse is latched until next impulse is calculated. For simplicity, the sampling interval Δt_s is taken to be equal to the control interval Δt_c. Fig. 3.7 shows the behavior of a discrete controller that samples a continuous measured signal and holds a discrete command signal. In the simulation, a condition syntax is used to keep the previous x value to simulate a *hold device*.

Closed loop with a PI-Controller ODE function: ClosedLoopODEPI.m

```
function [dlevel] = ClosedLoopODEPI(time,level,Kc,Tau_I,DeltaT,l_sp,x_bar,Cv,
Fi,A)
    % Calculate the error signal
    e = l_sp - level;

    % load the previous manipulated variable, the error signal and the controller
    % output. If these variables are empty, initialize them.
    global e_previous t_previous x_previous;

    if isempty(e_previous)
        e_previous = e;
    end
```

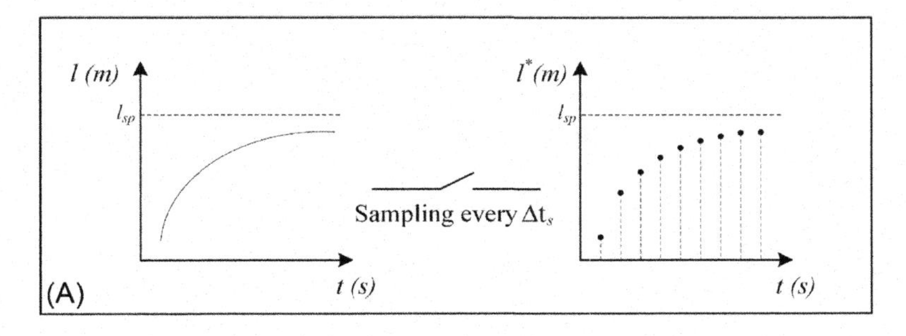

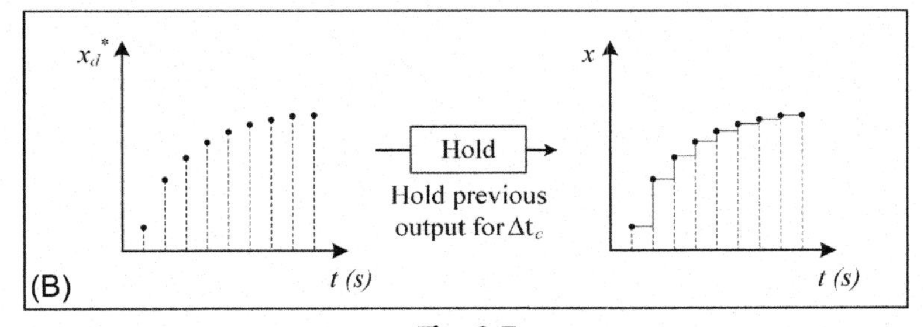

Fig. 3.7
(A) Sampling of a continuous measured signal and (B) D/A conversion and holding of the manipulated variable.

```
if isempty(t_previous)
   t_previous = 0;
end
if isempty(x_previous)
   x_previous = x_bar;
end

% Use the control law to update the manipulated variable
% Ensure the sampling/control interval is 'DeltaT'.
if time - t_previous > DeltaT
% If the elapsed time since last sampling/control is greater than DeltaT,
% measure and calculate new x,
% triple dots (...) is the line-continuation syntax.
    x = x_previous ...
           - Kc * (e - e_previous) ...
           - Kc / Tau_I * e * DeltaT;
% Set saturation for the valve position. (Anti reset-windup)
if x > 1
    x = 1;
end
```

```
        if x < 0
            x = 0;
        end

        % Store the time, the error signal, and the controller output to global var.
        % so that these variable can be reloaded next time entering this ODE function.
            t_previous = time;
            e_previous = e;
            x_previous = x;
        else
            % The interval since last sampling/control is still within DeltaT.
            % Controller does not sample or update the valve position.
            % The last valve position is maintained to simulate a zero order hold device.
            x = x_previous;
        end

        % Call the open loop function to calculate the derivative of the level.
        dlevel = OpenLoopODE(time, level, Cv, x, Fi, A);
end
```

A fixed-step Euler ODE solver is provided later to simulate this case, instead of using an embedded ODE solver within MATLAB. Make sure the following code ode1.m is in the same directory as the ClosedLoopODEPI.m, OpenLoopODE.m.

Fixed Step Euler ODE Solver: ode1.m

```
function [t,y] = ode1(dydt,tspan,y0,h,varargin)
    ti = tspan(1);
    tf = tspan(2);
    t = (ti:h:tf )';
    n = length(t);

    if t(n)<tf
      t(n+1) = tf;
      n = n+1;
    end

    for i=1:length(y0)
        y(i,:) = y0(i)*ones(1,n);
    end

    for i = 1:n-1
        y(:,i+1) = y(:,i) + dydt(t(i),y(:,i),varargin{:})*(t(i+1)-t(i));
    end
    y=y';
end
```

The script for simulating the PI controller with the Euler algorithm is provided as follows.

Script for closed loop simulation with the PI-controller:

```
%Clear the global variables
clear global;

%Assign parameters
A = 3;
Cv = 0.4;
x = 0.5;
Fi = 0.2 ;
l_init = 1;

%Assign controller parameters
Kc = 0.2;
l_sp = 5;
x_bar = 0.5;
Tau_I = 10;
DeltaT = 1;

Options = [];

%ODE Solver computational step size
Step = 0.1;
%Simulation time span
TimeSpan = [0 300];

%Simulate and plot the results. Note that Euler ODE1 solver is used.
[t,y]=ode1(@ClosedLoopODEPI,TimeSpan,l_init ,Step ,Kc, Tau_I, DeltaT, l_sp,
x_bar , Cv, Fi, A);
plot(t,y);
xlabel('Time (second)');
ylabel('Level (m)');
```

Figs. 3.8 and 3.9 show the closed-loop response using the P and PI controllers with different controller parameters.

In the P controller simulation, the set point level is 5 m. However, in Fig. 3.8 there is always an *offset* from the set point. The offset is defined as the difference between the set point and the ultimate value (the steady-state value) of the controlled variable. The offset decreases as the proportional gain K_c of the controller increases but cannot be eliminated using a pure P controller. The three curves overlap because of the saturation of the valve (fully open or fully closed condition). If there is no saturation, greater K_c will lead to faster dynamics.

In Fig. 3.9, the level approaches the set point (5 m) at steady state, i.e., in the presence of integral action the offset is eliminated. One may notice that there is a large

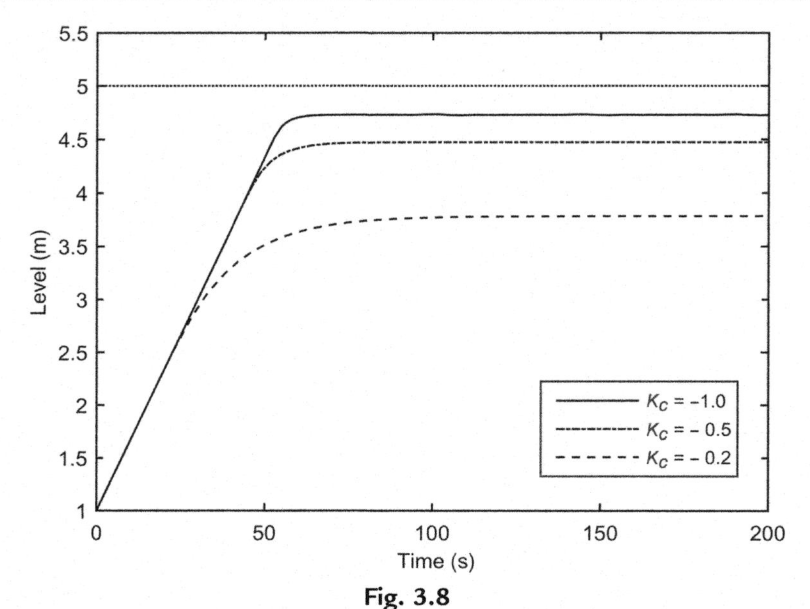

Fig. 3.8
Simulation results of surge tank with a P controller.

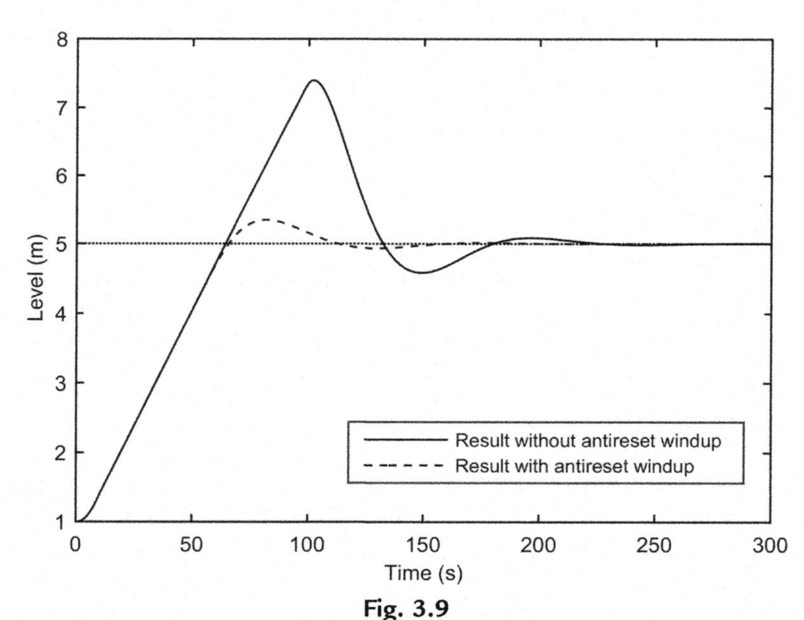

Fig. 3.9
Simulation results of the surge tank with a PI controller.

overshoot in the level around 100 s. This overshoot is caused by the reset windup, which causes the controller action to hit the saturation in the presence of a lasting error signal. Ultimately, the reset windup causes valve saturation. The industrial controllers are often equipped with an antireset windup mechanism which may be implemented using a hardware (in analog controllers) module or a software code (in digital controllers). In the code, a digital "antirest windup" is incorporated in file "ClosedLoopODEPI.m." With an antireset windup the overshoot will be lower. The integral- or reset-windup and antireset windup will be discussed in more detail in Chapter 4.

(c) Simulation of the open-loop surge tank on Simulink
The governing dynamic equation including the auxiliary equation describing the process is given by Eq. (3.12).

$$A\frac{dl(t)}{dt}=F_i(t)-C_v x(t)\sqrt{l(t)} \tag{3.12}$$

Fig. 3.10 represents the open-loop (for a fixed value of $x(t)=0.5$) simulation of the system on Simulink.

(d) Simulation of the closed-loop surge tank on Simulink
For the closed-loop simulation (in which the value of $x(t)$ is calculated by the controller), we use a "Subsystem" from the "Commonly Used Blocks" in the Simulink library to represent the surge tank, as is shown in Figs. 3.11 and 3.14, for the P- and PI-controller simulations, respectively. One of the inputs to the Subsystem, $x(t)$, which is the valve position, changing between 0 (closed) and 1 (fully open position), is used as the manipulated variable. The P- or PI-block receives the measured liquid level, $l(t)$, as the controlled variable, and calculates the valve position, $x(t)$, as the manipulated variable. Please note that the "Subsystem" in Fig. 3.12 is the same as described in Fig. 3.10 except for the value of $x(t)$ which in the closed-loop simulation is calculated by the P- or PI-controller subsystems shown in Figs. 3.13 and 3.15, instead of having a fixed value as was the case in the open-loop simulation.
Simulink provides an embedded PID block with adjustable parameters, saturation, and antiwindup configurations. However, the PID block is a transfer function which

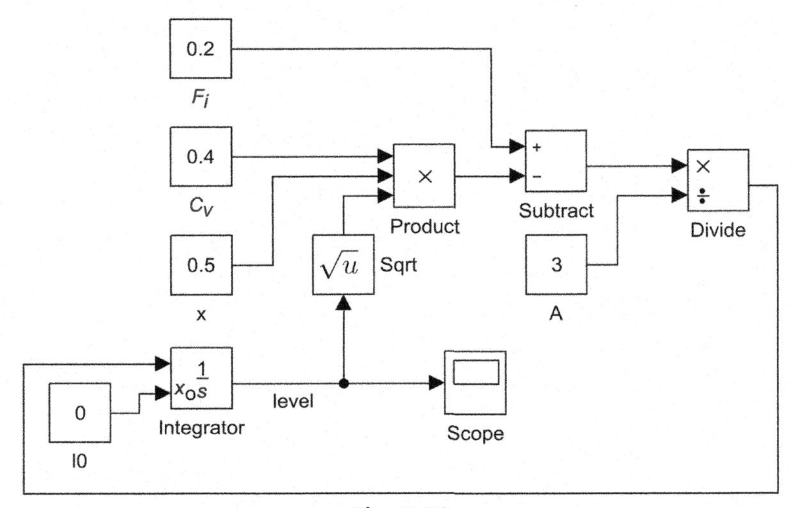

Fig. 3.10
The open-loop simulation of the surge tank on Simulink.

uses deviations variables to remove constant terms including the controller bias. Figs. 3.12 and 3.15 depict two subsystems, simulating a P- and a PI-controller. The simulation results with the P- and PI-controllers given in Figs. 3.11 and 3.14 are identical with those presented in Figs. 3.8 and 3.9 generated by MATLAB codes.
P-controller simulation:

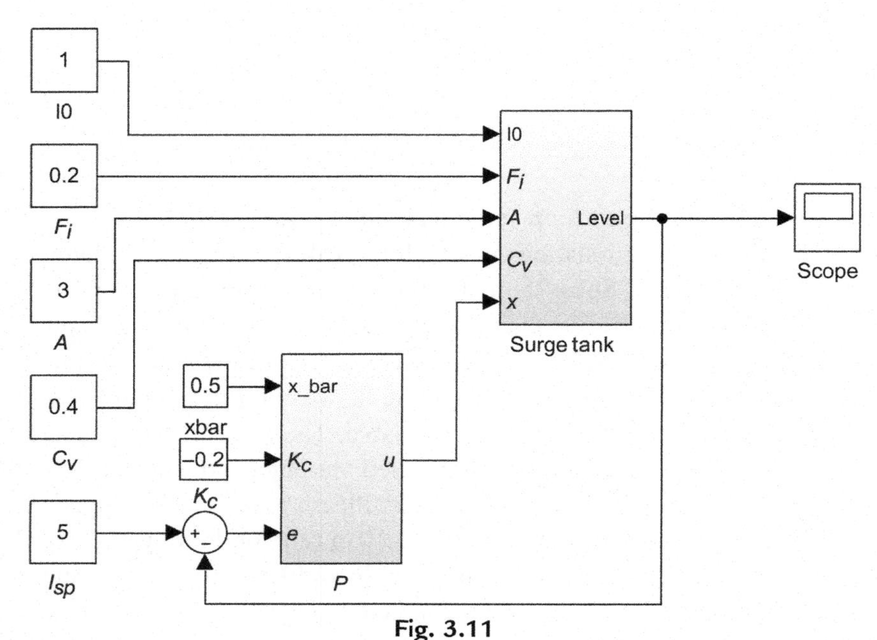

Fig. 3.11
Overall Simulink block diagram for simulation with a P controller.

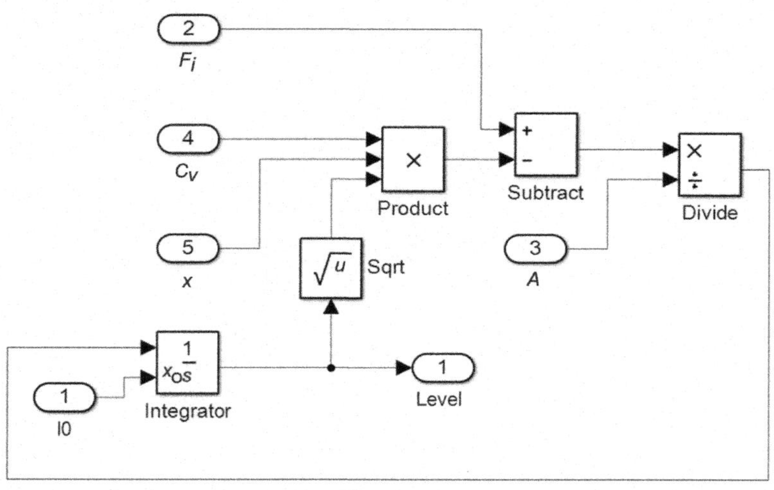

Fig. 3.12
Subsystem surge tank.

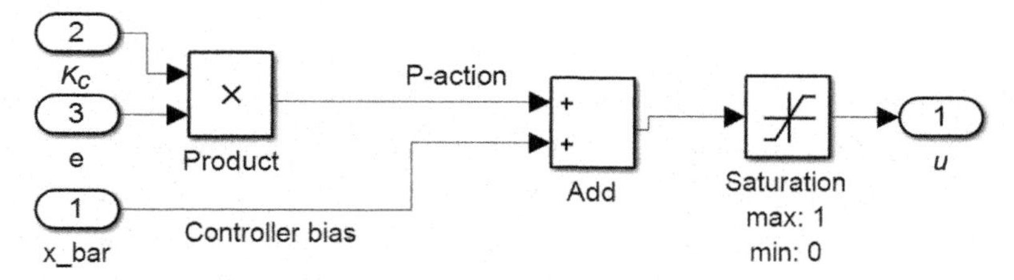

Fig. 3.13
Subsystem P controller.

PI-controller simulation:

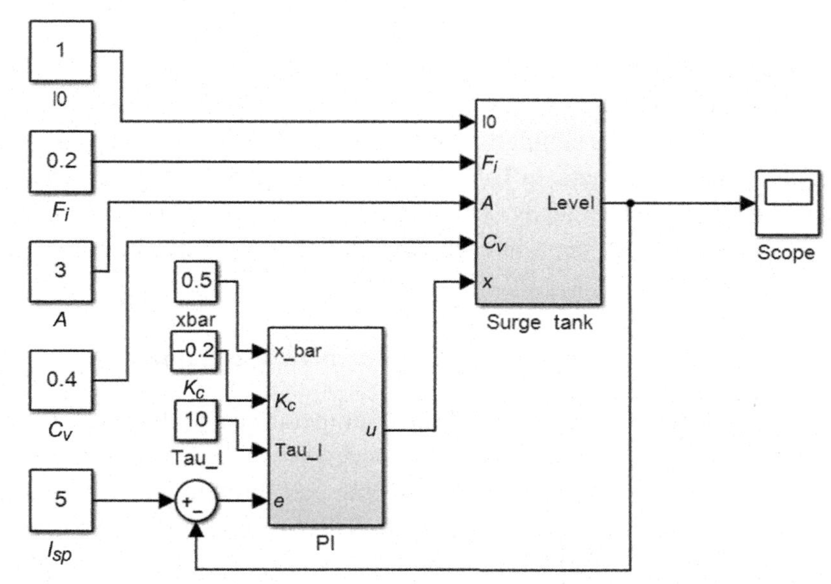

Fig. 3.14
Overall Simulink block diagram for simulation with a PI controller.

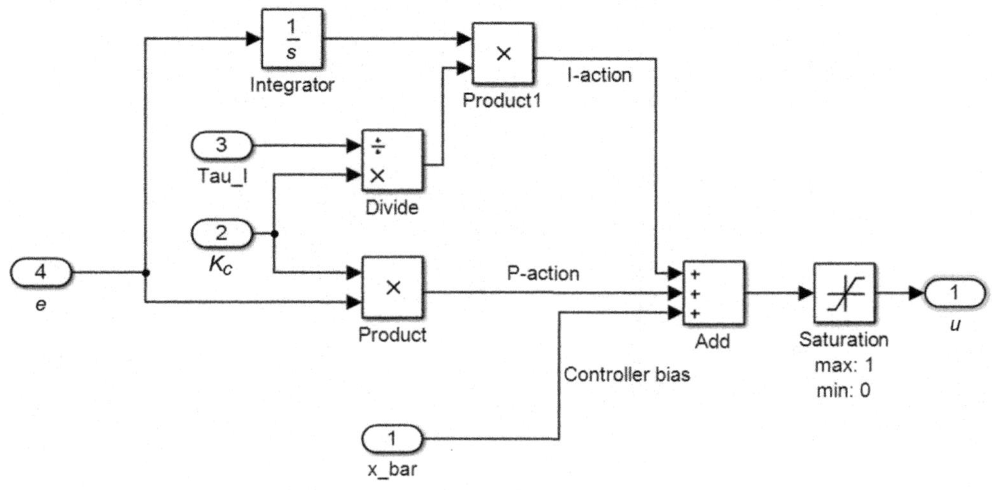

Fig. 3.15
Subsystem PI controller.

Step 8: Model validation requires reliable experimental data. In the absence of experimental data, one can perform a steady-state check by running the dynamic simulation to converge to the steady-state condition for a given set of operating conditions. The steady-state condition obtained from the dynamic simulation is checked against the calculated value from the steady-state equation. In the present example, one can find the expected steady-state head in the tank, $\bar{l}$; for a given valve size, C_v; a valve opening, $\bar{x}$; and an inlet flow rate, $\bar{F}_i$, from the steady-state equation given as follows:

$$\bar{F}_i = C_v \cdot \bar{x}\sqrt{\bar{l}} \tag{3.13}$$

The open-loop simulation result discussed earlier must converge to the calculated steady-state level, $\bar{l}$. Table 3.2 is set up with a fixed valve position $\bar{x} = 0.5$ and fixed cross-sectional area $A = 3\ \text{m}^3$. The simulation time span for Fig. 3.6 is 200 s, which is too short for approaching the steady state. In Table 3.2 the time span is 3600 s to ensure reaching the steady-state liquid level. The steady-state liquid level from the dynamic simulation and the steady-state equation is the same in each case, validating the dynamic model and simulation results.

3.3.2 A Stirred Tank Heater With Level and Temperature Control

In this example, it is required to control the effluent temperature and the liquid level in a well-mixed stirred tank heater (STH) that continuously receives water at a flow rate of (m^3/s) at a temperature T_i(K). The tank is equipped with a heating coil through which saturated steam at 101 kPa is supplied at a mass flow rate of q_s (kg/s). The steam leaves the heating coil as saturated condensed liquid at 101 kPa. The heat of condensation is given by λ_c (kJ/kg). Write down a complete dynamic model for this control system and specify the required control equations to maintain the effluent water temperature at T_{sp} and the liquid level or volume of the liquid in the tank, at V_{sp}. The tank has a constant cross-sectional area. Use proportional controllers only (Fig. 3.16).

Table 3.2 The validation table of the dynamic model and simulation

Parameters Set		Calculated Steady-State Level $\bar{l}$(m)	Steady State $\bar{l}$(m) From Dynamic Model
C_v	F_i (m^3/s)		
0.2	0.1	1.0	0.9999
	0.2	4.0	4.0000
	0.4	16.0	16.0000
0.4	0.1	0.25	0.2500
	0.2	1.0	0.9998
	0.4	4.0	3.9995
0.8	0.1	0.0625	0.0625
	0.2	0.25	0.2499
	0.4	1.0	1.0000

Step 1: The system is defined as the STH excluding the heating coil.
Step 2: The simplifying assumptions include

- The density and specific heat of water are assumed constant.
- The contents of the tank are well mixed that means there is no temperature variations with position and the temperature of the liquid inside the tank is only a function of time, $T = T(t)$.
- No heat losses from the tank. It is assumed that the heat of condensation of saturated steam is all transferred to the liquid content in the tank.
- The steam is saturated and the condensate is not subcooled, that is, only heat of condensation is released by the steam and there is no sensible heat involved.
- No change in the potential and kinetic energies of the liquid content in the tank and the incoming and outgoing liquid streams.

Step 3: Based on the previous assumptions, the liquid content in the tank comprises a lumped parameter system.
Step 4: The conservation principles include the total mass balance and energy balance on the liquid content in the tank.

Total mass balance

$$\frac{d[V(t)]}{dt} = F_i(t) - F(t) \tag{3.14}$$

Energy balance

$$\underbrace{\frac{d[U(t)]}{dt}}_{\text{Total internal energy of the liquid in tank}} = \underbrace{\dot{E}_i(t) - \dot{E}_o(t)}_{\text{Rate of energy entering and leaving the tank in the inlet and outlet streams}} = \dot{H}_i(t) + \underbrace{Q_s(t)}_{\text{Rate of steam condensation}} - \underbrace{\dot{H}_o(t)}_{\text{Rate of energy leaving the tank in the effluent stream}} \tag{3.15}$$

where $\dot{E}_i$ and $\dot{E}_o$ are the rate of energy input and output from all sources, $\dot{H}_i$ and $\dot{H}_o$ are the rate of enthalpy entering and leaving the tank with the inlet and effluent streams,

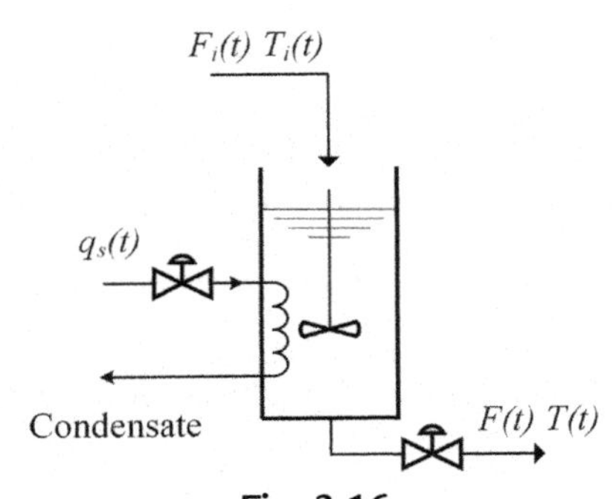

Fig. 3.16
Schematics of a STH (stirred tank heater).

and Q_s is the rate of heat released by the condensation of the steam in the steam coil which is transferred to the liquid in the tank.
Step 5: The supplementary equations comprise the internal energy of the liquid in the tank, the rate of enthalpy entering and leaving the tank in the inlet and outlet streams, and the rate of heat transfer from the steam to the liquid in the tank.

$$U=\rho VC(T-T_{ref}) \tag{3.16}$$

$$\dot{H}_i=\rho CF_i(T-T_{ref}) \tag{3.17}$$

$$Q_s=q_s\lambda_c \tag{3.18}$$

$$\dot{H}_o=\rho CF(T-T_{ref}) \tag{3.19}$$

where C is the liquid specific heat in kJ/(kg K) which is assumed to be constant. Inserting Eqs. (3.16)–(3.19) in Eq. (3.15) results in

$$\begin{aligned}\frac{d[\rho VC(T-T_{ref})]}{dt}&=\rho C\frac{d[V\cdot(T-T_{ref})]}{dt}=\rho C\left[T\frac{dV}{dt}+V\frac{dT}{dt}\right]-\rho C\left[T_{ref}\frac{dV}{dt}\right]\\&=\rho F_iC(T_i-T_{ref})+q_s\lambda_c-\rho FC(T-T_{ref})\end{aligned} \tag{3.20}$$

Substituting for dV/dt from Eq. (3.14), will further simplify the previous equation and get rid of terms involving T_{ref}.

$$\rho CV(t)\frac{d[T(t)]}{dt}=\rho CF_i\ (t)\ [T_i(t)-T(t)]+q_s(t)\lambda_c \tag{3.21}$$

It is clear from the previous equation that the temperature in the tank, T, does not depend on the effluent rate since the content of the tank is assumed to be well mixed.
Step 6: Check the unit consistency in Eqs. (3.14), (3.15). Each term in Eq. (3.14) has the unit of m^3/s. In Eq. (3.15), each term is in kJ/s or kW.
In order to check the degrees of freedom, the number of variables, disturbances, and state and auxiliary equations must be counted.
Degrees of freedom analysis:

Number of variables	10	$V,T,T_i,F_i,F,q_s,Q_s,U,\dot{H}_i,\dot{H}_o$
State equations	2	Eqs. (3.14), (3.15)
Auxiliary equations	4	Eqs. (3.16)–(3.19)
Disturbance	2	$F_i(t)$, T_i
Fixed parameters		$\rho,\lambda_c,C,T_{sp},V_{sp},K_{c1},K_{c2},\overline{F},\overline{q_s}$, initial conditions, $V(0)$, $T(0)$

Number of degrees of freedom $=10-2-4-2=2$.
Therefore two controller equations are needed to reduce the degrees of freedom of the dynamic model to zero.

$$F=\overline{F}-K_{c1}\left(V_{sp}-V\right) \tag{3.22}$$

$$q_s=\overline{q_s}+K_{c2}\left(T_{sp}-T\right) \tag{3.23}$$

Step 7: Numerical solution of the state equations, auxiliary equations, and control equations for the specified parameters can be performed using MATLAB.

STH simulation ODE function: STH.m

```
function dy = STH(t,y,Fi,Ti,density,C,Latent,Fbar,qsbar,Vsp,Tsp,Kc1,Kc2,Fmax,qsmax)
    %Enable P-Controller after 10 second.
    if (t>10)
        F = Fbar - Kc1 * (Vsp - y(1));
        qs = qsbar + Kc2 * (Tsp - y(2));
    else
        F = Fbar;
        qs = qsbar;
    end

    %Constraint on flowrate
    if (F<0)
        F=0;
    end
    if (F>Fmax)
         F=Fmax;
    end
    if (qs<0)
        qs=0;
    end
    if (qs > qsmax)
        qs = qsmax;
    end

    %y(1): volume of liquid; dy(1): time derivative of volume;
    %y(2): temperature of liquid; dy(2): time derivative temperature;

    %Mass balance
    dy(1) = Fi - F;
    %Energy balance
    dy(2) = 1/(density*C * y(1)) * (density * C * Fi * (Ti - y(2)) + qs* Latent);
    %transpose dy to make it a column vector.
    dy = dy';
end
```

STH Simulation script:

```
%Stream parameter setup
Fi = 0.5;     %m^3/s
Ti = 50;      %Celsius
density = 980;    %kg/m^3
C  = 4180;    % J/(kg*Celsius)
Latent = 2260e3;      % J/kg

%Controller setup
Kc1 = 5;  %controller gain for volume control
Kc2 = 5;  %controller gain for temperature control
Vsp = 15; %m^3
Tsp = 60; %Celsius

%Initial values
Vbar = 3; %m^3
Tbar = 55; %Celsius
InitVal = [Vbar,Tbar];

%Controller bias setup
Fbar = Fi;
qsbar = Fi*(Tbar-Ti)*density*C/Latent;

%Actuator constraint setup
Fmax = 1; %m^3/s
qsmax = 10;
Timespan = [0 40];
Options = [];

[t,y]=ode45(@STH,Timespan,InitVal,Options,Fi,Ti,density,C,Latent,...
             Fbar,qsbar,Vsp,Tsp,Kc1,Kc2,Fmax,qsmax);
plot(t,y);
```

The block diagram of the process is shown in Fig. 3.17. Fig. 3.18 reveals an important point that in the presence of a change in the disturbance, F_i, an *offset* appears because in the absence of a change in the *set point* the error remains zero, and the controller output does not change unless the controller bias is also changed.

Step 8: Again in the absence of experimental data, we can check if the dynamic model converges to the steady-state solution given by

$$\overline{F}_i\left[\overline{T} - T_i\right] = \frac{\lambda_c \overline{q}_s}{\rho C} \tag{3.24}$$

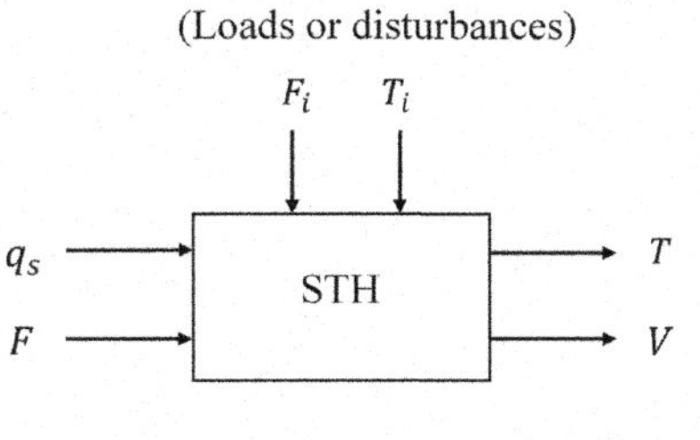

Fig. 3.17
Process block diagram.

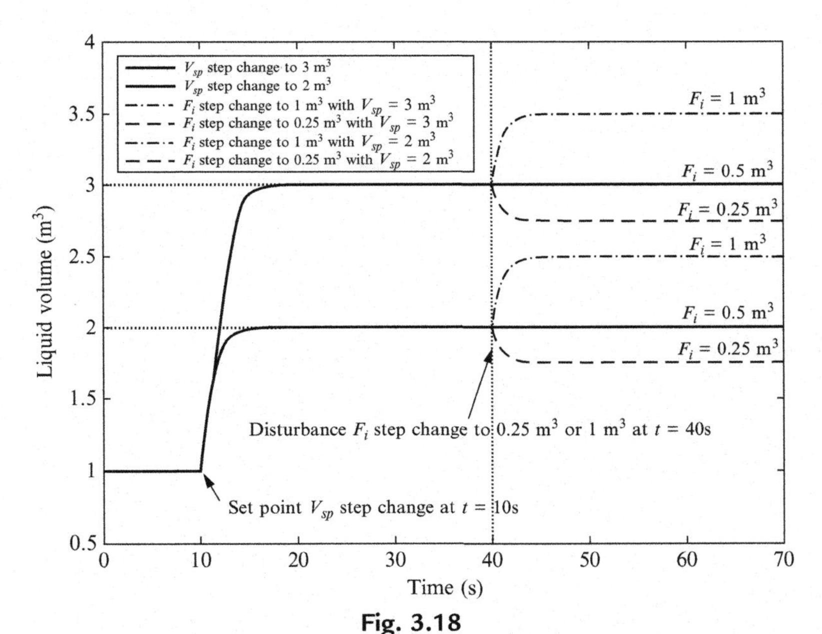

Fig. 3.18
The STH closed-loop simulation results.

3.3.3 A Nonisothermal Continuous Stirred Tank Reactor

A nonisothermal CSTR (Adopted in part from Process Modeling, Simulation and Control, Luyben[3]).

An irreversible isothermal reaction $A \rightarrow B$ (with $k = \alpha \cdot e^{-\frac{E}{RT}}$) is carried out in a well-mixed CSTR as shown in Fig. 3.19. Negligible heat losses and constant densities are assumed. The reaction is assumed to be first order in the concentration of component A. The heat of reaction is assumed to be λ (Btu/mole of A reacted). The reactor is surrounded by a cooling jacket. Cooling water feed rate is F_J(ft^3/s) and the inlet temperature of the feed is T_{Ji}(°R). The volume of water in the jacket, V_J(ft^3), is assumed constant. The jacket water is assumed to be perfectly mixed at all times. The mass of the reactor metal wall is considered to be negligible. Hence the thermal

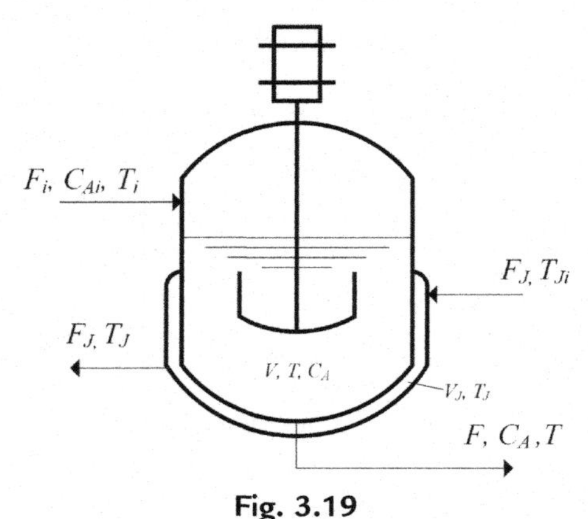

Fig. 3.19
A nonisothermal continuous stirred tank reactor (CSTR).

inertia of the metal need not be considered. The heat transfer between the process and the cooling water jacket may be defined in terms of an overall heat transfer coefficient:

$$Q = U \cdot A_H (T - T_J) \tag{3.25}$$

where Q is the heat transfer rate, Btu/h, U is the overall heat transfer coefficient, Btu/(h ft^2 °R), and A_H is the heat transfer area, ft^2

Step 1: The system to be modeled consists of two subsystems, the reactor and the cooling jacket. Therefore both subsystems must be modeled. The reactor holdup, $\overline{V}$, and jacket holdup, $\overline{V}_J$, are assumed constant.
Step 2: Many of the relevant assumptions are stated in the problem definition and the set of parameters given in the following table.

$\overline{F} = 40\,\text{ft}^3/\text{h}$	$U = 2700\,\text{Btu}/(\text{h ft}^2\,°\text{R})$
$\overline{F}_J = 49.9\,\text{ft}^3/\text{h}$	$A_H = 250\,\text{ft}^2$
$\overline{V} = 48\,\text{ft}^3$	$\lambda = -30{,}000\,\text{Btu/lbmol}$
$\overline{V}_J = 3.85\,\text{ft}^3$	$c = 0.75\,\text{Btu}/(\text{lb}_\text{m}\,°\text{R})$
$C_{Ai} = 0.50\,\text{lbmol}/\text{ft}^3$	$c_J = 1.0\,\text{Btu}/(\text{lb}_\text{m}\,°\text{R})$
$T_{Ji} = 530\,°\text{R}$	$\rho = 50\,\text{lb}_\text{m}/\text{ft}^3$
$T_i = 530\,°\text{R}$	$\rho_j = 62.3\,\text{lb}_\text{m}/\text{ft}^3$
$\alpha = 7.08 \times 10^{10}\,\text{h}^{-1}$	$K_c = 4.0\,\text{ft}^3/(\text{h}\,°\text{R})$
$E = 30{,}000\,\text{Btu/lbmol}$	$T_{sp} = 600\,°\text{R}$
$R = 1.99\,\text{Btu}/(\text{lbmol}\,°\text{R})$	

Step 3: Since the reactor content and the jacket are assumed to be well mixed, both subsystems are assumed to be lumped parameter.

Step 4: The reactor holdup is assumed to remain constant, therefore there is no need to consider the total mass balance. The component mass balance on A takes the following form:

$$\overline{V}\frac{d[C_A(t)]}{dt} = F_i(t)C_{Ai}(t) - F(t)C_A(t) - \overline{V}\cdot C_A(t)\cdot\alpha\cdot e^{-\frac{E}{R\cdot T(t)}} \tag{3.26}$$

The energy balance on the reactor content is

$$\begin{aligned}\rho c\overline{V}\frac{d\left[T(t)-T_{ref}\right]}{dt} &= \rho cF_i(t)\left[T_i(t)-T_{ref}\right] - \rho cF(t)\left[T(t)-T_{ref}\right] \\ &\quad - UA_H[T(t)-T_J(t)] - \lambda\overline{V}k(T)C_A(t)\end{aligned} \tag{3.27}$$

Similarly, the jacket holdup is assumed constant and hence no need to consider the total mass balance for the jacket content. The energy balance on the jacket, however, has to be considered:

$$\begin{aligned}\rho_J c_J\overline{V}_J\frac{d\left[T_J(t)-T_{ref}\right]}{dt} &= \rho_J c_J F_J(t)\left[T_{J,i}(t)-T_{ref}\right] \\ &\quad - \rho_J c_J F_J(t)\left[T_J(t)-T_{ref}\right] + UA_H[T(t)-T_J(t)]\end{aligned} \tag{3.28}$$

Step 5: The auxiliary equations include the reaction kinetics $r_A = \alpha\cdot e^{-\frac{E}{RT(t)}}\cdot C_A(t)$, and the heat transfer rate between the contents of the jacket and the reactor given by Eq. (3.25), and the internal energy and rate of enthalpy that have all been incorporated in Eqs. (3.27), (3.28).

Step 6: Each term in Eq. (3.26) has the units of mol/ft^3, and in Eqs. (3.27), (3.28), Btu/h. Degrees of freedom analysis:

Number of variables	8	T_i, T, T_{Ji}, T_J, F_J, F, C_A, C_{Ai}
State equations	3	Eqs. (3.26)–(3.28)
Auxiliary equations		All have been incorporated in the state equations
Disturbance	4	$F(t)$, T_i, T_{Ji}, C_{Ai}
Fixed parameters		ρ, ρ_J, c, c_J, U, A_H, α, E, R, T_{sp}, K_c, $\overline{F}_J$, $\overline{V}$, $\overline{V}_J$, T_{ref} initial conditions, $C_A(0)$, $T(0)$, $T_J(0)$

Therefore only one controller equation is needed to make the degrees of the freedom zero. A proportional controller measures the reactor temperature and adjusts the cooling water flow rate supplied to the reactor jacket to maintain the reactor temperature at T_{sp}.

$$F_J(t) = \overline{F}_J + K_C\left[T_{sp} - T(t)\right] \tag{3.29}$$

Step 7: Use MATLAB to simulate the closed-loop model. Fig. 3.20 shows the results of closed-loop simulation using the P controller given in Eq. (3.29). The reactor temperature

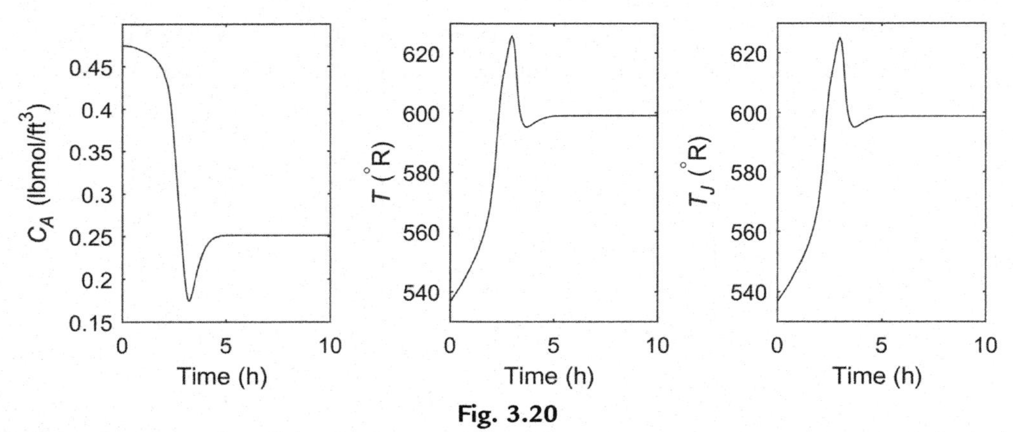

Fig. 3.20
CSTR simulation results of the reactant concentration, the reactor temperature, and the jacket temperature.

overshoots to 627°R before settling close to the set point value at steady state. The reactor jacket temperature and the reactant concentration are also depicted in Fig. 3.20.

CSTR model ODE function: CSTRODE.m

```
function [ dy ] = CSTRODE( t,y,FJ,CAi,Ti,TJi,F)
    %Derivative function of the main program
    %State variables
    CA = y(1); T = y(2); TJ = y(3);

    %Constant parameters
    V=48; Lambda=-3e4; Capacity=0.75; Density=50; U=0.75*3600; A=250; E=3e4;
    R=1.99; VJ=3.85; CapacityJ=1; DensityJ=62.3; alpha =7.06e10;
    %Intermediate variable
    k = alpha*exp(-E./(R*T));

    %Derivatives
    dCA = F/V*CAi - F/V*CA - CA*k;
    if ( CA <=0 && dCA<0)
        dCA = 0 ;
    end

    dT = F/V*Ti - F/V*T - Lambda/(Capacity*Density)*CA.*k ...
        -U*A/(Capacity*V*Density)*(T-TJ);
    dTJ= FJ*(TJi-TJ)/VJ + U*A/(CapacityJ*VJ*DensityJ)*(T-TJ);
    dy=[dCA;dT;dTJ];
end
```

CSTR Closed loop model ODE function: CSTRODE_ClosedLoop.m

```
function dy = CSTRODE_ClosedLoop( t,y,FJ_bar,CAi,Ti,TJi,F,Kc,Tsp )
    %Use P-controller to control the temperature
    FJ = FJ_bar + Kc*(Tsp - y(2));
    %Set constraint for flow rate.
    Fmax=95;
    if (FJ > Fmax)
        FJ =Fmax;
    end
    if (FJ<0)
        FJ = 0;
    end
    dy = CSTRODE(t,y,FJ,CAi,Ti,TJi,F);
end
```

Run the following script to simulate closed loop.

CSTR Closed loop simulation script:

```
% Parameters
FJ= 49.9; CAi = 0.5; Ti = 530;
F  = 40; TJi = 530;
% Initial conditions, arbitrary values
CA_init = 0.50; T_init = 530; TJ_init = 530;
y0 = [CA_init T_init TJ_init];
tspan = [0 10];%Hours
Options = [];
% Run open loop model to obtain steady state value
[t,y] = ode15s(@CSTRODE,tspan,y0,Options,FJ,CAi,Ti,TJi,F);
y0=y(end,:);

% Closed loop simulation
Kc = -4;
Tsp = 600;
[t,y]=ode15s(@CSTRODE_ClosedLoop,tspan,y0,Options,FJ,CAi,Ti,TJi,F,Kc,Tsp);
CA=y(:,1);
T=y(:,2);
TJ = y(:,3);

% Plot concentration response
subplot(1,3,1);
plot(t,CA);
```

```
% Plot reactor temperature response
subplot(1,3,2);
plot(t,T);

% Plot jacket temperature response
subplot(1,3,3);
plot(t,TJ);
```

Step 8: In the absence of experimental data, in some simple cases, the validity of the dynamic simulation is checked against the steady-state solution. Calculation of the steady-state temperature in the present example is somewhat convoluted but calculable.

3.3.4 A CSTR With Liquid Phase Endothermic Chemical Reactions

Dynamic modeling of a CSTR with liquid phase endothermic chemical reactions (Fig. 3.21):

$$A + R \xrightarrow{k_1} P_1$$

$$B + 2R \xrightarrow{k_2} P_2$$

Control objectives:

- maintain T at T_{sp}
- maintain V or l at V_{sp} or l_{sp}
- maintain C_{p1} at $C_{p1,sp}$
- maintain C_{p2} at $C_{p2,sp}$

Step 1: The system under consideration consists of the CSTR and the heating coil.
Step 2: Assumptions:

- The reactant streams and the product stream are dilute, therefore there is no change in the density and the specific heats.
- No heat losses from the reactor.
- Saturated steam with total condensation and no subcooling of the condensate.
- The reacting mixture is well mixed.

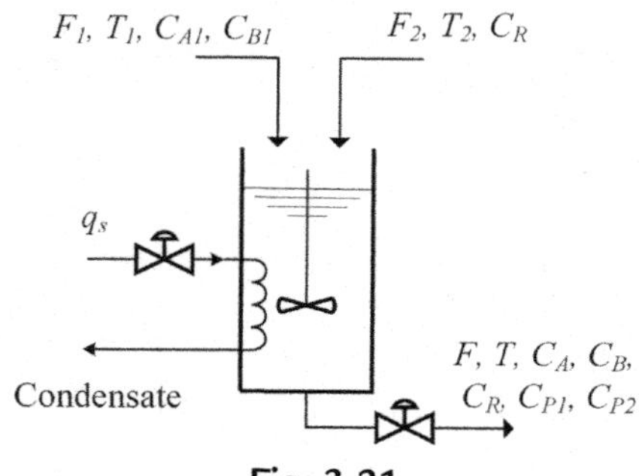

Fig. 3.21
The CSTR schematics.

Step 3: The previous assumptions imply that both the reactor and the heating coil can be considered as lumped parameter systems.
Step 4: Conservation principles lead to the following state equations:
Total mass balance on the reactor:

$$\frac{d(V\rho)}{dt} = \rho F_1 + \rho F_2 - \rho F$$
$$\frac{dV(t)}{dt} = F_1 + F_2 - F \quad (\text{constant } \rho) \tag{3.30}$$

Component mass balance:

$$\frac{d(VC_A)}{dt} = F_1 C_{A1} - FC_B - V \cdot r_1 \tag{3.31}$$

$$\frac{d(VC_B)}{dt} = F_1 C_{B1} - FC_B - V \cdot r_2 \tag{3.32}$$

$$\frac{d(VC_R)}{dt} = F_2 C_{R2} - FC_R - [V \cdot r_1 + V \cdot (2r_2)] \tag{3.33}$$

Where r_1 and r_2 are the reaction rates of the first and second reactions, respectively.

$$\frac{d(VC_{P1})}{dt} = 0 - FC_{P1} + V \cdot r_1 \tag{3.34}$$

$$\frac{d(VC_{P2})}{dt} = 0 - FC_{P2} + V \cdot r_2 \tag{3.35}$$

Instead of Eq. (3.35), an overall mole balance could be used.
Energy balance on the reactor:

$$\begin{aligned} \rho c \frac{d\{V(t)[T(t) - T_{ref}]\}}{dt} = & \rho c F_1(t)[T_1(t) - T_{ref}] + \rho c F_2(t)[T_2(t) - T_{ref}] \\ & - \rho c F(t)[T(t) - T_{ref}] + V(t) \cdot r_1(t) \cdot (-\Delta H_1) \\ & + V(t) \cdot r_2(t) \cdot (-\Delta H_2) + q_{st}(t) \end{aligned} \tag{3.36}$$

where q_{st} is the rate of heat transfer from steam to the reacting mixture, ΔH_1 is the heat of reaction 1 (Btu/lbmol), and ΔH_2 is the heat of reaction 2 (Btu/lbmol).
We assume no accumulation of mass and energy in the heating coil, therefore there is no need to consider the mass and energy balance in the heating coil. The assumption is that the heat of condensation, q_{st}, is completely and instantaneously transferred to the reactor content.
Step 5: Supplementary equations:

$$r_1 = k_1 C_A C_R (\text{assuming a 2nd reaction}) \tag{3.37}$$

$$r_2 = k_2 C_B C_R^2 (\text{assuming a 3rd reaction}) \tag{3.38}$$

$$q_{st} = F_{st} \cdot \lambda (F_{st}\,\text{lbm/h};\ \ \lambda\,\text{is the heat of condensation, Btu/lbmol}) \tag{3.39}$$

Step 6: Consistency check:
Unit of each term in Eq. (3.30) is ft^3/h, in Eqs. (3.31)–(3.35) is lbmol/h, and in Eq. (3.36) is Btu/h.
Degree of freedom analysis:

Number of variables	19	$T_1, T_2, T, q_{st}, F_1, F_2, F, C_A, C_B, C_R, C_{P1}, C_{P2}, C_{A1}, C_{B1}, C_{R2}, V(or\ l), r_1, r_2, F_{st}$
State equations	7	Eqs. (3.30)–(3.36)
Auxiliary equations	3	Eqs. (3.37)–(3.39)
Disturbances	5	$T_1, T_2, C_{A1}, C_{B1}, C_{R2}$
Parameters		$\rho, \lambda, T_{sp}, l_{sp}, C_{P1,sp}, C_{P2,sp}, K_{c1}, K_{c2}, K_{c3}, K_{c4}, \overline{F_1}, \overline{F_2}, \overline{F_{st}}, \overline{F}, C_A(0), C_B(0), C_R(0), C_{P1}(0), C_{P2}(0), T(0), \Delta H_1, \Delta H_2$

The number of required control equations: $19 - (7 + 3 + 5) = 4$.

$$\left.\begin{aligned} F &= \overline{F} - K_{c1}\left(l_{sp} - l\right) \\ F_{st} &= \overline{F_{st}} + K_{c2}\left(T_{sp} - T\right) \\ F_1 &= \overline{F_1} - K_{c3}\left(C_{P1,sp} - C_{P1}\right) \\ F_2 &= \overline{F_2} + K_{c4}\left(C_{P2,sp} - C_{P2}\right) \end{aligned}\right\}\ \text{Highly interactive loops} \tag{3.40}$$

The block diagram of the process is shown in Fig. 3.22.
Step 7: The numerical solution of this problem is left as an exercise to the reader.

3.4 Examples of Stage-Wise Systems

Stage-wise processes are common in the chemical industry. Examples include a series of CSTRs, multistage evaporators, and tray distillation columns. Such processes may be assumed as a series of homogeneous lumped parameter systems.

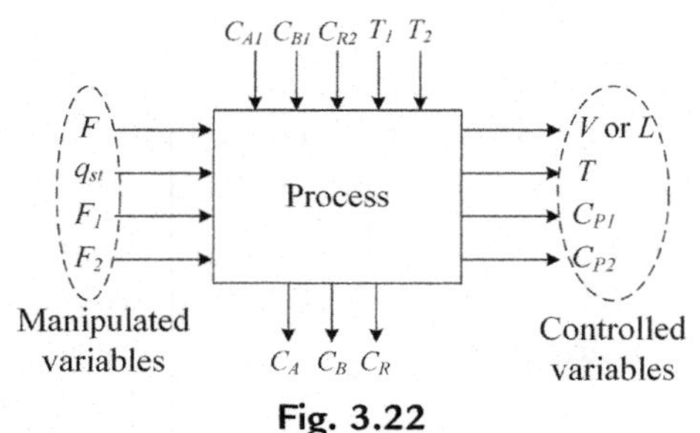

Fig. 3.22
Process block diagram for the CSTR.

3.4.1 A Binary Tray Distillation Column

A binary tray distillation column is considered for the separation of two components A and B. Fig. 3.23 depicts a schematic diagram of a binary distillation column. The necessary data for the simulation of the process are given in Table 3.3. The distillation column is assumed to run isothermally and the heat input in the reboiler and heat output in the condenser correspond to the latent heats of vaporization and condensation of the distillation mixture.

Given information as tabulated in Table 3.3.

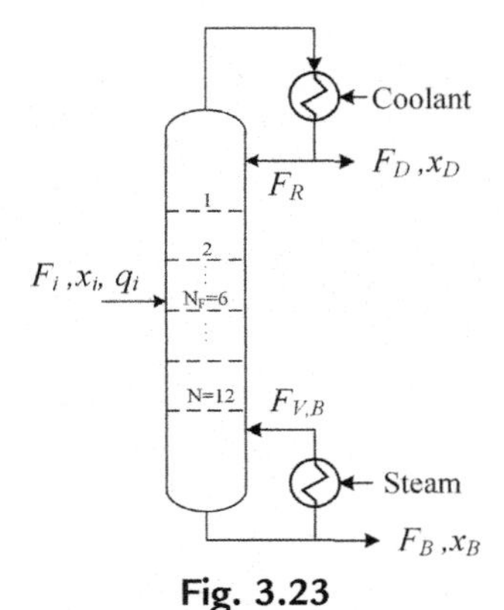

Fig. 3.23
The schematics of the binary distillation column.

Table 3.3 The required data for the distillation column

F_i	Inlet molar flow rate	50 mol/s
q_i	Feed quality (liquid fraction)	0.9
x_i	Inlet mole fraction	0.6
α	Relative volatility	2.2
N	Number of stages	12
R_c	Condenser reflux ratio	2.0
R_r	Reboiler ratio	4.69
ρ_A	Density of component A	800 kg/m^3
ρ_B	Density of component B	760 kg/m^3
MW_A	Molar weight of A	0.046 g/mol
MW_B	Molar weight of B	0.078 g/mol
h_w	Weir height	0.05 m
l_w	Weir length	0.375 m
A_t	Tray active area	0.7854 m^2

Additional information

- Francis weir equation: The relation of downcoming liquid flow rate and liquid holdup on tray k.

$$n_k = \frac{\overline{\rho_k}}{\overline{MW_k}} \cdot A_t \cdot \left[h_w + 1.41 \left(\frac{L_k \overline{MW_k}}{\sqrt{g} \cdot \overline{\rho} \cdot l_w} \right)^{\frac{2}{3}} \right] \tag{3.41}$$

n_k is the molar holdup on the kth tray (mol)

$\overline{\rho_k}$ is the average density of liquid holdup on the kth tray $\left(\frac{\text{kg}}{\text{m}^3}\right)$

$\overline{MW_k}$ is the average molar weight on the kth tray $\left(\frac{\text{kg}}{\text{mol}}\right)$

A_t is the active area of each tray m^2

h_w is the weir height (m)

L_k is the flow rate of liquid leaving the kth tray $\left(\frac{\text{mol}}{\text{s}}\right)$

l_w is the weir length (m)

g is the gravitational acceleration $g = 9.81 \frac{\text{m}}{\text{s}^2}$

- Equilibrium

The relative volatility is given by: $\alpha_{A,B} = \alpha = \dfrac{K_A}{K_B} = \dfrac{\frac{y_A}{x_A}}{\frac{y_B}{x_B}} = 2.2$

In a binary system $y_B = 1 - y_A$ and $x_B = 1 - x_A$

$$y_A = \frac{\alpha_{A,B} x_A}{1 + x_A(\alpha_{A,B} - 1)} \tag{3.42}$$

- There is no holdup in the condenser and reboiler and they are at steady state.

The dynamic model for the distillation column can be developed following the systematic 8-step procedure discussed earlier.

Step 1: The system is a stage-wise process consisting of 12 trays, a condenser, and a reboiler. Reflux ratio and reboiler ratios are manipulated to control the composition of distillate and bottoms product.

Step 2: Assumptions include

(a) The vapor holdup on each tray is negligible.

(b) Constant molar overflow, the molar heats of vaporization of the feed components are equal, for every mole of liquid vaporized, a mole of vapor is condensed, and the heat effects such as heats of solution are negligible.

(c) Vapor composition and liquid composition on each tray are in equilibrium.

(d) The vapor flow rate is constant in the rectifying and stripping sections.
(e) Both condenser and reboiler operate at steady state.
(f) No heat losses from the column, the operation is isothermal.
(g) Pressure drop along the column is negligible.
(h) Relative volatility remains constant.

Step 3: The system is a stage-wise process consisting of individual lumped parameter systems.

Step 4: The dynamic molar balances for each component in each phase on all trays and in the reboiler and condenser are considered.

Molar balance on the feed tray:

$$\frac{dn_{NF}}{dt} = F_i + L_{NF-1} + G_{NF+1} - L_{NF} - G_{NF} \tag{3.43}$$

Component mole balance on the feed tray:

$$\frac{dx_{NF}n_{NF}}{dt} = F_i x_i + L_{NF-1}x_{NF-1} + G_{NF+1}\,y_{NF+1} - L_{NF}x_{NF} - G_{NF}y_{NF} \tag{3.44}$$

$$\frac{dx_{NF}}{dt} = \frac{1}{n_{NF}} \cdot \left[F_i x_i + L_{NF-1}x_{NF-1} + G_{NF+1}\,y_{NF+1} - L_{NF}x_{NF} - G_{NF}y_{NF} - x_{NF}\frac{dn_{nf}}{dt}\right] \tag{3.45}$$

On tray k ($k = 1$ to N except for the feed tray NF):

Molar balance

$$\frac{dn_k}{dt} = L_{k-1} + G_{k+1} - L_k - G_k \tag{3.46}$$

Component mole balance

$$\frac{dn_k x_k}{dt} = L_{k-1}x_{k-1} + G_{k+1}\,y_{k+1} - L_k x_k - G_k y_k \tag{3.47}$$

$$\frac{dx_k}{dt} = \frac{1}{n_k} \cdot \left[L_{k-1}x_{k-1} + G_{k+1}\,y_{k+1} - L_k x_k - G_k y_k - x_k\frac{dn_k}{dt}\right] \tag{3.48}$$

Step 5: Auxiliary equations:

(a) Francis weir equation:

$$n_k = \frac{\overline{\rho}_k}{\overline{MW}_k} \cdot A_t \cdot \left[h_w + 1.41\left(\frac{L_k\,\overline{MW}_k}{\sqrt{g}\cdot\overline{\rho}\cdot l_w}\right)^{\frac{2}{3}}\right] \tag{3.49}$$

Rearrange to obtain L_k explicitly:

$$L_k = \begin{cases} \dfrac{0.59727\cdot l_w}{\overline{MW}_k}\left[\dfrac{\left(\overline{MW}_k\cdot n_k - A_t\cdot h_w\cdot\overline{\rho_k}\right)\cdot g^{\frac{1}{3}}}{A_t\cdot\overline{\rho_k}^{\frac{1}{3}}}\right]^{\frac{3}{2}} & \left(\overline{MW}_k\cdot n_k - A_t h_w\overline{\rho_k} > 0\right) \\ 0 & \left(\overline{MW}_k\cdot n_k - A_t h_w\overline{\rho_k} \le 0\right) \end{cases} \tag{3.50}$$

Equilibrium equation:

$$y_{A,k} = \frac{\alpha_{A,B} x_{A,k}}{1 + x_{A,k}(\alpha_{A,B} - 1)} \tag{3.51}$$

(b) In the condenser steady state holds. All vapor is condensed so the outlet liquid composition is equal to the inlet vapor composition. The reflux ratio R_c is:

$$R_c = \frac{F_R}{F_D} \tag{3.52}$$

And,

$$F_R + F_D = F_{V,A} \tag{3.53}$$

Therefore

$$F_R = \frac{F_{V,A} R_c}{R_c + 1} \tag{3.54}$$

The composition in stream F_R is assumed to be the same as in the inlet to the condenser.

$$x_D = y_1 \tag{3.55}$$

(c) In the reboiler steady state holds. All liquid is vaporized so vapor composition is the same as inlet composition. Similar to the condenser the reboiler ratio

$$R_r = \frac{F_{V,B}}{F_B} \tag{3.56}$$

And

$$F_B + F_{V,B} = L_N \tag{3.57}$$

Therefore

$$F_{V,B} = \frac{L_N R_r}{R_r + 1} \tag{3.58}$$

The composition in stream $F_{V,B}$ is assumed to be the same as in the inlet of the reboiler.

$$y_B = x_N \tag{3.59}$$

(d) The feed tray vapor flow rate changes due to the vapor content of the feed stream

$$F_{V,A} = F_{V,B} + (1 - q_i)F_i \tag{3.60}$$

On any tray other than the feed tray

$$L_{k-1} = \begin{cases} F_R & [k=1] \text{ tray below condenser} \\ L_{k-1} & [k \neq 1] \text{ calculate with Eq. (3.50)} \end{cases} \tag{3.61}$$

$$G_k = \begin{cases} F_{V,A} & 1 \leq k \leq NF \text{ tray above feed tray (include feed tray)} \\ F_{V,B} & NF < k \leq N \text{ Below feed tray} \end{cases} \tag{3.62}$$

Step 6: Units of each term in each equation should be checked for consistency. Overall degrees of freedom analysis:

Number of variables	• Condenser: R_c, F_D, $F_{V,A}$, F_R, x_D • Reboiler: R_r, F_B, $F_{V,B}$, y_B • Feed tray: F_i, q_i, x_i • All trays including the feed tray: ($\times N$) L_k, G_k, x_k, n_k, y_k	$12+5N=72$
State equations	• Condenser: Eqs. (3.54), (3.55) • Reboiler: Eqs. (3.58), (3.59) All trays including the feed tray: ($\times N$) • Molar balance: Eq. (3.43) or (3.46) • Component balance: Eq. (3.45) or (3.48)	$4+2N=28$
Auxiliary equations	• Distillate flow rate: Eq. (3.53) • Bottom flow rate: Eq. (3.57) • Feed tray vapor flow rate change: Eq. (3.60) All trays including the feed tray: ($\times N$) • Liquid flow rate: Eq. (3.50) • Equilibrium: Equation (3.51) o Excluding top and bottom trays: the compositions are same as condenser or reboiler • Vapor flow rate: Eq. (3.62)	$5+3N-2=39$
Disturbances	F_i, x_i, q_i	3
Degree of freedom		2

Two controller equations are required to make the degrees of freedom zero as is shown in Fig. 3.24. The control target is to maintain both the distillate composition and bottoms composition at their set points. One possible pair of controllers is

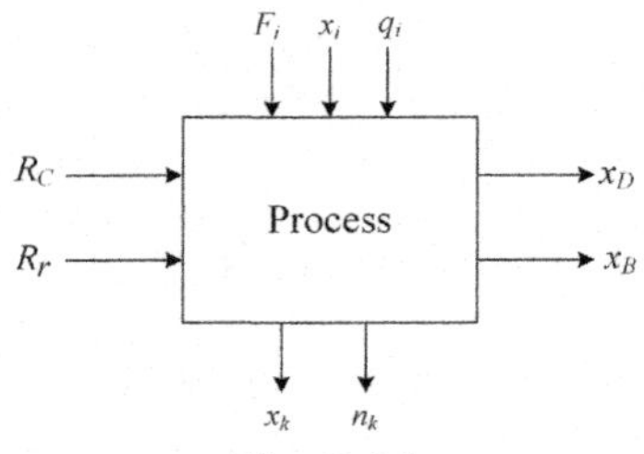

Fig. 3.24
Process block diagram for the binary distillation column.

$$\begin{cases} R_c = \overline{R_c} + K_c\left(x_{D,sp} - x_D\right) \\ R_r = \overline{R_r} + K_r\left(x_{B,sp} - x_B\right) \end{cases} \tag{3.63}$$

One may also use the distillate composition as the controlled variable. The optimum pairing of the manipulated variables and controlled variables in a MIMO system will be discussed in Chapter 5.

Step 7: The solution of the dynamic model presented previously in MATLAB code is given as follows.

Simulation of the binary distillation ODE code: BinaryDistillation.m

```
function [ dy ] = BinaryDistillation(t,y,p)
    % p is a structure variable. For example, access Fi by p.Fi.
    % dy(1:n) liquid molar holdup
    % dy(n+1:2n) liquid composition
    % N stages, excluding the condenser and the reboiler
    N = p.N;
    NF = p.NF;
    MOLAR_HEAD = 1;
    COMPOSITION_HEAD = N+1;
    if (t>100 && p.ClosedLoop == 1);
    % After 100 second, a P controller kicks in.
        p.Rc = p.Rc + p.Kc1 *(p.xtsp - y(COMPOSITION_HEAD));
        p.Rr = p.Rr + p.Kc2 *(p.xbsp - y(COMPOSITION_HEAD + N - 1));
    end

    %% Auxiliary: Reboiler
    % XN: liquid composition leaving stage N (bottom)
    XN = y(COMPOSITION_HEAD + N - 1);

    % LN: downcoming liquid flowrate from stage N (bottom). Pass p into
    % WeirEquation function and the required data will be pulled on demanded.
```

```
LN = WeirEquation(XN,y(MOLAR_HEAD + N - 1),p);
%FVB: Vapor flowrate below feed stage. All comes from reboiler.
FVB = LN * p.Rr /( p.Rr + 1 );

%% Auxiliary: Feed tray
% FVA: Vapor flowrate above feed stage.
FVA = FVB + p.Fi * (1-p.qi);


%% Auxiliary: condenser
% FR: Flowrate of reflux stream from condenser
FR = FVA* p.Rc/(p.Rc+1);
%% Tray: Feed tray (k = NF)
% Liquid composition of NF-1, NF and NF+1 trays
XNF_1 = y(COMPOSITION_HEAD + NF - 2);
XNF = y(COMPOSITION_HEAD + NF - 1);
XNFp1 = y(COMPOSITION_HEAD + NF);
% Vapor composition of NF and NF+1
YNF = ToGasComposition(XNF,p.alpha);
YNFp1 = ToGasComposition(XNFp1,p.alpha);
% Liquid holdup
nF_1 = y(MOLAR_HEAD + NF -2);
nF = y(MOLAR_HEAD + NF -1);
% Liquid flow rate from tray NF-1 and tray NF.
LNF_1 = WeirEquation(XNF_1,nF_1,p);
LNF = WeirEquation(XNF,nF,p);

% molar balance
dy(MOLAR_HEAD + NF - 1) = p.Fi* p.qi +LNF_1 - LNF;

% composition balance
dy(COMPOSITION_HEAD + NF - 1) = 1/y(MOLAR_HEAD + NF - 1) * ( ...
    + p.Fi * p.xi + FVB* YNFp1 + LNF_1* XNF_1 ...
    - FVA * YNF - LNF * XNF ...
    - XNF* dy(MOLAR_HEAD + NF - 1));

% Tray : Any trays excluding k = NF
for k = 1 : N
    % when k = NF, skip to next k.
    if (k == NF)
        continue;
    end;
```

```
  % Below feed tray the vapor flow rate is FVB, or it is FVA.
      if (k < NF)
          FV = FVA;
      else
          FV = FVB;
      end
% Composition of liquid and vapor on kth (current) tray.
    Xk = y(COMPOSITION_HEAD + k - 1);
    Yk = ToGasComposition(Xk,p.alpha);

% Calculate variables below current tray.
% The boundary is k=N at which the next tray is the reboiler
    if (k==N)
% At bottom tray, the incoming vapor is from reboiler
% and its composition is equal to the bottoms composition.
        Ykp1 = Xk;
    else

% The composition of the tray below current tray (k+1) is defined
% as y(COMPOSITION_HEAD + k)
        Xkp1 = y(COMPOSITION_HEAD + k);
        Ykp1 = ToGasComposition(Xkp1,p.alpha);
    end

% Calculate variables above current tray.
% The boundary is k=1 at which the previous tray is the condenser.

% On the top tray, the previous stage is the condenser, the liquid flow rate
% is FR and composition is the same as vapor composition of the top tray.
    if (k == 1)
        Xk_1 = Yk;
        Lk_1 = FR;
    else
        Xk_1 = y(COMPOSITION_HEAD + k - 2);
        Nk_1 = y(MOLAR_HEAD + k - 2);
        Lk_1 = WeirEquation(Xk_1,Nk_1,p);
    end

% Calculate variables in current tray.
% Molar holdup in kth (current) tray
    Nk = y(MOLAR_HEAD + k - 1);
% Outcoming liquid flowrate from kth (current) tray.
    Lk = WeirEquation(Xk,Nk,p);
```

```
    % Molar balance
        dy(MOLAR_HEAD + k - 1) = Lk_1 - Lk;
    % Composition balance
        dy(COMPOSITION_HEAD + k - 1) = 1/Nk * ( Lk_1*Xk_1 + FV*Ykp1 - FV*Yk ...
            - Lk*Xk - Xk*dy(MOLAR_HEAD + k - 1));
    end %for loop
    % convert to column vector
    dy = dy';
    end

    function [y] = ToGasComposition(comp,alpha)
    % comp: composition of A
      y = alpha * comp/( 1 + comp * (alpha -1 ));
    end

    function [L] = WeirEquation(comp, n, p)
    mA = p.MwA * comp;
    mB = p.MwB * (1-comp);
    mt = mA+mB;

    density = mA/mt * p.densityA + mB/mt * p.densityB;
    Mw = p.MwA *comp + p.MwB * (1-comp);
    g=9.81;

    if p.At*p.hw*density - Mw*n > 0
      L = 0;
    else
      L=0.59727/Mw*((-(p.At*p.hw*density-Mw*n)*g.^(1/3)/(p.At*density.^(1/3))).^(3/2)
      *p.lw);
    end
end
```

Binary distillation simulation script

```
p = {};
p.Fi = 50;          % Inlet volumetric flowrate
p.xi = 0.6;         % Inlet composition
p.qi = 0.9;         % Inlet quality (liquid fraction)
p.Rc = 2.0;         % Reflux ratio in condenser
p.Rr = 4.5;         % Reboiler ratio in reboiler
p.alpha = 2.2;      % Relative volatility
p.N = 12;           % Number of trays
p.NF = 6;           % The feed tray number
```

```
p.densityA = 800; % Density of pure A
p.densityB = 760; % Density of pure B
p.MwA = 0.046;    % Molar weight of pure A
p.MwB = 0.078;    % Molar weight of pure B
p.At = 0.7854;    % Active area of tray
p.hw = 0.05;      % Weir height
p.lw = 0.375;     % Weir length

p.ClosedLoop=0;   % Switch closed loop (1) or open loop (0);
p.Kc1 = 50;       % Proportional gain for distillate composition control
p.Kc2 = -50;      % Proportional gain for bottom composition control
p.xtsp = 0.95;    % Set point for distillate composition
p.xbsp = 0.05;    % Set point for bottom composition

% Set arbitrary initial holdup of 1000 mol in each tray
holdup = ones (p.N,1) * 1000;
% Set arbitrary initial composition for each tray with a linear distribution
xss = linspace(1,0,p.N)';
InitVal = [holdup;xss];

Timespan = [0 2000];
Options = [];
[t,y]=ode45(@BinaryDistillation,Timespan,InitVal,[],p);
% Update real hold up and composition for each stage
holdup = y(end,1:N)';
xss = y(end,N+1:2*N)';
InitVal = [holdup;xss]; %New Initial value.
p.ClosedLoop=1;  %Enable controller;
[t,y]=ode45(@BinaryDistillation,Timespan,InitVal,[],p);

subplot(1,2,1);
plot(t,y(:,1:N));
xlabel('Time (s)');
ylabel(' Liquid Holdup (mol)');
subplot(1,2,2);
plot(t,y(:,N+1:2*N));
xlabel('Time (s)');
ylabel(' Liquid Composition');
```

Step 8: Model validation requires experimental data. In the absence of such data, the dynamic model is checked at the steady-state condition.
The steady-state holdups and compositions on each tray can be obtained using the steady-state versions of the open-loop dynamic model discussed earlier.

Table 3.4 The steady-state values of the binary distillation column

Tray #	Holdup (mol)	Composition
1	1017.43	0.8797
2	978.41	0.8042
3	939.53	0.7203
4	904.26	0.6366
5	875.57	0.5618
6 (feed tray)	1012.3	0.5010
7	978.23	0.4055
8	943.63	0.2991
9	914.50	0.2003
10	893.16	0.1226
11	879.55	0.0695
12	871.00	0.0363

In the script file, arbitrary initial values are given and `p.ClosedLoop` is set to 0 to disable the controllers. With 2000-s time span, the final value of the holdups and compositions are the required steady-state values. The results are given in Table 3.4.

This example provides a good method to predict the steady-state values. At the steady state, there are 2 unknowns (holdup, composition) and 2 equations (molar balance, component balance) on each tray. Therefore, theoretically, a solution must exist. However, the vapor composition introduces nonlinearity and an iterative method has to be used. It is always preferred to use the dynamic model to obtain the steady-state data.

The closed-loop response of the entire system using two controller equations is obtained using two P-controllers after 100 s. The results are shown in Fig. 3.25. There is an overshoot

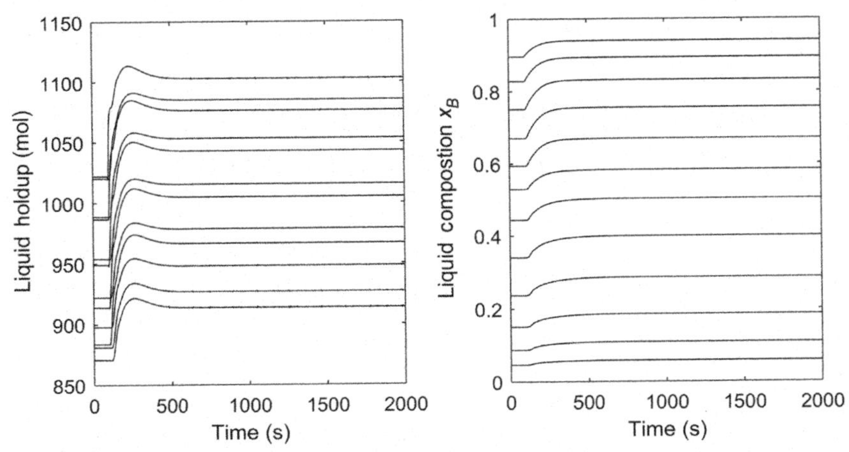

Fig. 3.25
The closed-loop simulation result for binary distillation column. The P controllers kick in at $t = 100$ s. From top to bottom the curves represent holdup/composition response of stage 1 (condenser) to tray N (reboiler).

in the holdup after the controllers kick in. At the peak holdup, the downcoming liquid may cause a flooding flow condition. The tray parameters should be adjusted to prevent such a problem.

3.5 Examples of Distributed Parameter Systems

The process variables in such systems depend on the spatial position in the system as well as time. Examples of such systems include heat exchangers, conduction in a metal bar, tubular reactors, fixed-bed reactors, packed absorption, distillation and extraction columns, etc. The dynamic models of such systems are expressed by PDEs. Depending on the flow patterns in a system, attempts are made to minimize the number of independent spatial variables. For example, only the axial variations of temperature in a shell-and-tube heat exchanger or the axial variations of a species concentration in a tubular reactor are considered and the radial and angular variations are neglected to simplify the dynamic model.

3.5.1 A Plug Flow Reactor

Develop a dynamic model for the plug flow reactor (PFR) (a tubular reactor with only axial variations in the velocity and concentration) shown in Fig. 3.26. It is assumed that a first-order liquid phase reaction occurs in the reactor.

$$A \xrightarrow{k} B$$

The reactor cross-sectional area is constant and the liquid density does not change. We also assume that the velocity profile in the reactor is flat, which means that the velocity and concentrations depend on time, t, and the axial position, z, but independent of the radial and angular positions.

$$\left.\begin{aligned} u_b &= u_b(t) \\ C_A &= C_A(t,z) \\ C_B &= C_B(t,z) \end{aligned}\right\} \text{Flat velocity profile}$$

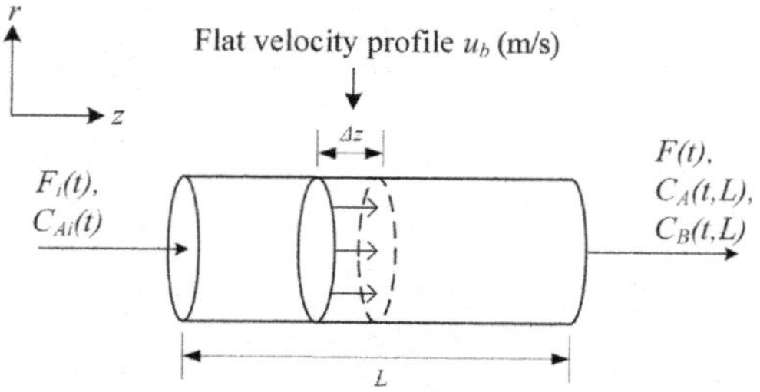

Fig. 3.26
Schematics of a tubular plug flow reactor (PFR).

Control objective: Keep the exit concentration of B, $C_B(t, L)$, at $C_{B,sp}$

Step 1: The system is the PFR.
Step 2: Assumptions:

- Isothermal reaction that suggests no energy balance is needed.
- Flat velocity profile leading to a plug flow behavior.
- Incompressible fluid with constant density is assumed.
- Reactor has a constant cross-sectional area.

Step 3: The system is a distributed parameter system. The control volume is selected to be a disk with thickness, Δz.
Step 4: Conservation principles must be applied to the control volume and NOT to the entire reactor due to the distributed parameter nature of the system.
The total mass balance on the control volume is redundant for a liquid phase reaction with constant density.

$$u_b \frac{d}{dt}[(A\Delta z)\rho] = F(t,z)\rho - F(t,z+\Delta z)\rho = F(t)\rho - F(t)\rho = 0 \tag{3.64}$$

The component molar balance in the control volume is

$$\frac{\partial}{\partial t}[(A\Delta z)\cdot C_A(t,z)] = F(t)\,C_A(t,z) - F(t)C_A(t,z+\Delta z) - r_A(t,z)A\Delta z \tag{3.65}$$

Where the term $[(A\Delta z)\cdot C_A(t,z)]$ represents the number of moles of component A in the control volume. Divide both sides of the previous equation by $(A\Delta z)$ and take the limit as $\Delta z \to 0$

$$\begin{aligned}\frac{\partial}{\partial t}[C_A(t,z)] &= \frac{F(t)}{A}\cdot\left[\lim_{\Delta z\to 0}\frac{C_A(t,z)-C_A(t,z+\Delta z)}{\Delta z}\right] - r_A(t,z)\\ &= \frac{F(t)}{A}\cdot\left[-\frac{\partial C_A(t,z)}{\partial z}\right] - r_A(t,z)\\ &= -u_b(t)\cdot\left[\frac{\partial C_A(t,z)}{\partial z}\right] - r_A(t,z)\end{aligned} \tag{3.66}$$

Similarly, a mole balance on component B is given by

$$\frac{\partial}{\partial t}[C_B(t,z)] = -u_b(t)\cdot\left[\frac{\partial C_B(t,z)}{\partial z}\right] + r_A(t,z) \tag{3.67}$$

The dynamic model consists of two first-order linear hyperbolic PDEs. In order to solve the PDEs, we require two *Initial Conditions,* which can be derived from the steady-state solution of the dynamic model. Setting Eqs. (3.66), (3.67) to zero results in the steady-state solution.

$$0 = -u_b(t) \cdot \left[\frac{\partial C_A(t,z)}{\partial z}\right] - r_A(t,z)$$

$$\begin{aligned} u_b \cdot \frac{dC_A(z)}{dz} &= -k \cdot C_A(z) \\ \int_{C_{Ai}}^{C_A(z)} \frac{dC_A(z)}{C_A(z)} &= -\frac{k}{u_b}\int_0^z dz \\ C_A(z) &= C_{Ai} \cdot e^{-\frac{k}{u_b}z} \end{aligned} \tag{3.68}$$

$$0 = -u_b(t) \cdot \left[\frac{\partial C_B(t,z)}{\partial z}\right] + r_A(t,z)$$

$$\begin{aligned} u_b \cdot \frac{dC_B(z)}{dz} &= -k \cdot C_A(z) \\ \int_{C_{Bi=0}}^{C_B(z)} dC_B(z) &= -\frac{k}{u_b}\int_0^z C_A(z)dz = -\frac{k \cdot C_{Ai}}{u_b}\int_0^z e^{-\frac{k}{u_b}z}dz \\ C_B(z) &= C_{Ai} \cdot \left[1 - e^{-\frac{k}{u_b} \cdot z}\right] \end{aligned} \tag{3.69}$$

Therefore the initial conditions are

$$C_A(0,z) = C_A(z) = C_{Ai} \cdot e^{-\frac{k}{u_b}z} \tag{3.70}$$

$$C_B(0,z) = C_B(z) = C_{Ai} \cdot \left[1 - e^{-\frac{k}{u_b}z}\right] \tag{3.71}$$

Note that the initial conditions use the parameters of previous steady-state iteration. In the simulation, a change has to occur to move either C_{Ai} or u_b from the steady-state condition, otherwise, the previous steady state will be maintained.
Besides the *Initial Conditions*, we also need two *Boundary Conditions*. In this example, the inlet concentrations are assumed constant in the feed stream:

$$C_A(t,0) = C_{Ai} \tag{3.72}$$

$$C_B(t,0) = 0 \tag{3.73}$$

Step 5: Supplementary equations include the reaction kinetics with a constant reaction rate k

$$r_A(t,z) = k\,C_A(t,z) \tag{3.74}$$

The reacting mixture bulk velocity is given by

$$u_b(t) = \frac{F(t)}{A} \tag{3.75}$$

Step 6: Consistency check:
Units: Each term in Eqs. (3.66), (3.67) has the unit of $\left(\frac{\text{mol/m}^3}{\text{s}}\right)$.
Degrees of freedom analysis:

Number of variables	6	$F(t)$, $C_A(t,z)$, $C_B(t,z)$, $C_{Ai}(t)$, $k(t), u_b(t)$
State equations	2	Eqs. (3.66), (3.67)
Auxiliary equations	2	Eqs. (3.74), (3.75)
Disturbances	1	$C_{A,i}$

There is one degree of freedom in this problem. One PI controller equation in velocity form can be written according to the block diagram shown in Fig. 3.27.

$$F_j = F_{j-1} - K_c \cdot [e_j - e_{j-1}] - \frac{K_c}{\tau_I} \cdot e_j; \quad \text{where } [e_j = C_{B,sp} - C_{B,j}(L)] \tag{3.76}$$

Increasing $F(t)$ decreases the residence time in the reactor and reduces the conversion or $C_B(t, L)$. One can determine the trend of C_B with changing $F(t)$. Since $F(t)$ is proportional to $u_b(t)$, changing trend can be predicted by determining the sign of $\frac{dC_B(L)}{d\,u_b}$. Eq. (3.77) indicates that the derivative is always negative. Therefore increasing u_b decreases the outlet concentration of product B and the sign of the controller gain K_c is thus negative.

$$\frac{dC_B(L)}{d\,u_b} = \frac{-e^{\frac{-k\cdot z}{u_b}} \cdot z \cdot k \cdot C_{Ai}}{u_b^2} < 0 \tag{3.77}$$

Step 7: Solution of the PDE is often performed numerically with finite difference method or finite element method. There are several methods to discretize Eqs. (3.66), (3.67) with different orders of local truncation error. In this example, a simple upwind scheme of discretization is used.

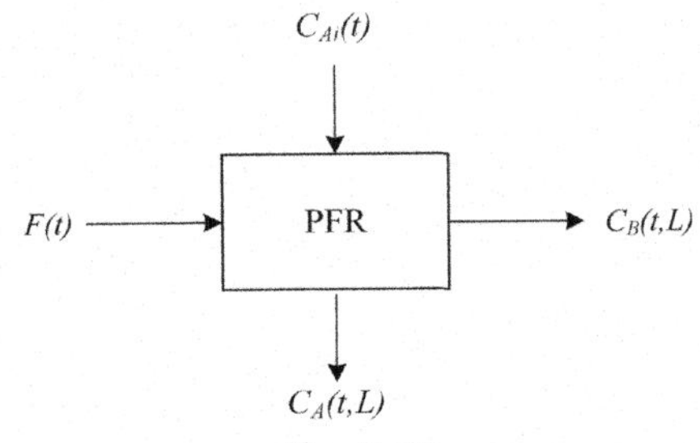

Fig. 3.27
Process block diagram for the PFR.

$$\frac{\partial C_A(t,z)}{\partial z} = \frac{1}{\Delta z} \cdot [C_A(t,z) - C_A(t,z-\Delta z)] + \mathcal{O}(\Delta z) \approx \frac{1}{\Delta z} \cdot \left[C_{A_j}^{i} - C_{A_j}^{i-1}\right] \tag{3.78}$$

$$\frac{\partial C_A(t,z)}{\partial t} = \frac{1}{\Delta t} \cdot [C_A(t+\Delta t,z) - C_A(t,z)] + \mathcal{O}(\Delta t) \approx \frac{1}{\Delta t} \cdot \left[C_{A_{j+1}}^{i} - C_{A_j}^{i}\right] \tag{3.79}$$

where the superscript i stands for the discretized position (z-axis) and the subscript j stands for the discretized time. For example, $u_j^i = u[(j-1)\cdot\Delta t, (i-1)\cdot\Delta z]$. The $\mathcal{O}(\Delta z)$ represents the local truncation error order. In the upwind scheme both time and space error orders are first order. There are methods with second-order accuracy, e.g., Lax-Friedrichs method.[4]

Substitute Eqs. (3.78), (3.79) in Eq. (3.66)

$$\frac{\partial}{\partial t}[C_A(t,z)] = -u_b(t) \cdot \left[\frac{\partial C_A(t,z)}{\partial z}\right] - k \cdot C_A(t,z)$$

$$\frac{1}{\Delta t} \cdot \left[C_{A_{j+1}}^{i} - C_{A_j}^{i}\right] \approx -u_b(t) \cdot \frac{1}{\Delta z} \cdot \left[C_{A_j}^{i} - C_{A_j}^{i-1}\right] - k \cdot C_{A_j}^{i} \tag{3.80}$$

$$C_{A_{j+1}}^{i} \approx C_{A_j}^{i} - u_b \cdot \frac{\Delta t}{\Delta z}\left(C_{A_j}^{i} - C_{A_j}^{i-1}\right) - k \cdot C_{A_j}^{i} \cdot \Delta t$$

Similarly, for component B, the discretized equation is

$$C_{B_{j+1}}^{i} \approx C_{B_j}^{i} - u_b \cdot \frac{\Delta t}{\Delta z}\left(C_{B_j}^{i} - C_{B_j}^{i-1}\right) + k \cdot C_{A_j}^{i} \cdot \Delta t \tag{3.81}$$

The numerical solution involves

Start from initial condition, where $C_{A_{j=1}}^{i}$ and $C_{B_{j=1}}^{i}$ are known. Iterate i from 2 to the maximum i. Then sweep all the nodes except for $i=1$ at the next time increment $j=2$.
Apply the boundary condition to the missing points, $i=1$. $C_{A_{j=2}}^{i=1} = C_{Ai}$ and $C_{B_{j=2}}^{i=1} = 0$
Repeat step 1 for time $j=1$ to $j_{max}-1$.

The selection of the time (Δt) and space (Δz) increments must be made to satisfy the Von Neumann stability condition. For the upwind scheme, the relation is

$$0 < \Delta t < \frac{\Delta z}{\Delta z \cdot k + u_b} \tag{3.82}$$

If the previous condition is not satisfied, the solution will be oscillating or unstable. This can be verified by setting the `Delta_t` ,(Δt), and `Delta_z` ,(Δz), ratio in the following MATLAB code.

The parameters used for the simulation of the PFR are given in the following table:

Variables	Value	Unit
$C_{A,i}$	3	mol/m^3
k	0.01	s^{-1}
A	0.1	m^2
$F(0^-)$	0.005	m^3/s
$F(0^+)$	0.008	m^3/s
K_c	−0.1	–
τ_I	5	–
$C_{B,sp}$	1.5	mol/m^3
L	1	m

PFR simulation script

```
clear;
%% Parameters Setup
k = 0.01;            % Kinetic constant k
A = 0.1;             % PFR cross-sectional area
CAi = 3;             % Inlet concentration of A
L = 1;               % PFR total length
F_previous = 0.005;  % Flowrate before t=0, which is the previous steady state value.
F_current = 0.008;   % Flowrate after t=0, which is for current simulation.
ub_previous = F_previous/A; % Velocity for the previous steady state.
FlagClosedLoop = 0;  % Set to 1 to enable PI-controller, 0 for open loop simulation.

%% Controller setup.
F_bar = F_previous;  % Set controller bias to the previous steady state value.
Kc = 0.1;            % Controller proportional gain.
tau_I = 5;           % Controller integral time.
CBsp = 1.5;          % Set point of concentration of B.
%% Problem scale in geometry and time.
tspan = 200;
zspan = L;
% Solver step size, change the ratio to see the unstable solution.
Delta_t = 0.05;
Delta_z = 0.1;
```

```
% Find the number of discretized time and space. Decimal part is discarded.
% For example, for a pipe length of 1m and 0.1m Delta_z, there are 10 pieces and 11% points.
jmax = fix(tspan/Delta_t) + 1;
imax = fix(zspan/Delta_z) + 1;

%% Initialization
% Allocate solution buffer. Let t represent the row and z the column.
CA = zeros(jmax,imax);
CB = zeros(jmax,imax);
% Set Initial Condition for solution. Real position start from 0, therefore
% it is (i-1)*Delta_z.
i = 1:imax;
Position = (i-1)*Delta_z;
% Apply the initial conditions.
CA(1,:)=CAi*exp(-k/ub_previous*Position);
CB(1,:)=CAi*(1 - exp(-k/ub_previous*Position));
%Initial the PI controller.
e_last = CBsp - CA(1,end);
F_last = F_bar;

%% Simulation
for j=1:jmax-1
% Code for PI-controller in velocity form
  if(FlagClosedLoop == 1)
      e_current = CBsp - CB(j,end);
      F_current = F_last - Kc * (e_current-e_last) - Kc/tau_I * e_current * Delta_t;
      F_current = max (0,F_current); % Flowrate cannot be negative
      F_last = F_current;
      e_last = CBsp - CB(j,end);
  end
 ub_current = F_current/A;
  for i=2:imax
      alpha = ub_current * Delta_t/Delta_z;
      CA(j+1,i)= CA(j,i) - alpha *( CA(j,i) - CA (j,i-1)) - k*CA(j,i)*Delta_t;
      CB(j+1,i)= CB(j,i) - alpha *( CB(j,i) - CB(j,i-1)) + k*CA(j,i)*Delta_t;
  end
% Apply boundary condition as we progress in time
  CA(j+1,1) = CAi;
  CB(j+1,1) = 0;
end
```

```
% Post-processing and plot the results
trange = Delta_t * ((1:jmax)-1);
zrange = Delta_z * ((1:imax)-1);
% Meshgrid(xvector,yvector) to generate a 2-D mesh grid for surf function
[z,t]=meshgrid(zrange,trange);
surf(z,t,CB,'edgecolor','none');

xlabel('z (m)');
ylabel('Time (s)');
zlabel('C_B (mol/m^3)');
```

The open-loop simulation results are presented in Fig. 3.28. It matches the expected trend of a decrease in $C_B(t,L)$ as the reactant flow rate decreases.

With the `FlagClosedLoop` set to 1, one can obtain the closed-loop simulation results with a PI controller shown in Fig. 3.29. The set point tracking takes about 70 s and eventually settles at $C_B = 1.5\,\text{mol/m}^3$ without offset, as expected with a PI controller.

Step 8: Model validation.

The product B concentration at the reactor effluent in open-loop simulation should be consistent with the model given in Eq. (3.71). The validation is presented in Table 3.5. To decrease the error of dynamic simulation, the step size Δt and Δz should be made smaller. However, this will increase the computational time and requires more memory space. It is recommended to select the step sizes wisely to fulfill both efficiency and accuracy.

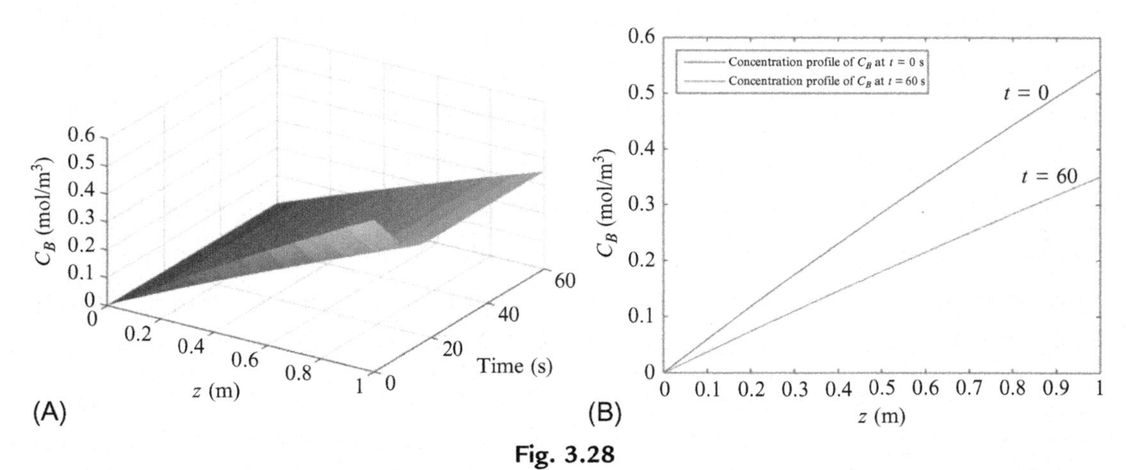

Fig. 3.28

Open-loop simulation results of the PFR. The volumetric flow rate changes from $0.005\,\text{m}^3/\text{s}$ to $0.008\,\text{m}^3/\text{s}$. (A) The surface plot for C_B representing the step change response. (B) The concentration distribution at $t = 0\,\text{s}$ and $t = 60\,\text{s}$.

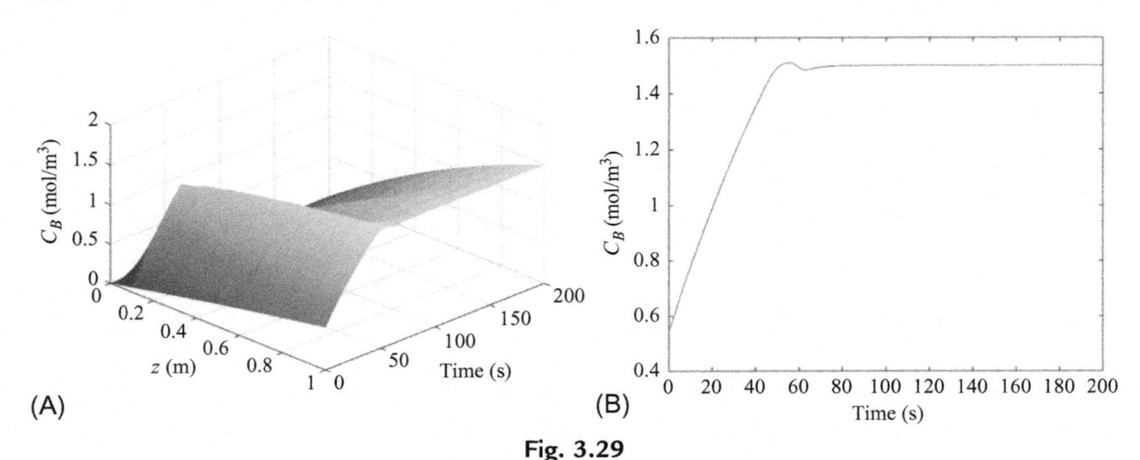

Fig. 3.29
Closed-loop simulation results of the PFR. The controller parameters are as follows: $K_c = 0.1$, $\tau_I = 5$, $C_{B,sp} = 1.5\,\text{mol/m}^3$. (A) The 3D surface plot of the concentration. (B) Concentration response at outlet ($z = L$).

Table 3.5 Steady-state validations of the PFR model, where the tube length $L=1\,\text{m}$, tube cross-sectional area $A=0.1\ \text{m}^2$, inlet concentration $C_{Ai}=3\ \text{mol/m}^3$, and volumetric reactant flow rate $F=0.005\,\text{m}^3/\text{s}$

Parameters k	Calculated Steady-State Outlet Concentration (mol/m^3)	Steady-State Outlet Concentration (mol/m^3) From Dynamic Model
0.001	0.0594	0.05934
0.01	0.5438	0.5390
0.1	2.594	2.5155

The computational time interval and space interval $\Delta t = 0.05\,\text{s}$ and $\Delta z = 0.1\,\text{m}$.

3.6 Problems

For the STH shown in Fig. 3.30, develop a dynamic mathematical model. Carry out a complete degrees of freedom analysis. Use the flow rates of the hot and cold water as the two manipulated variables. Write down the necessary mass and energy balances and the control equations. Use two proportional controllers to maintain the liquid level and the temperature in the tank at set points. Identify the disturbances. The flow rate out of the tank is proportional to the square root of the liquid level in the tank, i.e., $F(t) = \alpha\sqrt{l(t)}$, where α is a constant. Assume constant physical properties (constant density and specific heat). The cross-sectional area of the tank, A, is also assumed to be constant. There are no heat losses from the tank and the content of the tank is well mixed.

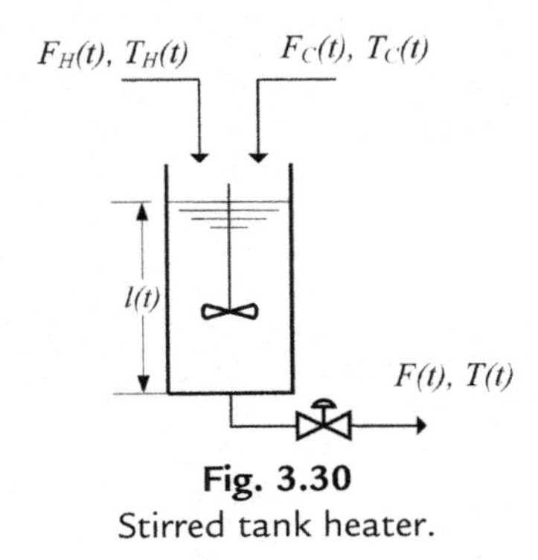

Fig. 3.30
Stirred tank heater.

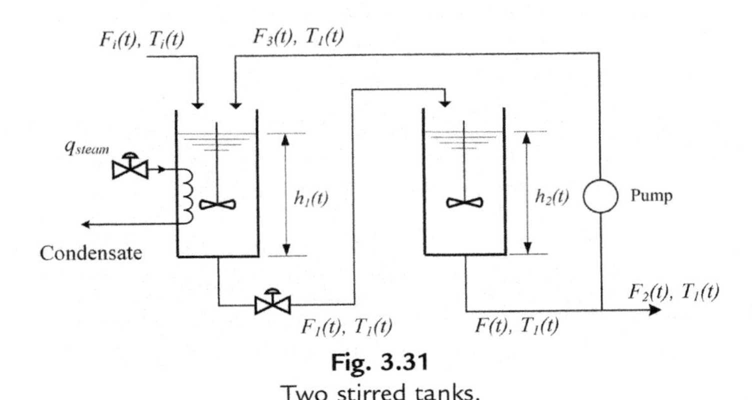

Fig. 3.31
Two stirred tanks.

Develop a dynamic mathematical model (in the time domain, use the 8-step approach discussed in this chapter) for the two stirred tanks shown in Fig. 3.31. Do not include the model of the steam heating coil. F_i, F, F_1, F_2, F_3 are volumetric flow rates (m^3/s) of various streams, h_1, and h_2 are the liquid levels in (m) of the tanks with cross-sectional area A_1 and A_2 (m^2).
The density and specific heats of all the liquid streams are constant and represented by ρ (kg/m^3) and c [kJ/(kg K)], respectively. The saturated steam mass flow rate is q_{steam} in (kg/s) and the heat of condensation is given by λ (kJ/kg). There is no subcooling of the condensate, that is, the total rate of heat released by condensation of the steam is $\lambda \times q_{steam}$ (kW). Both tanks are well mixed and the entire system (the tanks and the pipelines) is perfectly insulted, that is, heat losses are negligible. State any additional simplifying assumption that you make.

Discuss the possibility of controlling the levels of liquid and the temperatures in both tanks. How many manipulated variables do you need to maintain the levels and temperatures at set points? Please note that the temperatures in both tanks are the same, $T_1(t)$. Write down the required controller equations. Use only P controllers and make sure to use the correct sign for the controller proportional gains.

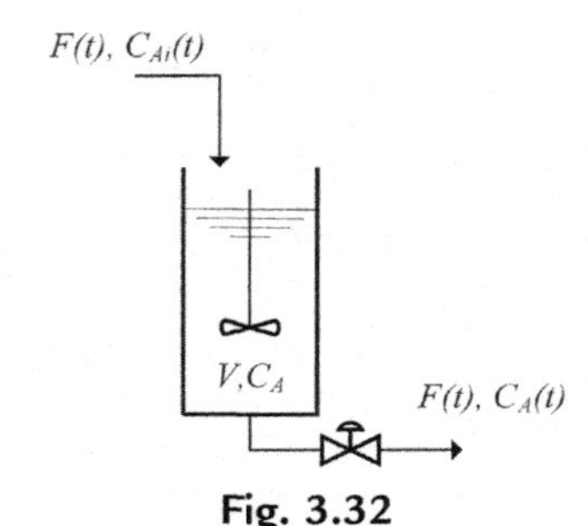

Fig. 3.32
A constant volume isothermal CSTR.

Write down the dynamic equation of a constant volume isothermal continuous stirred tank reactor (CSTR) in which a first-order reaction takes place, $A \rightarrow B$, where A is the reactant molecule and B is the product molecule. The reaction rate per unit volume is given by $r_A = -0.08\,C_A$, where C_A (mole/L) is the concentration of reactant in the reactor which is controlled by manipulating the reactant flow rate into the reactor, $F\,(\mathrm{L/min})$. Make sure that the degree of freedom is zero by incorporating a proportional controller relating the reactant flow rate, F, to the measured reactant concentration in the reactor, C_A.

Write down a MATLAB program to simulate the open loop and closed loop (using a P controller) of the CSTR. Develop the open-loop response of the reactant concentration in the reactor as the inlet reactant flow rate varies. Also investigate the closed-loop response of the system as a function of increasing the controller proportional gain (Fig. 3.32).

References

1. Jordan DW, Smith P. *Nonlinear ordinary differential equations: an introduction for scientists and engineers.* Oxford, NY: Oxford University Press; 2007.
2. Schesser WE, Griffiths GW. *A compendium of partial differential equation methods.* Cambridge, NY: Cambridge University Press; 2009.
3. Luyben WL. *Process modeling, simulation, and control for chemical engineers.* Tokyo: McGraw-Hill; 1973.
4. Bennett M, Rohani S. Solution of population balance equation with a new combined Lax-Wendroff Cranck-Nicholson method. *Chem Engng Sci* 2001;**56**:6623–33.

CHAPTER 4

Development of Linear State-Space Models and Transfer Functions for Chemical Processes

Sohrab Rohani, Yuanyi Wu
Western University, London, ON, Canada

Part A—Theoretical Development of Linear Models

In Chapter 2, we developed theoretical detailed models for typical chemical processes by considering the 8-step procedure involving the principles of conservation of mass, energy, and momentum. We noted that such models in a dynamic manner comprised of nonlinear differential equations, which were difficult to solve analytically and their solutions required numerical techniques. Moreover, the nonlinear dynamic models are not appropriate for the design of linear controllers such as the PID controllers. Therefore there is a need to develop linear dynamic models. There are two approaches to achieve this objective. The first approach is theoretical, while the second approach is empirical based on the analysis of the input-output data series of an existing process. In Part A of this chapter we shall develop two types of linear models by linearizing the nonlinear differential equations. The linearized differential equations in the time domain render the linear state-space models. Laplace transformation of the linearized differential equations, after some mathematical manipulations, results in linear algebraic equations in the Laplace domain or s-domain which are called transfer function models. The tools necessary for the development of the theoretical linear models, namely, the Taylor series expansion for the linearization of the nonlinear terms and Laplace/inverse Laplace transformation of the linearized equations, will be reviewed first. The transfer functions of the first order, second order, and some other simple systems will be developed in detail. In Part B of this chapter, various empirical approaches referred to as process identification methods will be introduced to obtain the linear transfer functions of existing processes.

Coulson and Richardson's Chemical Engineering. http://dx.doi.org/10.1016/B978-0-08-101095-2.00004-7

4.1 Tools to Develop Continuous Linear State-Space and Transfer Function Dynamic Models

4.1.1 Linearization of Nonlinear Differential Equations

Linearization of the nonlinear differential equations can be accomplished by approximating all the nonlinear terms appearing in a given equation by linear terms using the Taylor series expansion formula. A nonlinear function, $f(x)$, which is a function of one single independent variable, x, can be expanded in terms of an infinite series around a given point of linearization, x_0, by the Taylor series, according to the following equation:

$$f(x) = f(x_0) + \left.\frac{df}{dx}\right|_{x=x_0}(x - x_0) + \left.\frac{d^2f}{dx^2}\right|_{x=x_0}\frac{(x - x_0)^2}{2!} + \cdots \tag{4.1}$$

By discarding all the nonlinear terms in Eq. (4.1), a linear approximation is obtained for $f(x)$.

$$f(x) \approx f(x_0) + \left.\frac{df}{dx}\right|_{x=x_0}(x - x_0) = a + b(x - x_0) \tag{4.2}$$

where a and b are constants. The term $(x - x_0)$ is called the **deviation** or **perturbation** from the point of linearization which is often taken as the controller set point in a feedback control system. The approximation is ONLY valid in the vicinity of point x_0, as is shown in Fig. 4.1. If the linearization is performed around the set point value of the controlled variable, the approximation is justified provided that the controller performs its task effectively to maintain the process variable close to its set point.

4.1.1.1 Linearization of nonlinear terms involving more than one independent variable

Similarly, if the nonlinear terms appearing in the dynamic model of a system are functions of more than one independent variable, e.g., x and y, the linearization should be performed using the following equation:

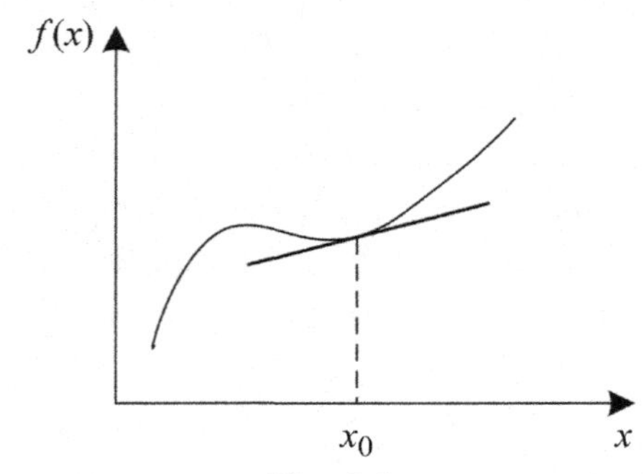

Fig. 4.1
Linearization of function, $f(x)$, around x_0.

$$f(x,y) = f(x_0,y_0) + \left.\frac{\partial f}{\partial x}\right|_{x_0,y_0}(x-x_0) + \left.\frac{\partial f}{\partial y}\right|_{x_0,y_0}(y-y_0) + \left.\frac{\partial^2 f}{\partial x^2}\right|_{x_0,y_0}\frac{(x-x_0)^2}{2!} + \left.\frac{\partial^2 f}{\partial y^2}\right|_{x_0,y_0}\frac{(y-y_0)^2}{2!} + \cdots \tag{4.3}$$

$$f(x,y) \approx f(x_0,y_0) + \left.\frac{\partial f}{\partial x}\right|_{x_0,y_0}(x-x_0) + \left.\frac{\partial f}{\partial y}\right|_{x_0,y_0}(y-y_0) = a + b(x-x_0) + c(y-y_0) \tag{4.4}$$

where a, b, and c are constants and the terms $(x-x_0)$ and $(y-y_0)$ are **deviations** or **perturbations** from the point of linearization or set point values in a feedback control system.

Example 4.1

Consider a liquid storage tank with a control valve installed at the outlet stream, as shown in Fig. 4.2.

The outlet flow rate depends on the valve's opening, $x(t)$, which can be considered as the manipulated variable in a feedback control system, and the pressure drop across the valve that depends on the hydrostatic head of the liquid in the tank, $l(t)$, i.e., $F(t) = \alpha x(t)\sqrt{l(t)}$, where α is the valve's constant. Therefore the mass balance on a constant cross-sectional area (A) cylindrical tank is given by:

$$A\frac{dl(t)}{dt} = F_i(t) - \alpha x(t)\sqrt{l(t)} \tag{4.5}$$

We desire to derive two one-to-one linear equations relating the controlled variable, $l(t)$, to the disturbance, $F_i(t)$, and the manipulated variable, $x(t)$. The nonlinear term $\alpha x(t)\sqrt{l(t)}$ must be linearized.

$$\alpha x(t)\sqrt{l(t)} \approx \alpha\bar{x}\sqrt{\bar{l}} + \alpha\sqrt{\bar{l}}(x-\bar{x}) + \frac{\alpha\bar{x}}{2\sqrt{\bar{l}}}\cdot(l-\bar{l}) \tag{4.6}$$

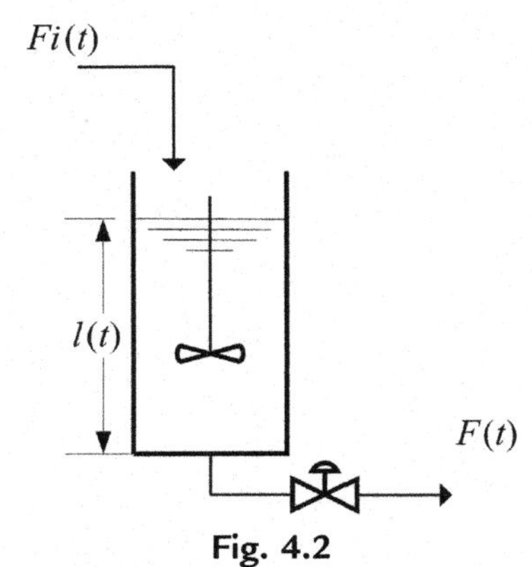

Fig. 4.2
Schematic of a liquid storage tank.

$$A\frac{dl}{dt} = F_i - \alpha\bar{x}\sqrt{\bar{l}} - \alpha\sqrt{\bar{l}}(x-\bar{x}) - \frac{\alpha\bar{x}}{2\sqrt{\bar{l}}}\cdot(l-\bar{l}) \tag{4.7}$$

where $\bar{l}$, $\bar{F}_i$, and $\bar{x}$ are the steady-state values of these variables. Subtracting the steady-state equation from the linearized equation will get rid of the constant term:

$$A\frac{d(l-\bar{l})}{dt} = (F_i - \bar{F}_i) - \alpha\sqrt{\bar{l}}(x-\bar{x}) - \frac{\alpha\bar{x}}{2\sqrt{\bar{l}}}(l-\bar{l}) \tag{4.8}$$

The deviations from the steady-state values of each variable is represented by placing a (′) on the variable:

$$A\frac{dl'}{dt} = F_i' - \alpha\sqrt{\bar{l}}x' - \frac{\alpha\bar{x}}{2\sqrt{\bar{l}}}\cdot l' \tag{4.9}$$

Rearranging the previous equation will render a first-order linearized ODE relating the controlled variable to the manipulated variable and the disturbance, all in terms of the deviation variables.

$$\frac{dl'}{dt} = -\frac{\alpha\bar{x}}{2A\sqrt{\bar{l}}}\cdot l' - \frac{\alpha\sqrt{\bar{l}}}{A}x' + \frac{1}{A}F_i' \tag{4.10}$$

Example 4.2

Let us consider the mass and energy balances in a cylindrical stirred tank heater (STH) shown in Fig. 4.3. It is assumed that the STH is insulated, its content is well mixed, and the density and specific heats of all streams are the same and constant.

The mass and energy balances are:

$$\frac{dV(t)}{dt} = F_H(t) + F_C(t) - F(t) \tag{4.11}$$

$$\frac{d\{V(t)\cdot[T(t)-T_{ref}]\}}{dt} = F_H(t)\cdot[T_H(t)-T_{ref}] + F_C(t)\cdot[T_C(t)-T_{ref}] - F(t)\cdot[T(t)-T_{ref}] \tag{4.12}$$

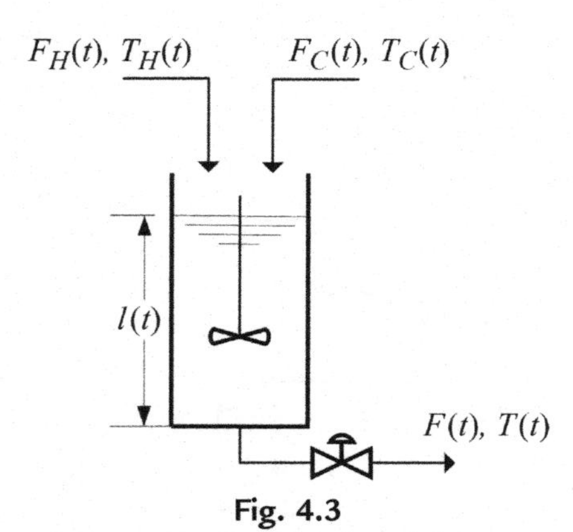

Fig. 4.3
Schematic of a stirred tank heater.

Combining Eqs. (4.11), (4.12) simplifies Eq. (4.12) to:

$$\frac{d[V(t)T(t)]}{dt} = F_H(t)T_H(t) + F_C(t)T_C(t) - F(t)T(t) \tag{4.13}$$

If the inlet and outlet flow rates, as well as their temperatures are time dependent, Eq. (4.13) will involve many nonlinear terms including, $[V(t)\cdot T(t)]$, $[F_H(t)\cdot T_H(t)]$, $[F_C(t)\cdot T_C(t)]$, and $[F(t)\cdot T(t)]$. In addition, $F(t)$ is a nonlinear function of the hydrostatic head, $l(t)$, and the valve's opening $x(t)$, given by $F(t) = \alpha x(t)\sqrt{l(t)}$. In order to develop 10 individual input-output equations that relate the two output variables, $l(t)$ or $V(t)$ and $T(t)$ to the five input variables; $F_H(t)$, $F(t)$ or $x(t)$ (the manipulated variables); and $F_C(t)$, $T_H(t)$, and $T_C(t)$ (the disturbances), we must first linearize Eq. (4.12) to obtain the linear state-space model. Subsequently, Laplace transform the linearized equations to obtain the required transfer functions in the s-domain.

One of the controlled variables can be either the total liquid volume in the tank, $V(t)$, or the liquid level (for a constant cross-sectional area), $l(t)$. The other controlled variable is the liquid temperature in the stirred tank, $T(t)$. Figs. 4.4 and 4.5 show the input and output variables of this system.

Linearization of the nonlinear term $V(t)T(t)$ results in:

$$V(t)T(t) \approx \overline{V}\cdot\overline{T} + \left.\frac{\partial(V\cdot T)}{\partial T}\right|_{\overline{V},\overline{T}}(T-\overline{T}) + \left.\frac{\partial(V\cdot T)}{\partial V}\right|_{\overline{V},\overline{T}}(V-\overline{V}) = \overline{V}\overline{T} + \overline{V}T' + \overline{T}V' \tag{4.14}$$

Similarly, other nonlinear terms can be linearized resulting in:

$$\overline{V}\frac{dT'}{dt} + \overline{T}\frac{dV'}{dt} = \overline{F}_H\overline{T}_H + \overline{F}_H T'_H + \overline{T_H}F'_H + \overline{F}_C\overline{T}_C + \overline{F}_C T'_C + \overline{T'_C}F'_C - \overline{F}\,\overline{T} - \overline{F}T' - \overline{T}F' \tag{4.15}$$

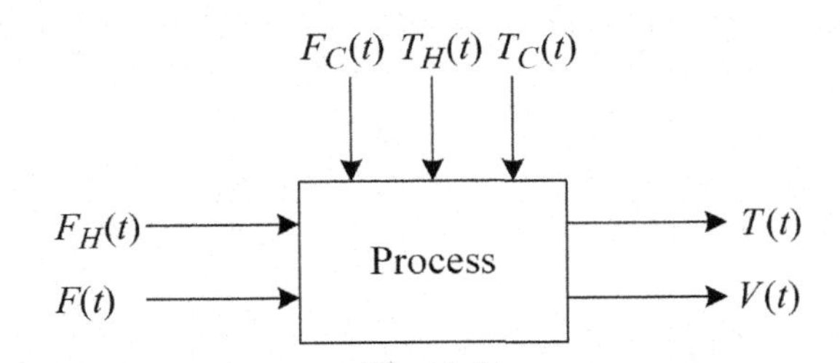

Fig. 4.4
The block diagram of the STH.

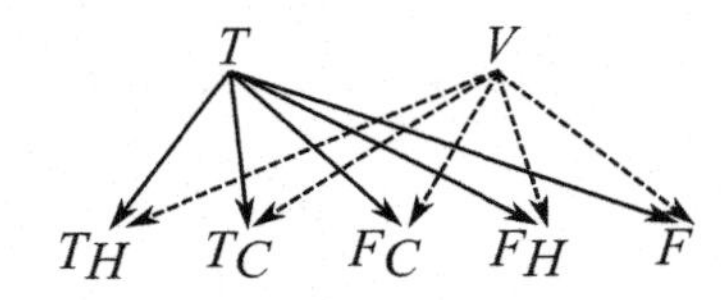

Fig. 4.5
Relations of input and output variables.

4.1.2 The Linear State-Space Models

The general form of a deterministic linear state-space model for both a SISO and a MIMO system is written in the form of:

$$\begin{cases} \dfrac{dx(t)}{dt} = \boldsymbol{A}\,x(t) + \boldsymbol{B}\,u(t) + \boldsymbol{D}\,d(t) \\ y(t) = \boldsymbol{C}\,x(t) \end{cases} \tag{4.16}$$

where $x(t)$ is a column vector which represents the system states; $u(t)$ and $d(t)$ are column vectors for a MIMO system and scalars for a SISO system representing the manipulated variables and disturbances, respectively; and $y(t)$ represents the measured controlled variable(s) column vector (a scalar for a SISO system). $\boldsymbol{A}$, $\boldsymbol{B}$, $\boldsymbol{C}$, and $\boldsymbol{D}$ are matrices of appropriate dimensions which are constant for a linear time-invariant (LTI) stationary system.

Note: A point of warning is in order, the symbol x is used to represent three different variables: the independent variable of a function, $f(x)$; the stem position of a control valve or the valve's opening; and the system states.

Eq. (4.10) representing the linearized dynamic model of a liquid storage tank discussed in Example 4.1 can be presented in the standard state-space form given by Eq. (4.16), by defining:

$$\boldsymbol{A} = -\frac{\alpha\bar{x}}{2\mathrm{A}\sqrt{\bar{l}}}, \quad \boldsymbol{B} = -\frac{\alpha\sqrt{\bar{l}}}{\mathrm{A}}, \quad \boldsymbol{D} = \frac{1}{\mathrm{A}}, \quad \text{and} \quad \boldsymbol{C} = 1 \tag{4.17}$$

Similarly the state-space representation of the process discussed in Example 4.2 is obtained by mathematical manipulation of Eqs. (4.11), (4.15) and use the steady-state equation given by, $\bar{F}_H\bar{T}_H + \bar{F}_C\bar{T}_C - \bar{F}\bar{T} = 0$:

$$\begin{cases} \dfrac{dV'}{dt} = F'_H + F'_C - F' \\ \dfrac{dT'}{dt} = \dfrac{\bar{F}_H}{\bar{V}}T'_H + \dfrac{\bar{F}_C}{\bar{V}}T'_C - \dfrac{\bar{F}}{\bar{V}}T' + \dfrac{\bar{T}_H - \bar{T}}{\bar{V}}F'_H + \dfrac{\bar{T}_C - \bar{T}}{\bar{V}}F'_C \end{cases} \tag{4.18}$$

Note that the variables are in deviation forms and the state vector is defined by $x = [V'\,T']^{\mathrm{T}}$; the manipulated variable is defined as $u = [F'_H F']^{\mathrm{T}}$, the disturbance vector is $d = [F'_C T'_H T'_C]^{\mathrm{T}}$, and the superscript T represents the transpose of a vector. The required matrices are given by:

$$\boldsymbol{A} = \begin{bmatrix} 0 & 0 \\ 0 & -\dfrac{\bar{F}}{\bar{V}} \end{bmatrix}; \quad \boldsymbol{B} = \begin{bmatrix} -1 & 1 \\ 0 & \dfrac{\bar{T}_H - \bar{T}}{\bar{V}} \end{bmatrix}; \quad \boldsymbol{C} = \begin{bmatrix} 1 & 0 \\ 0 & 1 \end{bmatrix}; \quad \boldsymbol{D} = \begin{bmatrix} 1 & 0 & 0 \\ \dfrac{\bar{T}_C - \bar{T}}{\bar{V}} & \dfrac{\bar{F}_H}{\bar{V}} & \dfrac{\bar{F}_C}{\bar{V}} \end{bmatrix} \tag{4.19}$$

4.1.3 Developing Transfer Function Models (T.F.)

A transfer function is defined as the "linear" relationship between the output and the input of a system in "deviation variables" in the Laplace domain by:

$$G(s) = \frac{\text{Laplace transform of the output variable in deviation form}}{\text{Laplace transform of the input variable in deviation form}} \tag{4.20}$$

In order to develop the linear input-output models in the s-domain, a preliminary knowledge of the Laplace transformation is necessary. Therefore, in what follows, we shall review the basic definitions and properties of Laplace transformation and inverse Laplace transformation.

4.1.3.1 Review of Laplace transform (L.T.)

Definition: The Laplace transform of a bounded function $f(t)$ is defined by:

$$\mathcal{L}[f(t)] = F(s) = \int_0^\infty f(t)e^{-st}dt \tag{4.21}$$

where s is the Laplace operator.

4.1.3.2 Laplace transform of simple functions

(1) $f(t) = S(t) = \begin{cases} 0 & t < 0 \\ 1 & t \geq 0 \end{cases}$ (where S is a unit step function shown in Fig. 4.8A)

$$\mathcal{L}[S(t)] = \int_0^\infty S(t)e^{-st}dt = \int_0^\infty 1 \cdot e^{-st}dt = \frac{1}{s} \tag{4.22}$$

(2) $f(t) = a = constant$, or a step function with magnitude, a, that can be presented by $a \cdot S(t)$

$$\mathcal{L}[a.S(t)] = \int_0^\infty a \cdot e^{-st}dt = -\frac{a \cdot e^{-st}}{s}\bigg|_0^\infty = \frac{a}{s} \tag{4.23}$$

(3) Laplace transform of a delayed function $f(t-\theta)$ due to a "time delay" or a "transportation lag," θ.

A delayed function of a nondelayed function $f(t)$ can be mathematically represented by:

$$\text{a delayed function represented as } f(t-\theta) \text{ or } f(t)S(t-\theta) \tag{4.24}$$

where S is the unit step function. The L.T. of the delayed function is:

$$\mathcal{L}[f(t-\theta)] = \int_0^\infty f(t-\theta)e^{-st}dt \tag{4.25}$$

Introducing:

$$\begin{cases} t^* = t - \theta \\ t = t^* + \theta \\ dt = dt^* \end{cases} \tag{4.26}$$

Eq. (4.25) becomes

$$\begin{aligned} \mathcal{L}[f(t-\theta)] &= \mathcal{L}[f(t^*)] = \int_0^\infty f(t^*)e^{-(t^*+\theta)s}dt^* \\ &= e^{-\theta s}\int_0^\infty f(t^*)e^{-t^*s}dt^* \\ &= e^{-\theta s}F(s) \end{aligned} \tag{4.27}$$

where $F(s) = \mathcal{L}[f(t)]$ and θ is the delay time.

An example of a delayed function is the temperature variations of a fluid in a perfectly insulated pipe in a plug flow regime shown in Fig. 4.6. The temperature of the fluid at the exit of the pipe is a delayed function of the inlet temperature. The delay or dead time is equal to the residence time of the liquid in the pipe.

In this case, θ = mean residence time of liquid in the pipe = time delay = $\dfrac{\text{pipe length, } L}{\text{bulk velocity of the liquid, } u_b}$ and the dynamic model of the pipe in the Laplace domain can be represented by $e^{-\theta s}$ which is the ratio of the L.T. of the delayed exit temperature to the L.T. of the inlet temperature given in Figs. 4.7 and 4.8E.

(4) A rectangular pulse function (shown in Fig. 4.8B) given by: $f(t) = \begin{cases} 0 & t < 0 \\ h & 0 < t < t_w \\ 0 & t \geq t_w \end{cases}$

$$\mathcal{L}[f(t)] = \int_0^\infty f(t)e^{-st}dt = \int_0^{t_w} h\,e^{-st}dt = -\frac{h\cdot e^{-st}}{s}\Big|_0^{t_w} = \frac{h}{s}\left(1 - e^{-st_w}\right) \tag{4.28}$$

(5) A Dirac delta function or a unit impulse function $f(t) = \delta(t)$ (shown in Fig. 4.8C), where $\int_{-\infty}^{\infty} \delta(t)dt = 1$ with a "strength" or an "area" of unity, a width of zero, and a height

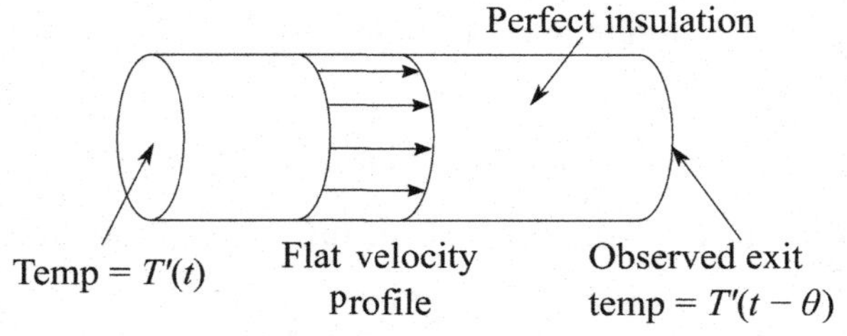

Fig. 4.6
Transportation delay of temperature in a perfectly insulated tube.

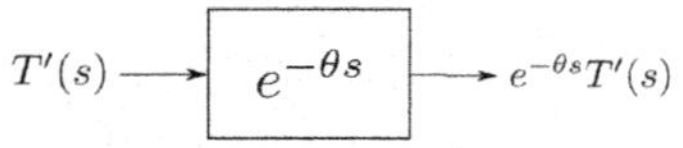

$$T'(s) \longrightarrow \boxed{e^{-\theta s}} \longrightarrow e^{-\theta s}T'(s)$$

Fig. 4.7
A pure time delay process block diagram.

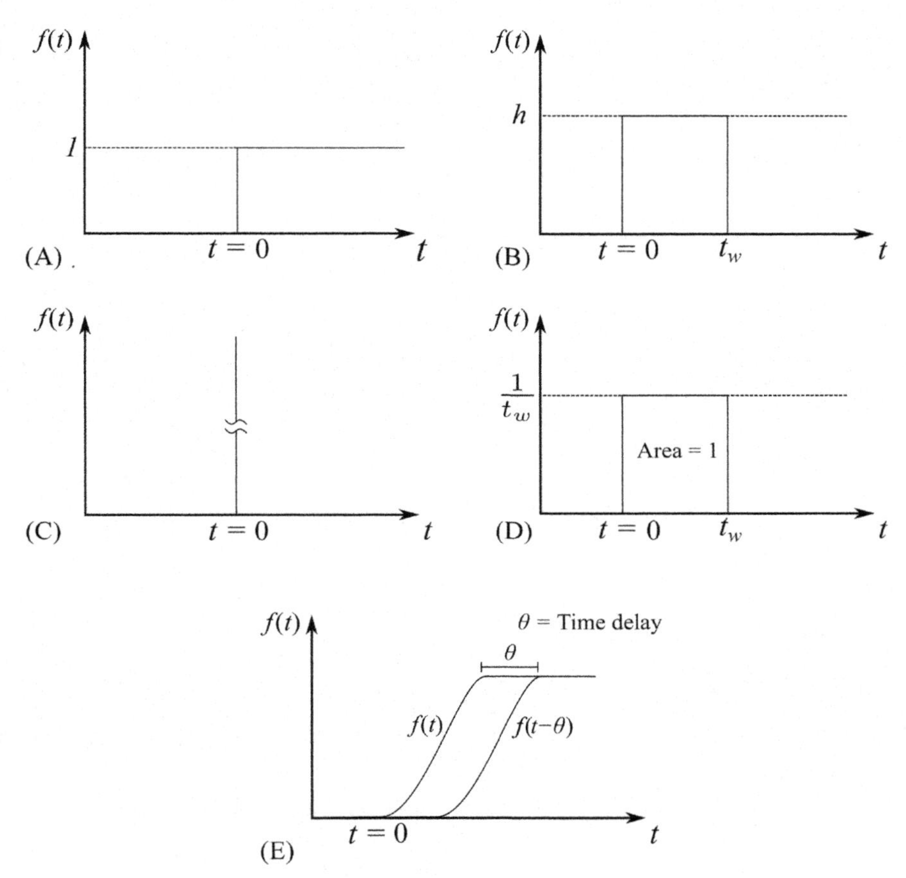

Fig. 4.8
Time-domain shapes of several functions: (A) a step change, (B) a rectangular impulse, (C) a Dirac function (unit impulse), (D) derivation of Dirac function, and (E) a time delayed function.

of infinity appearing at $t = 0$. Therefore $\delta(t)$ can be defined as the limit of a rectangular pulse function with area = 1, as the width goes to zero (Fig. 4.8D).

$$\mathcal{L}[\delta(t)] = \lim_{t_w \to 0} \frac{h}{s}\left(1 - e^{-t_w s}\right) \quad \text{where } h = \frac{1}{t_w} \tag{4.29}$$

Use L'Hôpital's rule and differentiate with respect to t_w,

$$\mathcal{L}[\delta(t)] = \lim_{t_w \to 0} \frac{\frac{d}{dt_w}(1 - e^{-t_w s})}{\frac{d}{dt_w}(s \cdot t_w)} = \lim_{t_w \to 0} \frac{e^{-t_w \cdot s} \cdot s}{s} = 1 \tag{4.30}$$

(6) Laplace transform of the derivative of a function defined as,

$$\mathcal{L}\left[\frac{df(t)}{dt}\right] = \int_0^\infty \frac{df(t)}{dt} e^{-st} dt \tag{4.31}$$

can be obtained using integration by parts:

$$\begin{aligned} u &= e^{-st} & dv &= df(t) \\ du &= -s \cdot e^{-st} \cdot dt & v &= f(t) \end{aligned} \tag{4.32}$$

Therefore

$$\begin{aligned} \mathcal{L}\left[\frac{df(t)}{dt}\right] &= \int_0^\infty \frac{df(t)}{dt} e^{-st} dt = \int_{t=0}^{t\to\infty} u\,dv = \int_{t=0}^{t\to\infty} [d(u \cdot v) - v\,du] \\ &= [u \cdot v]|_{t=0}^{t\to\infty} - \int_{t=0}^{t\to\infty} v\,du \\ &= [0 - f(0)] + s \cdot \int_0^\infty f(t) \cdot e^{-st} \cdot dt \end{aligned} \tag{4.33}$$

Resulting in:

$$\mathcal{L}\left[\frac{df(t)}{dt}\right] = s \cdot F(s) - f(0) \tag{4.34}$$

For the case where $f(0) = 0$, the pictorial presentation of the previous equation in the time domain and Laplace domain is given in Fig. 4.9.
Similarly, for the nth-order derivative of a function, the Laplace transform is

$$\mathcal{L}\left[\frac{d^n f(t)}{dt^n}\right] = s^n F(s) - s^{n-1} f(0) - s^{n-2} \left.\frac{df}{dt}\right|_{t=0} \cdots - \left.\frac{d^{n-1} f(t)}{dt^{n-1}}\right|_{t=0} \tag{4.35}$$

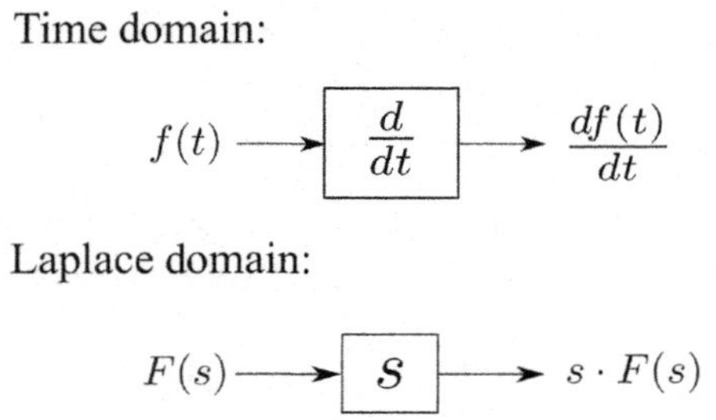

Time domain:

$$f(t) \longrightarrow \boxed{\frac{d}{dt}} \longrightarrow \frac{df(t)}{dt}$$

Laplace domain:

$$F(s) \longrightarrow \boxed{s} \longrightarrow s \cdot F(s)$$

Fig. 4.9
The block diagrams of the derivative operation in the time domain and Laplace domain.

(7) Laplace transform of the integral of a bounded function defined as,

$$\mathcal{L}\left[\int_0^t f(t)dt\right] = \int_0^\infty \left[\int_0^t f(t)dt\right] e^{-st}dt$$

can be obtained by integration by parts:

$$\begin{aligned} u &= \int_0^t f(t)dt \quad dv = e^{-st}dt \\ du &= f(t)dt \quad v = -\frac{e^{-st}}{s} \end{aligned} \tag{4.36}$$

$$\begin{aligned} \mathcal{L}\left[\int_0^t f(t)dt\right] &= \int_{t=0}^{t\to\infty} u\,dv = [u\cdot v]\Big|_{t=0}^{t\to\infty} - \int_{t=0}^{t\to\infty} v\,du \\ &= 0 + \frac{1}{s}\cdot\int_0^t f(t)e^{-st}\,dt \end{aligned} \tag{4.37}$$

Resulting in the following equation and pictorial presentation given in Fig. 4.10.

$$\mathcal{L}\left[\int_0^t f(t)dt\right] = \frac{1}{s}\cdot F(s) \tag{4.38}$$

(8) Laplace transform of an exponential function

$$f(t) = e^{-at} \tag{4.39}$$

$$\mathcal{L}[e^{-at}] = \int_0^\infty e^{-at}e^{-st}dt = \int_0^\infty e^{-(s+a)}dt = \frac{1}{s+a} \tag{4.40}$$

In general, if we have Laplace transform of a function $f(t)$ as $F(s)$, the Laplace transform of the function multiplied by an exponential function is given by the Laplace transform of the function in which s is replaced by $s+a$:

$$\mathcal{L}[e^{-at}f(t)] = F(s+a) \tag{4.41}$$

For example, if $f(t) = 1$, then $F(s) = \mathcal{L}[f(t)] = \frac{1}{s}$ and $\mathcal{L}[e^{-at}f(t)] = \frac{1}{s+a}$.

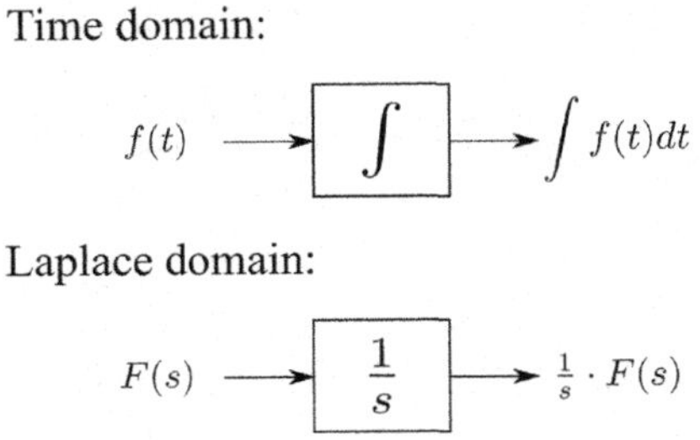

Time domain:

$f(t) \longrightarrow \boxed{\int} \longrightarrow \int f(t)dt$

Laplace domain:

$F(s) \longrightarrow \boxed{\frac{1}{s}} \longrightarrow \frac{1}{s}\cdot F(s)$

Fig. 4.10
Pictorial presentation of the integral operation in the time and Laplace domains.

(9) The initial value theorem

$$\lim_{t \to 0} f(t) = \lim_{s \to \infty} F(s) \tag{4.42}$$

(10) The final value theorem of Laplace transformation
The steady-state value or the final value of a function may be obtained from its Laplace transform, without having to take its inverse Laplace.

$$\begin{aligned} f(t \to \infty) &= \lim_{t \to \infty} f(t) \\ &= \lim_{s \to 0} [s \cdot F(s)] \end{aligned} \tag{4.43}$$

Proof: Recall that L.T. of a derivative of a function is given by:

$$\begin{aligned} &\mathcal{L}\left[\frac{df(t)}{dt}\right] = s \cdot F(s) - f(0) \\ &\lim_{s \to 0}\left[\int_0^\infty \frac{df(t)}{dt} e^{-st} dt\right] = s \cdot F(s) - f(0) \\ &\int_0^\infty \frac{df(t)}{dt} \cdot \lim_{s \to 0} e^{-st} dt = \lim_{s \to 0} s \cdot F(s) - f(0) \\ &f(\infty) - f(0) = \lim_{s \to 0} [s \cdot F(s)] - f(0) \\ &f(\infty) = \lim_{s \to 0} [s \cdot F(s)] \end{aligned} \tag{4.44}$$

Which proves Eq. (4.43).

(11) Superimposition theory. Since L.T. is a linear operator, we have:

$$\mathcal{L}[a \cdot f_1(t) + b \cdot f_2(t)] = a \cdot F_1(s) + b \cdot F_2(s) \tag{4.45}$$

Table 4.1 lists the Laplace transform of a few simple functions.

4.1.3.3 Inverse Laplace transform

In general, the input-output relations of a given process in Laplace domain can be written as:

$$\underbrace{Y(s)}_{\text{Controlled variable}} = \underbrace{G_p(s)}_{\text{Process transfer function}} \cdot \underbrace{U(s)}_{\text{Manipulated variable}} + \underbrace{G_d(s)}_{\text{Load transfer function}} \cdot \underbrace{D(s)}_{\text{Load}} \tag{4.46}$$

where $G_p(s)$ and $G_d(s)$ are the process and load transfer functions that are the ratios of the L.T. of the output variable to the manipulated variable and disturbance, respectively.

Table 4.1 Laplace transform of common functions

Time Domain $f(t)$	Laplace Domain $F(s)$	Remark
$\delta(t)$	1	Dirac unit impulse
$\delta(t-a)$	$e^{-a \cdot s}$	Shifted unit impulse
1	$\frac{1}{s}$	Unit step change
e^{at}	$\frac{1}{s-a}$	a is a constant
t^n	$\frac{n!}{s^{n+1}}$	$n=1,2,3\ldots$
$t^n e^{at}$	$\frac{n!}{(s-a)^{n+1}}$	$n=1,2,3\ldots$ and a is a constant
$\sin(at)$	$\frac{a}{s^2+a^2}$	a is a constant
$\cos(at)$	$\frac{s}{s^2+a^2}$	a is a constant
$\sinh(at)$	$\frac{a}{s^2-a^2}$	a is a constant
$\cosh(at)$	$\frac{s}{s^2-a^2}$	a is a constant
$e^{at}\sin(b\,t)$	$\frac{b}{(s-a)^2+b^2}$	a,b are constants
$e^{at}\cos(b\,t)$	$\frac{s-a}{(s-a)^2+b^2}$	a,b are constants
$e^{at}\sinh(b\,t)$	$\frac{b}{(s-a)^2-b^2}$	a,b are constants
$e^{at}\cosh(b\,t)$	$\frac{s-a}{(s-a)^2-b^2}$	a,b are constants

The product of $G_p(s)\cdot U(s)$ or $G_d(s)\cdot D(s)$ can be written in terms of the ratio of two general polynomials, for example, $G_p(s)\cdot U(s)=\frac{Q_1(s)}{P_1(s)}$ and $G_d(s)\cdot D(s)=\frac{Q_2(s)}{P_2(s)}$

$$\begin{aligned} y(t) &= \mathcal{L}^{-1}[Y(s)] = \mathcal{L}^{-1}\left[G_p(s)\cdot U(s)\right] + \mathcal{L}^{-1}[G_d(s)\cdot D(s)] \\ &= \mathcal{L}^{-1}\left[\frac{Q_1(s)}{P_1(s)}\right] + \mathcal{L}^{-1}\left[\frac{Q_2(s)}{P_2(s)}\right] \end{aligned} \tag{4.47}$$

To obtain the inverse L.T., $y(t)$, $P_1(s)$ and $P_2(s)$, or in general $P(s)$, must be factored into first- or second-degree polynomials for which the inverse L.T.s are available in the Laplace transform table (e.g., Table 4.1). Depending on the nature of the roots of $P_1(s)$ and $P_2(s)$ or $P(s)$, the denominator polynomial, three cases may arise.

Case I: Roots of $P(s)$ are all real and distinct

An example of this case is discussed as follows.

Example 4.3

Find the inverse Laplace transform of

$$Y(s) = G_p(s)U(s) = \frac{s^2 - s - 6}{s^3 - 2s^2 - s + 2} \tag{4.48}$$

Roots of $P(s) = s^3 - 2s^2 - s + 2 = 0$ are:

$$p_1 = -1; p_2 = +1; p_3 = +2$$

Rewrite $P(s) = s^3 - 2s^2 - s + 2 = (s+1)(s-1)(s-2)$.

Expand Eq. (4.48) in the partial fraction form,

$$Y(s) = \frac{s^2 - s - 6}{s^3 - 2s^2 - s + 2} = \frac{s^2 - s - 6}{(s+1)(s-1)(s-2)} = \frac{a_1}{s+1} + \frac{a_2}{s-1} + \frac{a_3}{s-2} \tag{4.49}$$

where a_i is a constant whose value can be obtained by either of the following two methods:

Method 1: Multiply both sides of Eq. (3.48) by $P(s)$ and equate the coefficients of like orders of "s."

$$s^2 - s - 6 = a_1(s-1)(s-2) + a_2(s+1)(s-2) + a_3(s+1)(s-1) \tag{4.50}$$

$$\begin{cases} s^2: \ 1 = a_1 + a_2 + a_3 \\ s^1: \ -1 = -3a_1 - a_2 \\ s^0: \ -6 = 2a_1 - 2a_2 - a_3 \end{cases} \tag{4.51}$$

Solving the previous three algebraic equations results in:

$$a_1 = -\frac{2}{3}; \ a_2 = 3; \ a_3 = -\frac{4}{3} \tag{4.52}$$

Method 2: This method uses the Heaviside's theorem. Multiplying both sides of Eq. (4.49) by the denominator of a_1, $(s+1)$, and taking the limit as $s \to p_1 = -1$, results in:

$$a_1 = \lim_{s \to -1} \frac{s^2 - s - 6}{(s-1)(s-2)} = -\frac{2}{3} \tag{4.53}$$

Similarly,

$$a_2 = \lim_{s \to 1} \frac{s^2 - s - 6}{(s+1)(s-2)} = 3 \tag{4.54}$$

$$a_3 = \lim_{s \to 2} \frac{s^2 - s - 6}{(s+1)(s-1)} = -\frac{4}{3} \tag{4.55}$$

Method 2 is quicker and easier to use. After all the constants are calculated, the inverse L.T. of Eq. (4.49) can be obtained using the table of Laplace Transform (Table 4.1)

$$\begin{aligned} y(t) &= \mathcal{L}^{-1}[Y(s)] = \mathcal{L}^{-1}\left[\frac{-2/3}{s+1} + \frac{3}{s-1} + \frac{-4/3}{s-2}\right] \\ &= -\frac{2}{3}\cdot\mathcal{L}^{-1}\left[\frac{1}{s+1}\right] + 3\cdot\mathcal{L}^{-1}\left[\frac{1}{s-1}\right] - \frac{4}{3}\mathcal{L}^{-1}\left[\frac{1}{s-2}\right] \end{aligned} \tag{4.56}$$

$$y(t) = -\frac{2}{3}\cdot e^{-t} + 3\cdot e^{t} - \frac{4}{3}\cdot e^{2t} \tag{4.57}$$

Let us analyze the stability of $y(t)$ in the time domain. As $t \rightarrow +\infty$, the three terms in Eq. (4.58) approach:

$$\begin{cases} -\frac{2}{3}\cdot e^{-t} \rightarrow 0 & \\ 3\cdot e^{t} \rightarrow \infty & \text{(goes to plus infinity, unstable)} \\ -\frac{4}{3}\cdot e^{2t} \rightarrow -\infty & \text{(goes to negative infinity faster)} \end{cases} \tag{4.58}$$

The function $y(t)$ is unstable (see Fig. 4.11, as the last two terms are unbounded). Therefore if one or more roots of $P(s)$ polynomial are positive, the process becomes unstable.

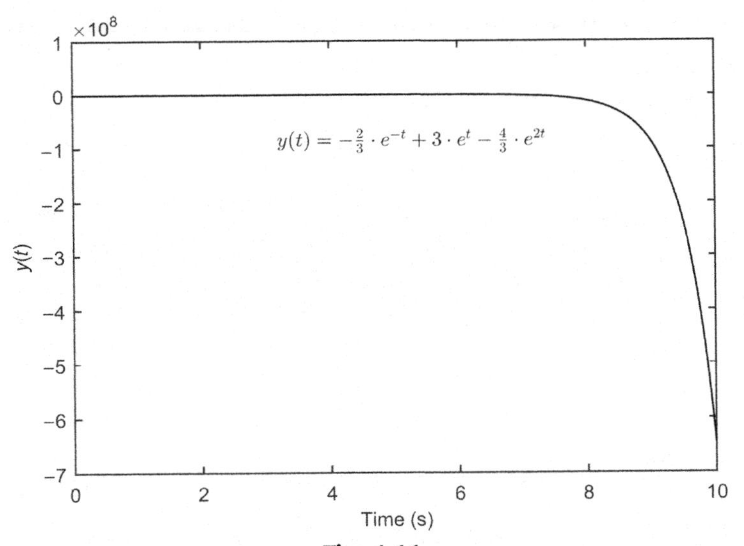

Fig. 4.11
Time-domain response for Example 4.3.

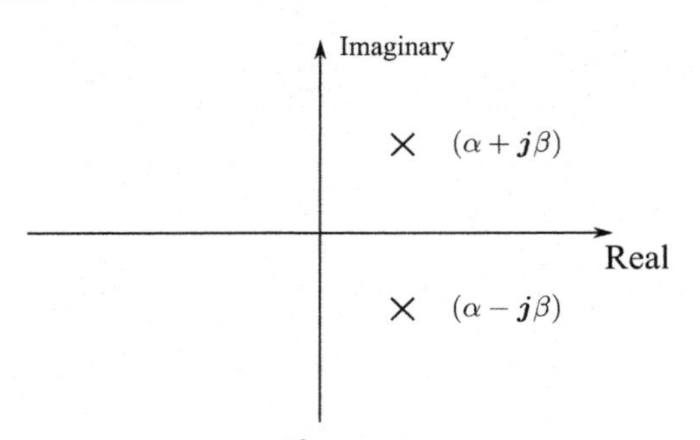

Fig. 4.12
Complex plane (s-plane).

Case II: Roots of P(s) are complex conjugates

Fig. 4.12 depicts the complex s-plane with the roots of the polynomial $P(s)$ indicated by two crosses X, at

$$p_{1,2} = +\alpha \pm j\beta \quad \text{where } j = \sqrt{-1} \tag{4.59}$$

Example 4.4

Find the inverse Laplace transform of

$$Y(s) = \frac{Q(s)}{P(s)} = \frac{s+1}{s^2 - 2s + 5} \tag{4.60}$$

The roots of $P(s) = s^2 - 2s + 5 = 0$ are:

$$p_{1,2} = 1 \pm 2j \tag{4.61}$$

Partial fraction yields,

$$\frac{s+1}{s^2 - 2s + 5} = \frac{a_1}{s - 1 - 2j} + \frac{a_2}{s - 1 + 2j} \tag{4.62}$$

Using Heaviside's theorem,

$$a_1 = \lim_{s \to 1 + 2j} \frac{s+1}{s - 1 + 2j} = \frac{2(1+j)}{4j} = \frac{1+j}{2j} \tag{4.63}$$

Multiplying both the numerator and the denominator by j to obtain a_1

$$a_1 = 0.5 - 0.5j \tag{4.64}$$

$$a_2 = \lim_{s \to 1 - 2j} \frac{s+1}{s - 1 - 2j} = \frac{2(1-j)}{-4j} \times \frac{j}{j} = 0.5 + 0.5j \tag{4.65}$$

Note that a_1 and a_2 are always complex conjugate pairs.

$$
\begin{aligned}
y(t) &= \mathcal{L}^{-1}[Y(s)] = \mathcal{L}^{-1}\left[\frac{0.5-0.5j}{s-1-2j}\right] + \mathcal{L}^{-1}\left[\frac{0.5+0.5j}{s-1+2j}\right] \\
&= (0.5-0.5j)\,e^{(1+2j)t} + (0.5+0.5j)\,e^{(1-2j)t} \\
&= 0.5\,e^{t}\left[(1-j)e^{2jt} + (1+j)e^{-2jt}\right]
\end{aligned}
\tag{4.66}
$$

From trigonometry,

$$
e^{\pm j\alpha} = \cos\alpha \pm j\sin\alpha \tag{4.67}
$$

$$
\begin{aligned}
y(t) &= 0.5e^{t}\cdot[(1-j)(\cos 2t + j\sin 2t) + (1+j)(\cos 2t - j\sin 2t)] \\
&= 0.5e^{t}\cdot[\cos 2t + j\sin 2t - j\cos 2t + \sin 2t + \cos 2t - j\sin 2t + j\cos 2t + \sin 2t] \\
&= e^{t}\cdot[\cos 2t + \sin 2t]
\end{aligned}
\tag{4.68}
$$

Using the trigonometric equation: $r\sin(\alpha+\beta) = r\sin\alpha\cos\beta + r\cos\alpha\sin\beta$, results in:

$$
\cos 2t + \sin 2t = \sqrt{2}\sin\left(2t + \frac{\pi}{4}\right) \tag{4.69}
$$

$$
\therefore y(t) = \sqrt{2}\,e^{t}\sin\left(2t + \frac{\pi}{4}\right) \tag{4.70}
$$

Recall that the two complex roots were at $p_{1,2} = +1 \pm 2j$. The exponent of the exponential part ($e^{1\times t}$) of the response originates from the real part of the complex roots, and the frequency (=2) of the sinusoidal part of the response $\sin\left(2t + \frac{\pi}{4}\right)$, is the imaginary part of the complex roots. This example shows that if the real part of the complex roots of the denominator polynomial is positive, the system becomes unstable, in an oscillatory manner with an ever-increasing amplitude (see Fig. 4.13).

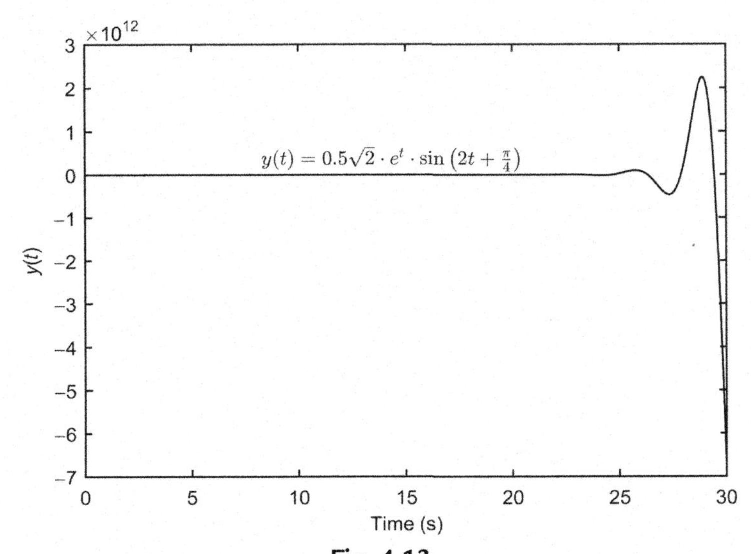

Fig. 4.13
Time-domain response for Example 4.4.

Case III: P(s) has multiple roots

The denominator polynomial may have a multiple real root. An example of such a case is discussed as follows.

Example 4.5

Find the inverse Laplace transform of

$$Y(s)=G_p(s)U(s)=\frac{Q(s)}{P(s)}=\frac{1}{(s+1)^3(s+2)} \tag{4.71}$$

Note that $P(s)$ has four roots, one at -2, and three multiple roots at -1. Expand Eq. (4.71) with partial fraction,

$$Y(s)=\frac{a_1}{s+1}+\frac{a_2}{(s+1)^2}+\frac{a_3}{(s+1)^3}+\frac{a_4}{s+2} \tag{4.72}$$

Use Heaviside's theorem,

$$a_4=\lim_{s\to-2}\frac{1}{(s+1)^3}=-1 \tag{4.73}$$

$$a_3=\lim_{s\to-1}\frac{1}{s+2}=1 \tag{4.74}$$

$$a_2=\lim_{s\to-1}\frac{1}{(s+1)(s+2)} \text{ and } a_1=\lim_{s\to-1}\frac{1}{(s+1)^2(s+2)} \tag{4.75}$$

$$a_2=\lim_{s\to-1}\frac{1}{(s+1)(s+2)} \text{ and } a_1=\lim_{s\to-1}\frac{1}{(s+1)^2(s+2)}$$

Note that the usual Heaviside's theorem does not work for the calculation of a_2 and a_1, therefore, we take the first and the second derivatives of the numerator and denominator to find a_2 and a_1.

$$a_2=\lim_{s\to-1}\frac{d}{ds}\left(\frac{1}{s+2}\right)=\frac{-1}{(s+2)^2}=-1 \tag{4.76}$$

$$a_1=\lim_{s\to-1}\frac{d^2}{ds^2}\frac{1}{2!}\left(\frac{1}{s+2}\right)=1 \tag{4.77}$$

In general, with n multiple roots, the ith multiple root is:

$$a_i=\lim_{s\to i^{th}\text{ root}}\left\{\frac{d^{n-i}}{ds^{n-i}}\frac{1}{(n-i)!}[Y(s)(s-p_i)^n]\right\} \tag{4.78}$$

Substitute the coefficients in Eq. (4.72),

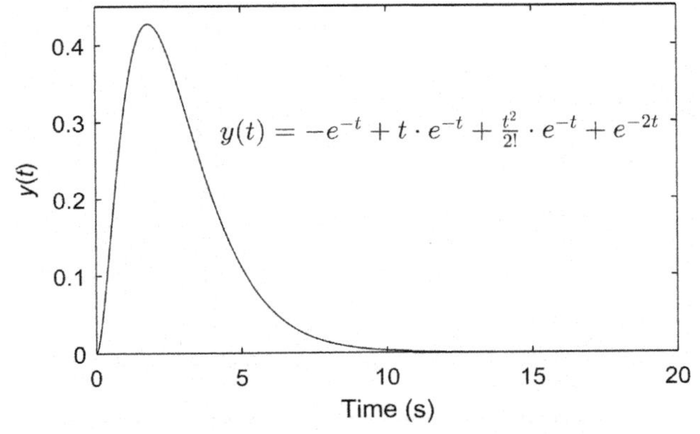

Fig. 4.14
Time-domain response for Example 4.5.

$$Y(s) = \frac{-1}{s+1} + \frac{1}{(s+1)^2} - \frac{1}{(s+1)^3} + \frac{1}{s+2} \tag{4.79}$$

$$y(t) = \mathcal{L}^{-1}[Y(s)] = -e^{-t} + te^{-t} + \frac{t^2}{2!}e^{-t} + e^{-2t} \tag{4.80}$$

Note that the process output, $y(t)$, approaches zero at steady state as is shown in Fig. 4.14, therefore the system is stable.

4.1.3.4 *Use of Laplace transform to solve differential equations*

One major application of the Laplace transformation is the solution of linear ordinary differential equations. Consider the following examples.

Example 4.6

A first-order linear ordinary differential equation (ODE) can be written as:

$$a_1 \frac{dy(t)}{dt} + y(t) = b_1 u(t) \quad \text{Assume } y(0) = 0 \tag{4.81}$$

Presenting $\mathcal{L}[y(t)]$ by $Y(s)$ and $\mathcal{L}[u(t)]$ by $U(s)$, we can L.T. Eq. (4.81):

$$\mathcal{L}\left[a_1 \frac{dy(t)}{dt} + y(t)\right] = \mathcal{L}[b_1 u(t)] \tag{4.82}$$

Use the superimposition theory,

$$a_1 \mathcal{L}\left[\frac{dy(t)}{dt}\right] + \mathcal{L}[y(t)] = \mathcal{L}[b_1 u(t)]; \; Y(s) = \frac{b_1}{a_1 s + 1} U(s) \tag{4.83}$$

$$a_1 [s \cdot Y(s) - y(0)] + Y(s) = b_1 U(s) \tag{4.84}$$

Assuming $y(0) = 0$, we have:

$$y(t) = \mathcal{L}^{-1}[Y(s)] = \mathcal{L}^{-1}\left[\frac{b_1}{a_1 s + 1} U(s)\right] \tag{4.85}$$

For example, if the input function $u(t)$ were an impulse function at $t = 0$, $U(s)$ would be 1. Insert in Eq. (4.85),

$$y(t) = \mathcal{L}^{-1}[Y(s)] = \mathcal{L}^{-1}\left[\frac{b_1}{a_1 s + 1}\right] = \frac{b_1}{a_1} \cdot \mathcal{L}^{-1}\left[\frac{1}{s + \frac{1}{a_1}}\right] = \frac{b_1}{a_1} \cdot e^{-\frac{t}{a_1}} \tag{4.86}$$

Example 4.7

Solving a second-order linear ODE by Laplace transformation.

$$a_1 \frac{dy^2}{dt^2} + a_2 \frac{dy}{dt} + a_3 y(t) = b_1\, u(t) \quad \left[\text{Assume}\, y(0) = \left.\frac{dy}{dt}\right|_{t=0} = \left.\frac{dy^2}{dt^2}\right|_{t=0} = 0\right] \tag{4.87}$$

Laplace transform,

$$a_1 s^2\, Y(s) + a_2\, s\, Y(s) + a_3\, Y(s) = b_1\, U(s) \tag{4.88}$$

$$Y(s) = \frac{b_1}{a_1 s^2 + a_2 s + a_3} U(s) \tag{4.89}$$

Inverse Laplace transform,

$$y(t) = \mathcal{L}^{-1}[Y(s)] = \mathcal{L}^{-1}\left[\frac{b_1}{a_1 s^2 + a_2 s + a_3} U(s)\right] \tag{4.90}$$

Example 4.8

Solve two coupled differential equations that may represent a 2×2 process (2-input2-output). Consider

$$\begin{cases} \dfrac{dy_1(t)}{dt} = a_{11} y_1(t) + a_{12} y_2(t) + b_{11}\, u_1(t) + b_{12}\, u_2(t) \\ \dfrac{dy_2(t)}{dt} = a_{21} y_1(t) + a_{22} y_2(t) + b_{21}\, u_1(t) + b_{22}\, u_2(t) \end{cases} \tag{4.91}$$

which fits the state-space representation of the form:

$$\begin{aligned} \frac{dx(t)}{dt} &= \boldsymbol{A}\, x(t) + \boldsymbol{B}\, u(t) \\ y(t) &= \mathbf{C}\, x(t) \end{aligned} \tag{4.92}$$

With

$$\boldsymbol{A} = \begin{bmatrix} a_{11} & a_{12} \\ a_{21} & a_{22} \end{bmatrix}, \quad \boldsymbol{B} = \begin{bmatrix} b_{11} & b_{12} \\ b_{21} & b_{22} \end{bmatrix}, \quad \text{and } \mathbf{C} = \begin{bmatrix} 1 & 0 \\ 0 & 1 \end{bmatrix} \tag{4.93}$$

Laplace transform both equations and rearrange (assuming that the initial conditions are zero),

$$\begin{cases} sY_1(s) = a_{11}Y_1(s) + a_{12}Y_2(s) + b_{11}U_1(s) + b_{12}U_2(s) \\ sY_2(s) = a_{21}Y_1(s) + a_{22}Y_2(s) + b_{21}U_1(s) + b_{22}U_2(s) \end{cases} \tag{4.94}$$

Convert Eq. (4.94) into matrix form,

$$\begin{bmatrix} s-a_{11} & -a_{12} \\ -a_{21} & s-a_{22} \end{bmatrix} \underbrace{\begin{bmatrix} Y_1(s) \\ Y_2(s) \end{bmatrix}}_{\boldsymbol{Y}(s)} = \underbrace{\begin{bmatrix} b_{11} & b_{12} \\ b_{21} & b_{22} \end{bmatrix}}_{\boldsymbol{B}} \underbrace{\begin{bmatrix} U_1(s) \\ U_2(s) \end{bmatrix}}_{\boldsymbol{U}(s)} \tag{4.95}$$

Denote the identity matrix $\boldsymbol{I} = \begin{bmatrix} 1 & 0 \\ 0 & 1 \end{bmatrix}$

$$(s \cdot \boldsymbol{I} - \boldsymbol{A}) \cdot \boldsymbol{Y}(s) = \boldsymbol{B} \cdot \boldsymbol{U}(s) \tag{4.96}$$

$$\boldsymbol{Y}(s) = (s \cdot \boldsymbol{I} - \boldsymbol{A})^{-1} \cdot \boldsymbol{B} \cdot \boldsymbol{U}(s) \tag{4.97}$$

$\boldsymbol{Y}(s)$ is a column vector of the controlled variables. Applying the inverse Laplace transform to both elements, we obtain $y_1(t)$ and $y_2(t)$.

4.1.3.5 Summary of Laplace transform and inverse Laplace transform

Based on the earlier review discussion, the following conclusions can be drawn:

- The dynamic behavior and stability of a transfer function representing the dynamics of a system depend on the roots of $P(s)$, the denominator or the characteristic polynomial of the transfer function.
- If $P(s)$ contains real positive roots, the time-domain response will consist of exponential terms with positive exponents which render the system unstable.
- If all roots of $P(s)$ are real and negative, the time-domain response will contain exponential terms with negative exponents and the system is stable.
- If $P(s)$ contains complex conjugate roots, the time-domain response will contain sinusoidal terms. In this case, if the real part of the roots is negative the system is stable with damped oscillations, and if the real part of the roots is positive, the system is unstable.

4.1.3.6 MATLAB commands for the calculation of Laplace transform and the inverse Laplace transform

To perform Laplace transform by MATLAB, first the variables s and t must be declared as symbolic variables, then the function is introduced in the time domain:

```
Run commands:

>> syms s t
>> f = 3 * t^4
      f =

      3*t^4
```

A call to `laplace` function will render the Laplace transform of the function. The first parameter is the function in the time domain, and the second parameter is optional and represents the Laplace operator symbol, s.

```
Run commands:

>> laplace(f,s)
       ans =

       72/s^5
```

The partial fraction of, for example, $\frac{s^2-s-6}{s^3-2s^2-s+2}$, can be found by the `residue` command:

```
Run commands:

>> [r,p,k]=residue([1 -1 -6], [1 -2 -1+2])
         r =
            -1.3333
             3.0000
            -0.6667
         p =
             2.0000
             1.0000
            -1.0000
         k =
              []
```

Note the first argument contains the coefficients vector of the numerator (a or r values) and the second argument contains the roots of the denominator polynomial, the p values. The return values `r,p,k` are defined as

$$\frac{r_1}{s-p_1}+\frac{r_2}{s-p_2}+\cdots+\frac{r_n}{s-p_n}+k(s)=\frac{Q(s)}{P(s)} \tag{4.98}$$

Detailed documentation is available in MATLAB by typing the function name `residue` and pressing the "F1" key.

To get the inverse Laplace directly, use the `ilaplace` command:

```
Run commands:

>> syms s t
>> fs=(s^2-s-6)/(s^3-2*s^2+2)
       fs =
       -(- s^2 + s + 6)/(s^3 - 2*s^2 + 2)

>>ft=ilaplace(fs,t)
         ft =
         symsum((r3*exp(r3*t))/(3*r3 - 4), r3 in RootOf(s3^3 - 2*s3^2 + 2, s3))
```

To get an approximate result, use `vpa` function and the `digits` function to set the precision digits.

Run commands:

```
>> digits(4)
>> vpa(ft)
      ans =
      exp(t*(1.42 - 0.6063*i))*(0.9073 - 1.864*i) - 0.8146*exp(-0.8393*t) +
      exp(t*(1.42 + 0.6063*i))*(0.9073 + 1.864*i)
```

The return value consists of n (nominator coefficients) and d (denominator coefficients).

4.2 The Basic Procedure to Develop the Transfer Function of SISO and MIMO Systems

We now have all the necessary tools for the derivation of the transfer function, which is the "linear" relationship between the output and input variables, of a system in "deviation form" in the Laplace domain. Let us consider a few examples.

Example 4.9

Consider the system discussed in Example 4.1 and start with the following equation after having gone through the linearization steps

$$A\frac{dl'}{dt} = F_i' - \alpha\sqrt{\bar{l}}x' - \frac{\alpha}{2\sqrt{\bar{l}}}\cdot l' \tag{4.99}$$

Rearrange the previous equation to the standard form of a first-order linear ODE,

$$\frac{2A\sqrt{\bar{l}}}{\alpha}\frac{dl'}{dt} + l' = \frac{2\sqrt{\bar{l}}}{\alpha}F_i' - 2\bar{l}x' \tag{4.100}$$

Eq. (4.100) represents a simple input-output relationship between the disturbance F_i', the manipulated variable x', and the output variable l', involving the derivative operation, all in deviation variable form. This linearized ODE is cumbersome to solve in the time domain. To simplify the mathematics, we use Laplace transform to convert the ODE to an algebraic expression between the input and output variables.

$$\mathcal{L}[f'(t)] = F'(s) \tag{4.101}$$

$$\mathcal{L}\left[\frac{df'(t)}{dt}\right] = sF'(s) - f'(0) \tag{4.102}$$

Note that $f'(t) = f(t) - f(0)$, therefore $f'(0) = 0$. Laplace transform the linearized equation Eq. (4.100):

$$\frac{2A\sqrt{\bar{l}}}{\alpha}[sL'(s) - 0] + L'(s) = \frac{2\sqrt{\bar{l}}}{\alpha}F_i'(s) - 2X'(s) \tag{4.103}$$

If we assume that the system has been at steady state at $t = 0$ (we often do), then the initial value of the variable in deviation form is always zero, i.e., $l'(0) = 0$.

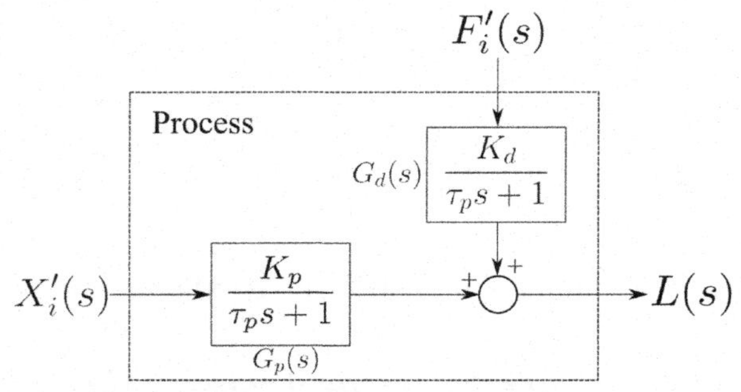

Fig. 4.15
The block diagram for the process in Example 4.9.

Note that working with the deviation variables has two advantages:

(1) All the constant terms in the linearized equations disappear.
(2) The initial condition of all equations becomes zero.

Eq. (4.103) becomes,

$$L'(s) = \frac{K_d}{\tau_p s + 1} F_i'(s) + \frac{K_p}{\tau_p s + 1} X'(s) \tag{4.104}$$

where $\tau_p = \dfrac{2A\sqrt{\bar{l}}}{\alpha}$, $K_d = \dfrac{2\sqrt{\bar{l}}}{\alpha}$, and $K_p = -2\bar{l}$. In addition, $L'(s)$ and $F_i'(s)$ and $X'(s)$ are the Laplace transforms of the output and the input variables in deviation form. From the definition of the transfer function (Fig. 4.15):

$$G_p(s) = \text{process transfer function} = \frac{L'(s)}{X_i'(s)} = \frac{K_p}{\tau_p s + 1} \tag{4.105}$$

$$G_d(s) = \text{load transfer function} = \frac{L'(s)}{F_i'(s)} = \frac{K_d}{\tau_p s + 1} \tag{4.106}$$

4.3 Steps to Derive the Transfer Function (T.F.) Models

(1) Using the Taylor series expansion, linearize all the nonlinear terms existing in the nonlinear dynamic model of a given system.
(2) Subtract the steady-state equations from the dynamic equations in order to express the variables and the equations in terms of the "deviation variables."
(3) Transfer all terms involving the output variables to the left-hand side of the equation and rearrange the equation to obtain a coefficient of 1 for the output variable.
(4) Laplace transform both sides of the equation.
(5) Derive the T.F. by dividing $\mathcal{L}$ {output} over $\mathcal{L}$ {input} in deviation form.

Example 4.10

Derive the transfer functions that describe the dynamics of the STH process discussed in Example 4.2.

Step 1: Linearization of the nonlinear mass and energy balances.

$$\textit{Mass balance}: \frac{dV}{dt} = F_H + F_C - F \tag{4.107}$$

$$\textit{Energy balance}: \frac{dV(t)T(t)}{dt} = F_H(t)T_H(t) + F_C(t)T_C(t) - F(t)T(t) \tag{4.108}$$

Eq. (4.107) is linear but Eq. (4.108) is nonlinear. Therefore

$$V(t)T(t) \approx \overline{V}\overline{T} + \overline{V}T' + \overline{T}V' \tag{4.109}$$

$$F_H(t)T_H(t) \approx \overline{F}_H\overline{T}_H + \overline{F}_H T'_H + \overline{T}_H F'_H \tag{4.110}$$

$$F_C(t)T_C(t) \approx \overline{F}_C\overline{T}_C + \overline{F}_C T'_C + \overline{T}_C F'_C \tag{4.111}$$

$$F(t)T(t) \approx \overline{F}\overline{T} + \overline{F}T' + \overline{T}F' \tag{4.112}$$

Substitute Eqs. (4.109)–(4.112) into Eq. (4.108). The argument term (t) is omitted for simplicity.

$$\frac{d[\overline{V}\overline{T} + \overline{V}T' + \overline{T}V']}{dt} = \overline{F}_H\overline{T}_H + \overline{F}_H T'_H + \overline{T}_H F'_H + \overline{F}_C\overline{T}_C + \overline{F}_C T'_C + \overline{T}_C F'_C - \overline{F}\overline{T} - \overline{F}T' - \overline{T}F' \tag{4.113}$$

Step 2: Express the equations in terms of deviation variables by subtracting the steady-state equations from the dynamic equations.

$$\frac{d\overline{V}}{dt} = \overline{F_H} + \overline{F_C} - \overline{F} = 0 \tag{4.114}$$

$$\frac{d\overline{V}\cdot\overline{T}}{dt} = \overline{F_H}\cdot\overline{T_H} + \overline{F_C}\cdot\overline{T_C} - \overline{F}\cdot\overline{T} = 0 \tag{4.115}$$

$$\frac{dV'}{dt} = F'_H + F'_C - F' \tag{4.116}$$

$$\overline{V}\frac{dT'}{dt} + \overline{T}\frac{dV'}{dt} = \overline{F_H}T'_H + \overline{T_H}F'_H + \overline{F_C}T'_C + \overline{T_C}F'_C - \overline{F}T' - \overline{T}F' \tag{4.117}$$

$V'(0) = V(0) - \overline{V} = 0$ (Assuming that at $t = 0$ the process has been at steady state). Simplify equations by substituting $\frac{dV'}{dt}$ from Eq. (4.116) into Eq. (4.117),

$$\overline{V}\frac{dT'}{dt} = (\overline{T_H} - \overline{T})F'_H + F'_c(\overline{T_C} - \overline{T}) + \overline{F_H}T'_H + \overline{F_C}T'_C - \overline{F}T' \tag{4.118}$$

Step 3: Simplify the equations and rearrange them to the standard form.

$$\frac{\overline{V}}{\overline{F}} = \frac{\textit{volume}}{\textit{flowrate}} = \text{mean residence time of liquid in the tank} = \tau_p \tag{4.119}$$

Eq. (4.118) becomes,

$$\tau_p \frac{dT'}{dt} + T' = \frac{(\overline{T_H} - \overline{T})}{\overline{F}} \cdot F'_H + \frac{(\overline{T_C} - \overline{T})}{\overline{F}} \cdot F'_C + \frac{\overline{F_H}}{\overline{F}} T'_H + \frac{\overline{F_C}}{\overline{F}} T'_C \tag{4.120}$$

Step 4: Laplace transform on Eq. (4.116),

$$s \cdot V'(s) - V'(0) = F'_H(s) + F'_C(s) - F'(s) \tag{4.121}$$

$$V'(s) = \frac{1}{s} \cdot F'_H(s) + \frac{1}{s} \cdot F'_C(s) - \frac{1}{s} \cdot F'(s) \tag{4.122}$$

Laplace transform of Eq. (4.120) results in,

$$(\tau_p s + 1) T'(s) - T'(0) = \frac{(\overline{T_H} - \overline{T})}{\overline{F}} \cdot F'_H(s) + \frac{(\overline{T_C} - \overline{T})}{\overline{F}} \cdot F'_C(s) + \frac{\overline{F_H}}{\overline{F}} \cdot T'_H(s) + \frac{\overline{F_C}}{\overline{F}} \cdot T'_C(s) \tag{4.123}$$

$$T'(s) = \frac{(\overline{T_H} - \overline{T})}{\overline{F}(\tau_p s + 1)} \cdot F'_H(s) + \frac{(\overline{T_C} - \overline{T})}{\overline{F}(\tau_p s + 1)} \cdot F'_C(s) + \frac{\overline{F_H}}{\overline{F}(\tau_p s + 1)} \cdot T'_H(s) + \frac{\overline{F_C}}{\overline{F}(\tau_p s + 1)} \cdot T'_C(s) \tag{4.124}$$

Note that $V'(s)$ is neither a function of $T'_H(s)$ nor $T_{C'}(s)$. Moreover, $T'(s)$ is not a function of $F'(s)$ because of the perfect mixing assumption. Two of the individual transfer functions are given in Eqs. (4.125), (4.126), while the transfer function matrix representation of the entire system is given in Eq. (4.127):

$$\frac{V'(s)}{F'_H(s)} = \frac{1}{s} \tag{4.125}$$

$$\frac{T'(s)}{T'_H(s)} = \frac{\overline{F_H}}{\overline{F}(\tau_p s + 1)} \tag{4.126}$$

$$\underbrace{\begin{bmatrix} V'(s) \\ T'(s) \end{bmatrix}}_{\text{Controlled variable vector}} = \underbrace{\begin{bmatrix} -\frac{1}{s} & \frac{1}{s} \\ 0 & \frac{(\overline{T_H} - \overline{T})}{\overline{F}(\tau_p s + 1)} \end{bmatrix}}_{\text{Process transfer function matrix}} \cdot \underbrace{\begin{bmatrix} F'(s) \\ F'_H(s) \end{bmatrix}}_{\text{Manipulated variable vector}} + \underbrace{\begin{bmatrix} 0 & 0 & \frac{1}{s} \\ \frac{\overline{F_H}}{\overline{F}(\tau_p s + 1)} & \frac{\overline{F_C}}{\overline{F}(\tau_p s + 1)} & \frac{(\overline{T_C} - \overline{T})}{\overline{F}(\tau_p s + 1)} \end{bmatrix}}_{\text{Load transfer function matrix}} \cdot \underbrace{\begin{bmatrix} T'_H(s) \\ T'_C(s) \\ F'_C(s) \end{bmatrix}}_{\text{Load or disturbance vector}} \tag{4.127}$$

In general, in a MIMO system, the process transfer function matrix, $\boldsymbol{G_p}(s)$, and the load transfer function matrix, $\boldsymbol{G_d}(s)$, are given by:

$$\boldsymbol{Y}'(s) = \boldsymbol{G_p}(s)\, \boldsymbol{U}'(s) + \boldsymbol{G_d}(s) \boldsymbol{D}'(s) \tag{4.128}$$

The block diagram representation of the earlier MIMO system is shown in Fig. 4.16.

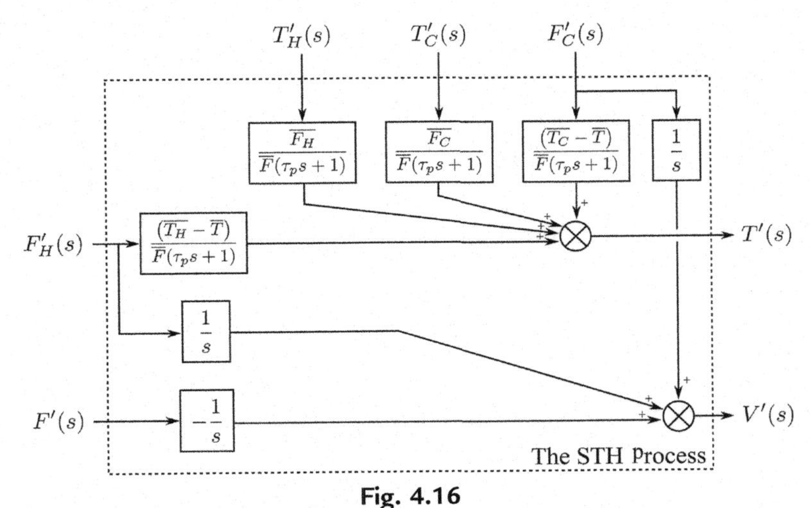

Fig. 4.16
The block diagram of the STH process.

Note: From now on, we shall drop the superscript (′) on all variables in the s-domain. However, in the time domain we shall maintain the superscript to emphasize that the variables are deviations from their steady-state values.

4.4 Transfer Function of Linear Systems

The dynamic behavior of the real processes is *often* approximated by a first-order or a second-order model with or without time delay. This provides an incentive to study these linear transfer functions in detail. Before doing that, however, it is instructive to look at the "simple" forms of process inputs that may occur in the manipulated variables or disturbances.

4.4.1 Simple Functional Forms of the Input Signals

A step input

Consider a step input of magnitude M. This function can be mathematically represented by multiplying M by a unit step function, i.e., $M \cdot S(t)$ which is given by:

$$M \cdot S(t) = \begin{cases} 0 & t<0 \\ M & t \geq 0 \end{cases} \tag{4.129}$$

The Laplace transformation of this function is:

$$\mathcal{L}[M \cdot S(t)] = \frac{M}{s} \tag{4.130}$$

A ramp input
A ramp input with a slope α

$$f(t) = \begin{cases} 0 & t < 0 \\ \alpha t & t \geq 0 \end{cases} \tag{4.131}$$

Laplace transformation of a ramp function with a slope α is given by:

$$\mathcal{L}[f(t)] = \frac{\alpha}{s^2} \tag{4.132}$$

A rectangular pulse input
A rectangular pulse with a magnitude of h and duration θ:

$$f(t) = \begin{cases} 0 & t < 0 \\ h & 0 \leq t < \theta \\ 0 & t \geq \theta \end{cases} \tag{4.133}$$

A rectangular pulse function can be obtained by subtracting a step function with magnitude h from the same step function delayed by θ:

$$f(t) = h \cdot [S(t) - S(t - \theta)] \tag{4.134}$$

Laplace transformation of the rectangular pulse function:

$$\mathcal{L}[f(t)] = \frac{h\left(1 - e^{-\theta s}\right)}{s} \tag{4.135}$$

A unit impulse input
This is a fictitious function that has a height of infinity and a duration of zero resulting in a strength (area) of unity occurring at $t = 0$ with Laplace transformation:

$$\mathcal{L}[\delta(t)] = 1 \tag{4.136}$$

A sinusoidal input
A sinusoidal input with amplitude A and frequency $\omega = \frac{2\pi}{P}$, where P is the period of oscillation.

$$f(t) = \begin{cases} 0 & t < 0 \\ A \sin \omega t & t \geq 0 \end{cases} \tag{4.137}$$

Laplace transformation of a sinusoidal function is:

$$\mathcal{L}[f(t)] = \frac{A\omega}{s^2 + \omega^2} \tag{4.138}$$

4.4.2 First-Order Transfer Function Models

A first-order ODE is described by the following equation

$$a_1 \frac{dy'(t)}{dt} + a_0 y'(t) = b\, u'(t) \quad (4.139)$$

$$\frac{a_1}{a_0} \cdot \frac{dy'(t)}{dt} + y'(t) = \frac{b}{a_0} u'(t) \quad (4.139)$$

Let $\tau_p = \frac{a_1}{a_0}, K_p = \frac{b}{a_0}$ and Laplace transformation of the previous equation results in:

$$\tau_p [\mathrm{Y}(s) - y'(0)] + \mathrm{Y}(s) = K_p\, U(s) \quad (4.140)$$

$$\frac{Y(s)}{U(s)} = G_p(s) = \frac{K_p}{\tau_p s + 1} \quad (4.141)$$

Eq. (4.141) represents the transfer function of a first-order system. This transfer function has two parameters:

- K_p is the steady-state gain of the process with a unit as the ratio of the unit of the process output to the unit of the process input.
- τ_p is the time constant of the process which is a dynamic parameter with the dimension of s or min.

4.4.2.1 The step response of a first-order system

Let us consider the response of a first-order system to a step change of magnitude M, i.e., $u'(t) = M \cdot S(t)$

$$U(s) = \frac{M}{s} \quad (4.142)$$

$$\frac{Y(s)}{U(s)} = G_p(s) = \frac{K_p}{\tau_p s + 1} \quad (4.143)$$

$$Y(s) = \frac{K_p}{\tau_p s + 1} \cdot \frac{M}{s} = \frac{a_1}{\tau_p s + 1} + \frac{a_2}{s} \quad (4.144)$$

Inverse Laplace transform:

$$y'(t) = M \cdot K_p \left(1 - e^{-\frac{t}{\tau_p}}\right) \quad (4.145)$$

As time goes to infinity, the output, $y'(t)$, approaches $M \cdot K_p$. Therefore $M \cdot K_p$ is the final value or the steady-state value of the output.

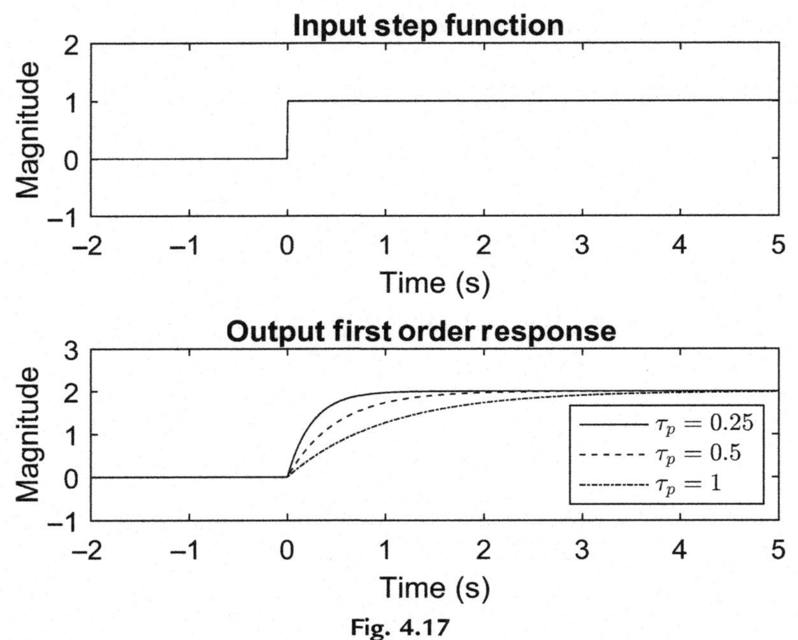

Fig. 4.17
The step response of a first-order system with a magnitude $M = 1$ and $K_p = 2$.

$$K_p = \text{steady} - \text{state gain of the process} = \frac{\text{ultimate change in the output}}{\text{introduced change in the input}} = \frac{M \cdot K_p}{M} \tag{4.146}$$

If $K_p > 1$, the process amplifies the input signal, and if $K_p < 1$, the process attenuates the input signal.

A positive steady-state gain demonstrates that the final value of the output increases with an increase in the input and a negative steady-state gain shows that the final value of the output decreases with an increase in the input.

τ_p represents the *dynamic* characteristic of the process and the speed of the process to reach its final steady-state value. Slow processes have a large τ_p and fast processes have a small τ_p as is shown in Fig. 4.17.

Note that at $t = \tau_p$, we have:

$$y'(t = \tau_p) = M \cdot K_p \left(1 - e^{-\frac{\tau_p}{\tau_p}}\right) = 0.632 M \cdot K_p \tag{4.147}$$

Therefore, at $t = \tau_p$, the output of a first-order process reaches 63.2% its final value. This property is useful to approximate the behavior of an existing process by a first-order transfer function using a simple graphical process identification technique, as will be discussed in Part B of this chapter.

4.4.2.2 Impulse response of a first-order process

$$\frac{Y(s)}{U(s)} = \frac{K_p}{\tau_p s + 1} \tag{4.148}$$

For an impulse input with a strength of B, we have:

$$\left.\begin{aligned} u(t) &= B\delta(t) \\ U(s) &= B \end{aligned}\right\} \text{ where } B \text{ is the strength of the impulse signal} \tag{4.149}$$

$$Y(s) = \frac{K_p}{\tau_p s + 1} \cdot B \tag{4.150}$$

Inverse Laplace transform,

$$y'(t) = \frac{K_p B}{\tau_p} e^{-t/\tau_p} \tag{4.151}$$

The impulse response of a first-order system is shown in Fig. 4.18. Note that the actual impulse does take a finite time to rise and drops back to its original value. Therefore the actual response would be the response to a rectangular pulse signal with the given area or strength, as is shown in Fig. 4.18.

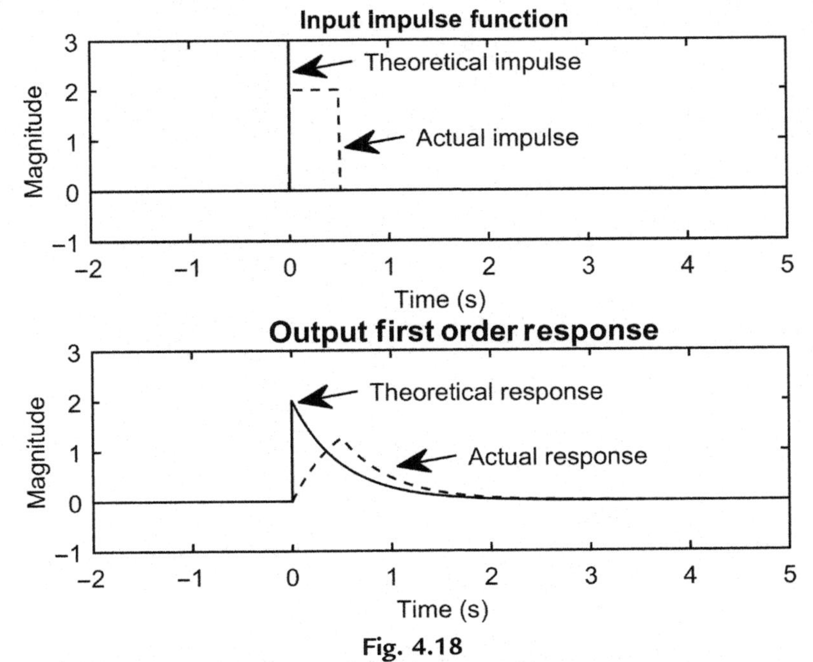

Fig. 4.18
Theoretical and actual impulse response of a first-order system.

4.4.2.3 MATLAB and Simulink commands

The command, `tf` is used to define a transfer function and obtain its response in MATLAB, e.g., given $g(s)=\dfrac{2}{0.5s+1}$, the MATLAB command is:

```
Run commands:
>> g = tf([2], [0.5 1]);
```

where the first argument includes the coefficients of the numerator polynomial and the second argument the coefficients of the denominator polynomial. Note that if the numerator is a first-order polynomial with a zero constant term, $2s$, the first argument should be given as `[2 0]`. Alternatively, `tf`, can be used to create a transfer function in the symbolic mode, which is useful to enter complex transfer functions.

```
Run commands:
>> s = tf('s');
>> g = 2/(0.5*s + 1)
```

The `step` and the `impulse` commands plot or return the point series for the `step` and `impulse` response of a given transfer function. When the return "value" is not specified, the `step` and `impulse` will generate the response curves.

```
Run commands:
>> [value,time] = step(g);   % Suppresses the plot
>> step(g);                  % Plots the step response
>> impulse(g);               % plots the impulse response
```

If the numerator or denominator polynomials contain the product of two polynomials, the function `conv` is used to multiply the two polynomials, for example

$$G(s)=\frac{s^2-s-6}{(s^3-2s^2-s+2)(s+1)} \tag{4.152}$$

```
Run commands:
>> den = conv([1 -2 -1 2] [1 1])
      den =
           1    -1    -3    1    2
```

The result is

$$\left(s^3-2s^2-s+2\right)(s+1)=s^4-s^3-3s^2+s+2. \tag{4.153}$$

The parallel function (`parallel`) adds two expressions:

$$G(s)=\frac{s^2-s-6}{s^3-2s^2-s+2}+\frac{s+1}{s^2+3s+1} \tag{4.154}$$

The result can be obtained with `parallel` function:

```
Run commands:

>> [n,d]=parallel([1 -1 -6],[1 -2 -1 2],[1 1],[1 3 1])
      n =
          0    2    1   -11   -18    -4
      d =
          1    1   -6    -3     5     2
```

4.4.2.4 Examples of real processes with first-order transfer functions

Although the first-order transfer function model is used to approximate the behavior of real complex processes, there are some physical processes whose dynamic behavior is actually in the form of a first-order transfer function. We shall consider a few such processes.

Example 4.11

A liquid-filled thermometer is shown in Fig. 4.19. Derive the transfer function between the temperature monitored by the thermometer, $T(t)$ (the output variable), and the surrounding temperature, $T_0(t)$ (the input variable). Assuming that, initially $T_0(t) > T(t)$, the relevant conservation principle on the liquid inside the thermometer bulb is the energy balance.

$$m \cdot c_p \frac{dT(t)}{dt} = h \cdot A[T_0(t) - T(t)] \tag{4.155}$$

where h is the heat transfer coefficient, A is the surface area of the thermometer bulb, and m is the mass of the liquid inside the bulb with a specific heat of c_p.

Expressing the previous equation in deviation variables by subtracting the steady-state energy balance and taking the Laplace transformation:

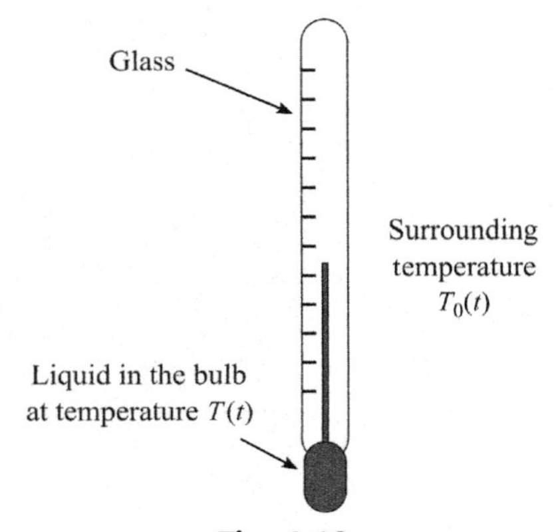

Fig. 4.19
A liquid-filled thermometer.

$$\left(\frac{m \cdot c_p}{h \cdot A} \cdot s + 1\right) T(s) = T_o(s) \tag{4.156}$$

$$G(s) = \frac{T(s)}{T_o(s)} = \frac{1}{\frac{m \cdot c_p}{h \cdot A} \cdot s + 1} \tag{4.157}$$

where $\tau_p = \frac{m \cdot c_p}{h \cdot A}$ is the time constant of the thermometer. Note that

$$\begin{cases} m \cdot c_p & \text{is the capacitance to store thermal energy} \\ \frac{1}{h \cdot A} & \text{is the resistance to the flow of thermal energy} \end{cases}$$

In general, the time constant is the product of the capacitance to store a quantity and the resistance to the flow of that quantity: $\tau_p = (\text{capacitance})(\text{resistance})$. It is instructive to note that as the mass of the liquid in the thermometer bulb decreases, and the area of the bulb and the heat transfer coefficient increase, the time constant decreases and the thermometer becomes a faster responding instrument, which is highly desirable.

A steady-state gain of $K_p = 1$ suggests that the measured temperature eventually reaches the surrounding temperature at steady state, which is to be expected.

Example 4.12

An electrical RC circuit consists of a power supply, e_i; a resistor, R; and a capacitor, C. If the switch is on B contact shown in Fig. 4.20, the capacitor charges up until e_o (output voltage) reaches e_i (input voltage). If the switch is on A contact, e_o discharges through the electrical resistor and reaches 0.

The electrical current flowing through the circuit is used to derive the transfer function representing this system:

$$i = C \cdot \frac{de_o}{dt} = \frac{e_i}{R} - \frac{e_o}{R} \tag{4.158}$$

Laplace transform,

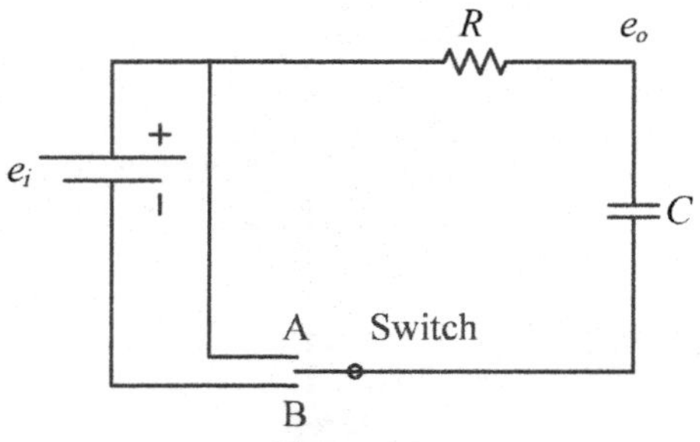

Fig. 4.20
The RC charge-discharge circuit.

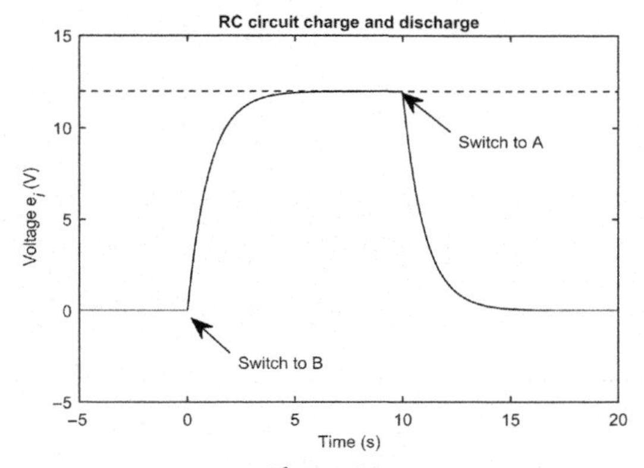

Fig. 4.21
The RC circuit charge and discharge response. $R = 1000\Omega$, $C = 1$ mF, $e_i = 12$V.

$$[(R\cdot C)s+1]\cdot e_o(s) = e_i(s) \tag{4.159}$$

$$\frac{e_o(s)}{e_i(s)} = \frac{1}{(RC)\cdot s+1} \tag{4.160}$$

A value of $K_p = 1$ suggests that the output voltage eventually reaches the input voltage at the steady state. The time constant is the product of the resistance to the flow of the electrical current and the capacitance to store the electric charge, $\tau_p = (\text{capacitance})(\text{resistance}) = RC$.

Fig. 4.21 shows the response of an RC circuit to a square pulse input signal.

Example 4.13

The time constant and the steady-state gain of a nonlinear process are time varying. Recall the model of a surge tank with no valve on the outlet stream includes a nonlinear term:

$$A\frac{dl(t)}{dt} = F_i(t) - \alpha\sqrt{l(t)} \tag{4.161}$$

where A is the cross-sectional area of the tank.

$$\alpha\sqrt{l(t)} \approx \sqrt{\bar{l}} + \frac{\alpha}{2\sqrt{\bar{l}}}\left[l(t) - \bar{l}\right] \tag{4.162}$$

$$A\frac{dl'(t)}{dt} = F_i'(t) - \frac{\alpha}{2\sqrt{\bar{l}}}\cdot l'(t) \tag{4.163}$$

Laplace transform,

$$A\cdot s\cdot L(s) = F_i(s) - \frac{\alpha}{2\sqrt{\bar{l}}}\cdot L(s) \tag{4.164}$$

$$G(s)=\frac{L(s)}{F_i(s)}=\frac{\frac{2\sqrt{\bar{l}}}{\alpha}}{\frac{2A\sqrt{\bar{l}}}{\alpha}s+1} \quad (4.165)$$

Note: Both τ_p and K_d depend on the steady-state value of the liquid level around which linearization is performed, i.e., $\bar{l}$. The implication of this in the design of a controller is that controller equation depends on the process models; online tuning or redesigning of the controller is needed for the nonlinear processes, as its set point (point of linearization) changes.

4.4.3 A Pure Capacitive or An Integrating Process

A process whose output signal is the time integral of its input signal is referred to as an "Integrating Process." Such a process only stores material or energy, and therefore it is also referred to as a "Pure Capacitive" process.

Example 4.14

A surge tank shown in Fig. 4.22 is equipped with a pump at its effluent stream to ensure a constant outlet flow rate irrespective of the liquid head in the tank represents a pure capacitive process. The mass balance on the tank with a constant outlet stream flow rate results in:

$$A\frac{dl(t)}{dt}=F_i(t)-\bar{F} \quad (4.166)$$

The steady-state equation is:

$$A\frac{d\bar{l}}{dt}=\bar{F}_i-\bar{F} \quad (4.167)$$

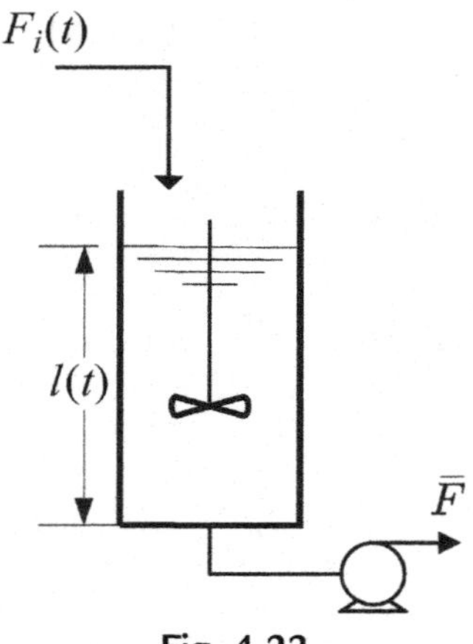

Fig. 4.22
Schematics of a pure capacitive process.

Subtract the steady-state from the total mass balance and Laplace transform,

$$\frac{L(s)}{F_i(s)} = \frac{1/A}{s} \tag{4.168}$$

Eq. (4.168) demonstrates that $l'(t)$ is proportional to the time integral of $F'_i(t)$. The control of such processes is often difficult, as they are nonself-regulatory. The response to a unit step change in $F'_i(t)$:

$$F_i(s) = \frac{1}{s} \tag{4.169}$$

$$Y(s) = \frac{1/A}{s^2} \tag{4.170}$$

Inverse Laplace transform,

$$y'(t) = \frac{1}{A}t \text{ (1/A is the slope of the ramp function)} \tag{4.171}$$

The step response is shown in Fig. 4.23. The tank overflows (unstable) as a result of a step increase in the inlet flow.

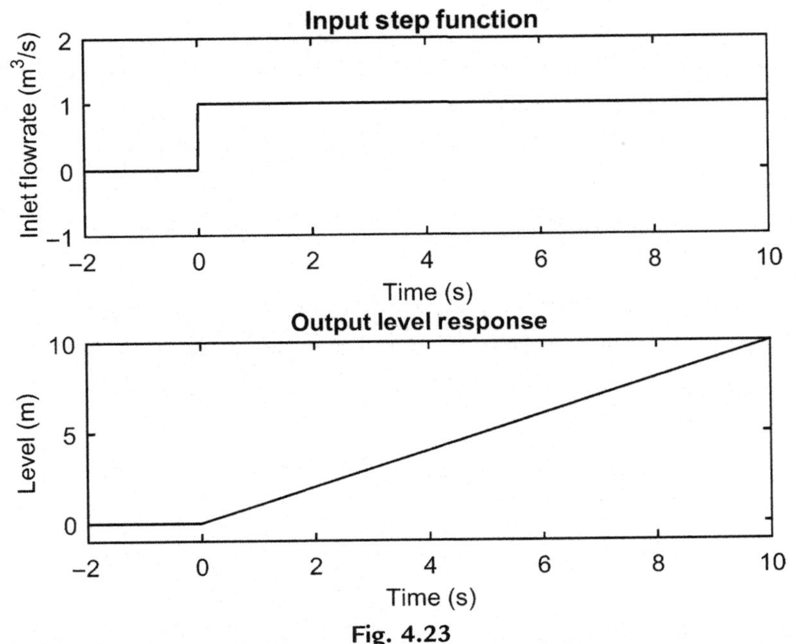

Fig. 4.23
Step response of a pure capacitive system.

4.4.4 Processes With Second-Order Dynamics

The dynamics of a second-order process is represented by a second-order ODE,

$$a_2 \frac{d^2y(t)}{dt^2} + a_1 \frac{dy(t)}{dt} + a_0 \cdot y(t) = b \cdot u(t) \tag{4.172}$$

Defining: $\frac{a_2}{a_0} = \tau^2$ where τ is the natural period of oscillation, representing the speed of the process,

$\frac{a_1}{a_0} = 2\xi\tau$ ξ is the damping coefficient, and

$\frac{b}{a_0} = K$ K is the steady-state gain of the process, and subtracting the steady-state equation, result in the general form of a dynamic model of a linear second-order system in deviation variables

$$\tau^2 \frac{d^2y'(t)}{dt^2} + 2\xi\tau \frac{dy'(t)}{dt} + y'(t) = K \cdot u'(t) \tag{4.173}$$

Laplace transforming Eq. (4.173) and assuming that the initial conditions are zero, renders the general transfer function of a second-order system.

$$\left(\tau^2 s^2 + 2\xi\tau s + 1\right) Y(s) = K \cdot U(t) \tag{4.174}$$

$$\frac{Y(s)}{U(s)} = G_p(s) = \frac{K}{\tau^2 s^2 + 2\xi\tau s + 1} \tag{4.175}$$

In order to understand the physical meanings of ξ and τ, we study the response, $y'(t)$, of a second-order transfer function to a step change in the input signal, $u'(t)$, with a magnitude M.

Input: $U(s) = \frac{M}{s}$

$$\begin{aligned} Y(s) &= \frac{K}{\tau^2 s^2 + 2\xi\tau s + 1} \cdot \frac{M}{s} \\ &= \frac{KM}{\tau^2} \left[\frac{a_1}{s} + \frac{a_2}{s + \frac{\xi}{\tau} - \frac{\sqrt{\xi^2 - 1}}{\tau}} + \frac{a_3}{s + \frac{\xi}{\tau} + \frac{\sqrt{\xi^2 - 1}}{\tau}} \right] \end{aligned} \tag{4.176}$$

The shape of the time-domain response of Eq. (4.176), $y'(t)$ depends on the roots of the characteristic polynomial, $s^2 + \frac{2\xi s}{\tau} + \frac{1}{\tau^2}$. The roots of this polynomial depend on the value of ξ, the damping coefficient of the system. Three cases are distinguishable:

$\xi > 1$ The system is stable and the response, $y'(t)$, will be smooth (nonoscillatory), there will be two negative real distinct roots, and the response is referred to as **overdamped**

$\xi = 1$ The system will be stable, there will be two multiple negative real roots, and the response will be smooth and **critically damped**

$\xi < 1$ The system will be stable; there will be a pair of complex, conjugate roots with negative real parts; and the response will be oscillatory **underdamped**

$\xi = 0$ There are two purely imaginary roots and the response will be pure sinusoidal **undamped**

$\xi < 0$ The system will be UNSTABLE in an oscillatory manner with two complex roots with positive real parts

4.4.4.1 *Case I: An overdamped system,* $\xi > 1$, two distinct real roots

For this case, the coefficients a_1, a_2, and a_3 in Eq. (4.176) are:

$$a_1 = \lim_{s \to 0} \frac{1}{s^2 + \frac{2\xi}{\tau}s + \frac{1}{\tau^2}} = \tau^2 \tag{4.177}$$

$$a_2 = \lim_{s \to -\frac{\xi}{\tau} + \frac{\sqrt{\xi^2-1}}{\tau}} \frac{1}{s\left(s + \frac{\xi}{\tau} + \frac{\sqrt{\xi^2-1}}{\tau}\right)} = \frac{\tau^2}{2\sqrt{\xi^2-1} \cdot \left(\sqrt{\xi^2-1} - \xi\right)} \tag{4.178}$$

$$a_3 = \lim_{s \to -\frac{\xi}{\tau} - \frac{\sqrt{\xi^2-1}}{\tau}} \frac{1}{s\left(s + \frac{\xi}{\tau} - \frac{\sqrt{\xi^2-1}}{\tau}\right)} = \frac{\tau^2}{2\sqrt{\xi^2-1} \cdot \left(\sqrt{\xi^2-1} + \xi\right)} \tag{4.179}$$

Substitute in Eq. (4.176),

$$\begin{aligned}
y'(t) &= \frac{MK_p}{\tau^2}\left\{\tau^2 + \frac{\tau^2}{2\sqrt{\xi^2-1}\cdot\left(\sqrt{\xi^2-1}-\xi\right)} \cdot e^{\left(-\frac{\xi}{\tau} + \frac{\sqrt{\xi^2-1}}{\tau}\right)t} + \frac{\tau^2}{2\sqrt{\xi^2-1}\cdot\left(\sqrt{\xi^2-1}+\xi\right)} \cdot e^{\left(-\frac{\xi}{\tau} + \frac{\sqrt{\xi^2-1}}{\tau}\right)t}\right\} \\
&= MK_p\left\{1 - e^{-\frac{\xi}{\tau}t}\left[\frac{2\left[(\xi^2-1) + \xi\sqrt{\xi^2-1}\right] \cdot e^{\frac{\sqrt{\xi^2-1}}{\tau}t} + 2\left[(\xi^2-1) - \xi\sqrt{\xi^2-1}\right] \cdot e^{-\frac{\sqrt{\xi^2-1}}{\tau}t}}{4(\xi^2-1)^2 - 4\xi^2(\xi^2-1)}\right]\right\} \\
&= MK_p\left\{1 - e^{-\frac{\xi}{\tau}t}\left[\frac{e^{\frac{\sqrt{\xi^2-1}}{\tau}\cdot t} + e^{-\frac{\sqrt{\xi^2-1}}{\tau}\cdot t}}{2} + \frac{\xi}{\sqrt{\xi^2-1}} \cdot \frac{e^{\frac{\sqrt{\xi^2-1}}{\tau}} - e^{-\frac{\sqrt{\xi^2-1}}{\tau}}}{2}\right]\right\}
\end{aligned} \tag{4.180}$$

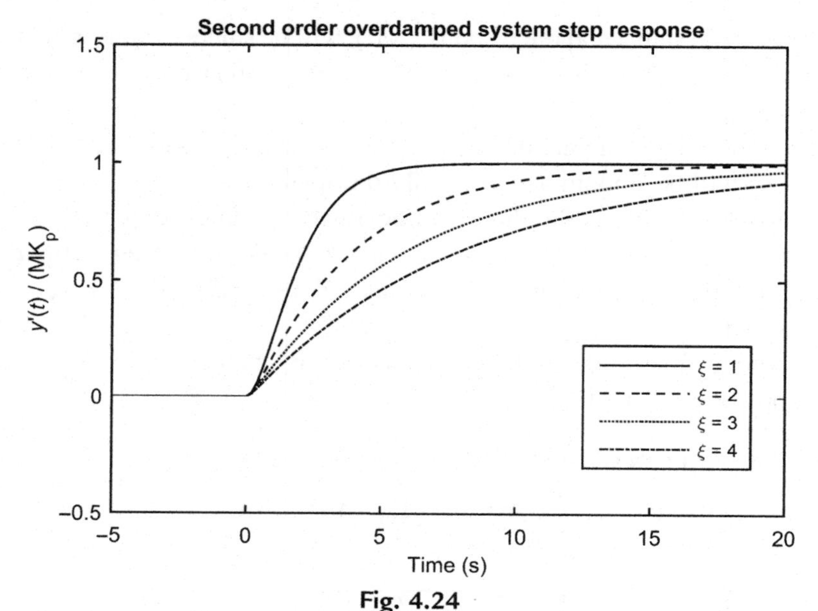

Fig. 4.24
The unit step response of a second-order overdamped system with $\tau = 1$ s.

Recall: $\sinh\alpha = \dfrac{e^{\alpha} - e^{-\alpha}}{2}$ and $\cosh\alpha = \dfrac{e^{\alpha} + e^{-\alpha}}{2}$

$$y'(t) = MK_p\left\{1 - e^{-\frac{\xi}{\tau}t}\left[\cosh\left(\frac{\sqrt{\xi^2 - 1}}{\tau}\cdot t\right) + \frac{\xi}{\sqrt{\xi^2 - 1}}\sinh\left(\frac{\sqrt{\xi^2 - 1}}{\tau}\cdot t\right)\right]\right\} \tag{4.181}$$

4.4.4.2 *Case II: A critically damped system,* $\xi = 1$, multiple real roots

The roots of the characteristic polynomial $s^2 + \dfrac{2\xi}{\tau} + \dfrac{1}{\tau^2}$ are $p_{1,2} = -\dfrac{1}{\tau}$. Therefore

$$y'(s) = \frac{MK_p}{\tau^2}\left\{\frac{1}{s^2 + \frac{2\xi s}{\tau} + \frac{1}{\tau^2}}\cdot\frac{1}{s}\right\} = \frac{MK_p}{\tau^2}\left\{\frac{a_1}{s} + \frac{a_2}{\left(s + \frac{1}{\tau}\right)^2} + \frac{a_3}{s + \frac{1}{\tau}}\right\} \tag{4.182}$$

$$a_1 = \lim_{s \to 0}\frac{1}{s^2 + \frac{2\xi}{\tau} + \frac{1}{\tau^2}} = \tau^2 \tag{4.183}$$

$$a_2 = \lim_{s \to -\frac{1}{\tau}} \left(\frac{1}{s}\right) = -\tau \tag{4.184}$$

$$a_3 = \lim_{s \to -\frac{1}{\tau}} \left[\frac{d}{ds}\left(\frac{1}{s}\right)\right] = -\tau^2 \tag{4.185}$$

$$y'(t) = MK_p\left\{1 - \left(\frac{t}{\tau} + 1\right)e^{-\frac{t}{\tau}}\right\} \tag{4.186}$$

A critically damped system has the fastest overdamped response. As ξ increases, the speed of the response decreases. See Fig. 4.24 for the shape of the step response of an overdamped and a critically damped system.

4.4.4.3 *Case III: An underdamped system, $\xi < 1$*, two distinct complex conjugate roots with negative real parts

$$p_{1,2} = -\frac{\xi}{\tau} \pm j \cdot \frac{\sqrt{1-\xi^2}}{\tau} \quad \text{where } j = \sqrt{-1} \tag{4.187}$$

$$y'(s) = \frac{MK_p}{\tau^2}\left\{\frac{1}{s^2 + \frac{2\xi s}{\tau} + \frac{1}{\tau^2}} \cdot \frac{1}{s}\right\} = \frac{MK_p}{\tau^2}\left\{\frac{a_1}{s} + \frac{a_2}{s-p_1} + \frac{a_3}{s-p_2}\right\} \tag{4.188}$$

$$a_1 = \lim_{s \to 0} \frac{1}{s^2 + \frac{2\xi}{\tau} + \frac{1}{\tau^2}} = \tau^2 \tag{4.189}$$

$$\begin{aligned} a_2 &= \lim_{s \to p_1}\left[\frac{1}{s(s-p_2)}\right] = \frac{1}{\left(-\frac{\xi}{\tau} + j \cdot \frac{\sqrt{1-\xi^2}}{\tau}\right) \cdot \left(j \cdot \frac{2\sqrt{1-\xi^2}}{\tau}\right)} \\ &= \frac{\tau^2}{-2\left[j\xi\sqrt{1-\xi^2} + (1-\xi^2)\right]} = \frac{\tau^2\left[(1-\xi^2) - j\xi\sqrt{1-\xi^2}\right]}{-2\left[(1-\xi^2)^2 + \xi^2(1-\xi^2)\right]} \\ &= \frac{\tau^2\left[(1-\xi^2) - j\xi\sqrt{1-\xi^2}\right]}{-2(1-\xi^2)} = -\frac{\tau^2}{2}\left[1 - j \cdot \frac{\xi}{\sqrt{1-\xi^2}}\right] \end{aligned} \tag{4.190}$$

$$
\begin{aligned}
a_3 &= \lim_{s\to p_2}\left[\frac{1}{s(s-p_1)}\right] = \frac{1}{\left(-\frac{\xi}{\tau}-j\cdot\frac{\sqrt{1-\xi^2}}{\tau}\right)\cdot\left(-j\cdot\frac{2\sqrt{1-\xi^2}}{\tau}\right)} \\
&= \frac{\tau^2}{-2\left[j\xi\sqrt{1-\xi^2}-(1-\xi^2)\right]} = \frac{\tau^2\left[(1-\xi^2)+j\xi\sqrt{1-\xi^2}\right]}{-2\left[(1-\xi^2)^2+\xi^2(1-\xi^2)\right]} \\
&= \frac{\tau^2\left[(1-\xi^2)+j\xi\sqrt{1-\xi^2}\right]}{-2(1-\xi^2)} = -\frac{\tau^2}{2}\left[1+j\cdot\frac{\xi}{\sqrt{1-\xi^2}}\right]
\end{aligned}
\tag{4.191}
$$

Note: The complex roots and the coefficients a_2 and a_3 are complex conjugates.

Substitute a_1, a_2, a_3 into Eq. (4.188) and inverse Laplace transform,

$$
\begin{aligned}
y'(t) &= \frac{MK_p}{\tau^2}\{a_1+a_2\cdot e^{p_1\cdot t}+a_3\cdot e^{p_2\cdot t}\} \\
&= MK_p\cdot\left\{1-e^{-\frac{\xi}{\tau}\cdot t}\left[\frac{e^{j\cdot\frac{\sqrt{1-\xi^2}}{\tau}t}+e^{-j\cdot\frac{\sqrt{1-\xi^2}}{\tau}t}}{2}+\frac{\xi}{\sqrt{1-\xi^2}}\cdot\frac{e^{j\cdot\frac{\sqrt{1-\xi^2}}{\tau}t}-e^{-j\cdot\frac{\sqrt{1-\xi^2}}{\tau}t}}{2j}\right]\right\}
\end{aligned}
\tag{4.192}
$$

Recall that $\sin\alpha=\frac{e^{j\alpha}-e^{-j\alpha}}{2j}$ and $\cos\alpha=\frac{e^{j\alpha}+e^{-j\alpha}}{2}$, therefore

$$
y'(t)=MK_p\left\{1-e^{-\frac{\xi}{\tau}\cdot t}\left[\cos\left(\frac{\sqrt{1-\xi^2}}{\tau}t\right)+\frac{\xi}{\sqrt{1-\xi^2}}\cdot\sin\left(\frac{\sqrt{1-\xi^2}}{\tau}t\right)\right]\right\}
\tag{4.193}
$$

Using the trigonometric identity,

$$
r\cdot\sin(\phi+\beta)=r\cdot\sin\phi\cdot\cos\beta+r\cdot\cos\phi\cdot\sin\beta \tag{4.194}
$$

$$
r\cdot\sin\phi=1 \tag{4.195}
$$

$$
r\cdot\cos\phi=\frac{\xi}{\sqrt{1-\xi^2}} \tag{4.196}
$$

$$
r^2\cdot\left(\sin^2\phi+\cos^2\phi\right)=1+\frac{\xi^2}{1-\xi^2}=r^2 \tag{4.197}
$$

We get the final expression of the output response.

$$
r=\frac{1}{\sqrt{1-\xi^2}} \text{ and } \phi=\tan^{-1}\frac{\sqrt{1-\xi^2}}{\xi} \tag{4.198}
$$

$$y'(t) = MK_p\left\{1 - \frac{e^{-\frac{\xi}{\tau}\cdot t}}{\sqrt{1-\xi^2}}[\sin(\omega t + \phi)]\right\} \tag{4.199}$$

where $\omega = \frac{\sqrt{1-\xi^2}}{\tau}$ is the natural frequency of oscillation. The smaller ξ gets, the more oscillatory the response will be. If $\xi = 0$, the response is **undamped** and there will be no damping. Fig. 4.25 shows the step response of a second-order underdamped ($\xi < 0$) and undamped ($\xi = 0$) system.

Of special interest is the underdamped response as is shown in Fig. 4.26 because it often represents the desired shape of the response of a closed-loop feedback control system. The parameters of interest that are often used as design criteria for the feedback controllers are:

$$\frac{a}{b} = \text{overshoot} = \text{O.S.} = \exp\left(\frac{-\pi\xi}{\sqrt{1-\xi^2}}\right) \tag{4.200}$$

$$\frac{c}{a} = \text{decay ratio} = \text{D.R.} = (\text{O.S.})^2 = \exp\left(\frac{-2\pi\xi}{\sqrt{1-\xi^2}}\right) \tag{4.201}$$

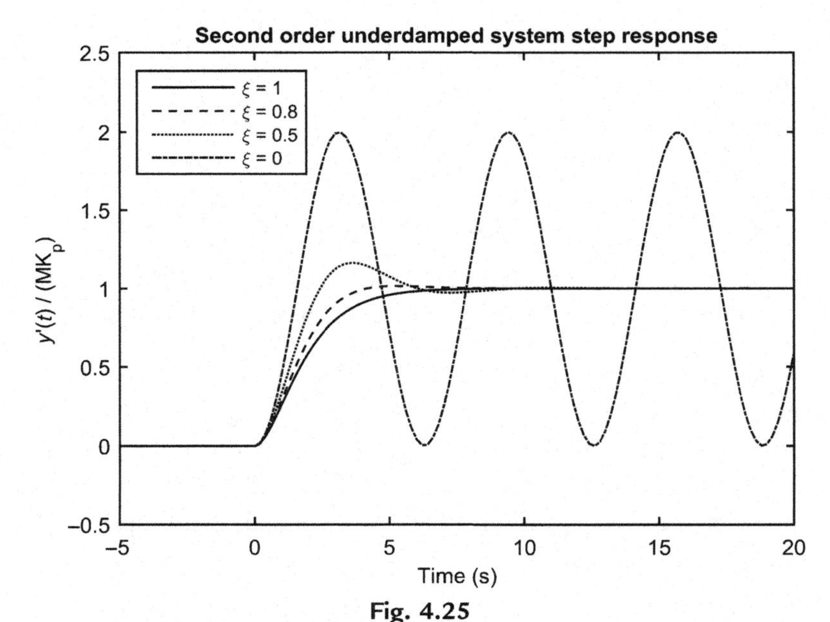

Fig. 4.25
The unit step response of a second-order underdamped system with $\tau = 1$ s.

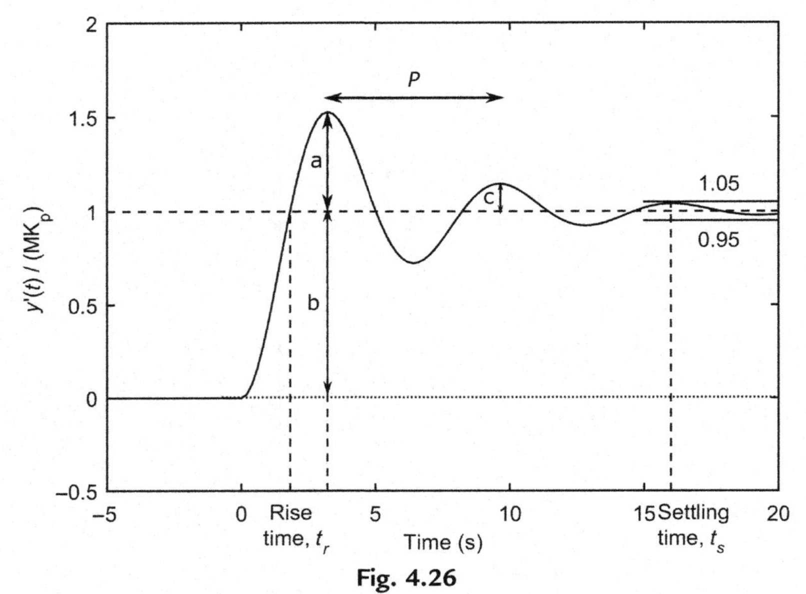

Fig. 4.26
The step response of a second-order underdamped system.

$$P = \text{period of oscillation} = \frac{2\pi\tau}{\sqrt{1-\xi^2}} \tag{4.202}$$

$$\omega = \text{frequency of oscillation} = \frac{2\pi}{P} = \frac{\sqrt{1-\xi^2}}{\tau} \tag{4.203}$$

4.4.4.4 Examples of real systems that have second-order dynamics (second-order transfer functions)

There are real systems that inherently have a second-order overdamped dynamics, and less frequently, underdamped dynamics. We shall consider the following examples.

(1) Two first-order processes in series.
(2) An inherently second-order underdamped process. Examples include the suspension system in a car, the movement of the liquid in a manometer arm as a result of the increase/decrease in the applied pressure to one of the arms, the movement of the control valve stem as a result of the increase/decrease in the control command signal applied to the diaphragm of a pneumatically actuated control valve.
(3) A feedback control system.

Example 4.15

Derive the transfer function of two isothermal continuous stirred tank reactors (CSTRs) in series (Fig. 4.27), in which a first-order chemical reaction takes place, with the following assumptions:

- $V_1 = \overline{V}_1 = \text{constant}$, $V_2 = \overline{V}_2 = \text{constant}$, $F = \overline{F} = \text{constant}$, the reaction rate is expressed by a first-order reaction $r_A = kC_A$
- Isothermal operation with constant properties

The component mole balance in the first tank is:

$$\overline{V}_1 \cdot \frac{dC_{A1}}{dt} = \overline{F} \cdot C_{Ai} - \overline{F} \cdot C_{A1} - k\overline{V}_1 C_{A1} \tag{4.204}$$

Laplace transform,

$$\left(\frac{\overline{V}_1}{\overline{F} + k\overline{V}_1} s + 1\right) C_{A1}(s) = \frac{\overline{F}}{\overline{F} + k\overline{V}_1} C_{Ai}(s) \tag{4.205}$$

$$\frac{C_{A1}(s)}{C_{Ai}(s)} = \frac{\dfrac{\overline{F}}{\overline{F} + k\overline{V}_1}}{\left(\dfrac{\overline{V}_1}{\overline{F} + k\overline{V}_1} s + 1\right)} = \frac{K_1}{\tau_1 s + 1} \tag{4.206}$$

The component mole balance in the second tank is:

$$\overline{V_2} \cdot \frac{dC_{A2}}{dt} = \overline{F} \cdot C_{A1} - \overline{F} \cdot C_{A2} - k\overline{V_2} C_{A2} \tag{4.207}$$

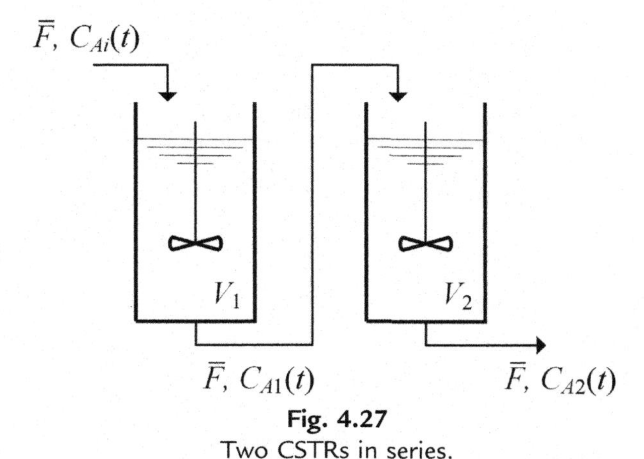

Fig. 4.27
Two CSTRs in series.

Laplace transform,

$$\left(\frac{\overline{V_2}}{\overline{F}+k\overline{V_2}}s+1\right)C_{A2}(s)=\frac{\overline{F}}{\overline{F}+k\overline{V_2}}C_{A1}(s) \quad (4.208)$$

$$\frac{C_{A2}(s)}{C_{A1}(s)}=\frac{\dfrac{\overline{F}}{\overline{F}+k\overline{V_2}}}{\left(\dfrac{\overline{V_2}}{\overline{F}+k\overline{V_2}}s+1\right)}=\frac{K_2}{\tau_2 s+1} \quad (4.209)$$

The overall transfer function relating the Laplace transform of the reactant concentration in the effluent stream of the second tank to the inlet reactant concentration to the first tank is the product of the previous two transfer functions:

$$\frac{C_{A2}(s)}{C_{Ai}(s)}=\frac{C_{A2}(s)}{C_{A1}(s)}\cdot\frac{C_{A1}(s)}{C_{Ai}(s)}=\frac{K_1K_2}{(\tau_1 s+1)(\tau_2 s+1)}=\frac{K_1K_2}{\tau_1\tau_2 s^2+(\tau_1+\tau_2)s+1} \quad (4.210)$$

Therefore the overall transfer function is a second-order transfer function with $\tau=\sqrt{\tau_1\tau_2}$, and a damping coefficient, $\xi=\frac{\tau_1+\tau_2}{2\sqrt{\tau_1\tau_2}}$. Note that for two first-order systems in series, the overall system response is always either overdamped or critically damped, and never underdamped.

$$\begin{cases}\xi>1 & (\tau_1\neq\tau_2)\\ \xi=1 & (\tau_1=\tau_2)\end{cases} \quad (4.211)$$

Example 4.16

In this example, it will be shown that a pneumatically actuated control valve has a second-order dynamics. The transfer function relating the change in the valve's "lift" or "position" x, to the applied pressure to the valve's diaphragm (the control command signal), P, can be developed by considering the forces acting on the valve stem. This can be achieved by writing a momentum balance, Newton's law of motion, on the valve stem movement.

Momentum balance: Newton's law states that the change in the momentum of a moving object is equal to the sum of the forces acting on the object.

$$\frac{d}{dt}\left(m\frac{dx}{dt}\right)=\text{rate of change of momentum} \quad (4.212)$$

where m is the mass of the valve stem, x is the valve stem displacement or position, and $\frac{dx}{dt}$ is the velocity of the valve stem movement. See Fig. 4.28.

$$\frac{d}{dt}\left(m\frac{dx}{dt}\right)=\sum\text{forces}=P_cA_d-mg-Kx-F_s+P_fA_p \quad (4.213)$$

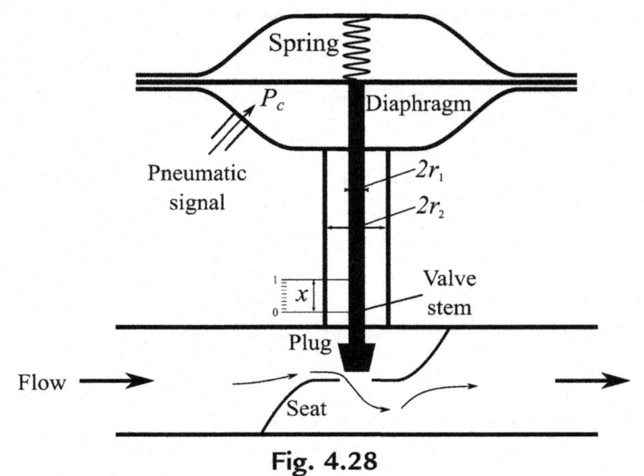

Fig. 4.28
An air-to-open control valve.

where A_d is the diaphragm area, mg is the gravitational force, Kx is the spring force with a spring constant K, F_s is the shear force acting on the stem (caused by the packing that keeps the valve stem in position), P_f is the fluid pressure, and A_p is the area of the valve's plug.

Assuming a Newtonian fluid for the packing, the shear force F_s acting on the area $(2\pi \cdot r_1 \cdot l)$ is

$$F_s = -\mu \frac{dV}{dr} \cdot (2\pi \cdot r_1 \cdot l) = \mu \frac{\left(\frac{dx}{dt} - 0\right)}{r_1 - r_2} (2\pi \cdot r_1 \cdot l) \tag{4.214}$$

where μ is the viscosity of the packing material that holds the valve stem in position and l is the length of the packing in contact with the valve stem. Substitute F_s in Eq. (4.213)

$$m \frac{d^2x}{dt^2} = P_c A_d - mg + P_f A_p - Kx - \mu \frac{(2\pi \cdot r_1 \cdot l)}{r_1 - r_2} \frac{dx}{dt} \tag{4.215}$$

Subtract the steady-state equation and assume negligible force due to the fluid pressure acting on the plug, we have:

$$\frac{m}{K} \frac{d^2x'}{dt^2} + \mu \frac{(2\pi \cdot r_1 \cdot l)}{(r_2 - r_1)K} \frac{dx'}{dt} + x' = \frac{A_d}{K} P'_c \tag{4.216}$$

Taking Laplace transform:

$$\frac{X(s)}{P_c(s)} = \frac{\frac{A_d}{K}}{\frac{m}{K} s^2 + \mu \frac{(2\pi \cdot r_1 \cdot l)}{(r_2 - r_1)K} s + 1} \tag{4.217}$$

Note that the previous equation is a second-order equation. Although, the actual dynamics of a control valve is in the form of a second-order transfer function, a first-order approximation is often used because m is small and the d^2x/dt^2 (or s^2) term drops out.

Example 4.17

The effluent flow rate from a tank is used to control the liquid level using a PI (proportional integral) controller, as is shown in Fig. 4.29.

Mass balance on the liquid contents in the tank:

$$A\frac{dl(t)}{dt} = F_i(t) - F(t) \tag{4.218}$$

$$F(t) = \overline{F} - K_c\left[l_{sp} - l(t)\right] - \frac{K_C}{\tau_I}\int_0^t \left[l_{sp} - l(t)\right] dt \tag{4.219}$$

Substitute for $F(t)$ in Eq. (4.218),

$$A\frac{dl(t)}{dt} = F_i(t) - \overline{F} + K_c\left[l_{sp} - l(t)\right] + \frac{K_C}{\tau_I}\int_0^t \left[l_{sp} - l(t)\right] dt \tag{4.220}$$

Subtract the steady-state equation:

$$A\frac{dl'}{dt} = F_i'(t) - K_c l' - \frac{K_c}{\tau_I}\int_0^t l' dt \tag{4.221}$$

Note that $l'(t) = l(t) - l_{sp}$. Laplace transform Eq. (4.221),

$$A \cdot s \cdot L(s) = F_i(s) - K_c \cdot L(s) - \frac{K_c}{\tau_I}\frac{L(s)}{s} \tag{4.222}$$

$$\frac{L(s)}{F_i(s)} = \frac{\tau_I \cdot s}{A\tau_I s^2 + \tau_I K_c s + K_c} \tag{4.223}$$

Note that ξ and τ are functions of the controller parameters K_c and τ_I. Depending on the choice of the controller parameters, an underdamped or an overdamped response can be achieved.

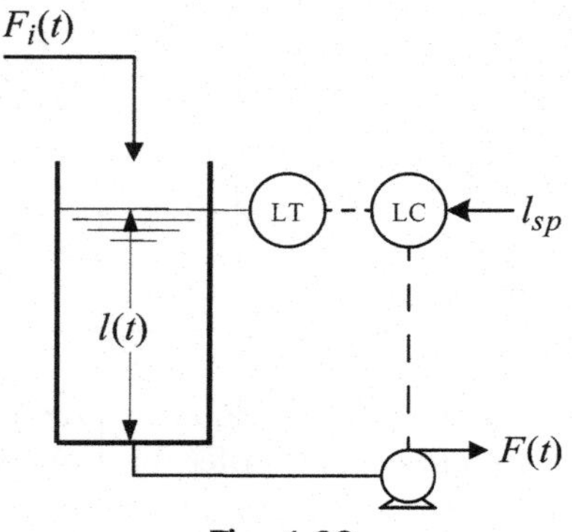

Fig. 4.29
A liquid tank with a PI level controller.

4.4.5 Significance of the Transfer Function Poles and Zeros

Based on the concepts introduced earlier, we know that the roots of the denominator polynomial of a transfer function—which is referred to as the process "characteristic equation or polynomial"—have significant effect on the process dynamic response. These roots are called the system **poles** in the s-plane and determine the system stability and the shape of the response curve (smooth or oscillatory, stable or unstable, etc.). Consider a general transfer function:

$$G(s)=\frac{Y(s)}{U(s)}=\frac{K}{s(\tau_1 s+1)(\tau_2 s^2+2\xi\tau_2 s+1)} \tag{4.224}$$

The previous transfer function, for $0\le\xi<1$, has four poles as shown in Fig. 4.30:

$$p_1=0;\ p_2=-\frac{1}{\tau_1};\ p_{3,4}=\ -\frac{1}{\tau_2}\pm j\frac{\sqrt{1-\xi^2}}{\tau_2} \tag{4.225}$$

The previous four poles contribute to the following terms in the overall time response of the output:

- A constant term due to the $\frac{1}{s}$ term,
- an exponential term $e^{-\frac{t}{\tau_1}}$ due to the $\frac{1}{\tau_1 s+1}$ term, and
- an exponentially decaying sinusoidal term due to the $\frac{1}{(\tau_2 s^2+2\xi\tau_2 s+1)}$.

$$\frac{e^{-\frac{\xi}{\tau_2}\cdot t}}{\sqrt{1-\xi^2}}\left[\sin\left(\frac{\sqrt{1-\xi^2}}{\tau_2}\cdot t+\tan^{-1}\frac{\sqrt{1-\xi^2}}{\xi}\right)\right] \tag{4.226}$$

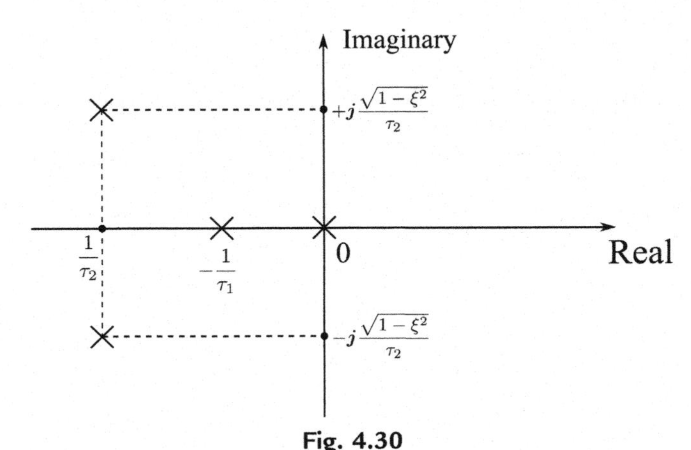

Fig. 4.30
The location of the roots on the complex s-plane.

Note that the complex roots lead to an oscillatory response and always appear in complex conjugate pairs. This is because they are the results of the solution of a quadratic equation in the characteristic polynomial. Since the real part of these complex roots is negative, the response will be damped (stable) and oscillatory. Also, it is of interest to note that the frequency of the sinusoidal term is given by the imaginary part of the complex roots.

The roots of the numerator polynomial of the transfer function are called the system **zeros**. The system zeros also influence the dynamic response of the system. In the s-plane, the system poles are shown by a cross ($\times$) and the system zeros by a circle ($\circ$). It will be shown that positive zeros of a transfer function lead to an inverse response (nonminimum phase behavior). The small poles (closer to the imaginary axis) are the **dominant** poles and exert a more influential effect on the system response. Similarly, the small negative zeros cause an overshoot in the response (in such a case, the time constant of the lead term in the transfer function is larger than the lag time constants). Note that the small poles correspond to a large **lag time constant** and small zeros correspond to a large **lead time constant**.

In general, the transfer function of a process can be represented either in a **pole-zero** or a **time constant** form:

$$G(s) = \frac{b_m s^m + b_{m-1} s^{m-1} + \cdots + b_0}{a_n s^n + a_{n-1} s^{n-1} + \cdots + a_0} = \frac{b_m (s - z_1)(s - z_2)\ldots(s - z_m)}{a_n (s - p_1)(s - p_2)\ldots(s - p_m)} \quad \text{(pole-zero form)} \qquad (4.227)$$

$$G(s) = \frac{K(\tau_{\text{lead},1} \cdot s + 1)(\tau_{\text{lead},2} \cdot s + 1)\ldots}{(\tau_{\text{lag},1} \cdot s + 1)(\tau_{\text{lag},2} \cdot s + 1)\ldots} \quad \text{(time constant form, lead-lag terms)} \qquad (4.228)$$

where $z_i = -\frac{1}{\tau_{\text{lead},i}}$ and $p_i = -\frac{1}{\tau_{\text{lag},i}}$.

Example 4.18

We have seen that the general transfer function of a second-order process with a time delay can be represented by Eq. (4.229), which represents the dynamics of two first-order systems in series (overdamped) or an underdamped system, with a time delay. The step response of such a system depending on the numerical value of the damping coefficient is shown in Fig. 4.31.

$$G_p(s) = \frac{K_p e^{-\theta s}}{\tau^2 s^2 + 2\xi \tau s + 1} \qquad (4.229)$$

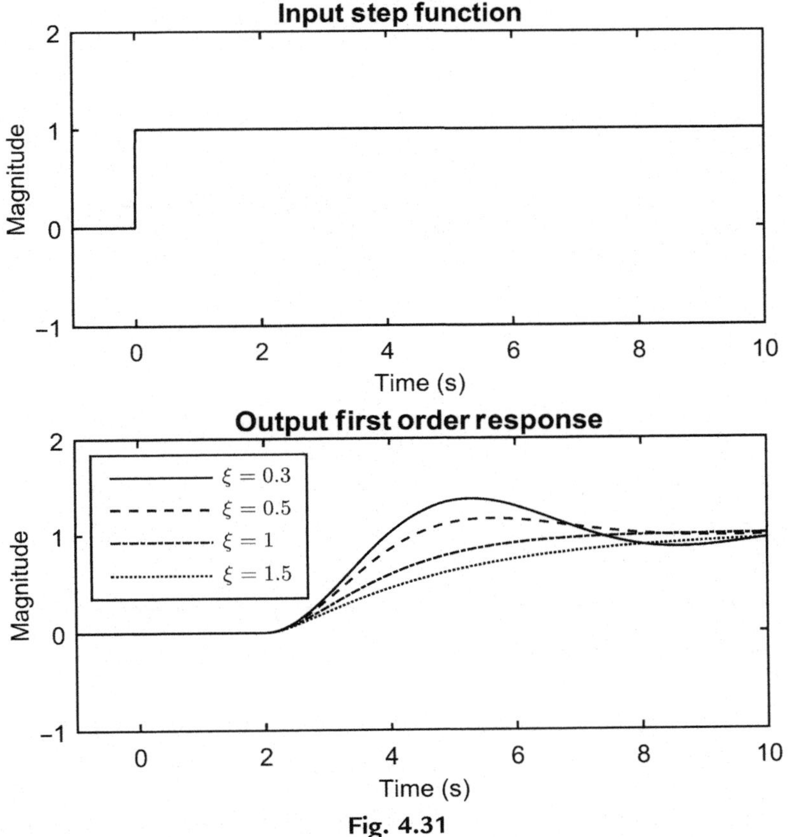

Fig. 4.31
Response of a second-order system with $\theta = 2s$ and $\tau = 1s$.

Similarly, the transfer function of a second-order process with a lead term is given by:

$$G_p(s) = \frac{Y(s)}{U(s)} = \frac{K(\tau_a s + 1)}{(\tau_1 s + 1)(\tau_2 s + 1)} \tag{4.230}$$

The step response of the transfer function given in Eq. (4.230) to a step change with magnitude M in the input is:

$$y'(t) = KM\left(1 + \frac{\tau_a - \tau_1}{\tau_1 - \tau_2} e^{-\frac{t}{\tau_1}} + \frac{\tau_a - \tau_2}{\tau_2 - \tau_1} e^{-\frac{t}{\tau_2}}\right) \tag{4.231}$$

Depending on the relative magnitude of the three time constants, the shape of the response curve changes. Note that if the lead time constant is larger than the lag time constants, the response will have an overshoot, and if the transfer function zero is positive (negative τ_a values), an inverse response is obtained (Fig. 4.32).

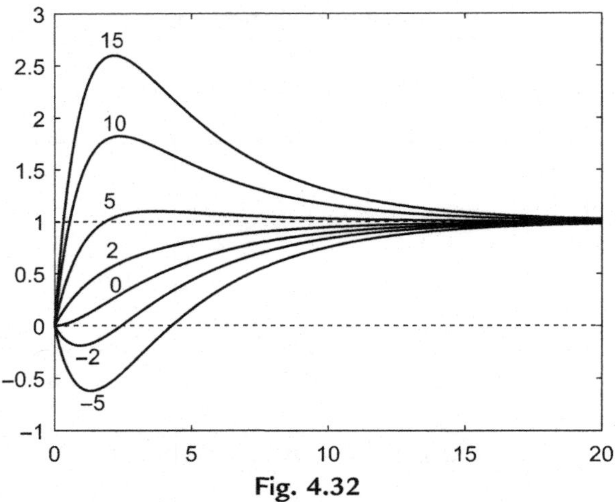

Fig. 4.32
The step response of an overdamped second-order system with a lead term given in Eq. (4.230), for different values of τ_a ($\tau_1 = 4$ and $\tau_2 = 1$).

4.4.5.1 *Summary of the significance of poles and zeros of a transfer function*

$$G(s) = \frac{Q(s)}{P(s)} \quad \begin{cases} \text{the roots of } Q(s) \text{ are } \textbf{zeros} \text{ of the transfer function} \\ \text{the roots of } P(s) \text{ are } \textbf{poles} \text{ of the transfer function} \end{cases} \tag{4.232}$$

The poles of a transfer function determine system stability. The zeros of a transfer function determine whether the system exhibits inverse response or an overshoot.

4.4.6 *Transfer Functions of More Complicated Processes—An Inverse Response (A Nonminimum Phase Process), A Higher Order Process and Processes With Time Delays*

The dynamics of the majority of real processes can be approximated by a first- or second-order transfer function with or without a delay term. There are, however, certain processes that exhibit inverse response. In what follows we shall consider more complicated forms of process transfer functions.

4.4.6.1 *Physical processes with inverse response*

If the ultimate response of a process to an input is in the opposite direction to its initial direction, the process is called an *inverse process* or a *nonminimum process*. The responses of a few nonminimum phase processes to a unit step increase in the input are shown in Fig. 4.33. Note that the output either increases initially and ultimately decreases, or vice versa. In general, if the process transfer function has a positive zero, the system exhibits an inverse response.

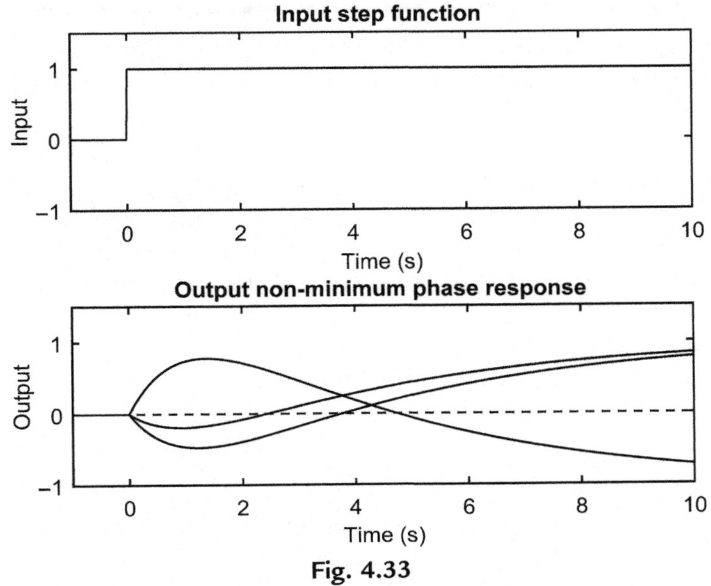

Fig. 4.33
An inverse response or a nonminimum phase response.

Example 4.19

The liquid level in a reboiler of a distillation column usually shows an inverse response to a step increase in the input steam pressure, in the absence of a level controller. The level in the reboiler increases initially due to the frothing and spill over from the trays immediately above the reboiler. Ultimately, because of the higher rate of boil-off of the liquid, the liquid level in the reboiler decreases (Fig. 4.34).

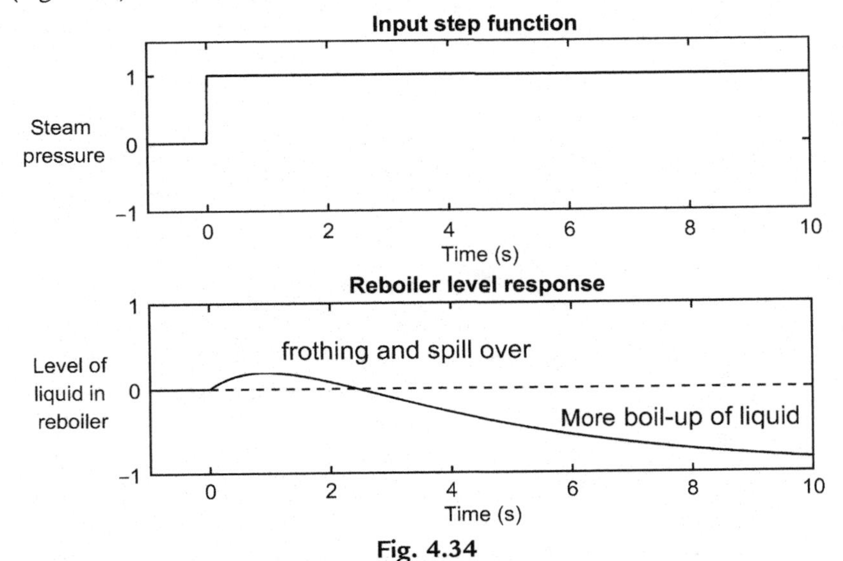

Fig. 4.34
The liquid level response in a reboiler with a step increase in the steam pressure.

Example 4.20

The exit temperature in a fixed-bed catalytic reactor with an exothermic reaction may show an inverse response (Fig. 4.35).

If the reactant inlet temperature, T_i, is increased by a step change, the reaction rate in the entrance region of the reactor increases and uses up the reactants. Since the reaction is assumed to be exothermic, there will be less heat of reaction generated due to the lower reactants concentration, and hence, initially we observe a drop in the exit temperature. However, since the inlet temperature has increased and is sustained by a step change, eventually the exit temperature increases as is shown in Fig. 4.36.

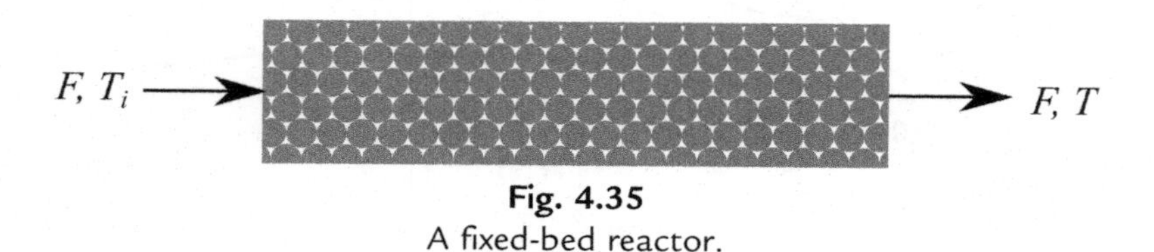

Fig. 4.35
A fixed-bed reactor.

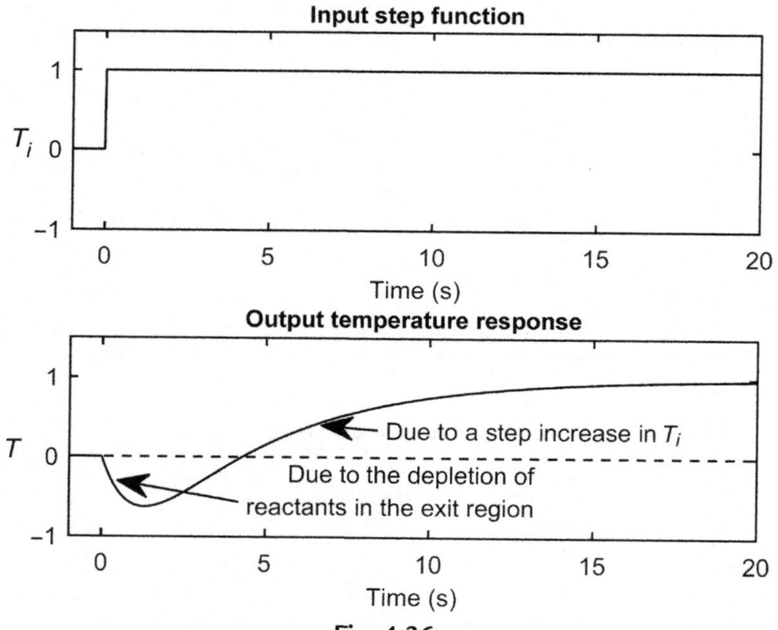

Fig. 4.36
Fixed-bed temperature response.

Example 4.21

Response in the liquid level of a steam boiler due to an increase in the flow rate of the inlet water shown in Fig. 4.37 may also show an inverse response.

As a result of higher cold water inflow, the steam bubbles in the boiler start to collapse. Let us assume that the reduction in the liquid level due to the collapse of the entrained bubbles can be represented by a first-order dynamics with a negative gain,

$$G_{\text{bubbles collapse}} = -\frac{K_{p1}}{\tau_p s + 1} \tag{4.233}$$

Similarly, it is reasonable to assume a pure capacitive transfer function between the liquid level in the boiler and the inlet water flow rate at a constant rate of steam generation:

$$G_{\text{capacitive}} = \frac{K_{p2}}{s} \tag{4.234}$$

The boiler dynamics is the sum of these two opposing subprocesses shown in Fig. 4.38:

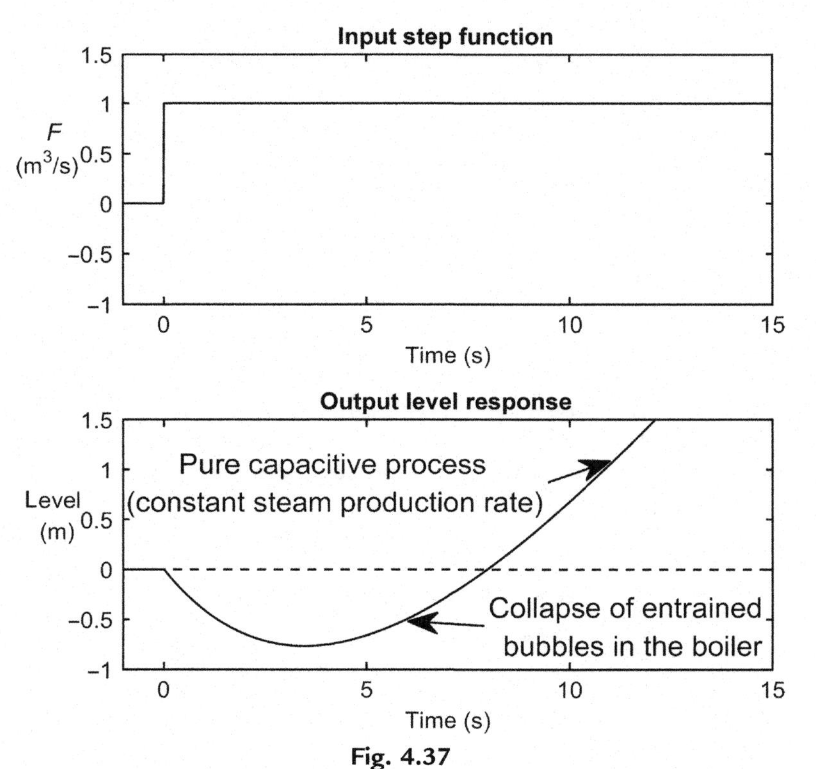

Fig. 4.37
Level response in a steam boiler with a step increase in the inlet flow rate.

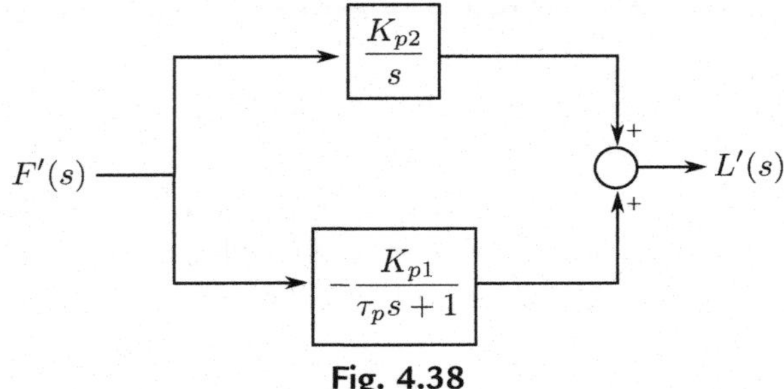

Fig. 4.38
The block diagram of a boiler.

$$\frac{L(s)}{F(s)}=\frac{K_{p2}}{s}+\left(-\frac{K_{p1}}{\tau_p s+1}\right)=\frac{(K_{p2}\tau_p-K_{p1})s+K_{p2}}{s(\tau_p s+1)} \tag{4.235}$$

Note that there is a first-order polynomial in "s" in the numerator with a root at:

$$z=\text{zero of transfer function}=-\frac{K_{p2}}{K_{p2}\tau_p-K_{p1}} \tag{4.236}$$

If $K_{p1}>K_{p2}\tau_p$, the transfer function zero is positive leading to an inverse response. However, if $K_{p1}<K_{p2}\tau_p$, the transfer function zero becomes negative and there will be no inverse response. For a step change in the inlet flow rate with magnitude M, we have: $F(s)=\frac{M}{s}$,

$$L(s)=\frac{MK_{p2}}{s^2}-\frac{MK_{p1}}{s(\tau_p s+1)} \tag{4.237}$$

Inverse Laplace transform,

$$l'(t)=MK_{p2}t-MK_{p1}\left(1-e^{-\frac{t}{\tau_p}}\right) \tag{4.238}$$

By considering the responses shown in Fig. 4.39, it becomes clear that the inverse response is caused when the transfer function zero becomes positive which happens if $K_{p1}>K_{p2}\tau_p$. The numerical examples for selected values of the process parameters confirm this point.

(a) $K_{p2}=2, K_{p1}=4, \tau_p=1$

$$K_{p2}<\frac{K_p}{\tau_p}\ \text{(expect inverse response)};\ G(s)=\frac{2(1-s)}{s(s+1)} \tag{4.239}$$

(b) $K_{p2}=2, K_{p1}=1, \tau_p=1$

$$K_{p2}>\frac{K_p}{\tau_p}\ \text{(do not expect inverse response)};\ G(s)=\frac{s+2}{s(s+1)} \tag{4.240}$$

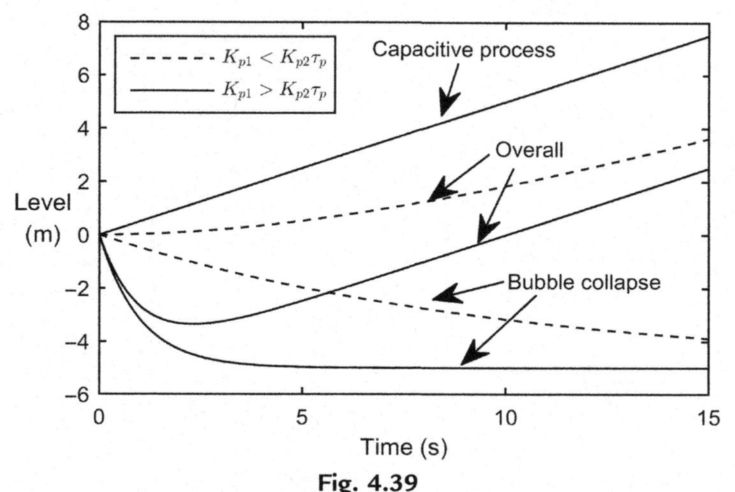

Fig. 4.39
The cause of inverse response in a steam boiler.

It appears that for a process to show an *inverse response*, there must be two opposing processes that work against each other.

4.4.7 Processes With Nth-Order Dynamics

The dynamic equation of an Nth-order system is represented by an Nth-order ODE of the general form:

$$a_N \frac{d^N y'(t)}{dt^N} + a_{N-1} \frac{d^{N-1} y'(t)}{dt^{N-1}} + \cdots + a_0 y'(t) = b \cdot u'(t) \tag{4.241}$$

That leads to an Nth-order transfer function.

$$\frac{Y(s)}{U(s)} = \frac{\dfrac{b}{a_0}}{\dfrac{a_N}{a_0} s^N + \dfrac{a_{N-1}}{a_0} s^{N-1} + \cdots + 1} \tag{4.242}$$

Example 4.22

Develop the transfer function of N isothermal CSTRs in series, shown in Fig. 4.40, with a first-order reaction, assuming that the volumes of all reactors remain constant.

The simplifying assumptions include, constant volume $\overline{V}_i$, constant flow rates $\overline{F}$, constant temperature, and density. The reaction kinetics is $r = kC_A$, where k is the kinetic constant.

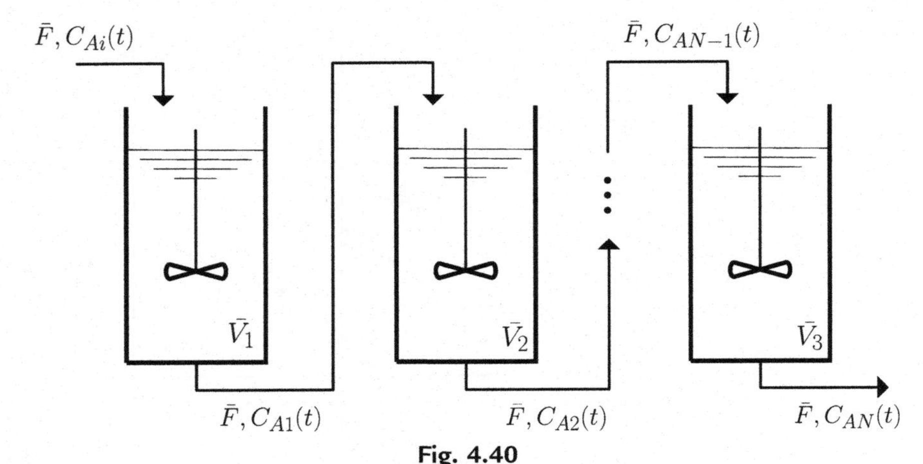

Fig. 4.40
The pictorial presentation of N isothermal CSTRs in series.

$$\frac{C_{AN}(s)}{C_{Ai}(s)} = \frac{K_{P1}}{\tau_{P1}s+1} \cdot \frac{K_{P2}}{\tau_{P2}s+1} \cdots \frac{K_{PN}}{\tau_{PN}s+1} = \prod_{i=1}^{i=N} \frac{K_{Pi}}{\tau_{Pi}s+1} \tag{4.243}$$

where $K_{Pi} = \frac{\overline{F}}{\overline{F}+k\overline{V}_i}$ and $\tau_{pi} = \frac{\overline{V}_i}{\overline{F}+k\overline{V}_i}$.

If all the reactors have equal volumes, the steady-state gains and time constants of all reactors will be the same:

$$\begin{aligned} K_{P1} &= K_{P2} = \cdots = K_{PN} = K_P \\ \tau_{P1} &= \tau_{P2} = \cdots = \tau_{PN} = \tau_P \end{aligned} \tag{4.244}$$

Therefore

$$\frac{C_{AN}(s)}{C_{Ai}(s)} = \frac{K_P^N}{(\tau_P s+1)^N} \tag{4.245}$$

The response of this system to a step change in the inlet reactant concentration with magnitude M (Eq. 4.246) is given by Eq. (4.247). In this case, there are N real multiple real roots. Using partial fractions, we obtain an expression for $C'_{AN}(t)$. The response is shown in Fig. 4.41.

$$C_{Ai}(s) = \frac{M}{s} \tag{4.246}$$

$$C'_{AN}(t) = MK_p^N \left\{ 1 - e^{-\frac{t}{\tau_p}} \left[1 + \frac{t/\tau_p}{1!} + \frac{(t/\tau_p)^2}{2!} + \cdots + \frac{(t/\tau_p)^{N-1}}{(N-1)!} \right] \right\} \tag{4.247}$$

Large N values slow down the process and increase the conversion. As N increases, the final output value decreases and the time at which the output starts changing (the delay time) increases. Therefore an Nth-order system may be represented by a second-order overdamped transfer function with a time delay. We can conclude that higher order dynamics can be

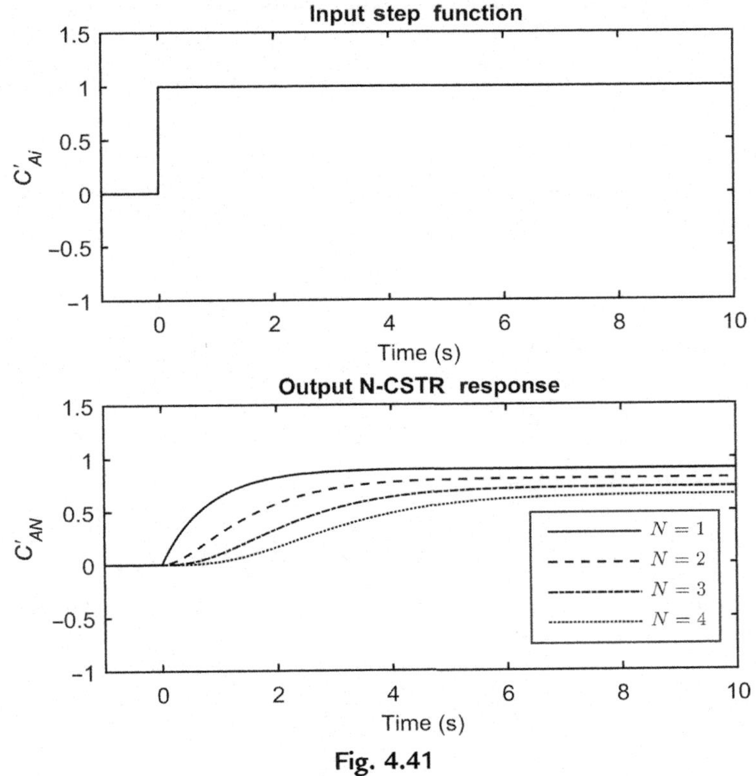

Fig. 4.41
The response of N isothermal CSTRs with different N ($K_p = 0.9$ and $\tau = 0.8$).

approximated by a second-order system (overdamped) with a time delay. The parameters τ, K_p, ξ, and θ should be determined experimentally using the "Process Identification" technique discussed later in this chapter.

4.4.8 Transfer Function of Distributed Parameter Systems

Dynamic models of the distributed parameter systems are in the form of partial differential equations (PDE) and are quite complex. Therefore they are often approximated by a first- or a second-order transfer function with a time delay.

The same steps used for the lumped parameter systems are followed to derive the transfer functions of the distributed parameter systems. After linearizing all the nonlinear terms in the equations resulting from the conservation principles, the linearized equations are Laplace transformed. The analysis and control of distributed control systems will be studied in detail in Chapter 13. Here, we consider an example of a simple distributed parameter system, namely, an isothermal plug flow reactor with a first-order reaction rate.

Example 4.23

Develop the transfer function of an isothermal tubular reactor shown in Fig. 4.42 in a plug flow regime with a first-order reaction, $r_A = kC_A(t,z)$.

Assumptions include plug flow regime, isothermal operation, constant fluid properties, a constant reactant flow rate, and a constant cross-sectional area of the tubular reactor.

The component mole balance on reactant A leads to:

$$\frac{\partial C_A(t,z)}{\partial t} = -\bar{u}_b \frac{\partial C_A(t,z)}{\partial z} - kC_A(t,z) \tag{4.248}$$

In terms of the deviation variables:

$$\frac{\partial C'_A(t,z)}{\partial t} = -\bar{u}_b \frac{\partial C'_A(t,z)}{\partial z} - kC'_A(t,z) \tag{4.249}$$

Laplace transform with respect to time,

$$s \cdot C'_A(sz) = -\bar{u}_b \frac{dC'_A(sz)}{dz} - kC'_A(sz) \tag{4.250}$$

Note that after Laplace transformation, the PDE becomes an ODE in the z domain. Rearranging the previous equation leads to:

$$\frac{dC_A(s,z)}{C_A(s,z)} = -\left(\frac{s+k}{\bar{u}_b}\right) dz \tag{4.251}$$

Integrating:

$$\int_{C_{Ai}(s)}^{C_A(s,L)} \frac{dC_A(s,z)}{C_A(s,z)} = -\left(\frac{s+k}{\bar{u}_b}\right) \int_0^L dz \tag{4.252}$$

$$\frac{C_A(s,L)}{C_{Ai}(s)} = \exp\left[-\left(\frac{s+k}{\bar{u}_b}\right) L\right] = e^{-\frac{L}{u_b} \cdot s} \cdot e^{-\frac{kL}{\bar{u}_b}} \tag{4.253}$$

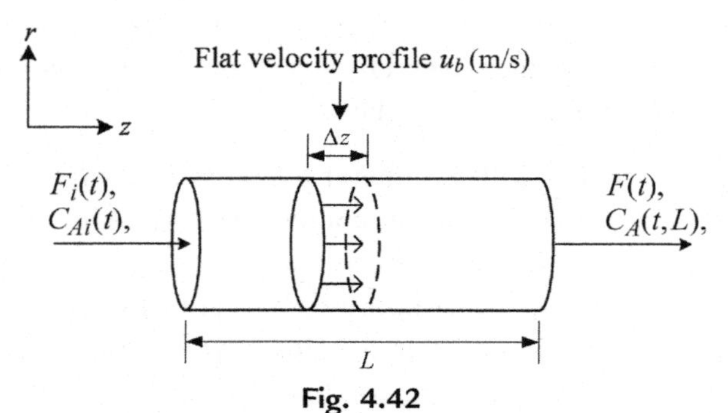

Fig. 4.42
A tubular reactor in a plug flow regime.

$$\frac{C_A(s, L)}{C_{Ai}(s)} = e^{-\theta \cdot s} \cdot e^{-\frac{kL}{\bar{u}_b}} \tag{4.254}$$

where θ is the delay time or residence time of the liquid droplets in the reactor $= \frac{L}{\bar{u}_b}$, and $\frac{kL}{\bar{u}_b}$ represents an attenuation factor that represents the extent of the conversion of the A component. As θ and k increase, the reactant concentration, C_A, at the reactor effluent decreases. Eq. (4.254) is for a plug flow condition, if there is dispersion due to a nonplug flow condition, the behavior of the reactor can be approximated by a first- or a second-order overdamped transfer function with a time delay. Fig. 4.43 shows the response in the outlet reactant concentration to a unit step change in the inlet reactant concentration. The output steady-state value is $e^{-\frac{kL}{\bar{u}_b}}$.

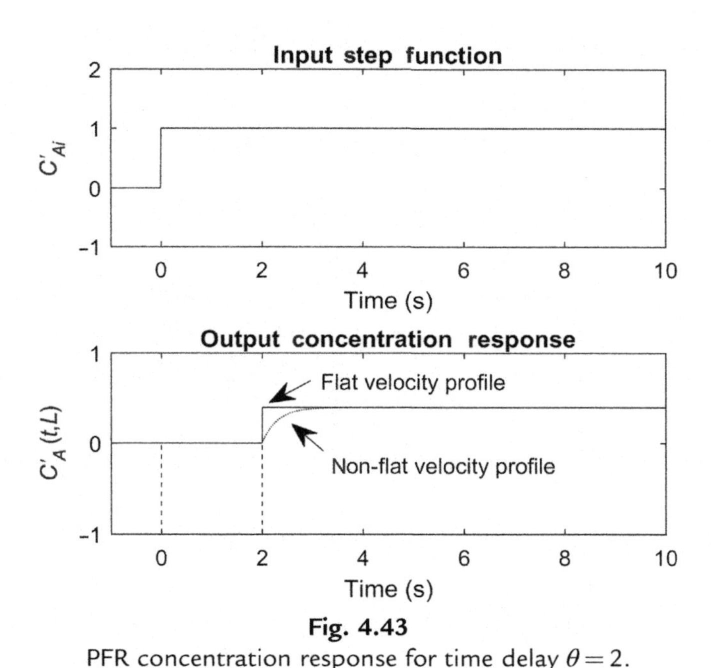

Fig. 4.43
PFR concentration response for time delay $\theta = 2$.

4.4.9 Processes With Significant Time Delays

Time delays, dead times, or transportation lags are common in chemical processes. They occur between the manipulated variable and the controlled variable (process time delay) or between the disturbance and the controlled variable (load time delay). The time delay may also be due to the measurement transducer using a batch-type analyzer, e.g., a GC (gas chromatograph) that requires a certain time interval for the analysis of a sample. Let us consider an example.

Example 4.24

Consider a well-mixed tank connected to a perfectly insulated pipe shown in Fig. 4.44. The temperature at the exit of the pipe is the same as the temperature at the pipe inlet with a dead time equal to the residence time of the liquid in the pipe.

Assumptions include constant fluid properties, constant volume ($\overline{V}$) and flow rate ($\overline{F}$) in the tank, plug flow regime in the pipe, a well-mixed tank, perfectly insulated pipe, and a constant rate of heat removal, $\overline{Q}$.

The energy balance on the tank renders a transfer function between the temperature at the exit and inlet of the tank,

$$\rho c \overline{V} \frac{dT}{dt} = \rho c \overline{F}[T_i(t) - T(t)] - \overline{Q} \tag{4.255}$$

Rearrange and convert to deviation form,

$$\frac{\overline{V}}{\overline{F}} \frac{dT'(t)}{dt} + T'(t) = T_i'(t) \tag{4.256}$$

Let $\tau_p = \frac{\overline{V}}{\overline{F}}$

$$\frac{T(s)}{T_i(s)} = \frac{K_p}{\tau_p s + 1} \tag{4.257}$$

Consider the transfer function of the long tube. Because of the assumption of plug flow and perfect insulation,

$$T_1(t) = T(t - \theta) \tag{4.258}$$

where $\theta =$ time delay in pipe $= \frac{L}{u_b}$

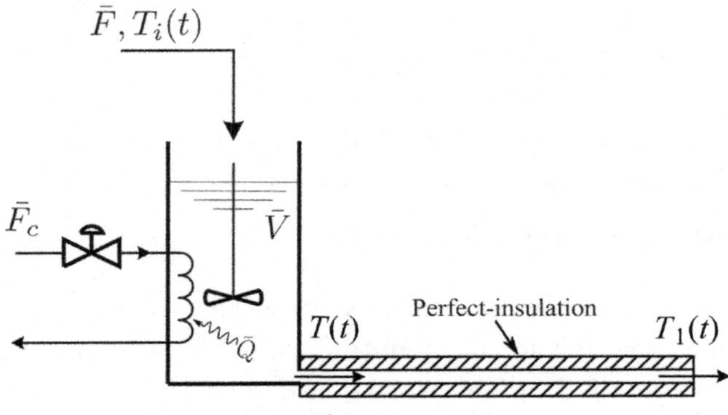

Fig. 4.44
A cooling tank with long insulated discharge tube.

Subtract the steady-state and Laplace transform,

$$T_1(s) = \mathcal{L}[T(t-\theta)] = e^{-\theta s}T(s) \tag{4.259}$$

The overall transfer function between the temperature at the pipe exit and the inlet temperature to the tank is:

$$G_p(s) = \frac{T_1(s)}{T_i(s)} = \frac{T_1(s)}{T(s)} \cdot \frac{T(s)}{T_i(s)} = e^{-\theta s} \cdot \frac{K_p}{\tau_p s + 1} = \frac{K_p e^{-\theta s}}{\tau_p s + 1} \tag{4.260}$$

The effect of time delay is simply to shift the response curve by θ time units.

4.4.9.1 *Effect of time delay in more detail*

The delay time $e^{-\theta s}$ can be expanded by the Taylor series,

$$e^{-\theta s} = 1 - \theta s + \frac{(\theta s)^2}{2!} - \frac{(\theta s)^3}{3!} + \cdots \tag{4.261}$$

This expansion suggests that the time delay introduces infinite number of roots and hence infinite number of positive and negative zeros in the transfer function, if the time delay appears in the numerator of the transfer function. Therefore one expects an inverse response in the presence of a time delay. Retaining all the infinite terms in the Taylor series is impractical, therefore, various approximations are used to replace the delay term. One such approximation is to retain only the first two terms of the time series expansion given in Eq. (4.261). Another approach is done using the first- or second-order Padé approximation:

$$\text{First-order Padé}: \; e^{-\theta s} \approx \frac{1 - \dfrac{\theta s}{2}}{1 + \dfrac{\theta s}{2}} \tag{4.262}$$

$$\text{Second-order Padé}: \; e^{-\theta s} \approx \frac{1 - \dfrac{\theta s}{2} + \dfrac{(\theta s)^2}{12}}{1 + \dfrac{\theta s}{2} + \dfrac{(\theta s)^2}{12}} \tag{4.263}$$

Fig. 4.45 shows the exact, first-order, and second-order Padé approximations of a time delay term.

Under certain conditions the time delay appears in the denominator of the transfer function (feedback control of a process with a time delay or having a recycle of energy or matter), and the only way to obtain the inverse transform is to approximate the delay term, $e^{-\theta s}$, by a Taylor series or a first- or a second-order Padé approximation.

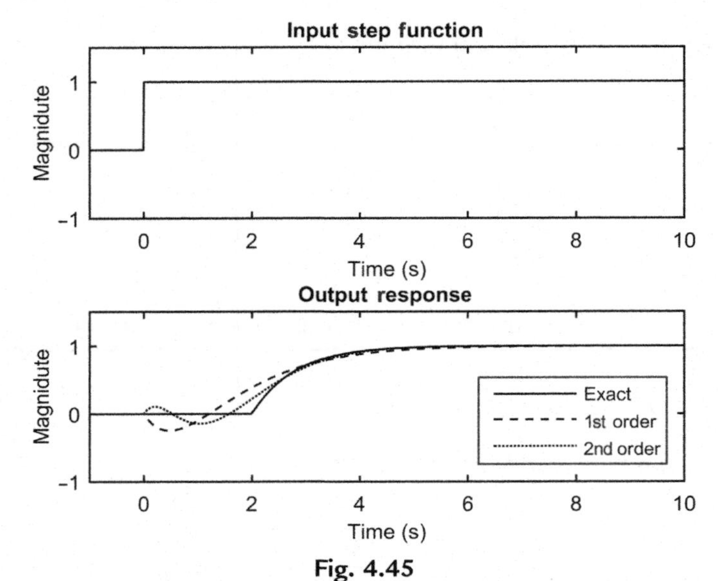

Fig. 4.45
The Padé approximation for a first-order system with time delay.

Example 4.25

Develop a transfer function for the mixing process in Fig. 4.46 with a recycle stream. The bulk velocity in the recycle pipe is $\bar{u}_b$.

In Fig. 4.46, θ represents the residence time in the pipe, $\theta = \frac{L}{\bar{u}_b}$. Writing the component mole balance on the reactor results in:

$$\overline{V}\frac{dC_{A1}(t)}{dt} = \overline{F}C_{Ai} + \overline{F}_1 C_{A1}(t-\theta) - (\overline{F} + \overline{F}_1)C_{A1}(t) \tag{4.264}$$

Laplace transform:

$$\overline{V}s\,C_{A1(s)} = \overline{F}C_{Ai}(s) + \overline{F}_1 C_{A1}(s)\,e^{-\theta s} - (\overline{F} + \overline{F}_1)C_{A1}(s) \tag{4.265}$$

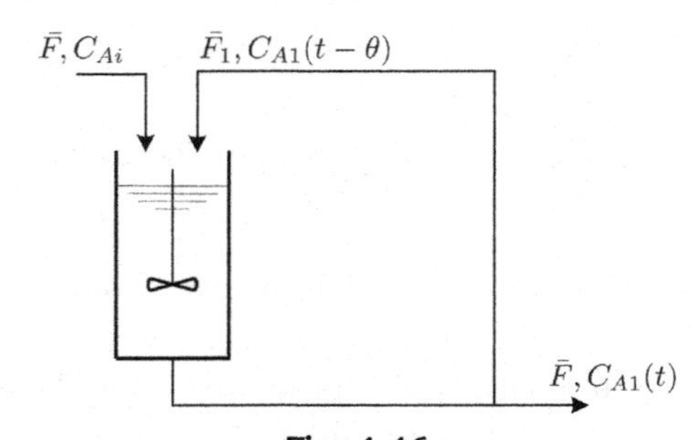

Fig. 4.46
A mixing process with recycle stream.

Rearrange and let $\tau_p = \overline{V}/\overline{F}$, $K = \overline{F}_1/\overline{F}$

$$\frac{\overline{V}}{\overline{F}} s\, C_{A1(s)} = C_{Ai}(s) + \frac{\overline{F}_1}{\overline{F}} C_{A1}(s)\, e^{-\theta s} - \frac{(\overline{F} + \overline{F}_1)}{\overline{F}} C_{A1}(s) \tag{4.266}$$

$$\left[\tau_p s + \left(-Ke^{-\theta s}\right) + (1 + K)\right] C_{A1}(s) = C_{Ai}(s) \tag{4.267}$$

The transfer function is

$$G(s) = \frac{C_{A1}(s)}{C_{Ai}(s)} = \frac{1}{\left[\tau_p s + 1 + K\left(1 - e^{-\theta s}\right)\right]} \tag{4.268}$$

To obtain the inverse Laplace transform or study the stability of the system, an approximation for the $e^{-\theta s}$ term is usually used.

4.4.9.2 *Approximation of higher order transfer functions by a first order plus time delay*

Using the Taylor series expansion and discarding higher terms, the following approximations can be written for the time delay and *time advance*:

$$\begin{aligned} e^{-\theta s} &\approx 1 - \theta s \\ e^{+\theta s} &\approx 1 + \theta s \end{aligned} \tag{4.269}$$

Therefore the lead and lag terms in a high-order transfer function can be approximated by a delay or advance time. Consider the following example:

$$G_p(s) = \frac{Y(s)}{U(s)} = \frac{3(-5s + 1)}{(6s + 1)(3s + 1)(2s + 1)} \tag{4.270}$$

Using the following approximations:

$$\begin{aligned} \frac{1}{3s + 1} &\approx \frac{1}{e^{3s}} = e^{-3s} \\ \frac{1}{2s + 1} &\approx \frac{1}{e^{2s}} = e^{-2s} \\ -5s + 1 &\approx e^{-5s} \end{aligned} \tag{4.271}$$

The original transfer function can then be approximated by:

$$G_p(s) = \frac{3(-5s + 1)}{(6s + 1)(3s + 1)(2s + 1)} \approx \frac{3e^{-5s}e^{-3s}e^{-2s}}{(6s + 1)} = \frac{3e^{-10s}}{6s + 1} \tag{4.272}$$

Therefore a first-orderlead-third-order lag transfer function given in Eq. (4.270) can be approximated by a first-order lag system with a time delay given by Eq. (4.272).

Example 4.26

Approximate the transfer function in Eq. (4.272) by a first-order plus time delay and use MATLAB to plot both the "actual" and the "approximate" responses of the system to a unit step change in the input.

The following MATLAB codes represent two methods to obtain the plots shown in Fig. 4.47.

Method 1: Use the inverse Laplace transform.

MATLAB script for inverse Laplace transform:

```
syms s t
exact_s = (3/s) * (1-5*s) / ((6*s+1) * (2*s + 1) * (3*s+1));
exact_t = ilaplace(exact_s,t);


approx_s = 3*exp(-10*s) / (s * (6*s + 1));
approx_t=ilaplace(approx_s,t);


hold on
% ezplot accepts a symbolic function and plot it within the range provided.
ezplot(exact_t,[0 100])
ezplot(approx_t,[0 100])
hold off
legend('exact response','approximation','Location','SouthEast')
```

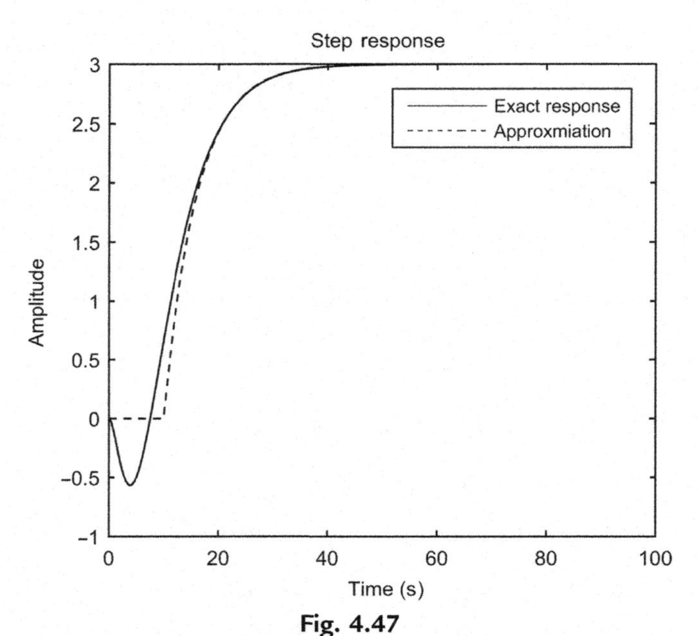

Fig. 4.47
The unit step response of the system in Example 4.26.

Method 2: Use transfer function.

```
MATLAB script for transfer function:

t=100;  % Simulation time range


exact_sys = tf([-15 3],conv([18 9 1], [2 1]));
approx_sys = tf(3, [6 1], 'inputdelay', 10);


hold on
step(exact_sys,t);
step(approx_sys,t);
hold off
```

Part B—The Empirical Approach to Develop Approximate Transfer Functions for Existing Processes

Real processes are complex and the theoretical derivation of transfer functions for such processes is a very time-consuming task if not impossible. An alternative to the theoretical approach to develop linear transfer functions is the use of process input-output data generated from an existing process and fitting them to low-order models. This approach may also be used to develop linear input-output models from the numerical simulation of nonlinear process models. The approach is primarily useful to develop transfer functions or linear state-space models of a process for the design of effective controllers. In this approach, the process is treated as a "black box" shown in Fig. 4.48.

There are two main methods to develop empirical transfer functions using the input-output data series:

(a) A graphical approach which is useful in the absence of process noise. It is based on the step response of a system, although other input signals such as an impulse, a square pulse, or a sinusoidal signal may also be used. Once the input-output data are obtained and plotted, an approximate graphical method is used to fit the input-output data to low-order transfer functions such as a first- or a second-order transfer function with or without a time delay.

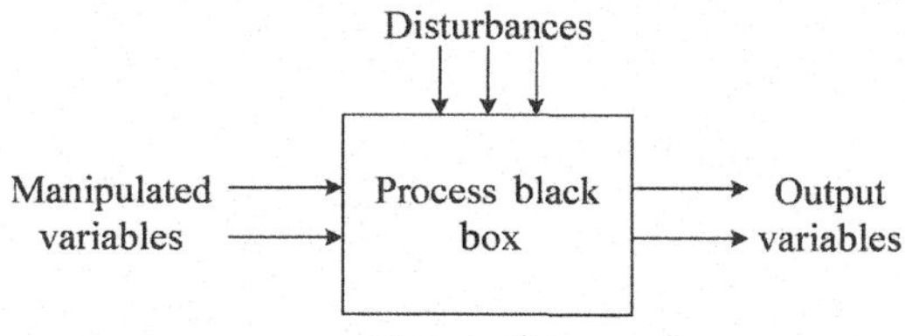

Fig. 4.48
The black box treatment of a process with unknown dynamics.

(b) A more rigorous approach is the use of a numerical technique such as the least squares, extended least squares, maximum likelihood, or instrumental variable analyses to obtain the transfer functions from the measured input-output data series. There are many advanced, software packages such as MATLAB System Identification Tool-Box that facilitate this approach.

4.5 The Graphical Methods for Process Identification

This method allows developing a transfer function for an existing process using the input-output historical graphs and involves the following steps:

Step 1: Bring the process to a steady state.
Step 2: Change one input (either a manipulated variable or a disturbance) at a time by a step change, while the other inputs are kept constant. An impulse change or a square pulse change can also be introduced in the input variable.
Step 3: Monitor and plot the effect of the input change on all the output variables.
Step 4: Approximate the input-output behavior by a simple transfer function.

The method is simple but suffers from the following shortcomings:

- Bringing the process to a steady state is time consuming.
- Interactions among input variables cannot be detected.
- Generation of an impulse change in the input is not easy, a step change is easier. The size of the step change should be large enough to excite the process but not too large to disturb the system significantly.
- The process noise and sensor noise make the analysis difficult.

Fig. 4.49 shows the real and ideal input-output. The noisy signals must be smoothed before a graphical identification approach can be implemented.

4.5.1 Approximation of the Unknown Process Dynamics by a First-Order Transfer Function With or Without a Time Delay

Assuming an ideal transfer function of a first-order system with time delay, the input (step change with magnitude M) and output responses are shown in Fig. 4.50.

With

$$\begin{aligned} G_p &= \frac{K_p e^{-\theta s}}{\tau_p s + 1} \\ U(s) &= \frac{M}{s} \end{aligned} \tag{4.273}$$

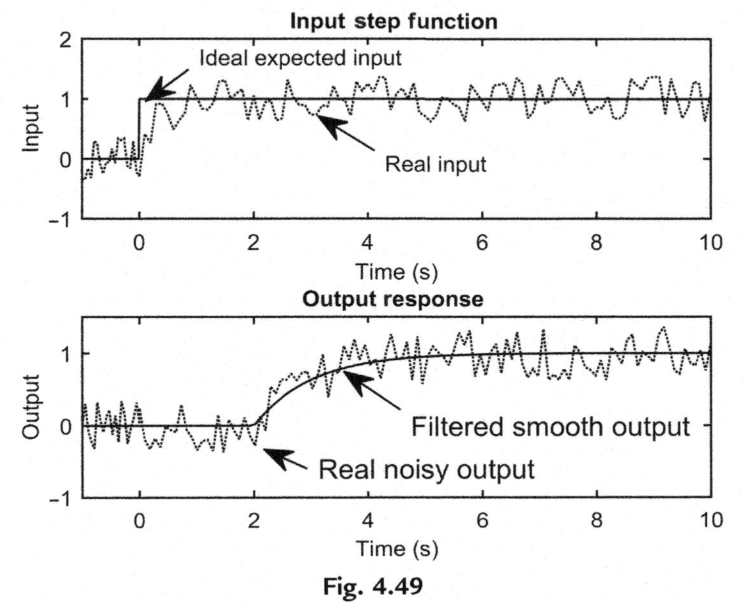

Fig. 4.49
A typical real and ideal process input and output data series.

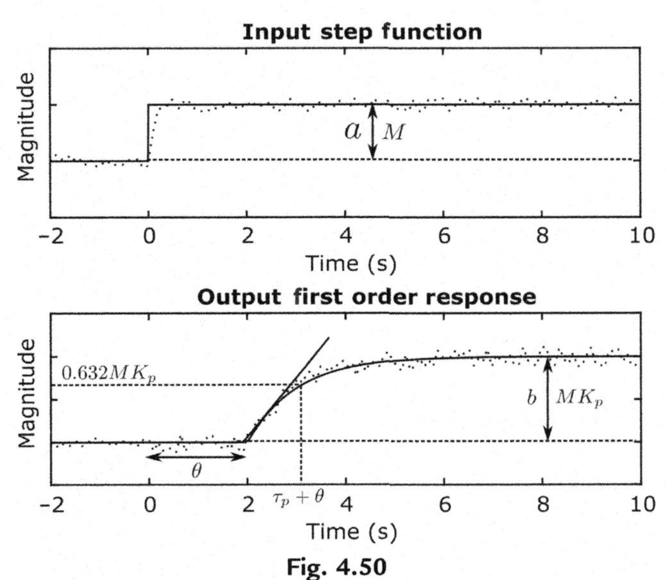

Fig. 4.50
The step response of a first-order system.

$$Y(s) = U(s) \cdot G(s)$$
$$y'(t) = MK_p\left[1 - e^{-\frac{t}{\tau_p}}\right] S(t-\theta) = M K_p\left[1 - e^{-\frac{t-\theta}{\tau_p}}\right] \tag{4.274}$$

where $S(t-\theta)$ is the unit step function or Heaviside function delayed by θ. Considering the smooth curve shown in Fig. 4.50, the following observations can be made:

- K_p is obtained by dividing $\frac{b}{a} = K_p$
- Drawing a tangent at the inflection point to the smooth curve intercepts the time axis and the steady-state value of the output.
- θ is obtained from the intercept of the tangent at the inflection point with the time axis.
- τ_p is obtained noting that the output reaches 63.2% of its final steady-state value, i.e., M, K_P, at $t = \tau_p + \theta$.

4.5.1.1 The Sundaresan and Krishnaswamy method[1]

It is difficult to draw a tangent at the inflection point of the output response curve. Instead, in this method, the times at which the response reaches 35.3% and 85.3% its final value ($t_{0.853}$ and $t_{0.353}$, respectively) are read from the experimental response curve and the time constant and time delay are calculated using the following two equations.

$$\tau = 0.67\,(t_{0.853} - t_{0.353}) \tag{4.275}$$

$$\theta = 1.3t_{0.353} - 0.29t_{0.853} \tag{4.276}$$

K_p is obtained by dividing the steady-state value of the output to the step input size.

4.5.2 Approximation by a Second-Order Transfer Function With a Time Delay

The experimental output response can be fitted to an overdamped second-order model with time delay

$$G_p(s) = \frac{K_p e^{-\theta s}}{(\tau_1 s + 1)(\tau_2 s + 1)} \tag{4.277}$$

The time-domain response is given by:

$$y'(t) = MK_p\left[1 - \frac{1}{\tau_1 - \tau_2}\left(\tau_1 e^{-\frac{t-\theta}{\tau_1}} - \tau_2 e^{-\frac{t-\theta}{\tau_2}}\right)\right] \tag{4.278}$$

The unknown parameters are K_p, θ, τ_1, τ_2. We consider two different methods to determine the unknown parameters.

4.5.2.1 The Smith's method[2]

This method works for both the underdamped and overdamped systems given by Eq. (4.229). From the response curve, the times at which the output reaches 20% and 60% of its final steady-state value are read and designated by t_{20} and t_{60}, respectively. Use Fig. 4.51 to read ξ and τ. K and θ are estimated as discussed earlier.

$$G_p(s) = \frac{K_p e^{-\theta s}}{\tau^2 s^2 + 2\xi\tau s + 1} \tag{4.229}$$

4.5.2.2 Fitting to an underdamped second-order model

The transfer function of a second underdamped system with time delay is given by

$$G_p(s) = \frac{K_p e^{-\theta s}}{\tau^2 s^2 + 2\xi\tau s + 1} \tag{4.229}$$

where $0 < \xi < 1$. The typical step response of this transfer function is shown in Fig. 4.52. The natural period of oscillation, τ, and the damping coefficient, ξ, can be obtained from the parameters obtained from the response curve, namely, a, b, c, and P.

From Fig. 4.52, calculate the decay ratio by dividing c by a to estimate. Measure the period P to estimate τ.

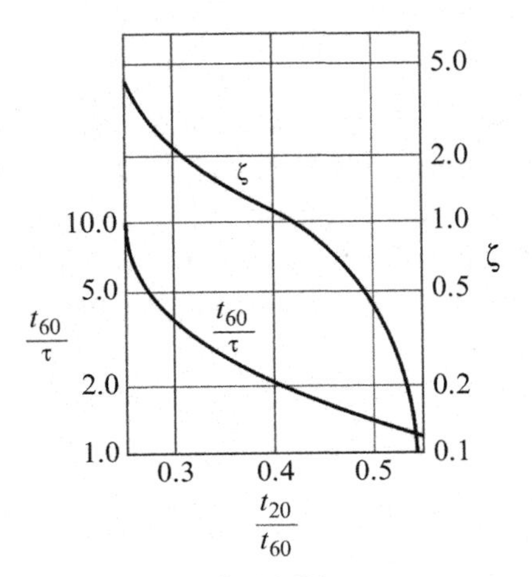

Fig. 4.51
The Smith method to estimate ξ and τ to approximate process input-output data with a second-order transfer function.[2]

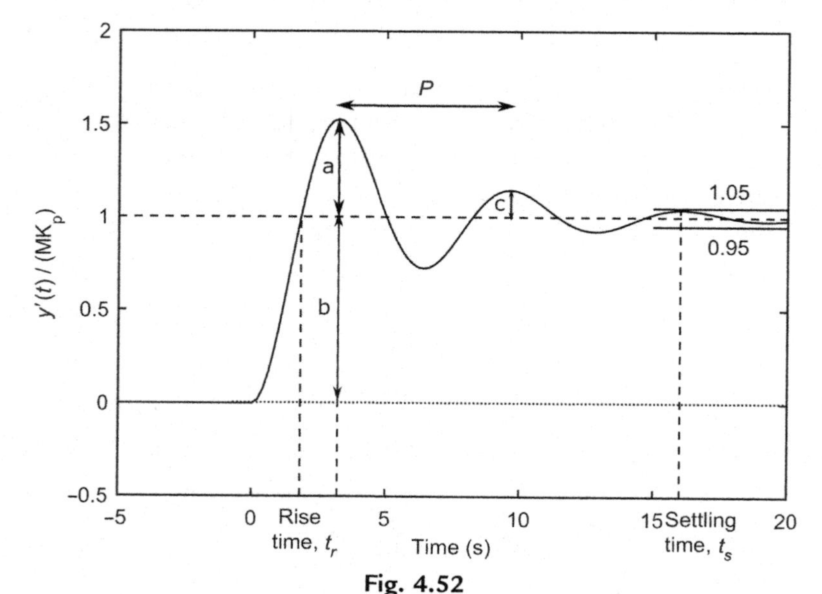

Fig. 4.52
Second-order system step response (underdamped).

$$\text{decay ratio} = \frac{c}{a} = \exp\left(\frac{-2\pi\xi}{\sqrt{1-\xi^2}}\right) \quad \text{for the estimation of } \xi$$
$$P = \text{period} = \frac{2\pi\tau}{\sqrt{1-\xi^2}} \quad \text{for the estimation of } \tau \tag{4.279}$$

There are other graphical identification techniques based on impulse, pulse, and sinusoidal inputs. We shall discuss the process identification in the frequency domain in Chapter 7.

In summary, the transfer functions of a first-order process with time delay (FOPTD), a second-order process with time delay (SOPTD), and a SOPTD with a lead term are presented. The time-domain expressions to a step response are also provided. The time-domain expressions can be used for graphical and numerical process identification using the experimental input-output data series collected from an existing plant.

FOPTD: $g_1(s) = \frac{Y_1(s)}{U(s)} = \frac{Ke^{-\theta s}}{\tau s + 1}$ Exponential

SOPTD: $g_2(s) = \frac{Y_2(s)}{U(s)} = \frac{Ke^{-\theta s}}{(\tau_1 s + 1)(\tau_2 s + 1)}$ S-shaped

SOPTD +a lead term: $g_3(s) = \frac{Y_3(s)}{U(s)} = \frac{K(\tau_3 s + 1)e^{-\theta s}}{(\tau_1 s + 1)(\tau_2 s + 1)}$ Overshoot or inverse response, depending on the magnitude and the sign of τ_3 with respect to τ_1 and τ_2

With an input signal $U(s)=\frac{M}{s}$, the inverse Laplace transform of the output of the previous transfer functions is:

$$y_1'(t)=KM\left\{1-e^{-\frac{(t-\theta)}{\tau}}\right\} \qquad \text{3 unknowns: } \theta,\ \tau,\ K$$

$$y_2'(t)=KM\left\{1-\frac{\tau_1}{\tau_1-\tau_2}e^{-\frac{t-\theta}{\tau_1}}-\frac{\tau_2}{\tau_2-\tau_1}e^{-\frac{t-\theta}{\tau_2}}\right\} \qquad \text{4 unknowns: } \theta,\ \tau_1,\ \tau_2,\ K$$

$$y_3'(t)=KM\left\{1-\frac{\tau_1-\tau_3}{\tau_1-\tau_2}e^{-\frac{t-\theta}{\tau_1}}-\frac{\tau_2-\tau_3}{\tau_2-\tau_1}e^{-\frac{t-\theta}{\tau_2}}\right\} \qquad \text{5 unknowns: } \theta,\ \tau_1,\ \tau_2,\ \tau_3,\ K$$

4.6 Process Identification Using Numerical Methods

There are many numerical techniques for the development of low-order transfer functions from the input-output data series of an existing process. The numerical process identification can be performed in an off-line or an online fashion. In the *off-line process identification*, the input-output data series are collected and a numerical optimization technique such as the minimization of the least squares of the errors between the plant data and the model predictions is used to obtain the unknown process model parameters. The least squares minimization can be performed by a host of tools including the Microsoft-Excel and the Process Identification toolbox of MATLAB. The same procedure can be implemented in an online fashion rendering the *online process identification* which is particularly useful for the design of *adaptive controllers* presented pictorially in Fig. 4.53. Since the numerical identification is very powerful, we are no longer restricted to simple *impulse* and *step* inputs. In fact, other types of signals that are statistically independent of each other are preferred. One such type of a signal is the *pseudo-random binary signal* (PRBS) which switches between a high and a low level at each user-specified switching interval.

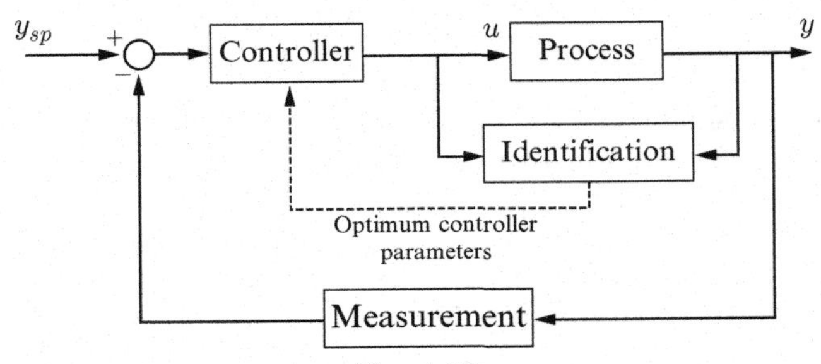

Fig. 4.53
A self-tuning adaptive controller using an online process identification.

4.6.1 The Least Squares Method

The sampled input (u) and output (y) signals are collected at time $t = k\Delta t$ for $k = 0, 1, 2, \ldots$, where Δt is the sampling interval. Note that $u(k)$ and $y(k)$ are the latest input and output data points. Assuming that $y(k)$ is a linear combination of the present and past input and output data series related by the following equation with the unknown parameters $a_0, a_1, \ldots, b_1, b_2, \ldots$

$$y(k) = a_0 u(k) + a_1 u(k-1) + \cdots + a_n u(k-n) + b_1 y(k-1) + \cdots + b_m y(k-m) \tag{4.280}$$

Collecting N samples for $k = 1, \ldots, N$ we get:

$$\underbrace{\begin{bmatrix} y(1) \\ y(2) \\ \vdots \\ y(N) \end{bmatrix}}_{\boldsymbol{Y}} \underbrace{\begin{bmatrix} u(1) & \cdots & u(1-n) & y(0) & \cdots & y(1-m) \\ \vdots & \vdots & \vdots & \vdots & \vdots & \vdots \\ u(N) & \cdots & u(N-n) & y(N-1) & \cdots & y(N-m) \end{bmatrix}}_{\boldsymbol{X}} \underbrace{\begin{bmatrix} a_0 \\ \vdots \\ a_n \\ b_1 \\ \vdots \\ b_m \end{bmatrix}}_{\boldsymbol{\theta}} \tag{4.281}$$

Putting Eq. (4.281) in a compact form:

$$\boldsymbol{Y} = \boldsymbol{X} \cdot \boldsymbol{\theta}$$

where $\boldsymbol{Y}$ and $\boldsymbol{X}$ are known from the measured input-output data, and $\boldsymbol{\theta}$ is the unknown parameter vector. Let $\hat{\boldsymbol{\theta}}$ represent the estimate of $\boldsymbol{\theta}$.

$$y_i = \boldsymbol{X}^T \hat{\boldsymbol{\theta}} + \varepsilon_i \tag{4.282}$$

$$\boldsymbol{Y} = \boldsymbol{X}^T \hat{\boldsymbol{\theta}} + \boldsymbol{E} \tag{4.283}$$

where ε_i is the estimation error at each interval and $\boldsymbol{E}$ is the estimation error vector. The least squares method minimizes the sum of the squares of the estimation errors:

$$S = \sum_{i=1}^{N} \varepsilon_i^2 = \boldsymbol{E}^T \boldsymbol{E} \tag{4.284}$$

To estimate the unknown parameter vector the first derivative of S with respect to $\boldsymbol{\theta}$ must be set to zero and the second derivative must be positive.

$$\frac{dS}{d\hat{\theta}} = 0 = \frac{d}{d\hat{\theta}} \left\{ \sum_{i=1}^{N} \varepsilon_i^2 \right\} \Rightarrow \hat{\theta} = \left(\boldsymbol{X}^T \boldsymbol{X}\right)^{-1} \boldsymbol{X}^T \boldsymbol{Y} \tag{4.285}$$

4.6.2 Using the "Solver" Function of Excel for the Estimation of the Parameter Vector in System Identification

In order to use the "Solver" function of Excel for optimization, the latter must be installed on the computer. Install the "solver" by going to "Excel Options," choose "Excel Add-Ins," press "Go," select the "Excel Solver," and press "Ok."

In an Excel Workbook, create the data (the experimental data (*y*-measured) at a given time, and the model prediction (*y*-model)). In Table 4.2, the model predictions are generated by a first-order model (Fig. 4.54).

Specify the objective function of the optimization as the sum of the squares between the measured and model data $(y_{measured} - y_{model})^2$, and the parameters as the model parameters, K and τ. Call "solver" by going to the "Data" tab to find the best set of parameters. Obviously, apart from the first-order model, higher order models can also be used.

4.6.3 A MATLAB Program for Parameter Estimation

Alternatively, a MATLAB program such as the one listed as follows may be used to fit the plant data to a given transfer function model. The details are given in the following program comment statements.

Table 4.2 Excel sheet for model identification

Time, t	y-Measured	y-Model	Sum of Squares	Parameters	
				K	τ
0	0.010	0	0.00010924	0.168945	2.126307
1	0.089	0.063386	0.000708328		
2	0.115	0.10299	0.000144245		
3	0.111	0.127735	0.000261959		
4	0.129	0.143196	0.000187592		
5	0.141	0.152857	0.000138224		
6	0.154	0.158893	1.51546E-05		
7	0.158	0.162664	1.34269E-05		
8	0.162	0.165021	6.3539E-06		
9	0.169	0.166493	1.22989E-05		
10	0.1909	0.167413	0.000556348		
		SUM	**0.002153172**		

Model:
$y\text{-model} = K(1 - \exp(-t/\tau))$

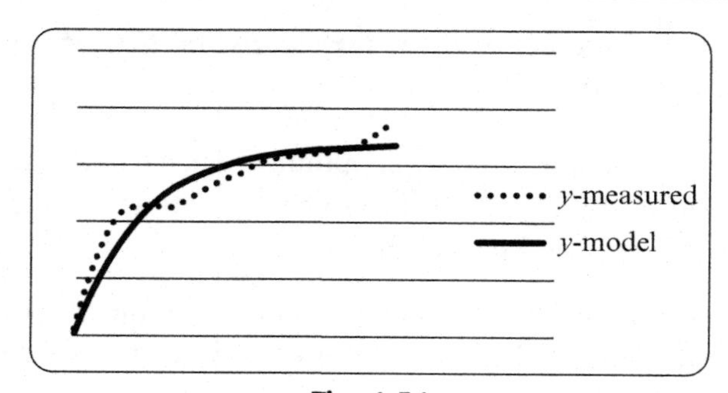

Fig. 4.54
The Excel model identification results.

MATLAB Script:

```
% This example uses the non-linear least squares technique to fit a
% first order plus time delay model (find K, tau, and theta) to a set of
% input - output data. The data are generated by a third order plus
% time delay model. Two cases are considered, case 1: with no noise
% added to the process output, case 2: random noise is added to the
% process output.
% This example illustrates a numerical method for Process Identification.
% Initially, generate a third order plus time delay transfer function to represent %
the actual experimental data.
sys = tf(2.5,[1, 4, 4.96, 1.92], I'InputDelay', 1.5)
t =0:0.5:10; % sampling time at which data are collected
% generate the output data at sampling instants defined by 't'
[y1,t1]=step(sys,t); % This set of data [y1 t1], represents the measured data
%% Case 1: No noise

% figure 1: plot of the measured data
figure(1)
plot(t1,y1,'*')

% Define a parameter vector, para0, and assign an initial guess of the
% unknown parameters tau, K, and theta to it
para0 = [2 1 1]; % [tau0 K0 theta0]
% Call lsqcurvefit MATLAB function, a non-linear optimization that finds
% K, tau, and theta values that will minimize the sum of the squares of
% the errors between the output data generated by the third order plus
% time delay transfer function (measured data) and the model (first order plus time
% delay), Minimize the Sum {yhat-y1}^2 by adjusting K, tau and theta and thereby
% finding the best values of these parameters that would minimize the
% sum of squares of error.
para1=lsqcurvefit('yt',para0,t1,y1);
disp('Time Constant, Gain,    Time Delay')
disp(para1)
```

```
% check the model fit
% simulate a set of data using the model obtained
tt  = 0:0.1:10;
yy1=yt(para1,tt);
figure(2)
plot(t1,y1,'*',tt,yy1,'-')
xlabel('time')
ylabel('y')
legend('measured data','model fit')
```

%% case 2: Output with noise

```
% figure  1: plot of the measurement data
y2 = y1  +  randn(length(y1),1)*0.05;
figure(3)
plot(t1,y2,'*')
t2 = t1;
% initial guess
para0 = [2 1 1];  % [tau0 K0 theta0]
para2=lsqcurvefit('yt',para0,t2,y2);
disp('Time Constant, Gain,     Time Delay')
disp(para2)
% check the model fit
% simulate a set of data using the model obtained
tt = 0:0.1:10;
yy2=yt(para2,tt);
figure(4)
plot(t2,y2,'*',tt,yy2,'-')
xlabel('time')
ylabel('y')
legend('measured data','model fit')
```

MATLAB file: yt.m

```
% This example uses the non-linear least squares technique to fit a
% first order plus time delay model (find K, tau, and theta) to a set of
% input - output data. The data are generated by a third order plus
function yhat = yt(para,t)
% The step response of a first order system in the time domain
%    para - parameters that describes the system
%    para = [tau K theta]
%    tau: time constant
%    K: gain
%    theta: time delay
%    t: time

tau = para(1);
K = para(2);
theta = para(3);

i = find(t<theta);% for example, if theta=6, then i=1,2,3,4,5
j = find(t>=theta);% for theta=6, j=6, 7, 8,. . ., 100
```

```
t1 = t(i);
t2 = t(j);
yhat1 = zeros(length(t1),1);
yhat2 = K*(1-exp(-(t2-theta)/tau));
yhat = [yhat1(:); yhat2(:)];
% set both yhat1 and yhat2 to be column vectors
```

Results:

Case 1: Noise free

Time Constant	Gain	Time Delay
2.0422	1.3503	2.3134

Case 2: With noise (Fig. 4.55)

Time Constant	Gain	Time Delay
1.8945	1.3248	2.3229

4.6.4 Using System Identification Toolbox of MATLAB

The System Identification toolbox of MATLAB provides a powerful technique for the process identification. There are many useful examples and demonstrations for the use of the system identification toolbox on MathWorks site. One simple approach is to use the graphical user interface (GUI) program by typing the following command on the MATLAB Command Window:

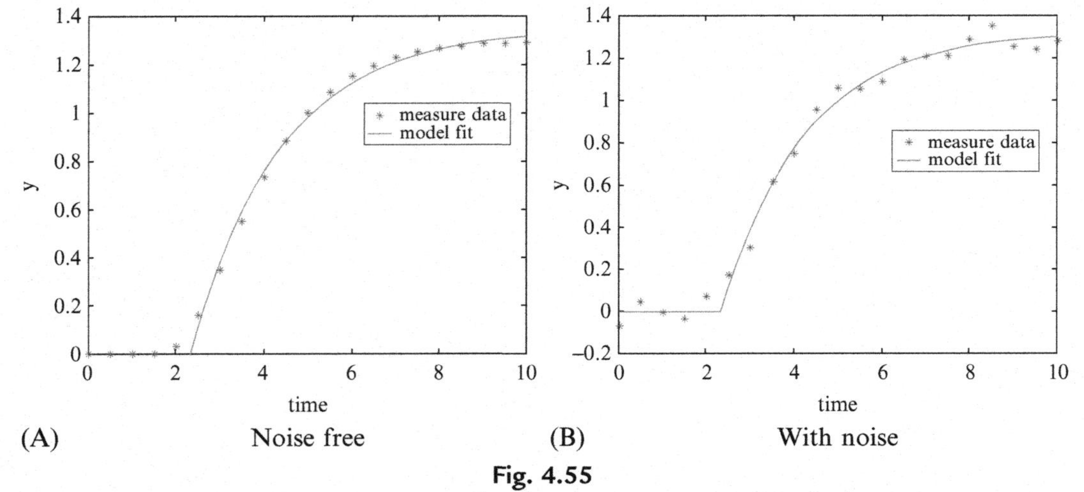

Fig. 4.55
The results of the model identification using the previous MATLAB codes.

Run Command:

```
>> ident
```

Once the GUI is opened, refer to the help and follow the demos listed for the use of the "ident" GUI. In what follows, we generate a unit step input-output data series of a fifth-order transfer function $G(s) = \frac{0.3}{(5s+1)^5}$. We then generate the input and output data sets with a sampling interval of 1 sec. We use the "ident" toolbox to estimate a first order plus time delay model to best fit the generated input-output data set.

MATLAB Script:

```
g = tf(0.3,[6 2 1]);
[y,t] = step(g);   % generate the output vector, y.
u=ones(size(y));  % generate the input vector, u.
delta_t = t(2) - t(1); % sampling interval. 'step' generates evenly spaced time
```

In the "ident" window, click Import menu indicated in Fig. 4.56 and select "Time domain data" as the generated series is in the time domain. Fill the fields as shown in Fig. 4.57 and then click "Import" button.

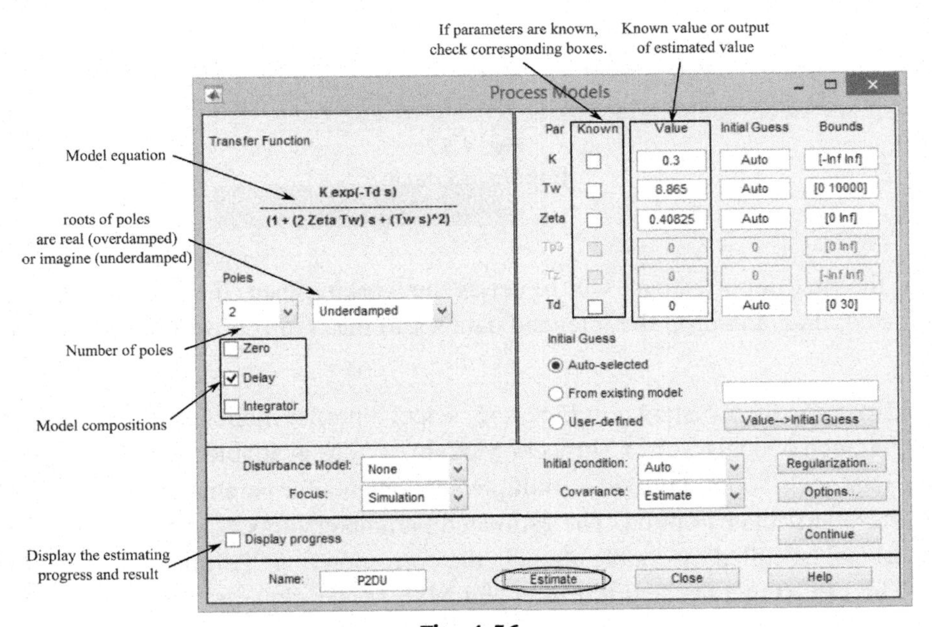

Fig. 4.56
User interface of identification toolbox.

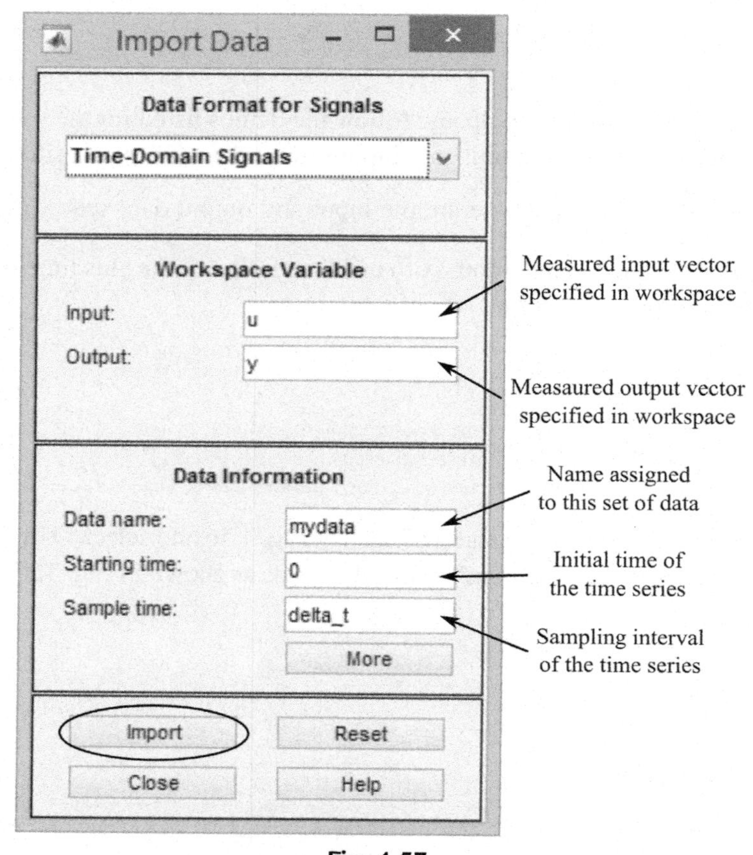

Fig. 4.57
Import interface.

By default, the first imported data will be set as the working data. If multiple data sets are to be estimated, drag and drop the intended data set to the "Current working data" block in Fig. 4.56.

Press "Identification menu" shown in Fig. 4.56, select a proper model type and continue. The "Transfer Function Models ..." or "Process Models ..." are available options in this case. The "Process Models" option provides multiple adjustments for parameter ranges as shown in Fig. 4.58. Press "Estimate" button. The estimated parameters appear in the Value field. The results report can be called by double clicking the corresponding "Estimated Model" object in Fig. 4.56. The FPE (Final Prediction Error) and MSE (Mean Squared Error) can be used to assess the model accuracy. If the error is out of the acceptable range, the model should be modified.

The results can also be plotted to assess the model fit (Fig. 4.56).

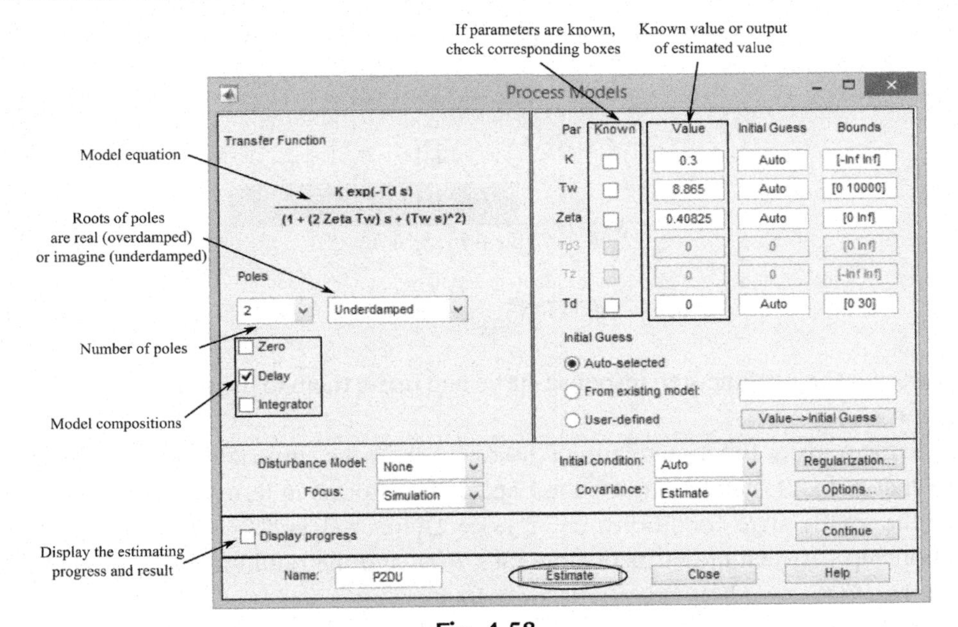

Fig. 4.58
The interface of Process Model in the Ident toolbox.

4.7 Problems

(1) The reaction rate constant in the Arrhenius equation is a function of temperature in the form of

$$k(T) = k_0 \cdot \exp\left(-\frac{E}{RT}\right) \tag{4.286}$$

Linearize the previous equation in terms of temperature.

(2) Derive the Laplace transform of the following functions by hand and check the results using MATLAB's "`laplace`" command in the symbolic mode.

(a)
$$f(t) = 3t^2 \exp(-3t) \tag{4.287}$$

(b)
$$g(t) = 5\cos(2t) \tag{4.288}$$

(3) Derive the inverse Laplace transform of the following functions by hand and check the results using MATLAB's "residue" and "ilaplace" commands in symbolic mode.

(a)
$$F_1(s) = \frac{3(s+1)}{(6s^2+8s+1)(3s+1)} \tag{4.289}$$

(b)
$$F_2(s) = \frac{0.5}{(8s+1)(s+1)} \tag{4.290}$$

(4) For the following transfer functions, plot the unit step and unit impulse responses using MATLAB commands.

$$g_1(s) = \frac{0.3}{3s+1} \tag{4.291}$$

$$g_2(s) = \frac{(4s+1)}{(3s+1)(5s+1)} \tag{4.292}$$

$$g_3(s) = \frac{0.3}{(3s^2+s+1)} \tag{4.293}$$

Discuss the shape of the step response curve and relate them to the transfer functions poles and zeros.

Derive the transfer function relating $L'(s)$ to $F_i'(s)$ for the surge tank depicted in Fig. 4.59. Note that both $L'(s)$ and $F_i'(s)$ are the Laplace transforms in terms of deviation variables from the steady-state conditions, i.e., $L'(s) = \mathcal{L}\{l(t) - \bar{l}\}$ and $F_i' = \mathcal{L}\{F_i(t) - \bar{F}_i\}$. The relationship between the effluent flow rate, $F(t)$, and the liquid level in the tank, $l(t)$, is given by: $F(t) = \alpha\sqrt{l(t)}$, where α is a constant.

(5) Derive the inverse Laplace transform of the following function

$$Y(s) = \frac{(3s+1)}{(4s^2+8s+1)} \tag{4.294}$$

(6) Using the Laplace transform table, determine the Laplace transform of the following function

$$f(t) = 8\exp(-2t)\sin(4t) \tag{4.295}$$

(7) Without taking the inverse Laplace transform, specify whether the time-domain response of the following systems is stable or unstable, oscillatory or smooth, and why? Sketch an approximate time-domain response of these systems to a unit step change in the input variables. Make sure to specify the final steady-state value of the output response.

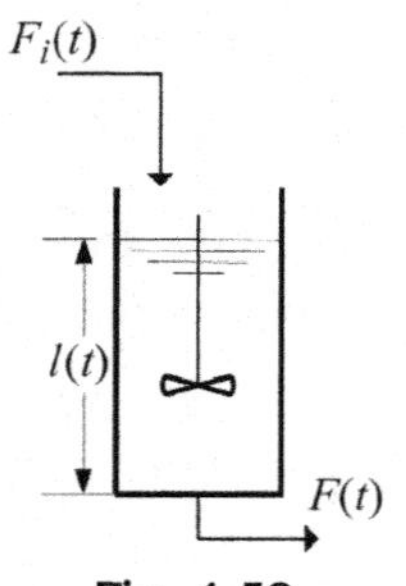

Fig. 4.59
A surge tank.

$$g_1(s) = \frac{2}{(3s^2+s+2)};\quad g_2(s) = \frac{0.2}{(s+5)(s+2)} \tag{4.296}$$

A thermometer has a first-order dynamic with a time constant of 1 sec and a steady-state gain of 1. It is placed in a temperature bath at 120°C. After the thermometer reaches steady state, it is suddenly placed in a bath at 140 °C for $0 \leq t \leq 10$s. Then it is returned to the bath at 100 °C.

Derive an expression for the measure temperature, $T_m(t)$, as a function of time.

(a) Plot using MATLAB, the variations of the measured temperature with time. Calculate $T_m(t)$ at $t = 0.5$s and at $t = 15$s.

Using MATLAB and Simulink, plot the unit step and impulse response of the system given by

$G(s) = \frac{K}{\tau s+1}$ for $K = 0.1$, 1, 10; and $\tau = 0.1$, 1, 10 min.

Plot the unit step and impulse response of g_1 to g_{12} given later. In addition, plot the location of the transfer function poles and zeros on the s-plane. Discuss the shape of each step response curve and relate them to the transfer function's poles and zeros.

$$\begin{aligned}
&g_1 = \frac{0.3}{3s+1} && g_2 = \frac{0.3\,e^{-5s}}{3s+1} && g_3 = \frac{0.3}{3s^2+s+1}\\
&g_4 = \frac{0.3\,e^{-9s}}{3s^2+8s+1} && g_5 = \frac{18s+1}{(3s+1)(5s+1)} && g_6 = \frac{4s+1}{(3s+1)(5s+1)}\\
&g_7 = \frac{0.1s+1}{(3s+1)(5s+1)} && g_8 = \frac{0.03}{s} && g_9 = \frac{-3s+1}{12s^2+2s+1}\\
&g_{10} = \frac{-3s+1}{s^2+2s+1} && g_{11} = \frac{3s-1}{12s^2+2s+1} && g_{12} = \frac{(-3s+1)(2s-1)}{12s^2+2s+1}
\end{aligned} \tag{4.297}$$

(8) Write down the general form of the transfer functions of the following systems and sketch the approximate unit step response of each system. On each sketch provide as much details as possible, i.e., the final value of the output, the approximate shape of the response curve, i.e., exponential or S-shape, etc.

- A first-order lag plus time delay with a pole at −4, a delay time equal to 20 s, and a steady-state gain of 0.8.
- A second-order underdamped system with a steady-state gain of 0.6, a damping coefficient of 0.6, and a natural period of oscillation of 2 s.
- A transfer function with two poles at −1 and −4, a zero at +1, and a steady-state gain of 0.9.

Write down the general forms of the transfer functions $g_1(s)$ through $g_4(s)$ and include estimates of the transfer function's steady-state gains and time delays in each case. Also discuss the relative magnitude of the lag time constant(s) to the lead time constant (if any) in each case (Fig. 4.60).

(9) The pole-zero map, pzmap, of the following systems is shown in Fig. 4.61. Specify whether the time-domain response of the following systems is stable or

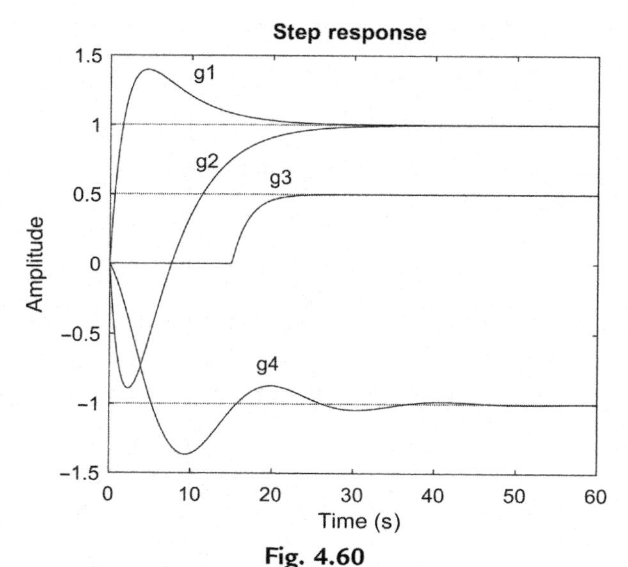

Fig. 4.60
Responses of unknown transfer functions.

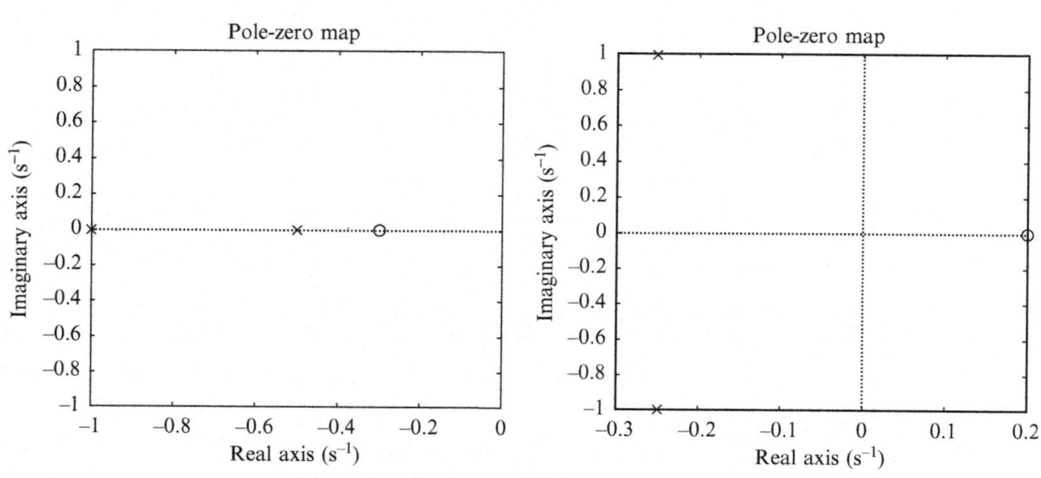

Fig. 4.61
Pole-zero maps of unknown systems.

unstable, oscillatory or smooth, and why? Do the systems show inverse response? Sketch an approximate time-domain response of these systems to a unit step change in the input variables. You are not required to specify the steady state of the outputs.
The unit step responses of two unknown processes are shown in Fig. 4.62. Use a graphical approach to obtain the approximate transfer function for each system. For the first system use a first order plus time delay approximation and for the second system use a second-order underdamped ($\xi < 1$) transfer function without time delay.

(10) The following unit step response shown in Fig. 4.63 has been generated by a third-order system

```
>> g=tf(0.4,[1 12 6 1])
```

Approximate the response by a first order plus time delay transfer function using:

- A graphical approach
- MATLAB Ident tool box
- Excel Solver

Plot the response of the original transfer function and the three approximations on the same graph and discuss the results.

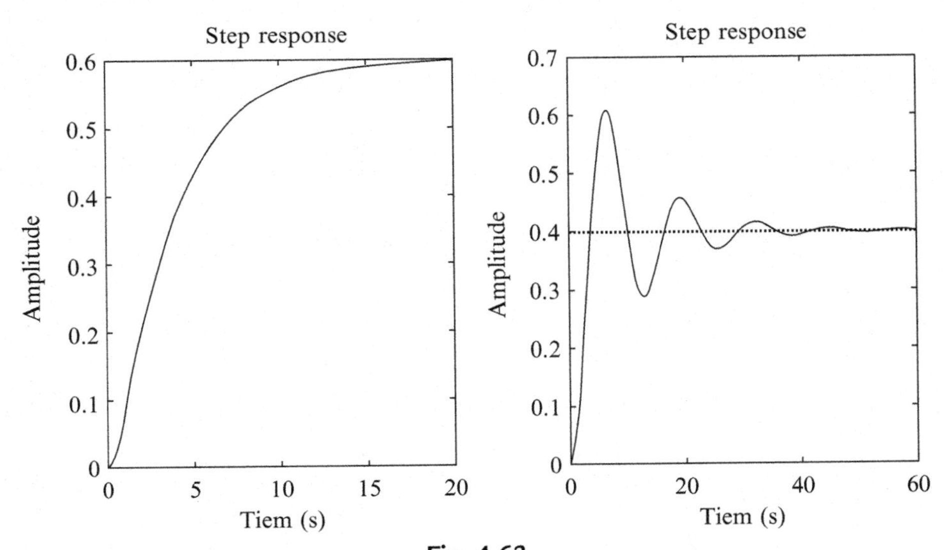

Fig. 4.62
Step responses of unknown systems.

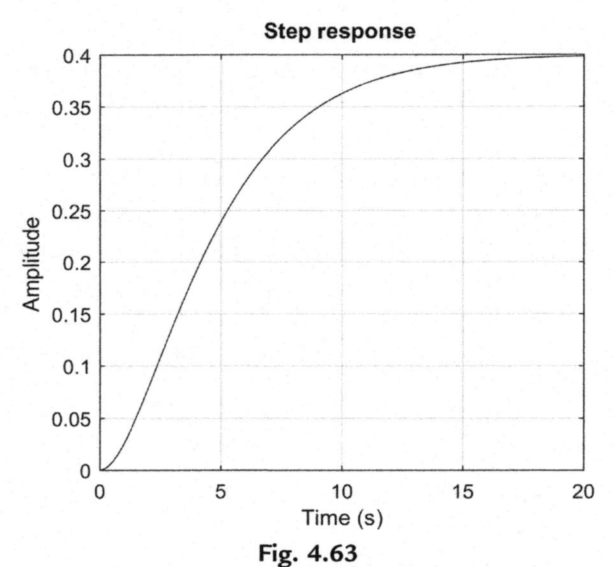

Fig. 4.63
Step response of unknown system.

References

1. Sundaresan KR, Krishnaswamy RR. Estimation of time delay, time constant parameters in time, frequency and Laplace domains. *Can J Chem Eng* 1978;**56**:257.
2. Smith CL. *Digital process control.* Scranton, PA: Intext; 1972.

CHAPTER 5

Dynamic Behavior and Stability of Closed-Loop Control Systems—Controller Design in the Laplace Domain

Sohrab Rohani, Yuanyi Wu

Western University, London, ON, Canada

The objectives of this chapter are to

- Develop the closed-loop transfer functions (CLTF) (regulatory and servo) of a single-input, single-output (SISO) feedback control system.
- Analyze the closed-loop behavior of a feedback control system equipped with a PID controller.
- Determine the effect of the P-, I-, and D-actions of a PID controller by simulating the closed-loop systems in the MATLAB and Simulink environments.
- Investigate the "stability" of a closed-loop system.
- Design of a PID controller and procedures for the controller tuning.
- Introduce enhanced feedback control systems such as cascade, override, selective, and adaptive controllers.
- Discuss the design and implementation of a feedforward controller.
- Discuss the implementation of the feedback controllers on the multiinput, multioutput (MIMO) systems.

5.1 The Closed-Loop Transfer Function of a Single-Input, Single-Output (SISO) Feedback Control System

The block diagram of a typical feedback control system is shown in Fig. 5.1, where $G_p(s)$ and $G_d(s)$ are the process and disturbance transfer functions of the system, $G_m(s)$, $G_v(s)$, and $G_c(s)$ are the transfer functions of the measuring element, the control valve, and the controller, respectively.

Often the valve's dynamics is combined with the process dynamics resulting in $G(s) = G_v(s)G_p(s)$ and a simplified block diagram shown in Fig. 5.2.

Coulson and Richardson's Chemical Engineering. http://dx.doi.org/10.1016/B978-0-08-101095-2.00005-9

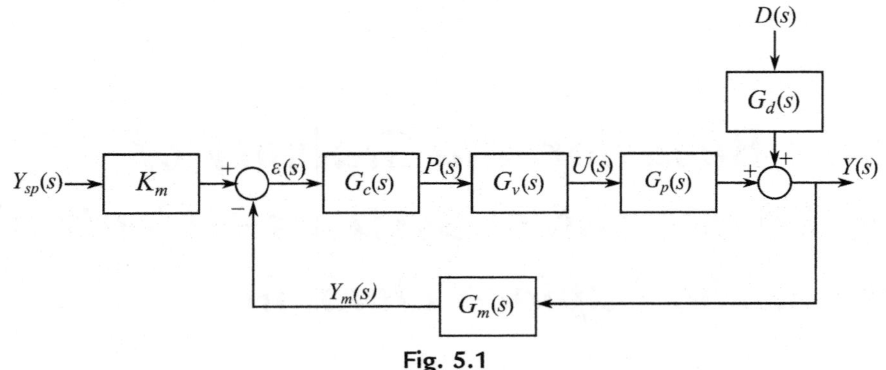

Fig. 5.1
The closed-loop system block diagram.

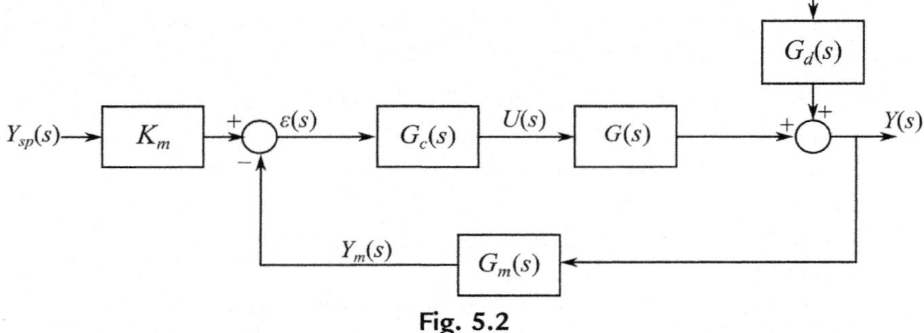

Fig. 5.2
The simplified closed-loop block diagram, where $G(s) = G_v(s)G_p(s)$.

The closed-loop transfer function can be derived by following the signal flow around the control loop.

$$\begin{aligned} Y(s) &= G_d(s)D(s) + G(s)U(s) \\ &= G_d(s)D(s) + \lfloor G_c(s)G_v(s)G_p(s) \rfloor E(s) \end{aligned} \tag{5.1}$$

$$\begin{aligned} E(s) &= Y_{sp}(s)K_m - Y_m(s) \\ &= Y_{sp}(s)K_m - Y(s)G_m(s) \end{aligned} \tag{5.2}$$

$$\begin{aligned} &Y(s) = D(s)G_d(s) + \lfloor K_m Y_{sp}(s) - Y(s)G_m(s) \rfloor G_c(s)G_v(s)G_p(s) \\ &Y(s)\left[1 + G_m(s)G_c(s)G_v(s)G_p(s)\right] = D(s)G_d(s) + Y_{sp}(s)K_m G_c(s)G_v(s)G_p(s) \end{aligned} \tag{5.3}$$

$$Y(s) = \underbrace{\frac{K_m G_c(s)G_v(s)G_p(s)}{1 + G_c(s)G_v(s)G_p(s)G_m(s)}}_{\text{Servo or command following closed loop function}} Y_{sp}(s) + \underbrace{\frac{G_d(s)}{1 + G_c(s)G_v(s)G_p(s)G_m(s)}}_{\text{Regulatory closed loop transfer function}} D(s) \tag{5.4}$$

The closed-loop transfer function relating the controlled variable to the controller set point is referred to as the "servo" or "command-following" closed-loop transfer function, while the closed-loop transfer function relating the controlled variable to the disturbance is referred to as the "regulatory" transfer function. Note that the denominator polynomial is common to both the servo and regulatory closed-loop transfer functions and is referred to as the closed-loop characteristic equation (*CLCE*) of the closed-loop system. The stability of the closed-loop system depends on the location of the roots of the *CLCE* on the s-plane. The numerator polynomial, in each case, is the product of the transfer functions of the components separating the two signals, namely, $Y(s)$ and $Y_{sp}(s)$, or $Y(s)$ and $D(s)$, respectively. Similar to the open-loop systems, we may define the closed-loop "servo" and "regulatory" steady-state gains as:

$$\frac{K_m G_c(0) G_v(0) G_p(0)}{1+G_c(0)G_v(0)G_p(0)G_m(0)} \quad \text{servo steady-state gain} \tag{5.5}$$

$$\frac{G_d(0)}{1+G_c(0)G_v(0)G_p(0)G_m(0)} \quad \text{regulatory steady-state gain} \tag{5.6}$$

5.2 Analysis of a Feedback Control System

In order to analyze the behavior of a closed-loop system, let us assume simple transfer functions for the various elements existing in the loop.

$$G_p(s)=\frac{K}{\tau s+1};\ G_d(s)=\frac{K_d}{\tau_d s+1};\ G_v(s)=1;\ G_m(s)=1 \tag{5.7}$$

5.2.1 A Proportional Controller

a. Let us consider the closed-loop response to a unit step change in the set point $Y_{sp}(s)=\frac{1}{s}$, while maintaining the disturbance unchanged, $D(s)=0$. For a proportional controller, the controller transfer function is $G_c(s)=K_c$.

$$Y(s)=\frac{K_c\left(\frac{K}{\tau s+1}\right)}{1+K_c\left(\frac{K}{\tau s+1}\right)}\left(\frac{1}{s}\right)=\frac{\left(\frac{KK_c}{\tau s+1}\right)}{1+s\left(\frac{\tau}{1+KK_c}\right)}\left(\frac{1}{s}\right)=\frac{\left(\frac{KK_c}{KK_c+1}\right)}{\left(\frac{\tau}{1+KK_c}\right)s+1}\left(\frac{1}{s}\right) \tag{5.8a}$$

$$=\left(\frac{K^*}{\tau^* s+1}\right)\left(\frac{1}{s}\right)$$

The closed-loop "offset" or the steady-state error is defined as the difference between the controlled variable and its set point at steady state and can be obtained using the final value theorem of Laplace:

$$\text{Offset} = \text{steady–state error} = y'_{sp}(\infty) - y'(\infty) = \lim_{s\to 0}\{s[Y_{sp}(s) - Y(s)]\}$$
$$= \lim_{s\to 0}\left\{s\left[\frac{1}{s} - \left(\frac{K^*}{\tau^* s+1}\right)\left(\frac{1}{s}\right)\right]\right\} = 1 - K^* = \frac{1}{KK_c+1} \neq 0 \tag{5.9a}$$

b. Let us now consider the regulatory closed-loop response to a unit step change in the load, $D(s) = \frac{1}{s}$ while keeping the set point unchanged, $Y_{sp}(s) = 0$.

$$Y(s) = \frac{G_d(s)}{1+G_c(s)G_v(s)G_p(s)G_m(s)}D(s) = \frac{\frac{K_d}{\tau_d s+1}}{1+\frac{KK_c}{\tau s+1}}\left(\frac{1}{s}\right) = \frac{K_d}{\tau s+1+KK_c}\left(\frac{1}{s}\right)$$
$$= \frac{\frac{K_d}{1+KK_c}}{\left(\frac{\tau}{1+KK_c}\right)s+1}\left(\frac{1}{s}\right) = \frac{K^*}{\tau^* s+1}\left(\frac{1}{s}\right) \tag{5.8b}$$

$$\text{Offset} = y'_{sp}(\infty) - y'(\infty) = \lim_{s\to 0}\{s[Y_{sp}(s) - Y(s)]\} = \lim_{s\to 0}\left\{s\left[0 - \left(\frac{K^*}{\tau^* s+1}\right)\frac{1}{s}\right]\right\}$$
$$= -K^* = -\frac{K_d}{1+KKc} \neq 0 \tag{5.9b}$$

Note that both in the servo and regulatory cases with a P controller, the offset is not zero. As K_c increases, the offset decreases and the closed-loop response becomes faster (τ^* decrease).

5.2.2 A Proportional-Integral (PI) Controller

Again, in order to study the behavior of the control system with an ideal PI controller, let us assume simple transfer functions for the process, control valve, and the transducer:

$$G_p(s) = \frac{K}{\tau s+1};\ G_v(s) = 1;\ G_m(s) = 1;\ G_c(s) = K_c\left(1+\frac{1}{\tau_I s}\right) = \frac{K_c(\tau_I s+1)}{\tau_I s} \tag{5.10}$$

The closed-loop servo transfer function for a unit step change in the set point and in the absence of a change in the disturbance, $Y_{sp}(s) = \frac{1}{s}$; $D(s) = 0$; is

$$Y(s) = \frac{K_m G_c(s) G_v(s) G_p(s)}{1 + G_c(s) G_v(s) G_p(s) G_m(s)} Y_{sp}(s) = \frac{\left[\frac{K_c(\tau_I s + 1)}{\tau_I s}\right]\left(\frac{K}{\tau s + 1}\right)}{1 + \left[\frac{K_c(\tau_I s + 1)}{\tau_I s}\right]\left(\frac{K}{\tau s + 1}\right)} \frac{1}{s}$$

$$= \frac{K_c K \tau_I s + K_c K}{\tau_I \tau s^2 + (\tau_I + K_c K \tau_I)s + K_c K} \frac{1}{s} = \frac{\tau_I s + 1}{\left(\frac{\tau_I \tau}{K_c K}\right)s^2 + \left(\frac{\tau_I + K_c K \tau_I}{K_c K}\right)s + 1}\left(\frac{1}{s}\right) \tag{5.11}$$

$$= \frac{\tau_I s + 1}{\tau^{*2} s^2 + 2\xi \tau^* s + 1}\left(\frac{1}{s}\right)$$

Note that the I-action adds one zero and one pole to the closed-loop transfer function. The presence of the additional pole will slow down the system response and the additional zero may cause an overshoot in the closed-loop response. Also with an I-action, the offset is always eliminated.

$$\text{Offset} = y'_{sp}(\infty) - y'(\infty) = \lim_{s \to 0} \{s[Y_{sp}(s) - Y(s)]\} = \lim_{s \to 0}\left\{s\left[\frac{1}{s} - Y(s)\right]\right\} = 0 \tag{5.12}$$

Summary. The P-action speeds up the closed-loop response (at high K_c values) but always results in an offset. The I-action eliminates the offset but adds a pole and a zero to the closed-loop transfer function. Therefore it slows down the closed-loop response and may lead to instability at high K_c/τ_I values.

Similarly, it can be shown that the D-action adds one zero to the closed-loop transfer function; and therefore it speeds up the response. The offset, however, remains nonzero with a D-action alone.

Example 5.1

For the following system, simulate the closed-loop response of the feedback control system in MATLAB with the various controller parameters given as follows.

$$G_p(s) = \frac{0.3}{4s + 1};\; G_d(s) = \frac{0.1}{3s + 1};\; G_v(s) = 1;\; G_m(s) = 1;\; G_c(s) = K_c \tag{5.13}$$

with $K_c = 0.5, 5, 20, 50$

```
clear; %Clear variables and functions from memory
gp=tf(0.3,[ 4 1]);gd=tf(0.1,[ 3 1]); gv=1; gm=tf(1); Km =1;
% Implementing a P-Controller with different parameters
Kc = [0.5 5 20 50];
for i=1:4;
    gc=tf(Kc(i));
    gclosedServo =Km*gc*gv*gp/(1+gc*gv*gp*gm);
    step (gclosedServo,15);
    hold on;
end;
```

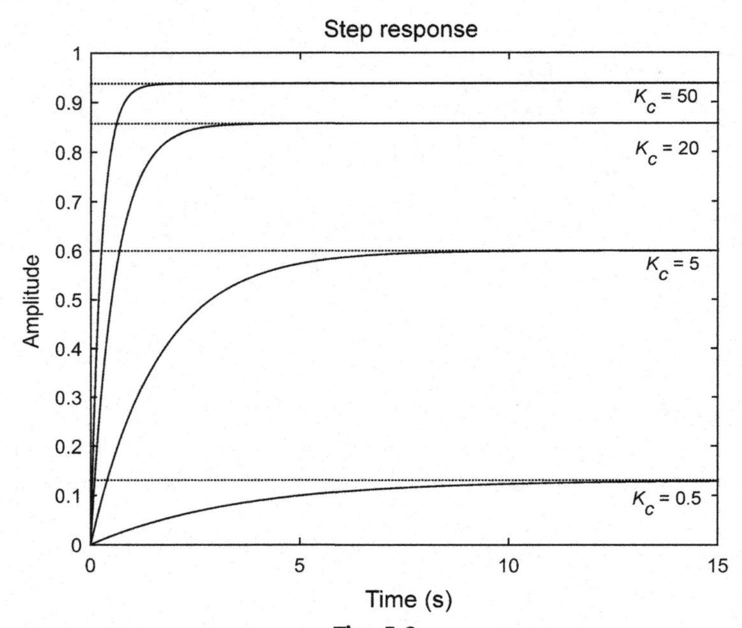

Fig. 5.3
The simulation results of Example 5.1.

The closed-loop response of the system to a unit step change in the set point using a P-controller is shown in Fig. 5.3. Note that as the controller proportional gain increases the offset decreases but is never completely eliminated.

Example 5.2

Repeat Example 5.1 with a PI controller in both MATLAB and Simulink environments.

$$G_c(s) = K_c\left(1 + \frac{1}{\tau_I s}\right) = \frac{K_c(\tau_I s + 1)}{\tau_I s} \tag{5.14}$$

With $K_c = 5$ and $\tau_I = 2, 10, 20$

```
clear;
gp=tf(0.3,[ 4 1]);gd=tf(0.1,[ 3 1]); gv=1; gm=tf(1); Km =1;
% Implementing a PI Controller
Kc1 =5;
TauI =[2 10 20];
for i=1:3;
      gc = (Kc1/TauI(i))*tf([TauI(i) 1],[1 0]);
      gclosedServo =Km*gc*gv*gp/(1+gc*gv*gp*gm);
      step (gclosedServo,50);
      hold on;
end;
```

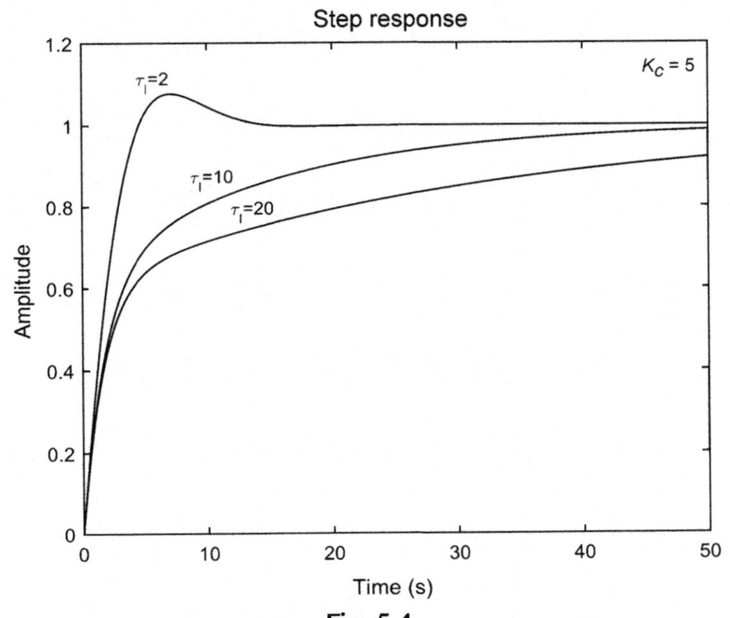

Fig. 5.4
The simulation results of Example 5.2.

The closed-loop response of the system to a unit step change in the set point using a PI-controller is shown in Fig. 5.4. Note that in all cases, the offset approaches zero. The simulation in the Simulink environment is shown in Fig. 5.5.

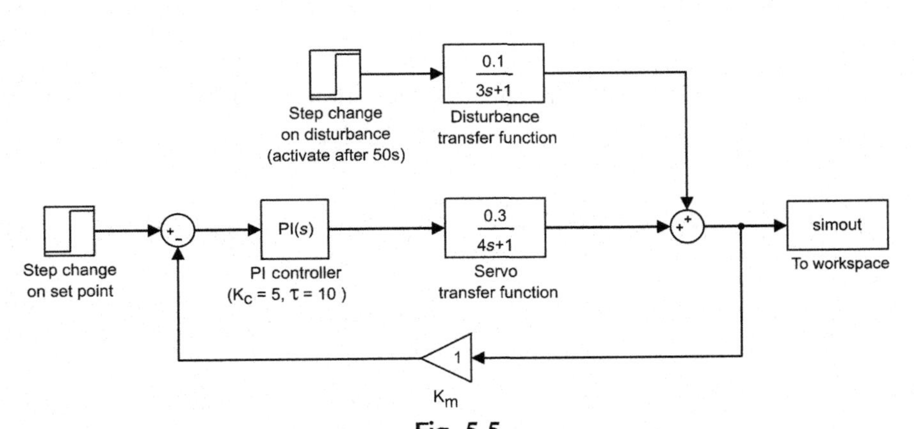

Fig. 5.5
The Simulink program of Example 5.2.

```
>>[t,x]=sim('Example11_2',100); % The time span is set at 100
>>plot(t,y)
```

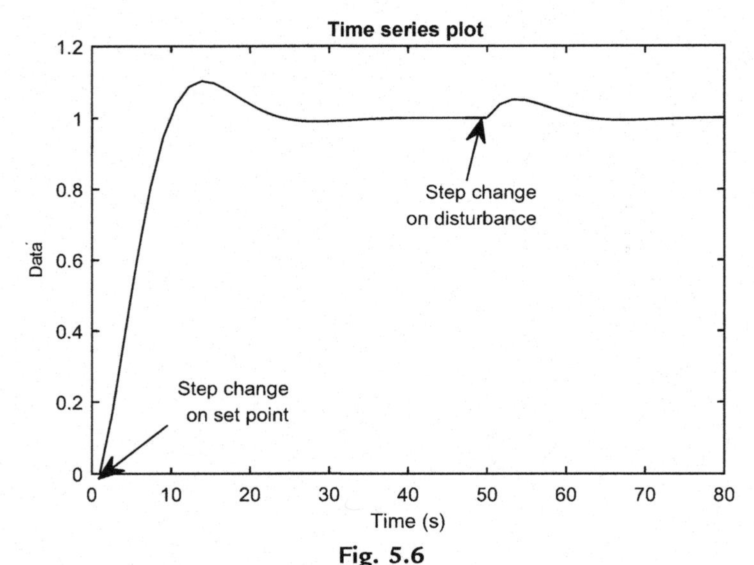

Fig. 5.6
The Simulink simulation results of Example 5.2.

Fig. 5.6 depicts the closed-loop response of the previous system in the presence of a step change in the set point (at $t=0$) and in the disturbance (at $t=50s$).

5.3 The Block Diagram Algebra

Block diagram manipulation can simplify and help derive the closed-loop transfer function of complex control systems such as a cascade control system to be discussed later in this chapter and shown in Fig. 5.7.

The inner loop can be replaced by an equivalent transfer function $G^*(s)$ as is shown in Fig. 5.8.

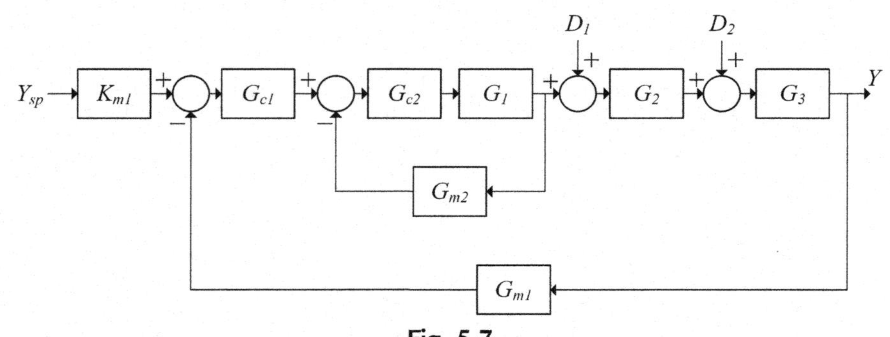

Fig. 5.7
A cascade system with an inner loop.

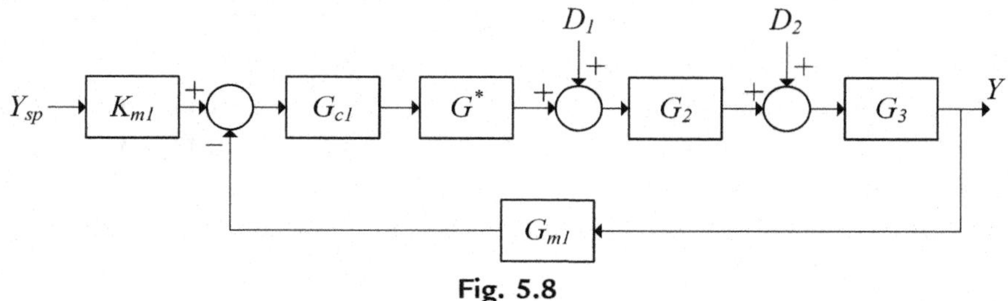

Fig. 5.8
Replacement of the inner loop with an equivalent transfer function $G^*(s)$.

$$G^* \triangleq \frac{G_{c2}G_1}{1+G_{c2}G_1G_{m2}}$$

After the complex block diagram is simplified to a standard feedback control block diagram, the usual approach can be used to derive the closed-loop servo and regulatory block diagrams.

$$Y(s) = \frac{K_{m1}G_{c1}G^*G_2G_3}{1+G_{c1}G^*G_2G_3G_{m1}}Y_{sp}(s) + \frac{G_2G_3}{1+G_{c1}G^*G_2G_3G_{m1}}D_1(s) + \frac{G_3}{1+G_{c1}G^*G_2G_3G_{m1}}D_2(s)$$

5.4 The Stability of the Closed-Loop Control Systems

The stability of a closed-loop control system is one of the main concerns for the safe operation of a chemical process/plant. In a physical sense, the stability refers to maintaining the process variables within the safe limits. If a control system experiences stability problems and departs from the safe operating limits, the final control elements, the control valves, saturate either at a fully open or a fully closed limit for a long period of time. This may then result in the controlled variable experiencing large and unsafe deviations from the safe operating conditions. Traditionally, a chemical process is designed using the nominal steady-state values of the operating conditions. In a more realistic situation, process design should accommodate departures from the steady-state conditions. The concept of closed-loop stability requires that the process variables remain bounded for bounded changes in the input variables. Obviously, if the input variables are unbounded, any system will eventually depart from stability. For example, if the input flow rate to a surge tank is increased by a large amount, the feedback controller takes a corrective action by opening the control valve on the effluent stream of the tank. However, if the valve has not been properly sized for such a large change in the input flow rate, irrespective of the precision of the measuring element and the controller, the system will become unstable and the liquid level rises until it spills over. Similarly, for a highly exothermic reaction, if the cooling duty is not properly designed, the temperature control in the reactor will not be able to prevent possible runaway conditions.

Mathematically, however, the stability of a closed-loop system is determined by the location of the roots of the denominator polynomial of the closed-loop transfer function (Eq. 5.4) referred to as the closed-loop characteristic equation (*CLCE*). The roots of the closed-loop characteristic equation are the *poles* of the closed-loop transfer function and determine the stability of the closed-loop system. If all the closed-loop system *poles* are on the left-hand side on the s-plane, the exponential terms in the time domain due to these negative poles will have negative exponents that die out with the progress of time, and therefore the closed-loop system will be *stable*. If even one of the roots of the characteristic equation (system closed-loop *poles*) is on the right-hand side of the s-plane (a positive exponential term in the time domain), the closed-loop system becomes unstable. Fig. 5.9 shows the location of possible real and complex closed-loop poles in the s-plane.

- p_1 A stable pole, resulting in a smooth overdamped response
- p_2 A pair of complex stable poles, resulting in an oscillatory underdamped response
- p_3 A pair of purely imaginary poles, resulting in purely sinusoidal behavior
- p_4 A pair of complex unstable poles with positive real part, resulting in an oscillatory unstable response
- p_5 An unstable pole, without oscillations

$$\text{Closed loop characteristic equation} = CLCE = 1 + G_c(s)G_v(s)G_p(s)G_m(s) \tag{5.15}$$

If the $CLCE$ is in the form of a polynomial in the s operator, the roots can be obtained by a polynomial solver such as the MATLAB function, `roots`

$$CLCE = a_n s^n + a_{n-1} s^{n-1} + \cdots + a_0 = 0 \tag{5.16}$$

```
>>clce = [anan-1... a0]
>>roots(clce)
```

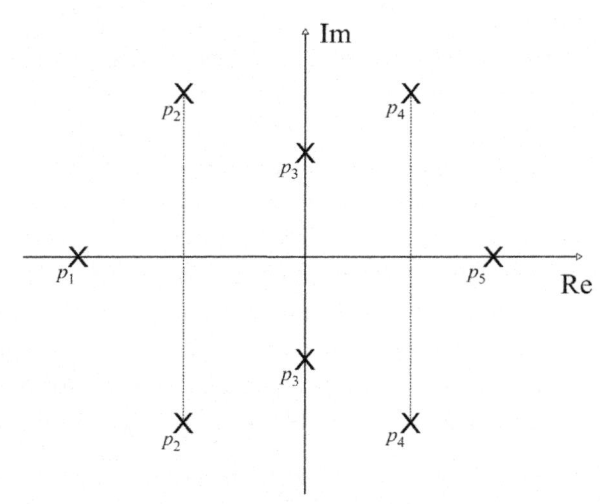

Fig. 5.9
The location of the closed-loop poles and zeros on the s-plane.

However, there are many stability tests that determine the system stability without having to find the roots of the characteristic equation. The main objective in the design of a control system is to choose the parameters of the controller such that the system is closed-loop stable.

5.5 Stability Tests

There are a few stability tests in the Laplace domain that we shall study in this chapter and one stability test in the frequency domain that will be covered in Chapter 7.

- Routh Test
- Direct Substitution
- Root Locus Diagram

5.5.1 Routh Test

The Routh stability test is used if the *CLCE* is a polynomial. It consists of two steps. In the first step, the *CLCE* is arranged in the decreasing orders of the Laplace operator, *s*. The necessary condition for the system stability is that there should be no sign change of the coefficients of the *CLCE*.

$$CLCE = a_n s^n + a_{n-1} s^{n-1} + \cdots + a_0 = 0 \tag{5.16}$$

If any of the coefficients a_{n-1} to a_0 is negative when a_n is positive, the system is unstable. The first test is a "necessary" but not "sufficient" condition for the system stability. If the first test is satisfied, the second step of the Routh test must also be satisfied to ensure the "sufficient" condition for system stability. In order to perform the second step, the Routh array is constructed. For an *n*th-order *CLCE* polynomial, there will be *n+1* rows arranged in the following manner.

$$\begin{array}{cccc} a_n & a_{n-2} & a_{n-4} & \cdots \\ a_{n-1} & a_{n-3} & a_{n-5} & \cdots \\ b_1 & b_2 & b_3 & \cdots \\ c_1 & c_2 & c_3 & \cdots \\ \vdots & \vdots & \vdots & \ddots \end{array} \tag{5.17}$$

The coefficients "b_i" and "c_i" are calculated by

$$\begin{aligned} b_1 &= \frac{[a_{n-1}a_{n-2} - a_n a_{n-3}]}{a_{n-1}}; \quad b_2 = \frac{[a_{n-1}a_{n-4} - a_n a_{n-5}]}{a_{n-1}} \\ c_1 &= \frac{[b_1 a_{n-3} - a_{n-1} b_2]}{b_1}; \quad c_2 = \frac{[b_1 a_{n-5} - a_{n-1} b_3]}{b_1} \end{aligned} \tag{5.18}$$

For the system to be stable, all the coefficients in the 1st column of the Routh array given in Eq. (5.17) must be positive.

Example 5.3

Find the maximum value of the controller gain that results in a marginally stable behavior, $K_{c,\max}$. Note that this value of the controller parameter results in two purely imaginary roots on the imaginary axis. The transfer functions of the various components in the loop are given as follows:

$$G_p(s) = \frac{5}{s^2+2s+1}; \; G_v(s) = 1; \; G_m(s) = \frac{6}{s+1}; \text{ and } G_c(s) = K_c.$$

The closed-loop characteristic equation is

$$CLCE = 1 + G_c(s)G_v(s)G_p(s)G_m(s) \tag{5.19}$$

$$\begin{aligned} 1 + \left(\frac{5}{s^2+2s+1}\right)(K_c)(1)\left(\frac{6}{s+1}\right) &= 0 \\ s^3 + 3s^2 + 3s + (30K_c + 1) &= 0 \end{aligned} \tag{5.20}$$

The first test (the necessary condition) is satisfied since in the context of stability tests, the magnitude of K_c is important and not its sign, therefore, K_c is always taken positive in studying the system stability. Let us check the second test, i.e., the sufficient condition.

Arrange the Routh row (there are $n+1=4$ rows).

$$\begin{array}{ccc} 1 & 3 & 0 \\ 3 & 1+30K_c & 0 \\ \dfrac{(9-1-30K_c)}{3} & 0 & 0 \\ 1+30K_c & 0 & 0 \\ 0 & 0 & 0 \end{array} \tag{5.21}$$

The entries in the first column of the Routh array must all be positive. This will help find $K_{c,\max}$.

$$\begin{aligned} \frac{(9-1-30K_c)}{3} &> 0 \\ 9-(1+30K_c) &> 0 \\ 1+30K_c &< 9 \\ 30K_c &< 8 \\ K_{c,\max} &= \frac{8}{30} \end{aligned} \tag{5.22}$$

At $K_{c,\max} = 8/30$, the system will oscillate as a sine wave with constant amplitude (2 pure imaginary poles). To show that for this value of $K_{c,\max} = 8/30$, the roots of the *CLCE* are pure imaginary, let us calculate the roots:

$$\begin{aligned} s^3 + 3s^2 + 3s + (30K_{c,\max} + 1) &= 0 \\ s^3 + 3s^2 + 3s + (8+1) &= 0 \\ s^3 + 3s^2 + 3s + 9 &= 0 \end{aligned} \tag{5.23}$$

The roots are

$$s_{1,2} = 0 \pm 1.73j; \; s_3 = -3 \tag{5.24}$$

This confirms that for the maximum controller gain, 2 of the 3 system *poles* lie on the imaginary axis.

5.5.2 Direct Substitution Method

For a marginally stable closed-loop system, the closed-loop *CLCE* has a pair of pure imaginary roots. Therefore if s is replaced by $j\omega$ in the *CLCE* shown in Fig. 5.10, the corresponding controller proportional gain will be the maximum or the ultimate gain $(K_{c,}\max = K_{ult})$.

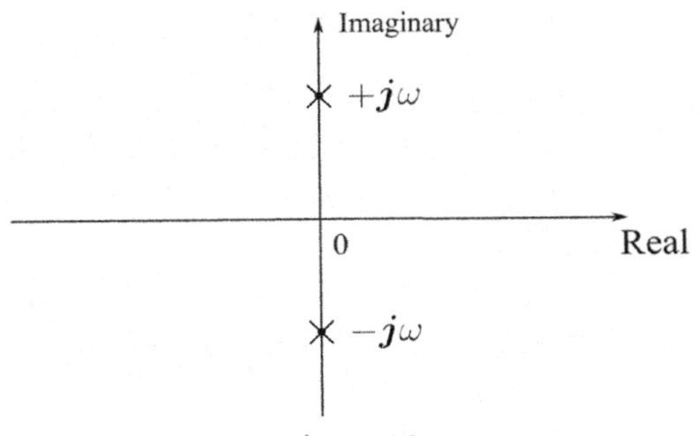

Fig. 5.10
The pictorial presentation of the direct substitution method with the two pure imaginary *CLCE* poles.

Example 5.4

Using the direct substitution method, find the maximum or the ultimate controller gain for the system whose *CLCE* is given by

$$CLCE = s^3 + 3s^2 + 3s + (1 + 30K_c) = 0 \tag{5.25}$$

$$CLCE = (j\omega)^3 + 3(j\omega)^2 + 3(j\omega) + (1 + 30K_c) = -j\omega^3 - 3\omega^2 + 3(j\omega) + (1 + 30K_c) = 0 + 0j$$

The imaginary part of the previous polynomial identity yields the value of $\omega = \pm\sqrt{3}$, once substituted in the real part of the equation, it renders the maximum controller gain.

$$\begin{gathered} -3(3) + 1 + 30K_{c,}\max = 0 \\ K_{c,}\max = K_{ult} = \frac{8}{30} \end{gathered} \tag{5.26}$$

As expected, the result is the same obtained by the Routh test. The advantage of this method is that it can be used in a *CLCE* which has a "time delay," without having to approximate the delay term with a Padé or Taylor series approximation.

Example 5.5

Using the direct substitution method, find the maximum or ultimate controller gain for the system that has a time delay given as follows:

$$G_p(s) = \frac{2e^{-4s}}{3s+1}; \quad G_v(s) = G_m(s) = 1; \quad G_c(s) = K_c \tag{5.27}$$

$$\begin{gathered} CLCE = 1 + \left(\frac{2e^{-4s}}{3s+1}\right)K_c = 0 \\ 3s + 1 + 2K_c e^{-4s} = 0 \end{gathered} \tag{5.28}$$

Use the direct substitution method:

$$3(j\omega) + 1 + 2K_{c,\max} e^{-4(j\omega)} = 0 \tag{5.29}$$

Using the Euler equation, $e^{-j\theta} = \cos\theta - j\sin\theta$, we have

$$3j\omega + 1 + 2K_{c,\max}[\cos(4\omega) - j\sin(4\omega)] = 0 + 0j \tag{5.30}$$

$$\begin{aligned} \text{Real} &\rightarrow 1 + 2K_{c,\max}\cos(4\omega) = 0 \\ \text{Imag} &\rightarrow 3\omega - 2K_{c,\max}\sin(4\omega) = 0 \end{aligned}$$

Divide the imaginary part by the real part,

$$\frac{\sin(4\omega)}{\cos(4\omega)} = -3\omega \tag{5.31}$$

Use trial and error to find $\omega = 0.532$ rad/s. If we substitute this value in the real part of the *CLCE*,

$$2K_{c,\max}\cos(4\omega) = 2K_{c,\max}\cos(4 \times 0.532) = -2K_{c,\max}(-0.5288) = -1 \tag{5.32}$$

$$K_{c,\max} = 9.45$$

5.5.3 *The Root Locus Diagram*

This is a powerful method for designing controllers. The root locus diagram is the locus of the roots of the *CLCE* (or the *poles* of the closed-loop system) as K_c varies from $0 \rightarrow \infty$. There are two methods for the construction of the root locus diagram.

5.5.3.1 The numerical method

In this method, the roots of the *CLCE* for different values of K_c over the entire range of 0 to ∞ are found numerically (e.g., using MATLAB) and are plotted on the s-plane.

$$CLCE = 1 + K_c G_v(s) G_p(s) G_m(s) \tag{5.33}$$

$$CLCE = 1 + G_{OL}(s) = 0 \tag{5.34}$$

where $G_{OL}(s)$ is referred to as the "overall" open-loop transfer function.

Example 5.6

Plot the root locus diagram for the system given as follows:

$$G_p(s) = \frac{50}{30s+1};\ G_v(s) = \frac{0.016}{3s+1};\ G_m(s) = \frac{1}{10s+1};\ G_c(s) = K_c \tag{5.35}$$

$$CLCE = 1 + K_c\left(\frac{50}{30s+1}\right)\left(\frac{0.016}{3s+1}\right)\left(\frac{1}{10s+1}\right) = 0 \tag{5.36}$$

$$900s^3 + 420s^2 + 43s + (1 + 0.8K_c) = 0 \tag{5.37}$$

The roots of the *CLLE* as K_c changes 0 to ∞, are shown in Table 5.1. These roots can be plotted on the *s*-plane to construct the root locus diagram as the controller gain changes over 0 to ∞,

Table 5.1 Roots of the *CLCE* to construct the root locus diagram

K_c	p_1	p_2 and p_3
0	−33	−0.10, −0.03
1	−0.345	0.061 ± j(0.046)
10	−0.41	0.028 ± j(0.154)
⋮		

Alternatively, a simple command in MATLAB, `rlocus`, constructs the root locus diagram shown in Fig. 5.11. In this case, however, the transfer function of the "overall" open-loop system should be supplied. Once the diagram is constructed, the command "`rlocfind`" determines the poles of the closed-loop system and the corresponding controller gain at any point, by clicking the cursor on the root locus plot.

```
MATLAB Script
% enter the OLTF, GOL(s) without Kc in the rlocus command
n=0.8; d=[900 420 43 1];
rlocus(n,d);
[k,poles]=rlocfind(n,d);
% The rlocfind determines the closed loop poles and the controller
% gain k (Kc) by placing the cursor on the diagram generated by the rlocus command
```

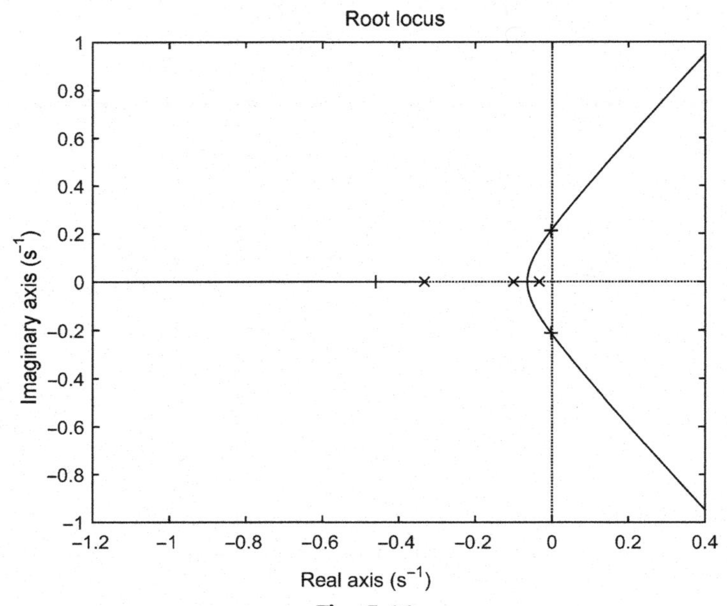

Fig. 5.11
The root locus diagram.

```
selected_point =
   0.2919 + 0.7609i

k =    784.6745

poles =
  -1.0522
   0.2928 + 0.7604i
   0.2928 - 0.7604i
```

5.5.3.2 The graphical method

Prior to the advent of powerful computers, determination of roots of a high-order polynomial was not a simple task. Therefore certain rules based on the mathematical properties of the polynomials and complex numbers were devised to construct the root locus diagram without having to know the numerical values of the individual roots at a given K_c value. Let us rearrange the *CLCE* in the following form:

$$CLCE = 1 + G_{OL}(s) = 1 + K_c \frac{N(s)}{P(s)} \tag{5.38}$$

where $N(s)$ has n roots (zeros of the "overall" open-loop transfer function $G_{OL}(s)$), and $P(s)$ has p roots (poles of the "overall" open-loop transfer function $G_{OL}(s)$.

Graphical rules for the construction of the approximate root locus diagram

- There will be p branches of the root locus diagram.
- Branches start from the open-loop poles of the $G_{OL}(s)$ and end at the open-loop zeros of $G_{OL}(s)$, OR at ∞.
- The real axis is a part of the root locus diagram if the total number of the $G_{OL}(s)$ poles and zeros to the right of that point is an odd number (1, 3, 5, ...).
- The breakaway point of the branches from the real axis can be found by solving the following equation.

$$\sum_{i=1}^{p}\frac{1}{s-p_i}=\sum_{j=1}^{n}\frac{1}{s-z_i} \tag{5.39}$$

- The angle of the asymptotes of the branches breaking away from the real axis and approaching infinity can be found from the following equation

$$\frac{\pi(2k+1)}{p-n}\rightarrow k=0,\ldots,p-n-1 \tag{5.40}$$

Example 5.7

Using the previous "rules," construct the *approximate* root locus diagram of the system in Example 5.6.

$$CLCE=1+K_c\left(\frac{50}{30s+1}\right)\left(\frac{0.016}{3s+1}\right)\left(\frac{1}{10s+1}\right)=0 \tag{5.41}$$

There are three open-loop *poles* ($p_1=-1/3$, $p_2=-0.1$, $p_3=-1/30$), since the denominator of the open-loop transfer function, $G_{OL}(s)$, is a third-order polynomial. Therefore the root locus diagram will have 3 branches. There are no open-loop *zeros*, therefore, all three branches end at infinity. The real axis to the left of p_1 is a part of the root locus diagram, since the total number of the open-loop *poles* and *zeros* to the right of p_1 is 3 (an odd number). The real axis to the left of p_2 will NOT be a part of the root locus diagram, since the total number of the open-loop *poles* and *zeros* to the right of p_2 is 2 (an even number). And finally, the real axis to the left of p_3 will be a part of the root locus diagram, since the total number of open-loop *poles* and *zeros* to the right of p_3 is 1 (an odd number).

The *breakaway point* of the branches of the root locus diagram from the real axis can be found from the following equation.

$$\frac{1}{s+1/3}+\frac{1}{s+1/30}+\frac{1}{s+0.1}=0\rightarrow s=-0.065 \tag{5.42}$$

And the angle of the asymptotes of the branches going to infinity can be found by

$$\frac{\pi(2k+1)}{p-n}=\frac{\pi(2k+1)}{3-0} \tag{5.43}$$

With k taking values of 0, 1, 2 which would be $\frac{\pi}{3}$, π, and $\frac{5\pi}{3}$.

The approximate root locus diagram is shown in Fig. 5.12.

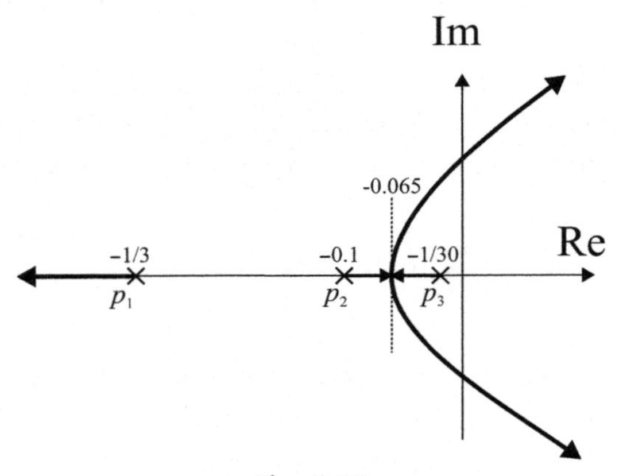

Fig. 5.12
Graphic method for root locus diagram.

5.5.3.3 Application of the root locus diagram to unstable processes

The root locus diagram can be used for the design of feedback controllers. The controller parameters are selected such that the closed-loop poles and zeros are placed on desirable locations in the s-plane.

Example 5.8

An unstable process has at least one positive pole. The following transfer function representing the open-loop dynamics of a system has a positive pole and a positive zero (the system is both nonminimum phase and open loop unstable). Assuming $G_m(s) = G_v(s) = 1$, the *CLCE* can be derived.

$$G_p(s) = \frac{\overbrace{(1-s)}^{\text{inverse response}}}{(1+s)\underbrace{(3s-1)}_{\text{instability}}} \tag{5.44}$$

$$\begin{aligned} CLCE &= 1 + K_c \frac{(1-s)}{(1+s)(3s-1)} \\ 0 &= (1+s)(3s-1) + K_c(1-s) \\ 0 &= 3s^2 + 2s - K_c s - 1 + K_c \\ 0 &= 3s^2 + (2-K_c)s + (K_c - 1) \end{aligned} \tag{5.45}$$

The closed-loop stability requires that:

$$2 - K_c > 0 \text{ and } K_c - 1 > 0 \tag{5.46}$$

Therefore to have a stable closed-loop system, the controller gain must have the following range:

$$1 < K_c < 2 \tag{5.47}$$

That is, for the closed-loop system to be stable, the controller gain must be larger than 1 and smaller than 2. Therefore there will be a minimum and a maximum controller gain, $K_{c,min}$ and $K_{c,max}$, to ensure closed-loop system stability.

There are two open-loop poles at $p_1 = +1/3$ and at $p_2 = -1$ where K_c is zero. Therefore there will be two branches of the root locus diagram, one of them ends at the open-loop zero at $z_1 = +1$, and the other one ends at infinity. Because of the positive zero, the real axis rule discussed earlier is reversed. Note that the two loci branch off from the real axis and converge back to the real axis after crossing the imaginary axis and entering the unstable region. One branch ends at the open-loop zero, while the other goes to infinity on the real axis.

The corresponding MATLAB commands for the construction of the root locus diagram shown in Fig. 5.13 are as follows:

```
MATLAB Script
n=[-1 1];d=conv([3 -1], [1 1]);rlocus(n,d)
```

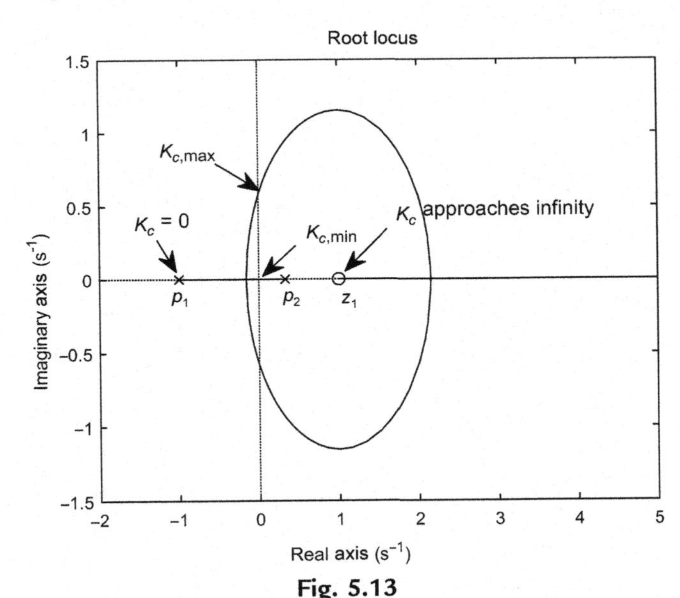

Fig. 5.13
Root locus diagram plotted in MATLAB.

5.6 Design and Tuning of the PID Controllers

Having discussed the main features of the PID controllers and developed certain stability analysis tools, we can now embark on the design of effective feedback controllers for the various chemical processes with given dynamics.

5.6.1 Controller Design Objectives

In the design of a PID controller, it is desired to have a fast and stable closed-loop response, often with no offset, with minimum changes in the manipulated variable (see Fig. 5.14). The design is usually carried out for the set point tracking; however, the controller must also perform well in the presence of the disturbances. The controller design is conducted using approximate models (transfer functions of the process, the valve, the measuring element, etc.), therefore the robustness of the controller in the presence of uncertainties and plant/model mismatch is also a desirable feature. Excessive variations in the manipulated variable cause wear and tear on the control valve and lead to greater energy consumption.

In the design of a PID controller for a given process, the following considerations should be borne in mind:

- Choosing the required control law (P, PI, or PID)
- Choosing the correct controller parameters (Controller Tuning)
- Choosing the correct controller action (Reverse or Direct action)

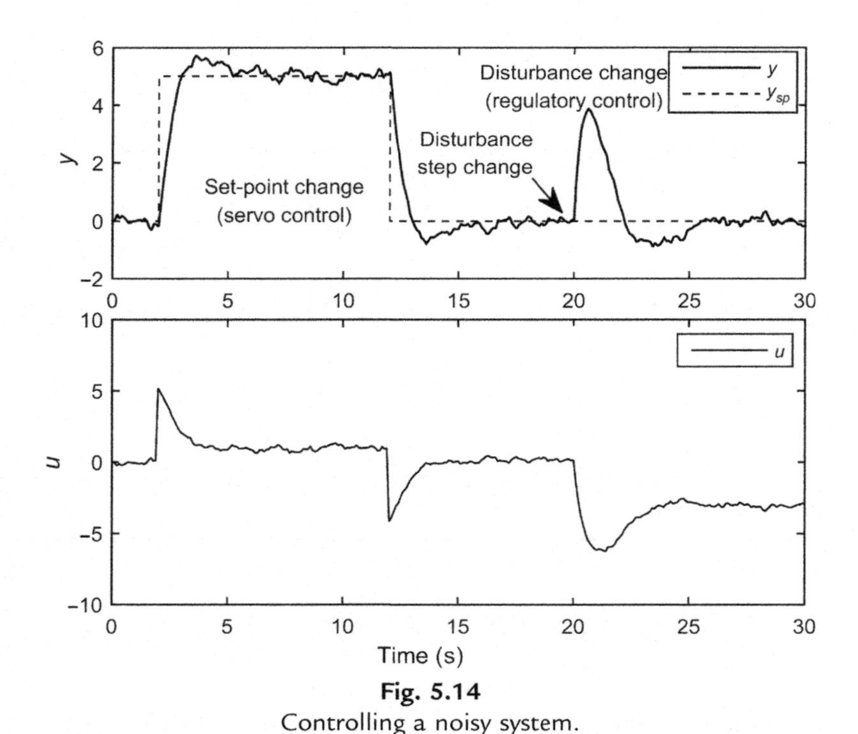

Fig. 5.14
Controlling a noisy system.

5.6.2 Choosing the Appropriate Control Law

In order to address the first consideration listed earlier, it is imperative to review the general characteristics of each of the individual controller actions in a PID controller:

P-action The P-action always introduces an offset (as K_c increases, the offset decreases, and the response becomes faster). The maximum magnitude of K_c is limited by the stability requirement of the closed-loop system. Therefore one cannot increase the controller proportional gain beyond a limit. The maximum or ultimate controller gain before the closed loop becomes marginally stable is shown by $K_{c,\ \max}$ or K_{ult} that can be obtained using the "stability tests" discussed previously. Therefore use the P-controller alone, if offset can be tolerated and a fast response is desirable

I-action The I-action slows down the process of closed-loop response as it adds a pole to the closed-loop transfer function. The I-action *eliminates* the offset. Therefore use the I-action if offset cannot be tolerated, e.g., in the control of temperature, pressure, flow rate, or concentration

D-action The D-action introduces some stability (robustness) in the loop due to its anticipatory nature (acts on the rate of change of the error signal). Therefore a larger K_c value can be employed in the presence of the D-action in a PD or a PID controller, without causing instability. However, the D-action should not be used if the signal is noisy. One may use the D-action if the noisy signal is "filtered" by passing it through a "low-pass" analog or a digital filter

5.6.3 Controller Tuning

The "Controller Tuning" involves finding the best or the optimum values of K_c, τ_I, τ_D that result in a "satisfactory" closed-loop response. In order to achieve this objective, certain performance criteria which quantitatively describe the best or the optimum closed-loop response (fast, no offset, stable, etc.) must be specified.

Fig. 5.15 shows a typical underdamped servo closed-loop response to a step change in the controller set point. A slightly underdamped closed-loop response as opposed to a slower overdamped response is often preferred. Based on this figure, a number of performance criteria may be defined. For example, the minimization of the "settling time" (the time at which the response reaches and remains within $\pm 5\%$ of the new set point), the minimization of the "rise time" (the time at which the response first reaches the new set point), or a one quarter (1/4) decay ratio (the ratio of the second peak to the first, c/a in Fig. 5.15) may be defined as the objective function.

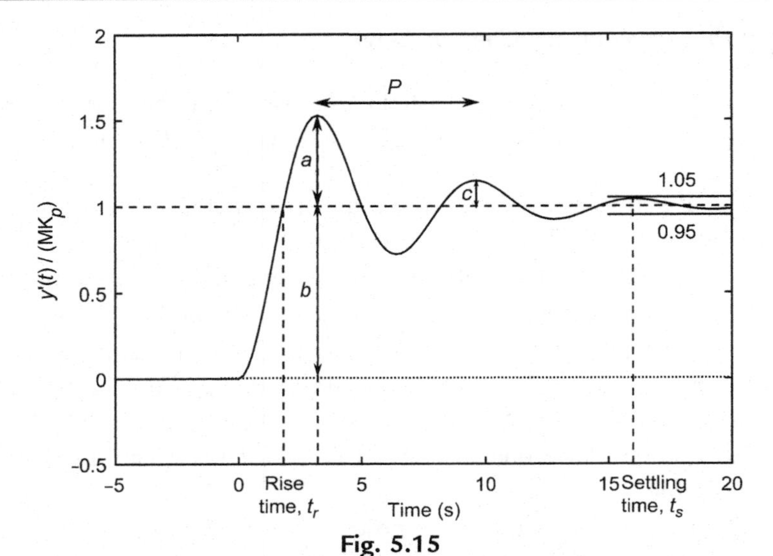

Fig. 5.15
The closed-loop response of a typical feedback control system to a step change in the set point.

$$\frac{c}{a} = \text{decay ratio} = D.R. = \exp\left(\frac{-2\pi\xi}{\sqrt{1-\xi^2}}\right) \tag{5.48}$$

Other performance criteria include the minimization of the integral of absolute value of error,

$$IAE = \int_0^t |\varepsilon|\, dt \tag{5.49}$$

Minimization of the integral of the squared error,

$$ISE = \int_0^t \varepsilon^2 dt \tag{5.50}$$

Minimization of the integral of time multiplied by the absolute error, and

$$ITAE = \int_0^t t|\varepsilon|\, dt \tag{5.51}$$

Minimization of the integral of time multiplied by the square error.

$$ITSE = \int_0^t t\varepsilon^2\, dt \tag{5.52}$$

There are theoretical and empirical approaches for the tuning of a PID controller in order to meet one of the above-stated performance criteria.

5.6.4 The Use of Model-Based Controllers to Tune a PID Controller (Theoretical Method)

There is a host of theoretical approaches that use model-based controllers for the tuning of a PID controller. We shall consider a few.

5.6.4.1 The direct synthesis method

The servo closed-loop transfer function (in the s-domain) of a typical feedback control system is given by

$$\frac{Y}{Y_{sp}}=\frac{K_mG_cG_vG_p}{1+G_cG_vG_p} \tag{5.53}$$

Assume G_v, G_m, and K_m are 1, therefore:

$$\frac{Y}{Y_{sp}}=\frac{G_cG_p}{1+G_cG_p}\rightarrow G_c=\frac{1}{G_p}\left[\frac{\frac{Y}{Y_{sp}}}{1-\frac{Y}{Y_{sp}}}\right] \tag{5.54}$$

The controller equation can be obtained from Eq. (5.54) if the process transfer function and the desired closed-loop servo transfer function (*DCLTF* or desired Y/Y_{sp}) are known. The simplest and yet a realistic specification of the *DCLTF* is a first-order lag transfer function with a small time constant and a steady-state gain equal to 1. Obviously, if the process has time delay, the *DCLTF* must also include a time delay at least equal to the process time delay.

$$\frac{Y}{Y_{sp}}=\frac{1}{\tau_cs+1}\ \text{or}\ \frac{1\,e^{-\theta s}}{\tau_cs+1}\ \text{if the process has a time delay of}\ \theta \tag{5.55}$$

Then, the controller transfer function is found by

$$G_c(s)=\frac{1}{G_p(s)}\left[\frac{\frac{1}{\tau_cs+1}}{1-\left(\frac{1}{\tau_cs+1}\right)}\right]\ \text{or}\ \frac{1}{G_p(s)}\left[\frac{\frac{1\,e^{-\theta s}}{\tau_cs+1}}{1-\left(\frac{1\,e^{-\theta s}}{\tau_cs+1}\right)}\right] \tag{5.56}$$

Example 5.9

Design a PI controller (find K_c and τ_I) for a process whose transfer function is given by $G_p(s)=\frac{1}{6s+1}$ to ensure a closed desired closed-loop transfer function (*DCLTF*) given by $\frac{Y}{Y_{sp}}=\frac{1}{2s+1}$.

$$G_c(s) = (6s+1)\left[\frac{\left(\frac{1}{2s+1}\right)}{1-\left(\frac{1}{2s+1}\right)}\right] \tag{5.57}$$

$$G_c(s) = 3 + \frac{1}{2s} = 3\left(1+\frac{1}{6s}\right) \tag{5.58}$$

Note that the transfer function of a PI controller is given by

$$G_c(s) = K_c + \frac{K_c}{\tau_I s} = K_c\left(1+\frac{1}{\tau_I s}\right) \tag{5.59}$$

Comparing the previous two equations gives the desired tuning parameters of the PI controller to achieve the specified closed-loop servo behavior. Therefore $K_c = 3$ and $\tau_I = 6$s.

5.6.4.2 The internal model control (IMC)

Fig. 5.16 represents the block diagram of an internal model controller (IMC) algorithm. The IMC is a powerful controller on its own. However, it has been used to tune the PID controllers. For the derivation of the IMC controller equation, let $G(s)$ represent the actual dynamics of the combined control valve, the process, and the measuring element with a gain $K_m = 1$. The simplified block diagram of the IMC with an internal model $\widetilde{G}(s)$ parallel to the plant is shown in Fig. 5.17. The controller $C(s)$ can be obtained by considering the closed-loop transfer function of the system.

$$U(s) = C(s)\left\{Y_{sp}(s) - Y(s) + U(s)\widetilde{G}(s)\right\} \tag{5.60}$$

$$\begin{aligned} Y(s) &= D(s) + U(s)G(s) \\ &= D(s) + \frac{\{Y_{sp}(s) - Y(s)\}C(s)}{1 - C(s)\widetilde{G}(s)}G(s) \end{aligned} \tag{5.61}$$

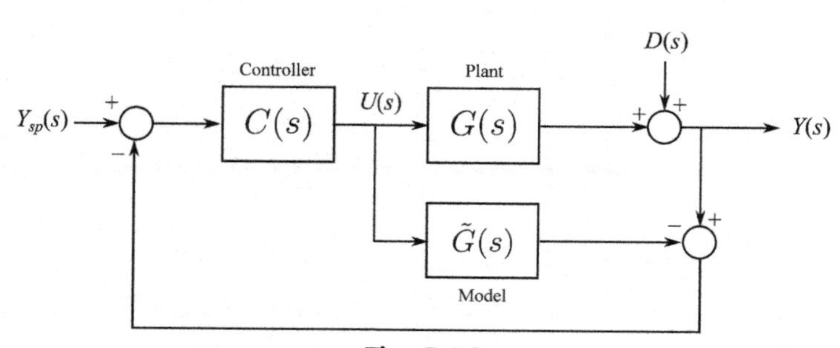

Fig. 5.16
The block diagram of the internal model control (IMC).

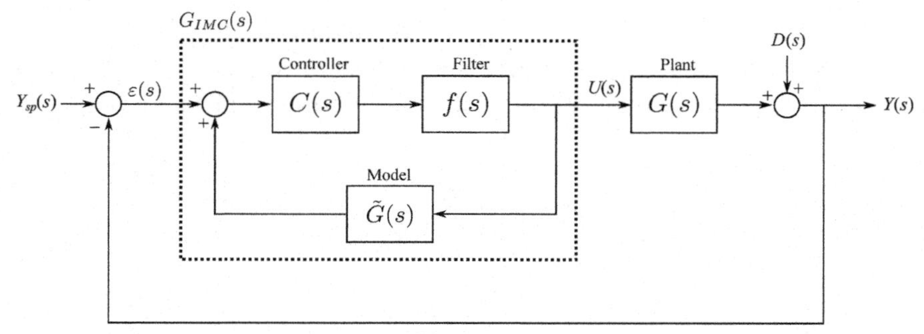

Fig. 5.17
The equivalent block diagram of the internal model control (IMC) shown in Fig. 5.16.

$$\left\{\frac{1-C(s)\widetilde{G}(s)+C(s)G(s)}{1-C(s)\widetilde{G}(s)}\right\}Y(s)=D(s)+\frac{C(s)G(s)}{1-C(s)\widetilde{G}(s)}Y_{sp}(s) \tag{5.62}$$

If $C(s)$ is chosen as $\frac{1}{\widetilde{G}(s)}$, "perfect" control can be achieved since $Y(s)$ becomes equal to $Y_{sp}(s)$ under all conditions.

$$Y(s)=\frac{1-\widetilde{G}(s)/\widetilde{G}(s)}{1-\widetilde{G}(s)/\widetilde{G}(s)+\widetilde{G}(s)/\widetilde{G}(s)}D(s)+\frac{G(s)/\widetilde{G}(s)}{1-\widetilde{G}(s)/\widetilde{G}(s)+G(s)/\widetilde{G}(s)}Y_{sp}(s) \tag{5.63}$$

The problem, however, is that $C(s)$ cannot always be chosen as the inverse of the plant model due to the possible presence of noninvertible terms, such as a positive zero or a time delay. Therefore $C(s)$ is chosen as the inverse of the invertible part of the plant model. For example, if the plant model is given by $\widetilde{G}(s)=\frac{0.25\,(s-2)e^{-4s}}{(s+2)(s+6)}$, the model contains a noninvertible part $\widetilde{G}(s)_+ = (s-2)\,e^{-4s}$ and an invertible part $\widetilde{G}(s)_- = \frac{0.25}{(s+2)(s+6)}$. This choice of $C(s)$, however, leads to an offset. To get rid of the offset, the steady-state part of $\widetilde{G}(0)_+$ is retained. Therefore

$$C(s)=\frac{1}{\widetilde{G}(s)_-\widetilde{G}(0)_+} \tag{5.64}$$

From the previous example, $C(s)$ can be derived as:

$$C(s)=\frac{1}{\left[\frac{0.25}{(s+2)(s+6)}\right](0-2)e^0}=\frac{-(s+2)(s+6)}{0.5} \tag{5.65}$$

Note that the transfer function of the controller $C(s)$ has a second-order lead term and a zeroth-order lag term, and therefore it is "improper." In order to make the controller "proper" and improve its performance, we add an rth order filter with a time constant τ_c and a steady-state gain of 1.

$$f(s) = \frac{1}{(\tau_c s + 1)^r} \tag{5.66}$$

The order of the filter, r, is chosen such that the product of $f(s)C(s)$ becomes "proper," i.e., having a denominator polynomial with at least the same order as the numerator polynomial. The filter time constant, τ_c, works as the IMC unique "tuning parameter."

The transfer function of the overall internal model controller is

$$G_{IMC}(s) = \frac{f(s)C(s)}{1 - f(s)C(s)\widetilde{G}(s)} \tag{5.67}$$

$$C(s) = \frac{1}{\widetilde{G}(s)_{-}\widetilde{G}(0)_{+}} \tag{5.68}$$

$$f(s) = \frac{1}{(\tau_c s + 1)^r} \tag{5.69}$$

Example 5.10

Design an IMC controller for a process whose model is given by

$$\widetilde{G}(s) = \frac{0.25\,(s-2)e^{-4s}}{(s+2)(s+6)} = \underbrace{\frac{0.25}{(s+2)(s+6)}}_{\widetilde{G}(s)_{-}}\underbrace{(s-2)e^{-4s}}_{\widetilde{G}(s)_{+}} \tag{5.70}$$

$$C(s) = -2(s+2)(s+6) \tag{5.71}$$

$$f(s) = \frac{1}{(\tau_c s+1)^2} = \frac{1}{(s+1)^2} \quad \text{Assume } \tau_c = 1 \tag{5.72}$$

$$G_{IMC}(s) = \frac{-2(s+2)(s+6)\left[\dfrac{1}{(s+1)^2}\right]}{1-(-2)(s+2)(s+6)\left[\dfrac{1}{(s+1)^2}\right]\left[\dfrac{0.25(s-2)e^{-4s}}{(s+2)(s+6)}\right]} \tag{5.73}$$

As the tuning parameter τ_c increases, the closed-loop response becomes slower.

In the following example, the IMC controller is used to tune a PI or a PID controller.

Example 5.11

Design an IMC controller for a process whose model is approximated by a first order plus time delay.

$$\widetilde{G}(s) = \frac{3e^{-4s}}{(6s+1)} \tag{5.74}$$

Then, rearrange the IMC controller equation as a PI or a PID controller and obtain the corresponding tuning parameters of the PI or the PID controller.

Initially, the delay term must be approximated either by a Padé approximation or a Taylor series approximation.

a. A 1/1 Padé approximation of the time delay results in:

$$e^{-4s} \approx \frac{1-2s}{1+2s} \tag{5.75}$$

With the Padé approximation, we have

$$\widetilde{G}(s) = \frac{3e^{-4s}}{(6s+1)} = \underbrace{\frac{3}{(6s+1)(1+2s)}}_{\widetilde{G}(s)_-}\underbrace{(1-2s)}_{\widetilde{G}(s)_+} = \underbrace{\frac{-3}{(6s+1)(1+2s)}}_{\widetilde{G}(s)_-}\underbrace{(2s-1)}_{\widetilde{G}(s)_+} \tag{5.76}$$

$$C(s) = \frac{1}{\widetilde{G}(s)_-\widetilde{G}(0)_+} = \frac{(6s+1)(2s+1)}{-3}\frac{1}{-1} = \frac{1}{3}(6s+1)(2s+1) \tag{5.77}$$

For a first-order filter ($r = 1$),

$$f(s) = \frac{1}{\tau_c s+1} \tag{5.78}$$

The IMC controller becomes

$$\begin{aligned} G_{IMC}(s) &= \frac{f(s)C(s)}{1-f(s)C(s)\widetilde{G}(s)} = \frac{\frac{1}{\tau_c s+1}\left[\frac{1}{3}(6s+1)(2s+1)\right]}{1-\frac{1}{\tau_c s+1}\left[\frac{1}{3}(6s+1)(2s+1)\right]\left[\frac{-3(2s-1)}{(6s+1)(1+2s)}\right]} \\ &= \frac{\frac{1}{3}(6s+1)(2s+1)}{\tau_c s+1+2s-1} = \frac{\frac{1}{3}(6s+1)(2s+1)}{(\tau_c+2)s} \\ &= \frac{1}{3(\tau_c+2)}\left[\frac{8s+1+12s^2}{s}\right] \end{aligned} \tag{5.79}$$

Once compared with the transfer function of an ideal PID controller, it renders the tuning parameters of the PID controller:

$$G_{PID}(s)=K_c+\frac{K_c}{\tau_I s}+K_c\tau_D s=\frac{K_c}{\tau_I}\left[\frac{\tau_I s+1+\tau_I\tau_D s^2}{s}\right] \tag{5.80}$$

Therefore

$$K_c=\frac{1}{3(\tau_c+2)};\quad \tau_I=8;\quad \tau_D=1.5$$

b. For a first-order Taylor series approximation of the time delay, we have

$$e^{-4s}\approx 1-4s \tag{5.81}$$

$$\tilde{G}(s)=\frac{3e^{-4s}}{(6s+1)}=\underbrace{\frac{3}{(6s+1)}}_{\tilde{G}(s)_-}\underbrace{(1-4s)}_{\tilde{G}(s)_+}=\underbrace{\frac{-3}{(6s+1)}}_{\tilde{G}(s)_-}\underbrace{(4s-1)}_{\tilde{G}(s)_+} \tag{5.82}$$

$$C(s)=\frac{1}{\tilde{G}(s)_-\tilde{G}(0)_+}=\frac{6s+1}{-3}\frac{1}{-1}=\frac{1}{3}(6s+1) \tag{5.83}$$

For a first-order filter ($r=1$),

$$f(s)=\frac{1}{\tau_c s+1} \tag{5.84}$$

the IMC controller is

$$\begin{aligned}G_{IMC}(s)&=\frac{f(s)C(s)}{1-f(s)C(s)\tilde{G}(s)}=\frac{\frac{1}{\tau_c s+1}\left[\frac{1}{3}(6s+1)\right]}{1-\frac{1}{\tau_c s+1}\left[\frac{1}{3}(6s+1)\right]\left[\frac{-3(4s-1)}{(6s+1)}\right]}\\&=\frac{\frac{1}{3}(6s+1)}{\tau_c s+1+4s-1}=\frac{\frac{1}{3}(6s+1)}{(\tau_c+4)s}\\&=\frac{1}{3(\tau_c+4)}\left[\frac{6s+1}{s}\right]\end{aligned} \tag{5.85}$$

Comparing the previous equation with the transfer function of an ideal PI controller renders the tuning parameters of the PI controller,

$$G_{PI}(s)=K_c+\frac{K_c}{\tau_I s}=\frac{K_c}{\tau_I}\left[\frac{\tau_I s+1}{s}\right] \tag{5.86}$$

results in, $K_c=\dfrac{1}{3(\tau_c+4)}$; $\tau_I=6$.

Seborg et al.[1] list the IMC-based PID controller settings for various process models (first order, second order with time delay, integrating, inverse response, etc.) in terms of the single filter time constant, τ_c.

5.6.5 Empirical Approaches to Tune a PID Controller

There are off-line (open-loop) and online (field tuning) empirical tuning procedures to obtain the optimum PID controller parameters.

5.6.5.1 The open-loop controller tuning or the process reaction curve approach

In this approach, the controller is switched to the "manual" mode and a step change is introduced in the controller output and its effect on the process variable is monitored. The "combined" transfer functions of the valve, process, and the measuring element is obtained as, for example, a first order plus time delay model.

$$G_v G_p G_m = \frac{Ke^{-\theta s}}{\tau s + 1} \tag{5.87}$$

There are numerous Tuning Relations developed in the literature[2] for the tuning of a PID controller in terms of the estimated parameters K, θ, τ. Table 5.2 lists the controller parameters K_c, τ_I, τ_D in terms of K, θ, τ, the process model parameters.

The Open-Loop Tuning Approach suffers from the following disadvantages:

- During the open-loop test to determine the process reaction curve and process model parameters (K, τ, θ), the controller is in the manual mode, i.e., the process is not under control. This may pose some stability risks.
- The procedure is time consuming.
- The identified process model is approximate and may not work well for the controller tuning, especially if the process is nonlinear.

5.6.5.2 Closed-loop controller tuning (field tuning), the continuous cycling, or the Ziegler-Nichols (Z-N) method

In this method, the controller is switched to the "automatic" mode as a P controller. For a given step change in the controller set point, the controller proportional gain is changed so that the closed-loop system is brought to the verge of instability, i.e., a continuous cycling in the

Table 5.2 Open-loop Ziegler-Nichols controller tuning based on the process reaction curve[2]

Controller Type	K_c	τ_I	τ_D
P	$\frac{1}{K}\left(\frac{\tau}{\theta}\right)$	—	—
PI	$\frac{0.9}{K}\left(\frac{\tau}{\theta}\right)$	$3.33\,\theta$	—
PID	$\frac{1.2}{K}\left(\frac{\tau}{\theta}\right)$	$2\,\theta$	$0.5\,\theta$

Table 5.3 Closed-loop (cycling) Ziegler-Nichols controller tuning[2]

Controller Type	K_c	τ_I	τ_D
P	$0.5\ K_{ult}$	—	—
PI	$0.45\ K_{ult}$	$P_{ult}/1.2$	—
PID	$0.6\ K_{ult}$	$P_{ult}/2$	$P_{ult}/8$

controlled variable is observed. The controller gain resulting in the continuous cycling of the closed-loop system is the ultimate controller gain, K_{ult}, and the period of oscillation is referred to as the ultimate period of oscillation, P_{ult}. These two parameters are used to obtain the PID parameters according to Table 5.3.

The advantage of the closed-loop tuning procedure is that the process remains under control during the controller tuning; however, the procedure may still be time consuming and the process must be brought to the verge of instability during the controller tuning that may lead to off-specification product quality.

No matter what method is used for the tuning of a PID controller, further FINE TUNING of the controller is always necessary, once the initial guesses of the controller parameters have been obtained by any of the above-mentioned procedures.

5.7 Enhanced Feedback and Feedforward Controllers

In what follows, we shall consider a few enhanced feedback control algorithms to improve the control quality and implement certain simple logics to ensure plant safety. A cascade controller that involves two or three feedback controllers in series improves the overall quality of the control by eliminating the disturbances in the inner loop(s). Selective control algorithms and override feedback controllers implement simple logics to improve the overall safety of the plant. Finally, a short introduction to the control of nonlinear processes and adaptive control architectures will be provided.

The feedforward controllers compensate for the effect of the measured disturbances and theoretically can achieve perfect control. However, due to the effect of nonmeasured disturbances, and the presence of model inaccuracies, the feedforward controllers do not perform ideally and are always implemented in conjunction with the feedback controllers. Complementing a feedback controller with a feedforward control either in a steady state or a dynamic form will improve the control quality.

We shall end the chapter by considering the control of multiinput, multioutput (MIMO) systems and derive simple controller equations for such processes in the s-domain.

5.7.1 Cascade Control

In order to improve the performance of the feedback controllers in the presence of frequent disturbances and large time delays, several improvements have been proposed and implemented in industry. One of the most industrially acceptable control strategies is the cascade control that involves more than one controller. Let us consider the temperature control in a CSTR equipped with a heating jacket. A simple feedback control of this system involves the measurement of the reactor temperature and the manipulation of the steam flow rate to the reactor jacket. The P&ID of the simple feedback control scheme is depicted in Fig. 5.18.

There are usually fluctuations in the steam supply pressure to the control valve since multiple users are on the same steam line. Therefore for a given valve opening, the actual steam supplied to the reactor jacket varies depending on the number of the users on the steam line at any given time. The changes in the actual supplied steam introduce disturbances affecting the "primary" controlled variable, the reactor temperature. In a more effective control configuration, the fluctuations in the steam supplied pressure can be corrected by a "secondary" or a "slave" pressure controller that receives its set point from the "primary" or "master" temperature controller (see Fig. 5.19A). The latter control configuration is referred to as a cascade controller since it involves more than one controller cascaded together. An alternative cascade control scheme is shown in Fig. 5.19B in which the "secondary" controller is another temperature controller that receives measurements of the reactor jacket temperature and attempts to minimize its fluctuations.

The corresponding block diagrams of the simple feedback and cascade temperature control of the previous reactor system are shown in Fig. 5.20A and B.

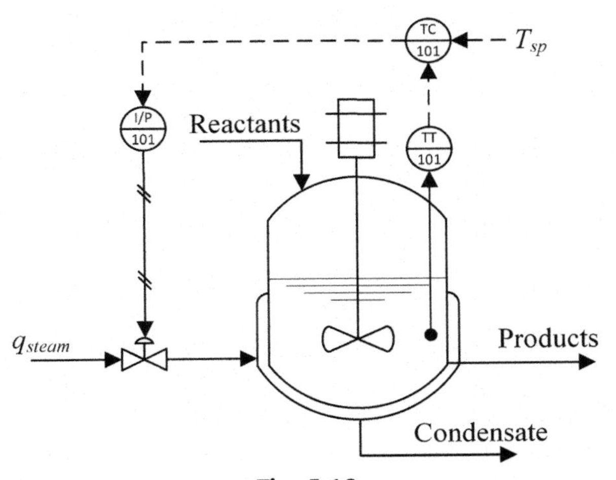

Fig. 5.18
A simple feedback control scheme for the temperature control of a reactor.

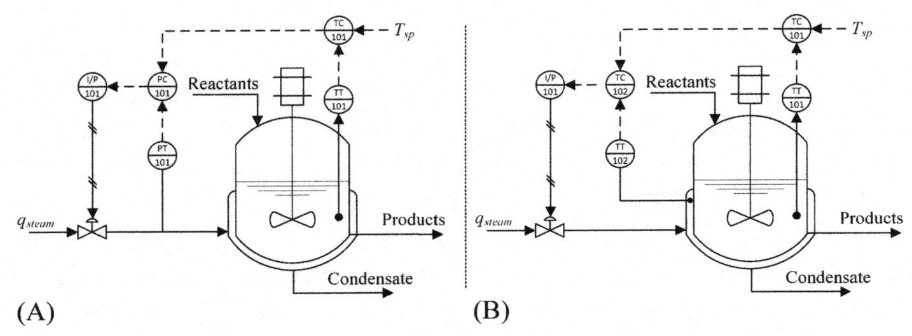

Fig. 5.19

Two possible cascade control configurations for the temperature control of a reactor. (A) Cascade control with the steam pressure controller as the slave controller; (B) cascade control with the jacket temperature controller as the slave controller.

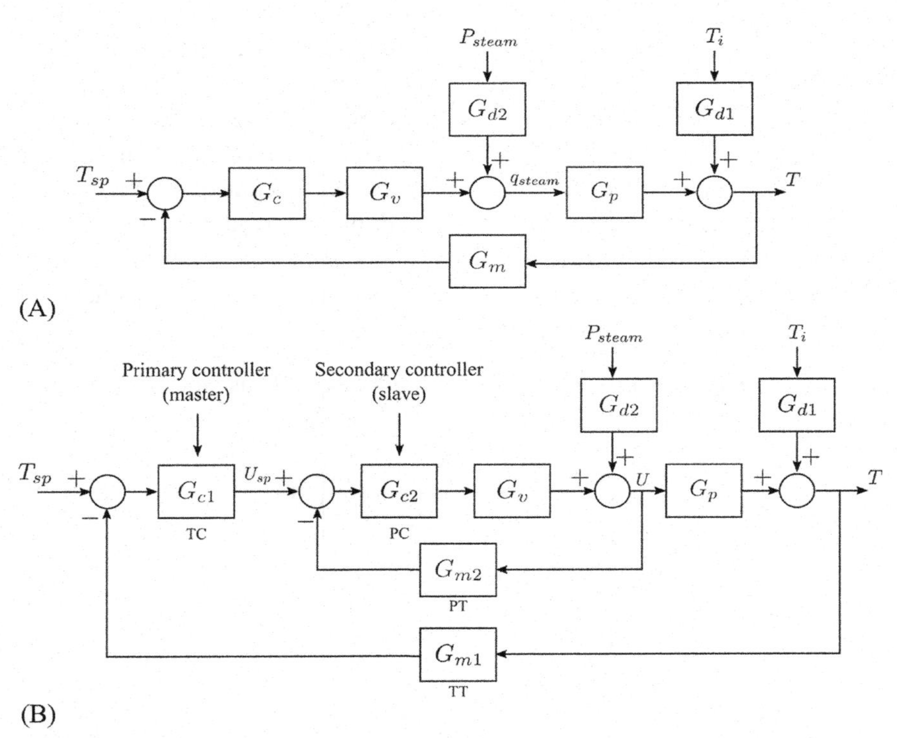

Fig. 5.20

The block diagrams of a simple feedback control (A) and a cascade control system (B).

5.7.1.1 *The closed-loop transfer function of a cascade control algorithm*

In order to derive the closed-loop transfer functions (servo and regulatory) of a cascade control system, the block diagram given in Fig. 5.21B) is simplified to that given in Fig. 5.22) using the following equations:

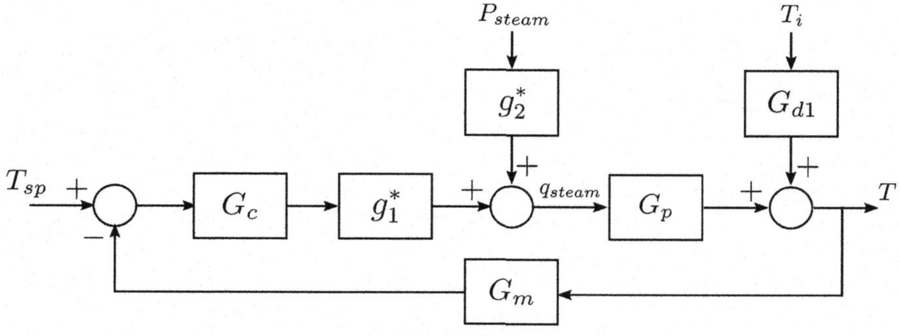

Fig. 5.21
The simplified block diagram of a cascade control system using Eq. (5.88).

$$U(s) = \underbrace{\frac{G_{c2}G_v}{1+G_{c2}G_vG_{m2}}}_{g_1^*}U_{sp}(s) + \underbrace{\frac{G_{d2}}{1+G_{c2}G_vG_{m2}}}_{g_2^*}P_{steam}(s)$$
$$= g_1^* U_{sp}(s) + g_2^* P_{steam}(s) \quad (5.88)$$

Note that in Eq. (5.88) if G_{c2} is very large compared to the other components, then $U(s)$ approaches $U_{sp}(s)$ and the effect of P_{steam} is nullified. The overall closed-loop transfer function of the cascade controller is

$$T(s) = \frac{G_{c1}g_1^*G_p}{1+G_{c1}g_1^*G_pG_{m1}}T_{sp}(s) + \frac{g_2^*G_p}{1+G_{c1}g_1^*G_pG_{m1}}P_{steam}(s) + \frac{G_{d1}}{1+G_{c1}g_1^*G_pG_{m1}}T_i(s) \quad (5.89)$$

In a cascade control system, the "secondary" controller is usually a P controller to increase the speed of the compensation in the inner loop. However, the "primary" controller is often a PI controller to eliminate the offset in the "primary" controlled variable. The choice is made since the "secondary" controlled variable does not have to be maintained at its set point, therefore, there is no need for the I-action in the "secondary" loop. The tuning of the "slave" controller is done first, followed by the "master" controller.

Example 5.12

For the control system given as follows, use MATLAB to obtain the closed-loop servo response to a step change in the set point, and compare the performance of the cascade controller with a simple feedback controller, when the manipulated variable is subjected to different disturbances.

Where G_v should be varied according to:

- $G_v = \dfrac{0.1}{4s+1}$, Normal steam pipeline status.

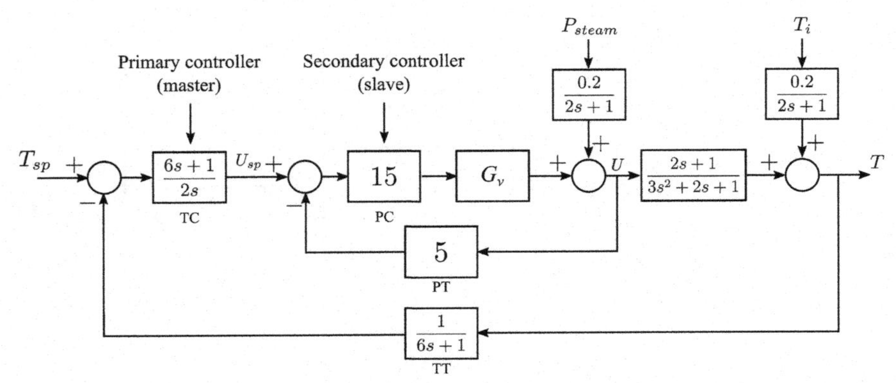

Fig. 5.22
The cascade control block diagram. *TC*, primary temperature controller, *PC*, secondary pressure controller, *TT*, temperature transducer, *PT*, pressure transducer.

- $G_v = \frac{0.07}{9s+1}$, Low pressure status. Too many devices are withdrawing steam. The pressure level (gain) is lower and the response is slower.
- $G_v = \frac{0.2}{s+1}$, High pressure status. Too much steam is being generated. The pressure level is higher and the response is faster.

```
MATLAB Script
% transfer function of the Master temperature controller
tc=tf([6 1], [2 0]);
% transfer function of the Slave pressure controller
pc=15;
% transfer functions of the the control valve
gv=[tf(0.1,[4 1]), tf(0.07,[9 1]), tf(0.2,[1 1])];
% transfer function of the process
gp=tf([2 1], [3 2 1]);
% transfer function of the pressure transmitter
pt=5;
% transfer function of the temperature transmitter
tt=tf(1,[6 1]);

for i=1:3
    % series transfer function of gv and PC
    gser1=pc*gv(i);
    % inner loop transfer function
    gstar=feedback(gser1,pt);
    gser2=tc*gstar*gp;
    goverall(i)=feedback(gser2,tt);

    % compute direct control transfer function
    g_direct(i)=feedback(tc*gv(i)*gp,tt);

end
```

```
subplot(1,2,1)
hold on
% plot cascade system response
for i=1:3; step(goverall(i),100); end;

subplot(1,2,2)
hold on
% plot non-cascade system response
for i=1:3; step(g_direct(i),100); end;
```

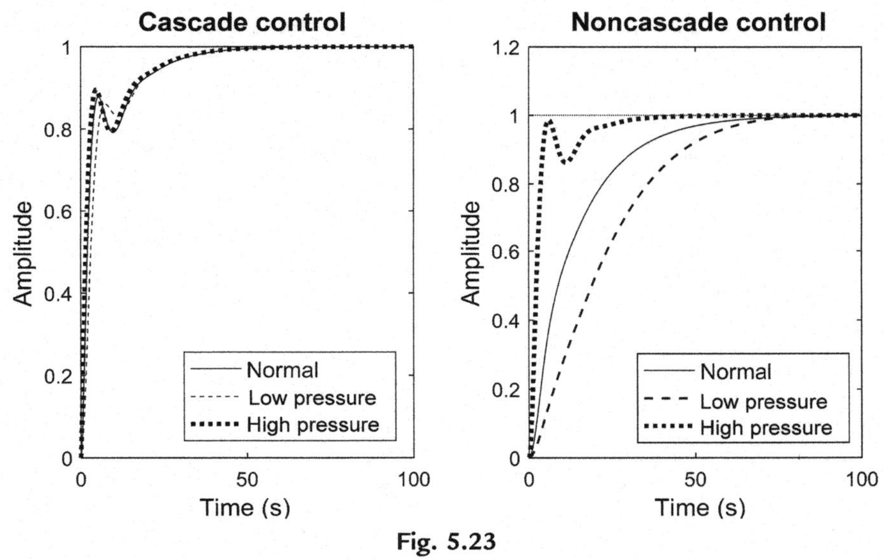

Fig. 5.23
The servo feedback response to a unit step change in the set-point temperature, using a cascade and a simple feedback controller.

Fig. 5.23 shows that, when the operating conditions vary (the valve's dynamics in this example), the cascade controller performs better than a simple feedback controller that requires frequent retuning.

5.7.2 Override Control

The override control is used as a "protective" strategy to ensure the safety of the personnel and equipment and improve the quality of the product. It is not as drastic as the "interlock" control which shuts down the plant or a part of a plant in the case of emergency. The override control switches the control of a manipulated variable from one controller to another in an abnormal condition. The override controller uses a High or a Low Selector switch to implement the logic of switching from the "normal" controller to an "abnormal" or "emergency" controller.

It is important to have auto-reset windup controllers for both the "selected" and the "nonselected" controllers so that neither of the controllers winds up (their outputs exceed 100%) while they are sitting idle. This is shown as Reset Feedback (RFB) in the following example.

Example 5.13

A hot liquid enters a tank from which it is pumped to the downstream processes (see Fig. 5.24). Under "normal" conditions, the level in the tank is controlled by the flow controller by adjusting the pump speed on the effluent stream of the tank. If the level, however, falls below h_2, the liquid level will not have enough net positive suction head (NPSH) and cavitation at the pump will occur. Therefore, under such conditions, the control of the pump speed must be switched to the level controller. Note that in order for the override controller to work, the level controller, LC-50, must be a "direct" acting and the flow controller, FC-50, must be a "reverse" acting controller. The Low Selector, LS-50, receives the controller outputs from LC-50 and FC-50 and chooses the lower of the two signals as its output that is sent to the pump.

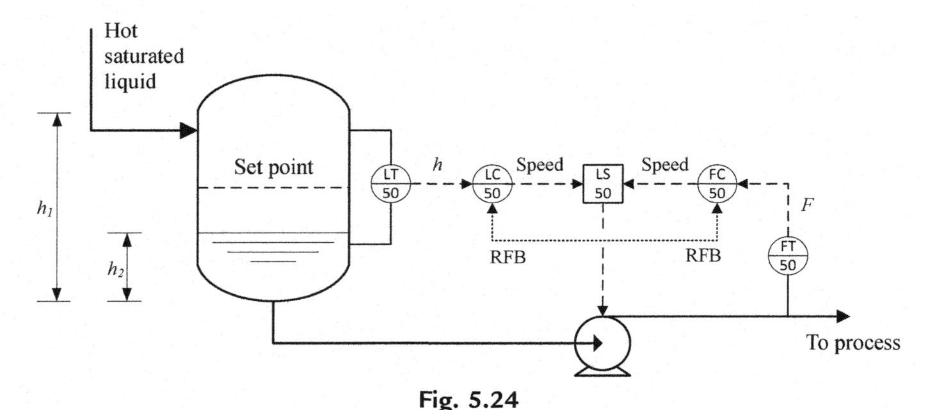

Fig. 5.24
An override control strategy.

Example 5.14

Control of a furnace shown in Fig. 5.25 under "normal" conditions is dictated by the temperature controller on the process stream, TC-12.

However, under "abnormal" conditions (higher pressure that can sustain a stable flame OR higher stack or tube temperatures, metallurgical temperature limit of the tube), the "selected" controller should be either the pressure controller, PC-14, or the temperature controller, TC-13, whichever has a lower output signal. Therefore the low selector, LS-11, receives three input signals and chooses the lowest of all as its output to be sent as the set point of the "slave" flow controller, FC-11.

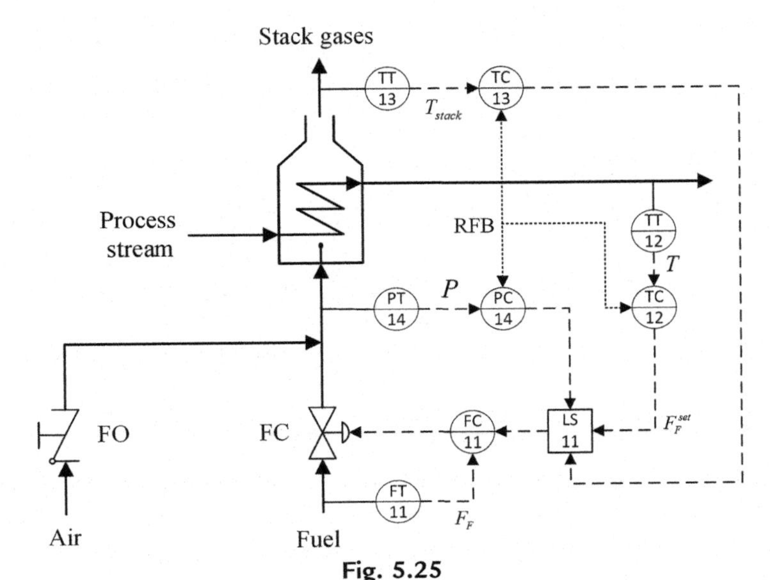

Fig. 5.25
Control of a furnace under "normal" conditions is dictated by the temperature controller on the process stream.

5.7.3 Selective Control

Selective controllers also implement logics for the "safe" and "optimal" plant operation. Selective controllers have a single manipulated variable and a number of measured process variables.

Example 5.15

A fixed-bed catalytic tubular reactor with an exothermic reaction may exhibit a "hot spot" along the length of the reactor (see Fig. 5.26). The actual location of the "hot spot" depends on the feed flow rate, catalyst activity, and the concentration of the reactants in the feed stream. The flow rate of the cooling medium in the annulus of the reactor jacket is adjusted using the temperature controller TC-15 to prevent excessive temperature rise at the hot spot in the reactor. Therefore the reactor temperature is measured in multiple locations along the length of the reactor by thermocouples (TT-15, TT-16, and TT-17). These measurements are sent to a

High Selector (HS-15) which chooses the highest measured temperature and sends it to the controller (TC-15) to avoid the occurrence of a hot spot.

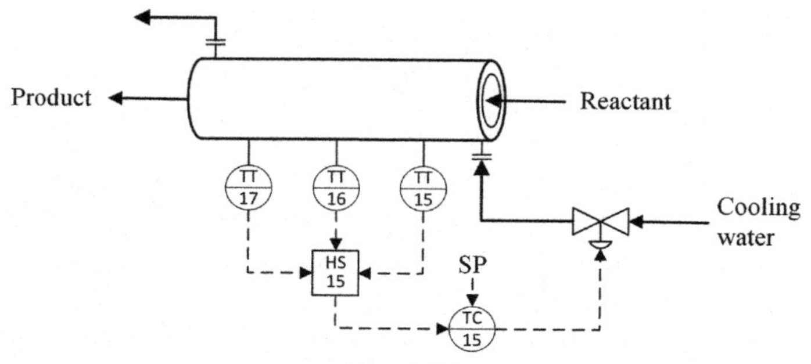

Fig. 5.26
A fixed-bed catalytic tubular reactor with an exothermic reaction may exhibit a "hot spot" along the length of the reactor.

Control of processes with abnormal dynamics. The majority of the chemical processes are self-regulatory and exhibit an overdamped or an underdamped response behavior in an open-loop condition. Control of such processes is often straightforward. However, processes with a significant time delay, processes with a nonminimum phase behavior, integrating processes, and nonlinear processes pose challenges. Fig. 5.27 shows the step response of a self-regulatory process (A); and a process with a time delay, a nonminimum phase process, and an open-loop unstable process (B); to a unit step change in the input signal, u. In addition, the location of the poles and zeros of such processes in the s-plane is depicted in Fig. 5.27.

5.7.4 Control of Processes With Large Time Delays

The delay time/dead time/transportation lag is detrimental to the closed-loop system stability. A controller that can handle a large dead time without causing closed-loop system stability is referred to as a "dead-time compensator," first proposed by Smith.[3] Let us consider a process with time delay whose actual dynamics is given by

$$G_p(s) = G_p^*(s)\, e^{-\alpha s} \tag{5.90}$$

where $G_p{}^*(s)$ represents the nondelayed part of the process transfer function. Note that the actual dynamics of a process is always "unknown"; however, for the simulation purposes, we assume that the actual dynamics is given by Eq. (5.90). The approximate model (transfer function) of the process with time delay is given by

$$\hat{G}_p(s) = \hat{G}_p^*(s)\, e^{-\hat{\alpha} s} \tag{5.91}$$

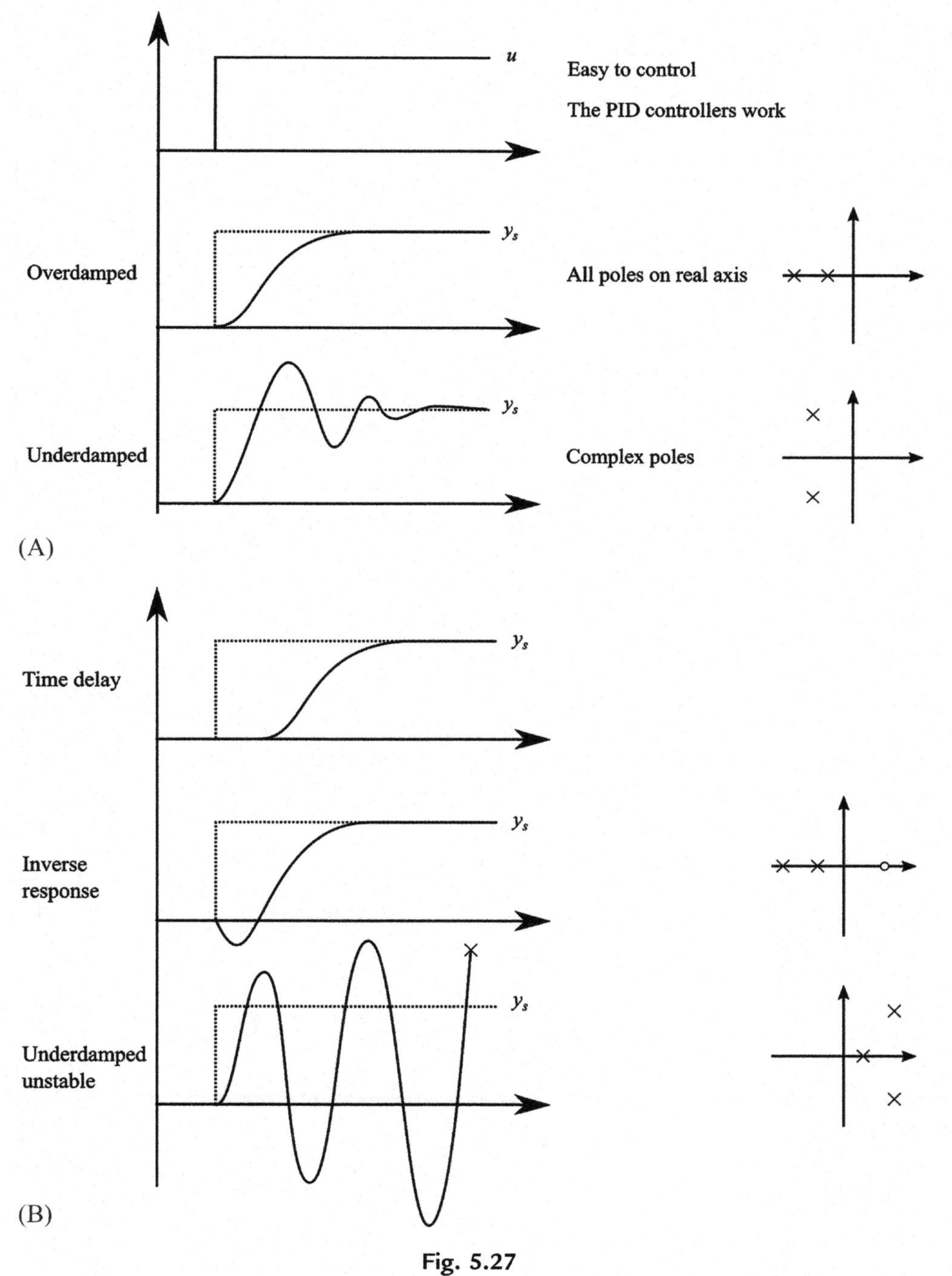

Fig. 5.27
Self-regulating processes (A) and processes with difficult dynamics (B).

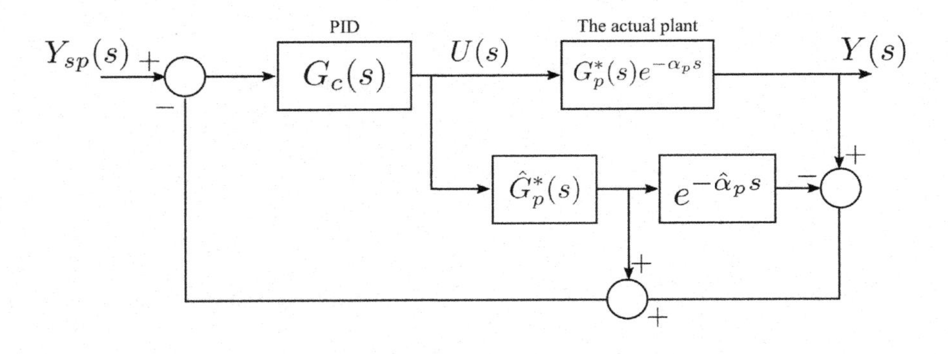

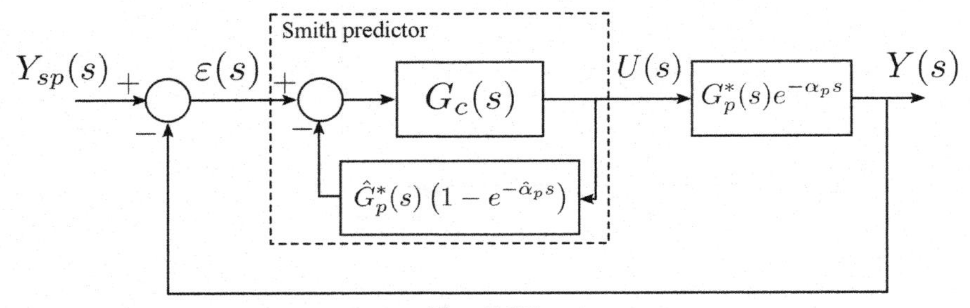

Fig. 5.28
The block diagram of the Smith dead-time compensator.

Consider the block diagram given in Fig. 5.28 in which the manipulated variable, U, is fed both to the actual process and the plant model. If the plant model is exact, the signal fed back to the controller would have no delay, and therefore, the controller will perform better. In the presence of a plant/model mismatch, it is expected that the dead-time compensator still performs better than the feedback controller alone.

Based on the equivalent block diagram shown in the lower part of Fig. 5.28, the equation of the Smith predictor/compensator is

$$G_{smith}(s) = \frac{U(s)}{\varepsilon(s)} = \frac{G_c(s)}{1 + G_c(s)\,\hat{G}_p^*\left(1 - e^{-\hat{\alpha}_p s}\right)} \tag{5.92}$$

Example 5.16

For the process given as follows, use Simulink to generate the closed-loop servo response of the system with and without a Smith dead-time compensator.

The "actual" process dynamics is,

$$G(s) = \frac{0.1\,e^{-3s}}{s^2 + 4s + 1} \tag{5.93}$$

and the approximate model is

$$\hat{G}(s) = \frac{0.3\, e^{-2s}}{s^2 + 8s + 1} \tag{5.94}$$

The Simulink program with the Smith predictor is given in Fig. 5.29, and the corresponding response and the controller output are shown in Fig. 5.30.

The Simulink block diagram and the response curves for a simple PI controller with the same settings are shown in Figs. 5.31 and 5.32. Clearly, the simple PI controller does not work and results in an unstable closed loop.

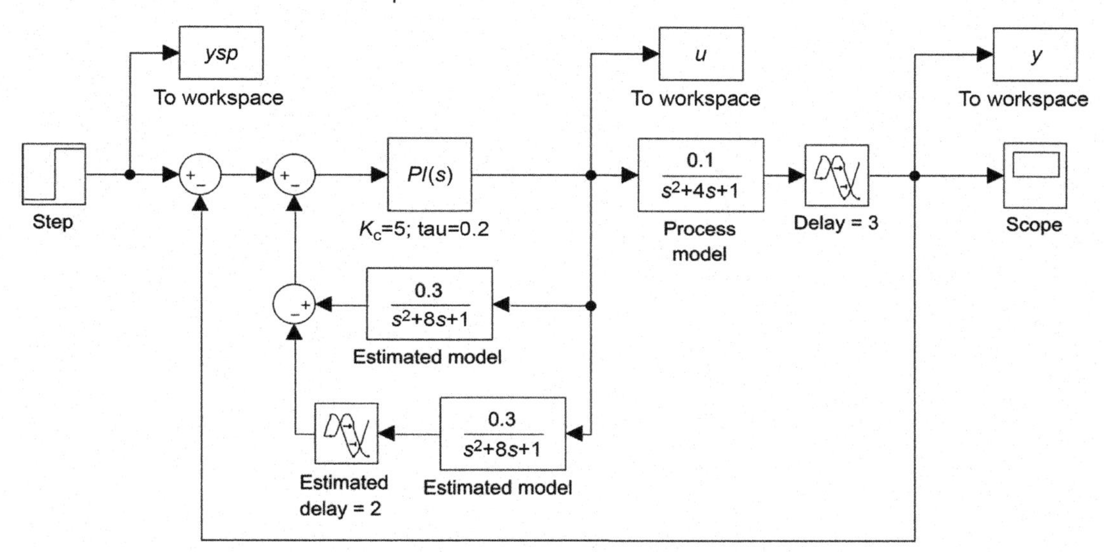

Fig. 5.29
The Simulink model for the Smith predictor.

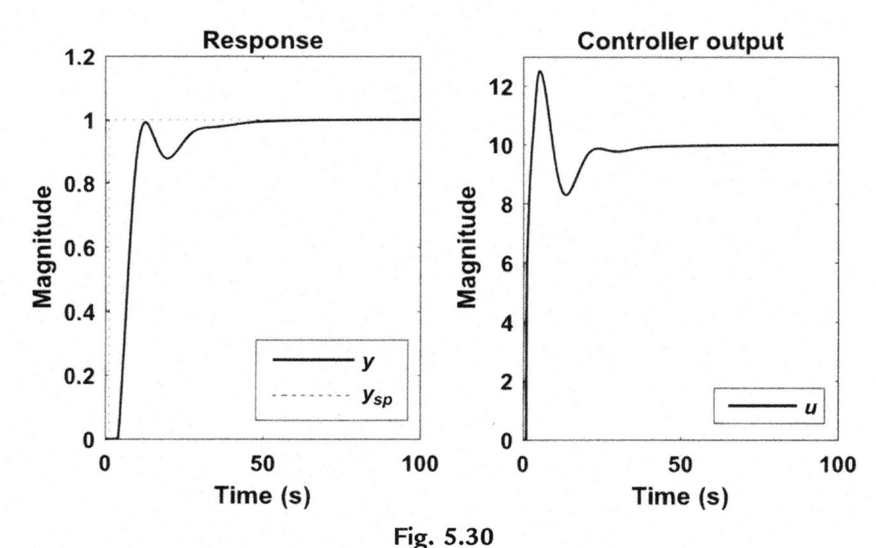

Fig. 5.30
The closed-loop response and the controller action with a dead-time compensator.

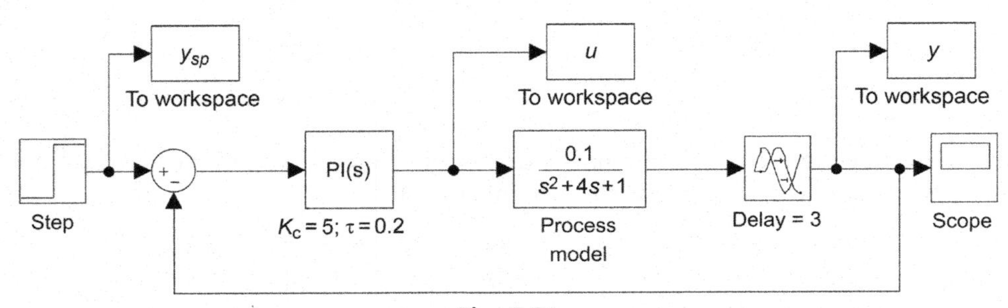

Fig. 5.31
A simple PI controller without a Smith predictor.

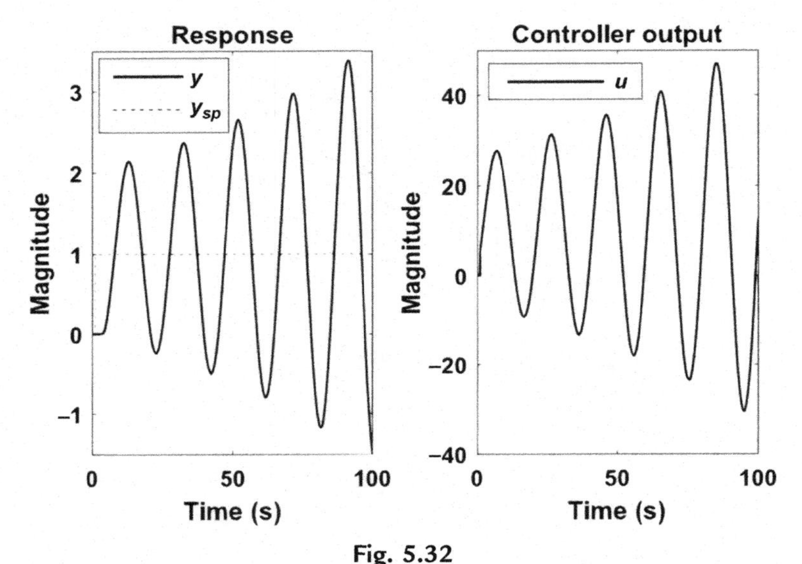

Fig. 5.32
The closed-loop response and the controller action with a PI controller.

5.7.5 Control of Nonlinear Processes

Fig. 5.33 depicts the relationship between the process input and output (u and y) in a linear and a nonlinear system. If the relationship between u and y is mildly nonlinear, a linear controller such as a PID controller with constant controller parameters would work fine. However, in the presence of strong nonlinearities, to achieve a satisfactory performance, the parameters of the linear controller must be retuned depending on the selected set point. The automatic regular tuning of the controller parameters can be performed by employing adaptive controllers. Adaptive controllers involve online system identification and controller design. There are many types of adaptive controllers:

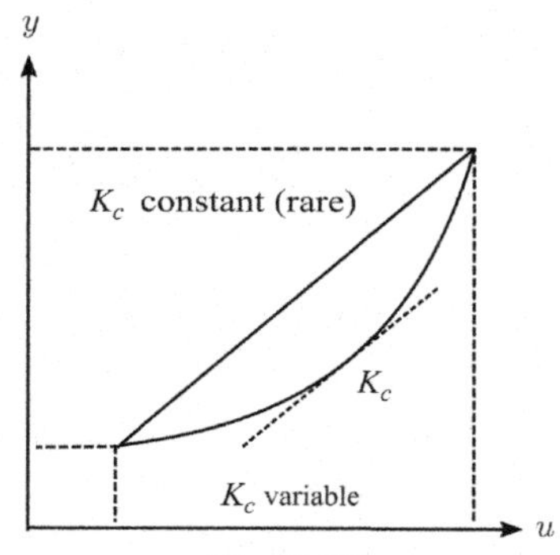

Fig. 5.33
The relationship between process input and output (u and y) in a linear and a nonlinear system.

a. Gain scheduling adaptive controllers
The overall gain of a feedback control system, namely, $K_{overall} = K_c K_v K_p K_m$, must be maintained constant for good performance. K_m and K_v are system dependent and undergo small changes. We may only change the controller gain in the presence of large changes in the process gain, K_p, in order that the overall steady-state gain, $K_{overall}$, is maintained constant. Therefore as the process gain changes, the controller gain (K_c) must be changed to maintain the overall gain constant. The implementation of this controller requires online estimation of the process gain $K_p = \Delta Y / \Delta U$, using an Estimation Block, as is shown in Fig. 5.34.

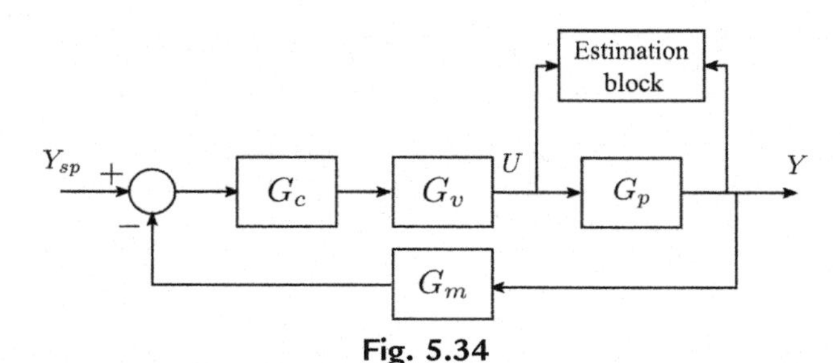

Fig. 5.34
The block diagram of a gain scheduling adaptive controller.

b. Model reference adaptive controllers
An alternative adaptive controller is the model reference architecture depicted in Fig. 5.35. In this algorithm, the system is controlled such that the closed-loop system behaves as a user-specified low-order model. The error between the actual closed-loop plant output and the output of the user-specified closed-loop model at each sampling interval is calculated. The controller parameters are adjusted to minimize the sum of the squares of these errors.

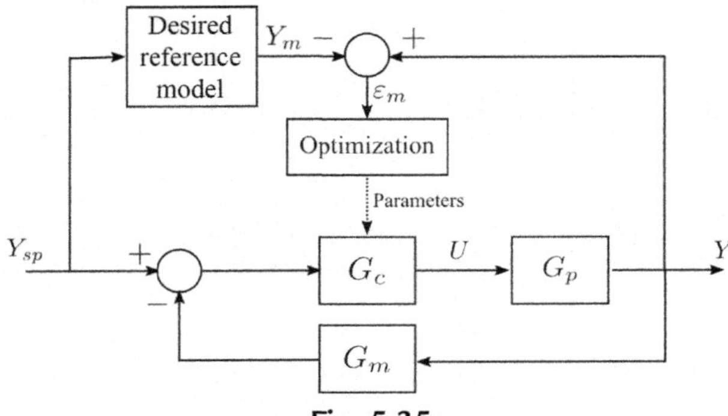

Fig. 5.35
The architecture of a model reference adaptive controller.

c. The self-tuning regulator (STR) adaptive controllers
The self-tuning regulators are the most common adaptive controllers used in industry. There are commercial self-tuning regulators, for example, the Foxboro's EXACT controller.[4] The input-output data are used to tune a controller either in an "explicit" or an "implicit" manner. In an explicit STR, the process model is identified using the input-output data, for example, as a first- or a second-order transfer function with or without time delay. The process identification block estimates the process model parameters, for example, K, θ, and τ of a model such as $G(s) = Ke^{-\theta s}/(\tau s + 1)$. The updated process model is then used to tune the controller.[5] In the implicit STR, the process model is not explicitly identified and the controller design step is carried out along with the parameter estimation (see Fig. 5.36).[6]

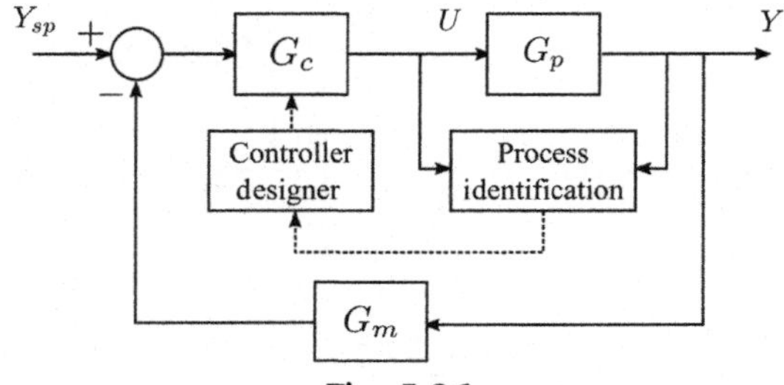

Fig. 5.36
The architecture of the self-tuning regulator (STR) adaptive controllers.

5.8 The Feedforward Controller (FFC)

The feedforward controller (FFC) is used in the presence of large and frequent disturbances affecting the controlled variable, especially in a slow process. A feedforward controller has a predictive nature to it and therefore can improve the overall control performance. However, since only the information on the major disturbances is provided to a feedforward controller, the feedforward controller is virtually unaware of the consequence of its action. The feedforward controller compensates for the adverse effects of the major measured disturbances. Therefore a feedforward controller should never be implemented on its own and must always be complemented by a feedback controller that does receive information about the controlled variable from the process. A feedforward controller requires accurate process and load transfer functions, and its performance very much depends on the model accuracy. It requires additional instrumentation and engineering effort, and therefore, is only used as an advanced control tool for those applications that a tight control of the process variables is needed.

Let us consider a distillation column whose distillate composition must be tightly controlled. If the feed flow rate and feed temperature fluctuate a lot, achieving a tight distillate composition control with a mere feedback controller would not be feasible. Addition of a feedforward controller to the feedback controller will improve the overall control performance. The feedforward controller, as is shown in Fig. 5.37, measures the major disturbances, namely, the feed flow rate and temperature, and calculates the reflux flow rate necessary to maintain the distillate composition at its set point, $X_{D,sp}$. The output of the feedforward controller is added to the output of the feedback controller and the combined corrective action is sent to the valve to manipulate the reflux flow rate.

The general block diagram of a combined FFC and a FB controller is shown in Fig. 5.38 in which D, the major disturbance is measured by a transmitter with a transfer function, G_t.

Assuming that $K_m = 1$, the closed-loop transfer function of the combined feedback plus feedforward control system is given by

$$Y = \left(G_d + G_t G_f G_v G_p\right) D + \left(G_{sp} G_v G_p\right) Y_{sp} + \left(Y_{sp} - G_m Y\right)\left(G_v G_c G_p\right) \tag{5.95}$$

$$Y = \underbrace{\frac{\left(G_{sp} + G_c\right) G_v G_p}{1 + G_c G_v G_p G_m}}_{g_1^*} Y_{sp} + \underbrace{\frac{G_d + G_t G_f G_v G_p}{1 + G_t G_f G_v G_p}}_{g_2^*} D \tag{5.96}$$

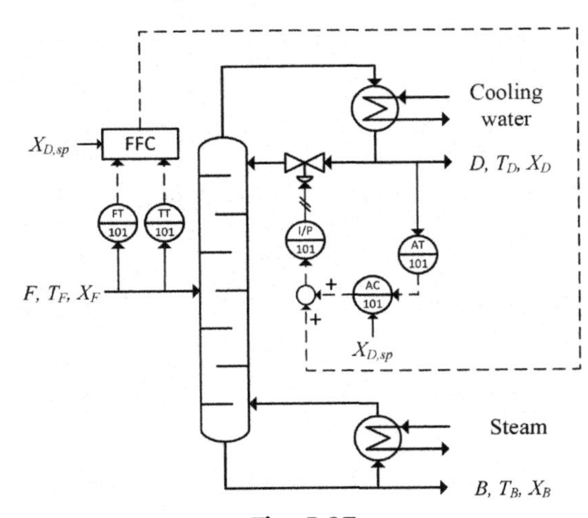

Fig. 5.37
The feedforward plus feedback control of the distillate composition.

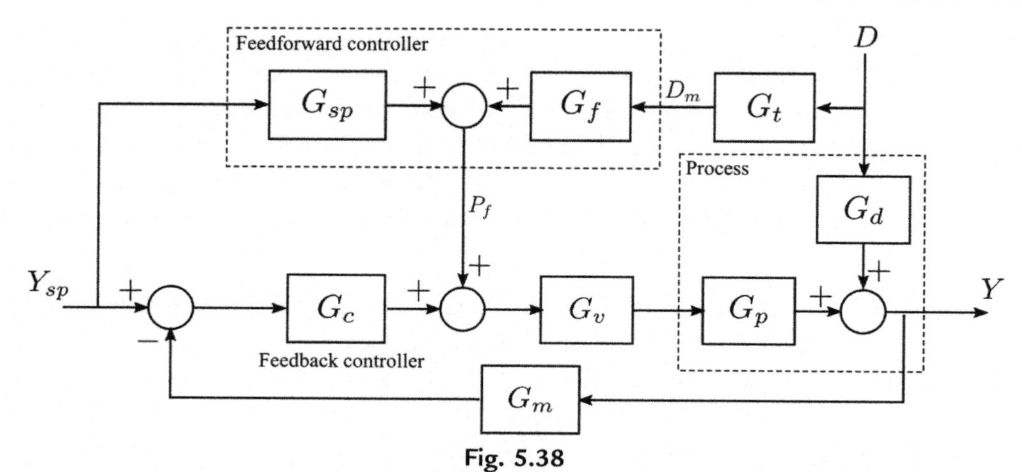

Fig. 5.38
The block diagram of a combined feedback plus feedforward control system.

The unknown transfer functions of the feedforward controller, namely, G_f and G_{sp}, can be derived using two design requirements of set point tracking, $g_1^* \rightarrow 1$, and disturbance rejection, $g_2^* \rightarrow 0$, to achieve perfect control (see Eq. 5.96). The disturbance rejection requires that $G_d + G_t G_f G_v G_p = 0$, and the set point tracking requires that $(G_{sp} + G_c) G_v G_p / (1 + G_c G_v G_p G_m) = 1$.

Assuming further, that $G_m = 1$ and $G_v = 1$, the required feedforward controller equation can be derived:

$$G_{sp} = \frac{1}{G_p} \tag{5.97}$$

$$G_f = -\frac{G_d}{G_p G_t} \tag{5.98}$$

Note that the closed-loop system stability only depends on the feedback controller as the feedforward controller equations do not appear in the closed-loop characteristic equation.

5.8.1 The Implementation of a Feedforward Controller

For the implementation of a dynamic feedforward controller, the process and load transfer functions and the transfer function of the measuring element on the disturbance must be known. For example, let us assume that both the process and load transfer functions are represented by a first order plus time delay transfer function:

$$G_p(s) = \frac{K e^{-\alpha s}}{\tau s + 1}; \quad G_d(s) = \frac{K_d e^{-\beta s}}{\tau_d s + 1} \tag{5.99}$$

If we also assume that $G_t = 1$, then:

$$G_f = -\frac{G_d}{G_p G_t} = \frac{K_d(\tau s + 1)e^{-(\beta-\alpha)s}}{K(\tau_d s + 1)} \tag{5.100}$$

If $\beta < \alpha$, there will be a time advance term which is not physically realizable. Using an analog system, the transfer function G_f excluding the time delay term can be implemented with a lead-lag unit that has adjustable parameters for the gain, the lead time constant, and the lag time constant:

$$G_{lead/lag} = \frac{K(\tau_{lead}s + 1)}{(\tau_{lag}s + 1)} \tag{5.101}$$

The new distributed control systems have numerical subroutines that mimic the function of a lead-lag unit with adjustable parameters.

For the digital implementation of a feedforward controller, one can derive the ordinary differential equation (ODE) between the feedforward controller output, $P_f(s)$, and the measured disturbance, $D_m(s)$, and then convert the ODE to a finite difference equation in terms of the controller output at each sampling instant. For example, let us assume that the controller equation is given by

$$G_f = \frac{P_f(s)}{D_m(s)} = \frac{2(3s+1)}{(4s+1)} \tag{5.102}$$

Converting the lead-lag term to a differential equation results in:

$$4\frac{dP_f}{dt} + P_f = 6\frac{dD_m}{dt} + 2D_m \tag{5.103}$$

Replacing the derivatives by a first-order backward difference equation gives:

$$4\frac{P_{f,n} - P_{f,n-1}}{\Delta t} + P_{f,n} = 6\frac{D_{m,n} - D_{m,n-1}}{\Delta t} + 2D_{m,n} \tag{5.104}$$

$$P_{f,n} = \frac{1}{4+\Delta t}\{4P_{f,n-1} + 6(D_{m,n} - D_{m,n-1}) + 2D_{m,n}\Delta t\} \tag{5.105}$$

In the previous example, the (lead-lag) term is a dynamic term and represents the implementation of the dynamic version of the feedforward controller. If the performance of the control system using the dynamic feedforward controller does not improve the overall control performance, a "steady-state" version of the feedforward controller can be implemented by replacing $s = 0$ in the dynamic feedforward controller equation.

$$G_f(s \to 0) = \frac{K_d(0+1)e^{-0}}{K(0+1)} = -\frac{K_d}{K} \tag{5.106}$$

5.8.2 The Ratio Control

The ratio control is a special steady-state feedforward controller that is used to control the ratio of the flow rates of two streams. One of the two streams is the wild stream, and the ratio controller is to control the flow rate of the other stream so that the ratio of the flow rates of the two streams remains close to the set point. There are many examples in industry in which such a ratio controller is needed. For example, the ratio of the air to fuel flow rate in a furnace, the ratio of the flow rates of two reactants fed to a reactor, or the reflux ratio which is the ratio of the reflux flow rate to the flow rate of the distillate in a distillation column.

A ratio controller can be implemented using two different configurations. In the first configuration shown in Fig. 5.39, the flow rates of both streams are measured and their ratio is calculated by dividing the measured flow rates. In the second configuration, the measured flow rate of the wild stream is multiplied by the desired ratio and the product is used as the set point of the flow controller of the controlled stream (see Fig. 5.40). Since the latter configuration avoids using a divider, it involves less rounding error and therefore is the preferred configuration.

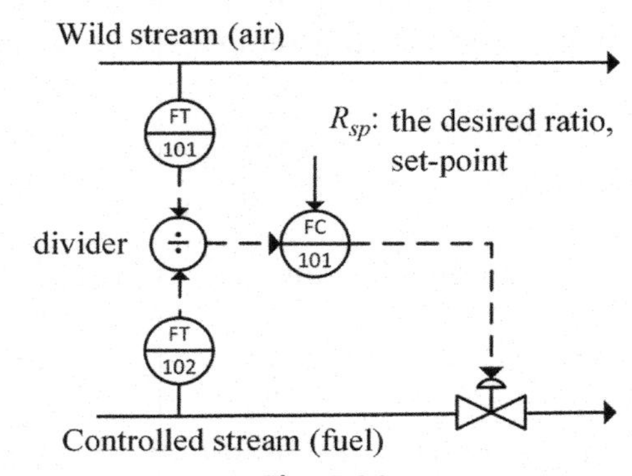

Fig. 5.39
The configuration using a divider for implementing of a ratio controller.

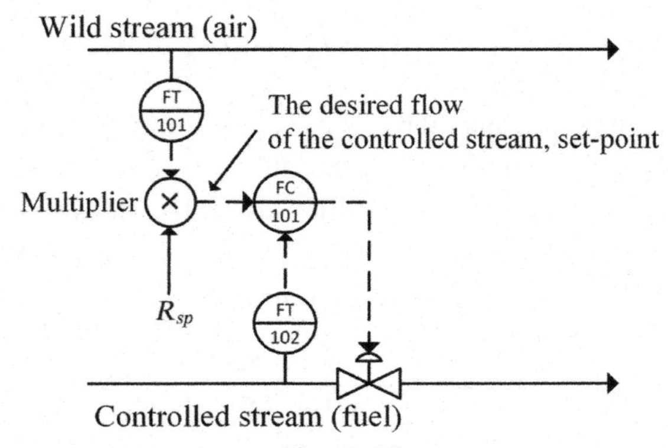

Fig. 5.40
The preferred configuration using a multiplier for implementing a ratio controller.

5.9 Control of Multiinput, Multioutput (MIMO) Processes

The multiinput, multioutput (MIMO) processes have more than one controlled variable. Most of the actual chemical processes belong to this category. Fig. 5.41 depicts a MIMO system with n manipulated and controlled variables and l disturbances.

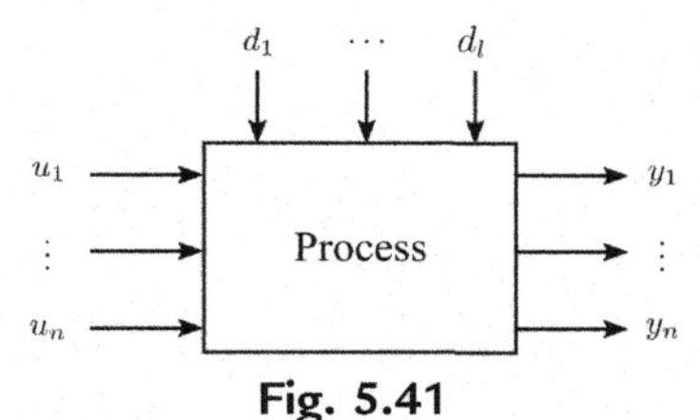

Fig. 5.41
The block diagram of a MIMO system.

The process model of a MIMO process can be represented by a transfer function matrix or by a state-space representation.

$$\begin{bmatrix} Y_1(s) \\ \vdots \\ Y_n(s) \end{bmatrix} = \begin{bmatrix} G_{11}(s) & G_{12}(s) & \cdots & G_n(s) \\ \vdots & \vdots & \ddots & \vdots \\ G_{n1}(s) & G_{n2}(s) & \cdots & G_{nn}(s) \end{bmatrix} \begin{bmatrix} U_1(s) \\ \vdots \\ U_n(s) \end{bmatrix} + \begin{bmatrix} G_{d11}(s) & G_{d12}(s) & \cdots & G_{d1l}(s) \\ \vdots & \vdots & \ddots & \vdots \\ G_{dn1}(s) & G_{dn2}(s) & \cdots & G_{dnl}(s) \end{bmatrix} \begin{bmatrix} D_1(s) \\ \vdots \\ D_l(s) \end{bmatrix} \tag{5.107}$$

$$\boldsymbol{Y}(s) = \boldsymbol{G}(s)\boldsymbol{U}(s) + \boldsymbol{G_d}(s)\boldsymbol{D}(s) \tag{5.108}$$

$$\begin{aligned} Y_1(s) = {} & G_{11}(s)U_1(s) + G_{12}(s)U_2(s) + \cdots + G_{1n}(s)U_n(s) \\ & + G_{d11}(s)D_1(s) + G_{d12}(s)D_2(s) + \cdots + G_{d1l}(s)D_l(s) \end{aligned} \tag{5.109}$$

For a noninteracting MIMO process (a process in which the controlled variable i is only affected by the manipulated variable i), the off-diagonal terms of the transfer function matrix, $\boldsymbol{G}(s)$, given in Eq. (5.108) are zero. In such a noninteracting MIMO system, each output is affected only by one input. However, this is a highly unlikely situation. Usually, there is interaction in the MIMO systems, and therefore, the SISO controllers do not perform well.

Example 5.17

Consider a binary distillation column with 5 controlled variables:

- the column pressure
- the level in the condenser

- the level in the bottom of the column
- the purity in the distillate, X_D
- the purity in the bottoms, X_B

There must also be five manipulated variables, for example:

- the cooling water flow rate
- the steam flow rate
- the reflux flow rate
- the bottoms flow rate
- the distillate flow rate

The question is how to pair the manipulated variables with the controlled variables, i.e., what is the best control configuration? Even if we use engineering judgment and common sense, there are many more than one potential control configuration. In general, for an $n \times n$ system (n manipulated and n controlled variables) there will be $n!$ ways to pair the manipulated variables and the controlled variables.

To find the best possible control configuration, we can identify the manipulated variables that have the largest, quickest, and the most direct effect on the controlled variables. However, where the engineering judgment fails, we need an analytical tool, such as the Bristol Relative Gain Array to choose a control configuration that results in the least amount of interaction. In what follows we discuss how to determine the level of interaction between the manipulated variables and the controlled variables, at steady state, and use this information to find the best control configuration. Note that the Bristol Relative Gain Array provides information only on the steady-state level of interaction in the system.

Once the proper pairing of $u \rightarrow y$ has been determined, the next step is to control the system effectively. There are two general approaches to control a MIMO system:

a. Use multiple SISO controllers such as the PID controllers in a multiloop architecture. In such an arrangement, the controllers work independent of each other. In order to minimize the interaction between the manipulated variables and controlled variables, in such a case, we can
   - either detune the controllers that control the less important variables, using a small K_c value
   - or, use decouplers that are essentially feedforward controllers to decouple the interacting control loop
b. Use multivariable controllers. In this approach, multivariate controllers such as the DMC (Dynamic Matrix Control) or MPC (Model Predictive Control) are used. In this case the multivariate controllers receive information on the error signals of all process variables and take the appropriate actions to minimize the error signals of all loops.

Example 5.18

Consider a 2×2 MIMO process with the block diagram shown in Fig. 5.42.

$$\begin{bmatrix} Y_1(s) \\ Y_2(s) \end{bmatrix} = \begin{bmatrix} G_{11}(s) & G_{12}(s) \\ G_{21}(s) & G_{22}(s) \end{bmatrix} \begin{bmatrix} U_1(s) \\ U_2(s) \end{bmatrix} \tag{5.110}$$

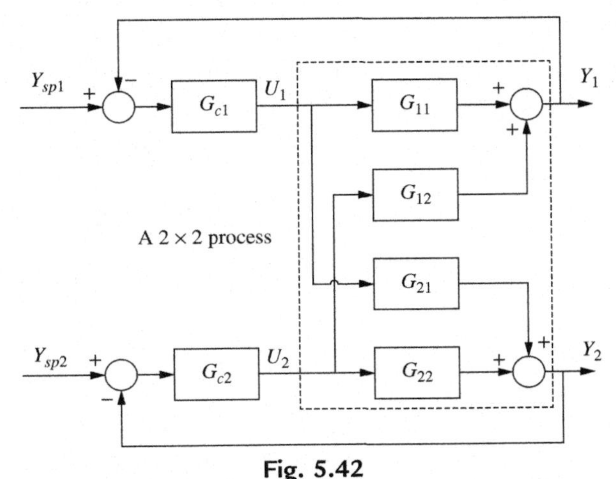

Fig. 5.42
A 2 × 2 process block diagram.

Note that if $G_{12}(s)$ and $G_{21}(s)$ are zero, then the two loops are independent and there would be no interactions in the system. Otherwise, Y_1 will be influenced by both U_1 and U_2. The same will be true for Y_2.

5.9.1 The Bristol Relative Gain Array (RGA) Matrix

The Bristol RGA matrix is a measure of the steady-state level of interaction in a MIMO system. For example, for a 2 × 2 system, it is a symmetric 2 × 2 matrix of the form:

$$\Lambda = \begin{bmatrix} \lambda_{11} & \lambda_{12} \\ \lambda_{21} & \lambda_{22} \end{bmatrix} \tag{5.111}$$

The individual entries of the RGA are defined:

$$\lambda_{11} = \frac{\text{the effect of } U_1 \text{ on } Y_1 \text{ assuming there is no interaction in the system}}{\text{the effect of } U_1 \text{ on } Y_1 \text{ in the presence of interaction}} \tag{5.112}$$

$$\lambda_{11} = \frac{(\Delta Y_1/\Delta U_1) \text{ all loops open}}{(\Delta Y_1/\Delta U_1) \text{ while } U_2 \leftrightarrow Y_2 \text{ loop is closed using a controller with I-action}} \tag{5.113}$$

The numerical value of λ_{11} determines if there is any interaction between the two loops.

$$\lambda_{11} = \begin{cases} 1 & \rightarrow \quad \text{no interaction or no effect of } U_2 \text{ on } Y_1 \\ 0.5 & \rightarrow \quad \text{maximum interaction meaning that } U_2 \text{ and } U_1 \text{have equal effect on } Y_1 \\ <0 & \rightarrow \quad U_2 \text{ has a deregulating effect on } Y_1 \text{ (loops fight each other)} \end{cases} \tag{5.114}$$

In general, it can be shown that the elements of each column or each row of Λ matrix add up to 1. Let us consider several cases with different numerical values of λ_{11}.

$$\text{If } \lambda_{11}=1, \quad \Lambda=\begin{bmatrix} 1 & 0 \\ 0 & 1 \end{bmatrix} \quad \text{no interaction}$$

$$\text{If } \lambda_{11}=0.7, \quad \Lambda=\begin{bmatrix} 0.7 & 0.3 \\ 0.3 & 0.1 \end{bmatrix} \quad \text{some interaction, } U_1 \text{ is paired to } Y_1 \text{ and } U_2 \text{ to } Y_2 \tag{5.115}$$

$$\text{If } \lambda_{11}=0.1, \quad \Lambda=\begin{bmatrix} 0.1 & 0.9 \\ 0.9 & 0.1 \end{bmatrix} \quad \text{some interaction, pair } U_1 \text{ to } Y_2 \text{ and } U_2 \text{ to } Y_1$$

There are two ways to determine Λ. Let us assume that we have a 2×2 system and we are to find λ_{11}.

a. Experimentally: One approach to obtain the elements of the RGA matrix is to perform two experiments. In the first experiment, both loops are left open (both controllers are set to manual). The system is allowed to reach steady state. Then, U_1 is changed by a step change, ΔU_1, and the corresponding change in Y_1 at steady state is measured. The ratio of the measured change in the output to the input at steady state, $\Delta Y_1/\Delta U_1$, will render the numerator of λ_{11}. In the second experiment, the second loop is closed, connecting U_2 to Y_2, with a PI controller and a step change in U_1 is introduced again, by an amount ΔU_1. Wait till the PI controller exerts its corrective action and eventually brings Y_2 back to its original set point. In the second experiment, U_1 will have a combined "direct" and an "indirect" effect on Y_1. The latter is a result of the possible interaction that U_1 may have on Y_2 causing Y_2 to deviate from its set point with a consequent corrective action by U_2 to bring Y_2 back to its original set point. The resulting change in U_2 may have an effect on Y_1, the "indirect" effect of U_1 on Y_1 through U_2. Again, we measure $\Delta Y_1/\Delta U_1$, this is now the result of the combined "direct" and "indirect" effect of U_1 and Y_1, the denominator of λ_{11}.

$$\lambda_{11} = \frac{\text{direct effect of } U_1 \text{on } Y_1}{\text{direct and indirect effect of } U_1 \text{ on } Y_1} \tag{5.116}$$

$$\Lambda = \begin{bmatrix} \lambda_{11} & 1-\lambda_{11} \\ 1-\lambda_{11} & \lambda_{22} \end{bmatrix} \tag{5.117}$$

b. Theoretically: If we have the individual steady-state gains of the system or the transfer function matrix of the system, the RGA matrix can easily be calculated. Consider the following example.

Example 5.19

Determine the correct pairing of the manipulated and controlled variables for the following system:

$$G=\begin{bmatrix}\dfrac{2.5e^{-2s}}{(s+0.5)} & \dfrac{2}{(s+1)}\\ \dfrac{3}{(2s+1)} & \dfrac{4}{(3s+1)}\end{bmatrix} \tag{5.118}$$

Which results in

$$Y_1=\left[\frac{2.5e^{-2s}}{(s+0.5)}\right]U_1+\left[\frac{2}{(s+1)}\right]U_2 \tag{5.119}$$

$$Y_1=\left[\frac{3}{(2s+1)}\right]U_1+\left[\frac{4}{(3s+1)}\right]U_2 \tag{5.120}$$

The steady-state gain matrix of the system can be obtained by setting $s=0$:

$$K=\begin{bmatrix}5 & 2\\ 3 & 4\end{bmatrix} \tag{5.121}$$

At steady state,

$$Y_1=5U_1+2U_2 \tag{5.122}$$

$$Y_1=3U_1+4U_2 \tag{5.123}$$

The first equation gives the "direct" effect of U_1 on Y_1, the numerator of λ_{11}

$$(\Delta Y_1/\Delta U_1)_{\text{all loops open}}=5 \tag{5.124}$$

If the second loop controller is on the automatic mode with a PI controller, at steady state, there will be no change in Y_2, i.e., no offset.

$$\begin{aligned}0&=3U_1+4U_2\\ U_2&=(-3/4)U_1\end{aligned} \tag{5.125}$$

$$Y_1=5U_1+2[(-3/4)\,U_1]$$

$$Y_1=5U_1-\left(\frac{6}{4}\right)U_1 \tag{5.126}$$

$$\Delta Y_1=(7/2)\Delta U_1$$

The combined "direct" and "indirect" effect of U_1 on Y_1 is given by

$$\left(\frac{\Delta Y_1}{\Delta U_1}\right)_{\text{the second loop } U2 \text{ to } Y2 \text{ is closed with a PI-controller}}=7/2 \tag{5.127}$$

$$\lambda_{11} = \frac{5}{7/2} = \frac{10}{7} \tag{5.128}$$

$$\Lambda = \begin{bmatrix} 10/7 & -3/7 \\ -3/7 & 10/7 \end{bmatrix} \tag{5.129}$$

The resulting Λ suggests the correct pairing, i.e., U_1 is paired to Y_1 while recognizing that U_2 has a deregulating effect on Y_1 (the two loops fight each other).

Another theoretical way to obtain the RGA is to use matrix operation: The transpose of the inverse of the steady-state gain matrix is obtained. The elements of the RGA are obtained by multiplying the corresponding elements of the steady-state gain by the elements of the transpose of the inverse of the steady-state gain matrix.

Example 5.20

Obtain the RGA matrix using the matrix operation:

$$\begin{cases} Y_1 = 5U_1 + 2U_2 \\ Y_2 = 3U_1 + 4U_2 \end{cases} \tag{5.130}$$

$$\begin{bmatrix} Y_1 \\ Y_2 \end{bmatrix} = \begin{bmatrix} 5 & 2 \\ 3 & 4 \end{bmatrix} \begin{bmatrix} U_1 \\ U_2 \end{bmatrix} \tag{5.131}$$

$$H = \left(K^{-1}\right)^T \tag{5.132}$$

$$\lambda_{ij} = k_{ij} h_{ig} \tag{5.133}$$

$$K^{-1} = \frac{\begin{bmatrix} 4 & -2 \\ -3 & 5 \end{bmatrix}}{(5 \times 4) - (2 \times 3)} = \begin{bmatrix} 4/14 & -2/14 \\ -3/14 & 5/14 \end{bmatrix} \tag{5.134}$$

$$\left(K^{-1}\right)^T = \begin{bmatrix} 4/14 & -3/14 \\ -2/14 & 5/14 \end{bmatrix} = H \tag{5.135}$$

$$\lambda_{11} = k_{11} h_{11} = 5\left(\frac{4}{14}\right) = \frac{20}{14} = \frac{10}{7} \tag{5.136}$$

Therefore U_1 is paired to Y_1, acknowledging that the two loops fight each other.

Example 5.21

The Bristol RGA of a process is given as follows suggesting the correct pairing of the manipulated and controlled variables.

$$\Lambda = \begin{bmatrix} \mathbf{0.8} & 0.1 & 0.1 \\ 0.05 & 0.05 & \mathbf{0.9} \\ 0.15 & \mathbf{0.85} & 0 \end{bmatrix} \tag{5.137}$$

The correct pairing will be

$$\begin{cases} U_1 \leftrightarrow Y_1 \\ U_2 \leftrightarrow Y_3 \\ U_3 \leftrightarrow Y_2 \end{cases} \tag{5.138}$$

5.9.2 Control of MIMO Processes in the Presence of Interaction Using Decouplers

A simple approach to control a MIMO process is to use multiloop single loop controllers such as PID controllers.

5.9.2.1 Design of decouplers

In the presence of interactions, in order to minimize the effect of the individual controllers on other loops, "decouplers," are added. The decouplers are essentially feedforward controllers that minimize the interactions between the single loop controllers. Their input is the output of the other controllers that are disturbance to the loop under consideration. Considering the block diagram of a two-by-two system shown in Fig. 5.43, one can write the following equations to derive the transfer functions of the unknown decouplers.

$$\begin{aligned} Y_1 &= G_{11}U_1 + G_{12}U_2 \\ Y_2 &= G_{21}U_1 + G_{22}U_2 \end{aligned} \tag{5.139}$$

$$\begin{aligned} U_1 &= v_1 + G_{I1}v_2 \\ U_2 &= v_2 + G_{I2}v_1 \end{aligned} \tag{5.140}$$

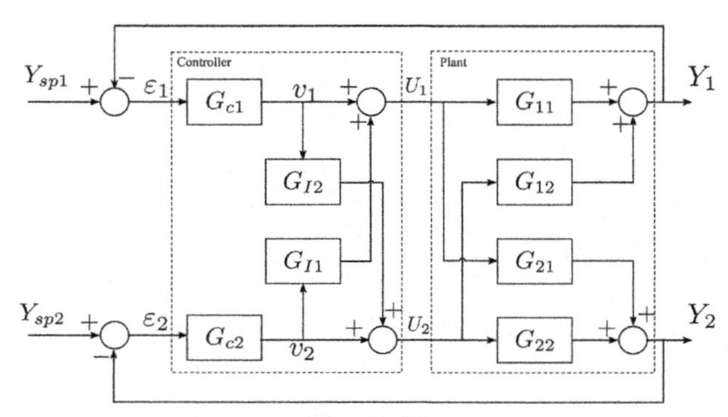

Fig. 5.43
Control of MIMO processes single loop controllers and decouplers.

Substitute Eq. (5.139) into Eq. (5.140) results in:

$$Y_1 = (G_{11} + G_{12}G_{I2})v_1 + (G_{12} + G_{11}G_{I1})v_2 \tag{5.141}$$

$$Y_2 = (G_{21} + G_{22}G_{I2})v_1 + (G_{22} + G_{21}G_{I1})v_2 \tag{5.142}$$

Ideally v_2 should not affect Y_1 and similarly v_1 should not affect Y_2.

$$G_{12} + G_{11}G_{I1} = 0 \rightarrow G_{I1} = -\frac{G_{12}}{G_{11}} \tag{5.143}$$

$$G_{21} + G_{22}G_{I2} = 0 \rightarrow G_{I2} = -\frac{G_{21}}{G_{22}} \tag{5.144}$$

Often, the dynamic transfer functions of the decouplers are physically unrealizable, therefore, only the steady-state compensators are implemented.

Example 5.22

Derive the decoupler equations for the following system.

$$\begin{bmatrix} Y_1(s) \\ Y_2(s) \end{bmatrix} = \begin{bmatrix} \dfrac{5e^{-2s}}{(s+0.5)} & \dfrac{2}{(s+1)} \\ \dfrac{3}{(2s+1)} & \dfrac{4}{(3s+1)} \end{bmatrix} \begin{bmatrix} U_1(s) \\ U_2(s) \end{bmatrix} \tag{5.145}$$

$$G_{I1}(s) = \frac{-2/(s+1)}{5e^{-2s}/(s+0.5)} = \frac{-2(s+0.5)\overbrace{e^{2s}}^{\text{not realizable}}}{5(s+1)} \tag{5.146}$$

$$G_{I2}(s) = \frac{-3/(2s+1)}{4/(3s+1)} = \frac{-0.75(3s+1)}{(2s+1)} \tag{5.147}$$

The steady-state interaction compensators (decouplers) are as follows:

$$G_{I1}(0) = -\frac{1}{5}, \quad G_{I2}(0) = -\frac{3}{4} \tag{5.148}$$

5.10 Problems

1. For the transfer functions shown as follows:

$$G_p(s) = \frac{0.3\,e^{-4s}}{4s+1}; \quad G_d(s) = \frac{K_d}{\tau_d s+1}; \quad G_v(s) = 1; \quad G_m(s) = 1 \tag{5.149}$$

Derive the closed-loop transfer functions of the servo and regulatory systems, and in each case obtain the closed-loop responses to a unit step change in the set point and in the disturbance. Use both MATLAB and Simulink for the simulation purposes.

2. Derive the closed-loop servo transfer function $Y(s)/Y_{sp}(s)$ of the system shown in Fig. 5.44 and obtain the response to a unit step change in $Y_{sp}(s)$. You may use MATLAB or the Simulink to obtain the unit step response of the system.

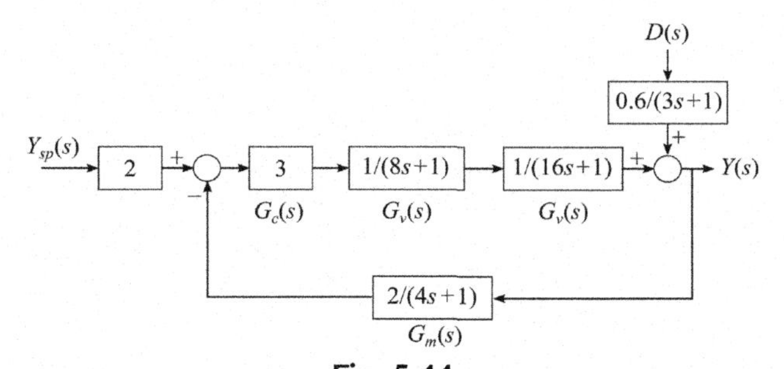

Fig. 5.44
Block diagram for problem 2.

3. The root locus diagram of a feedback control system is shown in Fig. 5.45. Specify the range of the controller gain, K_c, for which the closed-loop system response will be
   - oscillatory and stable
   - unstable
   - smooth and stable
   - oscillatory with constant amplitude, marginally stable

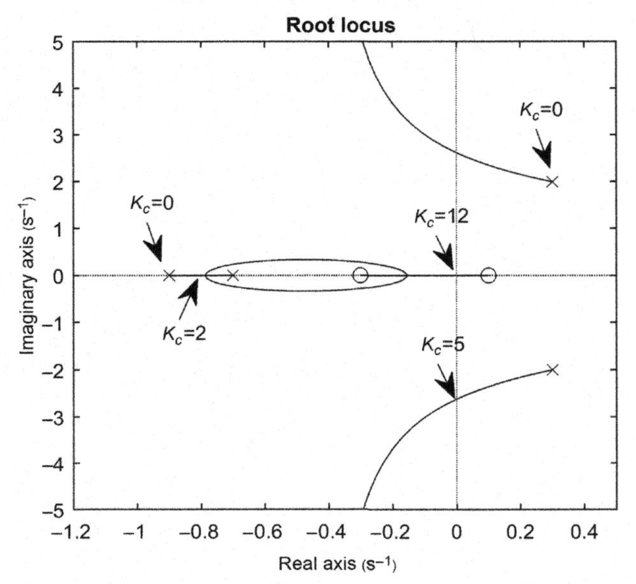

Fig. 5.45
Root locus diagram of problem 3.

4. The process transfer function, $G_p(s)$; load transfer function, $G_d(s)$; the transfer function of the measuring element, $G_m(s)$; the valve, $G_v(s)$; and the controller, $G_c(s)$ are given as follows:

$$G_p(s)=\frac{0.3}{40s+1};\ G_d(s)=\frac{0.8}{12s+1};\ G_m(s)=\frac{0.1}{3s+1};\ G_v(s)=\frac{1}{2s+1};\ G_c(s)=\frac{0.3(2s+1)}{2s} \tag{5.150}$$

Derive or write down the closed-loop servo and regulatory transfer functions. Do not simplify the resulting closed-loop transfer functions, only replace the given transfer functions in the expressions for the closed-loop transfer functions.
What is the closed-loop offsets (steady-state errors) of this control system for the servo and regulatory modes as a result of a unit step change in the controller set point or the disturbance?

5. The closed-loop characteristic equation ($CLCE$) of a control system is given by

$$CLCE=1+\frac{K_c}{(6s^3+5s^2+2s+1)} \tag{5.151}$$

Determine the maximum controller gain, $K_{c,max}$, above which the closed-loop system becomes unstable. You may use any stability test.

6. The closed-loop characteristic equation ($CLCE$) of a control system is given by

$$CLCE(s)=1+K_c\frac{0.3}{3s^3+2s^2+5s+1} \tag{5.152}$$

Using the Routh test and the direct substitution method, find the maximum controller gain, $K_{c,max}$, for a stable closed-loop system.
Also use MATLAB to obtain the root locus diagram of this system.

7. The actual dynamics of a process is given by

$$G_p(s)=\frac{0.3}{3s^3+2s^2+5s+1} \tag{5.153}$$

And a first order plus time delay approximation model of the process is given by

$$G_{p,model}(s)=\frac{0.3e^{-1.5s}}{4s+1} \tag{5.154}$$

Assuming that $G_m(s) = 1$, $K_m = 1$, $G_v(s) = 1$; calculate the controller parameters of a PI controller using the open-loop process reaction curve, and the Ziegler-Nichols continuous cycling method. In each case simulate the closed-loop response to a unit step change in the set point using MATLAB or Simulink.

8. For the following block diagram (Fig. 5.46)
   a. Derive the closed-loop servo and regulatory transfer functions, that is $\frac{Y(s)}{Y_{sp}(s)}$, and $\frac{Y(s)}{D(s)}$.

      Note, no simplification of the resulting closed-loop transfer functions is needed.
   b. Calculate the servo and regulatory closed-loop offsets (steady-state errors).
   c. Write down the necessary MATLAB commands to obtain the unit step servo response of the system.

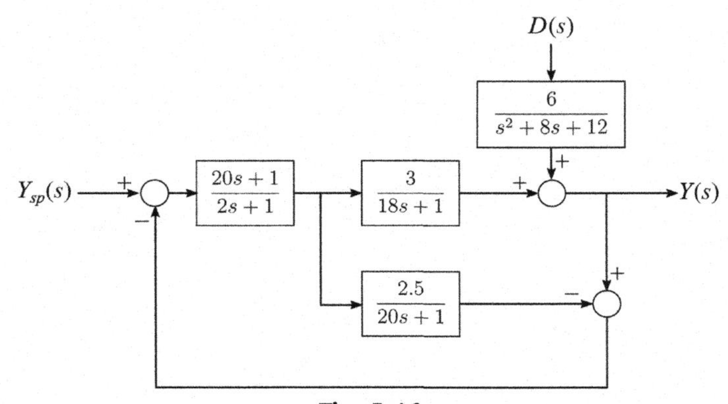

Fig. 5.46
Block diagram for problem 8.

9. The block diagram of a feedback control system is shown in Fig. 5.47.

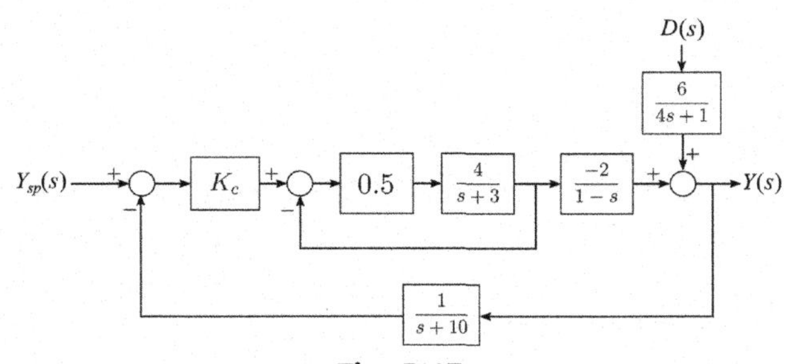

Fig. 5.47
Block diagram for problem 9.

Derive the closed-loop servo and regulatory transfer functions, i.e., $\frac{Y(s)}{Y_{sp}(s)}$ and $\frac{Y(s)}{D(s)}$ for this system.

Determine the maximum or minimum value of the controller gain, K_c, that results in a marginally stable closed-loop system. Use any stability tests that you may wish.

10. Smith and Corripio[3] suggest combining cascade control, ratio control, and selective control configurations for the control of a furnace/boiler system to generate steam at a desired pressure. In order to avoid the air-lean and air-rich conditions in the furnace, they propose several improvements shown in Fig. 5.48. Discuss the improvements suggested by referring to their book[3] and make any other suggestions to further improve the control quality.

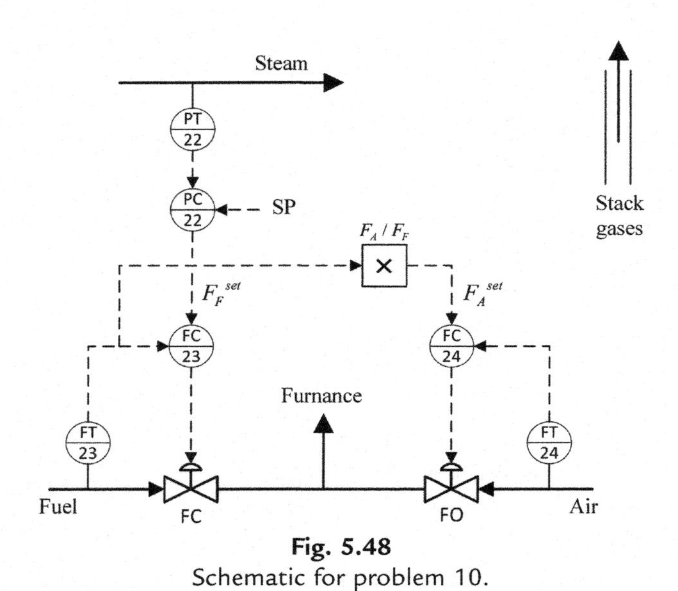

Fig. 5.48
Schematic for problem 10.

11. A digital controller is used to control the temperature of a highly exothermic reaction carried out in a continuous stirred tank reactor. There are significant variations in the composition and the flow rate of the reactants entering the reactor. Therefore it is recommended that a feedback plus feedforward controller be used for this system. The flow transmitter has an output of 0-5 V, the temperature transmitter has an output in the range of 4-20 mA, and reactant concentration measuring device has an output of 0-5 V. The "data acquisition board" accepts 0-5 V and 4-20 mA input signals and has an analog output module with 4-20 mA. The control valve supplying cooling water to the reactor jacket is pneumatically actuated accepting a signal in the range of 3-15 psig. Draw the block diagram and the Piping and Instrumentation Diagram (P&ID) of a combined feedback/feedforward controller for this system. Include all the hardware

components needed to implement the suggested control system. Clearly describe the nature of the signal (analog or digital; continuous or sampled), the unit and the range (°C; concentration; flow rate; 4-20 mA; 3-15 psig; binary) at each point around the control loop(s).

12. The "actual" dynamics of a process is given by

$$G_p(s) = \frac{0.3\,e^{-2s}}{2s^2 + 6s + 1} \tag{5.155}$$

And the approximate model is

$$G_m(s) = \frac{0.3\,e^{-4s}}{8s + 1} \tag{5.156}$$

Design a Smith predictor for this process using a well-tuned PI controller and simulate the closed-loop response of the process to a step change in the set point using Simulink. Assume a unity transfer function for the measuring element and the control valve.

13. Design a feedforward controller [obtain $G_f(s)$] for the distillation column shown in Fig. 5.49 that measures the feed composition, X_F with a transmitter whose transfer function is given by $G_t(s)$. Discuss the physical realizability and the implementation of the $G_f(s)$ on an analog and a digital controller.

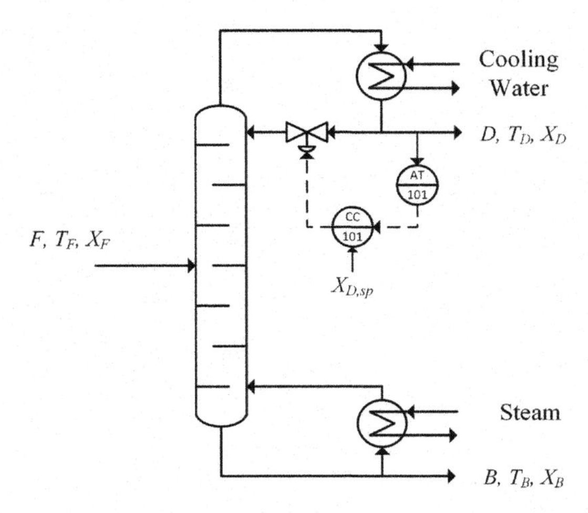

Fig. 5.49
Schematic for problem 13.

$$G_p(s) = \frac{X_D(s)}{R(s)} = \frac{0.3e^{-4s}}{12s+1};\ G_d(s) = \frac{X_D(s)}{X_F(s)} = \frac{0.8\,e^{-6s}}{7s+1};\ G_t(s) = \frac{\text{measured}\,X_F(s)}{X_F(s)} = 1.2\,e^{-3s} \tag{5.157}$$

14. The transfer function matrix of a two-input two-output system is given by

$$\begin{bmatrix} \dfrac{0.8\,e^{-5s}}{(8s+1)(5s+1)} & \dfrac{0.8}{(8s+1)} \\ \dfrac{0.7e^{-4s}}{(10s+1)(2s+1)} & \dfrac{1.4e^{-s}}{(20s+1)} \end{bmatrix} \tag{5.158}$$

Determine the Bristol relative gain array matrix for this system and comment on the steady-state interaction and pairing of the manipulated and the controlled variables.

15. The actual transfer function of a process is given by

$$G_p(s) = \frac{0.6s^{-3s}}{(s+1)(2s+1)(3s+1)} \tag{5.159}$$

And the approximate model is given by

$$\widetilde{G_p}(s) = \frac{0.6s^{-4.3s}}{(5s+1)} \tag{5.160}$$

Design an internal model controller (IMC) for this process. Discuss the controller performance by simulating the closed-loop response of the process in MATLAB or Simulink environment. The sampling/control interval in each is taken as 1 s.

References

1. Seborg DE, Edgar TF, Mellichamp DA, Doyle III FJ. *Process dynamics and control.* 3rd ed. Hoboken, NJ: John Wiley & Sons; 2011.
2. Ziegler JG, Nichols NB. Optimum setting for automatic controllers. *Trans ASME* 1942;**65**:759.
3. Smith CA, Corripio AB. *Principles and practice of automatic process control.* 3rd ed. Hoboken, NJ: John Wiley & Sons; 2006.
4. Astrom KJ, Wittenmark B. On self-tuning regulators. *Automatica* 1973;**9**(2):185–99.
5. Wittenmark B, Astrom KJ. Practical issues in the implementation of self-tuning control. *Automatica* 1984;**20** (5):595–605.
6. Butler H, Johansson R. Model reference adaptive control: from theory to practice. *Automatica* 1994;**30** (6):1073–5.

CHAPTER 6

Digital Sampling, Filtering, and Digital Control

Sohrab Rohani, Yuanyi Wu
Western University, London, ON, Canada

The objectives addressed in this chapter include:

- The definition of z-transform, its properties, and applications in the design of discrete controllers
- Analysis of the sampled-data systems
- The pulse transfer function of a continuous process and the discrete transfer function of a digital PID-controller in the z-domain
- The stability of the sampled-data systems
- The design of model-based single-input, single-output (SISO) and multi-input, multi-output (MIMO) digital controllers

Most of the controllers are digital, and therefore the design methodology must be carried out in the discrete or sampled environment. The equivalent tool to the Laplace transform for the design of digital controllers is the discrete Laplace transform or the z-transform. Once the discrete transfer function of the controller is derived, its implementation in the form of a discrete recursive difference equation in the time domain is straightforward. In view of the characteristics of chemical processes, it is evident that in the design of linear controllers, the plant-model mismatch must be incorporated in the design and the controllers must adapt their structures and/or their parameters to the time-varying nature of the plant. Moreover, state estimation and inference algorithms are necessary whenever unmeasurable states or controlled variables exist. Although, the design of linear controllers can be performed in the continuous state space and in the Laplace domain, the discrete state space or the z-domain offers far more flexibility. Feedback controllers may have a fixed structure with adjustable parameters which implicitly depend on the plant model. An example of such controllers is the commonly used PID (proportional-integral-derivative) controllers. On the other hand, the structure as well as the parameters of a controller may explicitly depend on the plant model.

Coulson and Richardson's Chemical Engineering. http://dx.doi.org/10.1016/B978-0-08-101095-2.00006-0

Such controllers are referred to as model-based controllers. The design and implementation of model-based controllers can be performed easier in the discrete domain.

6.1 Implementation of Digital Control Systems

Fig. 6.1 shows the block diagram of a general digital feedback control system. The measured controlled variable which may be corrupted by process noise (d) and sensor noise (d_s) is sampled every Δt_s (sampling interval) and then converted to a binary signal by an analog-to-digital converter (A/D) to be processed by the computer. The controller output is converted back to an analog signal by a digital-to-analog (D/A) converter and sent to a hold device at every control interval, Δt_c. For most applications, the control interval and the sampling interval are selected to be the same, that is, $\Delta t_s = \Delta t_c = \Delta t$. The function of the hold device is to reconstruct a pseudo-continuous signal from the discrete digital controller output before it is sent to the final control element. Usually, this is achieved by holding the signal during an interval at its value at the beginning of that interval. The device that performs this task is referred to as a zero-order hold (ZOH) device, which is a "capacitor" and is a part of the data acquisition system.

The final control element could be a pneumatically or a hydraulically actuated control valve, a variable speed pump, a stepping motor, a heating element, etc. The discrete or sampled signals in Fig. 6.1 are shown by a superscript "asterisk" and the binary signals by a subscript d.

Table 6.1 represents the analytical tools available to study continuous and discrete control systems in the time and Laplace or z-domains.

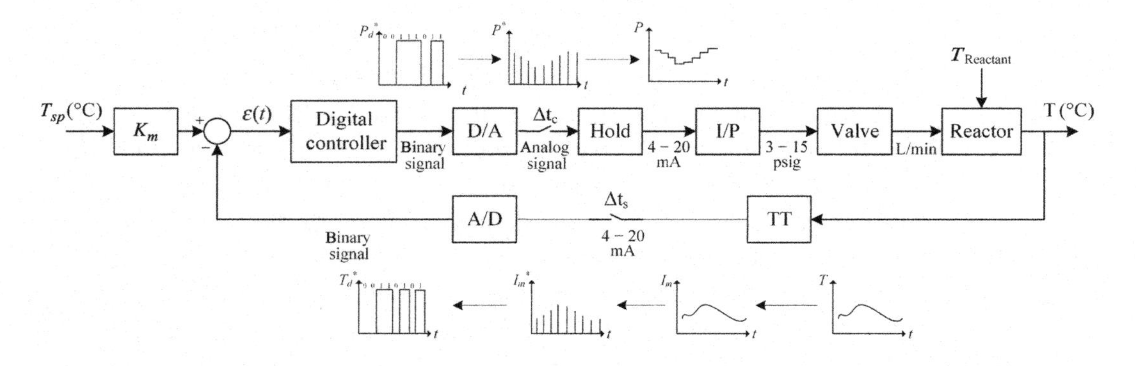

*The hold element is to maintain the controller output signal between the control intervals.
T_{sp}: Temperature set point
Δt_s: sampling interval in second.
Δt_c: control interval in second.
subscript d: represents a digital signal
superscript '*': represents a discrete signal

Fig. 6.1
Detailed block diagram of a digital feedback control system.

Table 6.1 Comparison of the continuous and discrete system expression

	Time Domain	Laplace Domain
Continuous system	Continuous state space	s-transfer function
Discrete system	Discrete state space	z-transfer function

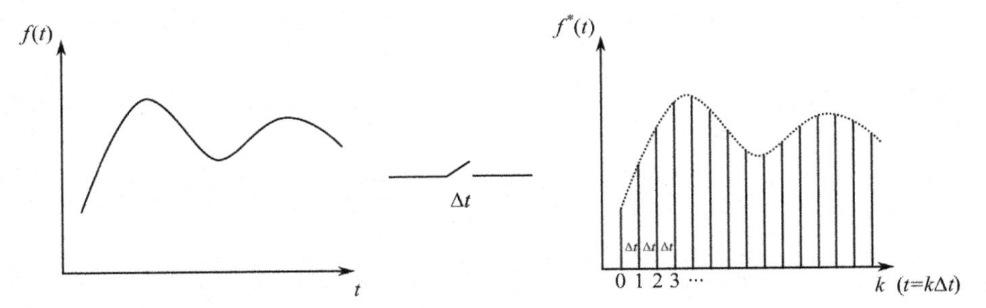

Fig. 6.2
Sampling operation.

The signal flow around the loop in Fig. 6.1 reveals that a continuous signal has to be sampled, converted to a digital signal, filtered to reduce the noise level, and the controller output must be reconstructed as a pseudo-continuous analog signal before it is fed to the final control element. In what follows, we shall discuss how these operations are performed physically and mathematically.

Sampling: Sampling is performed by an electrical switch, converting a continuous signal in the time domain to a discrete or sampled signal as shown in Fig. 6.2. The sampling interval is represented by Δt (s), its inverse is the sampling rate in hertz (Hz) ($cycles/s$) or the sampling frequency $2\pi/\Delta t$ (rad/s).

6.2 Mathematical Representation of a Sampled Signal

Sampling operation may be represented mathematically by:

$$I^*(t) = \sum_{k=0}^{\infty} \delta(t - k\Delta t) \tag{6.1}$$

where $\delta(t)$ is a unit impulse function $\int_{-\infty}^{+\infty} \delta(t)dt = 1$. Therefore the discrete representation of a continuous signal $f(t)$ after passing through a sampler is:

$$f^*(t) = f_k = f(t = k\Delta t) = \sum_{k=0}^{\infty} f(k\Delta t)\underbrace{\delta(t - k\Delta t)}_{\text{unit delayed impulse}} \tag{6.2}$$

Note that due to the presence of the delayed unit impulse function, the summation in Eq. (6.2) results in a single value corresponding to the value of the continuous signal at that sampling instant and takes a zero value in between the sampling intervals. Given

$$F(s) = \int_0^{\infty} f(t)e^{-st}dt \tag{6.3}$$

Laplace transforming the discrete function, $f^*(t)$, results in:

$$F^*(s) = \mathcal{L}\left\{\sum_{k=0}^{\infty} f(k\Delta t)\delta(t - k\Delta t)\right\} = \sum_{k=0}^{\infty} f(k\Delta t)e^{-k\Delta t_s} \tag{6.4}$$

Let $z = e^{\Delta t_s}$, then the z-transform of a discrete function $f^*(t) = f(k\Delta t) = f_k$ is

$$F(z) = Z\{f^*(t)\} = \sum_{k=0}^{\infty} f(k\Delta t)\, z^{-k} \tag{6.5}$$

6.3 z-Transform of a Few Simple Functions

Table 6.2 lists the z-transform and modified z-transform of a few simple functions.

6.3.1 A Discrete Unit Step Function (Fig. 6.3)

$$\begin{aligned} Z\{f^*(t)\} = F(z) &= \sum_{k=0}^{\infty} f(k\Delta t)\, z^{-k} = \sum_{k=0}^{\infty} (1) z^{-k} \\ &= 1 + z^{-1} + z^{-2} + \ldots \\ &= \frac{1}{1 - z^{-1}} \end{aligned} \tag{6.6}$$

Assuming $|z| \leq 1$, which is the case for a stable system as shall be discussed later.

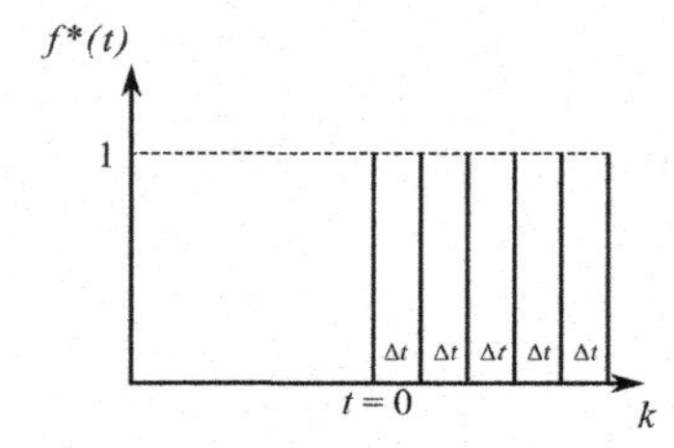

Fig. 6.3
A discrete unit step function.

6.3.2 A Unit Impulse Function

$$\delta(z) = \sum_{k=0}^{\infty} \delta(t=0)\, z^{-k} = 1 \tag{6.7}$$

6.3.3 A Discrete Exponential Function (Fig. 6.4)

$$\begin{aligned} Z\{f^*(t)\} &= Z\{e^{-at}\}^* = \sum_{k=0}^{\infty} \left(e^{-ak\Delta t}\right) z^{-k} \\ &= 1 + e^{-a\Delta t} z^{-1} + e^{-2a\Delta t} z^{-2} + \cdots \\ &= \frac{1}{1 - e^{-a\Delta t} z^{-1}} \end{aligned} \tag{6.8}$$

Assuming that $e^{-\alpha\Delta t}|z| \leq 1$ (Fig. 6.4).

6.3.4 A Discrete Delayed Function Where θ Is the Delay Time

Case 1: θ is a multiple integer of the sampling interval, i.e., $\theta = N\Delta t$

$$Z\{f(t-\theta)\}^* = \sum_{k=0}^{\infty} f(k\Delta t - N\Delta t) z^{-k} \tag{6.9}$$

Introducing a new dummy integer $P = k - N$, then

$$\begin{aligned} Z\{f(t-\theta)\}^* &= \sum_{P=0}^{\infty} f(P\Delta t) z^{-N} z^{-P} \\ &= z^{-N} \sum_{P=0}^{\infty} f(P\Delta t)\, z^{-P} \\ &= z^{-N} F(z) \end{aligned} \tag{6.10}$$

Case 2: $\theta = (N + r)t$, where r is a fraction of the sampling interval

$$Z\{f(t-\theta)\}^* = \sum_{k=0}^{\infty} f(k\Delta t - N\Delta t - r\Delta t) z^{-k} \tag{6.11}$$

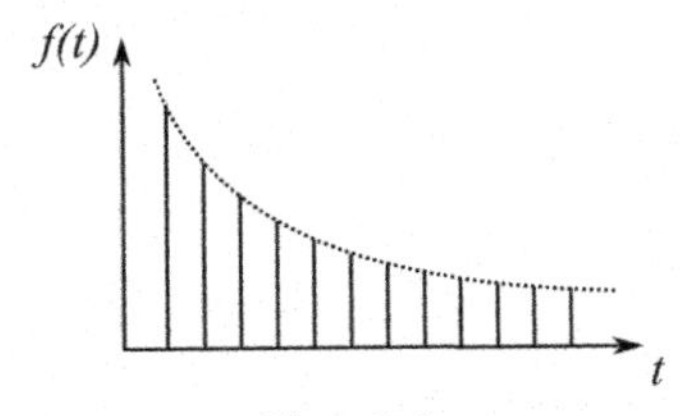

Fig. 6.4
A discrete exponential function.

Let $m=1-r$, Eq. (6.11) becomes

$$Z\{f(t-\theta)\}^* = \sum_{k=0}^{\infty} f(k\Delta t - N\Delta t - \Delta t + m\Delta t)z^{-k} \tag{6.12}$$

Let $P=k-N-1$,

$$\begin{aligned} Z\{f(t-\theta)\}^* &= \sum_{P=-N-1}^{\infty} f(P\Delta t + m\Delta t)z^{-N}z^{-P}z^{-1} \\ &= z^{-N}z^{-1}\sum_{P=0}^{\infty} f(P\Delta t + m\Delta t)z^{-P} \\ &= z^{-N}F(z,m) = z^{-N}F_m(z) \end{aligned} \tag{6.13}$$

Eq. (6.13) provides the definition of the modified z-transform of a discrete function:

$$F(z,m) = F_m(z) = z^{-1}\sum_{k=0}^{\infty} f(k\Delta t + m\Delta t)z^{-k} \tag{6.14}$$

Table 6.2 List of the z-transform and modified z-transform of a few simple functions

$f(t)$	$f(k\Delta t)$ or f_k	$F(z)$	$F_m(z)$
$\delta(t)$	$\delta(k\Delta t)$	1	z^{-1}
$\delta(t-\theta), \theta=k\Delta t$	$\delta(t-k\Delta t)$	z^{-k}	$z^{-(k+1)}$
$S(t)$	$S(k\Delta t)$	$\frac{1}{1-z^{-1}}$	$\frac{z^{-1}}{1-z^{-1}}$
e^{-at}	$e^{-ak\Delta t}$	$\frac{1}{1-e^{-a\Delta t}z^{-1}}$	$\frac{e^{-am\Delta t}z^{-1}}{1-e^{-a\Delta t}z^{-1}}$
t	$k\Delta t$	$\frac{\Delta t z^{-1}}{(1-z^{-1})^2}$	$\frac{m\Delta t(1-z^{-1})+\Delta t z^{-1}}{(1-z^{-1})^2}z^{-1}$

6.4 Some Useful Properties of the z-Transform

(a) Linearity

$$Z\{af^*(t)+bg^*(t)\} = aF(z)+bG(z) \tag{6.15}$$

(b) Initial value theorem

$$\lim_{k\to 0} f(k\Delta t) = \lim_{z\to\infty} F(z) \tag{6.16}$$

(c) Final value theorem

$$\lim_{k\to\infty} f(k\Delta t) = \lim_{z^{-1}\to 1} (1-z^{-1})F(z) \tag{6.17}$$

6.5 Inverse z-Transform

There are two methods to obtain the inverse of $F(z)$, which is the discrete function $f^*(t)$. The first method is exact and uses partial fraction similar to the inverse Laplace transform.

(1) Partial fraction. This method is tedious but provides the exact solution.

Example 6.1

Derive the inverse of $F(z)$

$$\begin{aligned} F(z) &= \frac{2z^{-1}}{1-1.2z^{-1}+0.2z^{-2}} \\ &= \frac{2z^{-1}}{(1-z^{-1})(1-0.2z^{-1})} \\ &= \frac{a}{1-z^{-1}} + \frac{b}{1-0.2z^{-1}} \end{aligned} \tag{6.18}$$

$$a = \lim_{z^{-1}\to 1} (1-z^{-1})f(z) = 2.5 \tag{6.19}$$

$$b = \lim_{z^{-1}\to 1/0.2} (1-0.2z^{-1})f(z) = -2.5 \tag{6.20}$$

$$\begin{aligned} f^*(t) = f(k\Delta t) &= 2.5S^*(t) - 2.5(0.2)^k \\ &= 2.5 - 2.5(0.2)^k \end{aligned} \tag{6.21}$$

where $S^*(t)$ is the discrete unit step function.

(2) The second method uses long division. This method is easier but leads to an approximate result.

Example 6.2

Derive the inverse of $F(z)$

$$\begin{aligned} F(z) &= \frac{1-0.3z^{-1}}{1-1.3z^{-1}+0.6z^{-2}} \\ &= \sum_{k=0}^{\infty} f(k\Delta t)z^{-k} = f(0) + f(\Delta t)z^{-1} + f(2\Delta t)z^{-2} + \cdots \end{aligned} \tag{6.22}$$

Long division:

$$
\begin{array}{r|l}
 & \underline{1 + z^{-1} + 0.7z^{-2} + \cdots} \\
 & 1 - 0.3z^{-1} \\
\left(1 - 1.3z^{-1} + 0.6z^{-2}\right) & \underline{1 - 1.3z^{-1} + 0.6z^{-2}} \qquad 0.7z^{-2} - 0.6z^{-3} \qquad \vdots \\
 & z^{-1} - 0.6z^{-2} \\
 & \underline{z^{-1} - 1.3z^{-2} + 0.6z^{-3}}
\end{array}
\tag{6.23}
$$

We get

$$F(z) = 1 + z^{-1} + 0.7z^{-2} + \ldots \tag{6.24}$$

Comparing Eq. (6.24) with the definition of z-transfer given in Eq. (6.5), we have:

$$f(0) = 1; f(\Delta t) = 1; f(2\Delta t) = 0.7; \ldots \tag{6.25}$$

6.6 Conversion of an Equation From the z-Domain to a Discrete Equation in the Time Domain

It is sometimes necessary to convert a discrete transfer function to a recursive difference equation in the time domain. An example is a digital controller equation that has to be converted from the z-domain to time domain before it is implemented.

Example 6.3

Given $D(z) = \dfrac{U(z)}{E(z)} = \dfrac{1 - 0.7z^{-1}}{1 - 1.2z^{-1} + 0.2z^{-2}}$, express u_k in terms of u_{k-1}, u_{k-2} e_k, e_{k-1}...

$$
\begin{aligned}
U(z)\left[1 - 1.2z^{-1} + 0.2z^{-2}\right] &= E(z)\left(1 - 0.7z^{-1}\right) \\
U(z) &= 1.2U(z)z^{-1} - 0.2U(z)z^{-2} + E(z) - 0.7E(z)z^{-1}
\end{aligned}
\tag{6.26}
$$

Eq. (6.26), in the time domain, results in:

$$u_k = 1.2u_{k-1} - 0.2u_{k-2} + e_k - 0.7e_{k-1} \tag{6.27}$$

6.7 Derivation of the Closed-Loop Transfer Function (CLTF) of a Digital Control System

In the closed-loop block diagram of a digitally controlled system shown in Fig. 6.5, three cases are distinguishable:

Case 1: A continuous process is disturbed by a continuous input, and its output is sampled (the continuous process $G_d(s)$ is disturbed by a continuous disturbance signal, $D(s)$,

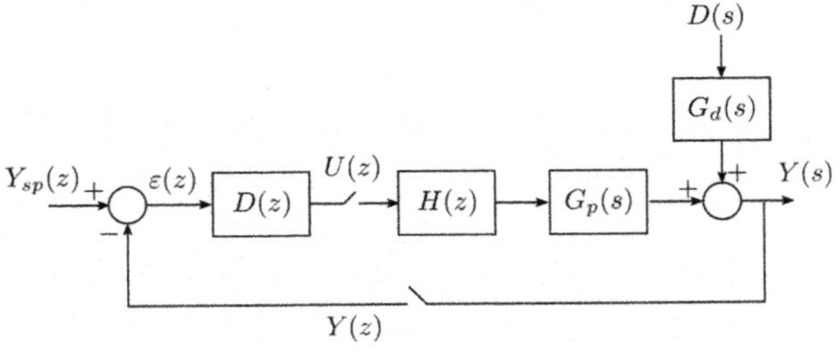

Fig. 6.5
The block diagram of a digital feedback control system.

shown in Fig. 6.5). This sampling operation is illustrated mathematically in the following example.

Example 6.4

Assuming $G_d(s) = \frac{0.6\,e^{-2.3s}}{2s+1}$; $D(s) = \frac{1}{s}$; $\Delta t = 1$ sec, derive an expression for $y^*(t)$

$$Y(z) = Z\{\mathcal{L}^{-1}[D(s)G_d(s)]\}^* = D\,G_d(z) \tag{6.28}$$

$$D(s)G_d(s) = \frac{1}{s}\frac{0.6e^{-2.3s}}{2s+1} \tag{6.29}$$

Note that the expression $DG_d(z)$ is the shorthand notation that involves a series of operations including the calculation of the inverse of the product of $D(s)G_d(s)$, sampling the product signal, and then taking the z-transform of the resulting discrete/sampled signal. Note that since the sampling interval is 1 s and the delay time is 2.3 s, 2 s of delay are expressed by z^{-2} that can be factored out as following. The remaining 0.3 *s* which is a fraction of the time delay results in the modified z-transformation.

$$\begin{aligned} Y(z) &= 0.6z^{-2}\,Z\left\{\mathcal{L}^{-1}\left[\frac{1}{s}\frac{0.6e^{-0.3s}}{2s+1}\right]\right\}^* \\ &= 0.6z^{-2}Z_m\left\{\mathcal{L}^{-1}\left[\frac{1}{s}\frac{1}{2s+1}\right]\right\}^* \text{ with } m = 1-0.3 = 0.7 \\ &= 0.6z^{-2}Z_m\left\{\mathcal{L}^{-1}\left[\frac{1}{s} + \frac{-2}{2s+1}\right]\right\}^* \\ &= 0.6z^{-2}\left[\frac{z^{-1}}{1-z^{-1}} - \frac{e^{-m\Delta t/2}z^{-1}}{1-e^{-\Delta t/2}z^{-1}}\right] \text{ with } \Delta t = 1 \\ &= 0.6z^{-3}\left[\frac{1}{1-z^{-1}} - \frac{0.705}{1-0.606z^{-1}}\right] \end{aligned} \tag{6.30}$$

$$y^*(t) = y(k\Delta t) = y_k = 0.6\left[S(k-3) - 0.705(0.606)^{k-3}\right] \tag{6.31}$$

Therefore $y_0 = 0$; $y_1 = 0$; $y_2 = 0$; $y_3 = 0.6[1-0.705]$, ...

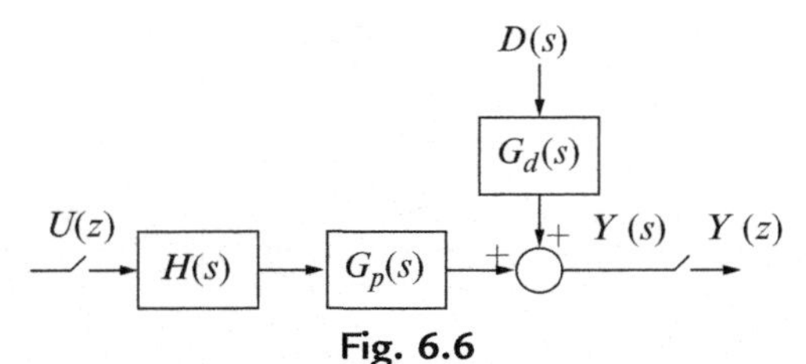

Fig. 6.6
The block diagram of a continuous system and a zero-order hold disturbed by a discrete input.

Case 2: A series of continuous systems disturbed by a discrete input signal and sampled at the output (refer to Fig. 6.6).where $H(s)$ is the transfer function of a zero-order hold element. $H(s)$ can be derived by assuming a unit impulse input that results in an output signal equal to 1 during one sampling interval.

$$H(s) = \frac{\text{Laplace}\{\text{output}\}}{\text{Laplace}\{\text{input}\}} = \frac{1/s - e^{-\Delta ts}/s}{1} = \frac{1-e^{-\Delta ts}}{s} \tag{6.32}$$

There are many methods to convert a plant transfer function from the s-domain to z-domain and obtain the pulse transfer function. One method is the use of a ZOH device.

$$Y(z) = Z\{\mathcal{L}^{-1}[H(s)G_p(s)]\}^* U(z) + Z\{\mathcal{L}^{-1}[D(s)G_d(s)]\}^* \tag{6.33}$$

The plant inputs consist of a discrete signal (controller output, $U(z)$) and a continuous signal (disturbance, $D(s)$) and the output of the plant is sampled before it is fed back to the controller. The derivation of the pulse transfer function of the process is discussed in the following example.

Example 6.5

Derive the pulse transfer functions of the following plant whose output is sampled every $\Delta t = 1$ sec

$$G_p(s) = \frac{0.3e^{-3s}}{4s+1};\ G_D(s) = \frac{0.6e^{-2s}}{2s+1};\ D(s) = \frac{1}{s} \tag{6.34}$$

$$\begin{aligned} Y(z) &= Z\left\{\mathcal{L}^{-1}\left[\frac{1-e^{-\Delta ts}}{s}\cdot\frac{0.3e^{-3s}}{4s+1}\right]\right\}^* U(z) + Z\left\{\mathcal{L}^{-1}\left[\frac{0.6s^{-2s}}{2s+1}\frac{1}{s}\right]\right\}^* \\ &= \left[0.3\left(1-z^{-1}\right)z^{-3}\right]Z\left\{\mathcal{L}^{-1}\left[\frac{1}{s}-\frac{1}{s+0.25}\right]\right\}^* U(z) + 0.6z^{-2}Z\left\{\mathcal{L}^{-1}\left[\frac{1}{s}-\frac{1}{s+0.5}\right]\right\} \\ &= 0.3\left(1-z^{-1}\right)z^{-3}\left[\frac{1}{1-z^{-1}}-\frac{1}{1-0.779z^{-1}}\right]U(z) + 0.6z^{-2}\left[\frac{1}{1-z^{-1}}-\frac{1}{1-0.606z^{-1}}\right] \\ &= \frac{0.066z^{-4}}{1-0.779z^{-1}}U(z) + \frac{0.236z^{-3}}{1-1.606z^{-1}+0.606z^{-2}} \end{aligned} \tag{6.35}$$

The shorthand notation for the previous series of operations is

$$Y(z) = HG_p(z)U(z) + DG_d(z) \tag{6.36}$$

where $HG_p(z)$ is called the *pulse transfer function* of the plant. The MATLAB command for this operation is defined by the *c2d*, which stands for continuous to discrete conversion. For example, for the system $G_p(s) = \frac{0.2e^{-5s}}{3s+1}$, the following commands will render the pulse transfer function of the process.

```
>> sys=tf(0.2,[3 1],'inputdelay',5);
>> sysd=c2d(sys,1,'zoh') % or even without 'zoh', as it is the default.
   Transfer function:
                    0.05669
          z^(-5) * ----------
                   z - 0.7165
>> step(sysd)
```

Case 3: A discrete process with a discrete input and a sampled output. The *discrete transform function* of a digital controller or a digital filter can often be expressed as

$$D(z) = \frac{U(z)}{\varepsilon(z)} = \frac{b_0 + b_1 z^{-1} + \ldots + b_m z^{-m}}{1 + a_1 z^{-1} + \ldots + a_n z^{-n}} \quad (6.37)$$

where the order of the controller/digital filter is given by m and n, and their parameters by $a_1, \ldots, a_n, b_0, \ldots, b_m$.

For a discrete system, the location of the poles and zeros can be determined by the *pzmap* command of MATLAB. For example, let us consider $D(z)$ given as follows, whose zeros (roots of the numerator polynomial) and poles (roots of the denominator polynomial) are all inside the Unit Circle on the z-plane:

$$D(z) = \frac{U(z)}{\varepsilon(z)} = \frac{0.3 - 0.2z^{-1}}{1 + 0.2z^{-1} + 0.6z^{-2}} \quad (6.38)$$

The following commands generate the pole-zero map of the system as shown in Fig. 6.7, which is the location of the system poles and zeros on the z-plane.

```
>> dz=tf([0.3 -0.2 0], [1 0.2 0.6],1,'variable','z^-1')
dz =

        0.3 - 0.2 z^-1
   ----------------------
   1 + 0.2 z^-1 + 0.6 z^-2

Sample time: 1 seconds
Discrete-time transfer function.
>> pzmap(dz)
```

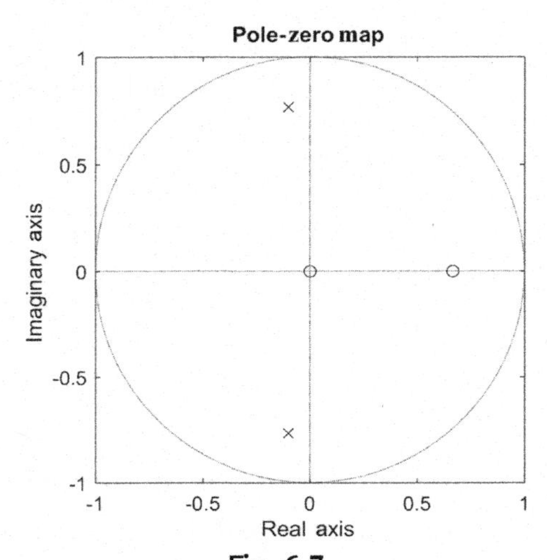

Fig. 6.7
Pole-zero map of a digital controller.

6.8 The Closed-Loop Pulse Transfer Function of a Digital Control System

Based on the discussion of the three distinct cases earlier and considering Fig. 6.6, one may write:

$$Y(z) = DG_d(z) + HG_p(z)U(z) \tag{6.39}$$

$$U(z) = D(z)E(z) + D(z)\left[Y_{sp}(z) - Y(z)\right] \tag{6.40}$$

Combining the previous equations and rearranging

$$Y(z) = \frac{D(z)HG_p(z)}{1 + D(z)HG_p(z)}Y_{sp}(z) + \frac{DG_d(z)}{1 + D(z)HG_p(z)} \tag{6.41}$$

Eq. (6.41) provides the closed-loop servo and regulatory transfer functions of a digitally controlled continuous process in the z-domain. Note that the terms $HG_p(z)$ and $DG_d(z)$ are shorthand notations of a series of mathematical operations as discussed earlier.

6.9 Selection of the Sampling Interval

There are several approaches to choose an appropriate sampling interval, Δt.

(a) The simplest approach is to choose the sampling interval as one-tenth of the dominant time constant of the process or the delay time, whichever is smaller.

$$\Delta t = 0.1 \times \begin{cases} \tau & \text{if } \theta > \tau \\ \theta & \text{if } \theta < \tau \end{cases} \tag{6.42}$$

(b) The second approach is to use a heuristic set of rules. The flow control loops are faster than the level and pressure loops and the slowest loops are temperature and concentration control loops. Therefore the choice of the sampling interval for the flow controllers is usually 1 s, for the level and pressure controllers is around 5 s, and for the temperature and concentration controllers is more than 20 s.

$$\begin{cases} \text{flow loop} & \Delta t = 1\,s \\ \text{level, pressure} & \Delta t = 5\,s \\ \text{Temp, concentration} & \Delta t > 20\,s \end{cases} \tag{6.43}$$

The third approach is to use Shannon's sampling theorem. This theorem sets a minimum limit for the sampling frequency $\omega = 2/\Delta t\ (\text{rad s}^{-1})$ or the sampling rate $f = 1/\Delta t\ (\text{s}^{-1})$, to at least twice the maximum frequency (ω_{max}) of the continuous signal, in order to allow separation of the signal from the high frequency noise.

6.10 Filtering

Most systems are corrupted with the process and sensor noise that have to be filtered before the signal is fed to the controller

- *Process noise*: This type of noise is caused by the medium to high frequency signals superimposed on the process variables due the presence of the turbulence and other disturbances in the process.
- *Sensor noise*: This type of noise consists of high frequency noise caused by the electrical motors, compressors, pumps, etc., in the vicinity of the sensors.

Noisy signals must be passed through filters to improve the signal-to-noise ratio. Filters are classified as analog filters and digital filters. Analog filters include passive filters and active filters. Digital filters are in the form of recursive equations.

A passive analog filter: An analog low-pass filter is a passive filter consisting of a resistor and a capacitor (an RC filter) arranged in the following manner (Fig. 6.8).

Where y and $\hat{y}$ are raw and filtered voltage signals. We have

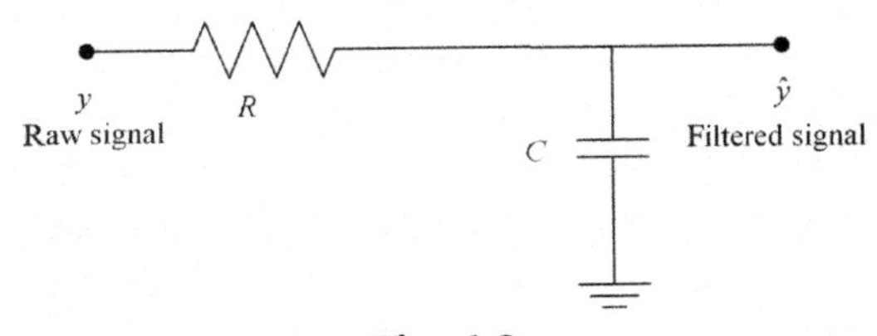

Fig. 6.8
Schematics of an RC filter.

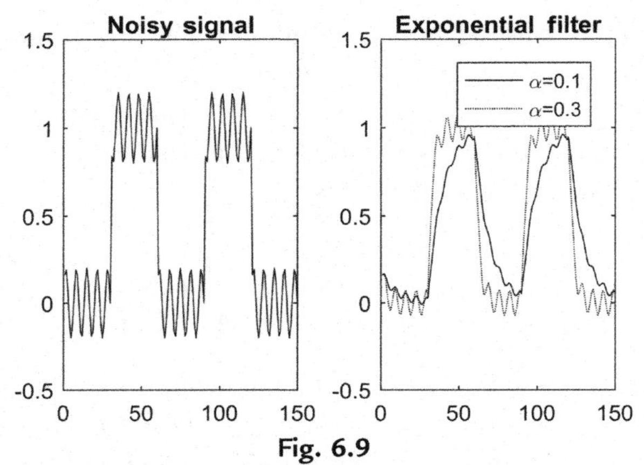

Fig. 6.9
A comparison of filter performance for additive sinusoidal noise.

$$\frac{y-\hat{y}}{R}=C\frac{d\hat{y}}{dt}\Rightarrow RC\frac{d\hat{y}}{dt}+\hat{y}=y \tag{6.44}$$

The product of the resistance and capacitance is the filter time constant that can be adjusted to change the cutting threshold of the filter, $RC=\tau_F$.

Digital filters: There are many types of digital filters. The simplest one is the first order or exponential filter which is similar to an RC analog filter and it can be derived by discretizing the equation of the equivalent analog filter given in Eq. (6.44).

$$\tau_F\frac{\hat{y}_k-\hat{y}_{k-1}}{\Delta t}+\hat{y}_{k-1}=y_k \;\Rightarrow\; \hat{y}_k=\alpha y_k+(1-\alpha)\hat{y}_{k-1} \tag{6.45}$$

where $\alpha=\frac{\Delta t}{\tau_F}$, $\alpha=\begin{cases}0 & \text{complete filtering or blocking of the noisy signal}\\ 1 & \text{no filtering}\end{cases}$

The discrete transfer function of a first-order digital filter is given in Eq. (6.46). Fig. 6.9 compares the performance of an exponential digital filter with different filter constants, α, to remove the noise from a sinusoidal noisy signal.

$$F(z)=\frac{\hat{Y}(z)}{Y(z)}=\frac{\alpha}{1-(1-\alpha)z^{-1}} \tag{6.46}$$

6.11 Mapping Between the s-Plane and the z-Plane

Recall that the relationship between the z and the s operators is described by $z=e^{\Delta ts}$. The z operator is a complex number with a magnitude $|z|$, which represents the distance from the origin, and an argument, θ, which is the angle it makes with the real axis on the z-plane. Similarly, s is a complex operator expressed in terms of its real and imaginary parts by $s=(\alpha+j\beta)$.

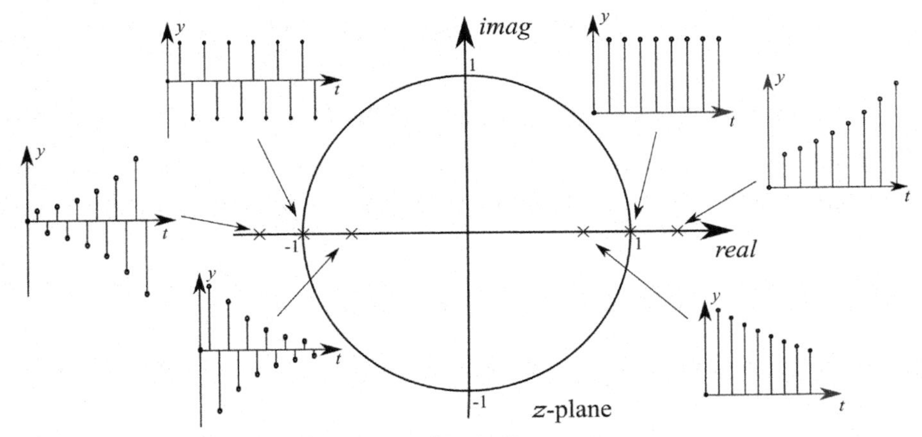

Fig. 6.10
Time-domain responses of a first-order lag transfer function with different locations of the pole, indicated by an x, to a unit impulse input at $k=0$.

$$z = e^{\Delta ts} = e^{(\alpha + j\beta)\Delta t} = e^{\alpha \Delta t} e^{j\beta \Delta t} = |z| e^{j\theta} \tag{6.47}$$

Therefore the magnitude of z is related to the real part of the s operator by $|z| = e^{\alpha \Delta t}$. Note that if α is positive the system is unstable in the s plane, if α is negative the system is stable and for a value of $\alpha = 0$, the system poles are pure imaginary and the system is marginally stable. Therefore the entire stable region of the s-plane maps onto the interior of a unit circle (UC) on the z-plane. Fig. 6.10 shows the region of stability and the expected unit impulse response of a first-order system with the designated pole locations on the z-plane.

$$|z| = e^{\alpha \Delta t} = \begin{cases} > 1 & \text{if } \alpha > 0 \quad \text{unstable} \\ = 1 & \text{if } \alpha = 0 \quad \text{marginally stable} \\ < 1 & \text{if } \alpha < 0 \quad \text{stable region} \end{cases} \tag{6.48}$$

6.11.1 The Concept of Stability in the z-Plane

Mapping from the s-plane to z-plane shows that a system, whether open loop or closed loop, is stable if the roots of its characteristic equation remain inside the unit circle (UC). This is further qualified by looking at a general servo discrete transfer function of a closed-loop system:

$$\frac{Y(z)}{Y_{sp}(z)} = \frac{D(z)HG_p(z)}{1 + D(z)HG_p(z)} = G_{CL}(z) \tag{6.49}$$

Which is, in general, the ratio of two polynomials, $Q(z)$ and $P(z)$

$$G_{CL}(z) = \frac{Q(z)}{P(z)} = \frac{a_1}{1 - p_1 z^{-1}} + \frac{a_2}{1 - p_2 z^{-1}} + \ldots + \frac{a_n}{1 - p_n z^{-1}} \tag{6.50}$$

where p_1, p_2, ..., p_n are poles that could be real or complex numbers. The output at the nth interval is

$$y_n = y(n\Delta t) = a_1(p_1)^n + a_2(p_2)^n + \cdots + a_n(p_n)^n + \text{contribution from the inputs} \quad (6.51)$$

If any of the system poles has a magnitude larger than 1 (lying outside the UC) the system will be unstable. Similar to the s-plane, there are a number of stability tests in the z-plane.

6.11.2 Routh-Hurwitz and Bilinear Transformation Test

The Routh test can be applied in the z-domain following a bilinear transformation.

$$\text{Let } z^{-1} = \frac{1-w}{1+w}, \text{ then } w = \frac{z-1}{z+1} \quad (6.52)$$

Note that the w-plane is the same as the s-plane.

Example 6.6

The closed-loop characteristics equation ($CLCE$) of a control system using a P controller is given as follows:

$$CLCE(z) = (1 - z^{-1})(1 - 0.779z^{-1}) + 0.199z^{-1}K_c(1.2 - z^{-1}) = 0 \quad (6.53)$$

If $K_c = 2$, roots of the $CLCE$ are at $p_1 = 0.856$ and $p_2 = 0.449$, which means the system is stable.

If $K_c = 9$, roots of the $CLCE$ are $p_1 = -1.21$ and $p_2 = 0.838$, which means the system is unstable.

Replacing z^{-1} by $\frac{1-w}{1+w}$ in the $CLCE$ results in:

$$CLCE(w) = \left(1 - \frac{1-w}{1+w}\right)\left(1 - 0.779\frac{1-w}{1+w}\right) + 0.199\left(\frac{1-w}{1+w}\right)K_c\left(1.2 - \frac{1-w}{1+w}\right) = 0 \quad (6.54)$$

The first condition of Routh test results in: $3.557 - 0.458K_c > 0 \Rightarrow K_c < 8.12$

The second condition of Routh test requires forming three arrays:

$$n = 2; n+1 = 3\ rows: \begin{cases} 3.557 - 0.458K_c & 0.039K_c \\ 0.442 + 0.398K_c & 0 \\ 0.039K_c & \end{cases} \quad (6.55)$$

All the element in the first column must be positive which results in $K_c < 8.12$. Therefore the maximum controller gain can be obtained as $K_{c,max} = 8.12$.

6.11.3 Jury's Stability Test in the z-Domain

This test can be directly applied in the z-plane to an open-loop characteristic equation ($OLCE$) or a closed-loop characteristic equation.

$$OLCE(z) \text{ or } CLCE(z) = a_n z^n + a_{n-1} z^{n-1} + \ldots + a_0 \quad (6.56)$$

Form the $(2n-3)$ Jury's rows:

$$rows\begin{cases} a_0 & a_1 & \cdots & a_n \\ a_n & a_{n-1} & \cdots & a_0 \\ b_0 & b_1 & \cdots & b_{n-1} \\ b_{n-1} & b_{n-2} & \cdots & b_0 \\ c_0 & c_1 & \cdots & c_{n-2} \\ c_{n-2} & c_{n-1} & \cdots & c_0 \\ \vdots & & & \end{cases} \quad (6.57)$$

Where $b_k = \begin{vmatrix} a_0 & a_{n-k} \\ a_n & a_k \end{vmatrix} = a_0a_k - a_na_{n-k}$, and $c_k = \begin{vmatrix} b_0 & b_{n-k-1} \\ b_{n-1} & b_k \end{vmatrix} = b_0b_k - b_nb_{n-k-1}$

For a system to be stable, all of the following conditions should be met. If one of them is violated, then the system is unstable.

(1) $CLCE(z=1) > 0$

(2) $(-1)^n CLCE(z=-1) > 0$

(3) $|a_0| < a_n$

(4) $|b_0| > |b_{n-1}|$

(5) $|c_0| > |c_{n-2}|$

Example 6.7

Determine the stability of the system having a *CLCE* given by:

$$CLCE(z) = z^2 + (0.239K_c - 1.779)z + (0.779 - 0.199K_c) = 0 \quad (6.58)$$

Note that $n=2$, therefore there will be $2n-3=1$ row:

$$(0.779 - 0.199K_c) \ (0.239K_c - 1.779) \ 1 \quad (6.59)$$

Check Jury's test's first condition:

$$\begin{aligned} CLCE(z=1) &= 1 + 0.238K_c - 1.779 + 0.779 - 0.199K_c > 0 \\ &\Rightarrow K_c > 0 \end{aligned} \quad (6.60)$$

Check the second condition:

$$\begin{aligned} (-1)^n CLCE(z=-1) &= 3.557 - 0.438K_c > 0 \\ &\Rightarrow K_c < 8.12 \end{aligned} \quad (6.61)$$

Check the third condition:

$$\begin{aligned} |0.779 - 0.199K_c| < 1 &\Rightarrow (0.779 - 0.199K_c)^2 < 1 \\ &\Rightarrow -1.104 < K_c < 8.93 \end{aligned} \quad (6.62)$$

So, when $0 < K_c < 8.12$, the system is stable.

6.12 Design of Digital Feedback Controllers for SISO Plants

PID Controllers: The controller output is proportional to the magnitude, duration, and time rate of change of the error signal. In the continuous time

$$u(t) = \bar{u} + K_c\varepsilon(t) + \frac{K_c}{\tau_I}\int_0^t \varepsilon(t)dt + K_c\tau_D\frac{d\varepsilon(t)}{dt} \tag{6.63}$$

where $\bar{u}$ is the controller output when $\varepsilon(t) = 0$, K_c, τ_I, and τ_D are the controller proportional gain, integral or reset time, and the derivative or rate time, respectively, and $\varepsilon(t) = y_{sp} - y(t)$ is the error signal. Using a finite difference approximation for the integral and first-order backward difference for the derivative term, we get:

$$u_n = \bar{u} + K_c\varepsilon_n + \frac{K_c\Delta t}{\tau_I}\sum_{i=0}^{n}\varepsilon_i + K_c\tau_D\frac{\varepsilon_n - \varepsilon_{n-1}}{\Delta t} \tag{6.64}$$

which is referred to as the *position form* of the PID controller. The recursive or *velocity form* of the PID controller is obtained by subtracting the controller output at the $(n-1)^{st}$ sampling interval from *un*.

$$u_n - u_{n-1} = K_c(\varepsilon_n - \varepsilon_{n-1}) + \frac{K_c\Delta t}{\tau_I}\varepsilon_n + \frac{K_c\tau_D}{\Delta t}(\varepsilon_n - 2\varepsilon_{n-1} + \varepsilon_{n-2}) \tag{6.65}$$

z-transformation of the previous equation results in the discrete transfer function of an ideal PID controller.

$$D_{PID}(z) = \frac{U(z)}{E(z)} = \frac{K_c\left(1 + \frac{\Delta t}{\tau_I} + \frac{\tau_D}{\Delta t}\right) - K_c\left(1 + \frac{2\tau_D}{\Delta t}\right)z^{-1} + \frac{K_c\tau_D}{\Delta t}z^{-2}}{1 - z^{-1}} \tag{6.66}$$

It should be noted that the term $(1 - z^{-1})$ in the denominator represents integral action which eliminates offset or the steady-state error.

Tuning of the Discrete PID controller: The discrete PID controller has 4 adjustable parameters: K_c, τ_I, τ_D, and Δt. There are several general guidelines as well as Shannon's sampling theorem for the selection of Δt. In addition, there are various tuning procedures to select K_c, τ_I, and τ_D. Some of these are open-loop procedures which require determination of the open-loop transfer function of the plant and its approximation as a first order plus time delay model. Tables have been constructed to suggest parameter settings for the discrete PID controllers using various performance criteria such as the 1/4 decay ratio, minimization of the integral of absolute error (IAE), and 5% overshoot.[1] In the closed-loop tuning[2] procedures such as the continuous cycling method due to Ziegler Nichols, the process is driven to the verge of instability under a P controller and the cycling period and the ultimate controller gain is obtained experimentally. Various tuning correlations are used to determine the controller parameters from the measured period of oscillation and the ultimate controller gain.

Example 6.8

Obtain the closed-loop response of a plant given by:

$$G_p(s) = \frac{0.3e^{-3s}}{4s+1} \tag{6.67}$$

To a unit step change in the controller set point using a discrete PI controller given by:

$$D_{PI}(z) = \frac{3.724 - 3.43z^{-1}}{1 - z^{-1}} \tag{6.68}$$

The closed-loop discrete transfer function is

$$Y(z) = \frac{D(z)HG_p(z)}{1 + D(z)\,HG_p(z)} Y_{sp}(z) \tag{6.69}$$

The closed-loop simulation can be performed easily in MATLAB or Simulink.

```
Gp=tf(0.3,[4 1],'inputdelay',3);
HGp=c2d(Gp,1);
D_PI=tf([3.724 -3.43], [1 -1],1,'variable','z^-1');
gcl=D_PI*HGp/(1+D_PI*HGp);
step(gcl)
```

The response is shown in Fig. 6.13.

Simulink

The process transfer function can be entered in the *s*-domain preceded by a zero-order hold element (Fig. 6.11), or discretized and the pulse transfer function be entered (Fig. 6.12). In both cases the sampling interval should be specified in all the blocks used, including the "Step" block.

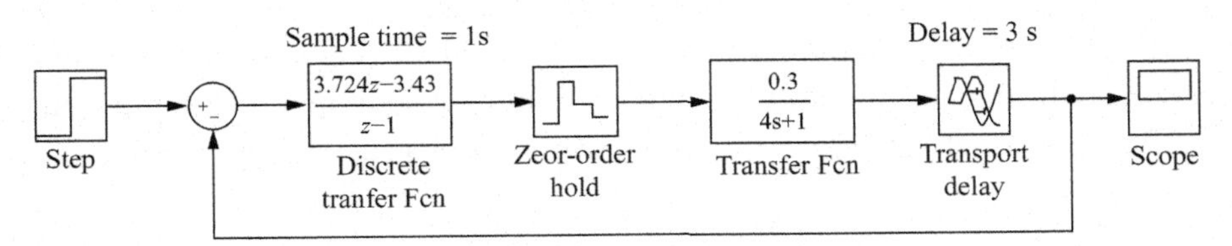

Fig. 6.11
Simulink method 1.

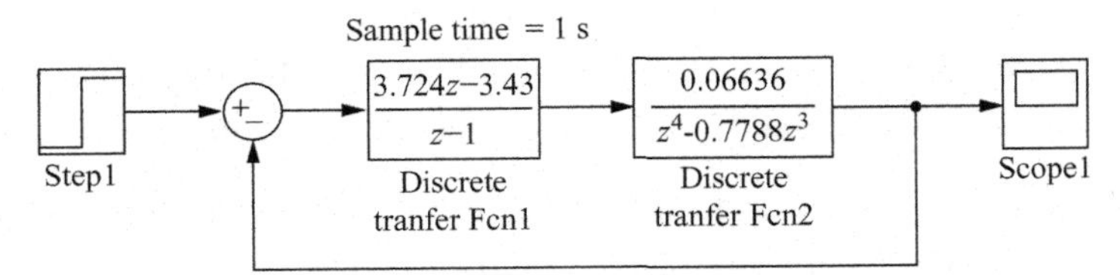

Fig. 6.12
Simulink method 2.

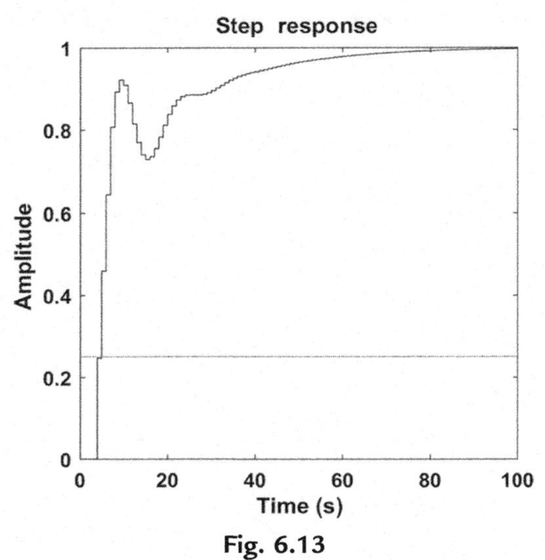

Fig. 6.13
Simulation with a discrete PI controller.

The simulation results in all three cases should be more or less the same and is shown in Fig. 6.13.

6.13 Design of Model-Based SISO Digital Controllers

Model-based digital controllers have generally a discrete transfer function of the form:

$$D(z)=\frac{U(z)}{E(z)}=\frac{a_0+a_1z^{-1}+\ldots+a_mz^{-m}}{1+b_1z^{-1}+\ldots+b_nz^{-n}} \tag{6.70}$$

where a_0, ..., a_m, $b_1,\ldots,b_n$ are the controller parameters and m and n are the controller orders which determine the structure of the controller. The roots of the polynomial in the numerator of $D(z)$ are called controller zeros and those in the denominator polynomial are referred to as the controller poles. Note that the PID controllers have a fixed structure $(m=2$ and $n=1)$ but adjustable parameters.

The structure of the majority of the model-based controllers depends on the inverse of the plant model. In the design of model-based controllers some general considerations regarding the process model must be borne in mind.

Nonminimum phase (NMP) models: If the plant model has a NMP structure with a zero outside the unit circle (UC), the controller, and consequently the closed-loop system becomes unstable.

Ringing poles: If the poles of the controller lie inside the left half of the UC, the controller output will exhibit intersample rippling effect. Such poles should be removed while their steady-state gain could be retained to avoid the closed-loop offset.

Plant–model mismatch: Perfect models are nonexistent. The performance of a model-based controller deteriorates as the mismatch between the actual plant and its model increases.

Constraints on inputs: There are always constraints on the controller outputs or their rate of change.

6.13.1 The Deadbeat Controllers (DB)

The deadbeat controller is a demanding controller and requires the controlled variable reach the set point in the minimum number of sampling intervals. Consider the closed-loop servo discrete transfer function of a closed-loop system. The equation of the deadbeat controller can be derived from the process model and the desired closed-loop transfer function.

$$\frac{Y(z)}{Y_{sp}(z)} = \frac{D(z)HG_p(z)}{1+D(z)HG_p(z)} \tag{6.71}$$

$$D_{DB}(z) = \frac{1}{\widetilde{HG_p}(z)} \frac{(Y/Y_{sp})_{\text{desired}}}{1-(Y/Y_{sp})_{\text{desired}}} \tag{6.72}$$

where $\widetilde{HG_p}(z)$ is the plant model and $(Y/Y_{sp})_{\text{desired}}$ is the desired closed-loop transfer function specified as

$$(Y/Y_{sp})_{\text{desired}} = z^{-(\widetilde{N}+1)} \tag{6.73}$$

for a plant model with $\widetilde{N}$ sampling intervals of delay. The additional one interval of delay is due to the zero-hold element.

Example 6.9

Design a DB controller for a plant given by the following model. Assume no plant–model mismatch.

$$\widetilde{HG_p}(z) = HG_p(z) = \frac{0.066z^{-4}}{1-0.779z^{-1}} \tag{6.74}$$

$$D_{DB}(z) = \frac{1-0.779z^{-1}}{0.066} \frac{z^{-1}}{1-z^{-5}} \tag{6.75}$$

```
z = tf('z', 1);
N = 4;
HGp= 0.066* z^-4/(1-0.779*z^-1);
D = 1/HGp *  z^-(N+1)/(1 - z^-(N+1));
G = feedback(HGp*D, 1);
step(G);
```

The closed-loop response to a unit step change in the set point is shown in Fig. 6.14 in the absence of plant–model mismatch.

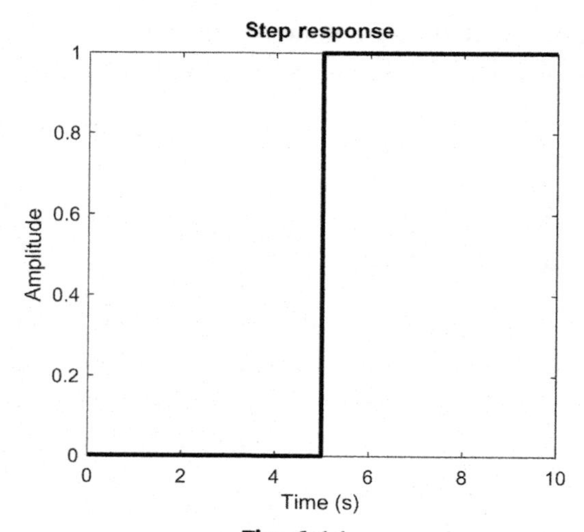

Fig. 6.14
Deadbeat controller step response $(\widetilde{N}=4)$.

Let us consider various cases of the previous example:

Case 1. The controller would be unstable if the plant model had a zero outside the UC, for example $(0.15-0.216z^{-1})$.

$$\widetilde{HG_p}(z)=\frac{(0.15-0.216z^{-1})z^{-4}}{1-0.779z^{-1}} \tag{6.76}$$

The controller is unstable as the model zero becomes the controller pole.

$$D_{DB}(z)=\frac{1-0.779z^{-1}}{0.15-0.216z^{-1}}\frac{z^{-1}}{1-z^{-5}} \tag{6.77}$$

Case 2. If the controller pole lies inside the left half of the UC, the controller will have a ringing pole. In order to avoid excessive controller action, the ringing pole has to be removed but its steady-state gain is retained to remove the offset. Consider the following example for the case where there is no plant–model mismatch:

$$\widetilde{HG_p}(z)=HG_p(z)=\frac{(0.035+0.031z^{-1})z^{-4}}{1-0.779z^{-1}} \tag{6.78}$$

$$D_{DB}(z)=\frac{1-0.779z^{-1}}{0.035+0.031z^{-1}}\frac{z^{-1}}{1-z^{-5}} \tag{6.79}$$

The closed-loop response to a unit step change in the set point with the DB controller with a ringing pole is depicted in Fig. 6.15. In order to reduce the excessive ringing of the controller output, the ringing pole is removed in the modified controller while its steady-state gain is retained. Fig. 6.15 was generated on Simulink, to obtain u in Matlab, lsim has to be used:

```
>>[y,t] = step(gcl)
>>Lsim(D,(1 - y),t) % D is the digital deadbeat controller, (1 - y) represents the error
signal and t is the time.
```

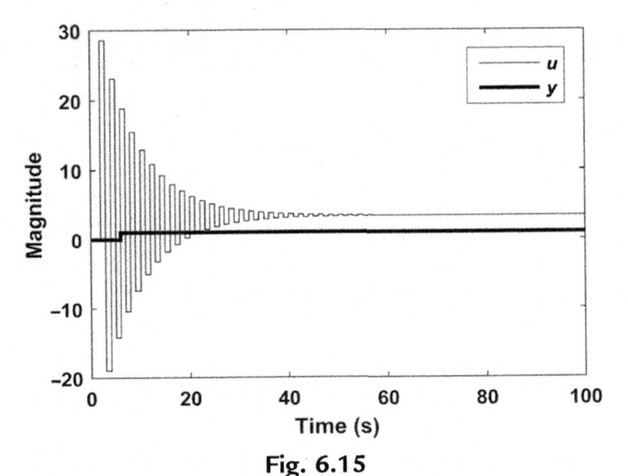

Fig. 6.15
Simulation with a DB controller with a ringing pole.

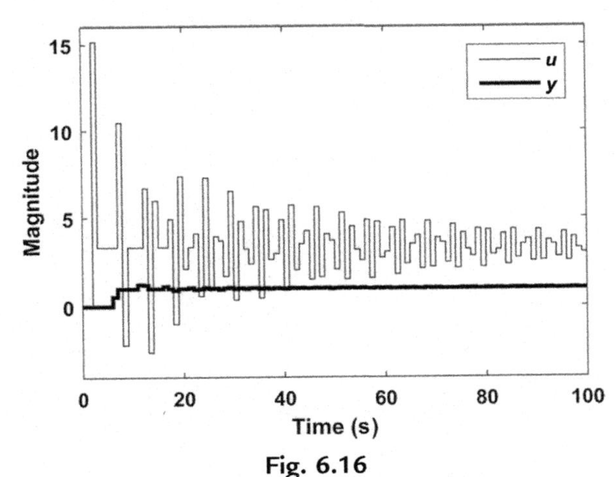

Fig. 6.16
Simulation with a modified DB controller.

$$D_{DB}(z) = \frac{1 - 0.779z^{-1}}{0.066} \frac{z^{-1}}{1 - z^{-5}} \tag{6.80}$$

The response with decreased amplitude and frequency of ringing with the modified controller after removing the ringing pole is shown in Fig. 6.16.

Case 3. Finally, if there is a plant–model mismatch, the controller performance deteriorates. Let us assume that the actual plant's dynamics is given by the following transfer function:

$$G_p(s) = \frac{0.6\, e^{-3s}}{(s+1)(2s+1)(3s+1)} \tag{6.81}$$

And use a first order plus time delay transfer function as the approximate model

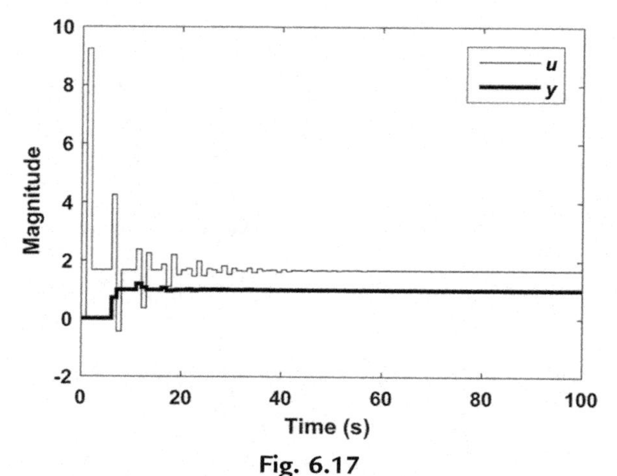

Fig. 6.17
Simulation with a DB controller in the presence of plant-model mismatch.

$$\widetilde{G_p}(s) = \frac{0.6\,e^{-4.3s}}{(5s+1)} \tag{6.82}$$

The pulse transfer function of the plant is obtained using

```
Gp=tf(0.6,[5 1],'inputdelay',4.3);
HGp=c2d(Gp,1)
```

The closed-loop response is shown in Fig. 6.17.

$$H\widetilde{G_p}(z) = \frac{0.078(1+0.387z^{-1})z^{-5}}{1-0.819z^{-1}} \tag{6.83}$$

The deadbeat controller is:

$$D_{DB}(z) = \frac{1-0.819z^{-1}}{0.078(1.387)}\frac{1}{1-z^{-5}} \tag{6.84}$$

The previous cases discussed for the deadbeat controller may happen with any model-based controller and similar conclusions can be drawn.

6.13.2 The Dahlin Controller

In the case of a Dahlin controller, the closed-loop system is to behave as a first-order system with a steady-state gain of 1 and a delay time at least equal to the delay time of the open-loop plant. The time constant of the desired closed-loop transfer function, λ, is used as the tuning parameter of the controller.

$$\left[\frac{Y(s)}{Y_{sp}(s)}\right]_{\text{desired}} = \frac{1\,e^{-\theta s}}{\lambda s + 1} \tag{6.85}$$

The desired servo closed-loop transfer function in the z-domain is:

$$\left[\frac{Y(s)}{Y_{sp}(s)}\right]_{\text{desired}} = \frac{1 - e^{-\Delta t/\lambda}}{1 - e^{-\Delta t/\lambda} z^{-1}} z^{-(N+1)} \tag{6.86}$$

where $\theta = N\Delta t$ is at least as large as plant delay and λ is the desired closed-loop time constant.

$$D_{Dah}(z) = \frac{1}{\widetilde{HG_p}(z)} \frac{\left(1 - e^{-\Delta t/\lambda}\right) z^{-(N+1)}}{1 - e^{-\Delta t/\lambda} z^{-1} - \left(1 - e^{-\Delta t/\lambda}\right) z^{-(N+1)}} \tag{6.87}$$

Note that the Dahlin controller should result in a zero offset due to the unit gain employed in the desired closed-loop transfer function.

Example 6.10

Design a Dahlin controller for a process with a perfect model given by:

$$HG_p(z) = \frac{0.066z^{-4}}{1 - 0.779z^{-1}} \tag{6.88}$$

Using a sampling interval of 1 s and $\lambda = 0.1$, 1 and 10.

$$\left[\frac{Y(s)}{Y_{sp}(s)}\right]_{\text{desired}} = \frac{\left(1 - e^{-1/\lambda}\right) z^{-4}}{1 - e^{-1/\lambda} z^{-1}} \tag{6.89}$$

For $\lambda = 0.1$, the Dahlin controller approaches the deadbeat controller.

$$D_{Dah,0.1}(z) = \frac{15.15 - 11.80z^{-1}}{1 - z^{-4}} \tag{6.90}$$

For $\lambda = 1$

$$D_{Dah,1}(z) = \frac{9.58 - 7.46z^{-1}}{1 - 0.37z^{-1} - 0.63z^{-4}} \tag{6.91}$$

For $\lambda = 10$, the closed-loop response is slow.

$$D_{Dah,10}(z) = \frac{1.44 - 1.121z^{-1}}{1 - 0.90z^{-1} - 0.095z^{-4}} \tag{6.92}$$

```
z = tf('z', 1);
HGp = 0.066*z^-4/(1-0.779*z^-1);
N=3;

Dah = @(lambda) (1/HGp*(1-exp(-1/lambda))*z^-(N+1)...
/(1-exp(-1/lambda)*z^-1 -(1-exp(-1/lambda))*z^-(N+1)) );
```

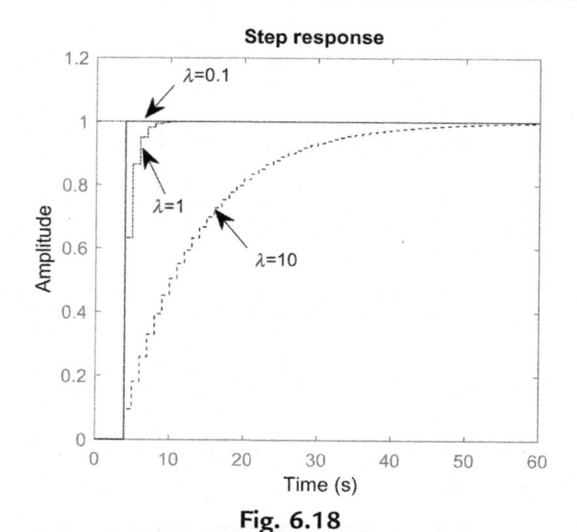

Fig. 6.18
Closed-loop simulation result of Dahlin controller.

```
hold on
step(feedback(Dah(0.1)*HGp,1));
step(feedback(Dah(1)*HGp,1));
step(feedback(Dah(10)*HGp,1));
```

The closed-loop response for different values of the controller tuning parameter, λ, is shown in Fig. 6.18.

6.13.3 The Smith Predictor[3]

The majority of chemical processes possess significant time delays. The dead time or time delay affects system stability and controller performance. The Smith predictor combines a conventional controller such as a PID controller with a dead time compensator to improve the overall closed-loop performance. Consider the block diagrams shown in Fig. 6.19. The plant dynamics is represented by a nondelayed portion multiplied by a delay time $HG_p(z)=G(z)\,z^{-N}$ whereas the model is represented by $\widetilde{HG_p}(z)=\widetilde{G}(z)z^{-\widetilde{N}}$.

Note that if $\widetilde{HG_p}(z)=HG_p(z)$, that is for a perfect model, the signal $Y_2(z)$ shown in Fig. 6.19 is zero and $Y_3(z)=Y_1(z)$ is a delay-free signal that is fed back to the controller. In the presence of plant–model mismatch, however, complete dead time compensation is not feasible, nevertheless an improvement in closed-loop behavior is expected. Note that the two block diagrams shown in Fig. 6.19 are identical and therefore the Smith predictor can be given by:

$$D_{Smith}(z)=\frac{U(z)}{E(z)}=\frac{D_{PID}(z)}{1+D_{PID}(z)\widetilde{G}(z)\left[1-z^{-\widetilde{N}}\right]} \tag{6.93}$$

where the PID controller, $D_{PID}(z)$, has to be properly tuned.

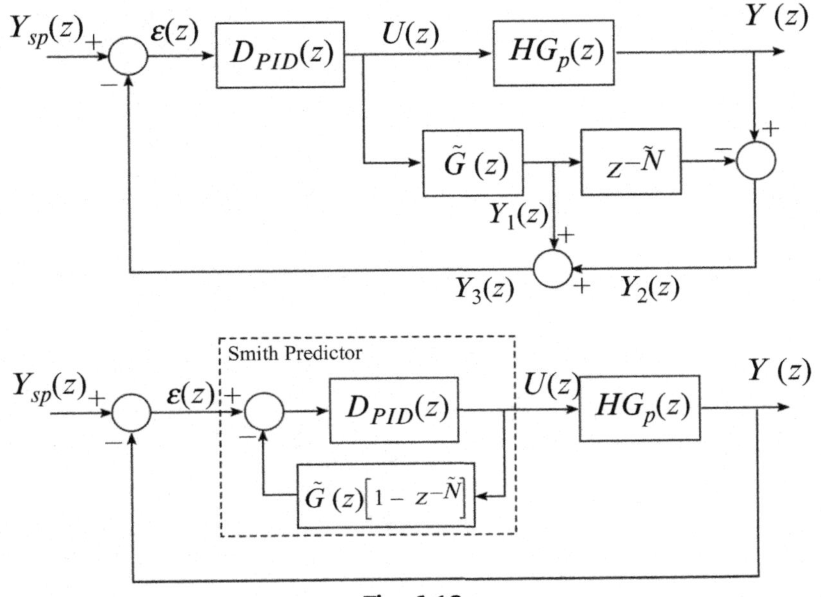

Fig. 6.19
Block diagram of the Smith Predictor.

Example 6.11

Derive the Smith predictor for a plant (Eq. 6.94) whose model is given by the transfer function in Eq. (6.95). Use a sampling interval of 1 s, a P controller with $K_c = 1$, and a PI controller with $K_c = 1$ and $\tau_I = 10$s.

$$G_p(s) = \frac{0.6e^{-3s}}{(s+1)(2s+1)(3s+1)} \tag{6.94}$$

$$\widetilde{G_p}(s) = \frac{0.6e^{-4.3s}}{5s+1}; \; \widetilde{HG_p}(z) = \frac{0.078z^{-1} + 0.03z^{-2}}{1 - 0.819z^{-1}} z^{-4} \tag{6.95}$$

$$D_{Smith}(z) = \frac{U(z)}{E(z)} = \frac{D_{PID}(z)}{1 + D_{PID}(z)\widetilde{G}(z)[1 - z^{-4}]} \tag{6.96}$$

Note that $\widetilde{G}(z) = \dfrac{0.078z^{-1} + 0.03z^{-2}}{1 - 0.819z^{-1}}$.

With a proportional controller , $D_{PID}(z) = 0.2$, the Smith predictor is:

$$D_{Smith}(z) = \frac{0.2(1 - 0.819z^{-1})}{1 - 0.819z^{-1} + (0.2)(0.078z^{-1} + 0.03z^{-2})(1 - z^{-4})} \tag{6.97}$$

With a proportional controller , $D_{PID}(z) = 1$, the Smith predictor is:

$$D_{Smith}(z) = \frac{1 - 0.819z^{-1}}{1 - 0.741z^{-1} + 0.03z^{-2} - 0.078z^{-5} - 0.03z^{-6}} \tag{6.98}$$

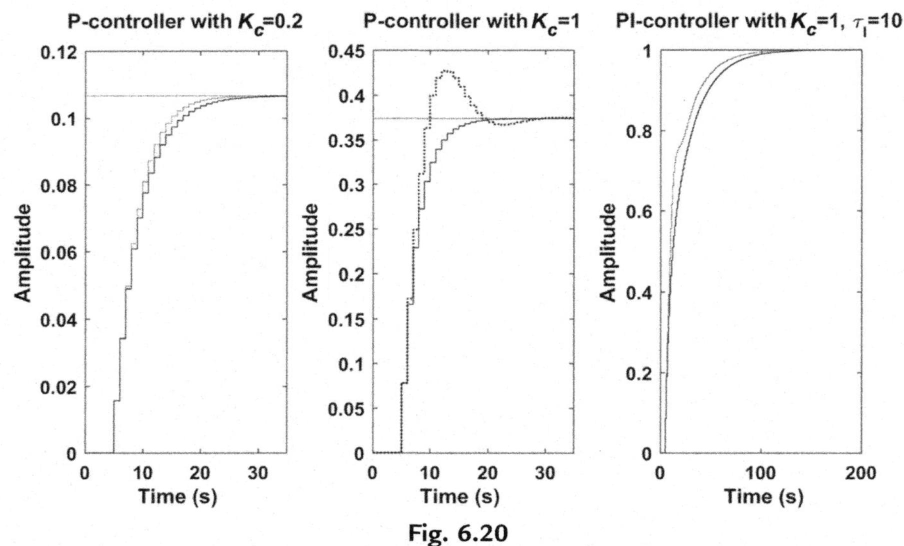

Fig. 6.20
Simulation with the Smith Predictor *(solid curve)* and without the Smith Predictor *(dotted curve)*.

With a PI controller , $D_{PID}(z) = \dfrac{1.1 - z^{-1}}{1 - z^{-1}}$

$$D_{Smith}(z) = \frac{1.013 - 1.749z^{-1} + 0.754z^{-2}}{1 - 1.732z^{-1} + 0.773z^{-2} - 0.086z^{-5} - 0.045z^{-6} + 0.03z^{-7}} \tag{6.99}$$

The closed-loop responses to a unit step change in the controller set point for each controller with and without the Smith predictor are shown in Fig. 6.20. The superiority of the Smith predictor in each case is evidenced by a faster response.

```
z = tf('z', 1);

HGp = (0.078*z^-1 + 0.03*z^-2)/(1-0.819*z^-1)*z^-4;
Gz = (0.078*z^-1 + 0.03*z^-2)/(1-0.819*z^-1);

subplot(1,3,1);
DPID =  (0.2);
Dsmith = DPID/(1 + DPID*Gz*(1-z^-4));
hold on
step(feedback(Dsmith*HGp,1));
step(feedback(DPID*HGp,1));
title('P-controller with K_c = 0.2')

subplot(1,3,2);
DPID =  (1);
Dsmith = DPID/(1 + DPID*Gz*(1-z^-4));
hold on
step(feedback(Dsmith*HGp,1));
step(feedback(DPID*HGp,1));
title('P-controller with K_c = 1')
subplot(1,3,3);
DPID = (1.1-z^-1)/(1-z^-1);
```

```
Dsmith=DPID/(1+DPID*Gz*(1-z^-4));
hold on
step(feedback(Dsmith*HGp,1));
step(feedback(DPID*HGp,1));
title('PI-controller with K_c=1, \tau_I=10')
```

6.13.4 The Kalman Controller[3]

The Kalman controller may be derived noting that the discrete transfer function of a digital controller can be arranged in the form of $\frac{P(z)}{1-Q(z)}$.

$$D(z)=\frac{P(z)}{1-Q(z)}=\frac{u(z)}{\varepsilon(z)}=\frac{u(z)}{y_{sp}(z)-y(z)}=\frac{u(z)/y_{sp}(z)}{1-y(z)/y_{sp}(z)} \tag{6.100}$$

where $P(z)$and $Q(z)$ are to be determined, and $u(z)$ and $\varepsilon(z)$ are the controller output and the error signal, respectively. Based on Eq. (6.100), $u(z)/y_{sp}(z)$ can be taken as $P(z)$, and $y(z)/y_{sp}(z)$as $Q(z)$. Dividing $y(z)/y_{sp}(z)$ by $u(z)/y_{sp}(z)$ yields $\frac{y(z)/y_{sp}(z)}{u(z)/y_{sp}(z)}=\frac{y(z)}{u(z)}=\frac{Q(z)}{P(z)}=H\tilde{G}_p(z)$, which is the pulse transfer function of the process model. Accordingly the numerator and denominator of $\widetilde{HG_p}(z)$ render the two unknown polynomials $Q(z)$ and $P(z)$, respectively. In order to ensure zero offset, however, the limit of $y(z)/y_{sp}(z)$ must approach unity at the steady state, that is

$$\lim_{z\to 1}\frac{y(z)}{y_{sp}(z)}=Q(z=1)=1 \tag{6.101}$$

So $P(z)$ and $Q(z)$ in the Kalman controller are chosen by dividing the numerator and denominator polynomials in $\widetilde{HG_p}(z)$ by $Q(z=1)$:

$$D_{Kal}(z)=\frac{P(z)/Q(z=1)}{1-Q(z)/Q(z=1)} \tag{6.102}$$

Example 6.12

Derive the Kalman controller for the process given in Example 6.11, where the pulse transfer function of the process model is:

$$\begin{aligned}
\widetilde{HG_p}(z)&=\frac{0.078z^{-1}+0.03z^{-2}}{1-0.819z^{-1}}z^{-4}\\
P(z)&=1-0.819z^{-1}\\
Q(z)&=(0.078+0.03z^{-1})z^{-5}\\
Q(1)&=0.108
\end{aligned} \tag{6.103}$$

$$D_{Kal}(z)=\frac{(1-0.819z^{-1})/0.108}{1-[(0.078+0.03z^{-1})z^{-5}]/0.108} \tag{6.104}$$

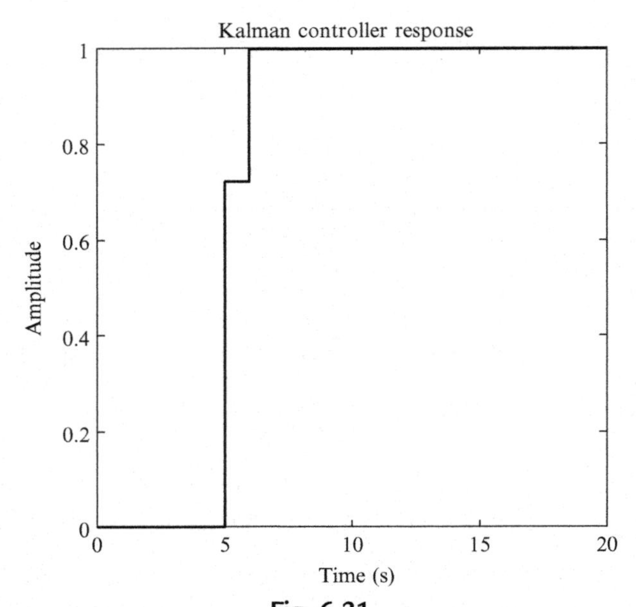

Fig. 6.21
The Kalman controller response.

The Kalman Controller has no apparent tuning parameter, however $Q(1)$ can be used as one, at the expense of creating some offset. Fig. 6.21 demonstrates the closed-loop response of the process to a unit step change in the set point.

```
z = tf('z', 1);

HGp = (0.078*z^-1+0.03*z^-2)/(1-0.819*z^-1)*z^-4;
DKal = ((1-0.819*z^-1)/0.108)/(1-((0.078+0.03*z^-1)*z^-5)/0.108);

step(feedback(DKal*HGp,1));
title('Kalman controller response')
```

6.13.5 Internal Model Controller (IMC)[4]

Consider the block diagram of a plant $G(z)$ and its model $\widetilde{G}(z)$ arranged as is shown in Fig. 6.22. The feedback signal, $Y_1(z)$, and the closed-loop transfer functions can be obtained from the block diagram.

$$Y_1(z) = \frac{C(z)\left[G(z) - \widetilde{G}(z)\right]}{1 + C(z)\left[G(z) - \widetilde{G}(z)\right]} Y_{sp}(z) + \frac{1}{1 + C(z)\left[G(z) - \widetilde{G}(z)\right]} d(z) \tag{6.105}$$

$$Y(z) = d(z) + C(z)\left[Y_{sp}(z) - Y_1(z)\right]G(z) \tag{6.106}$$

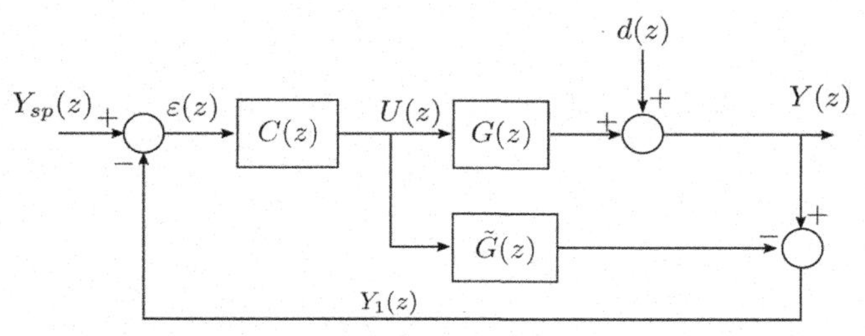

Fig. 6.22
The block diagram of the IMC structure.

where $C(z)$ is the internal model controller. Combining the previous two equations results in:

$$Y(z) = \frac{C(z)G(z)}{1 + C(z)\left[G(z) - \tilde{G}(z)\right]} Y_{sp}(z) + \frac{1 - C(z)\tilde{G}(z)}{1 + C(z)\left[G(z) - \tilde{G}(z)\right]} d(z) \tag{6.107}$$

If $C(z)$ is chosen to be the inverse of the plant model, $C(z) = \frac{1}{\tilde{G}(z)}$, then perfect control would always be achievable, as is shown as follows.

$$Y(z) = \frac{\frac{1}{\tilde{G}(z)} G(z)}{1 + \frac{1}{\tilde{G}(z)}\left[G(z) - \tilde{G}(z)\right]} Y_{sp}(z) + \frac{1 - \frac{1}{\tilde{G}(z)} \tilde{G}(z)}{1 + \frac{1}{\tilde{G}(z)}\left[G(z) - \tilde{G}(z)\right]} d(z) = Y_{sp}(z) + 0 \tag{6.108}$$

However, such a selection is not feasible due to the time delay, NMP zeros, and ringing zeros of the plant model. We can, however, factor out the plant model into an invertible part, $\tilde{G}_1(z)$, and a noninvertible part, $\tilde{G}_2(z)$, and select $C(z)$ as follows:

$$C(z) = \frac{1}{\tilde{G}_1(z)\tilde{G}_2(z = 1)} \tag{6.109}$$

The division by the steady-state gain of the noninvertible part of the transfer function, $\tilde{G}_2(z = 1)$, is necessary to ensure zero offset.

In order to improve robustness of the controller, an exponential filter $F(z)$ is added with an adjustable filter parameter, β, as shown in Fig. 6.23.

$$F(z) = \frac{1 - \beta}{1 - \beta z^{-1}} \tag{6.110}$$

where $0 < \beta < 1$. Higher speeds of closed-loop response can be achieved at small values of β. The overall IMC controller can be reduced to:

$$D_{IMC}(z) = \frac{u(z)}{\varepsilon(z)} = \frac{F(z)C(z)}{1 - F(z)C(z)\tilde{G}(z)} \tag{6.111}$$

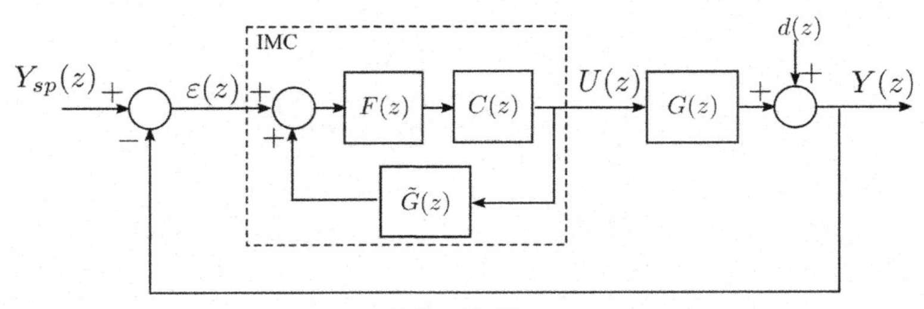

Fig. 6.23
IMC structure with an exponential filter.

Example 6.13

Derive the IMC controller for the plant model given by

$$\widetilde{HG_p}(z) = \frac{0.078(1+0.387z^{-1})z^{-5}}{1-0.819z^{-1}} \tag{6.112}$$

$$\widetilde{G_1}(z) = \frac{0.078}{1-0.819z^{-1}}; \ \widetilde{G_2}(z) = (1+0.387z^{-1})z^{-5}; \ \widetilde{G_2}(1) = 1.387 \tag{6.113}$$

$$C(z) = \frac{1}{\widetilde{G_1}(z)\widetilde{G_2}(1)} = \frac{1-0.819z^{-1}}{(0.0784)(1.387)} \tag{6.114}$$

$$F(z) = \frac{1-\beta}{1-\beta z^{-1}} \tag{6.115}$$

$$D_{IMC}(z) = \frac{F(z)C(z)}{1-F(z)C(z)\widetilde{G}(z)} \tag{6.116}$$

For $\beta = 0.1$:

$$D_{IMC}(z) = \frac{8.277-6.779z^{-1}}{1-0.1z^{-1}-0.6456z^{-5}-0.2498z^{-6}} \tag{6.117}$$

For $\beta = 0.5$:

$$D_{IMC}(z) = \frac{4.598-3.766z^{-1}}{1-0.5z^{-1}-0.3587z^{-5}-0.1388z^{-6}} \tag{6.118}$$

For $\beta = 0.9$:

$$D_{IMC}(z) = \frac{0.9196-1.506z^{-1}+0.6168z^{-2}}{1-1.719z^{-1}+0.7371z^{-2}-0.07173z^{-5}+0.03099z^{-6}+0.02274z^{-7}} \tag{6.119}$$

```
z = tf('z', 1);

HGp = (0.078*(1+0.387*z^-1)*z^-5)/(1-0.819*z^-1);
DIMC1 = (8.275-6.775*z^-1)/(1-0.1*z^-1-0.649*z^-5-0.251*z^-6);
```

```
hold on
beta = 0.1;
C =  (1-0.819*z^-1)/(0.0784*1.387);
F =  (1-beta)/(1-beta*z^-1);
dimc = F*C/(1-F*C*HGp);
step(feedback(dimc*HGp,1));

beta = 0.5;
C = (1-0.819*z^-1)/(0.0784*1.387);
F = (1-beta)/(1-beta*z^-1);
dimc = F*C/(1-F*C*HGp);
step(feedback(dimc*HGp,1));

beta = 0.9;
C = (1-0.819*z^-1)/(0.0784*1.387);
F = (1-beta)/(1-beta*z^-1);
dimc = F*C/(1-F*C*HGp);
step(feedback(dimc*HGp,1));
```

Fig. 6.24 depicts the closed-loop response to a unit step change in the set-point, in each case, in the absence of a plant-model mismatch. It is evident that β can be used as a tuning parameter, the larger it is, the more sluggish closed-loop response would become.

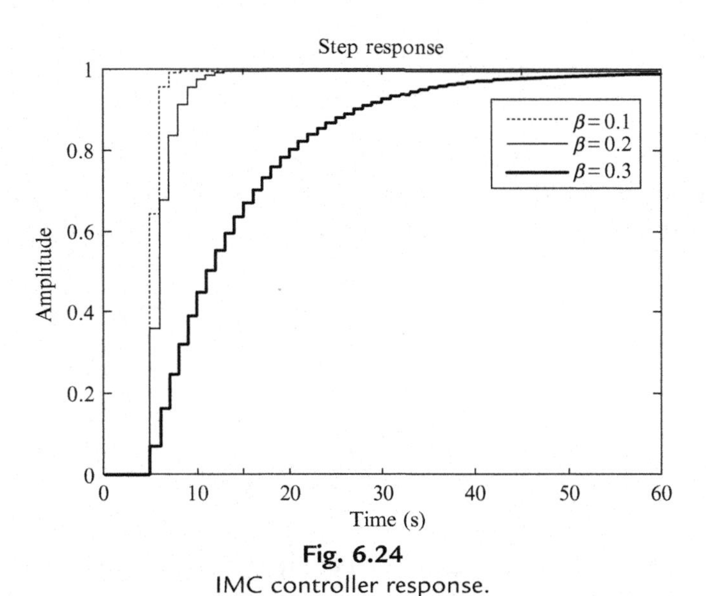

Fig. 6.24
IMC controller response.

6.13.6 The Pole Placement Controller[3]

This controller has a structure given by:

$$u(z) = \frac{H(z)}{F(z)} y_{sp}(z) - \frac{E(z)}{F(z)} y(z) \tag{6.120}$$

with unknown polynomials $H(z)$, $E(z)$ and $F(z)$. The desirable closed-loop characteristic equation is given by a known user-specified polynomial:

$$T(z) = 1 + t_1 z^{-1} + \ldots + t_{n_T} z^{-n_T} \tag{6.121}$$

The process model is given by:

$$\widetilde{G}(z) = \frac{y(z)}{u(z)} = \frac{B(z^{-1})}{A(z^{-1})} \tag{6.122}$$

where

$$A\left(z^{-1}\right) = 1 + a_1 z^{-1} + \ldots + a_{n_A} z^{-n_A} \tag{6.123}$$

$$B\left(z^{-1}\right) = b_1 z^{-1} + \ldots + b_{n_B} z^{-n_B} \tag{6.124}$$

The unknowns are the orders and the coefficients of the polynomials $E\left(z^{-1}\right)$, $F\left(z^{-1}\right)$, and $H\left(z^{-1}\right)$.

$$E\left(z^{-1}\right) = e_0 + e_1 z^{-1} + \ldots + e_{n_E} z^{-n_E} \tag{6.125}$$

$$F\left(z^{-1}\right) = 1 + f_1 z^{-1} + \ldots + f_{n_F} z^{-n_F} \tag{6.126}$$

$$H\left(z^{-1}\right) = h_0 + h_1 z^{-1} + \ldots + h_{n_H} z^{-n_H} \tag{6.127}$$

In order to determine the unknown polynomial orders and parameters, the closed-loop servo transfer function is derived by substituting the controller output, $u(z)$, from Eq. (6.120) in Eq. (6.122):

$$A\left(z^{-1}\right)y(z) = B\left(z^{-1}\right)u(z) = B\left(z^{-1}\right)\left[\frac{H(z^{-1})}{F(z^{-1})}y_{sp}(z) - \frac{E(z^{-1})}{F(z^{-1})}y(z)\right] \tag{6.128}$$

$$y(z) = \frac{B(z^{-1})H(z^{-1})}{A(z^{-1})F(z^{-1}) + B(z^{-1})E(z^{-1})} y_{sp}(z) \tag{6.129}$$

The closed-loop characteristic equation (*CLCE*) in Eq. (6.129) is equated to the user-specified desired *CLCE* and the solution of the following polynomial identity determines the unknown polynomials $E\left(z^{-1}\right)$ and $F\left(z^{-1}\right)$.

$$A\left(z^{-1}\right)F\left(z^{-1}\right) + B\left(z^{-1}\right)E\left(z^{-1}\right) = T\left(z^{-1}\right) \tag{6.130}$$

This polynomial identity can be written in a matrix form:

$$\overbrace{\begin{bmatrix} 1 & 0 & \cdots & 0 & b_1 & 0 & \cdots & 0 \\ a_1 & 1 & & \vdots & b_2 & b_1 & & \vdots \\ \vdots & a_1 & & 1 & \vdots & b_2 & & b_1 \\ a_{n_A} & \vdots & & a_1 & b_{n_B} & \vdots & & b_2 \\ 0 & a_{n_A} & & \vdots & 0 & b_{n_B} & & \vdots \\ \vdots & \vdots & & a_{n_A} & \vdots & 0 & & b_{n_B} \\ \multicolumn{4}{c}{\underbrace{\hphantom{xxxxxxxxxx}}_{n_F \text{ columns}}} & \multicolumn{4}{c}{\underbrace{\hphantom{xxxxxxxxxx}}_{n_E+1 \text{ columns}}} \end{bmatrix}}^{A} \overbrace{\begin{bmatrix} f_1 \\ \vdots \\ f_{n_F} \\ e_0 \\ \vdots \\ e_{n_G} \end{bmatrix}}^{\theta} = \overbrace{\begin{bmatrix} t_1 - a_1 \\ t_2 - a_2 \\ \\ \vdots \\ \\ t_n - a_n \end{bmatrix}}^{B} \tag{6.131}$$

where $n_E = n_A - 1$ and $n_F = n_B - 1$.

The unknown vector of parameters θ can be obtained by solving the following equation:

$$\theta = A^{-1}B \tag{6.132}$$

The remaining polynomial $H\left(z^{-1}\right)$ can be specified as a scalar quantity, h, to fulfill the zero offset requirement.

$$H\left(z^{-1}\right) = h = \frac{T(1)}{B(1)} \tag{6.133}$$

Example 6.14

Design a pole placement controller for a process having a model

$$\tilde{G}\left(z^{-1}\right) = \frac{z^{-1} - 0.5z^{-2}}{1 + 0.1z^{-2}} \tag{6.134}$$

which places the closed-loop poles given by

$$T\left(z^{-1}\right) = 1 - 0.3z^{-1} \tag{6.135}$$

The orders of the unknown polynomials $E\left(z^{-1}\right)$ and $F\left(z^{-1}\right)$ are as follows:

$$n_E = n_A - 1 = 2 - 1 = 1 \text{ and } n_F = n_B - 1 = 2 - 1 = 1 \tag{6.136}$$

$$\begin{bmatrix} 1 & 1 & 0 \\ 0 & -0.5 & 1 \\ 0.1 & 0 & -0.5 \end{bmatrix} \begin{bmatrix} f_1 \\ e_0 \\ e_1 \end{bmatrix} = \begin{bmatrix} -0.3 \\ -0.1 \\ 0 \end{bmatrix} \tag{6.137}$$

$$[f_1 \;\; e_0 \;\; e_1] = [-0.357 \;\; 0.057 \;\; -0.071] \tag{6.138}$$

$$H\left(z^{-1}\right) = \frac{T(1)}{B(1)} = \frac{0.7}{0.5} = 1.4 \tag{6.139}$$

The block diagram and the closed-loop response of this system are shown in Figs. 6.25 and 6.26.

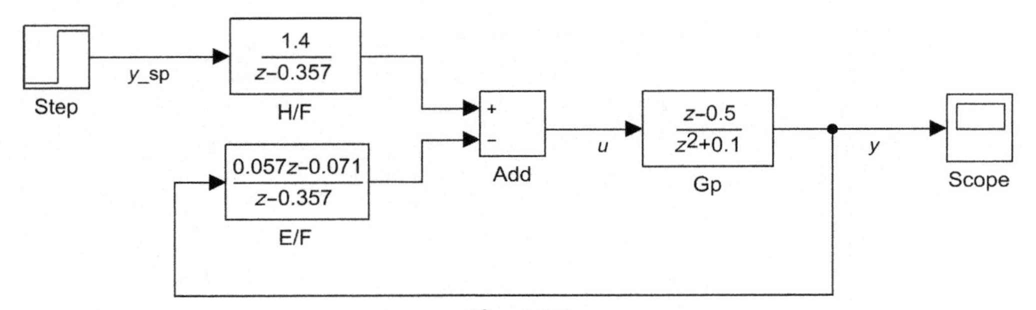

Fig. 6.25
Block diagram of the pole placement controller.

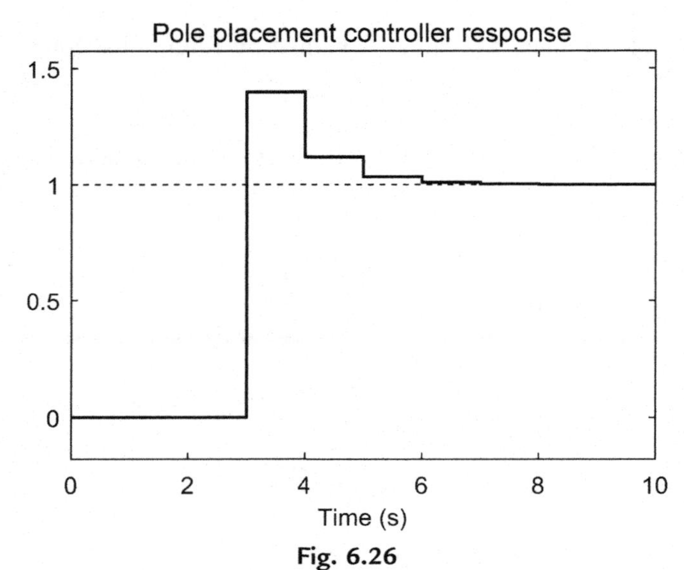

Fig. 6.26
Step response of the pole placement controller.

6.14 Design of Feedforward Controllers

A feedback controller is activated by a nonzero error signal irrespective of the source of the error signal. For slow processes with significant time delays, feedback controllers alone do not perform well since they start taking corrective action after the disturbances have entered and disturbed the process. The feedforward controllers, on the other hand, measure the major disturbances and take appropriate corrective action to compensate for their effects. Fig. 6.27 shows the block diagram of a combined feedback and feedforward control system for a SISO plant. Extension to MIMO plants is straightforward.

Considering the previous block diagram, the closed-loop transfer function is:

$$Y(z) = \frac{D(z)HG_p(z) + G_{sp}(z)G_F(z)HG_p(z)}{1 + D(z)HG_p(z)} Y_{sp}(z) + \frac{dG_d(z) + dG_{mF}(z)HG_p(z)G_F(z)}{1 + D(z)HG_p(z)} \quad (6.140)$$

where $G_{MF}(s)$ represents dynamics of the sensor that measures the selected disturbance. Note the usual shorthand notation applies to the definition of the pulse transfer function of the process, $HG_p(z)$, and $dG_d(z)$ and $dG_{mF}(z)$; related to the continuous disturbance whose effect is being sampled at the plant output. In order to have perfect control, the coefficient of $Y_{sp}(z)$ must be unity and the second term should be zero. These two requirements render the unknown transfer functions necessary for the design of the feedforward controller.

$$G_{sp}(z) = -\frac{dG_{mF}(z)}{dG_d(z)} = -\frac{HG_{mF}(z)}{HG_d(z)} \quad (6.141)$$

$$G_F(z) = -\frac{dG_d(z)}{dG_{mF}(z)HG_p(z)} = -\frac{HG_d(z)}{HG_{mF}(z)HG_p(z)} \quad (6.142)$$

Note that in order to calculate $dG_{mF}(z)$ and $dG_d(z)$, the nature of $d(s)$ should be known. In the absence of this knowledge the two terms are replaced by $HG_{mF}(z)$ and $HG_d(z)$, i.e., preceded by a zero-order hold device, that would render an exact solution if the disturbance were a step change or a series of step changes.

It should be noted that $G_{sp}(z)$ and $G_F(z)$ must be physically realizable. It is also worth noting that the closed-loop characteristic equation of the combined feedback and feedforward control is only influenced by the feedback controller. That is, the stability of the system is not influenced by the feedforward controller $G_F(z)$.

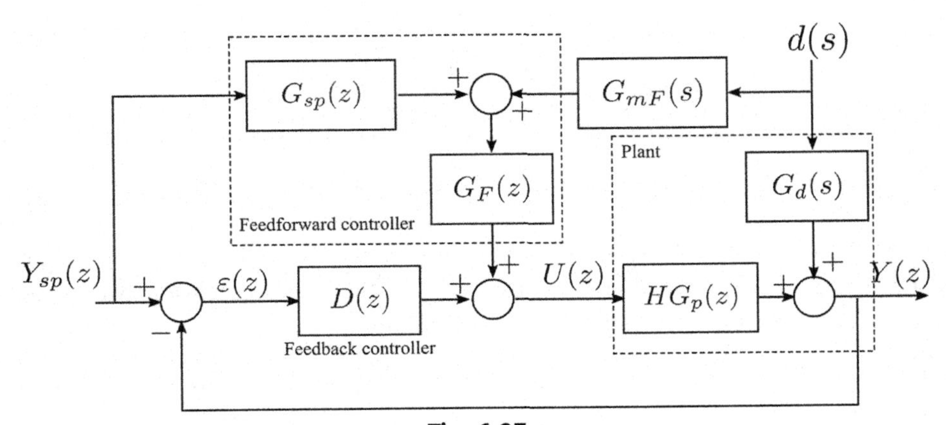

Fig. 6.27
Block diagram of feedforward feedback controller.

Since the performance of a feedforward controller depends heavily on the plant model and the feedforward controller can only compensate for the measured disturbances, it is imperative that the feedforward controller be used in conjunction with a feedback controller.

Example 6.15

Design a feedforward controller for a plant having the following models:

$$G_p(s) = \frac{0.6e^{-2s}}{4s+1} \tag{6.143}$$

$$G_d(s) = \frac{0.2e^{-4s}}{2s+1} \tag{6.144}$$

The dynamics of the disturbance sensor is given by

$$G_{MF}(s) = \frac{0.4}{s+1} \tag{6.145}$$

The sampling interval is 1 s.

Discretization of the transfer function driven by a ZOH results in the following pulse transfer functions:

$$HG_p(z) = \frac{0.133z^{-3}}{1-0.779z^{-1}} \tag{6.146}$$

$$HG_d(z) = \frac{0.079z^{-5}}{1-0.606z^{-1}} \tag{6.147}$$

$$HG_{mF}(z) = \frac{0.253z^{-1}}{1-0.368z^{-1}} \tag{6.148}$$

The feedforward controller transfer functions will be:

$$G_F(z) = -\frac{2.349(1-0.368z^{-1})(1-0.779z^{-1})z^{-1}}{(1-0.606z^{-1})} \tag{6.149}$$

$$G_{sp}(z) = -\frac{3.202(1-0.606z^{-1})z^{+4}}{1-0.368z^{-1}} \tag{6.150}$$

Obviously $G_{sp}(z)$ is not physically realizable due to the time advance term z^{+4}. This term is dropped in order to make the controller physical realizable.

6.15 Control of Multi-Input, Multi-Output (MIMO) Processes

The multi-input, multi-output (MIMO) processes have more than one controlled variable. Most of the actual chemical processes belong to this category. Fig. 6.28 depicts a MIMO system with n manipulated and controlled variables and l disturbances.

The model of a MIMO process can be represented by transfer function matrix or in the state space.

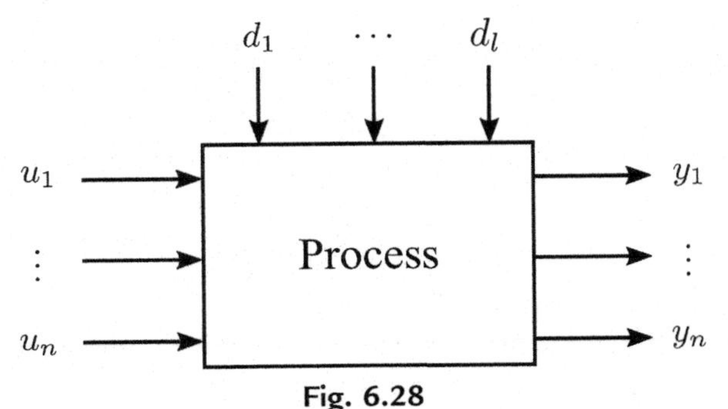

Fig. 6.28
MIMO system block diagram.

$$\begin{bmatrix} Y_1(s) \\ \vdots \\ Y_n(s) \end{bmatrix} = \begin{bmatrix} G_{11}(s) & G_{12}(s) & \cdots & G_n(s) \\ G_{n1}(s) & G_{n2}(s) & \cdots & G_{nn}(s) \end{bmatrix} \begin{bmatrix} U_1(s) \\ \vdots \\ U_n(s) \end{bmatrix} + \begin{bmatrix} G_{d11}(s) & G_{d12}(s) & \cdots & G_{d1l}(s) \\ G_{dn1}(s) & G_{dn2}(s) & \cdots & G_{dnl}(s) \end{bmatrix} \begin{bmatrix} D_1(s) \\ \vdots \\ D_l(s) \end{bmatrix} \tag{6.151}$$

$$Y(s) = \boldsymbol{G(s)}U(s) + \boldsymbol{G_d(s)}D(s) \tag{6.152}$$

$$Y_1(s) = G_{11}(s)U_1(s) + G_{12}(s)U_2(s) + \cdots + G_{1n}(s)U_n(s) + G_{d11}(s)D_1(s) + G_{d12}(s)D_2(s) + \cdots + G_{d1l}(s)D_l(s) \tag{6.153}$$

For a noninteracting MIMO process (a process in which the controlled variable i is only affected by the manipulated variable i), the off-diagonal terms of the transfer function matrix, $\boldsymbol{G}(\boldsymbol{s})$, will all be zero. Each input affects only one output in such a system. However, this is a highly unlikely situation. Usually, there is interaction in MIMO systems, and therefore the SISO controllers do not perform well.

Example 6.16

Determine the correct pairing of the manipulated and controlled variable for the following plant using the RGA (relative gain array method due to Bristol) method discussed in Chapter 5.

$$\begin{cases} \boldsymbol{x}_{k+1} = \boldsymbol{A}\boldsymbol{x}_k + \boldsymbol{B}\boldsymbol{u}_k \\ \boldsymbol{y}_k = \boldsymbol{C}\boldsymbol{x}_k \end{cases} \tag{6.154}$$

$$\boldsymbol{A} = \begin{bmatrix} 3 & 1 \\ 0.1 & 2 \end{bmatrix}, \boldsymbol{B} = \begin{bmatrix} 1 & 0 \\ 0 & 1 \end{bmatrix}, \text{ and } \boldsymbol{C} = \begin{bmatrix} 1 & 0 \\ 0 & 1 \end{bmatrix} \tag{6.155}$$

$$\boldsymbol{G_p}(z) = \boldsymbol{C}(z\boldsymbol{I} - \boldsymbol{A})^{-1}\boldsymbol{B} = \frac{1}{z^2 - 5z + 5.9} \begin{bmatrix} z-2 & 1 \\ 0.1 & z-3 \end{bmatrix} \tag{6.156}$$

$$K = G_p(z=1) = \frac{1}{1.9}\begin{bmatrix} -1 & 1 \\ 0.1 & -2 \end{bmatrix} \quad (6.157)$$

$$\left(K^{-1}\right)^T = \begin{bmatrix} -2 & -0.1 \\ -1 & -1 \end{bmatrix} \quad (6.158)$$

$$\Lambda = \begin{bmatrix} 1.05 & -0.05 \\ -0.05 & 1.05 \end{bmatrix} \quad (6.159)$$

which suggests u_1 should be used to control y_1 and u_2 to control y_2.

6.15.1 Singular Value Decomposition (SVD) and the Condition Number (CN)

The RGA method of Bristol may sometimes be misleading for the choice of the best control configuration. Consequently, a complementary property of the steady-state gain matrix which is referred to as the condition number (CN) is used in conjunction with the RGA. The CN of the steady-state gain matrix of a process is defined as the ratio of the largest (σ_l) to the smallest (σ_s) positive singular value of $\boldsymbol{K}^T\boldsymbol{K}$ matrix.

$$CN = \frac{\sigma_l}{\sigma_s} \quad (6.160)$$

The singular values of $\boldsymbol{K}^T\boldsymbol{K}$ are the square roots of the positive eigenvalues of $\boldsymbol{K}^T\boldsymbol{K}$. The eigenvalues of $\boldsymbol{K}^T\boldsymbol{K}$ are the roots of $\left(\boldsymbol{K}^T\boldsymbol{K} - \lambda \boldsymbol{I}\right) = 0$. A large CN indicates a poorly conditioned process or an ill-conditioned steady-state gain matrix which makes the control of a MIMO process challenging. If $\boldsymbol{K}$ is singular, the CN becomes infinite and the process is impossible to control no matter what pairing is selected.

Example 6.17

Using the condition number, confirm the pairing suggested by the RGA method for a system with the following steady-state gain matrix, K.

$$K = \begin{bmatrix} -0.526 & 0.526 \\ 0.053 & -1.053 \end{bmatrix} \quad (6.161)$$

$$\left|K^T K - \lambda I\right| = 0 \quad (6.162)$$

$$\lambda_1 = 1.478,\ \lambda_2 = 0.187 \quad (6.163)$$

$$\sigma_l = 1.216,\ \sigma_s = 0.432 \quad (6.164)$$

$$CN = \frac{\sigma_l}{\sigma_s} = 2.8 \quad (6.165)$$

which suggests that $\boldsymbol{K}$ is not ill conditioned and the recommended pairing by the RGA in Example 6.16 is acceptable.

6.15.2 Design of Multivariable Feedback Controllers for MIMO Plants

A MIMO plant with n manipulated variables and n controlled variables will have $n!$ possible different control configurations. The initial step in the design of feedback controllers for the MIMO processes involves the selection of the best pairing of the manipulated and controlled variables resulting in the smallest interaction among the various control loops. The next step is to use either multiloop SISO controllers such as the conventional PID controllers or truly multivariable MIMO controllers. In the former case, in order to improve the performance of the SISO controllers in the presence of interactions, an ad hoc approach is used by adding interaction compensators (decouplers) to each loop.

6.15.3 Dynamic and Steady-State Interaction Compensators (Decouplers) in the z-Domain

In the presence of interaction among different control loops, the performance of the multiloop SISO controllers is deteriorated. One way to improve their performance is to employ a dynamic or a steady-state compensator or decoupler as is shown in Fig. 6.29.

The closed-loop transfer function for the system shown in Fig. 6.29 can be developed recognizing that the input and output variables are all vectors, and the pulse transfer function of the process, $G(z)$, is a matrix of appropriate dimension. The multiloop PID controller, $\boldsymbol{D_{PID}}(z)$, consists of n individual controllers for an n by n system, and $\boldsymbol{D_c}(z)$ is the decoupler transfer function matrix that is to be determined.

$$\boldsymbol{y}(z) = [\boldsymbol{I} + \boldsymbol{G}(z)\boldsymbol{D_c}(z)\boldsymbol{D_{PID}}(z)]^{-1}\left[\boldsymbol{G}(z)\boldsymbol{D_c}(z)\boldsymbol{D_{PID}}(z)\boldsymbol{y_{sp}}(z) + \boldsymbol{d}(z)\right] \tag{6.166}$$

In order to eliminate interactions and realizing that $\boldsymbol{D_{PID}}(z)$ is a diagonal PID controller, the matrix $[\boldsymbol{I} + \boldsymbol{G}(z)\boldsymbol{D_c}(z)\boldsymbol{D_{PID}}(z)]^{-1}\boldsymbol{G}(z)\boldsymbol{D_c}(z)\boldsymbol{D_{PID}}(z)$ must also be diagonal to remove the system interactions. In order for this to happen $\boldsymbol{D_c}(z)$ must be chosen as:

$$\boldsymbol{D_c}(z) = \widetilde{\boldsymbol{G}}^{-1}(z)\, diag\left[\widetilde{\boldsymbol{G}}(z)\right] \tag{6.167}$$

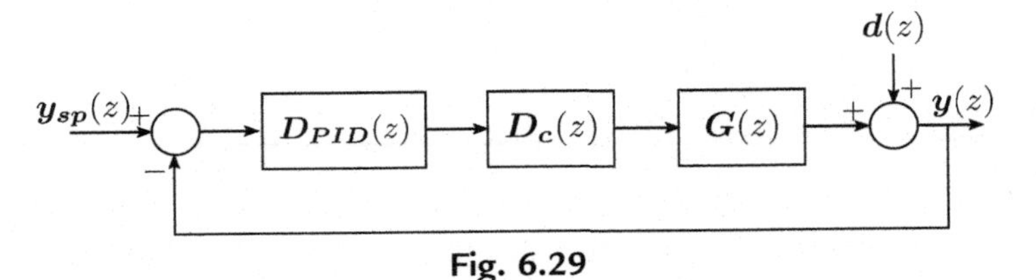

Fig. 6.29
Block diagram of the interaction compensator.

which results in the closed-loop transfer function matrix:

$$y(z) = \left[I + G(z)\widetilde{G}^{-1}(z)\,\text{diag}\left[\widetilde{G}(z)\right]D_{PID}(z)\right]^{-1}\left[G(z)\widetilde{G}^{-1}(z)\,\text{diag}[G(z)]D_{PID}(z)y_{sp}(z) + d(z)\right] \tag{6.168}$$

This choice of $D_c(z)$ would be feasible if $\widetilde{G}(z)$ were invertible. If $\widetilde{G}(z)$, however, contains time delays or zeros outside the UC, only the steady-state compensator can be employed:

$$D_c(z=1) = \lim_{z\to 1}\widetilde{G}^{-1}(z)\,\text{diag}\left[\widetilde{G}(z)\right] \tag{6.169}$$

This is an ad hoc design approach to implement single loop controllers with dynamic or steady-state compensators on a MIMO process. The overall performance of the system should improve provided that the compensator is realizable.

Example 6.18

Derive the dynamic and steady-state compensators for the following plant model:

$$\widetilde{G}(z) = \frac{1}{1-0.2z^{-1}}\begin{bmatrix} 0.2z^{-2} & 0.5z^{-3} \\ 0.3z^{-1} & 0.8z^{-4} \end{bmatrix} \tag{6.170}$$

$$\widetilde{G}^{-1}(z) = \frac{(1-0.2z^{-1})}{(0.16z^{-6}-0.15z^{-4})}\begin{bmatrix} 0.8z^{-4} & -0.5z^{-3} \\ -0.3z^{-1} & 0.2z^{-2} \end{bmatrix} \tag{6.171}$$

Note that $\widetilde{G}^{-1}(z)$ is not realizable as it contains prediction terms. In order to find $D_c(z)$, we need the diagonal matrix of the plant model.

$$\text{diag}\left[\widetilde{G}(z)\right] = \frac{1}{1-0.2z^{-1}}\begin{bmatrix} 0.2z^{-2} & 0 \\ 0 & 0.8z^{-4} \end{bmatrix} \tag{6.172}$$

$$D_c(z) = \widetilde{G}^{-1}(z)\,\text{diag}\left[\widetilde{G}(z)\right] = \frac{1}{0.16z^{-6}-0.15z^{-4}}\begin{bmatrix} 0.16z^{-6} & -0.4z^{-7} \\ -0.06z^{-3} & 0.16z^{-6} \end{bmatrix} \tag{6.173}$$

Note that $D_c(z)$ is unstable due to its poles outside the UC and it is also not physically realizable. So only the steady-state compensator can be implemented.

$$D_c(z=1) = \begin{bmatrix} 16 & -40 \\ -6 & 16 \end{bmatrix} \tag{6.174}$$

In what follows, instead of using multiloop PID controllers with compensators, we shall consider several MIMO controllers. The digital versions of a few single loop controllers discussed previously will be developed as potential MIMO controllers.

6.15.4 Multivariable Smith Predictor

The design procedure of the multivariable Smith predictor[1] follows closely its SISO version. Consider the block diagram shown in Fig. 6.30.

The closed-loop transfer function can be derived as:

$$\mathbf{y}(z) = [\mathbf{I} + \mathbf{G}(z)\mathbf{D}_{smith}(z)]^{-1}\left[\mathbf{G}(z)\mathbf{D}_{smith}(z)\mathbf{y}_{sp}(z) + \mathbf{d}(z)\right] \tag{6.175}$$

Where

$$\mathbf{D}_{smith}(z) = \left\{\mathbf{I} + \mathbf{D}_{PID}(z)\left[\widetilde{\mathbf{G}}^{*}(z) - \widetilde{\mathbf{G}}(z)\right]\right\}^{-1}\mathbf{D}_{PID}(z) \tag{6.176}$$

$\widetilde{\mathbf{G}}^{*}(z)$ is the plant transfer function matrix without the time delay. It is imperative to note that the closed-loop characteristic equation which determines the system stability does not contain any time delay. This can be shown for the servo closed-loop transfer function by:

$$\mathbf{y}(z) = \mathbf{G}(z)\left[\mathbf{I} + \mathbf{D}_{PID}(z)\widetilde{\mathbf{G}}^{*}(z)\right]^{-1}\mathbf{D}_{PID}(z)\mathbf{y}_{sp}(z) \tag{6.177}$$

in which the closed-loop poles are the delay-free roots of

$$\left[\mathbf{I} + \mathbf{D}_{PID}(z)\widetilde{\mathbf{G}}^{*}(z)\right] = 0 \tag{6.178}$$

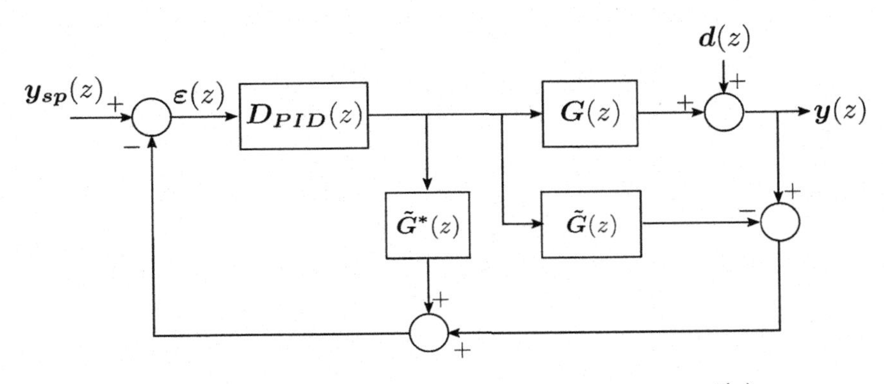

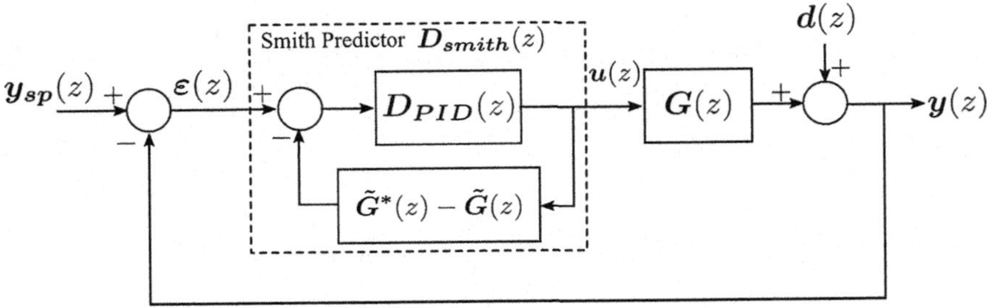

Fig. 6.30
The block diagrams of the MIMO Smith predictor.

Example 6.19

Derive a MIMO Smith predictor for the plant model given in Example 6.18

$$\tilde{G}^*(z) = \frac{1}{1-0.2z^{-1}} \begin{bmatrix} 0.2 & 0.5 \\ 0.3 & 0.8 \end{bmatrix} \tag{6.179}$$

$$\tilde{G}^*(z) - \tilde{G}(z) = \frac{1}{1-0.2z^{-1}} \begin{bmatrix} 0.2(1-z^{-2}) & 0.5(1-z^{-3}) \\ 0.3(1-z^{-1}) & 0.8(1-z^{-4}) \end{bmatrix} \tag{6.180}$$

The block diagram of the Smith predictor using a diagonal PI controller

$$D_{PI}(z) = \begin{bmatrix} \dfrac{1-0.3z^{-1}}{1-z^{-1}} & 0 \\ 0 & \dfrac{0.11-0.2z^{-1}}{1-z^{-1}} \end{bmatrix} \tag{6.181}$$

can be constructed. The corresponding closed-loop responses in y_1 and y_2 as well as the controller outputs can be obtained using Simulink.

6.15.5 Multivariable IMC Controller

Fig. 6.31 shows the IMC structure for a MIMO plant.[2,3] The signal $\boldsymbol{y}_1(z)$ and the closed-loop transfer function can be derived by following steps shown as follows.

$$\boldsymbol{y}_1(z) = \boldsymbol{d}(z) + \left[\boldsymbol{G}(z) - \tilde{\boldsymbol{G}}(z)\right]\boldsymbol{u}(z) \tag{6.182}$$

$$\boldsymbol{u}(z) = \boldsymbol{C}(z)\left[\boldsymbol{y}_{sp}(z) - \boldsymbol{y}_1(z)\right] \tag{6.183}$$

$$\boldsymbol{y}(z) = \boldsymbol{d}(z) + \boldsymbol{G}(z)\boldsymbol{u}(z) \tag{6.184}$$

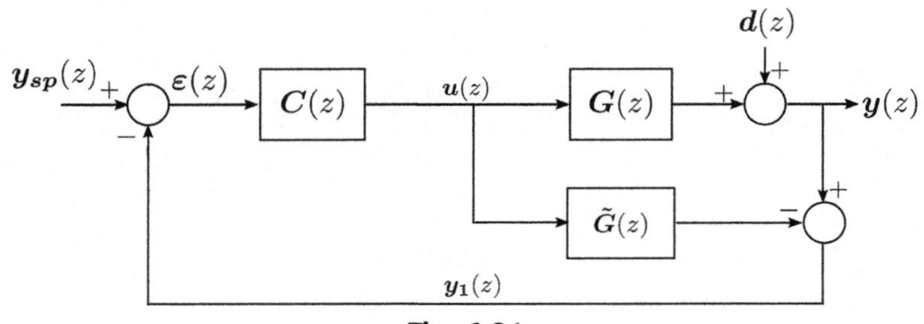

Fig. 6.31
Block diagram of the multivariable IMC.

Combining the previous three equations results in the closed-loop transfer function:

$$\boldsymbol{y}(z) = \boldsymbol{d}(z) + \boldsymbol{G}(z)\left\{\boldsymbol{I} + \boldsymbol{C}(z)\left[\boldsymbol{G}(z) - \widetilde{\boldsymbol{G}}(z)\right]\right\}^{-1}\boldsymbol{C}(z)\left[\boldsymbol{y}_{sp}(z) - \boldsymbol{d}(z)\right] \tag{6.185}$$

Note that if there is no plant–model mismatch, i.e., $\widetilde{\boldsymbol{G}}(z) = \boldsymbol{G}(z)$, the closed-loop stability would be guaranteed if $\boldsymbol{C}(z)$ and $\boldsymbol{G}(z)$ were stable. Moreover if $\boldsymbol{C}(z)$ is chosen as $\widetilde{\boldsymbol{G}}^{-1}(z)$ we can achieve perfect control:

$$\boldsymbol{y}(z) = \boldsymbol{G}(z)\left[\boldsymbol{I} + \widetilde{\boldsymbol{G}}^{-1}(z)\,\boldsymbol{G}(z) - \boldsymbol{I}\right]^{-1}\widetilde{\boldsymbol{G}}^{-1}(z)\left[\boldsymbol{y}_{sp}(z) - \boldsymbol{d}(z)\right] + \boldsymbol{d}(z) = \boldsymbol{y}_{sp}(z) \tag{6.186}$$

However, the inverse of the plant model transfer matrix is usually either unstable or not realizable. This is due to the existence of nonminimum phase (NMP) zeros and time delay terms in $\widetilde{\boldsymbol{G}}(z)$. Accordingly, $\widetilde{\boldsymbol{G}}(z)$ is factored into a part which is invertible $\widetilde{\boldsymbol{G}_1}(z)$ and a part which contains NMP zeros and time delay terms $\widetilde{\boldsymbol{G}_2}(z)$.

$$\widetilde{\boldsymbol{G}}(z) = \widetilde{\boldsymbol{G}}_1(z)\widetilde{\boldsymbol{G}}_2(z) \tag{6.187}$$

$\boldsymbol{C}(z)$ is then chosen as:

$$\boldsymbol{C}(z) = \widetilde{\boldsymbol{G}}_1^{-1}(z) = \widetilde{\boldsymbol{G}}^{-1}(z)\widetilde{\boldsymbol{G}}_2(z) \tag{6.188}$$

$\widetilde{\boldsymbol{G}}_2(z)$ is further factored into $\widetilde{\boldsymbol{G}}_{21}(z)$ and $\widetilde{\boldsymbol{G}}_{22}(z)$, where $\widetilde{\boldsymbol{G}}_{21}(z)$ is chosen as a diagonal matrix to make $\widetilde{\boldsymbol{G}}^{-1}(z)\,\widetilde{\boldsymbol{G}}_{21}(z)$ realizable and $\widetilde{\boldsymbol{G}}_{22}(z)$ is chosen such that the matrix product $\widetilde{\boldsymbol{G}}^{-1}(z)\,\widetilde{\boldsymbol{G}}_{21}(z)$ $\widetilde{\boldsymbol{G}}_{22}(z)$ is stable. Moreover, for zero offset the steady-state gain of $\widetilde{\boldsymbol{G}}_2(z=1)$ should be the identity matrix which means both $\widetilde{\boldsymbol{G}}_{21}(z)$ and $\widetilde{\boldsymbol{G}}_{22}(z)$ must approach the identity matrix at steady state.

If the time delays of individual transfer functions in $\widetilde{\boldsymbol{G}}(z)$ are equal, then $\widetilde{\boldsymbol{G}}_{21}(z)$ is chosen as $z^{-(N+1)}\boldsymbol{I}$ where N is the time delay of the individual transfer functions. However, if the time delays of the individual transfer functions are not equal, the choice of $\widetilde{\boldsymbol{G}}_{21}(z)$ follows:

$$\boldsymbol{G}_{21}(z) = \text{diag}\left[z^{-N_1}\cdots z^{-N_r}\right] \tag{6.189}$$

where

$$N_j = \max_i \max\left(0, \overline{N}_{ij}\right)\,\text{for}\,j = 1, \cdots, r \tag{6.190}$$

and $\overline{N}_{ij}$ is the prediction time in $\widetilde{\boldsymbol{G}}^{-1}(z)$ with r columns.

Similar to the SISO version of the IMC, a diagonal exponential filter is added to the system to improve the overall system robustness and its properness.

$$\boldsymbol{F}(z) = \text{diag}\left(\frac{1-\beta_j}{1-\beta_j z^{-1}}\right) \tag{6.191}$$

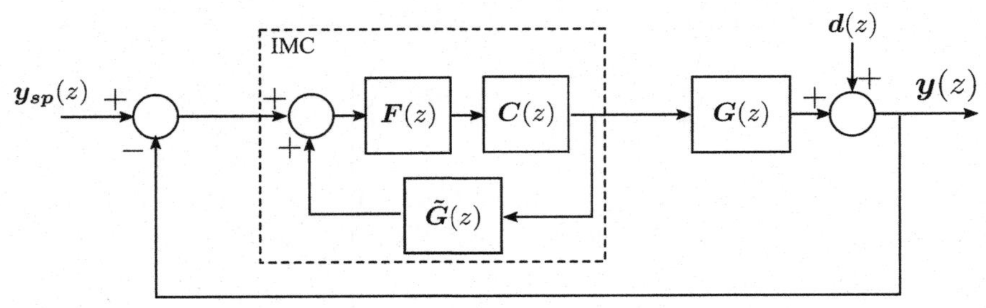

Fig. 6.32
The simplified block diagram of the multivariable IMC.

The block diagram of the entire system is shown in Fig. 6.32.

$$D_{IMC}(z) = \left[I - C(z)F(z)\tilde{G}(z)\right]^{-1} C(z)F(z) \tag{6.192}$$

Example 6.20

Derive the multivariable IMC controller for the plant model given in Example 6.18

$$\tilde{G}^{-1}(z) = \frac{1 - 0.2z^{-1}}{0.16z^{-2} - 0.15} \begin{bmatrix} 0.8 & -0.5z \\ -0.3z^3 & 0.2z^2 \end{bmatrix} \tag{6.193}$$

Accordingly $\tilde{G}_{21}(z)$ should be chosen as:

$$\tilde{G}_{21}(z) = \begin{bmatrix} z^{-3} & 0 \\ 0 & z^{-2} \end{bmatrix} \tag{6.194}$$

which makes $\tilde{G}^{-1}(z)\ \tilde{G}_{21}(z)$ realizable

$$\tilde{G}^{-1}(z)\tilde{G}_{21}(z) = \frac{1 - 0.2z^{-1}}{0.16z^{-2} - 0.15} \begin{bmatrix} 0.8z^{-3} & -0.5z^{-1} \\ -0.3 & 0.2 \end{bmatrix} \tag{6.195}$$

$\tilde{G}_{22}(z)$ is chosen such that $\tilde{G}^{-1}(z)\ \tilde{G}_2(z)$ is stable, while its steady-state gain is I. Considering $\tilde{G}^{-1}(z)$, a reasonable choice of $\tilde{G}_{22}(z)$ would be:

$$\tilde{G}_{22}(z) = \frac{0.16z^{-2} - 0.15}{0.16 - 0.15} I \tag{6.196}$$

which eliminates the unstable poles of $\tilde{G}^{-1}(z)$ and has a unity steady-state gain.

$$C(z) = \tilde{G}^{-1}(z)\tilde{G}_{21}(z)\tilde{G}_{22}(z) = \left(1 - 0.2z^{-1}\right) \begin{bmatrix} 80z^{-3} & -50z^{-1} \\ -30 & 20 \end{bmatrix} \tag{6.197}$$

$$\mathbf{F}(z) = \begin{bmatrix} \frac{1-\beta_1}{1-\beta_1 z^{-1}} & 0 \\ 0 & \frac{1-\beta_2}{1-\beta_2 z^{-1}} \end{bmatrix} \tag{6.198}$$

If we assume perfect model $\tilde{\mathbf{G}}(z) = \mathbf{G}(z)$

$$\mathbf{y}(z) = \mathbf{G}(z)\mathbf{C}(z)\mathbf{F}(z)\mathbf{y}_{sp}(z) \tag{6.199}$$

If we further assume $\mathbf{F}(z) = \mathbf{I}$ in view of no plant–model mismatch we shall have

$$\mathbf{y}(z) = \begin{bmatrix} 16z^{-4} - 15z^{-3} & 0 \\ 24z^{-3} - 24z^{-4} & -15z^{-2} + 16^{-4} \end{bmatrix} \mathbf{y}_{sp}(z) \tag{6.200}$$

Problems

1. The discrete transfer function of a process is given by

$$\frac{Y(z)}{U(z)} = \frac{5z^{-1} + 3z^{-2}}{1 + z^{-1} + 0.41z^{-2}} \tag{6.201}$$

Convert this transfer function to an equivalent difference equation. Calculate the response $y(k)$ to a discrete unit step change in u.

2. Using the MATLAB commands rlocus(gol) and rlocfind(gol), plot the root locus diagram of the following system and determine the range of K_c for which the system is stable, oscillatory, marginally stable, and unstable.

$$C.L.C.E.(s) = 3s^3 + 2s^2 + (3K_c + 1)s + 1 + K_c = 0 \tag{6.202}$$

What would be the corresponding range of K_c for which the system would be stable, oscillatory, marginally stable, and unstable, if a digital controller were to be used with a sampling interval of 1 s?

3. The actual transfer function of a process is:

$$g(s) = \frac{0.8(s-1)e^{-3s}}{4s^2 + 2s + 1} \tag{6.203}$$

And its unit step response is given as follows (Fig. 6.33):
Design an IMC controller for this process by first identifying a model for the process. Tune the IMC by changing the filter time constant. Simulate the closed-loop system on Simulink.

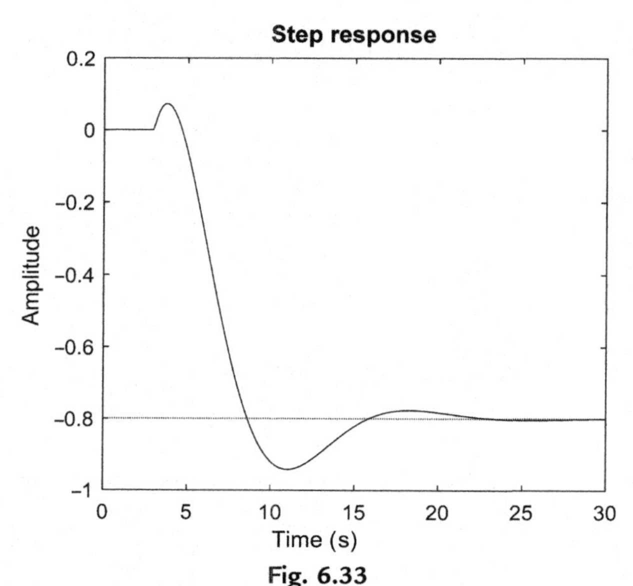

Fig. 6.33
Unit step response of the transfer function in Problem 16.6.3.

4. The open-loop pulse transfer function of a process is given by

$$HG(z) = \frac{(1.3 - z^{-1})z^{-2}}{(1 - 0.4z^{-1})(1 - 0.6z^{-1})} \tag{6.204}$$

Design various deterministic digital controllers discussed in this chapter such as a discrete PI controller (with appropriate tuning parameters), the Dahlin's algorithm, the Smith predictor (with the same PI controller parameter), and the IMC algorithm for this process. Discuss the controller performance by simulating the closed-loop response of the process in MATLAB or Simulink environment. The sampling/control interval in each is taken as 1 *s*.

5. The root locus diagram of a control system is shown in Fig. 6.34. Determine the range of controller gain, *K*, for which the closed-loop system is
   - stable,
   - unstable,
   - overdamped, and
   - underdamped.
6. Derive the time-domain controller output (u_k) using an internal model control (IMC) for a process whose transfer function is given by:

$$HG_p(z) = \frac{0.3(1.2 + z)}{z^3(0.8 - z)(0.3 - z)} \tag{6.205}$$

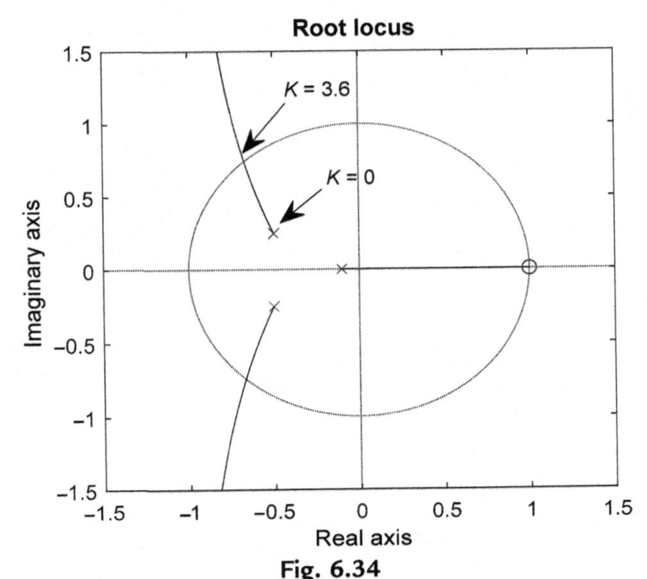

Fig. 6.34
Root locus diagram for Problem 16.6.6.

7. Discuss the type of the controller action (oscillatory or overdamped, ringing or nonringing, stable or unstable, …) of each of the following controllers.

$$D_1(z)=\frac{1-0.3z^{-1}}{1-1.8z^{-1}+z^{-2}}\quad D_2(z)=\frac{1+0.4z^{-1}}{1-1.6z^{-1}+0.6z^{-2}}$$
$$D_3(z)=\frac{z^2-0.9z}{z^2+0.1z-0.12}\qquad u_k=0.8u_{k-1}+0.9\varepsilon_k \tag{6.206}$$

where u_k and $_k$ are the controller action and the error signal at time $u_k=u(t=k\Delta t)$. Which controller results in offset and why?

8. Design a feedforward controller for a process with the following transfer functions

$$HG_p(z)=\frac{0.3z^3}{(z-0.3)(z-0.1)(z-0.4)};\quad HG_d(z)=\frac{0.3z}{z-0.3} \tag{6.207}$$

The transfer function of the measuring element on the disturbance signal is $0.8/(3s+1)$ and the sampling interval is 1 s. Sketch the block diagram and discuss the implementation of the feedforward controller in the time domain, i.e., write down the finite difference equation of the feedforward controller.

References

1. Cominos P, Munro N. PID controller: recent tuning methods and design to specification. *IEE Proc Control Theory Appl* 2002;**149**(1):46–53.
2. Lee J. On-line PID controller tuning from a single closed loop test. *AIChE J* 1989;**35**(2):329–31.
3. Smith CL. *Digital computer process control.* Hawaii: Intext Educational Publishers; 1972.
4. Garcia CE, Morari M. Internal model control. A unifying review and some new results. *Ind Eng Process Des Dev* 1982;**21**:308–23.

Further Reading

Zafirirou E, Morari M. Digital controllers for SISO systems: a review and a new algorithm. *Int J Control* 1985;**42** (4):855–76.

CHAPTER 7

Control System Design in the State Space and Frequency Domain

Sohrab Rohani, Yuanyi Wu
Western University, London, ON, Canada

7.1 State-Space Representation

The state-space representation of a continuous linear system is given by

$$\begin{cases} \frac{dx(t)}{dt} = \boldsymbol{A_c}x(t) + \boldsymbol{B_c}u(t) + \boldsymbol{D_c}d(t) \\ y(t) = \boldsymbol{C}x(t) \end{cases} \tag{7.1}$$

where x is the column vector of the system states with dimension (n); u is the manipulated variable vector of dimension (m); d is the disturbance vector of dimension (l); y is the vector of controlled variables of dimension (s); and A_c, B_c, C, and D_c are constant matrices for a linear time-invariant (LTI) stationary system with dimensions (n by n), (n by m), (n by l), and (s by n), respectively. For a SISO system $m = s = 1$. For a square MIMO system, that is a system in which the number of manipulated and controlled variables is the same, $m = s$. There are other variations of Eq. (4.16), for example, the state-space representation model used in MATLAB is of the form:

$$\begin{cases} \frac{dx(t)}{dt} = \boldsymbol{A_c}x(t) + \boldsymbol{B_c}u(t) \\ y(t) = \boldsymbol{C}x(t) + \boldsymbol{D_c}u(t) \end{cases} \tag{7.2}$$

Eq. (4.16) can be converted uniquely to Laplace domain by

$$y(s) = \boldsymbol{C}(s\boldsymbol{I} - \boldsymbol{A_c})^{-1}\boldsymbol{B_c}u(s) + \boldsymbol{C}(s\boldsymbol{I} - \boldsymbol{A_c})^{-1}\boldsymbol{D_c}d(s) \tag{7.3}$$

However, conversion from the Laplace domain (transfer function) to the state space is not unique. There are various conversion techniques such as the minimal realization and canonical presentation[1] for conversion from Laplace domain to the state space.

Coulson and Richardson's Chemical Engineering. http://dx.doi.org/10.1016/B978-0-08-101095-2.00007-2

Example 7.1

Given the state-space representation of a system by

$$A_c = \begin{bmatrix} -0.067 & 0.002 \\ -3.828 & -0.203 \end{bmatrix}; \quad B_c = \begin{bmatrix} 1.250 & 0 \\ -5.0 & 0.029 \end{bmatrix}$$
$$C = \begin{bmatrix} 1 & 0 \\ 0 & 1 \end{bmatrix}; \quad D_c = \begin{bmatrix} 0.025 & -0.2 & 0 & 0 \\ 0 & -2.0 & 0.025 & 0.005 \end{bmatrix} \tag{7.4}$$

Obtain the transfer function matrices of the system according to Eq. (7.3).

$$\begin{aligned} G_p(s) &= C(sI - A_c)^{-1}B_c \\ &= \begin{bmatrix} 1 & 0 \\ 0 & 1 \end{bmatrix} \begin{bmatrix} s+0.067 & -0.002 \\ +3.828 & s+0.203 \end{bmatrix}^{-1} \begin{bmatrix} 1.25 & 0 \\ -5.0 & 0.029 \end{bmatrix} \\ &= \frac{1}{s^2+0.37s+0.021} \begin{bmatrix} 1.25s+0.244 & 6\times 10^{-5} \\ -(5s+5.12) & 0.029s+0.002 \end{bmatrix} \end{aligned} \tag{7.5}$$

$$\begin{aligned} G_d(s) &= C(sI - A_c)^{-1}D_c \\ &= \frac{1}{s^2+0.37s+0.021} \begin{bmatrix} 0.025(s+0.203) & -0.2(s+0.403) & 5\times 10^{-5} & 1\times 10^{-5} \\ -0.096 & -2s+0.632 & 0.025(s+0.067) & 0.005(s+0.067) \end{bmatrix} \end{aligned} \tag{7.6}$$

7.1.1 The Minimal State-Space Realization

This method results in a state-space representation with minimum number of states. For simplicity let us assume that the poles of the process and load transfer function matrices are distinct and given by $-\lambda_i$ $(i=1, \ldots, p)$. Moreover, let us assume that the number of the controlled and manipulated variables is the same. So $G_p(s)$ would be a square matrix (m by m) and $G_d(s)$ a matrix (m by l). Using partial fraction

$$G_p(s) = C(sI - A_c)^{-1}B_c = \sum_{i=1}^{p} \frac{N_i}{s+\lambda_i} \tag{7.7}$$

where $-\lambda_i$ $(i=1, \ldots, p)$ are the distinct poles, N_i is matrix of residuals defined by

$$N_i = \lim_{s \to -\lambda_i} \left[(s+\lambda_i) G_p(s) \right] \tag{7.8}$$

The number of states, n, would be the sum of the ranks of the individual N_i, that is

$$n = \sum_{i=1}^{p} n_i \tag{7.9}$$

where n_i is the rank of N_i. The matrix A_c is chosen as follows:

$$A_c = \begin{bmatrix} a_{11} & \cdots & 0 \\ \vdots & \ddots & \vdots \\ 0 & \cdots & a_{pp} \end{bmatrix} \tag{7.10}$$

where $a_{ii} = -\lambda_i I_{n_i}$ with I_{n_i} being an identity matrix of dimension n_i. For the selection of B_c and C, note that N_i can be written as:

$$N_i = \sum_{j=1}^{n_i} c_{ij} b_{c,ij}^T \quad (i = 1,\ldots,p) \tag{7.11}$$

where c_{ij} and $b_{c,ij}^T$ are the column and row vectors of C and B_c, respectively.

$$B_c = \left[b_{c,11}\ldots b_{c,1n_1} | b_{c,21}\ldots b_{c,2n_2} | \ldots | b_{c,p_1}\ldots b_{c,pn_p}\right] \tag{7.12}$$

And

$$C = \left[c_{11}\ldots c_{1n_1} | c_{21}\ldots c_{2n_2} | \ldots | c_{p1}\ldots c_{pn_p}\right] \tag{7.13}$$

Similarly, for the load transfer function, we have

$$G_d(s) = C(sI - A_c)^{-1} D_c = \sum_{i=1}^{p} \frac{N_i^d}{s + \lambda_i} \tag{7.14}$$

where $N_i^d = \sum_{j=1}^{n_i} c_{ij} d_{c,ij}^T \quad i = 1,\ldots,p$. With

$$D_c = \left[d_{c\,11}\ldots d_{c\,1n_1} \;|\; d_{c\,21}\ldots d_{c\,2n_2} \;|\; \ldots \;|\; d_{c\,p1}\ldots d_{c\,pn_p}\right] \tag{7.15}$$

Example 7.2

The transfer function matrix of a MIMO system is given by

$$G_p(s) = \begin{bmatrix} \dfrac{3}{1+10s} & \dfrac{1}{1+10s} \\ \dfrac{2}{1+5s} & \dfrac{4}{1+5s} \end{bmatrix} \tag{7.16}$$

Obtain the minimal state-space representation of this plant.

Poles of $G_p(s)$ are at -0.1 and -0.2.

$$N_1 = \lim_{s \to -0.1} (s+0.1)\, G_p(s) = \begin{bmatrix} 0.3 & 0.1 \\ 0 & 0 \end{bmatrix} \tag{7.17}$$

$$N_2 = \lim_{s \to -0.2} (s+0.2)\, G_p(s) = \begin{bmatrix} 0 & 0 \\ 0.4 & 0.8 \end{bmatrix} \tag{7.18}$$

Ranks of N_1 and N_2 are both 1, that is, $n_1 = n_2 = 1$. So there will be 2 states $n_1 + n_2 = 2$.

$$A_c = \begin{bmatrix} -0.1 & 0 \\ 0 & -0.2 \end{bmatrix} \quad N_1 = \begin{bmatrix} 1 \\ 0 \end{bmatrix} [0.3 \ \ 0.1] \quad N_2 = \begin{bmatrix} 0 \\ 1 \end{bmatrix} [0.4 \ \ 0.8] \tag{7.19}$$

Therefore

$$\begin{cases} \dfrac{dx}{dt} = \begin{bmatrix} -0.1 & 0 \\ 0 & -0.2 \end{bmatrix} x + \begin{bmatrix} 0.3 & 0.1 \\ 0.4 & 0.8 \end{bmatrix} u \\ y = \begin{bmatrix} 1 & 0 \\ 0 & 1 \end{bmatrix} x \end{cases} \tag{7.20}$$

7.1.2 Canonical Form State-Space Realization

Consider a SISO system represented by the following transfer function:

$$\frac{y(s)}{u(s)} = \frac{Q(s)}{(s-p_1)(s-p_2)\ldots(s-p_n)} = \frac{a_1}{s-p_1} + \cdots + \frac{a_n}{s-p_n} \tag{7.21}$$

where all the transfer function poles are assumed to be distinct.

Define

$$\frac{x_i(s)}{u(s)} = \frac{1}{s-p_1} \tag{7.22}$$

which results in:

$$\begin{cases} \dfrac{dx_i}{dt} = p_i x_i + u \qquad \text{for } i = 1, \ldots, n \\ y = a_1 x_1 + \cdots + a_n x_n \end{cases} \tag{7.23}$$

The two matrices A_c and C are defined by:

$$A_c = \begin{bmatrix} p_1 & & & 0 \\ & p_2 & & \\ & & \ddots & \\ 0 & & & p_n \end{bmatrix} \quad B_c = \begin{bmatrix} 1 \\ \vdots \\ 1 \end{bmatrix} \quad C_c = [a_1 \ \ a_2 \ \ \ldots \ \ a_n] \tag{7.24}$$

The MATLAB commands for entering a matrix and conversion from the state space to transfer function (`ss2tf`) and the reverse (`tf2ss`) are as follows:

```
% entering matrices
A = [1 2; 3 2];     B = [2 1; 1 0]; C = [1 0; 0 1]; D = [1 4; 5 1];

% conversion from state space to transfer function,
% iu is the number of input (manipulated) variables.
[num, den] = ss2tf (A, B, C, D, iu)

% entering the transfer function
num = [3 1]; den = [2 2 1];

% conversion to state space (not unique)
[A, B, C, D] = tf2ss (num, den)

% 'method' refers to various techniques for the transformation
% such as 'modal' or 'companion'
[A, B, C, D, dt] = canon(a,b,c,d,'method')
```

7.1.3 Discretization of the Continuous State-Space Formulation

In order to design digital controllers for continuous plants, the plant dynamic models have to be discretized. This discretization can be performed either in the state space or in the Laplace domain. Following, the discretization from the continuous to discrete state space is discussed. The discrete state-space formulation is

$$\begin{cases} x_{k+1} = A\,x_k + B\,u_k \\ y_k = C\,x_k \end{cases} \tag{7.25}$$

where $x_{k+1} = x[(k+1)\Delta t]$ is the state of the plant at the $(k+1)$ th sampling interval. Laplace transform of Eq. (7.25) gives

$$x(s) = (s\boldsymbol{I} - \boldsymbol{A}_c)^{-1}x(0) + (s\boldsymbol{I} - \boldsymbol{A}_c)^{-1}\boldsymbol{B}_c\,u(s) \tag{7.26}$$

Inverse transforming of previous equation results in

$$x(t) = \boldsymbol{\phi}(t)\,x(0) + \int_0^t \boldsymbol{\phi}(t-\tau)\,\boldsymbol{B}_c\,u(\tau)d\tau \tag{7.27}$$

where $\boldsymbol{\phi}(t) = \mathcal{L}^{-1}\left[(s\boldsymbol{I} - \boldsymbol{A}_C)^{-1}\right]$ is called the 'state transition matrix' and $\mathcal{L}^{-1}$ stands for the inverse Laplace transform. For an arbitrary starting time t_0 we will have

$$x(t) = \boldsymbol{\phi}(t-t_0)\,x(t_0) + \int_{t_0}^t \boldsymbol{\phi}(t-\tau)\,\boldsymbol{B}_c\,u(\tau)d\tau \tag{7.28}$$

During $t = k\,\Delta t$ and $(k+1)\,\Delta t$, the control signal $u(t)$ is constant at $u(k\,\Delta t) = u_k$ in the presence of a zero-order hold (ZOH) device:

$$x_{k+1} = \boldsymbol{\phi}(\Delta t)\,x_k + \int_{k\Delta t}^{(k+1)\Delta t} \boldsymbol{\phi}[(k+1)\,\Delta t - \tau]\,d\tau\,\boldsymbol{B_c u}_k \tag{7.29}$$

Let $(k+1)\Delta t - \tau = \gamma,\;\; d\tau = -d\gamma$

$$x_{k+1} = \boldsymbol{\phi}(\Delta t)\,x_k + \boldsymbol{B_c u}_k \int_0^{\Delta t} \boldsymbol{\phi}(\gamma)d\gamma \tag{7.30}$$

So, the discrete state-space matrices are

$$\begin{cases} \boldsymbol{A} = \phi(\Delta t) = \mathcal{L}^{-1}\left[(s\boldsymbol{I} - \boldsymbol{A_C})^{-1}\right]\Big|_{t=\Delta t} \\ \boldsymbol{B} = \boldsymbol{B_c}\displaystyle\int_0^{\Delta t} \phi(\gamma)d\gamma \end{cases} \tag{7.31}$$

Example 7.3

Derive the discrete state-space representation of the following plant for a sampling interval of $\Delta t = 1\,\text{s}$.

$$\frac{d\boldsymbol{x}}{dt} = \begin{bmatrix} -1 & 1 \\ 0.7 & 0.4 \end{bmatrix}\boldsymbol{x} + \begin{bmatrix} 0 \\ 1 \end{bmatrix}u \tag{7.32}$$

$$\begin{aligned} \boldsymbol{\phi} &= \mathcal{L}^{-1}\left\{\begin{bmatrix} s & 0 \\ 0 & s \end{bmatrix} - \begin{bmatrix} -1 & 1 \\ 0.7 & 0.4 \end{bmatrix}\right\}^{-1} \\ &= \begin{bmatrix} 0.821\,e^{-1.391t} + 0.179\,e^{0.791t} & -0.458e^{-1.39t} + 0.458e^{0.791t} \\ -0.321e^{-1.391t} + 0.321e^{0.791t} & 0.179\,e^{-1.391t} + 0.821\,e^{0.791t} \end{bmatrix} \end{aligned} \tag{7.33}$$

$$\boldsymbol{A} = \phi(\Delta t) = \begin{bmatrix} 0.599 & 0.896 \\ 0.629 & 1.855 \end{bmatrix} \tag{7.34}$$

$$\boldsymbol{B} = \left[\int_0^1 \phi(\gamma)d\gamma\right]\begin{bmatrix} 0 \\ 1 \end{bmatrix} = \int_0^1 \begin{bmatrix} -0.458e^{-1.39t} + 0.458e^{0.791t} \\ 0.179e^{-1.391t} + 0.821e^{0.791t} \end{bmatrix} d\gamma = \begin{bmatrix} 0.451 \\ 1.347 \end{bmatrix} \tag{7.35}$$

The MATLAB command for conversion from the continuous state space to discrete state space is

```
[A, B] = c2d(Ac, Bc, dt)
```

7.1.4 Discretization of Continuous Transfer Functions

Another method to discretize plant models is to convert from the Laplace domain to its discrete z-domain which was discussed in Chapter 6. The mathematical definition of a sampler which is an electronic switch is given by

$$I=\sum_{n=0}^{\infty}\delta(t-n\,\Delta t) \tag{7.36}$$

Therefore the sampled version of a continuous signal, $f(t)$, can be written as

$$f(n\Delta t)=f_n=f^*(t)=\sum f(n\Delta t)\delta(t-n\Delta t)=\sum_{n=0}^{\infty} f(n\Delta t)e^{-n\,s\,\Delta t} \tag{7.37}$$

where the superscript (*) represents a discrete function. Taking Laplace transform of the discrete signal and representing $e^{\,s\Delta t}$ by the operator z, results in

$$f(z)=\sum_{n=0}^{\infty} f(n\Delta t)z^{-n} \tag{7.38}$$

There are many methods to convert plant transfer functions from the s-domain to z-domain. One method in which the plant is driven by a ZOH device is shown in Fig. 7.1:

$$y(z)=Z\{\mathcal{L}^{-1}[H(s)G_p(s)]\}^*u(z)+Z\{\mathcal{L}^{-1}[d(s)G_d(s)]\} \tag{7.39}$$

where the z operator represents the z-transform. The shorthand notation of the previous equation is

$$y(z)=HG_p(z)u(z)+dGd(z) \tag{7.40}$$

where $HG_p(z)$ is called pulse transfer function of the plant. The MATLAB command for this operation is

```
[numd,dend]=c2dm(num, den, dt, 'zoh')
```

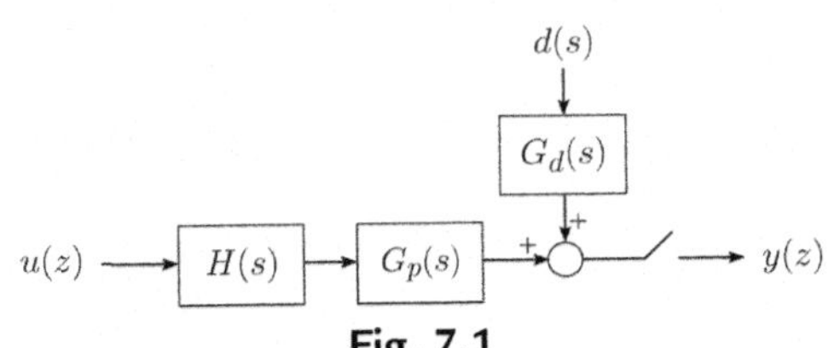

Fig. 7.1
Plant driven by a ZOH device.

The plant inputs consist of a discrete signal (controller output, u^*) and a continuous signal (disturbance, d) and the output of the plant is sampled before it is fed back to the controller. The transfer function of the zero-order hold device is

$$H(s) = \frac{1 - e^{-s\Delta t}}{s} \tag{7.41}$$

Example 7.4

Derive the pulse transfer function of the following plant whose output is sampled every $\Delta t = 1$ s.

$$G_p(s) = \frac{0.3e^{-3s}}{4s+1}, \quad G_d(s) = \frac{0.6e^{-2s}}{2s+1}, \quad d(s) = \frac{1}{s} \tag{7.42}$$

$$\begin{aligned} y(z) &= Z\left\{\mathcal{L}^{-1}\left[\frac{1-e^{-s\Delta t}}{s}\frac{0.3e^{-3s}}{4s+1}\right]\right\}u(z) + Z\left\{\mathcal{L}^{-1}\left[\frac{0.6e^{-2s}}{2s+1}\frac{1}{s}\right]\right\} \\ &= \left[0.3\left(1-z^{-1}\right)z^{-3}\right]Z\left\{\mathcal{L}^{-1}\left[\frac{1}{s} - \frac{1}{s+0.25}\right]\right\}^* u(z) + 0.6z^{-2}Z\left\{\mathcal{L}^{-1}\left[\frac{1}{s} - \frac{1}{s+0.5}\right]\right\}^* \\ &= 0.3\left(1-z^{-1}\right)z^{-3}\left[\frac{1}{1-z^{-1}} - \frac{1}{1-0.779z^{-1}}\right]u(z) + 0.6z^{-2}\left[\frac{1}{1-z^{-1}} - \frac{1}{1-0.606z^{-1}}\right] \end{aligned} \tag{7.43}$$

$$y(z) = \frac{0.066z^{-4}}{1-0.779z^{-1}}u(z) + \frac{0.236z^{-3}}{1-1.606z^{-1}+0.606z^{-2}} \tag{7.44}$$

7.1.5 Conversion of Plant Models From the Discrete State Space to the z-Domain

The conversion from the discrete state space to z-domain is unique. Consider the following plant:

$$\begin{cases} \boldsymbol{x}_{k+1} = \boldsymbol{A}\boldsymbol{x}_k + \boldsymbol{B}\boldsymbol{u}_k + \boldsymbol{D}\boldsymbol{d}_k \\ \boldsymbol{y}_k = \boldsymbol{C}\boldsymbol{x}_k \end{cases} \tag{7.45}$$

z-transforming of the equations and assuming $\boldsymbol{x}(0) = 0$, results in

$$\begin{cases} z\boldsymbol{x}(z) = \boldsymbol{A}\boldsymbol{x}(z) + \boldsymbol{B}\boldsymbol{u}(z) + \boldsymbol{D}\boldsymbol{d}(z) \\ \boldsymbol{y}(z) = \boldsymbol{C}\boldsymbol{x}(z) \end{cases} \tag{7.46}$$

which can be represented by

$$\boldsymbol{y}(z) = \boldsymbol{C}\left(z\boldsymbol{I} - \boldsymbol{A}\right)^{-1}\boldsymbol{B}\boldsymbol{u}(z) + \boldsymbol{C}\left(z\boldsymbol{I} - \boldsymbol{A}\right)^{-1}\boldsymbol{D}\boldsymbol{d}(z) \tag{7.47}$$

Therefore

$$G_p(z) = C(zI - A)^{-1}B = \begin{bmatrix} g_{p\,11} & \cdots & g_{p\,1m} \\ \vdots & \ddots & \\ g_{p\,s1} & & g_{p\,sm} \end{bmatrix} \tag{7.48}$$

$$G_d(z) = C(zI - A)^{-1}D = \begin{bmatrix} g_{d\,11} & \cdots & g_{d\,1l} \\ \vdots & \ddots & \\ g_{d\,s1} & & g_{d\,sl} \end{bmatrix} \tag{7.49}$$

Example 7.5

Derive the pulse transfer function of a plant given by

$$\begin{cases} x_{k+1} = \begin{bmatrix} 3 & 1 \\ 0.9 & -2 \end{bmatrix} x_k + \begin{bmatrix} 2 \\ 1 \end{bmatrix} u_k \\ y_k = \begin{bmatrix} 1 & 0 \end{bmatrix} x_k \end{cases} \tag{7.50}$$

Following the previous procedure, results in

$$G_p(z) = \begin{bmatrix} 1 & 0 \end{bmatrix} \begin{bmatrix} z-3 & -1 \\ -0.9 & z+2 \end{bmatrix}^{-1} \begin{bmatrix} 2 \\ 1 \end{bmatrix} = \frac{2z+5}{z^2 - z - 6.9} \tag{7.51}$$

7.1.6 Conversion From z-Domain to Discrete State Space

Conversion from z-domain to discrete state space is not unique, that is, there are different state-space representations for the same transfer function. We will consider the canonical realization for a SISO transfer function given by

$$HG_p(z) = \frac{y(z)}{u(z)} = \frac{b_0 + b_1 z^{-1} + \cdots + b_n z^{-n}}{1 + a_1 z^{-1} + \cdots + a_n z^{-n}} \tag{7.52}$$

If the order of the numerator polynomial is equal to that of the denominator, the transfer function is made "proper" by long division

$$\frac{y(z)}{u(z)} = b_0 + \frac{(b_1 - b_0 a_1) z^{n-1} + \cdots + (b_n - b_0 a_n)}{z^n + a_1 z^{n-1} + \cdots + a_n} \tag{7.53}$$

Case 1 Assuming that all the poles of $HG_p(z)$ are distinct:

$$\frac{y(z)}{u(z)} = b_0 + \frac{a_1}{z - p_1} + \cdots + \frac{a_n}{z - p_n} \tag{7.54}$$

Define:

$$\frac{x_i(z)}{u(z)} = \frac{1}{z - p_i} \quad \text{for } i = 1, \ldots, n \tag{7.55}$$

which results in

$$\begin{cases} x_i(k+1) = p_i x_i(k) + u(k) \quad \text{for } i = 1, \ldots, n \\ y(k) = b_0\, u(k) + \sum_{i=1}^{n} a_i x_i(k) \end{cases} \tag{7.56}$$

Therefore the state-space representation is given by

$$\begin{cases} \boldsymbol{x}_{k+1} = \boldsymbol{A}\,\boldsymbol{x}_k + \boldsymbol{b}\,\boldsymbol{u}_k \\ \boldsymbol{y}_k = \boldsymbol{C}\,\boldsymbol{x}_k + \boldsymbol{b}_0\,\boldsymbol{u}_k \end{cases} \tag{7.57}$$

where $\boldsymbol{A} = \begin{bmatrix} p_1 & & 0 \\ & \ddots & \\ 0 & & p_n \end{bmatrix}$, $\boldsymbol{b}^T = [1 \ \ldots \ 1]$, and $\boldsymbol{C} = [a_1 \ a_2 \ \ldots \ a_n]$

Example 7.6

Derive the canonical state-space realization of a plant given by

$$HG_p(z) = \frac{2z^2 + z + 6}{z^2 - 1} \tag{7.58}$$

Long division and partial fraction result in

$$HG_p(z) = \frac{y(z)}{u(z)} = 2 - \frac{3.5}{z+1} + \frac{4.5}{z-1} \tag{7.59}$$

$$\boldsymbol{A} = \begin{bmatrix} -1 & 0 \\ 0 & 1 \end{bmatrix} \quad \boldsymbol{b} = \begin{bmatrix} 1 \\ 1 \end{bmatrix} \quad \boldsymbol{C} = [-3.5 \ \ 4.5] \tag{7.60}$$

which results in

$$\begin{cases} \boldsymbol{x}_{k+1} = \boldsymbol{A}\,\boldsymbol{x}_k + \boldsymbol{b}\,\boldsymbol{u}_k \\ \boldsymbol{y}_k = \boldsymbol{C}\,\boldsymbol{x}_k + 2\,\boldsymbol{u}_k \end{cases} \tag{7.61}$$

Case 2 Let us assume that $HG_p(z)$ has a repeated pole, p_1, with multiplicity q.

$$\frac{y(z)}{u(z)} = b_0 + \frac{a_{1q}}{(z-p_1)^q} + \cdots + \frac{a_{11}}{(z-p_1)} + \frac{a_2}{(z-p_2)} + \cdots + \frac{a_{n-q+1}}{(z-p_{n-q+1})} \tag{7.62}$$

Define: $\dfrac{x_1(z)}{u(z)} = \dfrac{1}{z - p_1}$

$$\begin{gathered}
\frac{x_2(z)}{u(z)} = \frac{1}{(z-p_1)^2} = \frac{x_1(z)}{u(z)}\frac{1}{z-p_1} \\
\vdots \\
\frac{x_q(z)}{u(z)} = \frac{1}{(z-p_1)^q} = \frac{x_{q-1}(z)}{u(z)}\frac{1}{z-p_1} \\
\vdots \\
\frac{x_{q+1}(z)}{u(z)} = \frac{1}{z-p_2} \\
\vdots \\
\frac{x_n(z)}{u(z)} = \frac{1}{z-p_{n-q+1}}
\end{gathered} \tag{7.63}$$

which results in

$$\begin{cases}
x_1(k+1) = p_1 x_1(k) + u(k) \\
x_2(k+1) = p_1 x_2(k) + x_1(k) \\
\vdots \\
x_q(k+1) = p_1 x_q(k) + x_{q-1}(k) \\
x_{q+1}(k+1) = p_2 x_{q+1}(k) + u(k) \\
\vdots \\
x_n(k+1) = p_{n-q+1} x_n(k) + u(k)
\end{cases} \tag{7.64}$$

The state-space realization would be

$$\begin{bmatrix} x_1(k+1) \\ x_2(k+1) \\ \vdots \\ x_q(k+1) \\ x_{q+1}(k+1) \\ \vdots \\ x_n(k+1) \end{bmatrix} = \begin{bmatrix} p_1 & 0 & & & & & \\ 1 & p_1 & 0 & & & & \\ & 1 & \ddots & \ddots & & & \\ & & \ddots & p_2 & & & \\ & & & & p_2 & & \\ & & & & & \ddots & 0 \\ & & & & & 1 & p_{n-q+1} \end{bmatrix} \begin{bmatrix} x_1(k+1) \\ x_2(k+1) \\ \vdots \\ x_q(k+1) \\ x_{q+1}(k+1) \\ \vdots \\ x_n(k+1) \end{bmatrix} + \begin{bmatrix} 1 \\ 0 \\ \vdots \\ 0 \\ 1 \\ \vdots \\ 1 \end{bmatrix} u(k) \tag{7.65}$$

$$y(k) = [a_{11} \ \ldots \ a_{1q} \ a_2 \ \ldots \ a_{n-q+1}] x_k + b_0\, u(k) \tag{7.66}$$

Example 7.7

Derive the canonical state-space representation of the following plant.

$$HG_p(z) = \frac{y(z)}{u(z)} = \frac{2z+1}{(z-0.5)^2(z-0.6)} = \frac{-20}{(z-0.5)^2} - \frac{220}{z-0.5} + \frac{220}{z-0.6} \tag{7.67}$$

Following the previous procedure results in

$$\begin{cases} \boldsymbol{x}_{k+1} = \begin{bmatrix} 0.5 & 0 & 0 \\ 1 & 0.5 & 0 \\ 0 & 0 & 0.6 \end{bmatrix} \boldsymbol{x}_k + \begin{bmatrix} 1 \\ 0 \\ 1 \end{bmatrix} \boldsymbol{u}_k \\ \boldsymbol{y}_k = [-20 \quad 220 \quad -220] \boldsymbol{x}_k + 0\, \boldsymbol{u}_k \end{cases} \tag{7.68}$$

7.2 Design of Controllers in the State Space

In what follows simple design concepts in the state-space domain will be introduced. Design methodology in the state space has advanced tremendously in the last few decades. This chapter only provides the basic tools and methodologies to investigate the wealth of knowledge in this area in the literature.

7.2.1 Solution of the State-Space Equation

The simplest method of solving the state-space equation is the power series approach. Consider the system given as follows:

$$\begin{cases} \boldsymbol{x}_{k+1} = \boldsymbol{A}\,\boldsymbol{x}_k + \boldsymbol{B}\,\boldsymbol{u}_k \\ \boldsymbol{y}_k = \boldsymbol{C}\,\boldsymbol{x}_k \end{cases} \tag{7.69}$$

where $\boldsymbol{x}$ is an (n) column vector, $\boldsymbol{A}$ is an $(n \times n)$ matrix, $\boldsymbol{B}$ is an $(n \times m)$ matrix, $\boldsymbol{u}_i$ is an (m) column vector. Assuming that the state and manipulated variable vectors at $t=0$, i.e., x_0 and u_0, are known, the future state vector can be determined as follows:

$$\boldsymbol{x}_1 = \boldsymbol{A}\,\boldsymbol{x}_0 + \boldsymbol{B}\,\boldsymbol{u}_0 \tag{7.70}$$

$$\boldsymbol{x}_2 = \boldsymbol{A}\,\boldsymbol{x}_1 + \boldsymbol{B}\,\boldsymbol{u}_1 = \boldsymbol{A}^2\,\boldsymbol{x}_0 + \boldsymbol{AB}\,\boldsymbol{u}_0 + \boldsymbol{B}\,\boldsymbol{u}_1 \tag{7.71}$$

And in general

$$\boldsymbol{x}_n = \boldsymbol{A}^n\,\boldsymbol{x}_0 + \sum_{i=0}^{n-1} \boldsymbol{A}^{n-i-1}\,\boldsymbol{B}\,\boldsymbol{u}_i \tag{7.72}$$

The solution of the state equation given earlier is used to derive some useful concepts in the control theory which are discussed in the following sections.

7.2.2 Controllability

A system is said to be controllable if it can be brought from an arbitrary initial state x_0 to a final state x_n in n sampling intervals. Consider the following system.[1]

$$\begin{cases} \boldsymbol{x}_{k+1} = \boldsymbol{A}\,\boldsymbol{x}_k + \boldsymbol{B}\,\boldsymbol{u}_k \\ \boldsymbol{y}_k = \boldsymbol{C}\,\boldsymbol{x}_k \end{cases} \tag{7.73}$$

which has a solution given by

$$\boldsymbol{x}_n = \boldsymbol{A}^n\,\boldsymbol{x}_0 + \sum_{i=0}^{n-1} \boldsymbol{A}^{n-i-1}\,\boldsymbol{B}\,\boldsymbol{u}_i \tag{7.74}$$

Rearranging the last equation results in

$$\boldsymbol{x}_n - \boldsymbol{A}^n\,\boldsymbol{x}_0 = \left[\boldsymbol{A}^{n-1}\boldsymbol{B}\ \ \boldsymbol{A}^{n-2}\boldsymbol{B}\ \ \cdots\boldsymbol{B}\right] \begin{bmatrix} \boldsymbol{u}_0 \\ \vdots \\ \boldsymbol{u}_{n-1} \end{bmatrix} \tag{7.75}$$

The matrix $\left[\boldsymbol{A}^{n-1}\boldsymbol{B}\ \ \boldsymbol{A}^{n-2}\boldsymbol{B}\ \ \cdots\boldsymbol{B}\right]$ is referred to as the controllability matrix and is of $(n \times nm)$ dimension. In order for the system to be controllable the rank of the controllability matrix must be n.

Example 7.8

Determine the controllability of the following system:

$$\boldsymbol{x}_{k+1} = \begin{bmatrix} 3 & 1 \\ 0.9 & -2 \end{bmatrix} \boldsymbol{x}_k + \begin{bmatrix} 2 \\ 1 \end{bmatrix} \boldsymbol{u}_k \tag{7.76}$$

The controllability matrix would be

$$[\boldsymbol{AB}\ \ \boldsymbol{B}] = \left[\begin{bmatrix} 3 & 1 \\ 0.9 & -2 \end{bmatrix}\begin{bmatrix} 2 \\ 1 \end{bmatrix}\ \begin{bmatrix} 2 \\ 1 \end{bmatrix}\right] = \begin{bmatrix} 7 & 2 \\ -0.2 & 1 \end{bmatrix} \tag{7.77}$$

Which has a rank of 2, the same as the number of system states. Accordingly, the system is controllable.

7.2.3 Observability

The concept of observability determines whether the initial state of the system $\boldsymbol{x}_0$ can be found, given the sequence of measurements $\boldsymbol{y}_0, \boldsymbol{y}_1, \ldots, \boldsymbol{y}_{n-1}$.

The state-space representation of the system is given by

$$\begin{cases} \boldsymbol{x}_{k+1} = \boldsymbol{A}\boldsymbol{x}_k + \boldsymbol{B}\boldsymbol{u}_k \\ \boldsymbol{y}_k = \boldsymbol{C}\boldsymbol{x}_k \end{cases} \tag{7.78}$$

The solution of the state equation in terms of $\boldsymbol{y}_k$ is

$$\boldsymbol{y}_k = \boldsymbol{C}\boldsymbol{A}^n \boldsymbol{x}_0 + \sum_{i=0}^{k-1} \boldsymbol{A}^{k-i-1} \boldsymbol{B}\boldsymbol{u}_i \tag{7.79}$$

Which can be written as

$$\begin{bmatrix} [\boldsymbol{y}_0] \\ [\boldsymbol{y}_1 - \boldsymbol{C}\boldsymbol{B}\boldsymbol{u}_0] \\ \vdots \\ \left[\boldsymbol{y}_{n-1} - \boldsymbol{C}\sum_{i=0}^{n-2} \boldsymbol{A}^{n-2-i} \boldsymbol{B}\boldsymbol{u}_i\right] \end{bmatrix} = \begin{bmatrix} \boldsymbol{C} \\ \boldsymbol{C}\boldsymbol{A} \\ \vdots \\ \boldsymbol{C}\boldsymbol{A}^{n-1} \end{bmatrix} \boldsymbol{x}_0 \tag{7.80}$$

The term on the left-hand side of the previous equation is a column vector of dimension $(ns \times 1)$. The observability matrix is defined by

$$\begin{bmatrix} \boldsymbol{C} \\ \boldsymbol{C}\boldsymbol{A} \\ \vdots \\ \boldsymbol{C}\boldsymbol{A}^{n-1} \end{bmatrix} \tag{7.81}$$

Which has a dimension $(sn \times n)$ and $\boldsymbol{x}_0$ is an (n) column vector. The system would be observable if the observability matrix has a rank of n.

Example 7.9

Determine the observability of the system defined by

$$\begin{cases} \boldsymbol{x}_{k+1} = \begin{bmatrix} 3 & 1 \\ 0.9 & -2 \end{bmatrix} \boldsymbol{x}_k + \begin{bmatrix} 2 \\ 1 \end{bmatrix} \boldsymbol{u}_k \\ \boldsymbol{y}_k = [1 \;\; 0]\boldsymbol{x}_k \end{cases} \tag{7.82}$$

The observability matrix

$$\begin{bmatrix} C \\ CA \end{bmatrix} = \begin{bmatrix} 1 & 0 \\ 3 & 1 \end{bmatrix} \tag{7.83}$$

Which has a rank of 2 equal to the number of system states suggesting that the system is observable.

7.2.4 The State Feedback Regulator (SFR)[1]

Assuming that all system states are available (measurable), a state feedback regulator may be designed to control the output around the zero state:

$$\boldsymbol{u}_k = -\boldsymbol{K}\boldsymbol{x}_k \tag{7.84}$$

where $\boldsymbol{K}$ is the state feedback regulator (SFR) gain matrix. In order to determine the elements of $\boldsymbol{K}$, $\boldsymbol{u}_k$ is substituted in the state equation and z-transformed.

$$\boldsymbol{x}_{k+1} = \boldsymbol{A}\boldsymbol{x}_k - \boldsymbol{B}\boldsymbol{K}\boldsymbol{x}_k = (\boldsymbol{A} - \boldsymbol{B}\boldsymbol{K})\boldsymbol{x}_k \tag{7.85}$$

$$\boldsymbol{x}(z) = (z\boldsymbol{I} - \boldsymbol{A} + \boldsymbol{B}\boldsymbol{K})^{-1} z\boldsymbol{x}_0 \tag{7.86}$$

The stability of the closed-loop system depends on the roots of determinant $|z\boldsymbol{I} - \boldsymbol{A} + \boldsymbol{B}\boldsymbol{K}| = 0$ which is the closed-loop characteristic equation of the system. Note that for n states and m manipulated variables, the state feedback gain matrix $\boldsymbol{K}$ will have $n \times m$ unknown elements. One can specify n desired closed-loop poles and have $n \times (m - n)$ degrees of freedom. There are techniques to reduce the number of the degrees of freedom to zero.

Example 7.10

Design a state feedback regulator for the system given by

$$\begin{cases} \boldsymbol{x}_{k+1} = \begin{bmatrix} 3 & 1 \\ 0.9 & -2 \end{bmatrix} \boldsymbol{x}_k + \begin{bmatrix} 2 \\ 1 \end{bmatrix} \boldsymbol{u}_k \\ y_k = [1 \quad 0]\boldsymbol{x}_k \end{cases} \tag{7.87}$$

Which places two closed-loop poles at +0.5 and +0.6

$$|z\boldsymbol{I} - \boldsymbol{A} + \boldsymbol{B}\boldsymbol{K}| = (z - 0.5)(z - 0.6) \tag{7.88}$$

$$\left\| \begin{bmatrix} z-3 & -1 \\ -0.9 & z+2 \end{bmatrix} + \begin{bmatrix} 2k_1 & 2k_2 \\ k_1 & k_2 \end{bmatrix} \right\| = z^2 + (2k_1 + k_2 - 1)z + (5k_1 - 1.2k_2 - 6.9) \tag{7.89}$$

$$\boldsymbol{K} = [2.72 \quad -5.54]$$

7.2.5 The State Feedback Control With Incomplete State Information

Consider a plant given by

$$\begin{cases} \boldsymbol{x}_{k+1} = \boldsymbol{A}\boldsymbol{x}_k + \boldsymbol{B}\boldsymbol{u}_k \\ \boldsymbol{y}_k = \boldsymbol{C}\boldsymbol{x}_k \end{cases} \tag{7.90}$$

Let us assume that some of the states are not measurable, but the system is both controllable and observable. One possible way to control such a system is to use the available measured variables to construct the control law

$$u_k = -K y_k \tag{7.91}$$

Another approach is to design an estimator or an observer to estimate the unmeasurable states and construct the control law as a linear combination of the estimated states

$$u_k = -K \hat{x}_k \tag{7.92}$$

An obvious choice of the estimator model would be a model similar to the plant dynamics but corrected by the measured estimation error. So the observer model would be of the form:

$$x_{k+1} = A x_k + B u_k + L (y_k - C \hat{x}_k) \tag{7.93}$$

where L is the estimator gain matrix. If the estimator is stable, the estimation error approaches zero asymptotically. The estimator gain matrix can be determined by locating the poles of the error dynamics at some desired locations.

$$|zI - A + LC| \tag{7.94}$$

The dynamics of the estimator should be faster than those of the closed-loop system roughly by a factor of 2. If the output is a scalar (i.e., a single-input, single-output system), L would be a vector and its elements can be found uniquely by the specified pole locations of the estimator.

If the estimator is used in conjunction with a state feedback regulator, the "separation principle" suggests that the observer/estimator and the controller can be designed separately. After the elements of L are selected to give the desired closed-loop pole locations, then the elements of K are calculated.

$$\hat{x}_{k+1} = A \hat{x}_k + B K \hat{x}_k + L (y_k - C \hat{x}_k) \tag{7.95}$$

$$\hat{x}_{k+1} = [A - BK - LC] \hat{x}_k + L y_k \tag{7.96}$$

If u_k and y_k are both scalars, a transfer function of the observer–controller combination can be derived by z-transforming of the previous equation.

$$\hat{x}(z) = (zI - A + BK + LC)^{-1} + L y(z) \tag{7.97}$$

And using $u(z) = -K \hat{x}(z)$

$$D(z) = \frac{u(z)}{-y(z)} = -K(zI - A + BK + LC)^{-1} L \tag{7.98}$$

With a characteristic equation given by

$$|zI - A + BK + LC| \tag{7.99}$$

Example 7.11

Design an observer and a state feedback regulator for the following system

$$\begin{cases} x_{k+1} = \begin{bmatrix} 1 & 0.5 \\ 3 & 2 \end{bmatrix} x_k + \begin{bmatrix} 0.1 \\ 0.6 \end{bmatrix} u_k \\ y_k = [1 \quad 0] x_k \end{cases} \tag{7.100}$$

The observer poles are to be located at +0.3 and +0.4 while the regulator poles are at +0.8 and +0.9.

The observer dynamics is expressed by

$$\hat{x}_{k+1} = A\hat{x}_k + B u_k + L(y_k - C\hat{x}_k) \tag{7.101}$$

Taking the z-transform of this equation results in the observer characteristic equation given by

$$|zI - A + LC| = (z - 0.3)(z - 0.4) \tag{7.102}$$

$$\begin{vmatrix} z - 1 + l_1 & -0.5 \\ -3 + l_2 & z - 2 \end{vmatrix} = z^2 - 0.7z + 0.12 \tag{7.103}$$

$$L = \begin{bmatrix} 2.3 \\ 8.44 \end{bmatrix}$$

The SISO state feedback regulator would be $u_k = -K\hat{x}_k$

$$\hat{x}_{k+1} = A\hat{x}_k - BK\hat{x}_k + L(y_k - C\hat{x}_k) \tag{7.104}$$

With a characteristic equation given by

$$|zI - A + LC| = (z - 0.8)(z - 0.9) \tag{7.105}$$

$$\begin{vmatrix} z + 1.3 + 0.1k_1 & -0.5 + 0.1k_2 \\ 5.44 + 0.6k_1 & z - 2 + 0.6k_2 \end{vmatrix} = z^2 - 1.7z + 0.72 \tag{7.106}$$

$$K = [-16.59 \quad 1.1]$$

7.2.6 Time Optimal Control

Given the system,

$$\begin{cases} x_{k+1} = A x_k + B u_k \\ y_k = C x_k \end{cases} \tag{7.107}$$

it is desired to find out a sequence of control actions over a horizon, 0, 1, ..., $N-1$ that minimizes a performance index defined by

$$\min_{u_0, \cdots, u_{N-1}} J = \frac{1}{2} x_N^T H x_N + \sum_{k=0}^{N-1} \left[\frac{1}{2} x_k^T Q x_k + \frac{1}{2} u_k^T R u_k \right] \tag{7.108}$$

where $\boldsymbol{H}$ and $\boldsymbol{Q}$ are user-defined positive semidefinite symmetric matrices and $\boldsymbol{R}$ is a positive definite symmetric weighting matrix.[2]

In order to derive the optimal control sequence, we augment the performance index by

$$\boldsymbol{\lambda}_{k+1}^{T}\left(\boldsymbol{A}\boldsymbol{x}_{k}+\boldsymbol{B}\boldsymbol{u}_{k}-\boldsymbol{x}_{k+1}\right) \tag{7.109}$$

Which is virtually zero. In Eq. (7.109), λ is the Lagrangian multiplier vector. After augmentation of the performance index, the derivative of $\boldsymbol{J}$ is set to zero:

$$dJ=\boldsymbol{x}_{N}^{T}\boldsymbol{H}d\boldsymbol{x}_{N}+\sum_{k=0}^{N-1}\left[\boldsymbol{x}_{k}^{T}\boldsymbol{Q}\,d\boldsymbol{x}_{k}+\boldsymbol{u}_{k}^{T}\boldsymbol{R}\,d\boldsymbol{u}_{k}\right]+\lambda_{k+1}^{T}\left[\boldsymbol{A}\,d\boldsymbol{x}_{k}+\boldsymbol{B}\,d\boldsymbol{u}_{k}-d\boldsymbol{x}_{k+1}\right]=0 \tag{7.110}$$

Which results in

$$dJ=(\boldsymbol{x}_{N}^{T}\boldsymbol{H}-\boldsymbol{\lambda}_{N}^{T})d\boldsymbol{x}_{N}+\sum_{k=1}^{N-1}[\boldsymbol{x}_{k}^{T}\boldsymbol{Q}+\boldsymbol{\lambda}_{k+1}^{T}\boldsymbol{A}-\boldsymbol{\lambda}_{k}^{T}]\,d\boldsymbol{x}_{k}+\sum_{k=0}^{N-1}[\boldsymbol{u}_{k}^{T}\boldsymbol{R}+\boldsymbol{\lambda}_{k+1}^{T}\boldsymbol{B}]\,d\boldsymbol{u}_{k}=0 \tag{7.111}$$

The previous equation results in the following three equations:

$$\lambda_{k}=\boldsymbol{Q}\boldsymbol{x}_{k}+\boldsymbol{A}^{T}\lambda_{k+1} \tag{7.112}$$

$$u_{k}=-\boldsymbol{R}^{-1}\boldsymbol{B}^{T}\lambda_{k+1} \tag{7.113}$$

$$\lambda_{N}=\boldsymbol{H}\boldsymbol{x}_{N} \tag{7.114}$$

Substitution of Eq. (7.113) in Eq. (7.107) yields

$$\boldsymbol{x}_{k+1}=\boldsymbol{A}\boldsymbol{x}_{k}-\boldsymbol{B}\boldsymbol{R}^{-1}\boldsymbol{B}^{T}\lambda_{k+1} \tag{7.115}$$

Rearranging Eq. (7.112) gives

$$\lambda_{k+1}=\left(\boldsymbol{A}^{T}\right)^{-1}\left(\lambda_{k}-\boldsymbol{Q}\boldsymbol{x}_{k}\right) \tag{7.116}$$

Eq. (7.114) can be generalized as:

$$\lambda_{k}=\boldsymbol{P}_{k}\boldsymbol{x}_{k} \tag{7.117}$$

When substituted in Eqs. (7.117), (7.112), we have

$$\boldsymbol{x}_{k+1}=\left(\boldsymbol{I}+\boldsymbol{B}\boldsymbol{R}^{-1}\boldsymbol{B}^{T}\boldsymbol{P}_{k+1}\right)^{-1}\boldsymbol{A}\boldsymbol{x}_{k} \tag{7.118}$$

$$\boldsymbol{P}_{k}\boldsymbol{x}_{k}=\boldsymbol{Q}\boldsymbol{x}_{k}+\boldsymbol{A}^{T}\boldsymbol{P}_{k+1}\left(\boldsymbol{I}+\boldsymbol{B}\boldsymbol{R}^{-1}\boldsymbol{B}^{T}\boldsymbol{P}_{k+1}\right)^{-1}\boldsymbol{A}\boldsymbol{x}_{k} \tag{7.119}$$

Taking $\boldsymbol{P}_{k+1}$ inside the bracket and using the matrix equation $\left(\boldsymbol{C}\boldsymbol{M}\boldsymbol{N}^{-1}\right)=\left(\boldsymbol{M}\boldsymbol{N}\boldsymbol{C}^{-1}\right)$ where $\boldsymbol{M}=\left(\boldsymbol{B}\boldsymbol{R}^{-1}\boldsymbol{B}^{T}\right)$, $\boldsymbol{N}^{-1}=\boldsymbol{P}_{k+1}$, and $\boldsymbol{C}=\boldsymbol{P}_{k+1}^{-1}$, Eq. (7.119) can be written:

$$P_k = Q + A^T\left(P_{k+1}^{-1} + BR^{-1}B^T\right)^{-1}A \tag{7.120}$$

Eq. (7.120) can be expanded using the well-known matrix inversion lemma.

$$(M + NCD)^{-1} = M^{-1} - M^{-1}N\left(C^{-1} + DM^{-1}N\right)^{-1}DM^{-1} \tag{7.121}$$

$$P_k = A^T S_{k+1} A + Q \tag{7.122}$$

where

$$S_{k+1} = P_{k+1} - P_{k+1}B\left(B^T \mathbf{P}_{k+1}B + R\right)^{-1}B^T P_{k+1} \tag{7.123}$$

The last two recursive equations are known as the discrete Riccati equations. The optimal control sequence is given by

$$u_k = -R^{-1}B^T\left(A^T\right)^{-1}(P_k - Q)x_k \tag{7.124}$$

The algorithm starts with $P_N = H$, then S_N is calculated using Eq. (7.123), P_{N-1} is calculated using Eq. (7.122). The sequence is repeated until all the control sequence $u_N, u_{N-1}, \cdots, u_0$ is calculated.

Example 7.12

Derive the time optimal control sequence for the system given by the state equation

$$\begin{cases} x_{k+1} = \begin{bmatrix} 0.6 & 0 & 0 \\ 1 & 0.6 & 0 \\ 0 & 0 & 0.8 \end{bmatrix} x_k + \begin{bmatrix} 1 \\ 0 \\ 1 \end{bmatrix} u_k \\ y_k = \begin{bmatrix} -7.5 & -0.5 & 7.5 \end{bmatrix} x_k \end{cases} \tag{7.125}$$

And minimize:

$$J = \frac{1}{2}x_4 H x_4 + \frac{1}{2}\sum_{i=1}^{3}\left[x_i^T Q x_i + u_i^T R u_i\right] \tag{7.126}$$

Where $H = 3I$, $Q = 2I$, and $R = 1$

Starting with $P_4 = H$, the solution to the Riccati equation and the optimal control sequence may be obtained. The results are

$$S_4 = \begin{bmatrix} 1.714 & 0 & -1.29 \\ 0 & 3 & 0 \\ -1.29 & 0 & 1.71 \end{bmatrix} \quad P_3 = \begin{bmatrix} 5.62 & 1.8 & -0.62 \\ 1.80 & 3.08 & 0 \\ -0.62 & 0 & 3.10 \end{bmatrix} \tag{7.127}$$

$$u_3 = \begin{bmatrix} -0.257 & 0 & -0.343 \end{bmatrix} x_3 \tag{7.128}$$

$$S_3 = \begin{bmatrix} 2.67 & 0.739 & -2.08 \\ 0.739 & 2.7 & -0.526 \\ -2.08 & -0.526 & 2.37 \end{bmatrix} \quad P_2 = \begin{bmatrix} 6.55 & 1.88 & -1.42 \\ 1.88 & 2.97 & -0.253 \\ -1.42 & -0.253 & 3.52 \end{bmatrix} \tag{7.129}$$

$$u_2 = [-0.566 \;\; -0.127 \;\; -0.237]x_2 \tag{7.130}$$

$$S_2 = \begin{bmatrix} 3.35 & 0.868 & -2.73 \\ 0.868 & 2.65 & -0.669 \\ -2.73 & -0.669 & 2.98 \end{bmatrix} \quad P_1 = \begin{bmatrix} 6.89 & 1.90 & -1.84 \\ 1.90 & 2.95 & -0.321 \\ -1.84 & -0.34 & 3.91 \end{bmatrix} \tag{7.131}$$

$$u_1 = [-0.572 \;\; -0.119 \;\; -0.204]x_1 \tag{7.132}$$

$$S_1 = \begin{bmatrix} 3.75 & 0.918 & -3.13 \\ 0.918 & 2.65 & -0.723 \\ -3.13 & -0.723 & 3.30 \end{bmatrix} \quad P_0 = \begin{bmatrix} 7.10 & 1.92 & -2.08 \\ 1.92 & 2.95 & -0.347 \\ -2.08 & -0.347 & 4.17 \end{bmatrix} \tag{7.133}$$

$$u_0 = [-0.568 \;\; -0.117 \;\; -0.204]x_0 \tag{7.134}$$

7.3 Frequency Response of Linear Systems and the Design of PID Controllers in the Frequency Domain

The objectives of this section are to

- introduce the frequency response technique
- present a shortcut approach to derive the amplitude ratio (AR) and the phase difference (Φ) of a transfer function
- study the graphical and numerical methods (MATLAB) to draw the Bodé and Nyquist diagrams, and
- use the frequency response technique for the
  - system identification
  - controller design (stability test in the frequency domain)

7.3.1 Definition of the Amplitude Ratio and Phase Difference of a Linear System

If the input to a linear system is changed by a sine wave, the output at steady state (as $t \rightarrow \infty$) is also a sine wave with the same frequency but a different amplitude and a different phase difference (see Fig. 7.2). The ratio of the amplitude of the output signal to the amplitude of the input signal is called the amplitude ratio $AR(\omega) = \frac{B}{A}$ and the phase difference between the

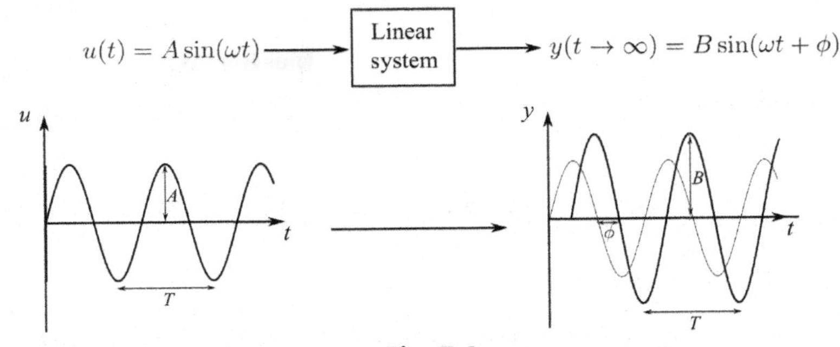

Fig. 7.2
Linear system frequency response.

output and input signals is designated by $\phi(\omega)$. Both AR and ϕ are functions of the frequency of the input sine wave.

To ascertain the earlier statement, let us consider a first-order system that is disturbed by a sinusoidal input.

$$G(s) = \frac{Y(s)}{U(s)} = \frac{K}{\tau s + 1} \tag{7.135}$$

$$U(s) = \frac{A\omega}{s^2 + \omega^2} \tag{7.136}$$

Therefore

$$Y(s) = \left(\frac{K}{\tau s + 1}\right)\left(\frac{A\omega}{s^2 + \omega^2}\right) \tag{7.137}$$

$$\begin{aligned} y'(t) &= \mathcal{L}^{-1}\{Y(s)\} = \mathcal{L}^{-1}\left\{\frac{a}{\tau s + 1} + \frac{b}{s + j\omega} + \frac{c}{s - j\omega}\right\} \\ &= \frac{KA}{(\tau\omega)^2 + 1} e^{-\frac{t}{\tau}} + \frac{KA}{\sqrt{(\tau\omega)^2 + 1}} \sin\left\{\omega t + \tan^{-1}(-\tau\omega)\right\} \end{aligned} \tag{7.138}$$

As $t \to \infty$, the first term approaches zero and the steady-state response becomes

$$y'(t \to \infty) = \underbrace{\frac{KA}{\sqrt{(\tau\omega)^2 + 1}}}_{B} \sin\left\{\omega t + \underbrace{\tan^{-1}(-\tau\omega)}_{\phi}\right\} \tag{7.139}$$

Note that the output signal at steady state is a sine wave with the same frequency as the input signal but with a different amplitude and a different phase difference. Also note that AR and ϕ are both functions of ω.

$$AR = \frac{B}{A} = \frac{K}{\sqrt{(\tau\omega)^2 + 1}} \tag{7.140}$$

$$\phi = \tan^{-1}(-\tau\omega) \tag{7.141}$$

7.3.2 Review of Complex Numbers

Replacing s by $j\omega$ in a general transfer function renders a complex number $G(j\omega)$ that has a real part and an imaginary part similar to a complex number $z = a + jb$ (see Fig. 7.3):

$$G(j\omega) = \text{Re}[G(j\omega)] + j\,\text{Im}[G(j\omega)] \tag{7.142}$$

The magnitude of a complex number z in the z-plane is the distance of that point from the origin and the argument is the angle it makes with the real axis.

$$\text{Magnitude of } z: \ |z| = \sqrt{a^2 + b^2} \tag{7.143}$$

$$\text{Argument of } z: \angle z = \tan^{-1}\left(\frac{b}{a}\right) \tag{7.144}$$

Note that a complex number can also be presented in terms of its magnitude and its argument in the polar coordinate:

$$a = |z|\cos(\angle z) \tag{7.145}$$

$$b = |z|\sin(\angle z) \tag{7.146}$$

$$\begin{aligned} z &= a + jb \\ &= |z|\cos(\angle z) + j|z|\sin(\angle z) \\ &= |z|\{\cos(\angle z) + j\sin(\angle z)\} \\ &= |z|\exp\{j\angle z\} \end{aligned} \tag{7.147}$$

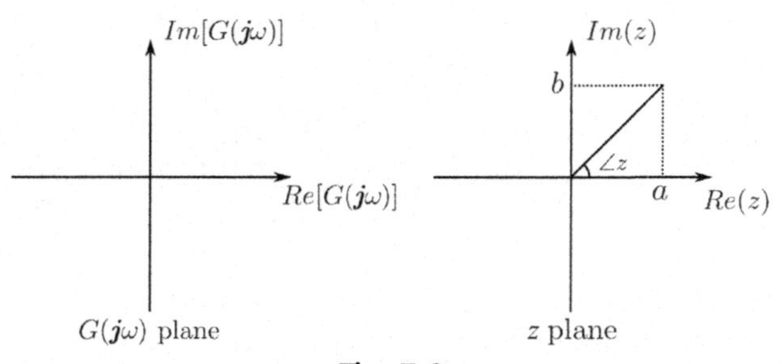

Fig. 7.3
Planes of imaginary numbers.

If z is the product or the ratio of two other complex numbers, z_1 and z_2, its magnitude and argument can be obtained in terms of the corresponding values of z_1 and z_2.

$$z = z_1 \times z_2 \Rightarrow \begin{matrix} |z| = |z_1| \times |z_2| \\ \angle z = \angle z_1 + \angle z_2 \end{matrix} \tag{7.148}$$

$$z = \frac{z_1}{z_2} \Rightarrow \begin{matrix} |z| = \dfrac{|z_1|}{|z_2|} \\ \angle z = \angle z_1 - \angle z_2 \end{matrix} \tag{7.149}$$

Based on the earlier properties of complex numbers, there is a shortcut method to determine the AR and ϕ of a given transfer function.

7.3.3 The Shortcut Method to Determine AR(ω) and ϕ(ω) of Linear Systems

Theorem *If s is replaced by j in the transfer function of a linear system, a complex number $G(j\omega)$ is obtained whose magnitude $|G(j\omega)|$ is equal to the amplitude ratio AR (ω) and its argument $\angle G(j)$ is the phase difference $\phi(\omega)$.*

$$AR = |G(j\omega)| \tag{7.150}$$

$$\phi = \angle G(j\omega) \tag{7.151}$$

Proof Let us consider a system that is being disturbed by a sine wave and has a general transfer function $G(s)$ as a ratio of two polynomials, $Q(s)$ and $P(s)$, with p poles.

$$G(s) = \frac{Y(s)}{U(s)} = \frac{Q(s)}{P(s)} \quad U(s) = \frac{A\omega}{s^2 + \omega^2} \tag{7.152}$$

$$Y(s) = \frac{Q(s)}{P(s)} \left(\frac{A\omega}{s^2 + \omega^2} \right) = \frac{a_1}{s - p_1} + \frac{a_2}{s - p_2} + \cdots + \frac{a_P}{s - p_P} + \frac{b_1}{s + j\omega} + \frac{b_2}{s - j\omega} \tag{7.153}$$

$$y'(t) = \mathcal{L}^{-1}\{Y(s)\} = a_1 e^{p_1 t} + a_2 e^{p_2 t} + \cdots + b_1 e^{-j\omega t} + b_2 e^{+j\omega t} \tag{7.154}$$

If the system is stable, system poles $p_1, \ldots, p_P$ will all be negative, therefore

$$y'(t \to \infty) = b_1 e^{-j\omega t} + b_2 e^{+j\omega t} \tag{7.155}$$

$$b_1 = \lim_{s \to -j\omega} G(s) \frac{A}{s - j\omega} = G(-j\omega) \frac{A}{-2j} \tag{7.156}$$

$$b_2 = \lim_{s \to j\omega} G(s) \frac{A}{s + j\omega} = G(j\omega) \frac{A}{2j} \tag{7.157}$$

$$
\begin{aligned}
y'(t\to\infty) &= -G(-j\omega)\frac{A}{2j}e^{-j\omega t}+G(j\omega)\frac{A}{2j}e^{j\omega t}\\
&= A\,|G(j\omega)|\frac{e^{j[\omega t+\angle G(j\omega)]}-e^{-j[\omega t+\angle G(j\omega)]}}{2j}\\
&= \underbrace{A\,|G(j\omega)|}_{B}\,\sin\left[\omega t+\underbrace{\angle G(j\omega)}_{\phi}\right]
\end{aligned}
\tag{7.158}
$$

Considering the polar presentation of $G(j\omega)=|G(j\omega)|\exp\{\angle G(j\omega\}$, proves that $AR=\frac{B}{A}=|G(j\omega)|$, and $\phi=\angle G(j\omega)$

In the following example the previous shortcut approach is used to calculate the amplitude ratio and phase difference of a few simple transfer functions.

Example 7.13

Let us consider a few simple transfer functions and calculate their corresponding AR and ϕ.

1. A First-Order System

$$G_1(s)=\frac{K}{\tau s+1}\ \Rightarrow\ G_1(j\omega)=\frac{K}{\tau(j\omega)+1} \tag{7.159}$$

$$AR=|G_1(j\omega)|=\frac{|K|}{|\tau(j\omega)+1|}=\frac{K}{\sqrt{\tau^2\omega^2+1}} \tag{7.160}$$

$$\phi=\angle G_1(j\omega)=\angle K-\angle(1+j\omega\tau)=0-tan^{-1}\left(\frac{\omega\tau}{1}\right)=-tan^{-1}(\omega\tau) \tag{7.161}$$

2. A Pure Capacitive or an Integrating System

$$G_2(s)=\frac{K^*}{s}\ \Rightarrow\ G_2(j\omega)=\frac{K^*}{j\omega} \tag{7.162}$$

$$AR=|G_2(j\omega)|=\frac{|K^*|}{|j\omega|}=\frac{K^*}{\omega} \tag{7.163}$$

$$\phi=\angle G_2(j\omega)=\angle K^*-\angle j\omega=0-\tan^{-1}\left(\frac{\omega}{0}\right)=0-\tan^{-1}(\infty)=-\frac{\pi}{2} \tag{7.164}$$

Note that a first-order lag transfer function results in $-\frac{\pi}{2}$ phase lag at high frequencies.

3. A Pure Time Delay

$$G_3(s)=e^{-\alpha s}\ \Rightarrow\ G_3(j\omega)=e^{-j\alpha\omega} \tag{7.165}$$

$$\text{Recall polar presentation, } z=|z|e^{j(\angle z)} \tag{7.147}$$

$$AR=|G_3(j\omega)|=1 \tag{7.166}$$

$$\phi=\angle G_3(j\omega)=-\alpha\omega \tag{7.167}$$

Note: A pure time delay causes a phase lag proportional to the delay time. The process output has the same amplitude as the input, i.e., $AR = 1$.

4. A Lead–Lag term

$$G_4(s) = \frac{K(\tau_1 s + 1)}{\tau_2 s + 1} \Rightarrow G_4(j\omega) = \frac{K(j\tau_1\omega + 1)}{j\omega\tau_2 + 1} \tag{7.168}$$

$$AR = \frac{K\sqrt{(\tau_1\omega)^2 + 1}}{\sqrt{(\tau_2\omega)^2 + 1}} \tag{7.169}$$

$$\phi = \tan^{-1}\left(\underbrace{\tau_1}_{\text{Lead}}\ \omega\right) - \tan^{-1}\left(\underbrace{\tau_2}_{\text{Lag}}\ \omega\right) \tag{7.170}$$

Note that a first-order lead transfer function results in $+\frac{\pi}{2}$ phase lead at high frequencies.

5. A Second-Order System

$$G_5(s) = \frac{K}{(\tau s)^2 + 2\xi\tau s + 1} \Rightarrow G_5(j\omega) = \frac{K}{\tau^2(j\omega)^2 + 2\xi\tau(j\omega) + 1} \tag{7.171}$$

$$G_5(j\omega) = \frac{K}{1 - \tau^2\omega^2 + 2j\xi\tau\omega} \tag{7.172}$$

$$AR = |G_5(j\omega)| = \frac{K}{\sqrt{(1 - \tau^2\omega^2)^2 + (2\xi\tau\omega)^2}} \tag{7.173}$$

$$\begin{aligned}\phi &= \angle G_5(j\omega) = \angle K - \angle\{(1 - \tau^2\omega^2) + (2\xi\tau\omega)j\} \\ &= 0 - tan^{-1}\left\{\frac{2\xi\tau\omega}{1 - \tau^2\omega^2}\right\}\end{aligned} \tag{7.174}$$

Note that a second-order transfer function results in $-\pi$ phase lag at high frequencies.

6. A general Transfer Function

$$G_6(s) = \frac{K(\tau_1 s + 1)e^{-\alpha s}}{(\tau_2 s + 1)(\tau_3 s + 1)} \quad \text{Recall: } z = |z|e^{j\angle z} \tag{7.175}$$

$$AR = |G_6(\boldsymbol{j}\omega)| = \frac{K\sqrt{1 + (\tau_1\omega)^2}\,(1)}{\sqrt{1 + (\tau_2\omega)^2}\sqrt{1 + (\tau_3\omega)^2}} \tag{7.176}$$

$$\phi = \{0 - \tan^{-1}(\tau_1\omega) - \alpha\omega\} - \{\tan^{-1}(\tau_2\omega) + \tan^{-1}(\tau_3\omega)\} \tag{7.177}$$

7.3.4 Graphical Representation of AR and ϕ and Their Applications

The AR and ϕ are used for the "system identification" and "controller design" using graphical methods in the form of the Bodé, Nyquist, and Nichols plots. We shall only consider the Bodé and Nyquist plots.

7.3.4.1 The Bodé diagram

The Bodé diagram consists of two subplots shown in Fig. 7.4, the first is a plot of the AR versus ω in a log-log scale, and the second is the plot of ϕ in degrees versus ω in a semilog plot.

7.3.4.2 The Nyquist diagram

The Nyquist diagram is a polar representation of AR and ϕ on the $G(j\omega)$ plane as ω changes from 0 to infinity shown in Fig. 7.5. For a given value of frequency, there will be a distinct point on the $G(j\omega)$-plane. The distance of this point from the origin is equal to the AR and its angle with the real axis is ϕ.

Fig. 7.4
The Bodé diagram coordination axes.

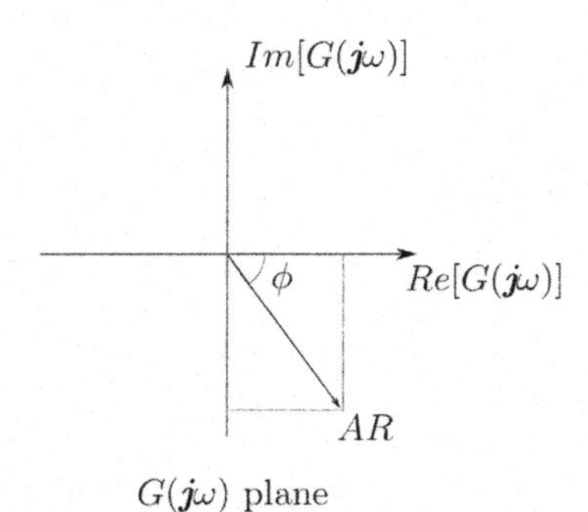

Fig. 7.5
The Nyquist diagram in the $G(j\omega)$ plane.

The Bodé and Nyquist diagrams can be plotted either graphically (which is approximate) or numerically.ss

7.3.5 Graphical Construction of the Approximate Bodé Plots

The following steps are to be taken to construct the approximate Bodé plots:

1. Obtain the expressions for the AR and ϕ as $\omega \to 0$, the low frequency asymptote (lfa).
2. Obtain expressions for the AR and ϕ as $\omega \to \infty$, the high frequency asymptote (hfa).
3. Using the "hfa," determine the slope of log(AR)versus log ω.
4. Obtain the intersection of the "lfa" with the "hfa" by equating the corresponding expressions and obtain the corner frequency, ω_n.

Example 7.14

Draw the "approximate" Bodé plots plot for a first-order system given by: $G(s) = \frac{2}{5s+1}$, and compare them with the MATLAB generated plots.

$$G(s) = \frac{K}{\tau s + 1}; \quad AR = \frac{K}{\sqrt{\tau^2\omega^2 + 1}}; \quad \phi = -\tan^{-1}(\tau\omega) \tag{7.178}$$

$$\text{The "lfa" asymptote is } \begin{array}{l} AR(\omega \to 0) = K \\ \phi(\omega \to 0) = 0 \end{array} \tag{7.179}$$

$$\text{The "hfa" asymptote is } \begin{array}{l} AR(\omega \to \infty) = 0 \\ \phi(\omega \to \infty) = -\frac{\pi}{2} \end{array} \tag{7.180}$$

The slope of the "hfa," as $\omega \to \infty$, is obtained from the AR expression

$$AR \to \frac{K}{\sqrt{\tau^2\omega^2}} = \frac{K}{\tau\omega} \tag{7.181}$$

$$\log AR \approx \log K - \log \tau - \log \omega \tag{7.182}$$

Therefore on a log-log scale, the amplitude ratio plot will have a slope of −1 cycle/cycle at high frequencies. The "lfa" and "hfa" intersect at the corner frequency, ω_n, that can be obtained by equating the "lfa" and "hfa" expressions.

$$K = \frac{K}{\tau\omega}; \quad \omega_n = \frac{1}{\tau} \tag{7.183}$$

It is shown as follows that for a first-order transfer function without time delay, the phase difference is −45 degrees at the corner frequency, $\omega_n = \frac{1}{\tau}$. Fig. 7.6 shows the Bodé plots of a first-order transfer function.

$$\phi(\omega_n) = -\tan^{-1}(\tau\omega) = -\tan^{-1}(1) = -\frac{\pi}{4} \tag{7.184}$$

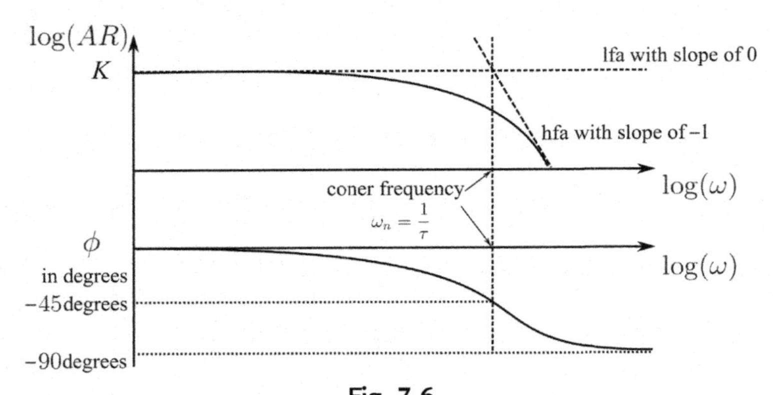

Fig. 7.6
Approximate Bodé plot.

7.3.6 Graphical Construction of the Approximate Nyquist Plot

Note that as ω goes to zero (*lfa*), the *AR* approaches *K* with an angle of approach $\phi = 0$. At high frequencies (*hfa*), the *AR* approaches the origin with an angle of approach (ϕ) equal to -90 degrees. Therefore the Nyquist diagram for a first-order lag process is a half circle which with its mirror image forms a complete circle, as is shown in Fig. 7.7. The Nyquist diagram is the location of the tip of a vector whose magnitude is the *AR* and its angle with the real axis is ϕ, as ω changes from 0 to infinity.

7.3.7 Numerical Construction of Bodé and Nyquist Plots

The Bodé and Nyquist plots can easily be generated in MATLAB. However, the Bodé *AR* plot is given in dB which is 20 $\log_{10}AR$. For a first-order system given by $G(s) = \frac{2}{5s+1}$, the Bodé and Nyquist plots shown in Figs. 7.8 and 7.9 can be easily constructed using the following MATLAB command.

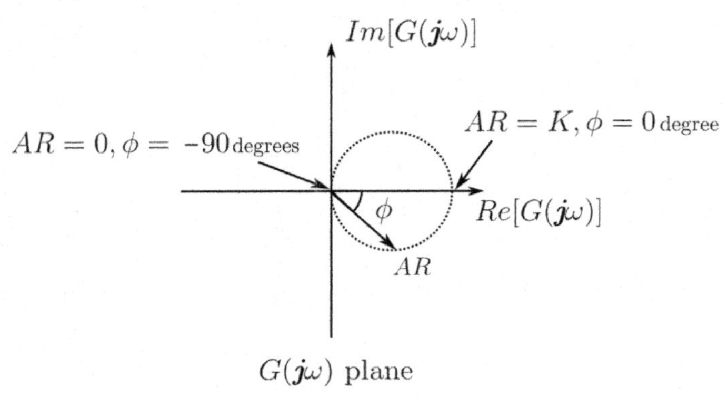

Fig. 7.7
The approximate Nyquist plot.

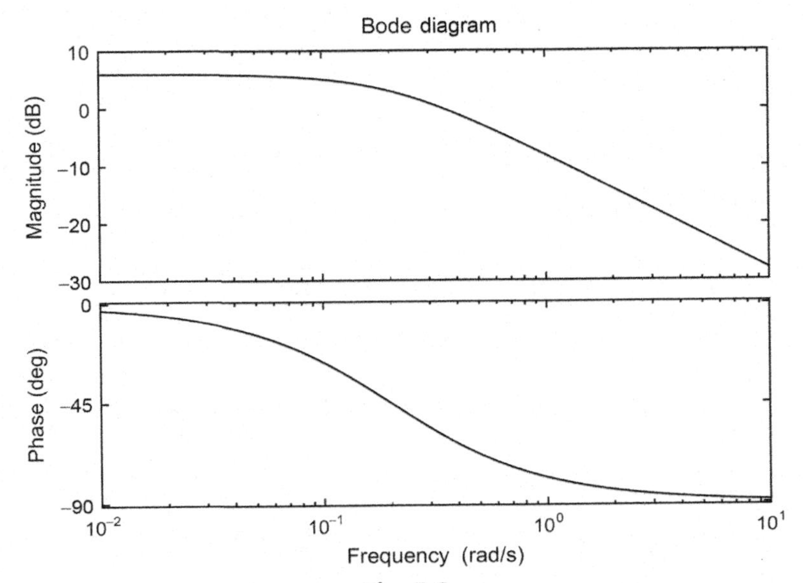

Fig. 7.8
Bodé plot generated in MATLAB.

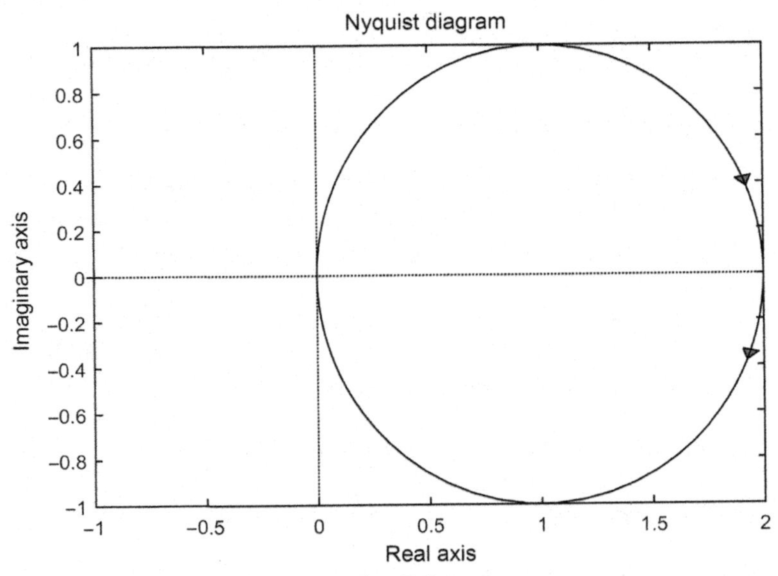

Fig. 7.9
Nyquist plot generated in MATLAB.

```
>> sys=tf(2,[5 1]);
>> bode(sys);
>> nyquist(sys);
```

In order to plot the log(AR) instead of dB of magnitude ratio and ϕ versus log of ω, the following commands can be used:

```
>> n=2;
>> d=[5 1];
>> [ar,phi,w]=bode(n,d);
>> subplot(2,1,1);
>> loglog(w,ar);
>> subplot(2,1,2);
>> semilogx(w,phi);
```

7.3.8 Applications of the Frequency Response Technique

The applications of frequency response are in

1. Process identification—empirically obtain AR and ϕ, compare them with the AR and ϕ of known low-order transfer functions, and thereby, approximate the unknown system with a low-order transfer function.
2. Controller design—a PID controller can be tuned using the stability criterion in the frequency domain, as it will be discussed in the following sections.

7.3.8.1 Process identification in the frequency domain

The AR and the phase difference of an existing process with unknown dynamics can be obtained empirically, either by disturbing the process with a sinusoidal input with different frequencies using a signal generator (Fig. 7.10), or with a single pulse input (Fig. 7.11).

In the latter case, the AR and Φ can be calculated from the measured input and output pulse functions.

$$G(s) = \frac{Y(s)}{U(s)} = \frac{\int_0^{\infty} y(t)e^{-st}dt}{\int_0^{\infty} u(t)e^{-st}dt} \tag{7.185}$$

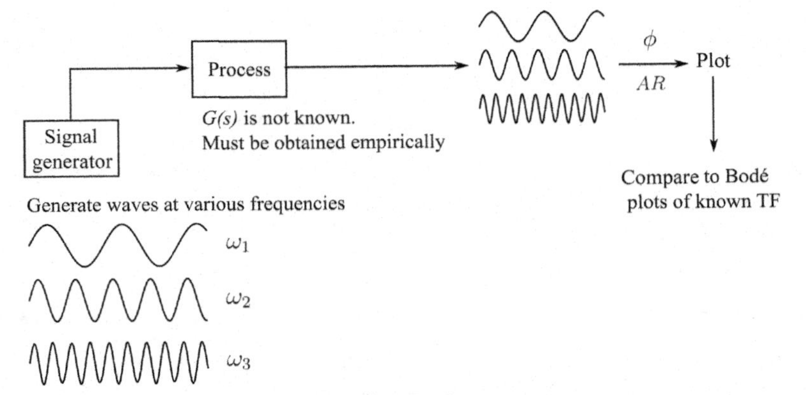

Fig. 7.10
Process identification in the frequency domain using a signal generator.

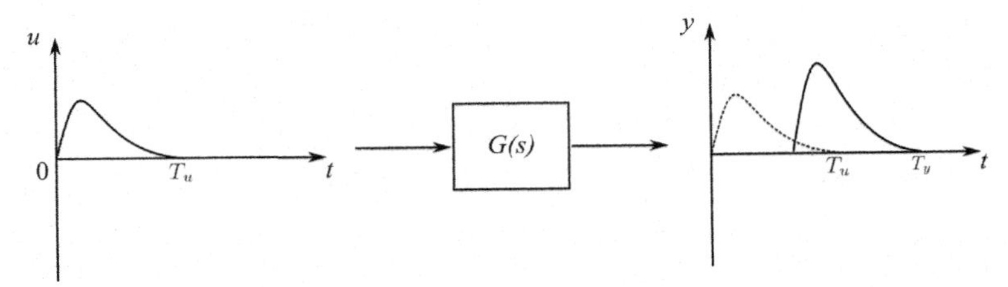

Fig. 7.11
Process identification in the frequency domain using a single pulse signal.

$$G(j\omega) = \frac{\int_0^{\infty} y(t)e^{-j\omega t}dt}{\int_0^{\infty} u(t)e^{-j\omega t}dt} \tag{7.186}$$

Recall: $e^{-j\theta} = \cos\theta - j\sin\theta$

$$G(j\omega) = \frac{\int_0^{T_y} y(t)\cos(\omega t)dt - j\int_0^{T_y} y(t)\sin(\omega t)dt}{\int_0^{T_u} u(t)\cos(\omega t)dt - j\int_0^{T_u} u(t)\sin(\omega t)dt} \tag{7.187}$$

$$G(j\omega) = \frac{A_1(\omega) - jA_2(\omega)}{B_1(\omega) - jB_2(\omega)} \tag{7.188}$$

At a given frequency ω, one can calculate A_1, A_2, B_1, and B_2 from the measured input and output pulses and obtain the corresponding values of AR and ϕ.

$$G(j\omega) = Re(\omega) \pm j\,Im(\omega) \tag{7.189}$$

$$AR = \sqrt{Re^2 + Im^2} \quad \phi = \tan^{-1}\left(\frac{Im}{Re}\right) \tag{7.190}$$

Therefore with one single experiment, $AR(\omega)$ and $\phi(\omega)$ of an unknown process can be calculated as a function of ω, and the Bodé diagram can be constructed. The Bodé plots can then be compared with the Bodé diagram of known low-order transfer functions to estimate the approximate transfer function of the unknown system and its parameters.

Example 7.15

Using the Bodé plots (Fig. 7.12) generated from a process with unknown dynamics, determine the transfer function (TF) of the unknown process.

Note that the slope of the *"hfa"* at high frequencies is −1 on a log-log scale or −20 db/cycle, and the slope of the *"lfa"* is zero. This suggests that the unknown process can be approximated by a first-order transfer function. Looking at the ϕ diagram, it is noted that the phase lag decreases at high frequencies and does not converge to −90 degrees which is typical of a first-order transfer function without a time delay. This suggests that the unknown process must have a time delay and be of the general form of $G(s) = \frac{K\,e^{-\theta s}}{\tau s + 1}$. Therefore there are three unknown parameters, K, τ, and θ that must be estimated from the Bodé plots. From the AR Bodé plot,

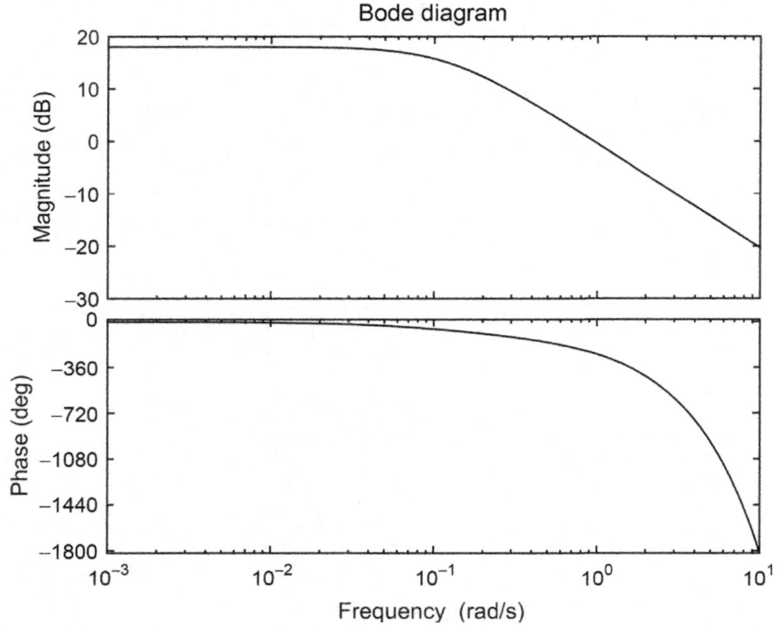

Fig. 7.12
Bodé plot of an unknown system.

$$20\log AR|_{\omega\to 0} = 20\log K \approx 18 \Rightarrow K = 7.9 \tag{7.191}$$

The corner frequency, ω_n, at the intersection of the "*lfa*" and the "*hfa*" is obtained to be equal to 0.12 The corner frequency is the inverse of the process time constant, i.e., $1/\tau$, resulting in $\tau = 8.3$ s. Finally, to determine the unknown time delay, from the given phase difference plot, for example, at a value of frequency equal to $\omega = 1$ rad/s, the phase difference is read from the experimental plot $\phi \approx -260° = 260/360 \times 2\pi = -4.53$ rad. The phase difference of a first-order transfer function with time delay is given by

$$\begin{aligned} \phi &= -\tan^{-1}(\tau\omega) - \theta\omega \\ -4.53 &= -\tan^{-1}(8.3 \times 1) - \theta \times 1 \end{aligned} \tag{7.192}$$

which renders the unknown delay time: $\theta = 3$ s. Therefore the unknown process can be approximated by the following transfer function:

$$G(s) \approx \frac{7.9\, e^{-3s}}{8.3s + 1} \tag{7.193}$$

7.3.8.2 *The stability analysis and design of feedback controllers using frequency response technique*

Recall that the closed-loop transfer function of a SISO system, assuming $G_m = K_m = 1$, is

$$Y(s) = \frac{G_c(s)G_v(s)G_P(s)}{1 + G_c(s)G_v(s)G_P(s)} Y_{sp}(s) + \frac{G_d(s)}{1 + G_c(s)G_v(s)G_P(s)} D(s) \tag{7.194}$$

With the closed-loop characteristic equation (*CLCE*) given by

$$CLCE = 1 + G_c(s)G_v(s)G_P(s) = 1 + G_{OL}(s) = 0 \tag{7.195}$$

Theorem *A system is closed-loop stable if the amplitude ratio of its open-loop* (AR_{OL}) *system is less than* 1 *at the critical frequency, i.e., the frequency at which the open-loop phase difference,* ϕ_{OL}, *is* -180 degrees.

Proof This is a qualitative proof of the theorem. Consider the control system shown in Fig. 7.13 in which the set point is perturbed by $A\sin(\omega t)$ and the loop is disconnected after the measuring element. Since the combination of a number of linear systems is also linear, the measured

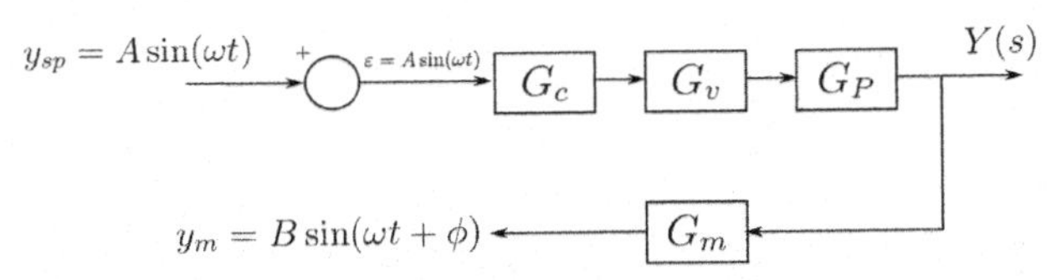

Fig. 7.13
The open-loop configuration by disconnecting the feedback signal from the controller.

output, after the transient portion is settled, will be a sinusoidal signal with the same frequency as the set point but a different amplitude and a phase lag, i.e., $y_m = B\sin(\omega t + \phi)$. The overall transfer function of the disconnected system referred to as the open-loop transfer function, $G_{OL}(s)$, and the corresponding amplitude ratio and phase difference are given by:

$$G_{OL}(s) = G_c(s)G_v(s)G_P(s)G_m(s) \tag{7.196}$$

$$AR_{OL} = \frac{B}{A} = |G_{OL}(j\omega)| \tag{7.197}$$

$$\phi_{OL} = \angle G_{OL}(j\omega) \tag{7.198}$$

If the frequency ω is varied such that the $AR_{OL} = 1$ and $\phi_{OL} = -180° = -\pi$, the measured output y_m will be:

$$y_m = A\sin(\omega t - \pi) = -A\sin(\omega t) \tag{7.199}$$

Under such a condition, let us reconnect the feedback signal and set the set point to zero, $y_{sp} = 0$. Note that as is shown in Fig. 7.14, the error signal is still the same as when the loop was open and y_{sp} was (A sin ωt).

Under these conditions, the closed-loop system oscillates continuously with a constant amplitude (i.e., the closed-loop system is at the verge of instability). If AR_{OL} is increased slightly (e.g., by increasing the controller gain) the closed-loop system becomes unstable (oscillates with an increasing amplitude). If AR_{OL} is decreased, the system will be closed-loop stable (oscillates with a decreasing amplitude until it settles at a steady-state value). Therefore if the open-loop transfer function of a control system has an AR_{OL} larger than 1 at the critical frequency, the system is closed-loop unstable. The pictorial presentation of the stability theorem in the form of Bodé and Nyquist diagrams is shown in Figs. 7.15 and 7.16.

7.3.8.3 The stability test in the frequency domain and the implementation of the theorem

In order to apply the stability test in the frequency domain, the following steps should be followed.

- Obtain the expressions for AR_{OL} and ϕ_{OL}.
- Find the critical frequency, ω_c, by letting $\phi_{OL} = -\pi$.

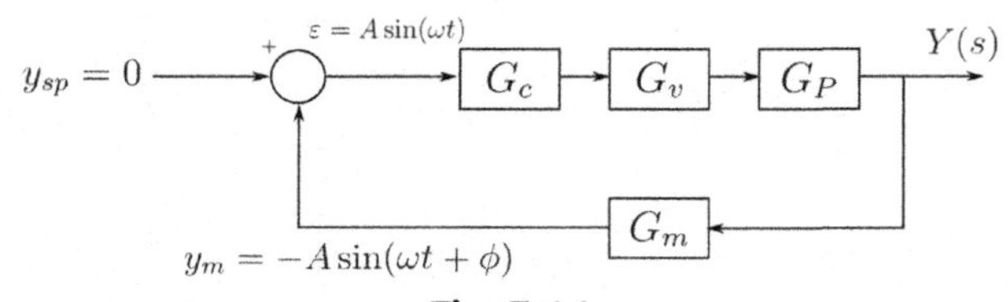

Fig. 7.14
The closed-loop configuration.

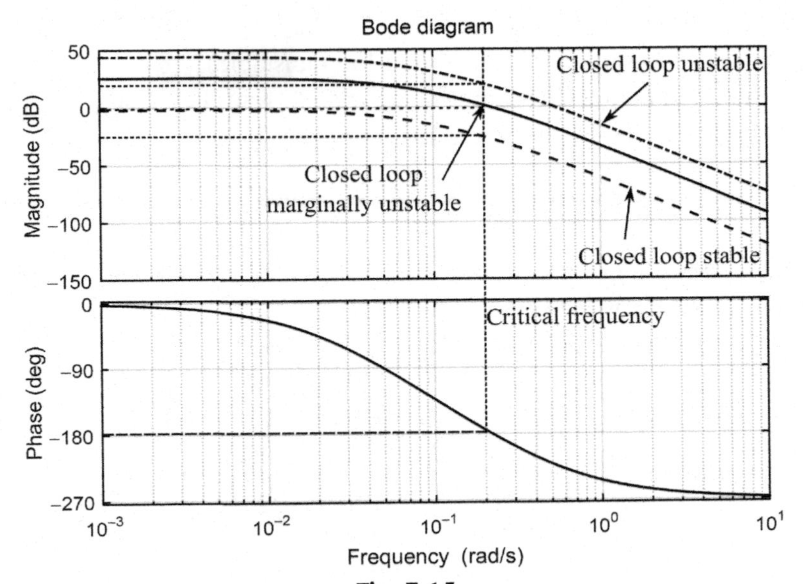

Fig. 7.15
Presentation of the stability criterion on Bodé plot.

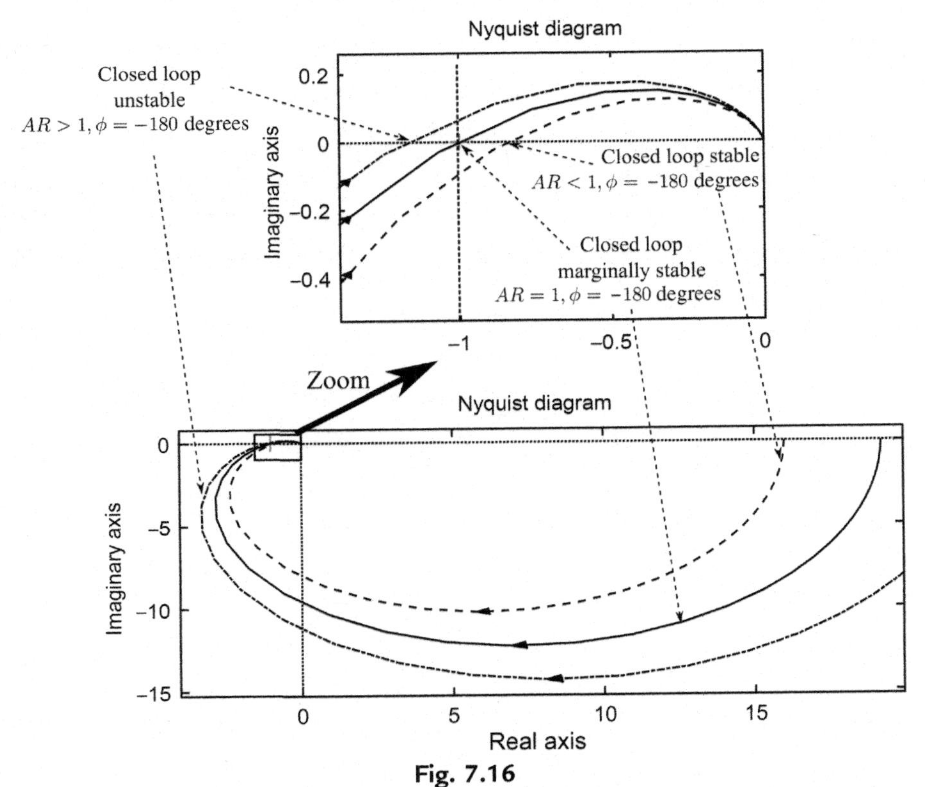

Fig. 7.16
Presentation of the stability theorem on Nyquist plot.

- For a given controller gain, K_c , substitute ω_c in the AR_{OL} expression and determine if $AR_{OL} < 1$ (closed-loop stable) or >1 (closed-loop unstable).
- To find the maximum controller gain ($K_{c,max} = K_{ult}$), find the value of K_c that results in $AR_{OL} = 1$ (marginally stable) at the critical frequency.

Example 7.16

For the control system given as follows, find the maximum or the ultimate controller gain, K_{ult}, that results in a marginally stable closed-loop system.

$$G_P = \frac{50}{30s+1} \quad G_v = \frac{0.016}{3s+1} \quad G_m = \frac{1}{10s+1} \quad G_c = K_c \tag{7.200}$$

$$G_{OL}(s) = \left(\frac{50}{30s+1}\right)\left(\frac{0.016}{3s+1}\right)\left(\frac{1}{10s+1}\right)(K_c) \tag{7.201}$$

$$AR_{OL} = K_c \frac{0.8}{\sqrt{900\omega^2+1}\sqrt{9\omega^2+1}\sqrt{100\omega^2+1}} \tag{7.202}$$

$$\phi_{OL} = -\tan^{-1}(3\omega) - \tan^{-1}(30\omega) - \tan^{-1}(10\omega) = -\pi \rightarrow \omega_c = 0.218\,\text{rad/s} \tag{7.203}$$

$$AR_{OL} = K_{ult} \frac{0.8}{\sqrt{900\omega_c^2+1}\sqrt{9\omega_c^2+1}\sqrt{100\omega_c^2+1}} = 1 \rightarrow K_{ult} = K_{c,max} = 23.8 \tag{7.204}$$

7.3.8.4 Definitions of the gain margin and the phase margin

Two measures of stability, i.e., the gain margin (GM) and phase margin (PM) determine how far the closed-loop system is from the "marginally stable" condition. The larger the gain margin and the phase margin are, the more conservative the controller will be. Moreover, at larger GM and PM, the closed-loop system will be slower. The pictorial presentations of the GM and PM are shown in Fig. 7.17.

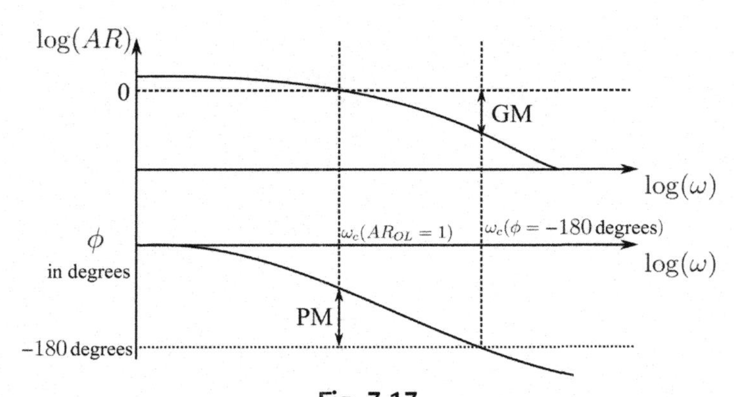

Fig. 7.17
Gain margin and phase margin.

$$GM = 1/AR_{OL} \tag{7.205}$$

$$PM = 180° + \phi_{OL} \tag{7.206}$$

A heuristic approach to design a feedback controller is to have a GM between 1.7 and 4 and a PM between 30 and 45 degrees.

Example 7.17

Design a feedback controller (obtain K_c) for the following system that ensures a GM of 2 and a PM of 30 degrees, whichever is more conservative.

$$G_P = \frac{50}{30s+1} \quad G_v = \frac{0.016}{3s+1} \quad G_m = \frac{1}{10s+1} G_c = K_c \tag{7.207}$$

Solution

$$G_{OL}(s) = \left(\frac{50}{30s+1}\right)\left(\frac{0.016}{3s+1}\right)\left(\frac{1}{10s+1}\right)(K_c) \tag{7.208}$$

$$AR_{OL} = K_c \frac{0.8}{\sqrt{900\omega^2+1}\sqrt{9\omega^2+1}\sqrt{100\omega^2+1}} \tag{7.209}$$

$$\phi_{OL}(\omega_c) = -\tan^{-1}(3\omega_c) - \tan^{-1}(30\omega_c) - \tan^{-1}(10\omega_c) = -\pi \rightarrow \omega_c = 0.218\,\text{rad/s} \tag{7.210}$$

$$GM = 2 = \frac{1}{AR_{OL}(\text{critical frequency})} \rightarrow AR_{OL}(\text{critical frequency}) = \frac{1}{2} \tag{7.211}$$

$$AR_{OL}(\text{critical frequency}) =$$

$$\frac{1}{2} = \frac{0.8K_c}{\sqrt{900\omega_c^2+1}\sqrt{9\omega_c^2+1}\sqrt{100\omega_c^2+1}} \tag{7.212}$$

$$\frac{1}{2} = \frac{0.8K_c}{\sqrt{900(0.218)^2+1}\sqrt{9(0.218)^2+1}\sqrt{100(0.218)^2+1}}$$

$$K_c = \frac{0.5(1.19)(6.54)(2.39)}{0.8} = 11.7 \tag{7.213}$$

Note that this value of K_c is smaller than the maximum value of $K_{c,max} = 23.8$ obtained for the system in Example 7.16.

Now, let us consider the stated design criterion on the phase margin.

$$PM = 30° = 180° + \phi_{OL}(AR_{OL} = 1) \tag{7.214}$$

$$\phi_{OL}(AR_{OL} = 1) = -180° + 30° = -150° = -\frac{5}{6}\pi = -2.616\text{ rad} \tag{7.215}$$

$$\phi_{OL}(AR_{OL} = 1) = -\tan^{-1}(3\omega) - \tan^{-1}(30\omega) - \tan^{-1}(10\omega) = -2.616\text{ rad} \tag{7.216}$$

Find ω by trial and error:

$$\omega \approx 0.14\text{ rad/s} \tag{7.217}$$

$$AR_{OL} = 1 = K_c \frac{0.8}{\sqrt{900(0.14)^2+1}\sqrt{9(0.14)^2+1}\sqrt{100(0.14)^2+1}}$$
$$= K_c \frac{0.8}{(1.08)(4.31)(1.72)} \tag{7.218}$$

$$K_c = \frac{(1.08)(4.31)(1.72)}{8} = 10.05 \tag{7.219}$$

Therefore the choice of the controller gain according to the given design specification in this problem is dictated by the PM which asks for a smaller controller gain, $K_c = 10.05$. This results in a more stable (slower) closed-loop response.

7.4 Problems

1. The discrete state-space representation of a SISO process is given by:

$$\begin{cases} x_{k+1} = \begin{bmatrix} 3 & 1 \\ -2 & 5 \end{bmatrix} x_k + \begin{bmatrix} 2 \\ 1 \end{bmatrix} u_k \\ y_k = \begin{bmatrix} 1 & 1 \end{bmatrix} x_k \end{cases} \tag{7.220}$$

   - Derive the discrete transfer function of this process.
   - Discuss the observability and the controllability of this system.
   - Derive a feedback regulator that places the closed-loop poles at $z_{1,2} = 0.6 \pm j0.3$.

2. Discuss the observability and controllability conditions for the following system:

$$\begin{cases} \dot{x} = Ax + Bu \\ y = Cx \end{cases} \tag{7.221}$$

   where $A = \begin{bmatrix} 2 & 1 \\ 4 & 3 \end{bmatrix}$, $B = \begin{bmatrix} 2 \\ 3 \end{bmatrix}$, $C = \begin{bmatrix} 3 & 2 \\ 1 & 2 \end{bmatrix}$

3. Derive a time optimal control sequence for the system given by the state equation

$$\begin{cases} \boldsymbol{x}_{k+1} = \begin{bmatrix} 0.1 & 0.1 & 0 \\ 1.2 & 0.3 & 0 \\ 0 & 0.3 & 0.4 \end{bmatrix} \boldsymbol{x}_k + \begin{bmatrix} 1 \\ 0.2 \\ 0.4 \end{bmatrix} \boldsymbol{u}_k \\ \boldsymbol{y}_k = \begin{bmatrix} -1.5 & 0.5 & 2.5 \end{bmatrix} \boldsymbol{x}_k \end{cases} \tag{7.222}$$

Which minimizes

$$J = \frac{1}{2} x_4 H x_4 + \frac{1}{2} \sum_{i=1}^{3} \left[x_i^T Q x_i + u_i^T R u_i \right] \tag{7.223}$$

where $H = 3I$, $Q = 2I$ and $R = 1$

4. For each of the following transfer functions, determine the expressions for the amplitude ratio and phase difference and sketch the approximate Bodé and Nyquist plots. Compare the plots with those generated by MATLAB.

$$g_1(s) = \frac{0.3e^{-4s}}{6s+1} \tag{7.224}$$

$$g_2(s) = \frac{0.3e^{-4s}}{(6s+1)(3s+1)} \tag{7.225}$$

$$g_3(s) = \frac{0.3e^{-4s}}{6s^2+3s+1} \tag{7.226}$$

5. The Bodé diagrams of a process with unknown dynamics are shown in Fig. 7.18. Discuss the major steady-state and dynamic characteristics of this process. That is, what is the order

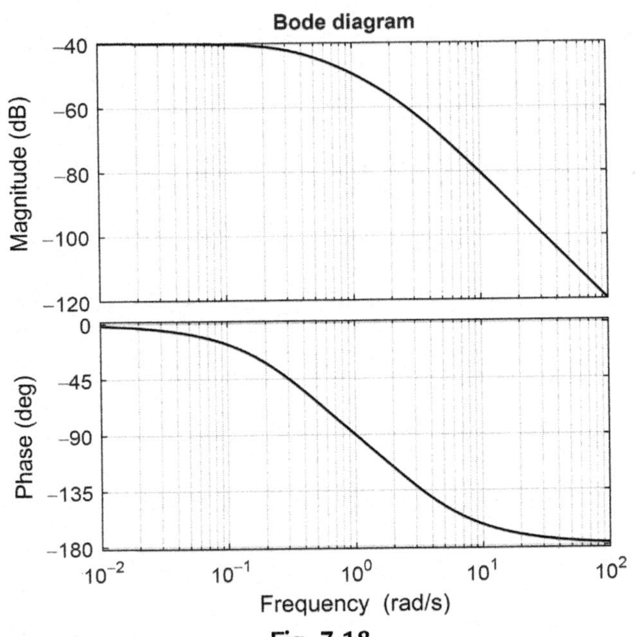

Fig. 7.18
Bode diagram of unknown system in Problem 5.

of the numerator and denominator polynomials in the transfer function? What is the steady-state gain of the process? Does the system have delay time and if so, estimate its value?

6. A conventional feedback control system has the following transfer functions:

$$g_p(s) = \frac{5}{10s+1}; \quad g_c(s) = K_c\frac{5s+1}{5s}; \quad g_v(s) = 1; \quad g_m(s) = \frac{1}{s+1} \tag{7.227}$$

- For what values of K_c is the system stable?
- Write down the expressions for the open-loop amplitude ratio and the phase difference for this system.

7. The closed-loop characteristic equation ($CLCE$) of a feedback control system is given by

$$CLCE(s) = 2s^2(3s+1) + K_c e^{-3s} \tag{7.228}$$

Using the Bodé (frequency response technique) stability criterion, determine the controller gain, if any, that ensures a gain margin of 2.

References

1. De Schutter B. Minimal state-space realization in linear system theory: an overview. *J Computational Appl Math* 2000;**121**(1–2):331–54.
2. Iserman R. *Digital control systems.* Berlin, Germany: Springer-Verlag; 1981.

Further Reading

Kuo BC. *Digital control systems.* New York: Reinhart and Winston; 1980.

CHAPTER 8

Modeling and Control of Stochastic Processes

Sohrab Rohani, Yuanyi Wu
Western University, London, ON, Canada

The majority of real processes are subject to uncertainties due to the process noise and/or the measuring element noise. The relationship between the measured output and the manipulated variable in such processes cannot be expressed by a deterministic equation. Such processes are referred to as stochastic processes. In what follows the modeling and control of stochastic processes will be discussed.

8.1 Modeling of Stochastic Processes

A stochastic process is subject to uncertainties and therefore its model includes the effect of the input variable (process model) and the noise (noise model).

8.1.1 Process and Noise Models

In a SISO stochastic process, the output at time t, y_t, can be expressed by:

$$y_t = \frac{B(z^{-1})}{A(z^{-1})} u_{t-b} + N_t \tag{8.1}$$

in which $A(z^{-1})$ and $B(z^{-1})$ are polynomials in the backshift operator z^{-1} (i.e., $z^{-b}y_t = y_{t-b}$) defined as:

$$A(z^{-1}) = 1 + a_1 z^{-1} + \cdots + a_{n_A} z^{-n_A} \tag{8.2}$$

$$B(z^{-1}) = b_0 + b_1 z^{-1} + \cdots + b_{n_B} z^{-n_B} \tag{8.3}$$

and b is the number of sampling intervals in the process time delay between the output y and input u. The first term on the right-hand side of Eq. (8.1) represents the effect of the manipulated variable on the controlled variable, while the second term, N_t, is the noise model that represents the effect of all other variables including the disturbances, the process noise, and the sensor noise, on y_t.

Coulson and Richardson's Chemical Engineering. http://dx.doi.org/10.1016/B978-0-08-101095-2.00008-4

It has been shown[1,2] that the noise term, N_t, can be modeled by passing a sequence of random numbers (white noise) through a linear filter:

$$N_t = \frac{C(z^{-1})}{D(z^{-1})\nabla^d} e_t \tag{8.4}$$

where $C(z^{-1})$ is referred to as the moving average (MA) operator, $D(z^{-1})$ is called the auto-regressive (AR) operator, and $\nabla = 1 - z^{-1}$ is a difference operator for modeling nonstationary processes.

$$C(z^{-1}) = 1 + c_1 z^{-1} + \cdots + c_{nC} z^{-nC} \tag{8.5}$$

$$D(z^{-1}) = 1 + d_1 z^{-1} + \cdots + d_{nD} z^{-nD} \tag{8.6}$$

Figs. 8.1–8.3 show the outputs of arbitrarily chosen AR(1), MA(1), ARMA(1,1), and ARIMA (1,1,1) processes generated on MATLAB.

```
% generates 100 random numbers
e=randn(100,1);
n=zeros(100,1);
n(1)=e(1);
d = 0.8;
for i =2 : 100
    % generates an AR(1) of the form n_t = (1/(1+0.8z^-1)) e_t
    n(i) = d*n(i-1) + e(i);
end
plot(n);
```

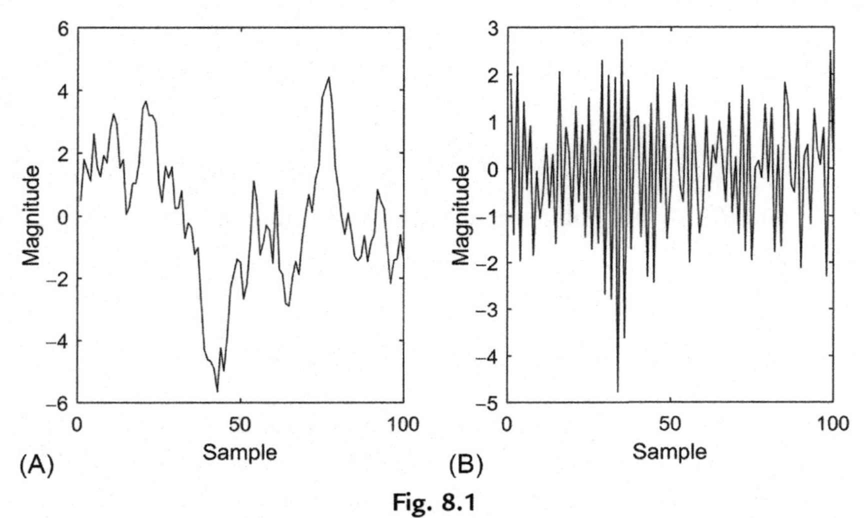

Fig. 8.1
The output of an arbitrarily chosen AR process of order 1. (A) AR(1) $n_t = e_t + 0.8n_{t-1}$. (B) AR(1) $n_t = e_t - 0.8n_{t-1}$.

```
e = randn(100,1);
c = 0.3;
n(1) = e(1);
for i = 2 : 100
    % generates an MA(1) of the form n_t = (1/(1+0.3z^-1)) e_t
    n(i) = c*e(i-1) + e(i);
end
plot(n);
```

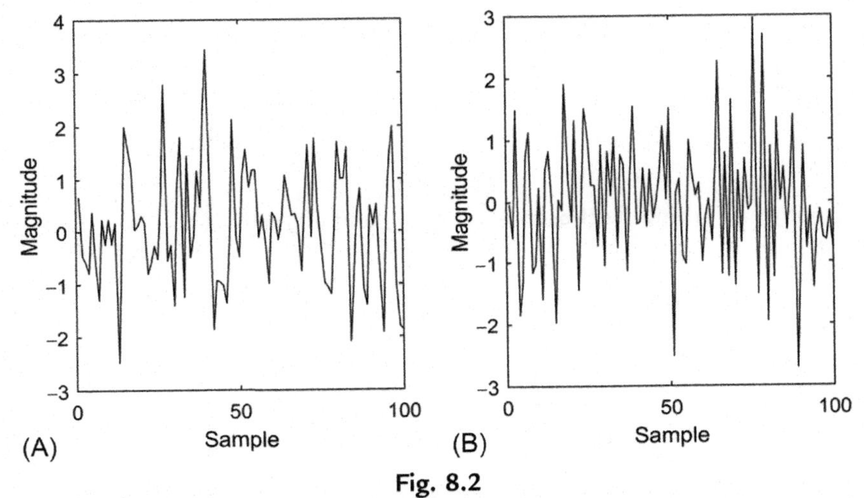

Fig. 8.2
The output of an arbitrarily chosen MA process of order 1. (A) MA(1) $n_t = e_t + 0.3e_{t-1}$. (B) MA(1) $n_t = e_t - 0.3e_{t-1}$.

```
e = randn(100,1);
n = zeros(100,1);
n(1) = e(1);
c = -0.3;
d = -0.8;
for i = 2 : 100
    % generates an ARMA(1,1)
    n(i) = c*e(i-1) + e(i) - d*n(i-1);
end
plot(n);
m(1:2) = e(1:2);
g = -0.2;
for i = 3 : 100
    % generates an ARIMA(1,1,1)
    m(i) = c*e(i-1) + e(i) - g*m(i-2) + d*m(i-1);
end
plot(m);
```

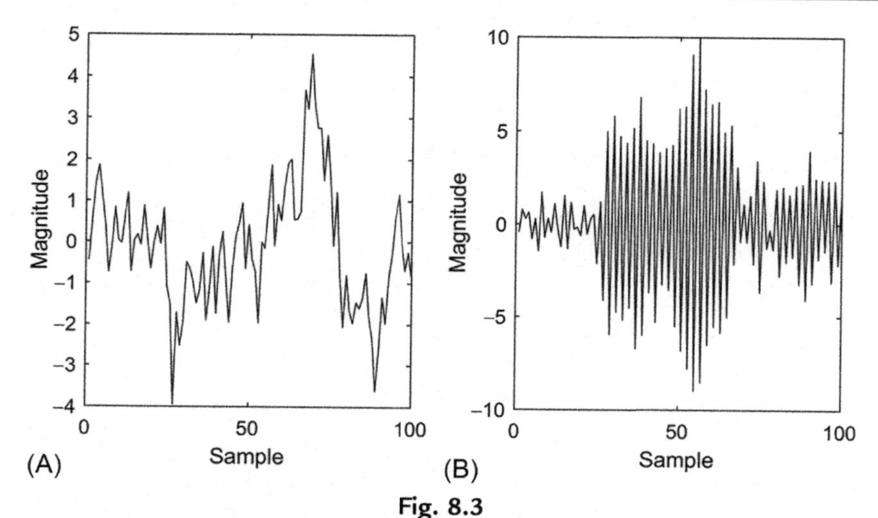

Fig. 8.3
The output of an arbitrarily chosen ARMA(1,1) and an ARIMA(1,1,1). (A) ARMA(1) $n_t = e_t - 0.3e_{t-1} + 0.8n_{t-1}$. (B) ARIMA(1) $n_t = e_t - 0.3e_{t-1} - 0.2n_{t-1} - 0.8n_{t-2}$.

The combined process and noise models can be written as

$$y_t = \frac{B(z^{-1})}{A(z^{-1})} u_{t-b} + \frac{C(z^{-1})}{D(z^{-1})\nabla^d} e_t \tag{8.7}$$

A simplified version of this model is obtained if $D(z^{-1})\nabla^d = A(z^{-1})$ in which case the model renders itself to the use of linear parameter estimation techniques.

$$y_t = \frac{B(z^{-1})}{A(z^{-1})} u_{t-b} + \frac{C(z^{-1})}{A(z^{-1})} e_t \tag{8.8}$$

Alternatively, the process model can be expressed as an impulse response model:

$$y_t = V(z^{-1}) u_t + N_t \tag{8.9}$$

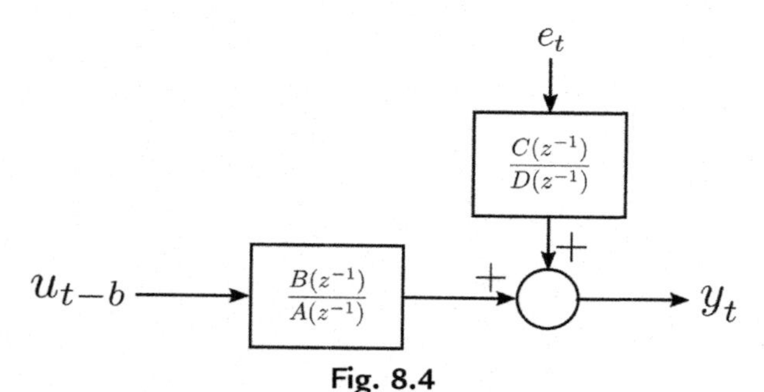

Fig. 8.4
The block diagram of a SISO stochastic process.

in which $V(z^{-1})$ is obtained by long division of $B(z^{-1})$ by $A(z^{-1})$.

$$V(z^{-1}) = \frac{B(z^{-1})}{A(z^{-1})} = v_0 + v_1 z^{-1} + \cdots \tag{8.10}$$

Note that $V(z^{-1})$ is an infinite series. The coefficients of $V(z^{-1})$ are the impulse weights of the process.

Fig. 8.4 shows the block diagram of a typical stochastic process.

8.1.2 Review of Some Useful Concepts in the Probability Theory

In the analysis of a stochastic process, one deals with uncertainties in the process variables and as such, the statistical properties of process variables such as their mean, variance, and probability density function are of significance.

Random variables (RV) and their statistical properties: The actual value of a random variable is unknown. Only the probability of the RV taking a specific value or a range of values can be defined. A random variable is either discrete or continuous. An example of a discrete RV is the outcome of flipping a coin which could either be heads or tails. The probability of a discrete RV to take a specific value is not zero. For example, the probability of getting a head in tossing a coin is 1/2. On the other hand, the probability of a continuous RV, for example, the yield of a chemical reactor taking a specific value would be zero. The probability of a continuous RV, X, falling in a given range of values can be expressed in terms of its probability density function (pdf):

$$Prob[x < X < x + dx] = p(x)\,dx \tag{8.11}$$

where $p(s)$ is the pdf of the continuous random variable, X.

$$Prob[x_1 < X < x_2] = \int_{x_1}^{x_2} p(x)dx \tag{8.12}$$

If the pdf of a RV is known, the useful statistical properties of that RV can be defined:

$$\text{statistical mean} = \text{expected value of } X = \overline{X} = E[X] = \int_{-\infty}^{+\infty} x\,p(x)\,dx \tag{8.13}$$

where $E[\,]$ is the expected value operator and $\overline{X}$ is the statistical or ensemble average of X which is the first moment of the pdf. The variance of X which determines the spread of X around its mean value is the second moment of the pdf around $\overline{X}$:

$$\sigma_X^2 = Var(X) = \int_{-\infty}^{+\infty} (x - \bar{x})^2 p(x)\,dx = E\left[(X - \overline{X})^2\right] \tag{8.14}$$

These statistical properties can be calculated for a random variable if its pdf is known. However, if the probability density function of a RV is not known, only estimates of the mean and the variance of the RV can be calculated:

$$\langle X \rangle = \text{estimated mean} = \lim_{N \to \infty} \frac{1}{2N+1} \sum_{i=-N}^{+N} x_i \tag{8.15}$$

$$\langle \sigma^2 \rangle = \text{estimated variance} = \lim_{N \to \infty} \frac{1}{2N+1} \sum_{i=-N}^{+N} [x_i - \langle x \rangle]^2 \tag{8.16}$$

Joint probability density function of two random variables: The joint pdf of two random variables, X and Y, is defined by:

$$p(x,y)dxdy = Prob[x < X < x + dx \text{ and } y < Y < y + dy] \tag{8.17}$$

If the two RV are dependent, there would be a cross-covariance between X and Y which is defined by:

$$E\left[(X - \overline{X})(Y - \overline{Y})\right] = \int_{-\infty}^{+\infty} \int_{-\infty}^{+\infty} (x - \overline{x})(y - \overline{y})\, p(x,y)\, dxdy = \rho \sigma_X \sigma_Y \tag{8.18}$$

where ρ is the correlation coefficient between X and Y that would be zero if X and Y were independent. σ_x and σ_y are the standard deviations of X and Y, respectively. The covariance matrix between X and Y is defined as:

$$\text{Covariance matrix} = P = E\left[\begin{pmatrix} X - \overline{X} \\ Y - \overline{Y} \end{pmatrix} \begin{pmatrix} X - \overline{X} \\ Y - \overline{Y} \end{pmatrix}^T\right] = \begin{bmatrix} \sigma_x^2 & \rho \sigma_X \sigma_Y \\ \rho \sigma_X \sigma_Y & \sigma_Y^2 \end{bmatrix} \tag{8.19}$$

The off-diagonal terms of P represent the interdependence of X and Y, while the diagonal terms represent the variances of X and Y from their mean values. The sum of the diagonal terms of P is known as the trace of the covariance matrix, $tr(P)$, whose magnitude represents the spread or uncertainty associated with X and Y. The previous definitions can be extended to n random variables in a straightforward manner.

Random signals: In the control theory we deal with random signals which also have temporal probability distributions associated with them. Fig. 8.5 depicts the input and output signals of a stochastic process over a given time period. If many identical experiments are carried out, the statistical or ensemble properties associated with either of the two random signals may be defined at a given time which would be different from the temporal properties obtained from a single experiment as shown in Fig. 8.5. However, if the random signals were stationary and the process were ergodic (i.e., a process which would return to its original state given sufficient

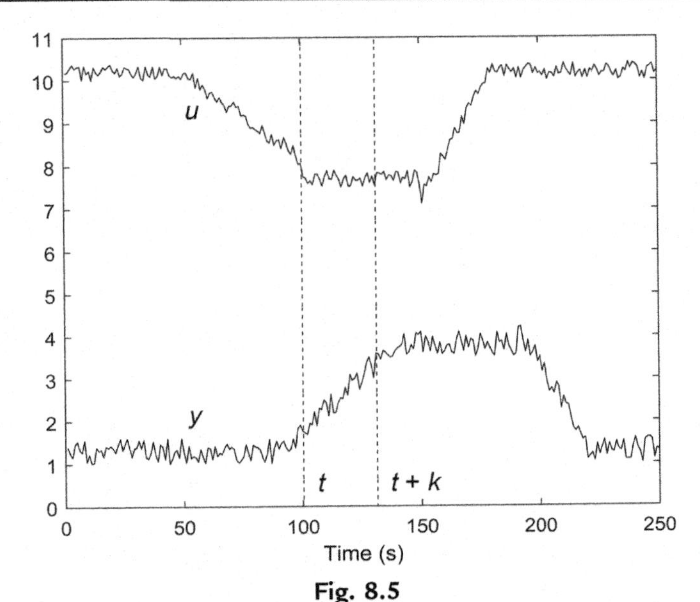

Fig. 8.5
The random input and output signals of a stochastic process.

time), the temporal and statistical properties may be taken to be equal. It should be noted that if the signals are not stationary, this condition can be induced by taking the first or second time difference of the random signals, i.e., to work with ∇u_t and ∇y_t or $\nabla^2 u_t$ and $\nabla^2 y_t$ instead of the original signals. This is a very important result which allows us to derive the required statistical properties of the random input-output signals from the outcome of a single experiment. The measured input-output data are used for the identification of the process and noise models of a stochastic process.

The auto- and cross-covariance matrices can be defined for the input and output signals separated by different lags:

$$\Gamma_{uu} = \text{auto-covariance matrix of } u = E\left[\begin{pmatrix} u_t - \bar{u} \\ u_{t+1} - \bar{u} \\ \vdots \\ u_{t+N+1} - \bar{u} \end{pmatrix}\begin{pmatrix} u_t - \bar{u} \\ u_{t+1} - \bar{u} \\ \vdots \\ u_{t+N+1} - \bar{u} \end{pmatrix}^T\right]$$

$$= \begin{bmatrix} \gamma_{uu}(0) & \gamma_{uu}(1) & \cdots & \gamma_{uu}(N-1) \\ \gamma_{uu}(1) & \ddots & & \\ \vdots & & & \\ \gamma_{uu}(N-1) & & & \gamma_{uu}(0) \end{bmatrix}$$

$$\Gamma_{uu} = \text{auto-covariance matrix of } u = E\left[\begin{pmatrix} u_t - \bar{u} \\ u_{t+1} - \bar{u} \\ \vdots \\ u_{t+N+1} - \bar{u} \end{pmatrix}\begin{pmatrix} u_t - \bar{u} \\ u_{t+1} - \bar{u} \\ \vdots \\ u_{t+N+1} - \bar{u} \end{pmatrix}^T\right]$$

$$= \begin{bmatrix} \gamma_{uu}(0) & \gamma_{uu}(1) & \cdots & \gamma_{uu}(N-1) \\ \gamma_{uu}(1) & \ddots & & \\ \vdots & & & \\ \gamma_{uu}(N-1) & & & \gamma_{uu}(0) \end{bmatrix} \tag{8.20}$$

$$\Gamma_{yy} = \text{auto-covariance matrix of } y = E\left[\begin{pmatrix} y_t - \bar{y} \\ y_{t+1} - \bar{y} \\ \vdots \\ y_{t+N+1} - \bar{y} \end{pmatrix}\begin{pmatrix} y_t - \bar{y} \\ y_{t+1} - \bar{y} \\ \vdots \\ y_{t+N+1} - \bar{y} \end{pmatrix}^T\right]$$

$$= \begin{bmatrix} \gamma_{yy}(0) & \gamma_{yy}(1) & \cdots & \gamma_{yy}(N-1) \\ \gamma_{yy}(1) & \ddots & & \\ \vdots & & & \\ \gamma_{yy}(N-1) & & & \gamma_{yy}(0) \end{bmatrix} \tag{8.21}$$

$$\Gamma_{uy} = \text{auto-covariance matrix of } y = E\left[\begin{pmatrix} u_t - \bar{u} \\ u_{t+1} - \bar{u} \\ \vdots \\ u_{t+N+1} - \bar{u} \end{pmatrix}\begin{pmatrix} y_t - \bar{y} \\ y_{t+1} - \bar{y} \\ \vdots \\ y_{t+N+1} - \bar{y} \end{pmatrix}^T\right]$$

$$= \begin{bmatrix} \gamma_{uy}(0) & \gamma_{uy}(1) & \cdots & \gamma_{uy}(N-1) \\ \gamma_{yu}(1) & \ddots & & \\ \vdots & & & \\ \gamma_{yu}(N-1) & & & \gamma_{uy}(0) \end{bmatrix} \tag{8.22}$$

Note that the auto-covariance matrix is symmetric while the cross-covariance matrix is not. This is due to the fact that $\gamma_{uy}(k) \neq \gamma_{yu}(k) = \gamma_{uy}(-k)$. The diagonal elements of the auto-covariance matrix represent the variance of the individual signal, and the off-diagonal terms represent the correlation of a signal with its past and future values.

The auto- and cross-correlation matrices can be obtained by dividing the elements of the auto-covariance matrices by their diagonal elements. The individual entries of the previous matrices can be estimated by the following equations:

$$\langle \gamma_{uu}(k) \rangle = \frac{1}{N}\sum_{t=1}^{N-k}(u_t - \langle u \rangle)(u_{t+k} - \langle u \rangle) \tag{8.23}$$

$$\langle \gamma_{yy}(k) \rangle = \frac{1}{N}\sum_{t=1}^{N-k}(y_t - \langle y \rangle)(y_{t+k} - \langle y \rangle) \tag{8.24}$$

$$\langle \gamma_{uy}(k) \rangle = \frac{1}{N}\sum_{t=1}^{N-k}(u_t - \langle u \rangle)(y_{t+k} - \langle y \rangle) \tag{8.25}$$

where N is the number of the observations which should be larger than 50, and $\langle u \rangle$ and $\langle y \rangle$ are the estimated mean values of u and y sequences.

$$\langle u \rangle = \frac{1}{2N+1}\sum_{i=-N}^{+N} u_{t+i} \tag{8.26}$$

$$\langle y \rangle = \frac{1}{2N+1}\sum_{i=-N}^{+N} y_{t+i} \tag{8.27}$$

k in Eqs. (8.23)–(8.25) is the number of lags for which the covariances are calculated

$$k = 0, 1, \cdots, \frac{N}{4} \tag{8.28}$$

8.2 Identification of Stochastic Processes

Design of controllers for stochastic processes requires models of the process and of the noise. Development of such models using a theoretical approach is almost an impossible task. Accordingly, an empirical approach is adopted to obtain models from the measured input-output sequences and their statistical properties. Process identification can be either carried out in an off-line mode using the collected input-output data, or in an online manner using the various recursive parameter estimation techniques. The latter is always used in conjunction with the adaptive controllers.

8.2.1 Off-line Process Identification

In this mode of process identification, a series of input-output data which contains dynamic information about the stochastic plant is collected and analyzed for the determination of the process and noise models. A single-input, single-output stochastic process is represented by:

$$y_t = \frac{B(z^{-1})}{A(z^{-1})}u_{t-b} + \frac{C(z^{-1})}{D(z^{-1})\nabla^d}e_t \tag{8.29}$$

whereas a multi-input, single-output (MISO) stochastic process is expressed by:

$$y_t = \frac{B_1(z^{-1})}{A_1(z^{-1})}u_{1,\, t-b_1} + \cdots + \frac{B_m(z^{-1})}{A_m(z^{-1})}u_{m,\, t-b_m} + \frac{C(z^{-1})}{D(z^{-1})\nabla^d}e_t \tag{8.30}$$

The corresponding model for a MIMO stochastic process is a direct extension of Eq. (8.30) which involves matrices of polynomials.

The objective of the off-line process identification is the determination of the orders and parameters of polynomials $A(z^{-1})$, $B(z^{-1})$, $C(z^{-1})$, and $D(z^{-1})$; the process time delay, b; and the power of the difference operator in the noise model, d.

There are four steps in an off-line process identification procedure.

8.2.1.1 Collection of the input-output data

The input-output data must contain enough dynamic information about the plant without seriously disturbing it. For a MIMO process with m input variables and n output variables, there would be $m \times n$ unknown process models and n unknown noise models. In order to reduce the data collection effort, all input variables are varied simultaneously and their effects on all output variables are monitored. The input signals must have certain characteristics for efficient process identification:

- they must be noncorrelated with one another,
- they must be noncorrelated with the process noise, and
- they must excite the plant sufficiently for retrieving useful dynamic information without causing significant disturbance to the plant operation.

These requirements are met by two types of signals:

- a sequence of pseudo random binary signal (PRBS), or
- an ARMA signal.

A PRBS is a signal which is randomly switched between a high and a low value. At each switching instant the probability of the PRBS taking the high or the low value is 1/2. There are two design parameters associated with a PRBS, its magnitude and its switching interval. The magnitude of the PRBS should be chosen such that it excites the plant sufficiently without introducing significant disturbance to the process operation. Its switching interval should be between one-tenth and one-half of the dominant time constant of the plant.

Collection of the input-output data for off-line process identification can be performed during the open-loop operation of the plant or while it is under feedback control. In the former case, the process inputs are varied using different PRBS or ARMA signals. In the feedback mode, the PRBS or ARMA signal (a dither signal) is added to the controller output or its set points as shown in Fig. 8.6.

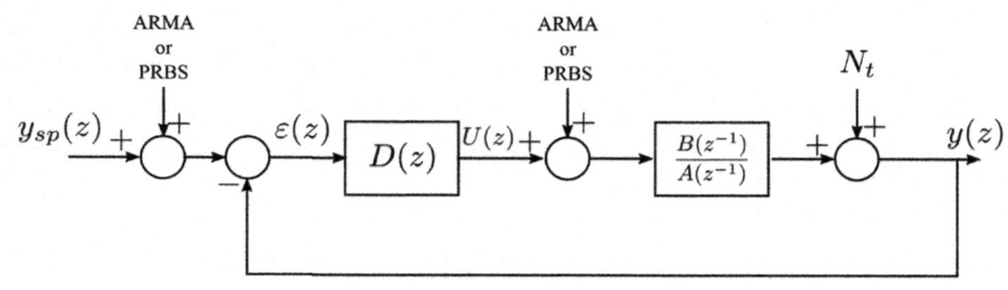

Fig. 8.6
Addition of PRBS or ARMA signal to the controller output or set point.

8.2.1.2 Identification of the structures of the process and noise models

In this step, the orders of $A(z^{-1})$, $B(z^{-1})$, $C(z^{-1})$, and $D(z^{-1})$ polynomials as well as the process time delay b and the power of the difference operator, d, are determined.

Determination of n_A, n_B, and b: Consider the impulse model of a stochastic process:

$$y_t = V(z^{-1})u_t + N_t = v_0 u_t + v_1 u_{t-1} + \cdots + N_t \tag{8.31}$$

Multiplying both sides of Eq. (8.31) by u_{t-k} and taking the expected values of each term results in:

$$E[y_t u_{t-k}] = v_0 E[u_t u_{t-k}] + v_1 E[u_{t-1} u_{t-k}] + \cdots E[N_t u_{t-k}] \tag{8.32}$$

The last term in Eq. (8.32) is zero since the input signal is assumed to be independent of the process noise, as was discussed in step 1. Eq. (8.32) can be written as:

$$\gamma_{uy} = v_0 \gamma_{uu}(k) + v_1 \gamma_{uu}(k-1) + \cdots \tag{8.33}$$

The estimated values of the auto- and cross-covariance terms appearing in Eq. (8.33), $\langle \gamma_{uu}(k) \rangle$ and $\langle \gamma_{uy}(k) \rangle$, can be calculated from the measured input-output data according to what was discussed earlier (Eqs. 8.23–8.27). The impulse weights of a stable finite impulse response (FIR) process, $v0$, $v1$, …, die out after say L sampling intervals. Consequently, the solution of a set of L equations represented by Eq. (8.33), using the estimated auto- and cross-covariance terms, renders the unknown impulse weights $v0$, $v1$, …, v_L. Therefore, the impulse and step responses of the unknown plant can be determined. The step response weights s_i can be determined in terms of the impulse response weights by:

$$s_i = \sum_{j=1}^{i} v_j \tag{8.34}$$

The calculated impulse and step response of the unknown plant are then compared with those of known low-order process models to determine n_A, n_B, and b.

Determination of n_C, n_D, and d: The identification of the structure of the noise model proceeds after the process model has been determined and verified. That is after n_A, n_B, and b and the parameters of $A(z^{-1})$ and $B(z^{-1})$ polynomials have been determined. The verification of the process model will be discussed in step 4. So, the residuals or the noise portion of the model can be written as:

$$N_t = y_t - \frac{\hat{B}(z^{-1})}{\hat{A}(z^{-1})} u_{t-\hat{b}} \tag{8.35}$$

where $\hat{A}$, $\hat{B}$, and $\hat{b}$ are the estimates of A, B, and b. Now the estimated auto-correlation, $\rho_{uu}(k)$, and the partial auto-correlation, $\phi_{NN}(k)$, of the residuals are calculated and compared with those of known ARIMA processes to determine n_C, n_D, and d. The partial auto-correlation at lag k is the remaining auto-correlation at lag k after subtracting from it the auto-correlations at lags $0, 1, \ldots, k-1$. For an AR process with order n_D, the partial auto-correlation cuts off after n_D lags, while for an MA process with order n_C, the auto-correlation cuts off at lag n_C. For an ARMA process, both the auto-correlation and the partial auto-correlation functions are infinite. And for an ARIMA process $\rho_{uu}(k)$ and $\phi_{NN}(k)$ are both 1 at lag 1 and have infinite extent.

8.2.1.3 Estimation of parameters

The unknown parameters are the coefficients of $A(z^{-1})$, $B(z^{-1})$, $C(z^{-1})$, and $D(z^{-1})$ polynomials. There are various parameter estimation methods such as the least squares (LS), the extended least squares (ELS), the maximum likelihood (ML), and the instrumental variable (IV) techniques.[1–4] The simplest method is the least squares algorithm which can only be used if $C(z^{-1}) = 1$ and $D(z^{-1})\nabla^d = A(z^{-1})$. Otherwise the LS method does not converge and leads to a biased estimate. In this section, we shall only consider the LS method. In the online process identification section, we will look at the ELS and the approximate ML approach in a recursive manner.

Consider the plant dynamics as:

$$y_t = \frac{B'(z^{-1})}{A(z^{-1})} u_t + \frac{1}{A(z^{-1})} e_t \tag{8.36}$$

Notice that in Eq. (8.36) the process time delay is incorporated in $B'(z^{-1}) = B(z^{-1})z^{-b}$ polynomial. This is a more general formulation and allows estimation of the process time delay as the number of the leading coefficients of $B'(z^{-1})$ which are zero. Eq. (8.36) can be expanded as

$$y_t = -a_1 y_{t-1} - \cdots - a_{n_A} y_{t-n_A} + b_0 u_t + \cdots b_{n_B} u_{t-n_b} + e_t \tag{8.37}$$

which then can be written as:

$$y_t = \phi_t^T \theta_t + e_t \tag{8.38}$$

where ϕ_t and θ_t are the measured input-output vector and the parameter vector.

$$\phi_t^T = [y_{t-1} \cdots y_{t-n_A} \; u_t \cdots u_{t-n_B}] \tag{8.39}$$

$$\theta_t^T = [-a_1 \cdots - a_{n_A} b_0 \cdots b_{n_B}] \tag{8.40}$$

There are $n_A + n_B + 1$ unknown parameters in Eq. (8.40). If the process were deterministic, a total of $n_A + n_B + 1$ data points would be sufficient to estimate these parameters. In the presence of uncertainties, however, the required number of data points N must be much larger than $n_A + n_B + 1$. Collecting N data points and arranging them in the following manner:

$$\begin{bmatrix} y_t \\ y_{t+1} \\ \vdots \\ y_{t+N-1} \end{bmatrix} = \begin{bmatrix} \phi_t^T \\ \phi_{t+1}^T \\ \vdots \\ \phi_{t+N-1}^T \end{bmatrix} \theta_t + \begin{bmatrix} e_t \\ e_{t+1} \\ \vdots \\ e_{t+N-1} \end{bmatrix} \tag{8.41}$$

results in a compact representation:

$$\boldsymbol{Y} = \boldsymbol{\Phi} \theta_t + \boldsymbol{e} \tag{8.42}$$

where $\boldsymbol{Y}$ is the measured output sequence, $\boldsymbol{\Phi}$ is the measured input-output matrix, θ_t is the unknown parameter vector, and $\boldsymbol{e}$ is a sequence of white noise.

If the parameter vector converges to its true value θ_t, the noise vector would be a sequence of random numbers according to Eq. (8.42). However, during the estimation process and before the parameter vector converges to its true value, the noise vector would not be a white noise and is referred to as residuals, $\boldsymbol{\eta}$.

$$\boldsymbol{Y} = \boldsymbol{\Phi} \hat{\theta}_t + \boldsymbol{\eta} \tag{8.43}$$

The objective of the least squares method is to estimate $\hat{\theta}_t$ by minimizing the sum of the squares of the residuals.

$$\min_{\hat{\boldsymbol{\theta}}_t} J = \boldsymbol{\eta}^T \boldsymbol{\eta} = \sum_{i=0}^{N-1} \eta_{t+i}^2 \tag{8.44}$$

where $\boldsymbol{\eta} = \boldsymbol{Y} - \boldsymbol{\Phi} \hat{\theta}_t$. Substitution of Eq. (8.43) in Eq. (8.44) results in:

$$J = \boldsymbol{Y}^T \boldsymbol{Y} - \hat{\theta}_t^T \boldsymbol{\Phi}^T \boldsymbol{Y} - \boldsymbol{Y}^T \boldsymbol{\Phi} \hat{\theta}_t^T + \hat{\theta}_t^T \boldsymbol{\Phi}^T \boldsymbol{\Phi} \hat{\theta}_t \tag{8.45}$$

It should be noted that all terms in Eq. (8.45) are scalar. Differentiating J with respect to $\hat{\theta}_t$ and setting it to zero will render the estimate of the parameter vector $\hat{\theta}_t$.

$$\frac{dJ}{d\hat{\theta}} = -2\boldsymbol{\Phi}^T\boldsymbol{Y} + 2\boldsymbol{\Phi}^T\boldsymbol{\Phi}\hat{\theta}_t = 0 \tag{8.46}$$

The second derivative of J with respect to $\hat{\theta}_t$ is $2\left(\boldsymbol{\Phi}^T\boldsymbol{\Phi}\right) \geq 0$, which is positive and confirms the minimization of J.

$$\hat{\theta}_t = \left[\boldsymbol{\Phi}^T\boldsymbol{\Phi}\right]^{-1}\boldsymbol{\Phi}^T\boldsymbol{Y} \tag{8.47}$$

Eq. (8.47) is the least squares estimate of the parameter vector.In order to show that the LS method renders the unbiased estimation of the parameter vector provided that $C\left(z^{-1}\right) = 1$ and $D\left(z^{-1}\right)\nabla^d = A\left(z^{-1}\right)$, let us substitute Eq. (8.42) in Eq. (8.47) for $\boldsymbol{Y}$:

$$\hat{\theta}_t = \left[\boldsymbol{\Phi}^T\boldsymbol{\Phi}\right]^{-1}\boldsymbol{\Phi}^T\left[\boldsymbol{\Phi}\theta_t + \boldsymbol{e}\right] = \theta_t + \left[\boldsymbol{\Phi}^T\boldsymbol{\Phi}\right]^{-1}\boldsymbol{\Phi}^T\boldsymbol{e} \tag{8.48}$$

So, the estimation bias $\hat{\theta}_t - \theta_t$ approaches zero only if $\boldsymbol{e}$ were a white noise sequence with a zero mean. This statement can be proven by taking the expected value of Eq. (8.48).

Often it is required to assign different weights to the past collected data. This can be achieved by incorporating a weighting matrix

$$\boldsymbol{W} = \text{diag}(w_i) \tag{8.49}$$

in the derivation of the LS equation. The weighted least squares (WLS) leads to the following estimate of the parameter vector:

$$\hat{\theta}_t = \left[\boldsymbol{\Phi}^T\boldsymbol{W}\boldsymbol{\Phi}\right]^{-1}\boldsymbol{\Phi}^T\boldsymbol{W}\boldsymbol{Y} \tag{8.50}$$

8.2.1.4 Test of model validity

It is important to note that in the identification of a stochastic process, slightly different models can be fitted to the same input-output data. Accordingly, the statistical tests must be employed to choose the best among the possible identified models. Consider the model of a stochastic process given by:

$$y_t = \frac{B(z^{-1})}{A(z^{-1})}u_{t-b} + \frac{C(z^{-1})}{D(z^{-1})\nabla^d}e_t \tag{8.51}$$

which can also be written as

$$y_t = V\left(z^{-1}\right)u_t + W\left(z^{-1}\right)e_t \tag{8.52}$$

where $V\left(z^{-1}\right)$ and $W\left(z^{-1}\right)$ are the exact models of the process and of the noise, respectively. The estimated models for the process and the noise are represented by $\widetilde{V}\left(z^{-1}\right)$ and $\widetilde{W}\left(z^{-1}\right)$.

Note that if the estimated process and noise models are not exact the residuals, η_t, will not be a white noise sequence, e_t:

$$y_t = \widetilde{V}(z^{-1})u_t + \widetilde{W}(z^{-1})\eta_t \tag{8.53}$$

Therefore in order that the estimated process and noise models be statistically acceptable, the residuals η_t should approach e_t. Let us assume that the process model is exact, i.e., $\widetilde{V}(z^{-1}) = V(z^{-1})$. Under this condition, the division of Eq. (8.52) by Eq. (8.53) yields:

$$\eta_t = \frac{W(z^{-1})}{\widetilde{W}(z^{-1})} e_t \tag{8.54}$$

which suggests that the residuals are only auto-correlated but not cross-correlated with the input signal, u_t, as the latter does not appear in Eq. (8.54). This is an important conclusion and suggests that in order to have a valid process model, there should be no cross-correlation between the residuals and the input signal. If, however, $\widetilde{V}(z^{-1}) \neq V(z^{-1})$, i.e., the process model were not exact, we would have,

$$\eta_t = \left[\frac{y_t - \widetilde{V}(z^{-1})u_t}{y_t - V(z^{-1})u_t}\right] \frac{W(z^{-1})}{\widetilde{W}(z^{-1})} e_t \tag{8.55}$$

which shows that the residuals would not only be auto-correlated but also cross-correlated with the input signal.

In the first step of model validation we have to ensure that there is no cross-correlation between the residuals and the input signal by repeatedly updating the process model to remove any statistical dependence between the residuals and the input signal. Once an acceptable process model is obtained for the process, the process model is frozen and an attempt is made to derive an acceptable noise model. Initially, the auto-correlation and partial auto-correlation of the residuals are calculated. The noise model will be updated until no auto-correlations and partial auto-correlations remain and the residuals approach a white noise.

Instead of considering the individual correlations at each lag, some cumulative measures of correlations over a range of sampling intervals may be calculated for model validation. Two such statistical measures are:

$$P_1 = N \sum_{k=1}^{L} \left\langle \rho_{\eta\eta}^2(k) \right\rangle \tag{8.56}$$

$$P_2 = N \sum_{k=0}^{L} \left\langle \rho_{\eta u}^2 \right\rangle \tag{8.57}$$

where N is the total number of data points, and L is the number of sampling intervals up to which the correlations are calculated. If the individual correlations are statistically insignificant, the sum of their squares, that is, P_1 and P_2 defined in Eqs. (8.56) and (8.57), must have Chi-squared distributions. Therefore, at any degrees of freedom ($L - n_D - n_C$ for P_1 and $L - n_A - n_B$ for P_2), P_1 and P_2 should have Chi-squared distributions if both the process and noise models are statistically acceptable.

Based on the previous discussion the iterative nature of the process identification method in model updating and parameter estimation becomes evident. MATLAB makes this otherwise tedious job a relatively simple task.

Example 8.1

Fig. 8.7 represents the input-output time series of a stochastic process with

$$A = [1 \ \ -1.3 \ \ 0.9]; \ \ B = [0 \ \ 1 \ \ 0.7]$$

and added random noise. The objective is to fit the given input-output data to an ARX(2,2,1) model and obtain estimates of A' and B' polynomials, the best estimates for A and B.

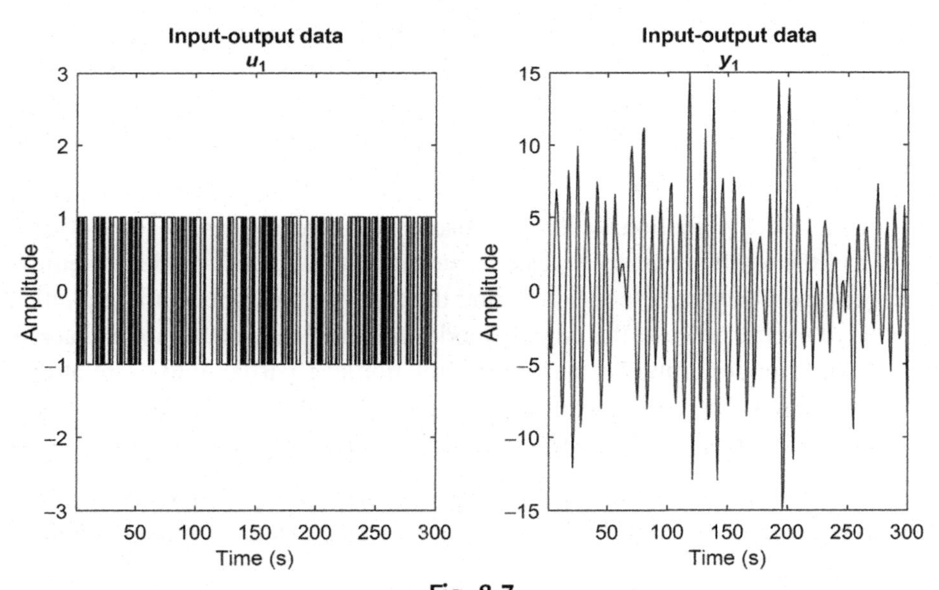

Fig. 8.7
The input-output data of a process to be estimated as an ARX(2,2,1) by MATLAB.

```
% enter the coefficients of A and B polynomials of the plant
A = [1 -1.3 0.9]; B = [0 1 0.7];
% convert the plant to a polynomial model with identifiable parameters
p0 = idpoly(A,B);
% create a data object to encapsulate the input/output data and their
% properties, rbs is the input in the form of a random binary signal
```

```
u = iddata([],idinput(300,'rbs'));
% e is the random noise
e = iddata([],randn(300,1));
% simulate the process and obtain the output y
% generate a data object for the subsequent model identification by the
% arx command
y = sim(p0,[u e]);
w = [y,u];
% compute the least squares estimate of an arx model
p = arx(w,[2 2 1])
```

The estimated model is:

```
p = Discrete-time ARX model: A'(z)y(t) = B'(z)u(t) + e(t)
  A'(z) = 1 - 1.303 z^-1 + 0.9247 z^-2
  B'(z) = 1.036 z^-1 + 0.7075 z^-2
Sample time: 1 seconds
Parameterization:
    Polynomial orders: na'=2    nb'=2    b'=1
    Number of free coefficients: 4
    Use "polydata", "getpvec", "getcov" for parameters and their uncertainties.
Status:
Estimated using ARX on time domain data "w".
Fit to estimation data: 86.87% (prediction focus)
FPE: 0.9223, MSE: 0.8862% the final precision error and the mean square error
```

Note that the estimated polynomials A' and B' are close to the actual polynomials A and B and the mean square error (MSE) is 0.886%. This is an open-loop off-line process identification example, in which the input-output data are collected and then analyzed. Note that the input signal is an RBS sequence with a switching interval of 1 s and a magnitude between -1 and $+1$.

In Example 8.1, the model structure, i.e., the orders of the polynomials A and B and the time delay, b, were all assumed to be known. In the absence of such knowledge, using the input-output signals, u_t and y_t, over say 300 sampling intervals ($N = 300$), the auto-correlations $\gamma_{uu}(k)$ and cross-correlations $\gamma_{uy}(k)$ for lags k are calculated. Solution of a set of equations using the estimated auto-correlations $\gamma_{uu}(k)$ and cross-correlations $\gamma_{uy}(k)$ renders the impulse weights of the plant. Knowing the impulse weights the step response of the plant is determined and compared with the impulse response and step response of low-order models rendering guesses for the model order, i.e., $n_A = 2$ and $n_B = 2$, and a time delay of one sampling interval ($b = 1$). Subsequently, the unknown parameters of the A and B polynomials are estimated. The residual sequence η_t is calculated for $t = 1, \ldots, N$, since the input and output signals are known and the process model has been identified. In order to validate the identified process model, the cross-correlations between the residuals and the input signal are calculated. After the identification of an acceptable process model, this model is frozen, and an appropriate ARX model for the residuals is developed.

The previous example demonstrates the power of the time series analysis of MATLAB for process identification. Many similar examples of ARMAX (auto-regressive moving average exogenous) processes can be found in the online MATLAB tutorials.

8.2.2 Online Process Identification

The online process identification is usually used in conjunction with adaptive controllers. Due to the limited time available for process identification which is followed by controller adaptation and implementation, the most computationally intensive step in the off-line identification, namely, the structural identification of the process and noise models, is not performed in the online identification. Instead, based on *a priori* information about the plant, n_A, n_B, n_C, n_D, d and the minimum time delay are assumed to be known, and only parameters $a_1,\ldots, a_{n_A}$, b_0, $b_1,\ldots,bn_B,c_1,\ldots,c_{n_C},d_1,\ldots,d_{n_D}$ are estimated online. Model validation step is also performed differently and not in an iterative manner as discussed in the off-line identification method. The three main steps in the online process identification are as follows:

1. *Collection of input-output data:* In order to collect useful input–output data which meet the closed-loop identifiability condition, the plant must be persistently excited during the identification process without being unacceptably disturbed. To fulfill these requirements, a PRBS or an ARMA signal, whose parameters (switching interval, magnitude, etc.) have been carefully selected, is either added to the controller output or to its set point as shown in Fig. 8.6. At each sampling interval, the newly collected input-output data pair is used to update the parameters of the process and noise models.
2. *Recursive Parameter Estimation Techniques:* Once a new pair of input-output data is received the plant models parameters are updated. Consequently, a recursive version of the parameter estimation methods is used. The recursive version of a few parameter estimation methods is discussed as follows.
   Recursive Least Squares (RLS): In the recursive least squares parameter estimation technique, the objective is to update the estimates of the parameter vector at each sampling interval, i.e., to express $\hat{\theta}_{t+1}$ in terms of $\hat{\theta}_t$. The least squares estimate of the parameter vector was derived earlier as:

$$\hat{\theta}_t = \left[\mathbf{\Phi}^T\mathbf{\Phi}_t\right]^{-1}\mathbf{\Phi}_t^T\mathbf{Y}_t \tag{8.58}$$

At time $t+1$ we receive a new set of input-output data, therefore, the new estimate of the parameter vector is:

$$\hat{\theta}_{t+1} = \left[\mathbf{\Phi}_{t+1}^T\mathbf{\Phi}_{t+1}\right]^{-1}\mathbf{\Phi}_{t+1}^T\mathbf{Y}_{t+1} \tag{8.59}$$

where

$$\mathbf{\Phi}_{t+1} = \begin{bmatrix}\mathbf{\Phi}_t \\ \phi_{t+1}^T\end{bmatrix} \quad \text{and} \quad \mathbf{Y}_{t+1} = \begin{bmatrix}\mathbf{Y}_t \\ y_{t+1}\end{bmatrix} \tag{8.60}$$

Introducing a new matrix as the covariance matrix, we have

$$\boldsymbol{P}_t = \left[\boldsymbol{\Phi}_t^T \boldsymbol{\Phi}_t\right]^{-1} \tag{8.61}$$

Eqs. (8.58) and (8.59) can be written as:

$$\hat{\theta}_t = \boldsymbol{P}_t \boldsymbol{\Phi}_t^T \boldsymbol{Y}_t \tag{8.62}$$

$$\hat{\theta}_{t+1} = \boldsymbol{P}_{t+1} \boldsymbol{\Phi}_{t+1}^T \boldsymbol{Y}_{t+1} \tag{8.63}$$

But

$$\begin{aligned} \boldsymbol{\Phi}_{t+1}^T \boldsymbol{Y}_{t+1} &= \boldsymbol{\Phi}_t^T Y_t + \phi_{t+1} y_{t+1} \\ &= \boldsymbol{P}_t^{-1} \hat{\theta}_t + \phi_{t+1} y_{t+1} \end{aligned} \tag{8.64}$$

And

$$\boldsymbol{\Phi}_{t+1}^T \boldsymbol{\Phi}_{t+1} = \boldsymbol{P}_{t+1}^{-1} = \boldsymbol{\Phi}_t^T \boldsymbol{\Phi}_t + \phi_{t+1} \phi_{t+1}^T \tag{8.65}$$

Eq. (8.65) can be written as:

$$\boldsymbol{P}_t^{-1} = \boldsymbol{P}_{t+1}^{-1} - \phi_{t+1} \phi_{t+1}^T \tag{8.66}$$

Eqs. (8.59)–(8.64) result in:

$$\hat{\theta}_{t+1} = \boldsymbol{P}_{t+1} \left[\boldsymbol{P}_t^{-1} \hat{\theta}_t + \phi_{t+1} y_{t+1}\right] \tag{8.67}$$

And Eqs. (8.66) and (8.67) result in:

$$\hat{\theta}_{t+1} = \boldsymbol{P}_{t+1} \left[\boldsymbol{P}_{t+1}^{-1} \hat{\theta}_t - \phi_{t+1} \phi_{t+1}^T \hat{\theta}_t + \phi_{t+1} y_{t+1}\right] \tag{8.68}$$

$$\hat{\theta}_{t+1} = \hat{\theta}_t + \boldsymbol{P}_{t+1} \phi_{t+1} \left[y_{t+1} - \phi_{t+1}^T \hat{\theta}_t\right] \tag{8.69}$$

On the other hand, Eq. (8.66) results in:

$$\boldsymbol{P}_{t+1}^{-1} = \boldsymbol{P}_t^{-1} + \phi_{t+1} \phi_{t+1}^T \tag{8.70}$$

Applying the well-known matrix inversion lemma

$$(\boldsymbol{A} + \boldsymbol{BCD})^{-1} = \boldsymbol{A}^{-1} - \boldsymbol{A}^{-1} \boldsymbol{B} \left(\boldsymbol{C}^{-1} + \boldsymbol{D} \boldsymbol{A}^{-1} \boldsymbol{B}\right)^{-1} \boldsymbol{D} \boldsymbol{A}^{-1} \tag{8.71}$$

With $\boldsymbol{P}_t^{-1} = \boldsymbol{A}$, $\phi_{t+1} = \boldsymbol{B}$, $\phi_{t+1}^T = \boldsymbol{D}$, and $\boldsymbol{C} = 1$ to Eq. (8.71), we get:

$$\boldsymbol{P}_{t+1} = \boldsymbol{P}_t - \boldsymbol{P}_t \phi_{t+1} \left[1 + \phi_{t+1}^T \boldsymbol{P}_t \phi_{t+1}\right]^{-1} \phi_{t+1}^T \boldsymbol{P}_t \tag{8.72}$$

$$\boldsymbol{P}_{t+1} = \boldsymbol{P}_t \left[I - \frac{\phi_{t+1} \phi_{t+1}^T \boldsymbol{P}_t}{1 + \phi_{t+1}^T \boldsymbol{P}_t \phi_{t+1}}\right] \tag{8.73}$$

Eq. (8.73) allows updating the covariance matrix $\boldsymbol{P}$ at time $t+1$ in terms of its value at time t without having to take the inverse of a matrix. It must be noted that the term appearing in the denominator of $\boldsymbol{P}_{t+1}$ in Eq. (8.73) is a scalar quantity. Therefore the entire RLS technique consists of the following steps as soon as u_{t+1} and y_{t+1} become available.

- Form

$$\phi_{t+1}^T = [y_t \cdots y_{t-n_A+1} \; u_{t+1} \cdots u_{t-n_B+1}] \tag{8.74}$$

- Form the innovation series at time $t+1$ which are the difference between the measured output and the predicted output

$$\varepsilon_{t+1} = y_{t+1} - \phi_{t+1}^T \theta_t \tag{8.75}$$

- Using Eq. (8.73) update the covariance matrix

$$\boldsymbol{P}_{t+1} = \boldsymbol{P}_t \left[\boldsymbol{I} - \frac{\phi_{t+1} \phi_{t+1}^T \boldsymbol{P}_t}{1 + \phi_{t+1}^T \boldsymbol{P}_t \phi_{t+1}} \right] \tag{8.76}$$

- Using Eq. (8.69) update the parameter vector

$$\hat{\theta}_{t+1} = \hat{\theta}_t + \boldsymbol{P}_{t+1} \phi_{t+1} \left[y_{t+1} - \phi_{t+1}^T \hat{\theta}_t \right] = \hat{\theta}_t + \boldsymbol{P}_{t+1} \phi_{t+1} e_{t+1} \tag{8.77}$$

In order to improve the convergence properties of the RLS technique, several modifications have been suggested to the basic algorithm. The most widely used approach is to use a forgetting factor $0.95 < \lambda < 1$ which places successively less weights on the old data. This is equivalent to having a weighting matrix in the least squares algorithm.

$$\Lambda = \begin{bmatrix} \lambda^n & & & 0 \\ & \lambda^{n-1} & & \\ & & \ddots & \\ 0 & & & \lambda^0 \end{bmatrix} \tag{8.78}$$

And in the RLS algorithm it can be implemented by dividing the covariance matrix (Eq. 8.76) by λ. It has been shown[4] that a variable forgetting factor further improves the convergence properties of the RLS algorithm.

Another improvement in the RLS algorithm can be introduced in updating the covariance matrix. Ljung and Soderstrom[3] have introduced the UD-factorization method which prevents the $\boldsymbol{P}$ matrix to become exceedingly large.

$$\boldsymbol{P} = \boldsymbol{U}\boldsymbol{D}\boldsymbol{U}^T \tag{8.79}$$

in which $\boldsymbol{U}$ is an upper triangular matrix with diagonal terms equal to 1 and $\boldsymbol{D}$ is a diagonal matrix.

The RLS algorithm must have some initial conditions $\boldsymbol{P}(0)$ and $\hat{\theta}(0)$. Initially the parameter vector is far from its actual value and consequently $\boldsymbol{P}(0)$, which represents the uncertainty involved in the estimation process, must be quite large. Therefore a reasonable choice of the initial values would be:

$$\boldsymbol{P}(0)=\alpha \boldsymbol{I},\hat{\theta}(0)=0 \tag{8.80}$$

where α is a large number such as a 1000.

Recursive Extended Least Squares (RELS): The LS algorithm can only render unbiased estimates of the parameter vector when the polynomial $C(z^{-1})=1$ and $D(z^{-1})\nabla^d=A(z^{-1})$. If, however, the noise model contains the $C(z^{-1})$ polynomial with multiple parameters, the input-output vector and the parameter vector take the following forms:

$$\phi_t^T=[y_{t-1}\cdots y_{t-n_A}\ u_t\cdots u_{t-n_B}\ e_{t-1}\cdots e_{t-n_C}] \tag{8.81}$$

$$\theta_t^T=[-a_1\cdots a_{n_A}\ b_0\ b_1\cdots b_{n_B}\ c_1\cdots c_{n_C}] \tag{8.82}$$

The process output can now be represented by

$$y_t=\phi_t^T\theta_t+e_t \tag{8.38}$$

In order to estimate the unknown parameters c_1, c_2, ..., c_{n_C} the white noise sequence $e_{t-1}\cdots e_{t-n_C}$ are required which are not measurable. Accordingly, the most one can do is to replace this nonmeasurable sequence of white noise with an approximate sequence which can be calculated. In the RELS algorithm the innovation series is used instead of the white noise sequence:

$$\varepsilon_{t+1}=y_{t+1}-\phi_{t+1}^T\hat{\theta}_t \tag{8.83}$$

Initially the innovation series is different from the white noise sequence, however as the parameter vector converges, ε_t approaches e_t. The following steps are involved in the implementation of the RESL algorithm at time $t+1$ as the new input-output data become available:

- Form ϕ_{t+1}^T using the measured u_{t+1} and y_{t+1}

$$\phi_{t+1}^T=[y_t\cdots y_{t-n_A+1}\ u_{t+1}\cdots u_{t-n_B+1}\ \varepsilon_t\cdots\varepsilon_{t-n_C+1}] \tag{8.84}$$

- Update $\boldsymbol{P}_{t+1}$ using Eq. (8.76)
- Update the parameter vector

$$\hat{\theta}_{t+1}=\hat{\theta}_t+\boldsymbol{P}_{t+1}\phi_{t+1}\varepsilon_{t+1} \tag{8.85}$$

where ε_{t+1} is given by Eq. (8.83) that depends on $\hat{\theta}_t$, available at time $t+1$.

Recursive Approximate Maximum Likelihood (RAML): Another approach is to substitute the residuals for the nonmeasurable white noise in the ϕ vector given in Eq. (8.81). The residuals are related to the innovation series. The scalar residual at time $t+1$ is defined by

$$\eta_{t+1}=y_{t+1}-\phi_{t+1}^T\hat{\theta}_{t+1} \tag{8.86}$$

Substituting for y_{t+1} in Eq. (8.86) from Eq. (8.83), results in:

$$\eta_{t+1} = \varepsilon_{t+1} - \phi_{t+1}^T \left[\hat{\theta}_{t+1} - \hat{\theta}_t\right] \tag{8.87}$$

Using Eq. (8.84), we have

$$\eta_{t+1} = \varepsilon_{t+1} - \phi_{t+1}^T \boldsymbol{P}_{t+1} \phi_{t+1} \varepsilon_{t+1} \tag{8.88}$$

$$\eta_{t+1} = \varepsilon_{t+1} \left[1 - \phi_{t+1}^T \boldsymbol{P}_{t+1} \phi_{t+1}\right]$$

Eqs. (8.85) and (8.86) demonstrate that in order to calculate η_{t+1}, only the estimate of the parameter vector at time t is required. The implementation of the RAML algorithm involves the following steps:

- Form ϕ_{t+1}^T using the measured u_{t+1} and y_{t+1}

$$\phi_{t+1}^T = \left[y_t \cdots y_{t-n_A+1} \;\; u_{t+1} \cdots u_{t-n_B+1} \;\; \eta_t \cdots \eta_{t-n_C+1}\right] \tag{8.89}$$

- Update the covariance matrix using Eq. (8.76)
- Update the parameter vector using Eq. (8.77)

8.2.3 Test of Convergence of Parameter Vector in the Online Model Identification

The iterative methods used for the validation of the estimated models in the off-line identification procedure are time consuming and cannot be implemented in the online identification algorithms. However, there are techniques by which the convergence properties of the algorithm may be tested. At each sampling interval, the absolute value of the difference between the estimated value of each parameter and its estimated value at the previous sampling interval can be calculated. For a converging algorithm, the absolute value of the difference should decrease as time proceeds. This test, however, fails if new disturbances enter the process. The absolute value of the residuals can also be checked online. Another test of convergence is the size of the trace of the covariance matrix. If $tr(\boldsymbol{P})$ decreases with time the parameter vector is converging to its true value. If the estimated parameters are to design an adaptive controller, the new set of parameters should be rejected if the convergence tests fail. Either the controller should be switched to a fixed parameter controller with reasonable performance or the old set of data which have proven to produce stable closed-loop behavior should be used.

8.3 Design of Stochastic Controllers

We have introduced the concept of the stochastic processes and used time series analysis technique to identify the process and noise models of such processes. These models can be used to design stochastic controllers either in an off-line or an online manner. The majority of real processes are subject to uncertainties and as such stochastic controllers offer potential in the control of industrial processes.

8.3.1 The Minimum Variance Controller (MVC)

The simplest stochastic controller is the MVC which minimizes the future variance of the output from its set point. For a SISO plant with b intervals of delay, the objective at time t is to find the controller action u_t which minimizes the variance of the output from its set point at $t+b$:

$$\min_{u_t} J = E\left[\left(Y_{t+b}-Y_{sp}\right)\left(Y_{t+b}-Y_{sp}\right)\right] = E\left[y_{t+b}^2\right] \tag{8.90}$$

Let us consider the general model of a stochastic process

$$y_t = \frac{B(z^{-1})}{A(z^{-1})}u_{t-b} + \frac{C(z^{-1})}{D(z^{-1})\nabla^d}e_t \tag{8.91}$$

which can be written as:

$$y_{t+b} = \frac{B(z^{-1})}{A(z^{-1})}u_t + \frac{C(z^{-1})}{D(z^{-1})\nabla^d}e_{t+b} \tag{8.92}$$

The second term on the right-hand side of Eq. (8.92) involves the past and future noise which may be separated using the following polynomial identity:

$$\frac{C(z^{-1})}{D(z^{-1})\nabla^d} = R\left(z^{-1}\right) + \frac{S(z^{-1})}{D(z^{-1})\nabla^d}z^{-b} \tag{8.93}$$

where

$$R\left(z^{-1}\right) = 1 + r_1 z^{-1} + \cdots + r_{b-1}z^{-(b-1)} \tag{8.94}$$

$$S\left(z^{-1}\right) = s_0 + s_1 z^{-1} + \cdots + s_{nD+d-1}z^{-(n_D+d-1)} \tag{8.95}$$

Substitution of Eq. (8.93) in (8.92) results in:

$$y_{t+b} = \left[\frac{B(z^{-1})}{A(z^{-1})}u_t + \frac{S(z^{-1})}{D(z^{-1})\nabla^d}e_t\right] + R\left(z^{-1}\right)e_{t+b} \tag{8.96}$$

The last term on the right-hand side of Eq. (8.96) involves only the future noise terms. Using Eq. (8.92), e_t can be expressed in terms of the measured input-output signals:

$$e_t = \frac{D(z^{-1})\nabla^d}{C(z^{-1})}y_t - \frac{D(z^{-1})\nabla^d B(z^{-1})}{C(z^{-1})\ A(z^{-1})}u_{t-b} \tag{8.97}$$

Substitute e_t from Eq. (8.97) into Eq. (8.96) results in:

$$y_{t+b} = \left[\frac{S(z^{-1})}{C(z^{-1})}y_t + \frac{B(z^{-1})R(z^{-1})D(z^{-1})\nabla^d}{A(z^{-1}) \quad C(z^{-1})}u_t\right] + R\left(z^{-1}\right)e_{t+b} \tag{8.98}$$

The variance of y_{t+b} assuming that the future noise is independent of u_t and y_t can be written:

$$J = E\left[\left(\frac{S(z^{-1})}{C(z^{-1})}y_t + \frac{B(z^{-1})R(z^{-1})D(z^{-1})\nabla^d}{A(z^{-1}) \quad C(z^{-1})}u_t\right)^2\right] + E\left[\left(R\left(z^{-1}\right)e_{t+b}\right)^2\right] \tag{8.99}$$

where E is the expectation operator. Minimization of J requires that the first term be zero because the second term represents the variance of the noise. Accordingly, the MVC can be derived as:

$$u_t = -\frac{A(z^{-1})}{B(z^{-1})}\frac{1}{R(z^{-1})}\frac{S(z^{-1})}{D(z^{-1})\,\nabla^d}y_t \tag{8.100}$$

There are three distinct terms in Eq. (8.100). The first term $\frac{A(z^{-1})}{B(z^{-1})}$ is the inverse of the process model without the time delay which indicates that the MVC does not work for NMP processes with a zero outside the unit circle. The second term $\frac{1}{R(z^{-1})}$ is a dead-time compensator which provides a memory to the controller to remember its past actions. The third term $\frac{S(z^{-1})}{D(z^{-1})\nabla^d}$ is the b-step ahead optimal noise predictor.

For a simpler plant model when $C\left(z^{-1}\right) = 1$ and $D\left(z^{-1}\right)\nabla^d = A\left(z^{-1}\right)$, the MVC given in Eq. (8.100) takes the following form:

$$u_t = -\frac{S(z^{-1})}{B(z^{-1})R(z^{-1})}y_t \tag{8.101}$$

Example 8.2

Derive a MVC controller for a plant given by:

$$y_t = -\frac{0.138 + 0.077z^{-1} + 0.035z^{-2}}{1 - 0.741z^{-1}}u_{t-3} + \frac{1}{1 - 0.643z^{-1}}e_t \tag{8.102}$$

Solution: The noise is stationary, that is, $d = 0$. The process time delay is $b = 3$. Long division of the noise model according to Eq. (8.91) results in:

$$\frac{1}{1 - 0.643z^{-1}} = \left(1 + 0.643z^{-1} + 0.413z^{-2}\right) + \frac{0.266}{1 - 0.643z^{-1}}z^{-3} \tag{8.103}$$

The MVC can be derived using Eq. (8.100):

$$u_t = -\frac{0.266(1-0.741z^{-1})}{(0.138+0.077z^{-1}+0.35z^{-2})(1+0.643z^{-1}+0.413z^{-2})(1-0.643z^{-1})}y_t \tag{8.104}$$

The closed-loop response will have an offset, the addition of an integral action to the controller eliminates the offset.

Example 8.3

Design a minimum variance controller (MVC) for the stochastic process given by:

$$y_t = \frac{1-0.7z^{-1}}{1+0.8z^{-1}+0.3z^{-2}}u_{t-1} + \frac{1-0.3z^{-1}}{1+0.8z^{-1}+0.3z^{-2}}e_t \tag{8.105}$$

Simulate the closed-loop control system and find the response to a unit step change in the set point using Simulink platform.

Solution: The MVC can be expressed as:

$$\frac{C(z^{-1})}{D(z^{-1})\nabla^d} = R(z^{-1}) + \frac{S(z^{-1})}{D(z^{-1})\nabla^d}z^{-b} \tag{8.93}$$

So, $b=1$ (delay), and $d=0$ (stationary noise)

Considering the definition of $R(z^{-1})$ and $S(z^{-1})$ as:

$$R(z^{-1}) = 1 + r_1 z^{-1} + \cdots + r_{(b-1)}z^{-(b-1)} \tag{8.94}$$

$$S(z^{-1}) = s_0 + s_1 z^{-1} + s_2 z^{-2} + \cdots + s_{n_D} z^{-(n_D+d-1)} \tag{8.95}$$

for $b=1$, $R(z^{-1})=1$; and for $d=0$, $S(z^{-1}) = s_0 + s_1 z^{-1} + s_2 z^{-2}$.

To solve for s_0 and s_1, consider the general expression:

$$\frac{C(z^{-1})}{D(z^{-1})\nabla^d} = \frac{(1)D(z^{-1})\nabla^d + (s_0 z^{-1} + s_1 z^{-2})}{D(z^{-1})\nabla^d} \tag{8.106}$$

$$1-0.3z^{-1} = 1 + 0.8z^{-1} + 0.3z^{-2} + (s_0 z^{-1} + s_1 z^{-2}) \tag{8.107}$$

Resulting in $s_0=0$, $s_1=-0.3$, and $s_2=-1.1$.

The closed-loop form of the MVC is:

$$u_t = -\frac{A(z^{-1})}{B(z^{-1})}\frac{1}{R(z^{-1})}\frac{S(z^{-1})}{D(z^{-1})\nabla^d}y' \tag{8.108}$$

Substituting from previous equations in Eq. (8.108) renders the MVC controller:

$$u_t = -\frac{(1+0.8z^{-1}+0.3z^{-2})}{(1-0.7z^{-1})}\frac{1}{(1)}\frac{(-1.1-0.3z^{-1})}{(1+0.8z^{-1}+0.3z^{-2})}y' \tag{8.109}$$

where $y' = y - y_{sp}$. The resulting Simulink model is shown in Fig. 8.8.

The system response to a step change in the set point (with white noise variance at 0.01) is shown in Fig. 8.9.

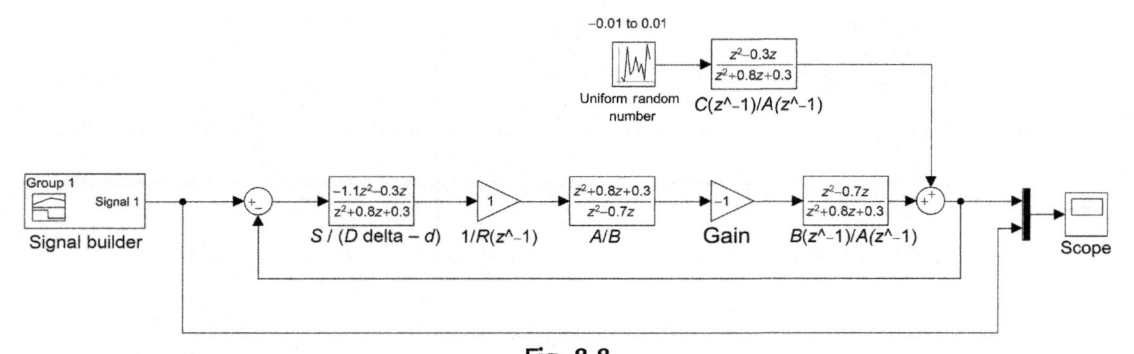

Fig. 8.8
The Simulink program of the MVC controller with a white noise.

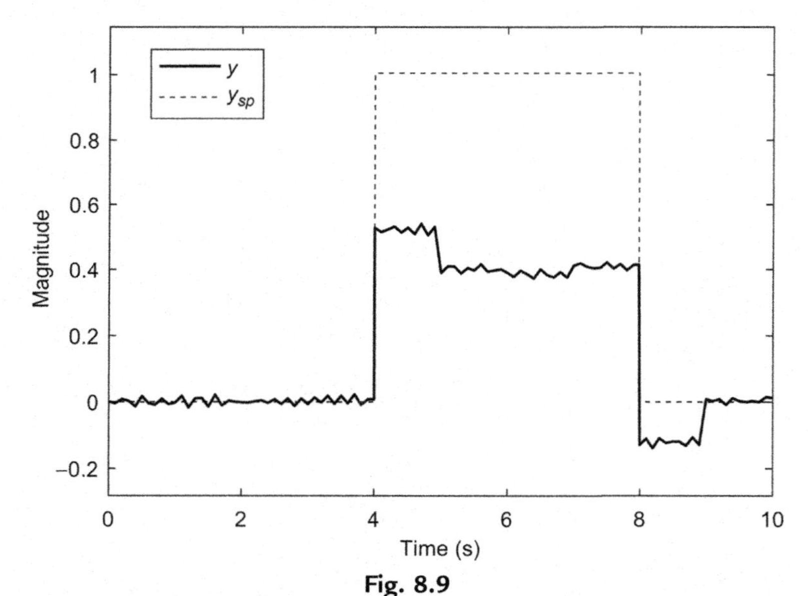

Fig. 8.9
The system closed-loop response with the MVC controller.

In order to get rid of the closed-loop offset, an integrator is added to the controller. The modified MVC Simulink program is shown in Fig. 8.10.

With the added integrator, the system response to a step change in the set point (with white noise variance at 0.01) is shown in Fig. 8.11.

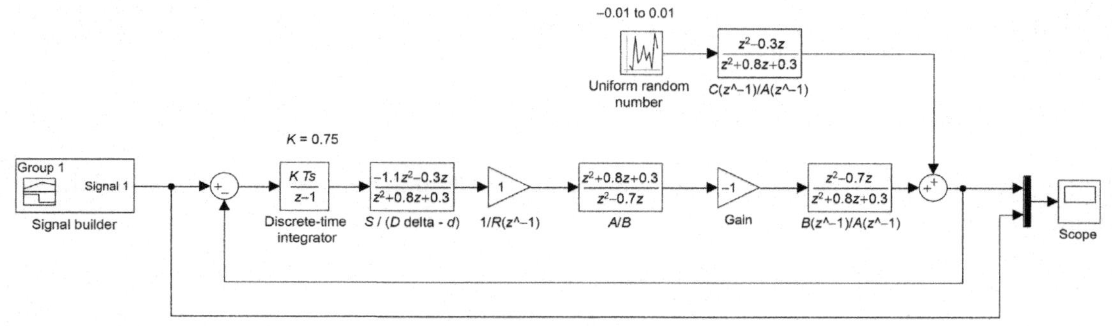

Fig. 8.10
The Simulink program of the MVC controller with an added integrator with a process with added white noise.

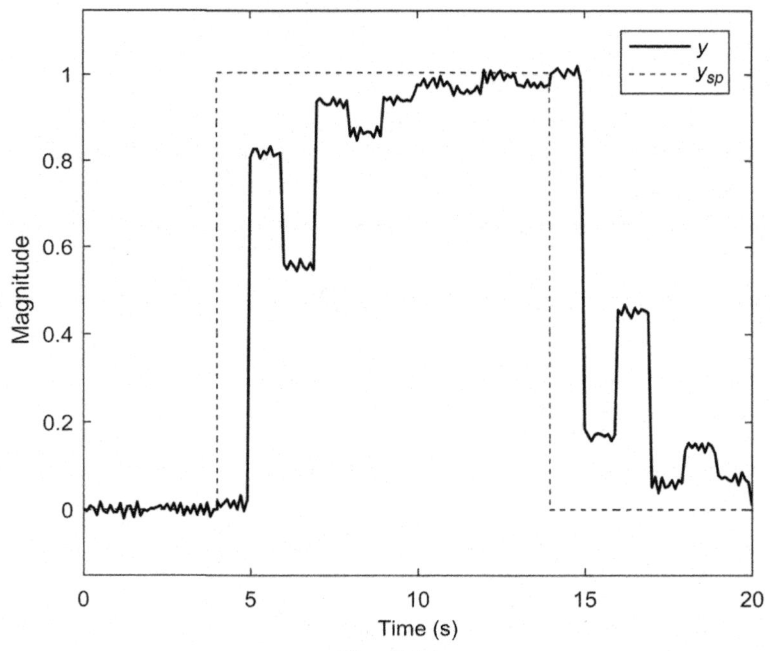

Fig. 8.11
The system closed-loop response with MVC controller with an added integrator to remove offset.

8.3.2 The Generalized Minimum Variance Controllers (GMVC)

The major problems with the minimum variance controller are its inability to handle nonminimum phase processes and that it can only minimize the variance of the future output variable. Accordingly, often it results in an excessive controller output variance. These problems may be alleviated if a general performance index is defined in terms of the variance of the output and input variables.

$$\min_{u_t} J = E\left[\left(P(z^{-1})y_{t+b} + Q(z^{-1})u_t - W(z^{-1})y_{sp}\right)^2\right] \tag{8.110}$$

where $P(z^{-1})$, $Q(z^{-1})$, and $W(z^{-1})$ are user-defined weighting polynomials on y_t, u_t and the set point. Considering a plant model of the form:

$$y_t = \frac{B(z^{-1})}{A(z^{-1})}u_{t-b} + \frac{C(z^{-1})}{A(z^{-1})}e_t \tag{8.111}$$

and following a similar approach to what was discussed in the derivation of the MVC, the generalized minimum variance controller (GMVC) controller action can be expressed by:

$$u_t = -\frac{S(z^{-1}) - C(z^{-1})W(z^{-1})y_{sp}}{C(z^{-1})Q(z^{-1}) + B(z^{-1})R(z^{-1})} \tag{8.112}$$

As it is evident from Eq. (8.112), the GMVC does not suffer from the problems associated with the MVC. Since it does not depend on the inverse of the process model, it can handle the nonminimum phase processes, also it reduces excessive changes in the input variable. Usually $P(z^{-1})$ is taken as 1 and $Q(z^{-1})$ as λ or $\lambda(1 - z^{-1})$, where λ is a tuning parameter.

Example 8.4

Design a generalized minimum variance controller (GMVC) for the stochastic process given by:

$$y_t = \frac{1 - 0.7z^{-1}}{1 + 0.8z^{-1} + 0.3z^{-2}}u_{t-1} + \frac{1 - 0.3z^{-1}}{1 + 0.8z^{-1} + 0.3z^{-2}}e_t \tag{8.113}$$

With $P(z^{-1}) = 1$, $Q(z^{-1}) = 0.4(1 - z^{-1})$, and $W(z^{-1}) = 1$.

Solution: Substitute the given process models in the general form of u_t:

$$u_t = -\frac{S(z^{-1})y_t - C(z^{-1})W(z^{-1})y_{sp}}{C(z^{-1})Q(z^{-1}) + B(z^{-1})R(z^{-1})} \tag{8.114}$$

Resulting in:

$$\begin{aligned} u_t &= -\frac{(-1.1 - 0.3z^{-1})y_t - (1 - 0.3z^{-1})(1)y_{sp}}{(1 - 0.3z^{-1})(0.4 - 0.4z^{-1}) + (1 - 0.7z^{-1})(1)} \\ &= -\frac{(-1.1 - 0.3z^{-1})y_t - (1 - 0.3z^{-1})y_{sp}}{1.4 - 1.23z^{-1} + -1.2z^{-2}} \end{aligned} \tag{8.115}$$

The Simulink model is given in Fig. 8.12

The system response to a unit step change in the set point with a white noise variance at 0.01 is shown in Fig. 8.13.

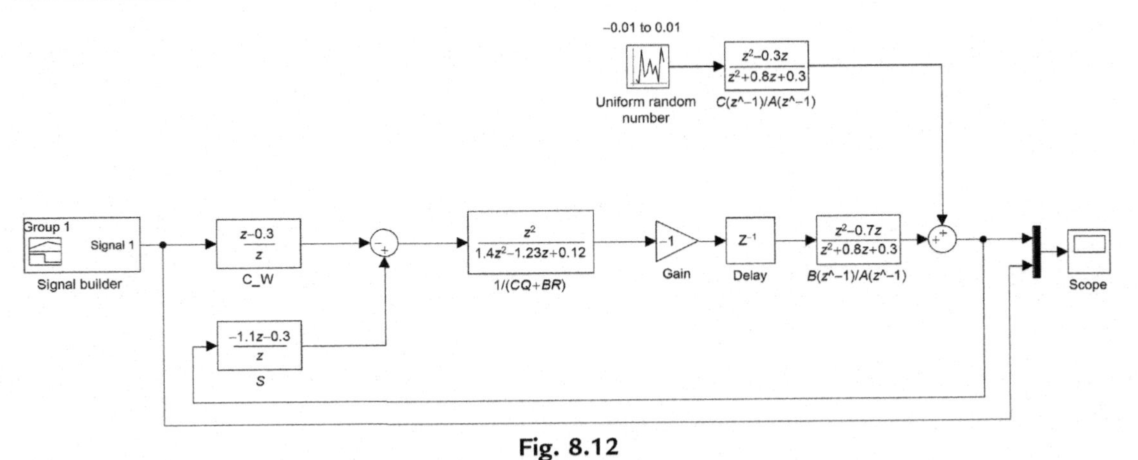

Fig. 8.12
The generalized minimum variance controller.

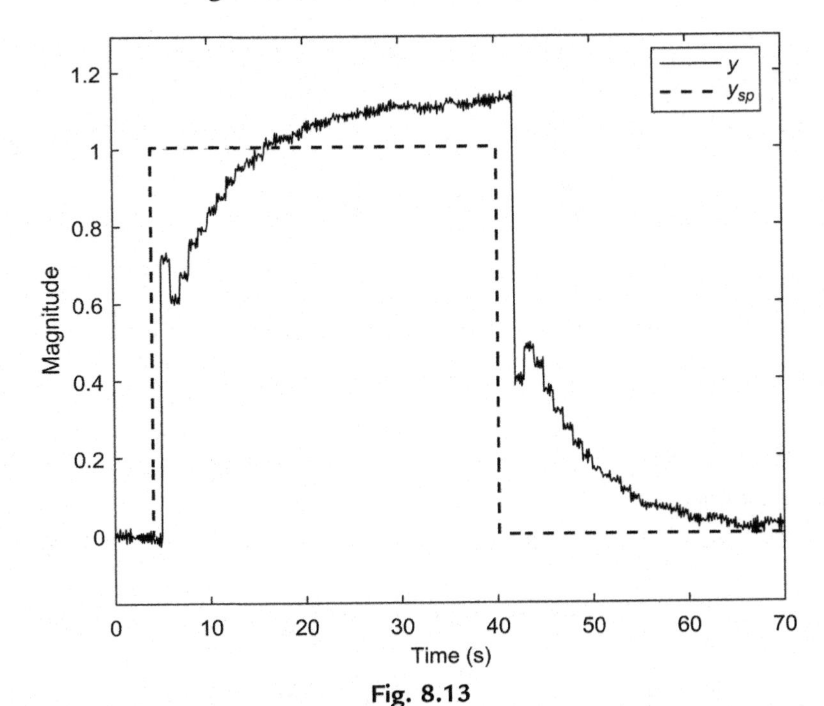

Fig. 8.13
The generalized minimum variance controller response to a unit step change in the set point.

8.3.3 The Pole Placement Controllers (PPC)

The stochastic version of the pole placement controller introduced in Chapter 6 can be derived for a process of the form

$$y_t = \frac{B(z^{-1})}{A(z^{-1})} u_{t-b} + \frac{C(z^{-1})}{A(z^{-1})} e_t \tag{8.116}$$

By substituting a controller of the type:

$$u_t = \frac{H(z^{-1})}{F(z^{-1})} y_{sp} + \frac{E(z^{-1})}{F(z^{-1})} y_t \tag{8.117}$$

in Eq. (8.116), the closed-loop transfer function can be obtained:

$$y_t = \frac{B(z^{-1})H(z^{-1})z^{-b}}{F(z^{-1})A(z^{-1}) + B(z^{-1})E(z^{-1})z^{-b}} y_{sp} + \frac{C(z^{-1})F(z^{-1})}{F(z^{-1})A(z^{-1}) + B(z^{-1})E(z^{-1})z^{-b}} e_t \tag{8.118}$$

Let us assume a desired closed-loop transfer function of the form

$$F(z^{-1})A(z^{-1}) + B(z^{-1})E(z^{-1})z^{-b} = T(z^{-1})C(z^{-1}) \tag{8.119}$$

where $T(z^{-1})$ contains the desired closed-loop poles. Solution of Eq. (8.119) renders the unknown polynomials $F(z^{-1})$ and $E(z^{-1})$. The remaining polynomial, $H(z^{-1})$ can be obtained from the zero offset requirement:

$$H(z^{-1}) = C(z^{-1}) \lim_{z \to 1} \frac{T(z^{-1})}{B(z^{-1})} \tag{8.120}$$

8.3.4 The Pole-Placement Minimum Variance Controller (PPMVC)

Combining the GMVC and the pole placement controller (PPC) results in the pole placement minimum variance controller (PPMVC). The controller equation is given by Eq. (8.112); however, the polynomials $P(z^{-1})$, $Q(z^{-1})$, and $W(z^{-1})$ which are user specified in the GMV controller are determined by specifying a desired closed-loop characteristic equation, $T(z^{-1})$. Substitution of Eq. (8.112) in Eq. (8.116) renders the closed-loop transfer function:

$$y_{t+b} = \frac{B(z^{-1})W(z^{-1})}{P(z^{-1})B(z^{-1}) + Q(z^{-1})A(z^{-1})} y_{sp} + \frac{B(z^{-1})R(z^{-1}) + C(z^{-1}) + Q(z^{-1})}{P(z^{-1})B(z^{-1}) + Q(z^{-1})A(z^{-1})} e_t \tag{8.121}$$

Polynomials $P(z^{-1})$ and $Q(z^{-1})$ are obtained from the solution of the following polynomial identity

$$P(z^{-1})B(z^{-1}) + Q(z^{-1})A(z^{-1}) = T(z^{-1}) \tag{8.122}$$

The remaining polynomial $W(z^{-1})$ is chosen as a scalar equal to $P(z=1)$ while $Q(z=1)$ is taken as zero to ensure zero offset. The expected value of y_{t+b} can be obtained from Eq. (8.121),

$$E[y_{t+b}] = \frac{B(1)W(1)}{P(1)B(1) + Q(1)A(1)} y_{sp} \tag{8.123}$$

which indicates that for such a choice of $W(z^{-1})$ and $Q(z=1)$, a zero offset can be achieved.

8.3.5 Self-Tuning Regulators (STR)

Controllers with fixed structures and fixed parameters perform well only for linear time-invariant plants. The majority of chemical processes are, however, time varying and nonlinear. Accordingly, in order to maintain the controller stability and performance, the controller has to be redesigned or its parameters retuned. The combination of an online parameter estimation algorithm with the online adjustment of the controller parameters results in the self-tuning and regulation of the controller which has gained success in industry. Fig. 8.14 The block diagram of a self-tuning regulator (STR) shows the structure of a STR which is a special type of an adaptive controller.

Any of the stochastic controllers discussed in this chapter can be combined with an appropriate online parameter estimation algorithm for the implementation of a STR. For example, consider the self-tuning version of the PPMVC controller which involves the following computational steps at each sampling instants:

- The structures of the process and noise models are assumed based on *a priori* information on the plant.
- An appropriate parameter estimation method, e.g., RELS or RAML algorithm is used to estimate the parameters of the process and the noise polynomials, $A(z^{-1})$, $B(z^{-1})$, and $C(z^{-1})$.
- For a given desired closed-loop characteristic equation $T(z^{-1})$, Eq. (8.122) is solved to obtain $P(z^{-1})$ and $Q(z^{-1})$.
- $W(z^{-1})$Z is taken as $P(1)$ and $Q(1)=0$ for zero offset.
- A controller equation such as Eq. (8.112) is used to calculate the controller action.

The previous steps are implemented at each sampling interval. However, convergence tests on the parameters must be performed before the controller is implemented. The stability of the

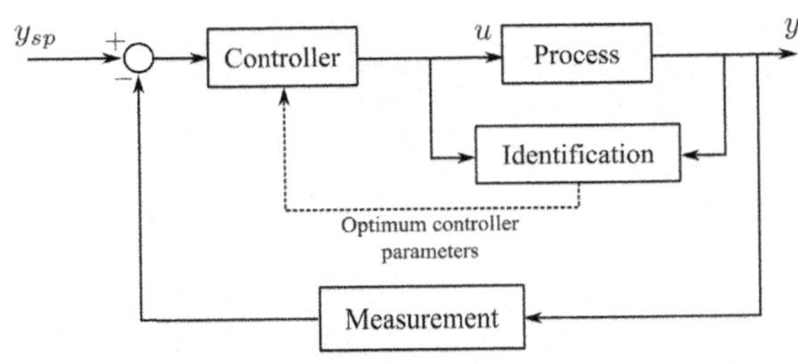

Fig. 8.14
The block diagram of a self-tuning regulator.

closed-loop system is of paramount significance and care must be exercised to reject unreasonable estimated parameters. The implementation of a self-tuning regulator on a tube wall reactor with a Fischer-Tropsch reaction is discussed by Rohani et al.[5]

8.4 Problems

8.4.1 For the following sequence, determine the cross-correlation between u and y signals at lag 3, i.e., $\gamma_{u,y}(3)$ (Table 8.1).

8.4.2 For the following system

$$y_t = \frac{1-0.7z^{-1}}{1+0.8z^{-1}+0.3z^{-2}}u_{t-1} \tag{8.124}$$

Generate the input (e.g., for a step change) and output data, form an object data `z = iddata(y,u)`, use various identification commands, such as `pem`, `arx`, `n4sid`, in MATLAB environment, to identify a model for the process and "compare" the identified model with the process actual output.

Add certain noise, for example:

```
n(1)=0;
for i=2:size(y)
n(i)=0.01*randn(1,1)-0.8*0.01*n(i-1);
end
```

and reidentify the process and noise models.

8.4.3 Design a minimum variance controller for the process given by:

$$y_t = \frac{1-0.5z^{-1}}{1+0.2z^{-1}-0.1z^{-2}}u_{t-2} + \frac{1-0.4z^{-1}}{1+0.2z^{-1}-0.1z^{-2}}e_t \tag{8.125}$$

And simulate the closed-loop control system and find the response to a unit step change in the set point using Simulink.

8.4.4 Repeat the previous problem and design a generalized minimum variance controller with

$$P(z^{-1})=2 \quad Q(z^{-1})=0.3(1-z^{-1}) \quad W(z^{-1})=1.2$$

Perform the closed-loop simulation on Simulink.

8.4.5 Derive the minimum variance controller for a stochastic process given by

$$y_t = \frac{B(z^{-1})}{A(z^{-1})}u_{t-3} + \frac{C(z^{-1})}{A(z^{-1})\Delta}e_t \tag{8.126}$$

Table 8.1 Data sequence used in Problem 8.4.1

i	0	1	2	3	4	5	6	7	8	9	10	11	12	13	14	15	16	17	18
u_{t+i}	2	2	2	−2	2	−2	−2	−2	2	2	−2	−2	2	−2	2	2	2	−2	−2
y_{t+i}	0.3	0.35	0.6	0.2	0.33	0.5	0.45	0.44	0.47	0.66	0.25	0.28	0.27	0.33	0.35	0.36	0.33	0.38	0.33
i	19	20	21	22	23	24	25	26	27	28	29	30	31	32	33	34	35	36	37
u_{t+i}	2	2	−2	−2	2	−2	2	2	−2	2	2	−2	−2	−2	−2	2	−2	2	2
y_{t+i}	0.55	0.54	0.52	0.53	0.35	0.66	0.34	0.25	0.36	0.55	0.45	0.44	0.49	0.42	0.34	0.46	0.51	0.53	0.46
i	38	39	40	41	42	43	44	45	46	47	48	49	50	51	52	53	54	55	56
u_{t+i}	2	2	−2	−2	−2	−2	2	−2	−2	2	2	2	−2	−2	2	−2	−2	2	2
y_{t+i}	0.14	0.22	0.25	0.26	0.36	0.25	0.25	0.27	0.29	0.35	0.55	0.56	0.25	0.25	0.44	0.46	0.46	0.42	0.43

where $\Delta = 1 - z^{-1}$. Discuss the significance of each term in the resulting controller. Does the controller work for a process whose $B\left(z^{-1}\right) = 1 - 1.2z^{-1}$? Does the controller exhibit offset and why?

8.4.6 Write down the general form of the equations for the following processes: an AR(2), a MA(2), an ARMA(1,2), and an ARIMA(2,2,2).

8.4.7 The actual transfer function of a process is given by

$$y_t = \frac{1 - 0.2z^{-1}}{1 + 0.1z^{-1} - 0.2z^{-2}} u_{t-2} + \frac{1 - 0.3z^{-1}}{1 + 0.3z^{-1} - 0.4z^{-2}} e_t \tag{8.127}$$

where e_t represents a sequence of random numbers. Assuming that with simple tests you have determined that the process and noise models have the following structures

$$y_t = \frac{b_0}{1 + a_1 z^{-1}} u_{t-2} + \frac{1 + c_1 z^{-1}}{1 + a_1 z^{-1}} e_t \tag{8.128}$$

Set up a recursive parameter estimation algorithm that allows you to estimate the unknown parameter vector $\theta^T = [b_0 \quad -a_1 \quad c_1]$. Describe in detail all the steps necessary to implement your suggested parameter estimation algorithm in an online fashion. Make sure to include the necessary equations.

Assuming that the best estimate of the parameter vector is given by $\theta^T = [0.3 \quad -0.6 \quad -0.2]$, design a minimum variance controller (MVC) for the process given by Eq. (8.128).

8.4.8 Derive the equation for the weighted least squares parameter estimation technique, i.e.,

$$\hat{\theta}_t = \left[\Phi_t^T W \Phi_t\right]^{-1} \Phi_t^T W Y_t \tag{8.129}$$

where $\hat{\theta}_t$ is the estimated unknown parameter vector, W_t is a diagonal weighting matrix, Φ_t is the measured input-output matrix, and Y_t is the measured output vector, at time t.

References

1. Box GEP, Jenkins M. Time series analysis forecasting and control. *Holden Day* 1970.
2. Goodwin GC, Payne RL. *Dynamic system identification: experimental design and analysis*. New York: Academic Press; 1977.
3. Ljung L, Soderstrom T. *Theory and practice of recursive identification*. Cambridge, MA: The MIT Press; 1983.
4. Astrom KJ, Wittenmark B. *Computer controlled systems: theory and design*. Englewood Cliffs, NJ: Prentice-Hall; 1984.
5. Rohani S, Quail D, Bakhshi NN. Self-tuning control of Fischer-Tropsch synthesis in a tube-wall reactor. *Can J Chem Eng* 1988;**66**:485–92.

CHAPTER 9

Model Predictive Control of Chemical Processes: A Tutorial

Victoria M. Ehlinger, Ali Mesbah
University of California, Berkeley, CA, United States

In this chapter, we will illustrate the ability of model predictive control (MPC) in dealing with the multivariable nature of process dynamics, process constraints, and multiple control objectives in chemical processes. The mathematical formulation of MPC is presented for a general class of processes. The receding-horizon implementation of MPC is demonstrated for a batch crystallization process and a continuous fermentation process. We will discuss the importance of state estimation for output-feedback implementation of MPC when the knowledge of the process state is not fully available. This chapter will conclude with a brief introduction to advanced topics in MPC that are of relevance to chemical process control.

9.1 Why MPC?

Chemical processes commonly exhibit nonlinear, multivariable dynamics subject to input and output constraints. Since its introduction in the 1920s, proportional-integral-derivative (PID) control has been the workhorse of process control.[1] Despite its widespread use, PID control lacks the ability to handle processes with multiple inputs and multiple outputs (i.e., multivariable process dynamics) and process constrais.[1] Hence, in practice, PID controllers are typically implemented in control structures such as cascade, split range, selectors, and overrides to deal with the multivariable nature of process dynamics, making the control structure selection a key consideration in PID control.[3,4] In the 1970s, MPC emerged in the process industry to address the shortcomings of PID control in coping with multivariable process dynamics and constraints.[5] Engineers at Shell published some of the first research papers on MPC and demonstrated its application to a fluid catalytic cracker. Their algorithm became known as dynamic matrix control (DMC), named for the dynamic matrix used to relate predicted future output changes to future inputs moves (i.e., a process model).[6] Although the DMC algorithm performed well for multivariable process control, output constraints could not

[1] Actuation constraints can be accounted for in PID control by adding antiwindup effect.[2]

Coulson and Richardson's Chemical Engineering. http://dx.doi.org/10.1016/B978-0-08-101095-2.00009-6

be handled systematically. Engineers at Shell improved upon DMC by explicitly incorporating the input and output constraints into the online optimization problem; this technique became known as quadratic dynamic model control.[7] These techniques were taken to academia in the early 1980s and triggered extensive research activities over the next three decades to develop MPC techniques for systems with complex dynamics (e.g., nonlinear, uncertain, distributed parameter, hybrid, etc.) and constraints in a wide range of applications. Nowadays in the process industry, MPC is widely used for advanced control of petrochemicals, specialty chemicals, pharmaceuticals, and biochemical processes.[5]

To describe the benefits of MPC, consider a crude unit in a petroleum refinery. The crude unit, the first process unit in a refinery, performs distillation of the crude oil into several fractions, which then undergo a series of reaction and separation processes to produce high value-added products (e.g., gasoline, diesel, and jet fuel).[8] Crude oil commonly contains dozens of chemical compounds, each of which with its own set of physical and thermodynamic properties, making the process dynamics complex. MPC can handle multiple control objectives, such as set point tracking for the several product streams in a crude unit (typically light gases, naphtha, kerosene, diesel, light oil, and heavy gas oil/bottoms). These set points are often adjusted based on the current pricing and market conditions for gasoline, diesel, and jet fuel.[9] While a PID controller can only control a single process output with a single process input, MPC, in theory, does not have any limitations to the number of inputs and outputs it can handle. This is a key advantage for processes with complex dynamics and multiple inputs and multiple outputs, and a primary motivator for the development of MPC in industry. In many chemical processes, maximizing the process productivity can be in conflict with maintaining the desired product quality. For example, maximum production from a crude unit occurs when the amount of light gas products is minimal; however, the light and heavy ends that are now part of the naphtha and diesel fractions result in lower quality products with a lower market value. In addition, a refinery may switch between different feeds with different compositions every few days, but the product specifications remain the same. MPC enables accounting for the range of acceptable product specifications as state or output constraints.[8] Other constraints due to the physical limitations of the crude unit, for example, environmental regulations and safety limitations, can also be accounted for in MPC. Additionally, MPC allows for operating the process close to the process constraints (e.g., operating conditions that correspond to the highest profitability of the process).[10] Overall, the improved process operation achieved by MPC in the process industry can be significant, while the payout time of MPC is typically less than a year.[9]

The key notion in MPC is to use a process model, which can be physics based or empirical, to optimize the process inputs over forecasts of process behavior made over a finite time horizon.[11] MPC offers several advantages over PID control including the ability to handle processes with multiple inputs and multiple outputs, input and output (similarly state) constraints, and multiple control objectives.[12] To illustrate the advantages of MPC, consider the

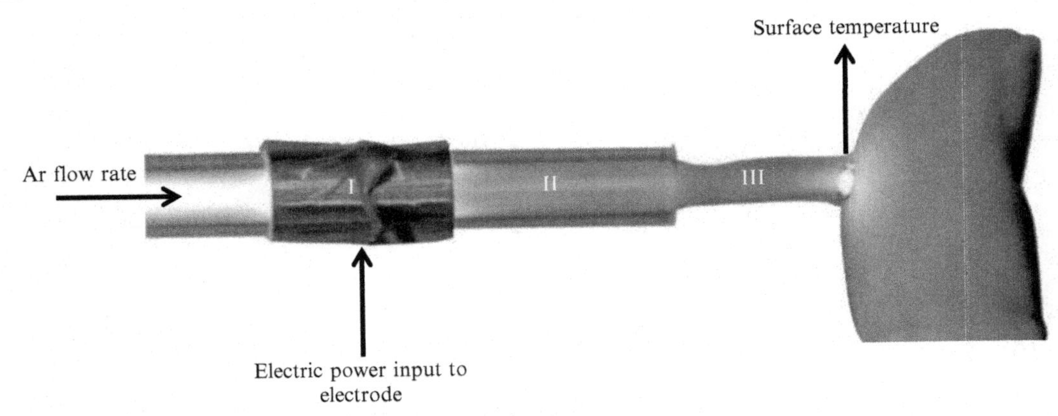

Fig. 9.1
Diagram of an atmospheric-pressure plasma jet (APPJ).[13]

Argon atmospheric-pressure plasma jet (APPJ) shown in Fig. 9.1 (see Ref. 13 for further details). APPJs are used in materials processing and biomedical applications for treatment of materials, such as biological tissues and organic materials, that are heat sensitive or unable to withstand vacuum pressures.[14] In the APPJ shown in Fig. 9.1, the ions and reactive species that are generated in Region I and II, as a result of applying electric field to Argon, extend through Region III to the target surface, which is cooled by the flow of a coolant (e.g., blood in a tissue) and by heat transfer with the surrounding air. An MPC controller is designed to regulate the thermal effects of the plasma on the target surface by driving the temperature of the target toward a desired set point. This is done by manipulating the two inputs of the APPJ device: the electric power input to the electrode and the inlet gas velocity. The APPJ under study is an example of a system with multiple inputs and multiple outputs (i.e., surface temperature and gas temperature in the jet). The ability of the MPC controller in handling the multivariable dynamics of the system alleviates the need to pair the different inputs and outputs of the system, as is the case in PID control.

The control objective of the APPJ is to maintain the surface temperature at a set point of 317 K while the APPJ is held at a prespecified distance from the surface. To ensure safe operation of the APPJ, the gas temperature in the plasma plume is constrained to be less than or equal to 319 K so that no significant changes occur in the plasma characteristics. In the MPC controller, both the velocity and power inputs are used as the system inputs to realize the control objective, whereas the PI controller can only adjust the electric power input to follow the desired surface temperature set point. In addition, the PI controller cannot enforce any constraint on the gas temperature. The closed-loop simulation results for both the MPC controller and the PI controller are shown in Fig. 9.2. At time 500 s, a step disturbance is applied to the distance of

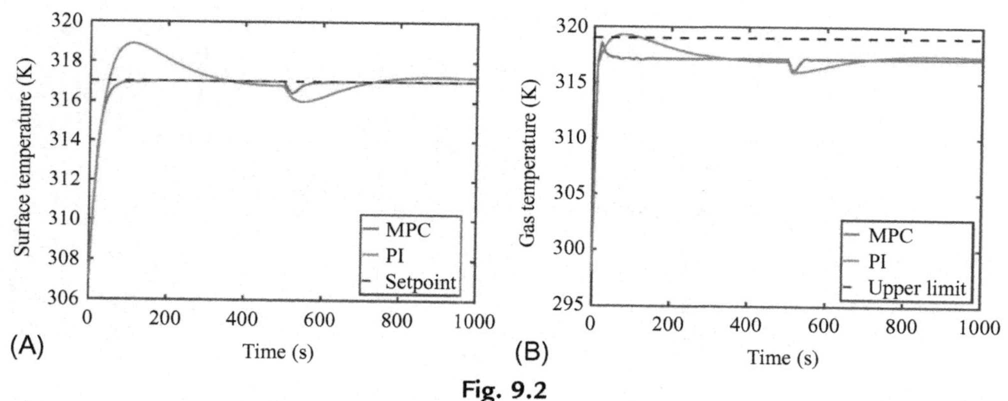

Fig. 9.2

Closed-loop simulation results of the MPC and PI control of the Argon atmospheric-pressure plasma jet shown in Fig. 9.1. The control objective is to follow a desired set point for the surface temperature, while maintaining the gas temperature below a certain limit. A step disturbance in the distance between the jet and the target surface is introduced at time 500 s. (A) Surface temperature set point tracking. The MPC controller shows minimal overshoot and faster set point tracking. (B) Gas temperature constraint handling. The MPC controller maintains the gas temperature within the constraint while the PI controller does not.

the APPJ from the target surface. Because the APPJ is now farther away from the target surface, the surface temperature decreases and the controller must adjust the inputs in order to bring the surface temperature back to its set point of 317 K. Fig. 9.2A shows that both the MPC and PI controllers are able to bring the surface temperature back to its set point after the disturbance is introduced. However, the MPC controller enables reaching the set point more quickly, without overshooting the set point. This is due to the fact that the MPC controller can systematically account for the multivariable nature of the jet dynamics and simultaneously actuate both system inputs toward achieving the control objective. Further, as shown in Fig. 9.2B, the PI controller violates the constraint imposed on the gas temperature, while the MPC controller stays within the constraint limit due to explicit incorporation of the gas temperature constraint in the control framework.

To demonstrate the flexibility of MPC in handling multiple control objectives, the objective function of the MPC controller in the previous example is modified to drive the surface temperature to the set point of 317 K while maintaining the input power level at the desired level of 12 W. The closed-loop simulation results are shown in Fig. 9.3. The MPC controller is able to simultaneously follow the set points for the surface temperature and power level, and therefore realize both control objectives. In contrast, when the PI controller is used for regulating the surface temperature, the (open-loop) power profile cannot meet its respective set point. This motivating example clearly shows the ability of MPC in dealing with the multivariable process dynamics, process constraints, and multiple control objectives.

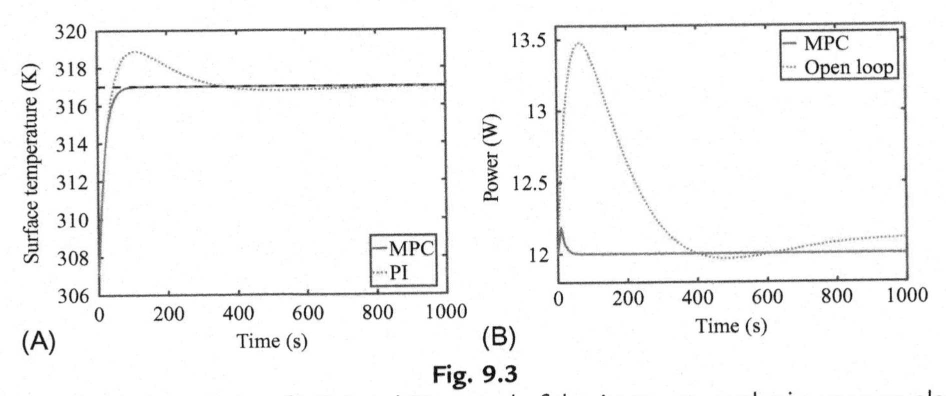

Fig. 9.3

Closed-loop simulation results of MPC and PI control of the Argon atmospheric-pressure plasma jet shown in Fig. 9.1. The MPC controller has multiple control objectives of maintaining the surface temperature and the power at the desired set points, whereas the PI controller can only be tasked to realize one control objective. (A) Surface temperature set point tracking. (B) Power set point tracking. The open-loop profile corresponds to the case in which the PI controller is used for regulating the surface temperature only.

9.2 Formulation of MPC

MPC involves solving an optimal control problem (OCP) in a receding-horizon fashion. The key components of an OCP, which include a process model, an objective function, and input and state constraints, as well as the receding-horizon implementation in MPC are discussed in this section.

9.2.1 Process Model

The first step in developing an MPC controller is to obtain a model of the process. Process models provide a description of the process dynamics in terms of process inputs, states, and outputs. Inputs, also known as manipulated variables, are process variables that can be adjusted by the controller in order to cause a change in the process dynamics. Examples of process inputs, depending on the process configuration, include the flow rate, temperature, and concentration of inlet streams to a unit operation. States are process variables that change in response to input changes and define the state of the process at different times. Examples of states include concentration, temperature, liquid level in a tank, and pressure in a gas drum. Outputs consist of the measured process variables that change in response to input variations. Process outputs are often defined as some function of states. In many chemical processes, process outputs are a subset of states because all states may not be measured due to various technical and economic limitations.

Process models are linear or nonlinear and can be described in discrete- or continuous-time form. The quality of a process model is critical to the performance of an MPC controller. Process models can be derived from data (i.e., empirical) or based on first principles. Empirical models usually have a limited range of validity, whereas first-principle models can typically predict the process behavior over a wider range of process conditions since they rely on mass, momentum, and energy conservation laws, along with constitutive equations (e.g., phase equilibria and chemical kinetics).[15] In its most general form, a process model can be represented by the following (continuous-time) nonlinear state-space representation

$$\begin{aligned} \frac{dx(t)}{dt} &= f(x(t), u(t), \theta), \quad x(t_0) = x_0 \\ y(t) &= h(x(t), u(t), \theta), \end{aligned} \tag{9.1}$$

where $x \in \mathbb{R}^n$ is the vector of states; $u \in \mathbb{R}^{n_u}$ is the vector of (manipulated) process inputs; $y \in \mathbb{R}^{n_y}$ is the vector of (measurable) process outputs; $\theta \in \mathbb{R}^{n_p}$ is the vector of model parameters; t is time; x_0 is the initial conditions of the states; and f and h are some nonlinear functions of the inputs, states, and parameters. $\mathbb{R}$ denotes the set of real numbers.

First-principle models of chemical processes are typically nonlinear due to the nonlinearity of conservation balances and constitutive equations.[16] Nonlinear first-principle models can seldom be solved analytically. Further, their numerical solution can be computationally expensive for real-time control applications. When a process model is intended to describe the system dynamics around a desired operating point (e.g., steady-state operating conditions), nonlinear first-principle models can be linearized around the operating point of interest. In this case, Eq. (9.1) will reduce to a linear state-space model

$$\begin{aligned} \frac{dx(t)}{dt} &= Ax(t) + Bu(t), \quad x(t_0) = x_0 \\ y(t) &= Cx(t) + Du(t), \end{aligned} \tag{9.2}$$

where $A \in \mathbb{R}^{n \times n}$, $B \in \mathbb{R}^{n \times n_u}$, $C \in \mathbb{R}^{n_y \times n}$, and $D \in \mathbb{R}^{n_y \times n_u}$, with $D = 0$ in many processes.

The state-space model (9.1) can also be written in the discrete-time form, which is convenient for programming of an MPC controller using digital computers. The discrete-time (nonlinear) state-space model takes the form

$$\begin{aligned} x(k+1) &= f_k(x(k), u(k), \theta), \quad x(k=0) = x_0 \\ y(k) &= h_k(x(k), u(k), \theta), \end{aligned} \tag{9.3}$$

where k is the time index; and the functions f_k and h_k are obtained based on the discretization of the functions f and h in Eq. (9.1).

Although first-principle models offer the advantage of a wider operating range and provide physics-based insights into the process dynamics, empirical models are frequently used for designing MPC controllers in industry because they are generally less expensive to develop and are often computationally more efficient than first-principle models for real-time control applications. Empirical models are derived by perturbing the process (e.g., using a series of step tests), collecting the output response of the process, and subsequently fitting the input-output data to a prespecified (often simple) model parameterization or a state-space model of the form Eq. (9.2).[17,17a] A critical consideration in empirical modeling is the choice of the input perturbations that must sufficiently excite the system dynamics to generate informative data for data-driven model development while being process friendly (e.g., avoid pushing the process outside of its safety limits or leading to an unacceptable amount of off-specification product).[18] Empirical models are only adequate for the operating range over which the data has been collected, whereas first-principle models can be used for a wider operating range. Generally speaking, the modeling approach to be adopted is application specific and critically depends on the process complexity as well as the computational requirements of the MPC application.

9.2.2 Objective Function

The objective function is some function of the process inputs and outputs (or states) that describes the control objective. Control objectives vary depending on the type of process (batch versus continuous) and various process considerations. Typical control objectives include disturbance rejection, set point tracking, batch time minimization, etc. In its general form, the objective function J consists of two parts

$$J(x(k),u(k)) = J_{\mathrm{f}}\left(x\left(N_{\mathrm{p}}\right)\right) + \sum_{i=0}^{N_{\mathrm{p}}-1} J_{\mathrm{c}}(x(i),u(i)), \tag{9.4}$$

where N_{p} is the prediction horizon, the time over which the model predicts the process states or outputs; J_{f} is the terminal cost term; and J_{c} is the running cost term. The running cost describes the control objective over the prediction horizon of the MPC controller, while the terminal cost describes the cost function at the end of the prediction horizon.

In MPC, the objective function is commonly defined in terms of a quadratic set point tracking function

$$J(x(k),u(k)) = \sum_{i=0}^{N_{\mathrm{p}}} \left(y_i - y_{\mathrm{sp}}\right)^{\top} Q \left(y_i - y_{\mathrm{sp}}\right) + \sum_{i=0}^{N_{\mathrm{c}}} \left(u_i - u_{\mathrm{sp}}\right)^{\top} R \left(u_i - u_{\mathrm{sp}}\right), \tag{9.5}$$

where y_{sp} and u_{sp} are the desired set points for the outputs and inputs, respectively; Q and R are symmetric and positive symmetric weight matrices, respectively; and N_{c} is the control horizon, the time over which the controller can adjust the inputs in order to optimize the objective function. Q, R, N_{p}, and N_{c} are generally the tuning parameters of an MPC controller and must be chosen by the user.

9.2.3 State and Input Constraints

One of the key features of MPC is the ability to systematically handle input and state constraints. Input constraints are defined by

$$u \in U, \tag{9.6}$$

which indicates that the process inputs u must lie in a compact set U. Linear input constraints are represented by

$$Du \leq d, \tag{9.7}$$

where $D \in \mathbb{R}^{n_u \times m}$ is the input constraint matrix; $d \in \mathbb{R}^m$ is a positive vector; and m denotes the number of input constraints. Input constraints represent physical limitations of the process such as valve positions and ranges of flow rates.

Like process models, state constraints are categorized as linear and nonlinear. Nonlinear state constraints can generally be defined by

$$g_k(x(k), u(k), \theta) \leq 0, \tag{9.8}$$

where g_k is the state constraint function. In the case of linear state constraints, Eq. (9.8) simplifies to

$$Hx \leq h, \tag{9.9}$$

where $H \in \mathbb{R}^{n \times p}$ is the constraint matrix; $h \in \mathbb{R}^p$ is a positive vector; and p denotes the number of state constraints. Output constraints are defined in a similar manner as state constraints. State or output constraints typically represent limits on species concentration or temperature and pressure of the process. These constraints can be used to systematically account for the safety, regulatory, and product quality requirements of the process.

9.2.4 Optimal Control Problem

The process model, objective function, and input and state constraints defined above can now be used to formulate the OCP, which is the cornerstone of an MPC controller. For the case of full-state feedback $x(k) = y(k)$ (i.e., all states are measured at the sampling time k), MPC solves the following OCP at each sampling time k

$$\min_{\boldsymbol{u}} \; J(x(k), \boldsymbol{u}) \tag{9.10}$$

subject to:

$$\begin{aligned}
&\bar{x}(i+1) = f_k(\bar{x}(i), u(i), \theta), && i = 0 \ldots N_\mathrm{p} - 1 \\
&u(i) \in U, && i = 0 \ldots N_\mathrm{c} - 1 \\
&g_k(\bar{x}(i), u(i), \theta) \leq 0, && i = 1 \ldots N_\mathrm{p} \\
&\bar{x}(0) = x(k),
\end{aligned}$$

where $u = [u(0), \ldots, u(N_c - 1)]^{\mathrm{T}}$ is the vector of process inputs over the control horizon N_c and $\bar{x}$ denotes the model predictions. $\boldsymbol{u}$ comprises the vector of decision variables of the optimization problem for which the OCP (Eq. 9.10) is solved. The optimal solution to the OCP is denoted by $\boldsymbol{u}^*$ and is referred to as the (open-loop) optimal control policy. When the process model used in the OCP (Eq. 9.10) is nonlinear, the control approach is commonly referred to as nonlinear MPC (NMPC).

9.2.5 Receding-Horizon Implementation

In practice, the performance of the open-loop optimal control policy $\boldsymbol{u}^*$ computed by Eq. (9.10) will degrade due to the inevitable plant-model mismatch (i.e., inaccuracies of the process model) and process disturbances. To mitigate the performance degradation of the (open-loop) optimal control policies, the OCP must be solved recursively when new process measurements $x(k)$ become available at every measurement sampling time instant k. This is known as the receding-horizon implementation of MPC[15] and is illustrated in Fig. 9.4. The idea is as follows. At the sampling time instant k, the OCP (Eq. 9.10) is solved over the control horizon N_c. The solution of the OCP is a sequence of optimal input values, for which the objective function is minimized over the prediction horizon N_p. The first element of the optimal input sequence, $u^*(0)$, is then applied to the process. Subsequently, the prediction horizon shifts one sample ahead and the procedure is repeated as soon as new measurements become available at the next sampling time instant $k+1$. In the case of full-state feedback, a closed-loop control system with an MPC controller is shown in Fig. 9.5. The optimal inputs from the MPC controller are applied to the plant and the measured outputs of the plant (in this case the full state

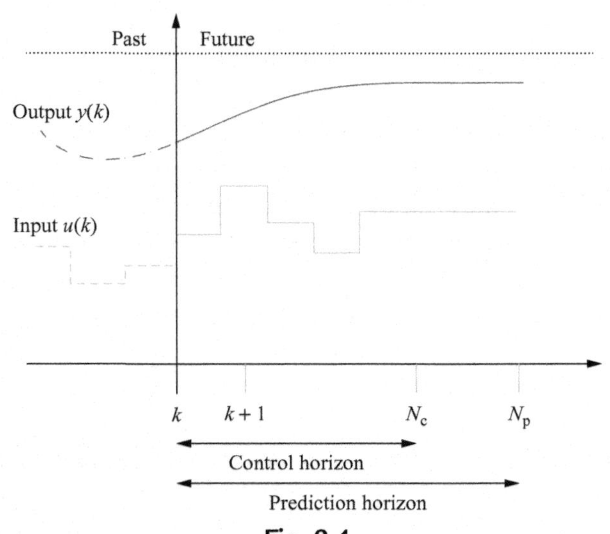

Fig. 9.4
The receding-horizon principle in MPC.

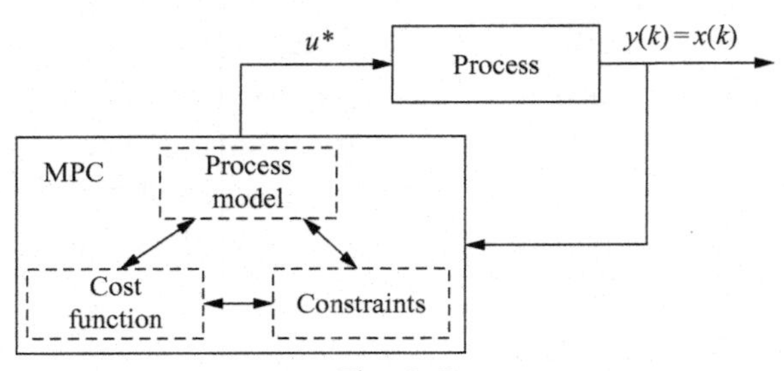

Fig. 9.5
Model predictive control in a closed-loop control system when all the state variables are measured (i.e., the case of full-state feedback).

vector) are fed to the MPC controller. The new measurements are used to reinitialize the process model at every sampling time k (see the OCP (Eq. 9.10)).

For processes with a fixed end time, the receding-horizon implementation of the OCP can be adapted to shrinking horizon. In this case, when the prediction horizon is shifted to the next sampling time instant, the prediction horizon shrinks as the process approaches the end time of the process.[19] The shrinking-horizon implementation is particularly useful for batch processes, whereas the receding-horizon implementation is typically used for continuous processes.

9.2.6 Optimization Solution Methods

The OCP (Eq. 9.10) is in fact a dynamic optimization problem. There exists various direct optimization strategies, namely, single shooting, multiple shooting, and simultaneous strategies, for solving the dynamic optimization problem. Generally speaking, these strategies convert the OCP into a finite dimensional nonlinear programming problem (NLP) through parameterizing the vector of decisions variables $\boldsymbol{u}$. In the single shooting strategy,[20] after reducing the infinitely many degrees of freedom of the control vector through parameterization, an initial value problem encompassing the model equations is numerically solved in each iteration step of the optimization procedure. The solution is obtained for the current values of the parameterized control vector. Hence, model simulation and optimization are carried out sequentially, guaranteeing the solution feasibility even in the case of premature optimization terminations. By contrast, the model equations are discretized along with the control vector in the simultaneous optimization strategy.[21] The discretized differential equations are included in the optimization problem as nonlinear constraints, typically leading to very large NLPs. Simultaneous model simulation and optimization can result in faster computations compared to the single shooting strategy. However, feasible state trajectories are obtained only after successful termination of the optimization since the discretized model equations are violated during the optimization procedure.

Table 9.1 Comparison of the direct optimization strategies[24]

	Single Shooting	Multiple Shooting	Simultaneous
Use of integration solvers	Yes	Yes	No
Size of nonlinear programming problem	Small	Intermediate	Large
Applicable to highly unstable systems	No	Yes	No
Model equations fulfilled in each iteration step	Yes	Partially	No

On the other hand, model simulation and optimization are not performed entirely sequentially or simultaneously in the multiple shooting strategy.[22,23] In this optimization strategy, the state trajectories and the control vector are parametrized over a predetermined number of intervals. The initial value problems are solved separately on each multiple shooting interval with a prespecified numerical accuracy. This makes the technique well suited for parallel computations. The relatively large number of variables often necessitates the use of tailored NLP algorithms, which exploit the special structure of the problem, e.g., sparsity, to yield faster convergence than for the single shooting strategy. However, the continuity of the state trajectories is only fulfilled after successful termination of the optimization procedure as with the simultaneous technique. Table 9.1 summarizes the direct optimization strategies.

9.3 MPC for Batch and Continuous Chemical Processes

This section demonstrates the NMPC for a batch crystallization process and a continuous fermentation process. The case studies include excerpts of code from MATLAB (MathWorks) to illustrate how a closed-loop control system, an optimization objective function, and process constraints are set up.

9.3.1 NMPC of a Batch Crystallization Process

Batch crystallization is prevalent in the specialty chemical, food, and pharmaceutical industries for manufacturing and purification of high value-added chemical substances. Crystallization processes are governed by several physicochemical phenomena such as nucleation, crystal growth, and agglomeration, which arise from the copresence of a continuous phase (i.e., solution) and a dispersed phase (i.e., particles).[25,26] Optimal operation of batch crystallization processes is particularly challenging due to the complexity of their models, uncertainty of the crystallization kinetics, and sensor limitations in reliably measuring the process variables.[27]

This case study demonstrates the NMPC for seeded batch crystallization of an ammonium sulfate-water system. The dynamics of crystal size distribution over the batch time can be described by the population balance equation (PBE), which is a nonlinear partial differential equation[28] (e.g., see Ref. 29 for an overview of various numerical solution methods for the PBE). In order to obtain a computationally efficient description of the process dynamics for the real-time control application at hand, the method of moments is applied to convert the PBE to a set of closed-form nonlinear ordinary differential equations of form (9.3) with the state vector

$$x = [m_0\, m_1\, m_2\, m_3\, m_4\, C]^\top,$$

where m_i is the ith moment of the crystal size distribution and C is the concentration of ammonium sulfate in solution. The input u consists of the heat input Q to the process, i.e., $u = [Q]$. A key process variable, which is closely related to the product quality and batch productivity, is the crystal growth rate that is defined in terms of the solute concentration

$$G = k_g (C - C^*)^g, \tag{9.11}$$

where k_g is the crystal growth rate constant; C^* is the saturation concentration; and g is the growth rate exponent. A detailed description of the batch crystallization process under study and the moment model can be found in Refs. 30,31. The objective of the NMPC controller is to maintain the crystal growth rate at a desired set point G_{max}, which provides a trade-off between the batch productivity and product quality. To this end, the OCP (Eq. 9.10) is defined as

$$\min_{u} \sum_{i=0}^{N_p} \left(\frac{G(i) - G_{max}}{G_{max}} \right)^2 \tag{9.12}$$

subject to:

$$\begin{aligned} &\bar{x}(i+1) = f\_k(\bar{x}(i), u(i), \theta), && i = 0 \ldots N_p - 1 \\ &9\text{kW} \leq u(i) \leq 13\text{kW}, && i = 0 \ldots N_c - 1 \\ &\bar{x}(0) = x(k), && \end{aligned}$$

where $G_{max} = 2.5 \times 10^{-8}\,\text{m/s}$.

To solve the OCP in Eq. (9.12), the MATLAB constrained optimization function `fmincon`,[32] which is a sequential quadratic programming solver, is used. For the case of full-state feedback, the excerpt of the code for shrinking-horizon implementation of the OCP is shown in Box 9.1. The code resembles the closed-loop control system depicted in Fig. 9.5. The optimizer `fmincon` relies on the objective function `objbatch`, in which the objective function of Eq. (9.12) is defined, and the constraint function `conbatch`. At each sampling time, the optimizer `fmincon` takes the initial guess of the decision variables `uopt_past`, their lower bound `lb` and upper bound `ub`, and the most recent observed states of the process `x_plant(k,:)` to compute the optimal process inputs `uopt`. The decision variables of the dynamic optimization problem are parameterized using a prespecified number, `ninter`, of piecewise-constant intervals over the prediction horizon `tp`. Since the NMPC controller is implemented in a shrinking-horizon

BOX 9.1

```
for k = 1:length(ts)-1,
    % The OCP
    [uopt] = fmincon(@objbatch,uopt_past,[],[],[],[],...
        lb,ub,@conbatch,[],x_plant(k,:),tp,ninter);

    % Adjusting the prediction horizon for shrinking-horizon implementation of
      the OCP
    tp = tb - ts(k);

    % Plant
    [t,x] = ode15s(@crystallizer,[ts(k) ts(k+1)],x_plant(k,:),[],uopt(1));
    x_plant(k+1,:) = x(end,:);
    uopt_past = uopt;
end
```

mode, the prediction horizon `tp` is shortened at every measurement sampling time, that is, `tp = tb - ts(k)` with `tb` being the fixed batch time. The optimizer computes the optimal inputs `uopt` over the control horizon, the first element of which is applied to the plant. The plant outputs `x_plant(k+1,:)` are then used to initialize the optimizer at the next sampling time instant `k+1`. See the Appendix for the code of the process model.

The first argument in `fmincon` is the objective function `objbatch`, shown in Box 9.2. The arguments of `objbatch` are the initial guess of the decision variables `u=uopt_past`, the initial conditions `x0=x_plant(k,:)`, the prediction horizon `tp`, and the number of intervals `ninter` over which the decision variables are parameterized. Note that the process model `crystallizer` is solved numerically using the MATLAB solver `ode15s` over the `ninter` intervals of the prediction horizon, in each of which the decision variable `u` takes a constant value. As defined in the OCP (Eq. 9.12), the objective function `objbatch` aims to drive the crystal growth rate to its set point G_{max} over the course of the batch run time. Since the OCP does not include any state constraints, the constraint function `conbatch` in the optimizer need not be used.

The simulation results are shown in Fig. 9.6. To illustrate the importance of the receding-horizon implementation of the NMPC controller, three scenarios have been considered: open-loop control under the assumption of perfect model as well as open-loop and closed-loop control in the presence of plant-model mismatch. In both open-loop control scenarios, the optimal heat input to the crystallizer is computed off-line by solving the OCP (Eq. 9.12) without reinitializing the model at every sampling time (i.e., the online process measurements are disregarded). The offline-computed optimal control inputs are then implemented on the crystallizer. In the hypothetical scenario of having a perfect process model, the open-loop optimal heat input and crystal growth rate profiles are shown in Figs. 9.6A and B,

BOX 9.2

```
function obj = objbatch(u,x0,tp,ninter)
IC(1,:) = x0;  % Initial state conditions for each interval of control input
                 parameterization
x1 = [];

% Model prediction over the prediction horizon using the parameterized control input
for i=1:ninter
    [t,x] = ode15s(@crystallizer,0:1:tp/ninter,IC(i,:),[],u(i));
    IC(i+1,:)=x(end,:);
    x1 = [x1;x];
end

% Calculate growth rate G using the predicted concentration
kg = 7.49567e-05;  % growth constant
g = 1;              % growth exponent
Cs = 0.456513;      % saturation concentration C*
Gmax = 2.5e-8;      % [m/s]
G= kg * (x1(:,6) - Cs).^g;

% Evaluate objective function
obj = sum(((G-Gmax)./Gmax).^2);
```

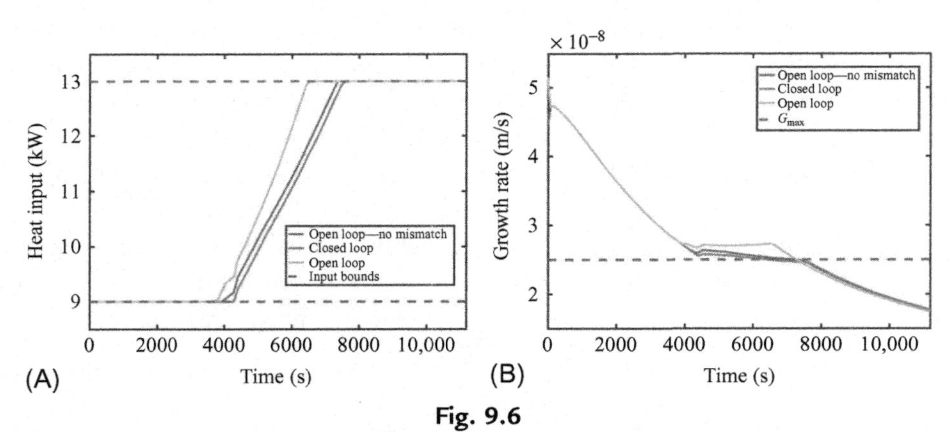

Fig. 9.6

The NMPC for the batch crystallization process. Comparison of three implementations of the OCP (Eq. 9.12): open-loop implementation under the assumption of perfect system model, open-loop and closed-loop implementations in the presence of plant-model mismatch. The open-loop implementation without plant-model mismatch is nearly identical to the closed-loop control in the presence of plant-model mismatch. This indicates the importance of the measurement feedback (i.e., receding-horizon implementation) in the NMPC controller to account for plant-model mismatch and process uncertainties. (A) Heat input to the crystallizer. (B) Crystal growth rate profile.

respectively. It is evident that the crystal growth rate can follow its set point only in a limited time period of 4200 s to 7800 s. This is due to the upper and lower bounds enforced on the process input. Initially, the controller must decrease the heat input below 9 kW to bring the crystal growth rate close to its set point. However, the lower heat input bound of 9 kW, imposed to ensure reproducible batch startups, prevents the controller from achieving its control objective. When the crystal growth rate naturally decays to its set point at approximately 4200 s, the controller starts increasing the heat input to maintain the growth rate at its desired set point. However, as soon as the heat input hits its upper bound of 13 kW due to the actuation limitations of the crystallizers, the controller can no longer effectively achieve its objective of G_{max} set point tracking. Fig. 9.6 clearly demonstrates the effect that the lack of process actuation (in this case due to the physical bounds on the heat input) can play in hampering effective process control.

The earlier discussed open-loop optimal control policy relied on the unrealistic assumption that a perfect process model was available. Due to process uncertainties and various modeling assumptions, there is always a certain degree of mismatch between the model predictions and the actual process behavior. Fig. 9.6 shows an open-loop optimal control policy that has been designed off-line with a process model that has parametric uncertainty in the crystallization kinetics (i.e., plant-model mismatch). Fig. 9.6B shows that the open-loop optimal control policy cannot closely follow the G_{max} set point even in the time period of 4200–7800 s where the controller can modulate the heat input to track the crystal growth set point. This is due to the fact that the OCP is solved off-line and, therefore, it cannot take corrective action using the online plant measurements to counteract the plant-model mismatch. Fig. 9.6 indicates that when the NMPC controller is implemented in closed loop by solving the OCP (Eq. 9.12) online, despite the plant-model mismatch, the optimal control inputs can fulfill the control objectives in the operating region that the process input can be adequately modulated (i.e., between 4200 s and 7800 s). As can be seen, the crystal growth rate profile in the case of the closed-loop control exhibits almost the same behavior as in the case of the open-loop control with no plant-model mismatch.

Notice that process constraints can be classified as either hard or soft constraints.[33] Hard constraints are those that are defined explicitly in the OCP and must be satisfied at all times. If a hard constraint is not satisfied, then the optimization problem will become infeasible. On the other hand, soft constraints are defined in the objective function. For example, in the OCP (Eq. 9.12), the crystal growth rate set point tracking can be interpreted as a soft constraint of the form $G \leq G_{max}$. Soft constraints can be violated to ensure feasibility of the optimization problem in the interest of satisfaction of the hard state constraints and/or input bounds. In the batch crystallization case study, the actuation constraints on the process input led to violation of the soft constraint on the crystal growth rate during the initial and final phase of the batch.

9.3.2 NMPC of a Continuous ABE Fermentation Process

Biobutanol derived from sustainable renewable resources such as lignocellulosic biomass has shown promise as a renewable drop-in fuel.[34,35] Biobutanol can be produced by bacteria of genus *Clostridium* in the so-called acetone-butanol-ethanol (ABE) fermentation process, which is a biphasic fermentation that converts sugars into acids (acetate, butyrate) and solvents (acetone, butanol, ethanol). During the first phase, known as acidogenesis, the primary products are the acidic metabolites. As the metabolism shifts to the second phase, solventogenesis, the acids are assimilated into the ABE solvents that comprise the main fermentation products.[36]

In this case study, an NMPC controller is designed for a continuous ABE fermentation process. The nonlinear process model used for the NMPC design consists of 12 states, which are the concentrations of various chemicals and enzymes in the fermentation culture.[37,38] The dynamics of the continuous fermentation process are described by the nonlinear state-space model (9.3) with the state vector defined by

$$x = [C_{\mathrm{AC}}\ C_{\mathrm{A}}\ C_{\mathrm{En}}\ C_{\mathrm{AaC}}\ C_{\mathrm{Aa}}\ C_{\mathrm{BC}}\ C_{\mathrm{B}}\ C_{\mathrm{An}}\ C_{\mathrm{Bn}}\ C_{\mathrm{Ad}}\ C_{\mathrm{Cf}}\ C_{\mathrm{Ah}}]^{\top},$$

where the abbreviations correspond to the species in the fermentation culture: AC is acetyl-CoA, A is acetate, En is ethanol, AaC is acetoacetate-CoA, Aa is acetoacetate, BC is butyryl-CoA, B is butyrate, An is acetone, Bn is butanol, Ad is adc, Cf is ctfA/B, and Ah is adhE (see Ref. 38). The inputs to the process are

$$u = [D\ G_0]^{\top},$$

where D (h^{-1}) is the dilution rate and G_0 is the inlet glucose concentration (mM). A detailed description of the metabolic pathway of *Clostridium acetobutylicum* can be found in Ref. 36. In the following, we denote each state variable as x_i, where the subscript i denotes the ith entry of the state vector.

The control objective is to track a set point for butanol concentration, while enforcing state constraints on the concentration of acids in order to retain the pH of the fermentation culture within a desired range. For this case study, the OCP (Eq. 9.10) takes the form

$$\min_{u} \sum_{i=0}^{N_p} (x_9(i) - C_{\mathrm{Bn}}^{\mathrm{sp}})^2 \tag{9.13}$$

subject to:

$$\begin{aligned}
&\hat{x}(i+1) = f_k(x(i), u(i), \theta), && i = 0 \ldots N_p - 1\\
&0.005\mathrm{h}^{-1} \le u_1(i) \le 0.145\mathrm{h}^{-1}, && i = 0 \ldots N_c - 1\\
&0\ \mathrm{mM} \le u_2(i) \le 80\mathrm{mM}, && i = 0 \ldots N_c - 1\\
&13.83\mathrm{mM} \le x_2(i) \le 15.68\mathrm{mM}, && i = 1 \ldots N_p\\
&10.55\ \mathrm{mM} \le x_7(i) \le 12.30\,\mathrm{mM}, && i = 1 \ldots N_p\\
&\bar{x}(0) = x(k),
\end{aligned}$$

where $C_{\mathrm{Bn}}^{\mathrm{sp}}$ denotes the butanol set point; u_1 denotes the dilution rate; u_2 denotes the inlet glucose concentration; x_2, x_7, and x_9 denote the acetate, butyrate, and butanol concentrations, respectively. The prediction and control horizons are chosen as $N_\mathrm{p} = N_\mathrm{c} = 150$ h. It is assumed that all the state variables are measured at every measurement sampling time. The sampling intervals are 0.33 h.

Like the previous case study, the MATLAB constrained optimization function `fmincon` is used to solve the OCP (Eq. 9.13). The code for the receding-horizon implementation of the OCP is given in Box 9.3. The optimizer `fmincon` calls the objective function `objfun` and the constraint function `constraints`. See the Appendix for the code of the process model.

In the objective function `objfun`, the process model is used to evaluate the cost function, which is defined as the sum of least squares between the predicted butanol concentration over the prediction horizon and the set point for butanol concentration. In the objective function `objfun`, `SP` $= C_{\mathrm{Bn}}^{\mathrm{sp}}$ is the set point for the butanol concentration and `Np` is the prediction horizon in hours (see Box 9.4). Here, `ode15s`, a stiff differential equation solver, is used to solve the system

BOX 9.3

```
for k = 1:N
    % The OCP
    u_opt(k+1,:)= fmincon(@objfun,u_opt(k,:),[], [], [],...
        [],lb,ub,@constraints,[],y(k,:),SP,Np);

    % Plant model
    [t,x]=ode15s(@ABE_model,meas_samp_time,y(k,:),[],...
        u_opt (k,:));

    % New measurements to reinitialize the OCP at the
      next sampling time
    y(k+1,:)=x(end,:);
end
```

BOX 9.4

```
function obj = objfun(u,x0,SP,Np)
     global A
     global B
     % model prediction
     [t, x]=ode15s(@ABE_model,0:1/3:Np,x0,[],u);
     A = x(:,2);  % acetate concentration
     B = x(:,7);  % butyrate concentration

     % objective function
     obj = sum((x(:,9)-SP).^2);
end
```

model ABE_model. In order to incorporate the state constraints into the optimization problem, the global variables A and B are declared and assigned as the predicted values of the acetate and butyrate concentrations over the prediction horizon. Through defining A and B as global variables, the predicted concentrations of acetate and butyrate are passed to the constraint function constraints, in which linear inequality constraints are defined in terms of the lower and upper bounds of each state (see Box 9.5). Notice that the arguments for the objective function objfun and constraint function constraints must be the same, regardless of whether or not each argument is used in both functions. The global variables A and B must be declared again inside of the constraints function. Once the predicted values of the states are evaluated in the objective function objfun, they are sent to the constraints function in order to impose the upper and lower bounds on the acetate and butyrate concentrations, as shown in Box 9.5.

The closed-loop simulation results of the NMPC controller are shown in Fig. 9.7. The process starts at steady state at time 0 h and a set point change is introduced at this time to increase the concentration of butanol by 10%. Once the fermentation process has reached the set point, another set point change is introduced at time 60 h to decrease the butanol concentration by 20%. The NMPC controller is tasked with accommodating the butanol set point changes while ensuring that the acetate and butyric concentrations remain within the process constraints. The closed-loop simulation results in Fig. 9.7 indicate that the NMPC controller can effectively retain the concentrations of acetate and butyrate within the specified limits, which is necessary to maintain the pH of the fernentation culture at a desired level and to remain within the validity range of the process model. At time 60 h, the acid concentrations decrease in response to the set point change in butanol concentration. The simulation results suggest that the NMPC controller enables effective transition of the process between the two operating conditions (i.e., butanol set points).

BOX 9.5

```
function [c,ceq] = constraints(u,x0,SP,time)
    global A
    global B
    Aub = 15.68; % upper bound on acetate concentration (mM)
    Alb = 13.83; % lower bound on acetate concentration (mM)
    Bub = 12.30; % upper bound on butyrate concentration(mM)
    Blb = 10.55; % lower bound on butyrate concentration(mM)

    % Inequality constraints
    c = [A - Aub; Alb - A; B - Bub; Blb - B];

    % Equality constraints
    ceq = [];
end
```

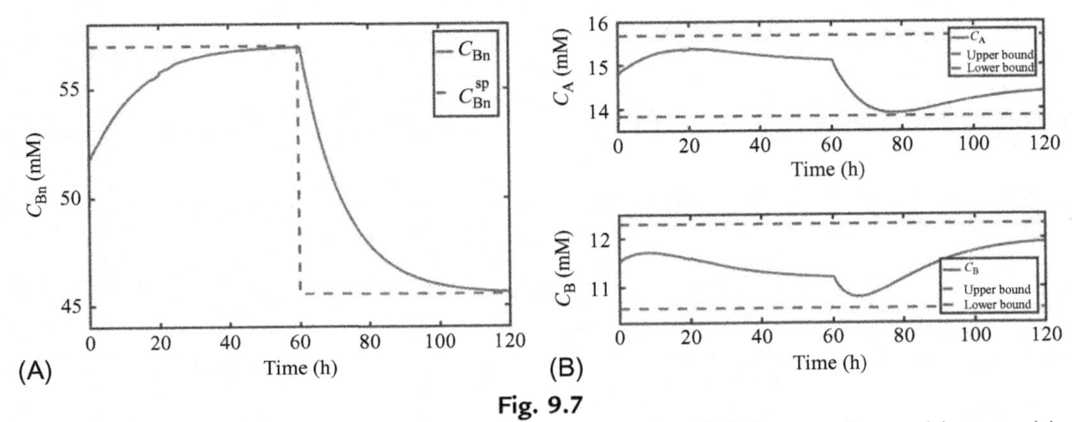

Fig. 9.7

The NMPC of the continuous ABE fermentation process. The NMPC controller enables transition from one operating condition to another (defined in terms of the butanol set point) without violating the state constraints on acetate and butyrate. (A) Set point tracking for butanol. (B) State constraints on acetate and butyrate.

The flexibility of using the NMPC controller over a wide operating range (i.e., changing the butanol set points) is due to the use of a nonlinear model in the NMPC controller. If the ABE fermentation process was to be controlled around a specific operating condition, a MPC controller could be designed using a linear model obtained via linearization of the nonlinear process model around the desired operating point. MPC based on linear models typically offers computational advantages over NMPC for real-time control applications.

9.4 Output-Feedback MPC

Up until this point, we have assumed that all of the process state variables can be measured and directly used for receding-horizon implementation of MPC in a full-state feedback fashion. In practice, however, it may not be feasible to measure all of the state variables due to various technical or economic limitations. This gives rise to the problem of state estimation. A state estimator uses a process model in combination with measurements from the process in order to reconstruct the state vector.[39,40] As shown in Fig. 9.8, the estimated states can then be used to reinitialize the process model in the MPC controller. This is known as the output-feedback implementation of MPC.

A wide range of methods have been developed for state estimation of nonlinear (bio)chemical processes.[41,42] A classification of nonlinear state estimation methods with exponential convergence properties is shown in Fig. 9.9. Broadly speaking, these methods can be categorized as methods developed under a deterministic framework and methods developed under the Bayesian estimation framework. The deterministic state estimation methods such as the extended Luenberger observer (ELO) neglect the stochastic process disturbances.[43,44]

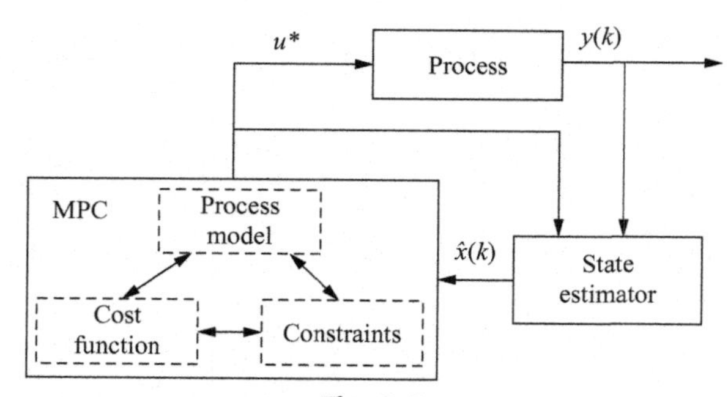

Fig. 9.8
Output-feedback implementation of MPC using a state estimator.

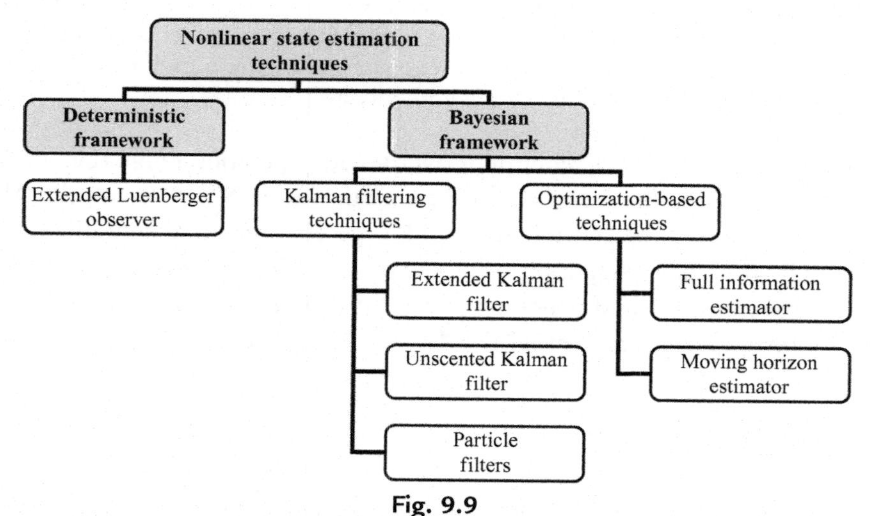

Fig. 9.9
A broad classification of nonlinear state estimation methods.

On the other hand, the Bayesian state estimation methods determine the probability density function (pdf) of the states under the effects of stochastic process and measurement noise. The most widely used nonlinear state estimation method is the extended Kalman filter (EKF), which is an adaptation of the Kalman filter for nonlinear systems.[45,46] EKF relies on linearization of the nonlinear process model around the current state estimates, and the assumption that the state variables and the process and measurement noise are described by Gaussian distributions. However, for highly nonlinear processes, the assumption that the state variables will retain their Gaussian distribution after propagation through the nonlinear process dynamics may not be valid. In addition, EKF may not be applicable to nondifferentiable systems since it requires linearization of the process model. These considerations have motivated the development of sample-based (derivative-free) methods for nonlinear state estimation. In the unscented Kalman filter (UKF), the linearization of the process model is avoided by an unscented transformation of a set of heuristically chosen samples through the nonlinear process model, thus eliminating the need for linearization.[47] However, UKF stills relies on the assumption that the state variables are described by a Gaussian distribution. On the other hand, particle filters do not make any assumptions about the process dynamics or the type of state distributions.[48] In particle filters, the pdfs of states are constructed using a series of random samples, called particles. Monte Carlo techniques are typically used to generate samples and then the pdfs of states are estimated based on Bayes rule. As the number of samples grows, the estimated pdfs of states must converge to their true pdfs.[49]

A different class of state estimation methods includes optimization-based estimation techniques.[50] Since (bio)chemical processes are often subject to state constraints due to various operational and economic considerations, optimization-based state estimation techniques can be used to explicitly include the state constraints into the nonlinear estimation problem. In contrast to the earlier discussed estimation methods that use the most recent process measurements only, optimization-based state estimators utilize the measurements obtained over a prespecified estimation horizon to minimize the difference between the model predictions and the measurements in a weighted least-squares sense. In general, nonlinear optimization-based estimation techniques can be classified as full information or moving horizon estimators.[51] The estimation horizon of the full information estimators grows as new process measurements become available, whereas in the moving horizon estimators the optimization is performed over a finite estimation horizon to avoid excessively large computational burdens.

Since we have adopted a deterministic setting for process modeling in this chapter (i.e., the process model (9.3) includes no stochastic noise), only the Luenberger observer and its extension for nonlinear processes are introduced in the remainder of this section.

9.4.1 Luenberger Observer

The linear state-space model (9.2) can be written in the discrete form

$$\begin{aligned} x(k+1) &= Ax(k) + Bu(k), \quad x(k=0) = x_0 \\ y(k) &= Cx(k), \end{aligned} \tag{9.14}$$

where $A \in \mathbb{R}^{n \times n}$, $B \in \mathbb{R}^{n \times n_u}$, and $C \in \mathbb{R}^{n_y \times n}$. A Luenberger observer is designed as[43]

$$\hat{x}(k+1) = Ax(k) + Bu(k) + K(k)(y(k) - C\hat{x}(k)), \tag{9.15}$$

where $\hat{x}(k)$ is an estimate of the state vector and $K(k)$ is the observer gain, which governs the accuracy and convergence properties of the state estimator. The right-hand side of Eq. (9.15) consists of a copy of the process model and a correction term that is defined as the difference between the estimated outputs and the measured process outputs multiplied by the observer gain.

The goal of the Luenberger observer is to provide an estimate of the state vector such that the estimation error

$$\begin{aligned} e(k+1) &= x(k+1) - \hat{x}(k+1) \\ &= A(\hat{x}(k) + e(k)) - A\hat{x}(k) - K(k)(C(\hat{x}(k) + e(k)) - C\hat{x}(k)), \end{aligned} \tag{9.16}$$

is zero. The estimation error is, by definition, the difference between the true states and the estimated states. The structure of the gain matrix $K(k)$ is defined on the basis of physical insights into the process, while its tuning parameters are typically obtained by running open-loop simulations and evaluating the evolution of the states.[52,53]

9.4.2 Extended Luenberger Observer

For nonlinear processes, the ELO is derived through linearization of the nonlinear process model. Using the nonlinear state-space model in Eq. (9.3), the ELO is written as

$$\hat{x}(k+1) = f_k(\hat{x}(k), u(k)) + K(k)(y(k) - h_k(\hat{x}(k))), \tag{9.17}$$

where the estimation error is of the form[44]

$$\begin{aligned} e(k+1) &= x(k+1) - \hat{x}(k+1) \\ &= f_k(\hat{x}(k) + e(k), u(k)) - f_k(\hat{x}(k), u(k)) - K(k)(h_k(\hat{x}(k) + e(k)) - h_k(\hat{x}(k))). \end{aligned} \tag{9.18}$$

Due to the nonlinear dynamics, the condition under which the error converges to zero cannot be readily determined from the error dynamics. Therefore, the observer must be designed based on a linearized version of the nonlinear process model. Linearization of the process model around the current state estimates $\hat{x}(k)$ yields

$$e(k+1) = (A(k) - K(k)C(k))e(k), \tag{9.19}$$

where $A(k) = \left[\frac{\partial f_k(x(k), u(k))}{\partial x}\right]\Big|_{x(k)=\hat{x}(k)}$ and $C(k) = \left[\frac{\partial h_k(x(k), u(k))}{\partial x}\right]\Big|_{x(k)=\hat{x}(k)}$. Generally, the estimation accuracy of the ELO depends on how well the linearized model represents the true process dynamics and how close the initial state estimates $\hat{x}(0)$ are to the true initial states (e.g., see Refs. 52,54).

9.4.3 NMPC of the Batch Crystallization Process Under Incomplete State Information

In the batch crystallization case study presented in Section 9.3, it was assumed that the full state vector could be measured. In many crystallization processes, however, obtaining reliable measurements for the solute concentration may be impractical. Revisiting this case study, we assume that the solute concentration measurement is not available. Thus, an ELO is designed to facilitate the output-feedback implementation of the NMPC controller. The ELO estimates the state variables, which are used for reinitializing the OCP (Eq. 9.12) at every measurement

BOX 9.6

```
for k = 1:length(ts)-1,
    % The OCP
    [uopt] = fmincon(@objbatch,uopt_past,[], [], [], [],...
        lb,ub,@conbatch,[],xhat(k,:),tp,ninter);

    % Adjusting the prediction horizon for shrinking-horizon implementation of the OCP
    tp = tb - ts(k);

    % Plant
    [t,x] = ode15s(@crystallizer,[ts(k) ts(k+1)],x_plant(k,:),[],uopt(1));
    x_plant(k+1,:) = x(end,:);
    uopt_past = uopt;
    % ELO (state estimation)

    [t,xobs] = ode15s(@ELO,[ts(k) ts(k+1)], xhat(k,:), [],uopt(1),
            x_plant(k,1:5), x_plant(k+1,1:5),ts(k),ts(k+1));
    xhat(k+1,:) = xobs(end,:);
end
```

sampling time. As demonstrated by the simulation results (see Fig. 9.6), the closed-loop implementation of the NMPC controller is essential for mitigating the performance degradation of the optimal control inputs due to plant-model mismatch and process uncertainties.

Box 9.6 gives the MATLAB excerpt for the output-feedback implementation of the NMPC, as depicted in Fig. 9.8. To this end, the feeback control loop described in Box 9.1 (for the case of full-state feedback) is modified by adding the ELO state estimator. At every sampling time `k`, the state estimator takes the plant outputs `x_plant` and constructs the vector of state estimates `xhat`. The state estimates `xhat` are then used for reinitializing the optimizer `fmincon`. Notice that the same process input `uopt(1)` applied to the process is also applied to the ELO. The state estimator `ELO` is given in Box 9.7. `ELO` involves solving the process model over the sampling interval. The model predictions are modified by the error terms `E` to account for the discrepancy between the model predictions and the process measurements. The observer gain `K` is computed off-line as discussed in Ref. 31. See Appendix for the MATLAB codes.

The closed-loop simulation results for the output-feedback NMPC controller are shown in Fig. 9.10. The optimal heat input and crystal growth rate profiles are similar to the case of the full-state feedback NMPC controller presented in Fig. 9.6 in spite of reconstructing the solute concentration using the available measurements. Fig. 9.10C shows the estimated solute concentration profile versus the true solute concentration profile, which is considered to be unmeasurable. The slight difference between the true concentration and the estimated concentration results from the plant-model mismatch that is due to the kinetic

BOX 9.7

```
function dxdt = ELO(t,x0,u,x1,x2,t1,t2)

% Initial state variables
m0 = x0(1); m1 = x0(2); m2 = x0(3); m3 = x0(4); m4 = x0(5); w = x0(6);

% Observer gain (computed offline)
K = [-0.00272458 84.5034 0.117009 0.441494 0.463708];

% Error signals
E(1) = x1(1) + (t - t1)*(x2(1) - x1(1))/(t2 - t1) - m0;
E(2) = x1(2) + (t - t1)*(x2(2) - x1(2))/(t2 - t1) - m1;
E(3) = x1(3) + (t - t1)*(x2(3) - x1(3))/(t2 - t1) - m2;
E(4) = x1(4) + (t - t1)*(x2(4) - x1(4))/(t2 - t1) - m3;
E(5) = x1(5) + (t - t1)*(x2(5) - x1(5))/(t2 - t1) - m4;

% Moment equations
G  = kg * (w - ws)^g;
B0 = kb * m3^km3 * G^kcg;

dm0dt =      B0 - m0*Qp/V + K(1)*E(4);
dm1dt =    G*m0 - m1*Qp/V + K(2)*E(5);
dm2dt = 2*G*m1 - m2*Qp/V + K(3)*E(3);
dm3dt = 3*G*m2 - m3*Qp/V + K(4)*E(4);
dm4dt = 4*G*m3 - m4*Qp/V + K(5)*E(5);

% Solute concentration balance
k1 = Hv*ws/(Hv-Hl)*(Rc/Rl-1+(Rl*Hl-Rc*Hc)/(Rl*Hv))-Rc/Rl;
k2 = ws/(V*Rl*(Hv-Hl));

dwdt = (Qp/V*(ws-w)+3*Kv*G*m2*(k1+w) + k2*Hin)/(1-Kv*m3);

dxdt=[dm0dt;dm1dt;dm2dt;dm3dt;dm4dt;dwdt];

end
```

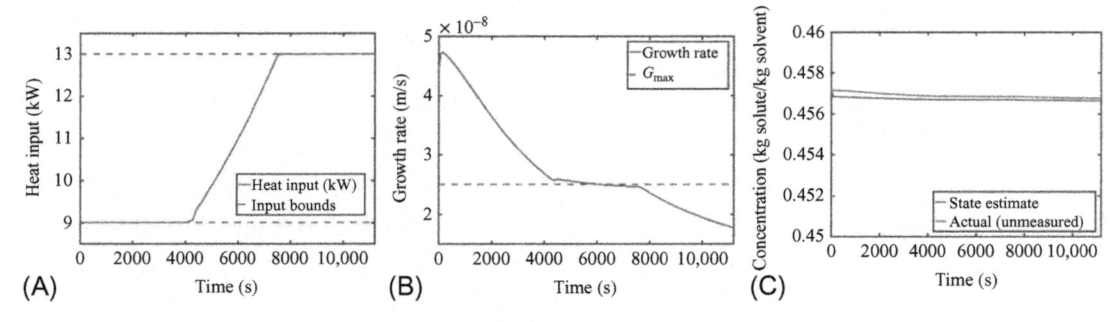

Fig. 9.10

The output-feedback NMPC of the batch crystallization process. The unmeasured solute concentration is estimated using the ELO. The ELO is able to effectively estimate the unmeasured solute concentration using the available process measurements, with a slight offset due to the plant-model mismatch in the underlying model of the ELO. (A) Heat input profile, (B) growth rate profile, and (C) a comparison between the (unmeasured) true and estimated profiles of the solute concentration.

parametric uncertainties and uncertainties in the initial process conditions. Despite the plant-model mismatch, updating the model predictions in the ELO based on the plant measurements enables obtaining accurate estimates of the state variables for the output-feedback implementation of the NMPC controller.

9.5 Advanced Process Control

In a chemical plant, MPC is typically incorporated into a hierarchical control scheme known as the advanced process control (APC)[55] scheme, which is illustrated in Fig. 9.11. The regulatory controllers, at the base of the APC hierarchy, are commonly made up of multi-loop PID controllers that provide improved stability and disturbance rejection. Constrained, multivariable optimization of the individual process units is performed in the MPC layer. As described earlier, MPC controllers predict the future outputs of a unit operation based on the current inputs and measurements and then determine the optimal control sequence to the unit. The first element of the optimal control sequence is applied to the process through the regulatory control layer. This implies that the set points of the PID controllers in the regulatory level are determined by the MPC controllers. In fact, the MPC layer above the regulatory layer enables optimal operation of the process close to the process constraints, typically where the process is the most profitable.

The production control layer (typically consisting of plant-wide optimization[56] and production planning and scheduling[57]) coordinates the operation of all of the process units across the plant. Running each unit at its local optimum may not lead to the most economical performance of the plant as a whole. The production control layer ensures that the plant operation meets the production demand, quality standards, plant utility constraints, and shipment requirements. The APC scheme shown in Fig. 9.11 offers a great advantage over traditional planning and scheduling approaches by streamlining of the entire control hierarchy into a single system.

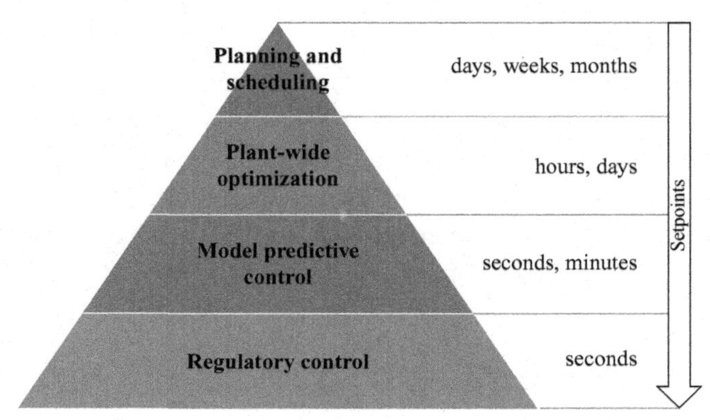

Fig. 9.11
Advanced process control scheme.[55]

9.6 Advanced Topics in MPC

9.6.1 Stability and Feasibility

Important theoretical properties of MPC consist of guaranteeing the feasibility and stability of the closed-loop control system. The feasibility requirement is to ensure that the OCP (Eq. 9.10) yields a feasible solution in light of the state constraints, whereas the stability requirement is to ensure that the optimal control inputs would not destabilize the control system. Generally speaking, conditions for establishing stability and feasibility are known for linear systems but often difficult to establish for nonlinear systems. Common approaches to enforcing closed-loop stability involve incorporating terminal equality constraints, terminal cost functions, and terminal constraint sets into the OCP. The addition of a terminal equality constraint forces the solution to the objective function to be a certain value at the end of the prediction horizon. When a terminal cost is used, an additional term (typically a quadratic function of the states) is added to the objective function. Although originally developed for linear systems, this method for establishing stability can be extended to nonlinear systems. A terminal constraint set is intended to drive the states to a constrained set within a finite length of time.

For further reading see Refs. 15,58–60.

9.6.2 MPC of Uncertain Systems

So far our discussion has focused on deterministic processes; however, all processes are subject to some form of uncertainty. These uncertainties may lead to degradation of the process model or cause constraint violation. Although the receding-horizon implementation of MPC improves the performance of the controller against uncertainties, it cannot systematically account for the process uncertainties. Robust MPC addresses uncertainties by utilizing a deterministic model with bounded, deterministic uncertainty descriptions. Robust MPC aims to ensure that certain process performance requirements and process constraints are met with respect to all possible process uncertainties. However, this may lead to conservative control. When the need for a process to remain within constraints is greater than that of achieving the performance objective, robust MPC offers an approach to guarantee that the process remains within the constraints. When probabilistic descriptions of uncertainties are available, a more natural approach to uncertainty handling will involve explicitly incorporating these probabilistic uncertainty descriptions into the OCP. Accounting for the probabilistic nature of uncertainties in the so-called stochastic MPC may allow for obtaining less conservative robust controllers.

For further reading see Refs. 61,62.

9.6.3 Distributed MPC

Complex, large-scale systems typically consist of a network of integrated subsystems that are connected to each other through mass, energy, and information streams. Examples of such large-scale processes include integrated chemical plants, power systems, and water distribution systems. Designing a centralized MPC for the system as a whole is typically impractical. However, complex systems can be divided into several subsystems, where each subsystem may only be coupled to few subsystems that share a few process variables in common. Distributed MPC involves designing MPC controllers for each individual subsystem such that some form of coordination takes place between the subsystems. Distributed control can address various practical and computational challenges associated with the MPC of large-scale processes.

For further reading see Refs. 63–65.

9.6.4 MPC With Integrated Model Adaptation

The performance of MPC controllers critically hinge on the quality of their underlying models. In most industrial process control applications, the inherent (gradual) time-varying nature of plant dynamics adversely impact the lifetime performance of MPC controllers. Changes in the plant dynamics over time increase the plant-model mismatch, which may eventually invalidate the model developed at the commissioning stage of a MPC controller. This necessitates regular maintenance of the process models and, consequently, the controller itself. An emerging area in the MPC of complex systems is MPC with integrated learning capability, where the control inputs must not only control the system dynamics in view of the control objectives but also gather informative process information for model adaptation. In MPC with integrated model adaptation, the OCP is modified to add some form of system perturbation capability to the control inputs. The addition of a process perturbation effect to the control inputs inevitably results in control performance loss due to the undesired system perturbations in the short run. However, the improved system learning facilitated by perturbations is expected to lead to better control performance over the future control stages and, as a result, decrease the overall performance loss that otherwise would be incurred due to large model uncertainty in the absence of learning.

For further reading see Refs. 66–69,69a,69b,69c.

9.6.5 Economic MPC

In chemical processes, achieving a desirable economic performance and control performance (e.g., in terms of set point tracking of a key process variable) may involve trade-offs. For an increasing number of processes, the hierarchical separation of MPC and (economic) plant-wide optimization is no longer optimal or desirable (see Fig. 9.11). An alternative to this hierarchical decomposition is to embed the economic objective of the process directly in the objective function of the MPC controller. In this approach, known as economic MPC, the controller optimizes directly the economic performance of the process in real time.

For further reading see Refs. 70–72.

Appendix

Batch Crystallization Case Study

Process model

```
function dxdt = crystallizer(t,x0,HI)

% State variables
m0 = x0(1);
m1 = x0(2);
m2 = x0(3);
m3 = x0(4);
m4 = x0(5);
w  = x0(6);

kg   = 7.49567e-05;
kb   = 1.02329e+14;

Hin = 0.001*HI; % [kW]
Qp  = 1.73120E-06; % Product flow (m3/s)

g   = 1;
kcg = 1;
km3 = 1;

Kv = 0.43;
V  = 75e-3; % [m^3]
Rc = 1767.3485; % [kg / m^3]
Rl = 1248.925; % [kg / m^3]
Hv = 2.590793151044623E+03; % [kJ / kg]
ws = 0.456513; % Solubility of AS at 50 degr. C. by Westhoff (2002) p. 159
Hl = 2.7945 * (50 - 25); % [kJ / kg]
Hc = 2.43   * (50 - 25); % [kJ / kg]
G  = kg * (w - ws)^g;
B0 = kb * m3^km3 * G^kcg;

% Moment equations
dm0dt =   B0   - m0*Qp/V;% + K1E(4);
dm1dt =   G*m0 - m1*Qp/V;% + K2*E(5);
dm2dt = 2*G*m1 - m2*Qp/V;% + K3*E(3);
dm3dt = 3*G*m2 - m3*Qp/V;% + K4*E(4);
dm4dt = 4*G*m3 - m4*Qp/V;% + K5*E(5);

% Solute concentration balance
k1 = Hv*ws/(Hv-Hl)*(Rc/Rl-1+(Rl*Hl-Rc*Hc)/(Rl*Hv))-Rc/Rl;
k2 = ws/(V*Rl*(Hv-Hl));

dwdt = (Qp/V*(ws-w)+3*Kv*G*m2*(k1+w) + k2*Hin)/(1-Kv*m3);

dxdt=[dm0dt;dm1dt;dm2dt;dm3dt;dm4dt;dwdt];

end
```

For full-state feedback, `crystallizer` and `crystallizer_model` have the same code. For the ELO case study:

```
kg  = 2*7.49567e-05;
kb  = 5*1.02329e+14;
```

in `crystallizer_model`, `objbatch`, and `ELO`.

Closed-loop MPC implementation with state estimator

```
ts = [0 70 144 214 288 352 426 494 584 656 728 838 960 1082 1194 1316...
        1438 1548 1666 1788 1908 2022 2142 2264 2382 2500 2616 2734 2852...
        2970 3092 3208 3328 3444 3562 3684 3800 3922 4036 4286 4406 4524...
        4642 4760 4878 4996 5116 5234 5354 5472 5590 5708 5826 5946 6064...
        6182 6300 6418 6538 6656 6774 6892 7010 7130 7248 7366 7486 7604...
        7722 7840 7960 8222 8340 8458 8696 8932 9050 9168 9288 9406 9524...
        9642 9762 9882 9998 10118 10236 10356 10474 10592 10712 10832...
        10948 11072 11186];

tb  = ts(end); % initial time horizon (s)
tp = tf;

x0(1,:) = [1.08457e10 342338 368.8758 0.0830 2.4860e-5 0.457106];
% initial values [m0 m1 m2 m3 m4 Cw]
xhat(1,:) = [1.08457e10 342338 368.8758 0.0830 2.4860e-5 0.4572];
% state estimates

x_plant(1,:) = x0; % plant measurements

ninter  = 3;
lb  = 9000*ones(1,ninter);
ub  = 13000*ones(1,ninter);

for k = 1:length(ts)-1,
    %Dynamic optimizer
    [uopt] = fmincon(@objbatch,uopt_past,[],[],[],[],...
        lb,ub,@conbatch, [], xhat(k,:),tp,ninter);
    % Adjusting the prediction horizon for shrinking-horizon implementation of
      the OCP
    tp = tb - ts(k+1);
    %Plant
    [t,x] = ode15s(@crystallizer,[ts(k) ts(k+1)],x_plant(k,:),[],uopt(1));
    uopt_past = uopt;
    x_plant(k+1,:) = x(end,:);
    % State  estimator
    [t,xobs]  = ode15s(@ELO,[ts(k) ts(k+1)],xhat(k,:),[],uopt(1),...
              x_plant(k,1:5), x_plant(k+1,1:5),ts(k),ts(k+1));
    xhat(k+1,:) = xobs(end,:);
end
```

Constraints function

```
function [conin,coneq] = conbatch(u,x0,tf,ninter)
conin  = [];
coneq  = [];
```

Extended Luenberger observer

```
function dxdt = ELO(t,x0,HI,x1,x2,t1,t2)

% State variables
m0 = x0(1);
m1 = x0(2);
m2 = x0(3);
m3 = x0(4);
m4 = x0(5);
w  = x0(6);

kg    = 2*7.49567e-05;
kb    = 5*1.02329e+14;

Hin = 0.001*HI; % [kW]
Qp  = 1.73120E-06; % Product flow (m3/s)

g   = 1;
kcg = 1;
km3 = 1;

Kv = 0.43;
V  = 75e-3; % [m^3]
Rc = 1767.3485; % [kg / m^3]
Rl = 1248.925; % [kg / m^3]
Hv = 2.590793151044623E+03; % [kJ / kg]
ws = 0.456513; % Solubility of AS at 50 degr. C. by Westhoff (2002) p. 159
Hl = 2.7945 * (50 - 25); % [kJ / kg]
Hc = 2.43   * (50 - 25); % [kJ / kg]
G  = kg * (w - ws)^g;
B0 = kb * m3^km3 * G^kcg;

K = [-0.00272458 84.5034 0.117009 0.441494 0.463708];

% Error signals
E(1) = x1(1) + (t - t1)*(x2(1) - x1(1))/(t2 - t1) - m0;
E(2) = x1(2) + (t - t1)*(x2(2) - x1(2))/(t2 - t1) - m1;
E(3) = x1(3) + (t - t1)*(x2(3) - x1(3))/(t2 - t1) - m2;
E(4) = x1(4) + (t - t1)*(x2(4) - x1(4))/(t2 - t1) - m3;
E(5) = x1(5) + (t - t1)*(x2(5) - x1(5))/(t2 - t1) - m4;

% Moment equations
dm0dt =    B0   - m0*Qp/V + K(1)*E(3);
dm1dt =    G*m0 - m1*Qp/V + K(2)*E(4);
dm2dt = 2*G*m1 - m2*Qp/V + K(3)*E(3);
dm3dt = 3*G*m2 - m3*Qp/V + K(4)*E(4);
dm4dt = 4*G*m3 - m4*Qp/V + K(5)*E(5);

% Solute concentration balance
k1 = Hv*ws/(Hv-Hl)*(Rc/Rl-1+(Rl*Hl-Rc*Hc)/(Rl*Hv))-Rc/Rl;
k2 = ws/(V*Rl*(Hv-Hl));

dwdt = (Qp/V*(ws-w)+3*Kv*G*m2*(k1+w) + k2*Hin)/(1-Kv*m3);

dxdt=[dm0dt;dm1dt;dm2dt;dm3dt;dm4dt;dwdt];

end
```

ABE Fermentation Case Study

Process model

```
function dxdt = ABE_model(t,x,u)

% Variables:
AC = x(1); %acetyl-CoA
A = x(2); %acetate
En = x(3); %ethanol
AaC = x(4); %acetoacetyl-CoA
Aa = x(5); %acetoacetate
BC = x(6); %butyryl-CoA
B = x(7); %butyrate
An = x(8); %acetone
Bn = x(9); %butanol
Ad = x(10); %adc
Cf = x(11); %ctfA/B
Ah = x(12); %adhE

%Kinetic parameters

V1 = 4.94;
V2 = 2.92;
V4 = 45.6;
V8 = 64.8;
V10 = 4.75;

K1 = 0.00158;
K2 = 0.00181;
K4 = 1.87;
K8 = 7.92E-6;
K10 = 1.40E-5;

a3 = 0.00517;
a5 = 0.0140;
a6 = 0.00537;
a7 = 4790;
a9 = 347000;

rad = 0.00547;
rcf = 0.000324;
rah = 0.289;

radp = 0.104;
rcfp = 1.06;
rahp = 2.56;

n = 485;
pstar = 4.50; %pH after switch

%inputs
D = u(1); %Dilution rate
G = u(2); %Glucose conc

p = 4.5; %keep pH constant

F = 1-tanh(n*(p-pstar)); %pH switch for enzymes
```

```
%Reactions
r1 = 2*V1*G/(K1+G);
r2 = V2*AC/(K2+AC);
r3 = a3*A*AaC*Cf;
r4 = V4*AC/(2*(K4+AC));
r5 = a5*AC*Ah;
r6 = a6*B*AaC*Cf;
r7 = a7*Aa*Ad;
r8 = V8*BC/(K8 + BC);
r9 = a9*BC*Ah;
r10 = V10*AaC/(K10+AaC);

%Species mass balances
dxdt(1) = r1 - r2 + r3 -r4 -r5 - D*AC;
dxdt(2) = r2 - r3 - D*A;
dxdt(3) = r5 - D*En; %ethanol
dxdt(4) = r4 - r3 - r6 - r10 - D*AaC;
dxdt(5) = r3 + r6 - r7 - D*Aa;
dxdt(6) = r10 - r8 + r6 - r9 - D*BC;
dxdt(7) = r8 - r6 - D*B;
dxdt(8) = r7 - D*An; %acetone
dxdt(9) = r9 - D*Bn; %butanol
dxdt(10) = rad + radp*F-D*Ad;
dxdt(11) = rcf + rcfp*F - D*Cf;
dxdt(12) = rah + rahp*F - D*Ah;

dxdt = dxdt';

end
```

Closed-loop MPC implementation with state constraints

```
Np = 60; %prediction horizon in hrs
meas_samp_time = 0:1/60:1/3; %measurement time in hrs (1/3 hr = 20 min)

%input bounds
lb =[0.005, 0]; % lower bound (hr^-1, mM)
ub = [0.145, 80]; % upper bound (hr^-1, mM)

%inputs, u = [D, G0]
u_opt(1,:) = [0.075 40];

%outputs, y = [AC, A, En, AaC, Aa, BC, B, An, Bn, Ad, Cf, Ah]
y0 = [0 13.73 3.02 0 0 0 8.81 1.82 0 0 0 0]; %initial condition

[t,x0]=ode15s(@ABE_model,0:1:150,y0,[],u(1,:));

y(1,:) = x0(end,:);

SP = 1.1*y(1,9);

global A; global B;
A=y(1,2); B=y(1,7);
```

```
for k = 1:(3*Np)
    % Optimizer
    u_opt(k+1,:)= fmincon(@objfun,u_opt(k,:),[], [], [], [],lb,ub,...
        @constraints,[],y(k,:),SP,Np);
    % Plant
    [t,x]=ode15s(@ABE_model,meas_samp_time,y(k,:),[],u_opt(k,:));
    % New measurements to reinitialize the optimizer at the next sampling time
    y(k+1,:)=x(end,:);
end

SP2 = 0.8*y(k,9);

for k = (3*Np)+1:(3*2*Np)
    % Optimizer
    u_opt(k+1,:)= fmincon(@objfun,u_opt(k,:),[],[],[],[],lb,ub,...
        @constraints,[],y(k,:),SP2,Np);
    % Plant
    [t,x]=ode15s(@ABE_model,meas_samp_time,y(k,:),[],u_opt(k,:));
    % New measurements to reinitialize the optimizer at the next sampling time
    y(k+1,:)=x(end,:);
end
```

Acknowledgments

Washington S. Reeder is acknowledged for his work on the ABE fermentation simulation case study.

References

1. Bennett S. A brief history of automatic control. *IEEE Control Syst* 1996;**16**(3):17–25.
2. Åström J, Hägglund T. *PID controllers: theory, design, and tuning research triangle park.* Raleigh, NC: Instrument Society of America; 1995.
3. Skogestad S, Postlethwaite I. *Multivariable feedback control: analysis and design.* 2nd ed. Southern Gate: John Wiley & Sons; 2005.
4. Ang KH, Chong G, Li Y. PID control system analysis, design and technology. *IEEE Trans Control Syst Technol* 2005;**13**(4):559–76.
5. Qin SJ, Badgwell TA. A survey of industrial model predictive control technology. *Control Eng Pract* 2003;**11** (7):733–64.
6. Cutler CR, Ramaker BL. Dynamic matrix control—a computer control algorithm. In: AIChE national meeting, Houston, TX; 1979.
7. Cutler C, Morshedi A, Haydel J. An industrial perspective on advanced control. In: AIChE national meeting, Washington, DC; 1983.
8. Basak K, Abhilash KS, Ganguly S, Saraf DN. On-line optimization of a crude distillation unit with constraints on product properties. *Ind Eng Chem Res* 2002;**41**(6):1557–68.
9. Richalet J. Industrial applications of model based predictive control. *Automatica* 1993;**29**(5):1251–74.
10. Garcia CE, Prett DM, Morari M. Model predictive control: theory and practice—a survey. *Automatica* 1989;**25** (3):335–48.
11. Richalet J, Rault A, Testud JL, Papon J. Algorithmic control of industrial processes. In: Proceedings of the 4th IFAC symposium on identification and system parameter estimation, Amsterdam; 1976. p. 1119–67.
12. Morari M, Lee JH. Model predictive control: past, present and future. *Comput Chem Eng* 1999;**23**(4):667–82.

13. Gidon D, Graves DB, Mesbah A. Model predictive control of thermal effects of an atmospheric plasma jet for biomedical applications. In: Proceedings of the American Control conference, Boston; 2016. p. 4889–94.
14. Schütze A, Jeong JY, Babyan SE, Park J, Selwyn GS, Hicks RF. The atmospheric-pressure plasma jet: a review and comparison to other plasma sources. *IEEE Trans Plasma Sci* 1998;**26**(6):1685–94.
15. Rawlings JB, Mayne DQ. *Model predictive control: theory and design*. Madison, WI: Nob Hill Publishing; 2015.
16. Allgöwer F, Findeisen R, Nagy ZK. Nonlinear model predictive control: from theory to application. *J Chin Inst Chem Eng* 2004;**35**(3):299–315.
17. Ljung L. *System identification: theory for the user*. Upper Saddle River, NJ: Prentice Hall PTR; 1999.
17a. Van Overschee P, De Moor BL. *Subspace identification for linear systems: Theory—Implementation—Applications*. Springer Science & Business Media; 2012.
18. Bequette BW. *Process control: modelling, design and simulation*. Upper Saddle River, NJ: Prentice Hall PTR; 2002.
19. Thomas MM, Kardos JL, Joseph B. Shrinking horizon model predictive control applied to autoclave curing of composite laminate materials. In: Proceedings of the American Control conference, Baltimore, MD; 1994. p. 505–9.
20. Biegler LT, Cuthrell JE. Improved infeasible path optimization for sequential modular simulators—II: the optimization algorithm. *Comput Chem Eng* 1985;**9**(3):257–67.
21. Cuthrell JE, Biegler LT. Simultaneous optimization and solution methods for batch reactor control profiles. *Comput Chem Eng* 1989;**13**(1):49–62.
22. Bock HG, Plitt KJ. A multiple shooting algorithm for direct solution of optimal control problems. In: Proceedings of the 9th IFAC World Congress, Budapest, Hungary; 1984. p. 242–7.
23. Diehl M, Bock HG, Schlöder JP, Findeisen R, Nagy Z, Allgöwer F. Real-time optimization and nonlinear model predictive control of processes governed by differential-algebraic equations. *J Process Control* 2002;**12**(4):577–85.
24. Binder T, Blank L, Bock HG, Bulirsch R, Dahmen W, Diehl M, et al. Introduction to model based optimization of chemical processes on moving horizons. In: *Online optimization of large scale systems*. Berlin, Heidelberg: Springer; 2001. p. 295–339.
25. Randolph A, Larson M. *Theory of particulate processes*. New York, NY: Academic Press; 1971.
26. Myerson AS. *Handbook of industrial crystallization*. Woburn, MA: Butterworth-Heinemann; 2002.
27. Nagy ZK, Braatz RD. Advances and new directions in crystallization control. *Annu Rev Chem Biomol Eng* 2012;**3**:55–75.
28. Hulbert HM, Katz S. Some problems in particle technology: a statistical mechanical formulation. *Chem Eng Sci* 1964;**19**(8):555–74.
29. Mesbah A, Kramer HJ, Huesman AE, Van den Hof PM. A control oriented study on the numerical solution of the population balance equation for crystallization processes. *Chem Eng Sci* 2009;**64**(20):4262–77.
30. Mesbah A, Nagy ZK, Huesman AE, Kramer HJ, Van den Hof PM. Nonlienar model-based control of a semi-industrial batch crystallizer using a population balance modeling framework. *IEEE Trans Control Syst Technol* 2012;**20**(5):1188–201.
31. Mesbah A, Landlust J, Huesman AE, Kramer HJ, Jansens PJ, Van den Hof PM. A model-based control framework for industrial batch crystallization processes. *Chem Eng Res Des* 2010;**88**(9):1223–33.
32. MathWorks. *fmincon*. Natick, MA: The Mathworks; 2016.
33. Kendall JW. Hard and soft constraints in linear programming. *Omega* 1975;**3**(6):709–15.
34. Dürre P. Biobutanol: an attractive biofuel. *Biotechnol J* 2007;**2**(12):1525–34.
35. Köpke M, Dürre P. The past, present, and future of biolfuels—biobutanol as promising alternative. In: Marco Aurelio Dos Santos Bernardes, editor. *Biofuel Production-Recent Developments and Prospects*. In Tech; 2011. pp. 451–86. http://dx.doi.org/10.5772/20113. Available from: https://www.intechopen.com/books/biofuel-production-recent-developments-and-prospects/the-past-present-and-future-of-biofuels-biobutanol-as-promising-alternative.
36. Jones DT, Woods DR. Acetone-butanol fermentation revisited. *Microbiol Rev* 1986;**50**(4):484–524.

37. Haus S, Jabbari S, Millat T, Janssen H, Fischer RJ, Bahl H, et al. A systems biology approach to investigate the effect of pH-induced gene regulation on solvent production by Clostridium acetobutylicum in continuous culture. *BMC Syst Biol* 2011;**5**(10).
38. Buehler EA, Mesbah A. Kinetic study of acetone-butanol-ethanol fermentation in continuous culture. *Public Libr Sci* 2016;**11**(8):1–21.
39. Soroush M. State and parameter estimations and their applications in process control. *Comput Chem Eng* 1998;**23**(2):229–45.
40. Dochain D. State and paramter estimation in chemical and biochemical processes: a tutorial. *J Process Control* 2003;**13**(8):801–18.
41. Soroush M. Nonlinear state-observer design with application to reactors. *Chem Eng Sci* 1997;**52** (3):387–404.
42. Patwardhan SC, Narasimhan S, Jagadeesan P, Gopaluni B, Shah SL. Nonlinear Bayesian state estimation: a review of recent developments. *Control Eng Pract* 2012;**20**(10):933–53.
43. Luenberger D. Observing the state of a linear system. *IEEE Trans Mil Electron* 1964;**8**(2):74–80.
44. Zeitz M. Extended Luenberger observer for nonlinear systems. *Syst Control Lett* 1987;**9**(2):149–56.
45. Kalman RE. A new approach to linear filtering and prediction problems. *J Basic Eng* 1960;**82**(1):35–45.
46. Kalman RE, Bucy RS. New results in linear filtering and prediction theory. *J Basic Eng* 1961;**83**(1):95–108.
47. Wan EA, Van Der Merwe R. The unscented Kalman filter for nonlinear estimation. In: Adaptive systems for signal processing, communications, and control symposium, Alberta, Canada; 2000. p. 153–8.
48. Arulampalam MS, Maskell S, Cordon N, Clapp T. A tutorial on particle filters for online nonlinear/non-Gaussian Bayesian tracking. *IEEE Trans Signal Process* 2002;**50**(2):174–88.
49. Crisan D, Doucet A. Survey of convergence results on particle filtering methods for practitioners. *IEEE Trans Signal Process* 2002;**50**(3):736–46.
50. Rawlings JB, Bakshi BR. Particle filtering and moving horizon estimation. *Comput Chem Eng* 2006;**30** (10):1529–41.
51. Rao CV, Rawlings JB, Mayne DQ. Constrained state estimation for nonlinear discrete-time systems: stability and moving horizon approximations. *IEEE Trans Autom Control* 2003;**48**(2):246–58.
52. Quintero-Marmol E, Luyben WL, Georgakis C. Application of an extended Luenberger observer to the control of multicomponent batch distillation. *Ind Eng Chem Res* 1991;**30**(8):1870–80.
53. Hulhoven X, Wouwever AV, Bogaerts P. Hybrid extended Luenberger-asymptotic observer for bioprocess state estimation. *Chem Eng Sci* 2006;**61**(21):7151–60.
54. Kazantzis N, Kravaris C, Wright RA. Nonlinear observer design for process monitoring. *Ind Eng Chem Res* 2000;**39**(2):408–19.
55. Robinson PR, Cima D. Advanced process control. In: *Practical advances in petroleum processing*. New York, NY: Springer; 2006. p. 695–703.
56. Tatjewski P. Advanced control and on-line optimization in multilayer structures. *Annu Rev Control* 2008;**32** (1):71–85.
57. Backx T, Bosgra O, Marquardt W. Integration of model predictive control and optimization of processes. In: IFAC symposium on advanced control of chemical processes, Pisa, Italy; 2000. p. 249–60.
58. Mayne DQ, Rawlings JB, Rao CV, Scokaert PO. Constrained model predictive control: stability and optimality. *Automatica* 2000;**36**(6):789–814.
59. Camacho EF, Bordons C. *Model predictive control*. London: Springer-Verlag; 1999.
60. Findeisen R, Imsland L, Allgöwer F, Foss BA. State and output feedback nonlinear model predictive control: an overview. *Eur J Control* 2003;**9**(2–3):190–206.
61. Bemporad A, Morari M. *Robust model predictive control: a survey. Robustness in identifiction and control*. London: Springer; 1999207–26.
62. Mesbah A. Stochastic model predictive control: an overview and perspectives for future research. *IEEE Control Syst Mag* 2016;**36**:30–44.
63. Stewart BT, Venkat AN, Rawlings JB, Wright SJ, Pannocchia G. Cooperative distributed model predictive control. *Syst Control Lett* 2010;**59**(8):460–9.

64. Venkat AN, Rawlings JB, Wright SJ. Stability and optimality of distributed model predictive control. In: Proceedings of the IEEE conference on decision and control, Seville, Spain; 2005. p. 6680–5.
65. Venkat AN, Rawlings JB, Wright SJ. Distributed model predictive control of large-scale systems. In: *Assessment and future directions of nonlinear model predictive control*. Berlin, Heidelberg: Springer; 2007. p. 591–605.
66. Larsson CA, Annergren M, Hjalmarsson H, Rojas CR, Bombois X, Mesbah A, et al. Model predictive control with integrated experiment design for output error systems. In: Proceedings of the European Control conference, Zurich, Switzerland; 2013. p. 3790–5.
67. Larsson CA, Rojas CR, Bombois X, Hjalmarsson H. Experimental evaluation of model predictive control with excitation (MPC-X) on an industrial depropanizer. *J Process Control* 2015;**31**:1–16.
68. Heirung TA, Foss B, Ydstie BE. MPC-based dual control with online experiment design. *J Process Control* 2015;**32**:64–76.
69. Marafioti G, Bitmead RR, Hovd M. Persistently exciting model predictive control. *Int J Adapt Control Signal Process* 2013;**28**:536–52.
69a. Bavdekar VA, Ehlinger V, Gidon D, Mesbah A. Stochastic predictive control with adaptive model maintenance. In: Proceedings of the 55th IEEE Conference on Decision and Control, Las Vegas; 2016. pp. 2745–50.
69b. Bavdekar V, Mesbah A. Stochastic model predictive control with integrated experiment design for nonlinear systems. In: Proceedings of the 11th IFAC Symposium on Dynamics and Control of Process Systems (DYCOPS), Trondheim; 2016. pp. 49–54.
69c. Heirung TAN, Mesbah A. Stochastic nonlinear model predictive control with active model discrimination for online fault detection. In: *Proceedings of the IFAC World Congress*, Toulouse; 2017. pp. 49–54, To Appear.
70. Ellis M, Durand H, Christofides PD. A tutorial review of economic model predictive control methods. *J Process Control* 2014;**24**(8):1156–78.
71. Ellis M, Liu J, Christofides PD. *Economic model predictive control: theory formulations and chemical process applications*. Switzerland: Springer International Publishing; 2017.
72. Rawlings JB, Angeli D, Bates CN. Fundamentals of economic model predictive control. In: IEEE conference on decision and control, Maui, HI; 2012. p. 3851–61.

CHAPTER 10

Optimal Control

Jean-Pierre Corriou
Lorraine University, Nancy Cedex, France

10.1 Introduction

Frequently, the engineer in charge of a process is faced with optimization problems. In fact, this may cover relatively different ideas, such as parameter identification or process optimization.

It is known that the reactive feed flow rate profile for a fed-batch reactor and the temperature or pressure profile to be followed for a batch reactor will have an influence on the yield, the selectivity or the product quality. To optimize production, one must then seek a time profile and perform a dynamic optimization with respect to the manipulated variables, while respecting the constraints of the system such as the bounds on temperature and temperature rise rate, the constraints related to the possible runaway of the reactor. Similarly, to optimize the conversion in a tubular reactor, one can seek the optimal temperature profile along the reactor. In the latter case, it is a spatial optimization very close to the dynamic optimization where the time is replaced by the abscissa along the reactor. The profile thus determined is calculated in open loop and will be applied as the set point in closed loop, which may lead to deviations between the effective result and the desired result. The direct closed-loop calculation of the profile in the nonlinear case is not studied here; on the contrary, the linear case is treated in linear quadratic (LQ) control and Gaussian LQ control.

In a continuous process, problems of dynamic optimization can also be considered with respect to the process changes from the nominal regime. For example, the quality of the raw petroleum feeding the refineries changes very often. The economic optimization realized offline imposes set point variations on the distillation columns. An objective can be to find the optimal profile to be followed during the change from one set of set points to another set.

In all cases, a dynamic model sufficiently representative of the behavior of the process is necessary, nevertheless of a reasonable complexity with respect to the difficulty of the mathematician and numerical task of solving.

Coulson and Richardson's Chemical Engineering. http://dx.doi.org/10.1016/B978-0-08-101095-2.00010-2

Among the criteria to be optimized, we often find the reaction yield or the selectivity, and also the end-time taken to reach a given yield, or any technical-economic criterion which simultaneously takes into account technical objectives, production, or investment costs.

Optimal control is the formulation of the dynamic optimization methods in the framework of a control problem.

10.2 Problem Statement

The optimal control problem is the first set in continuous time. The studied system is assumed to be nonlinear.

The fixed aim in this problem is the determination of the control $\boldsymbol{u}(t)$ minimizing a criterion $J(\boldsymbol{u})$ while verifying initial and final conditions and respecting constraints. The optimal control thus denoted by $\boldsymbol{u}^*(t)$ makes the state $\boldsymbol{x}(t)$ follow a trajectory $\boldsymbol{x}^*(t)$ which must belong to the set of admissible trajectories.

The formulation of the optimal control problem is given next.

Consider a system described in state space by the set of differential equations

$$\dot{\boldsymbol{x}}(t) = \boldsymbol{f}(\boldsymbol{x}(t), \boldsymbol{u}(t)); \quad t_0 \leq t \leq t_f \tag{10.1}$$

with $\boldsymbol{x}$ being a state vector of dimension n and $\boldsymbol{u}$ a control vector of dimension m. The system is subjected to initial and final conditions, called terminal (or at the boundaries)

$$\boldsymbol{k}(\boldsymbol{x}(t_0), t_0) = 0; \quad \boldsymbol{l}(\boldsymbol{x}(t_f), t_f) = 0 \tag{10.2}$$

Moreover, the system can be subjected to instantaneous inequality constraints

$$\boldsymbol{p}(\boldsymbol{x}(t), \boldsymbol{u}(t), t) \leq 0 \quad \forall t \tag{10.3}$$

or integral constraints (depending only on t_0 and t_f)

$$\int_{t_0}^{t_f} q(\boldsymbol{x}(t), \boldsymbol{u}(t), t)\, dt \leq 0 \tag{10.4}$$

The question is to find the set of the admissible controls $\boldsymbol{u}(t)$ which minimize a technical or economic performance criterion $J(\boldsymbol{u})$

$$J(\boldsymbol{u}) = G(\boldsymbol{x}(t_0), t_0, \boldsymbol{x}(t_f), t_f) + \int_{t_0}^{t_f} F(\boldsymbol{x}(t), \boldsymbol{u}(t), t)\, dt \tag{10.5}$$

G is called the algebraic part of the criterion. F is a functional. Frequently, the initial instant is taken as $t_0 = 0$.

In this general form, this problem makes use of the equality constraints corresponding to the state differential equations, the terminal equality constraints, possibly instantaneous or integral inequality constraints, and m independent functions, which are the controls $\boldsymbol{u}(t)$. The term $G(\boldsymbol{x}(t_0), t_0, \boldsymbol{x}(t_f), t_f)$ represents a contribution of the terminal conditions to the criterion whereas the integral term of Eq. (10.5) represents a time-accumulation contribution.

Several methods allow us to solve this type of problem: variational methods,[1] Pontryagin maximum principle,[2] and Bellman dynamic programming.[3] The books[4–10] propose compared approaches.

10.3 Optimal Control

10.3.1 Variational Methods

The basis of optimal control lies in variational calculus, which provides the fundamental principles in a mathematical framework.[11,12] The way to obtain the solution of the optimal control problem will be presented by first studying the variation of the criterion, then by three progressively more complete methods in continuous-time, that is, Euler conditions, Hamilton-Jacobi theory, and Pontryagin maximum principle. Finally, in discrete time, Bellman optimality principle will be presented.

The variables are divided in two types: state variables x_i $(1 \leq i \leq n)$ and control variables u_j $(1 \leq j \leq m)$, so that the optimal control problem is formulated as follows.

Given a criterion

$$J(\boldsymbol{u}) = G(\boldsymbol{x}(t_0), \boldsymbol{u}(t_0), \boldsymbol{x}(t_f), \boldsymbol{u}(t_f)) + \int_{t_0}^{t_f} F(\boldsymbol{x}(t), \boldsymbol{u}(t), t)\, dt \tag{10.6}$$

determine the optimal control trajectory $\boldsymbol{u}^*(t)$ that minimizes $J(\boldsymbol{u})$

$$\boldsymbol{u}^*(t) = \arg\left\{\min_{\boldsymbol{u}} J(\boldsymbol{u})\right\} \tag{10.7}$$

the state and control variables being subjected to the constraints

$$\text{Model:} \quad \phi_i = \dot{x}_i - f_i(\boldsymbol{x}, \boldsymbol{u}, t) = 0, \quad i = 1, \ldots, n \tag{10.8}$$

$$\text{Initial conditions:} \quad k_j(\boldsymbol{x}(t_0), \boldsymbol{u}(t_0), t_0) = 0, \quad j = 1, \ldots, n_0 \tag{10.9}$$

$$\text{Final conditions:} \quad l_j(\boldsymbol{x}(t_f), \boldsymbol{u}(t_f), t_f) = 0, \quad j = n_0 + 1, \ldots, n_0 + n_1 \leq 2n + 2 \tag{10.10}$$

In the criterion (10.6), the first term G is called the algebraic part and the second term is called the integral part. F is the functional. Note that the ordinary differential equations (10.8) represent the dynamic model of the process. Initial and final conditions, respectively, (10.9), (10.10) are algebraic equations.

10.3.2 Variation of the Criterion

Three general ideas, but different, will be evoked to describe the criterion variation.

- In the most general case, the criterion variation is equal to

$$
\begin{aligned}
\delta J = & \int_{t_0}^{t_f} \left\{ \left[\left(\frac{\partial F}{\partial \boldsymbol{x}} \right)^T - \boldsymbol{\psi}(t)^T \frac{\partial \boldsymbol{f}}{\partial \boldsymbol{x}} \right] \delta \boldsymbol{x} + \left[\left(\frac{\partial F}{\partial \boldsymbol{u}} \right)^T - \boldsymbol{\psi}(t)^T \frac{\partial \boldsymbol{f}}{\partial \boldsymbol{u}} \right] \delta \boldsymbol{u} \right\} dt \\
& + F(\boldsymbol{x}_f, \boldsymbol{u}_f, t_f)\, \delta t_f - F(\boldsymbol{x}_0, \boldsymbol{u}_0, t_0)\, \delta t_0 \\
& + \left[\left(\frac{\partial G}{\partial t_0} \right) \delta t_0 + \left(\frac{\partial G}{\partial \boldsymbol{x}_0} \right) \delta \boldsymbol{x}_0 + \left(\frac{\partial G}{\partial \boldsymbol{u}_0} \right) \delta \boldsymbol{u}_0 \right] \\
& + \left[\left(\frac{\partial G}{\partial t_f} \right) \delta t_f + \left(\frac{\partial G}{\partial \boldsymbol{x}_f} \right) \delta \boldsymbol{x}_f + \left(\frac{\partial G}{\partial \boldsymbol{u}_f} \right) \delta \boldsymbol{u}_f \right] \\
& + \boldsymbol{\psi}(t_f)^T \delta \boldsymbol{x}_f - \boldsymbol{\psi}(t_0)^T \delta \boldsymbol{x}_0 - \int_{t_0}^{t_f} \dot{\boldsymbol{\psi}}(t)^T \, \delta \boldsymbol{x} \, dt
\end{aligned}
\qquad (10.11)
$$

- If, according to Hamilton-Jacobi theory (Section 10.3.5), we furthermore introduce the Hamiltonian H equal to

$$
H(\boldsymbol{x}, \boldsymbol{u}, \boldsymbol{\psi}, t) = -F(\boldsymbol{x}, \boldsymbol{u}, t) + \boldsymbol{\psi}(t)^T \boldsymbol{f}(\boldsymbol{x}, \boldsymbol{u}, t) \qquad (10.12)
$$

the criterion (10.6) becomes

$$
J(\boldsymbol{u}) = G(\boldsymbol{x}(t_0), \boldsymbol{u}(t_0), \boldsymbol{x}(t_f), \boldsymbol{u}(t_f)) + \int_{t_0}^{t_f} \left[\boldsymbol{\psi}(t)^T \boldsymbol{f}(\boldsymbol{x}, \boldsymbol{u}, t) - H(\boldsymbol{x}, \boldsymbol{u}, \boldsymbol{\psi}, t) \right] dt \qquad (10.13)
$$

or

$$
J(\boldsymbol{u}) = G(\boldsymbol{x}(t_0), \boldsymbol{u}(t_0), \boldsymbol{x}(t_f), \boldsymbol{u}(t_f)) + \int_{t_0}^{t_f} \left[\boldsymbol{\psi}(t)^T \dot{\boldsymbol{x}}(t) - H(\boldsymbol{x}, \boldsymbol{u}, \boldsymbol{\psi}, t) \right] dt \qquad (10.14)
$$

Using the integration by parts, the variation of the criterion becomes

$$
\begin{aligned}
\delta J = & \int_{t_0}^{t_f} \left\{ - \left[\left(\frac{\partial H}{\partial \boldsymbol{x}} \right)^T + \dot{\boldsymbol{\psi}}(t)^T \right] \delta \boldsymbol{x} - \left[\left(\frac{\partial H}{\partial \boldsymbol{u}} \right)^T \right] \delta \boldsymbol{u} \right\} dt \\
& + \left[\left(\frac{\partial G}{\partial t_0} \right) + H(\boldsymbol{x}_0, \boldsymbol{u}_0, \boldsymbol{\psi}_0, t_0) \right] \delta t_0 + \left[\left(\frac{\partial G}{\partial \boldsymbol{x}_0} \right) - \boldsymbol{\psi}(t_0)^T \right] \delta \boldsymbol{x}_0 + \left(\frac{\partial G}{\partial \boldsymbol{u}_0} \right) \delta \boldsymbol{u}_0 \\
& + \left[\left(\frac{\partial G}{\partial t_f} \right) - H(\boldsymbol{x}_f, \boldsymbol{u}_f, \boldsymbol{\psi}_f, t_f) \right] \delta t_f + \left[\left(\frac{\partial G}{\partial \boldsymbol{x}_f} \right) + \boldsymbol{\psi}(t_f)^T \right] \delta \boldsymbol{x}_f + \left(\frac{\partial G}{\partial \boldsymbol{u}_f} \right) \delta \boldsymbol{u}_f
\end{aligned}
\qquad (10.15)
$$

This equation, giving the variation of the criterion, is necessary for understanding the origin of Hamilton-Jacobi equations (Section 10.3.5).

10.3.3 Euler Conditions

According to the performance index, the augmented function F^a is defined

$$F^a(\boldsymbol{x},\dot{\boldsymbol{x}},\boldsymbol{u},t) = F(\boldsymbol{x},\boldsymbol{u},t) + \sum_{i=1}^{n} \lambda_i \, \phi_i \tag{10.16}$$

Notice that the function G does not intervene in this augmented function, as G depends only on the terminal conditions. G would only intervene in F^a if the terminal conditions were varying.

The variables are the control vector $u(t)$, the state vector $x(t)$, and the Euler-Lagrange multipliers λ. Euler conditions give

$$\begin{aligned}
&\frac{\partial F^a}{\partial u_j} - \frac{d}{dt}\frac{\partial F^a}{\partial \dot{u}_j} = 0, \quad j = 1,\ldots,m \\
&\frac{\partial F^a}{\partial x_i} - \frac{d}{dt}\frac{\partial F^a}{\partial \dot{x}_i} = 0, \quad i = 1,\ldots,n \\
&\frac{\partial F^a}{\partial \lambda_i} - \frac{d}{dt}\frac{\partial F^a}{\partial \dot{\lambda}_i} = 0, \quad i = 1,\ldots,n
\end{aligned} \tag{10.17}$$

The third group of this system of equations corresponds to the constraints $\phi_i = 0$ which define the dynamic model, thus corresponds to a system of differential equations with respect to the state derivatives $\dot{x}$. The first group is a system of algebraic equations. The second group is a system of differential equations with respect to $\dot{\lambda}$.

If inequality constraints of the type (10.3) or (10.4) are present, the Valentine's method should be used to modify F^a consequently.

On the other hand, the terminal conditions (10.9), (10.10), which are transversality and discontinuity conditions, as well as the conditions relative to the second variations will have to be verified.

The transversality equations (refer to Eq. 10.15) are:

at initial time t_0

$$\begin{aligned}
&\left[-\frac{\partial G}{\partial t_0} + \left(F^a - \boldsymbol{\lambda}^T \dot{\boldsymbol{x}}\right)_0\right] \delta t_0 + \left[-\frac{\partial G}{\partial \boldsymbol{x}_0} + \boldsymbol{\lambda}(t_0)\right]^T \delta \boldsymbol{x}_0 = 0 \\
&\text{with } \left(\frac{\partial \boldsymbol{k}}{\partial t}\right)_0 \delta t_0 + \left(\frac{\partial \boldsymbol{k}}{\partial \boldsymbol{x}}\right)_0 \delta \boldsymbol{x}_0 = 0
\end{aligned} \tag{10.18}$$

at final time t_f

$$\left[\frac{\partial G}{\partial t_f} + \left(F^a - \boldsymbol{\lambda}^T \dot{\boldsymbol{x}}\right)_f\right] \delta t_f + \left[\frac{\partial G}{\partial \boldsymbol{x}_f} + \boldsymbol{\lambda}(t_f)\right]^T \delta \boldsymbol{x}_f = 0$$
$$\text{with} \quad \left(\frac{\partial l}{\partial t}\right)_f \delta t_f + \left(\frac{\partial l}{\partial \boldsymbol{x}}\right)_f \delta \boldsymbol{x}_f = 0 \qquad (10.19)$$

For a fixed final time, which is a frequently met condition, from Eq. (10.19), the following condition results

$$\boldsymbol{\lambda}(t_f) = -\frac{\partial G}{\partial \boldsymbol{x}_f} \qquad (10.20)$$

It must be underlined that, as such, Euler conditions provide only first-order conditions, which is not sufficient for an optimality problem. To determine the nature of the extremum, second-order conditions must be studied.

10.3.4 Weierstrass Condition and Hamiltonian Maximization

To complete the study of the extremum provided Euler conditions, it is necessary to make use of more advanced concepts and the use of Hamiltonian that will be presented later in Section 10.3.5.

Considering the augmented function

$$F^a(\boldsymbol{x},\dot{\boldsymbol{x}},\boldsymbol{u},t) = F(\boldsymbol{x},\boldsymbol{u},t) + \boldsymbol{\lambda}^T[\dot{\boldsymbol{x}} - \boldsymbol{f}(\boldsymbol{x},\boldsymbol{u},t)] \qquad (10.21)$$

the Weierstrass condition relative to second variations is applied to in the neighborhood of the optimum noted x^* obtained for the optimal control u^*, thus

$$F^a(\boldsymbol{x}^*,\dot{\boldsymbol{x}},\boldsymbol{u},t) - F^a(\boldsymbol{x}^*,\dot{\boldsymbol{x}}^*,\boldsymbol{u}^*,t) - (\dot{\boldsymbol{x}} - \dot{\boldsymbol{x}}^*)^T \left(\frac{\partial F^a}{\partial \dot{\boldsymbol{x}}}\right)_* \geq 0 \qquad (10.22)$$

By clarifying these terms and using the constraints

$$\dot{\boldsymbol{x}} = \boldsymbol{f}(\boldsymbol{x}^*,\boldsymbol{u},t)$$
$$\dot{\boldsymbol{x}}^* = \boldsymbol{f}(\boldsymbol{x}^*,\boldsymbol{u}^*,t) \qquad (10.23)$$

the Weierstrass condition is simplified as

$$F(\boldsymbol{x}^*,\boldsymbol{u},t) - F(\boldsymbol{x}^*,\boldsymbol{u}^*,t) - \boldsymbol{\lambda}^T(\boldsymbol{f}(\boldsymbol{x}^*,\boldsymbol{u},t) - \boldsymbol{f}(\boldsymbol{x}^*,\boldsymbol{u}^*,t)) \geq 0$$
$$\Leftrightarrow [\boldsymbol{\lambda}^T \boldsymbol{f}(\boldsymbol{x}^*,\boldsymbol{u}^*,t) - F(\boldsymbol{x}^*,\boldsymbol{u}^*,t)] - [\boldsymbol{\lambda}^T \boldsymbol{f}(\boldsymbol{x}^*,\boldsymbol{u},t) - F(\boldsymbol{x}^*,\boldsymbol{u},t)] \geq 0 \qquad (10.24)$$

in which the expression of the Hamiltonian (setting $\boldsymbol{\lambda} = \boldsymbol{\psi}$) can be recognized as

$$H(\boldsymbol{x}^*,\boldsymbol{u},\boldsymbol{\lambda},t) = -F(\boldsymbol{x}^*,\boldsymbol{u},t) + \boldsymbol{\lambda}^T \boldsymbol{f}(\boldsymbol{x}^*,\boldsymbol{u},t) \qquad (10.25)$$

It results the fundamental conclusion that the optimal control maximizes the Hamiltonian while respecting the constraints

$$H(\boldsymbol{x}^*,\boldsymbol{u}^*,\boldsymbol{\lambda},t) \geq H(\boldsymbol{x}^*,\boldsymbol{u},\boldsymbol{\lambda},t) \tag{10.26}$$

which will be generalized as Pontryagin's maximum principle.

Legendre-Clebsch condition for small variations would have allowed us to obtain the stationarity condition at the optimal trajectory, in the absence of constraints, as

$$\left(\frac{\partial H}{\partial u}\right)_* = 0 \tag{10.27}$$

and

$$\left(\frac{\partial^2 H}{\partial u^2}\right)_* \leq 0 \tag{10.28}$$

10.3.5 Hamilton-Jacobi Conditions and Equation

The Hamiltonian is deduced from the criterion (10.6) and from constraints (10.8); it is equal to

$$H(\boldsymbol{x}(t),\boldsymbol{u}(t),\boldsymbol{\psi}(t),t) = -F(\boldsymbol{x}(t),\boldsymbol{u}(t),t) + \boldsymbol{\psi}^T(t)\boldsymbol{f}(\boldsymbol{x}(t),\boldsymbol{u}(t),t) \tag{10.29}$$

Other authors use the definition of the Hamiltonian with an opposite sign before the functional, that is

$$H(\boldsymbol{x}(t),\boldsymbol{u}(t),\boldsymbol{\psi}(t),t) = F(\boldsymbol{x}(t),\boldsymbol{u}(t),t) + \boldsymbol{\psi}^T(t)\boldsymbol{f}(\boldsymbol{x}(t),\boldsymbol{u}(t),t)$$

which changes nothing, as long as we remain at the level of first-order conditions. However, the sign changes in condition (10.20).

The variation of the criterion has been expressed with respect to the Hamiltonian through Eq. (10.15). The canonical system of Hamilton conditions results as

$$\begin{aligned} \dot{\boldsymbol{x}} &= H_{\psi} \\ \dot{\boldsymbol{\psi}} &= -H_{x} \end{aligned} \tag{10.30}$$

which are equivalent to Euler conditions, to which the following equation must be added

$$H_t = -F_t \tag{10.31}$$

The second equation of Eq. (10.30) is, in fact, a system of equations called the costate equations, and $\boldsymbol{\psi}$ is called the costate or the vector of adjoint variables.

The derivative of the Hamiltonian is equal to

$$\frac{dH}{dt} = H_x^T \dot{\boldsymbol{x}} + H_u^T \dot{\boldsymbol{u}} + H_{\psi}^T \dot{\boldsymbol{\psi}} + H_t = H_u^T \dot{\boldsymbol{u}} + H_t \tag{10.32}$$

If $u(t)$ is an optimal control, one deduces

$$\dot{H} = H_t \tag{10.33}$$

Generally, the concerned physical system is time-invariant so that time does not intervene explicitly in f and also in the functional F, so that Eq. (10.33) becomes

$$\dot{H} = 0 \tag{10.34}$$

In this case, the Hamiltonian is constant along the optimal trajectory.

The transversality conditions (refer to Eq. 10.15) are:

at initial time t_0

$$\begin{aligned} &\left[\frac{\partial G}{\partial t_0} + H(t_0)\right] \delta t_0 + \left[\frac{\partial G}{\partial \boldsymbol{x}_0} - \boldsymbol{\psi}(t_0)\right]^T \delta \boldsymbol{x}_0 + \frac{\partial G}{\partial u_0} \delta u_0 = 0 \\ &\text{with } \left(\frac{\partial \boldsymbol{k}}{\partial t}\right)_0 \delta t_0 + \left(\frac{\partial \boldsymbol{k}}{\partial \boldsymbol{x}}\right)_0 \delta \boldsymbol{x}_0 = 0 \end{aligned} \tag{10.35}$$

at final time t_f

$$\begin{aligned} &\left[\frac{\partial G}{\partial t_f} - H(t_f)\right] \delta t_f + \left[\frac{\partial G}{\partial \boldsymbol{x}_f} + \boldsymbol{\psi}(t_f)\right]^T \delta \boldsymbol{x}_f + \frac{\partial G}{\partial u_f} \delta u_f = 0 \\ &\text{with } \left(\frac{\partial \boldsymbol{l}}{\partial t}\right)_f \delta t_f + \left(\frac{\partial \boldsymbol{l}}{\partial \boldsymbol{x}}\right)_f \delta \boldsymbol{x}_f = 0 \end{aligned} \tag{10.36}$$

It is possible to calculate the variation $\delta \mathcal{J}$ associated with the variation δt and with the trajectory change of $\delta \boldsymbol{x}$, the extremity $\boldsymbol{x}_f$ being fixed, for the time-dependent criterion $\mathcal{J}$ defined by

$$\mathcal{J}(\boldsymbol{x}^*, t) = G(\boldsymbol{x}^*(t_f), t_f) + \int_t^{t_f} F(\boldsymbol{x}^*, \boldsymbol{u}^*, \tau)\, d\tau \tag{10.37}$$

Note that

$$\mathcal{J}(\boldsymbol{x}^*, t_0) = J(\boldsymbol{u}^*) \tag{10.38}$$

The variation of the criterion can be expressed with respect to the Hamiltonian

$$\begin{aligned} \delta \mathcal{J}(\boldsymbol{x}^*, t) &= \mathcal{J}(\boldsymbol{x}^* + \delta \boldsymbol{x}(t), t + \delta t) - \mathcal{J}(\boldsymbol{x}^*, t) \\ &= H(\boldsymbol{x}^*, \boldsymbol{u}^*, \boldsymbol{\psi}, t)\, \delta t - \boldsymbol{\psi}^T(t)\, \delta \boldsymbol{x}(t) \end{aligned} \tag{10.39}$$

The optimal control corresponds to a maximum of the Hamiltonian. Frequently, the control vector is bounded in a domain U defined by $\boldsymbol{u}_{\min}$ and $\boldsymbol{u}_{\max}$. In this case, the condition that the Hamiltonian is maximum can be expressed in two different ways:

- When a constraint u_i is reached, the function H defined by Eq. (10.29) must be a maximum.
- When the control belongs strictly to the inner feasible domain U defined by $\boldsymbol{u}_{\min}$ and $\boldsymbol{u}_{\max}$, not reaching the bounds, the derivative of function H defined by Eq. (10.29) with respect to $\boldsymbol{u}$ is zero

$$\frac{\partial H}{\partial u} = 0 \tag{10.40}$$

This equation provides an implicit equation that allows us to express the optimal control with respect only to variables $\boldsymbol{x}, \boldsymbol{\psi}, t$: $\boldsymbol{u}^* = \boldsymbol{u}^*(\boldsymbol{x}, \boldsymbol{\psi}, t)$, hence the new expression of the criterion

$$\begin{aligned} \delta \mathcal{J}(\boldsymbol{x}^*, t) &= H(\boldsymbol{x}^*, \boldsymbol{u}^*(\boldsymbol{x}, \boldsymbol{\psi}, t), \boldsymbol{\psi}, t)\delta t - \boldsymbol{\psi}^T(t)\, \delta \boldsymbol{x}(t) \\ &= \mathcal{J}_t \delta t + \mathcal{J}_x^T \delta \boldsymbol{x} \end{aligned} \tag{10.41}$$

thus by identification

$$\begin{aligned} \mathcal{J}_t &= H(\boldsymbol{x}^*, \boldsymbol{u}^*(\boldsymbol{x}, \boldsymbol{\psi}, t), \boldsymbol{\psi}, t) \\ \mathcal{J}_x &= -\boldsymbol{\psi}(t) \end{aligned} \tag{10.42}$$

This equation shows that the optimal value of the Hamiltonian is equal to the derivative of criterion (10.37) with respect to time. The Hamilton-Jacobi equation results

$$\mathcal{J}_t - H(\boldsymbol{x}^*, \boldsymbol{u}^*(\boldsymbol{x}, -\mathcal{J}_x, t), -\mathcal{J}_x, t) = 0 \tag{10.43}$$

with boundary condition

$$\mathcal{J}(\boldsymbol{x}_f^*, t_f) = G(\boldsymbol{x}^*(t_f), t_f) \tag{10.44}$$

The Hamilton-Jacobi equation is a first-order partial derivative equation with respect to the sought function $\mathcal{J}$. Its solving is, in general, analytically impossible for a nonlinear system. In the case of a linear system such as Eq. (10.89), its solving is possible and leads to a Riccati differential equation (10.106). Thus it is possible to calculate the optimal control law by state feedback. Recall that the Hamilton-Jacobi equation (10.43) in discrete form corresponds to the Bellman optimality principle in dynamic programming (Section 10.4).

10.3.5.1 Case with constraints on control and state variables

Assume that general constraints of the form

$$g(x(t), u(t), t) = 0 \tag{10.45}$$

are to be respected in the considered problem. In that case, the augmented Hamiltonian is to be considered

$$H(x(t),u(t),\psi(t),t) = -F(x(t),u(t),t) + \psi^T(t) f(x(t),u(t),t) + \mu^T g(x(t),u(t),t) \qquad (10.46)$$

where μ is a vector of additional Lagrange multipliers. The Hamiltonian derivative yields

$$\frac{\partial H}{\partial u} = -\frac{\partial F}{\partial u} + \psi^T(t)\frac{\partial f}{\partial u} + \mu^T \frac{\partial g}{\partial u} = 0 \qquad (10.47)$$

together with Eq. (10.30) as

$$\dot{\psi} = -H_x = F_x - \psi^T(t) f_x - \mu^T g_x \qquad (10.48)$$

Particular cases of Eq. (10.45) are those where the constraints g depend only on the states or where a constraint on the state is valid only for a specific time t_1, such as

$$g(x(t_1),t_1) = 0 \qquad (10.49)$$

called interior-point constraints.[6] In that latter case, the state is continuous, but the Hamiltonian H and the adjoint variables ψ are no more continuous. Noting t_1^- and t_1^+ the times just before and after t_1, given the criterion J, they must verify the following relations

$$\psi^T(t_1^+) = \frac{\partial J}{\partial x(t_1)}; \quad H(t_1^+) = -\frac{\partial J}{\partial t_1} \qquad (10.50)$$

and

$$\psi^T(t_1^+) = \psi^T(t_1^-) - \nu^T \frac{\partial g}{\partial x(t_1)}; \quad H(t_1^+) = H(t_1^-) + \nu^T \frac{\partial g}{\partial t_1} \qquad (10.51)$$

where ν is Lagrange multipliers such that constraints (10.49) are satisfied.

10.3.5.2 Case with terminal constraints

A case frequently encountered in dynamic optimization is the one where terminal constraints are imposed

$$l_j(\boldsymbol{x}(t_f),\boldsymbol{u}(t_f),t_f) = 0 \qquad (10.52)$$

The transversality equation (10.36) becomes

$$\left[\frac{\partial G}{\partial t_f} - H(t_f) + \frac{\partial \boldsymbol{l}^T}{\partial t_f}\boldsymbol{\nu}\right]\delta t_f + \left[\frac{\partial G}{\partial \boldsymbol{x}_f} + \boldsymbol{\psi}(t_f) + \frac{\partial \boldsymbol{l}^T}{\partial \boldsymbol{x}_f}\boldsymbol{\nu}\right]^T \delta \boldsymbol{x}_f + \frac{\partial G}{\partial u_f}\delta u_f = 0 \qquad (10.53)$$

where $\boldsymbol{\nu}$ is a vector of Lagrange parameters. If the final time is fixed, the first term of Eq. (10.53) disappears. If the component $\boldsymbol{x}_i(t_f)$ is fixed at final time, that component disappears in Eq. (10.53).

10.3.6 Maximum Principle

In many articles, authors refer to the Minimum Principle, which simply results from the definition of the Hamiltonian H with an opposite sign of the functional. Comparing to definition (10.29), they define their Hamiltonian as

$$H(\boldsymbol{x}(t),\boldsymbol{u}(t),\boldsymbol{\psi}(t),t) = F(\boldsymbol{x}(t),\boldsymbol{u}(t),t) + \boldsymbol{\psi}^T(t)\boldsymbol{f}(\boldsymbol{x}(t),\boldsymbol{u}(t),t) \tag{10.54}$$

With that definition, the optimal control u^* minimizes the Hamiltonian. Furthermore, the so-called Minimum Principle presented in many articles is no more than Hamilton-Jacobi exposed in the previous section.

In the present section, the presentation follows the original publication by Pontryaguine et al.[2] and completes Hamilton-Jacobi theory. Now, let us examine briefly the Maximum Principle[2] about process optimal control. Pontryagin emphasizes several points:

- An important difference with respect to variational methods is that it is not necessary to consider two close controls in the admissible control domain.
- The control variables u_i are physical, thus they are constrained, for example, $|u_1| \le u_{\max}$, and they belong to a domain U. The admissible controls are piecewise continuous, that is, they are continuous nearly everywhere, except at some instants where they can undergo first-order discontinuities (jump from one value to another).
- Very frequently, the optimal control is composed by piecewise continuous functions: the control jumps from one summit of the polyhedron defined by U to another. These cases of control occupying only extreme positions cannot be solved by classical methods.

The process is described by a system of differential equations

$$\dot{x}^i(t) = f^i(\boldsymbol{x}(t),\boldsymbol{u}(t)), \quad i = 1,\ldots,n \tag{10.55}$$

An admissible control $\boldsymbol{u}$ is sought that transfers the system from point $\boldsymbol{x}_0$ in the phase space to point $\boldsymbol{x}_f$ and minimizes the criterion

$$J = G(\boldsymbol{x}_0, t_0, \boldsymbol{x}_f, t_f) + \int_{t_0}^{t_f} F(\boldsymbol{x}(t),\boldsymbol{u}(t))\, dt \tag{10.56}$$

To the n coordinates x^i in the phase space, we add the coordinate x^0 defined by

$$x^0 = G(\boldsymbol{x}_0, t_0, \boldsymbol{x}(t), t) + \int_{t_0}^{t} F(\boldsymbol{x}(\tau),\boldsymbol{u}(\tau))\, d\tau \tag{10.57}$$

so that if $\boldsymbol{x} = \boldsymbol{x}_f$ then $x^0(t_f) = J$. This notation is that of Pontryaguine et al.[2] The superscript corresponds to the rank i of the coordinate while the subscripts (0 and 1) or (0 and f), according to the authors, are reserved for the terminal conditions. The derivative of x^0 is equal to

$$\frac{dx^0}{dt} = G_x^T f + G_t + F(x(t), u(t)) \tag{10.58}$$

If time intervenes explicitly in the terminal conditions (10.2), or if it is not first fixed, we add the coordinate x^{n+1} to the state,[5] such that

$$\begin{aligned} x^{n+1} &= t \\ \dot{x}^{n+1} &= 1 \end{aligned} \tag{10.59}$$

The complete system of differential equations would then have dimension $n + 2$. In the following, in order not to make the notations cumbersome, we will only consider stationary problems of dimension $n + 1$ in the form

$$\dot{x}^i = f^i(x(t), u(t)), \quad i = 0, \ldots, n \tag{10.60}$$

by deducing f^0 from Eq. (10.57) by derivation (extended notation $\boldsymbol{f}$).

In the phase space of dimension $n + 1$, we define the initial point $\boldsymbol{x}_0$ and a straight line π parallel to the axis x^0 (i.e., the criterion), passing through the final point $\boldsymbol{x}_f$. The optimal control is, among the admissible controls such that the solution $\boldsymbol{x}(t)$, having as the initial condition $\boldsymbol{x}_0$, intersects the line π, the one which minimizes the coordinate x^0 at the intersection point with π.

The costate variables ψ are introduced such that

$$\dot{\boldsymbol{\psi}} = -\boldsymbol{f}_x^T \boldsymbol{\psi} \Leftrightarrow \dot{\psi}_i = -\sum_{j=0}^{n} \frac{\partial f^j(\boldsymbol{x}(t), \boldsymbol{u}(t))}{\partial x^i} \psi_j, \quad i = 0, \ldots, n \tag{10.61}$$

This system admits a unique solution $\boldsymbol{\psi}$ composed of piecewise continuous functions, corresponding to the control $\boldsymbol{u}$ and presenting the same discontinuity points.

In this setting, the Hamiltonian is equal to the scalar product of functions ψ and f

$$H(\boldsymbol{\psi}, \boldsymbol{x}, \boldsymbol{u}) = \boldsymbol{\psi}^T \boldsymbol{f} = \sum_{i=0}^{n} \psi_i f^i, \quad i = 0, \ldots, n \tag{10.62}$$

The systems can be written again in the Hamilton canonical form

$$\begin{aligned} \frac{dx^i}{dt} &= \frac{\partial H}{\partial \psi_i}, \quad i = 0, \ldots, n \\ \frac{d\psi_i}{dt} &= -\frac{\partial H}{\partial x^i}, \quad i = 0, \ldots, n \end{aligned} \tag{10.63}$$

When the solutions $\boldsymbol{x}$ and $\boldsymbol{\psi}$ are fixed, the Hamiltonian depends only on the admissible control $\boldsymbol{u}$, hence the notation

$$\mathcal{M}(\boldsymbol{\psi}, \boldsymbol{x}) = \sup_{u \in U} H(\boldsymbol{\psi}, \boldsymbol{x}, \boldsymbol{u}) \tag{10.64}$$

in order to mean that $\mathcal{M}$ is the maximum of H at fixed $\boldsymbol{x}$ and $\boldsymbol{\psi}$, or further

$$H(\boldsymbol{\psi}^*, \boldsymbol{x}^*, \boldsymbol{u}^*) \geq H(\boldsymbol{\psi}^*, \boldsymbol{x}^*, \boldsymbol{u}^* + \delta \boldsymbol{u}) \quad \forall \delta \boldsymbol{u} \tag{10.65}$$

We consider the admissible controls, defined on $[t_0, t_f]$, to be responding to the previous definition: the trajectory $\boldsymbol{x}(t)$ issued from $\boldsymbol{x}_0$ at t_0 intersects the straight line π at t_f. According to Pontryaguine et al.,[2] the first theorem of the Maximum Principle is expressed as follows:

So that the control $\boldsymbol{u}(t)$ and the trajectory $\boldsymbol{x}(t)$ are optimal, it is necessary that the continuous and nonzero vector, $\boldsymbol{\psi}(t) = [\psi_0(t), \psi_1(t), \ldots, \psi_n(t)]$ satisfying Hamilton canonical system (10.63), is such that:

1. The Hamiltonian $H[\boldsymbol{\psi}(t), \boldsymbol{x}(t), \boldsymbol{u}(t)]$ reaches its maximum at point $\boldsymbol{u} = \boldsymbol{u}(t)\ \forall t \in [t_0, t_f]$, thus

$$H[\boldsymbol{\psi}(t), \boldsymbol{x}(t), \boldsymbol{u}(t)] = \mathcal{M}[\boldsymbol{\psi}(t), \boldsymbol{x}(t)] \tag{10.66}$$

2. At the end-time t_f, the relations

$$\psi_0(t_f) \leq 0; \quad \mathcal{M}[\boldsymbol{\psi}(t_f), \boldsymbol{x}(t_f)] = 0 \tag{10.67}$$

are satisfied.

With Eq. (10.63) and condition (10.66) being verified, the time functions $\psi_0(t)$ and $\mathcal{M}[\boldsymbol{\psi}(t), \boldsymbol{x}(t)]$ are constant. In this case, the relation (10.67) is verified at any instant t included between t_0 and t_f.

10.3.7 Singular Arcs

In optimal control problems, it often occurs for some time intervals that the Maximum Principle does not give an explicit relation between the control and the state and costate variable: this is a singular optimal control problem which yields singular arcs.

Following Lamnabhi-Lagarrigue,[13] an extremal control has a singular arc $[a, b]$ in $[t_0, t_f]$ if and only if $H_u(\boldsymbol{\psi}^*, \boldsymbol{x}^*, \boldsymbol{u}^*) = 0$ and $H_{uu}(\boldsymbol{\psi}^*, \boldsymbol{x}^*, \boldsymbol{u}^*) = 0$, for all $t \in [a, b]$ and whatever ψ^* satisfies the Maximum Principle.

On the arcs corresponding to control constraints, it gives $H_u \neq 0$. Thus a transversality condition must be verified at the junctions between the arcs. Stengel[14] notes that, if a smooth transition of u is possible for some problems, in some cases, it is necessary to perform a Dirac impulse on the control to link the arcs.

Among problem of singular arcs, a frequently encountered case is the one where the Hamiltonian is linear with respect to the control $\boldsymbol{u}$

$$H(\boldsymbol{x}(t), \boldsymbol{\psi}(t), \boldsymbol{u}(t)) = \alpha(\boldsymbol{x}(t), \boldsymbol{\psi}(t), t)\, \boldsymbol{u}(t) \tag{10.68}$$

In that case, the condition

$$\frac{\partial H}{\partial \boldsymbol{u}} = 0 \tag{10.69}$$

depends on the sign of α and does not allow us to determine the control with respect to the state and the adjoint vector. To maximize $H(\boldsymbol{u})$, it results

$$\boldsymbol{u}(t) = \begin{cases} u_{\min} & \text{if } \alpha < 0 \\ \text{nondefined} & \text{if } \alpha = 0 \\ u_{\max} & \text{if } \alpha > 0 \end{cases} \tag{10.70}$$

The case where $\alpha = 0$ on a given time interval $[t_1, t_2]$ corresponds to a singular arc. It must then be imposed that the time derivatives of $\partial H / \partial u$ be zero along the singular arc. For a unique control u, the generalized Legendre-Clebsch conditions, also called Kelley conditions, which must be verified, are

$$(-1)^i \frac{\partial}{\partial u}\left(\frac{d^{2i}}{dt^{2i}} \frac{\partial H}{\partial u}\right) \geq 0, \quad i = 0, 1, \ldots \tag{10.71}$$

so that the singular arc be optimal.

10.3.8 Numerical Issues

In general, the dynamic optimization problem results in a set of two systems of first-order ordinary differential equations

$$\begin{aligned} \dot{\boldsymbol{x}} &= \dot{\boldsymbol{x}}(\boldsymbol{x}, \boldsymbol{\psi}, t) \text{ with } \boldsymbol{x}(t_0) = \boldsymbol{x}_0 \\ \dot{\boldsymbol{\psi}} &= \dot{\boldsymbol{\psi}}(\boldsymbol{x}, \boldsymbol{\psi}, t) \text{ with } \boldsymbol{\psi}(t_f) = \boldsymbol{\psi}_f \end{aligned} \tag{10.72}$$

where t_0 and t_f are initial and final time, respectively. Thus it is a two-point boundary-value problem. A criterion J is to be minimized with respect to a control vector. In general, in particular for nonlinear problems, there is no analytical solution.

The following general strategy is used to solve the two-point boundary-value problem (10.72): an initial vector $\boldsymbol{x}(t)$ or $\boldsymbol{\psi}(t)$ or $\boldsymbol{\psi}(t_0)$ or $\boldsymbol{u}(t)$ is chosen, then by an iterative procedure, the vectors are updated until all equations are respected, including in particular the initial and final conditions.

Different numerical techniques can be used to find the optimal control such as boundary condition iteration, multiple shooting, quasilinearization, invariant embedding, control vector iteration, control vector parameterization,[15,16] and collocation on finite elements or control and state parameterization,[17] iterative dynamic programming.[18–23] For more details, refer to Refs. 7,11,12.

10.4 Dynamic Programming

In Euler, Hamilton-Jacobi, and Pontryagin approaches, the system is defined in continuous time. In the dynamic programming approach by Bellman, the system is defined in discrete time.

10.4.1 Classical Dynamic Programming

Dynamic programming[3,24] has found many applications in chemical engineering,[25,26] in particular for economic optimization problems in refineries, and was frequently developed in the 1960s. Typical examples include the optimization of discontinuous reactors or reactors in series, catalyst replacement or regeneration, the optimization of the counter-current extraction process,[27] the optimal temperature profile of a tubular chemical reactor,[28] and the optimization of a cracking reaction.[29]

Optimality principle[3]

> *A policy is optimal if and only if, whatever the initial state and the initial decision, the decisions remaining to be taken constitute an optimal policy with respect to the state resulting from the first decision.*

Because of the principle of continuity, the optimal final value of the criterion is entirely determined by the initial condition and the number of stages. In fact, it is possible to start from any stage, even from the last one. For this reason, Kaufmann and Cruon[30] express the optimality principle in the following manner:

> *A policy is optimal if, at a given time, whatever the previous decisions, the decisions remaining to be taken constitute an optimal policy with respect to the result of the previous decisions,*

or further,

> *Any subpolicy (from* $\mathbf{x}_i$ *to* $\mathbf{x}_j$*) extracted from an optimal policy (from* $\mathbf{x}_0$ *to* $\mathbf{x}_N$*) is itself optimal from* $\mathbf{x}_i$ *to* $\mathbf{x}_j$*.*

At first, dynamic programming is discussed in the absence of constraints, which could be terminal constraints, constraints at any time (amplitude constraints) on the state $\boldsymbol{x}$ or on the control u, or inequality constraints. Moreover, we assume the absence of discontinuities.

In fact, as this is a numerical and not analytical solution, these particular cases previously mentioned would pose no problem and could be automatically considered.

In continuous form, the problem is the following:

Consider the state equation

$$\dot{\boldsymbol{x}} = f(\boldsymbol{x}, u) \quad \text{with } \boldsymbol{x}(0) = \boldsymbol{x}_0 \tag{10.73}$$

and the performance index to be minimized

$$J(u)=\int_0^{t_f} r(\boldsymbol{x},u)\,dt \tag{10.74}$$

where r represents an income or revenue.
In discrete form, the problem becomes:

Consider the state equation

$$\boldsymbol{x}_{n+1}=\boldsymbol{x}_n+f(\boldsymbol{x}_n,u_n)\Delta t \tag{10.75}$$

with $\Delta t = t_{n+1} - t_n$. The control u_n brings the system from the state $\boldsymbol{x}_n$ to the state $\boldsymbol{x}_{n+1}$ and results in an elementary income $r(\boldsymbol{x}_n, u_n)$ (integrating, in fact, the control period Δt, which will be omitted in the following).
According to the performance index in the integral form, define the performance index or total income at instant N (depending on the initial state $\boldsymbol{x}_0$ and the policy $\mathcal{U}_0^{N-1}$ followed from 0 to $N-1$, bringing from the state $\boldsymbol{x}_0$ to the state $\boldsymbol{x}_N$) as the sum of the elementary incomes $r(\boldsymbol{x}_i, u_i)$

$$J_0=\sum_{i=0}^{N-1} r(\boldsymbol{x}_i,u_i) \tag{10.76}$$

The values of the initial and final states are known

$$\boldsymbol{x}(t_0)=\boldsymbol{x}_0;\;\; \boldsymbol{x}(t_N)=\boldsymbol{x}_N \tag{10.77}$$

If the initial instant is n, note the performance index J_n.
The problem is to find the optimal policy $\mathcal{U}_0^{*,N-1}$ constituted by the succession of controls u_i^* ($i = 0, \ldots, N-1$) minimizing the performance index J_0. The optimal performance index $J^*(\boldsymbol{x}_0, 0)$ is defined as

$$J^*(\boldsymbol{x}_0,0)=\min_{u_i} J_0=\min_{u_i}\sum_{i=0}^{N-1} r(\boldsymbol{x}_i,u_i) \tag{10.78}$$

This performance index bears on the totality of the N stages and depends on the starting point $\boldsymbol{x}_0$. In fact, the optimality principle can be applied from any instant n, to which corresponds the optimal performance index $J^*(\boldsymbol{x}_n, n)$.

From the optimality principle, the following recurrent algorithm of search of the optimal policy is derived

$$J^*(\boldsymbol{x}_n,n)=\min_{u_n}\left[r(\boldsymbol{x}_n,u_n)+J^*(\boldsymbol{x}_n+f(\boldsymbol{x}_n,u_n),n+1)\right] \tag{10.79}$$

which allows us to calculate the series $J^*(\boldsymbol{x}_n, n)$, $J^*(\boldsymbol{x}_{n-1}, n-1)$, ..., $J^*(\boldsymbol{x}_0, 0)$ from the final state $\boldsymbol{x}_N$.

If the final state is free, choose $J^*(\boldsymbol{x}_N, N) = 0$. In the case where it is constrained, the last input u^*_{N-1} is calculated so as to satisfy the constraint.

The algorithm (10.79) could be written as

$$\begin{aligned} J^*(\boldsymbol{x}_n, n) &= \min_{u_n}\left[r(\boldsymbol{x}_n, u_n) + \min_{u_{n+1}}[r(\boldsymbol{x}_{n+1}, u_{n+1}) + J^*(\boldsymbol{x}_{n+2}, n+2)]\right] \\ &= \min_{u_n}\left[r(\boldsymbol{x}_n, u_n) + \min_{u_{n+1}}[r(\boldsymbol{x}_{n+1}, u_{n+1}) + \cdots]\right] \end{aligned} \tag{10.80}$$

However, a difficulty resides frequently in the formulation of a given problem in an adequate form for the solution by means of dynamic programming and, with the actual progress of numerical calculation and nonlinear constrained optimization methods, the latter are nowadays more employed. A variant of dynamic programming[31] called iterative dynamic programming can often provide good results with a lighter computational effort.[32–34]

10.4.2 Hamilton-Jacobi-Bellman Equation

Given the initial state $\boldsymbol{x}_0$ at time t_0, considering the state $\boldsymbol{x}$ and the control u, the optimal trajectory corresponds to the couple $(\boldsymbol{x}, u)$ such that

$$J^*(\boldsymbol{x}_0, t_0) = \min_{u(t)} J(\boldsymbol{x}_0, u, t_0) \tag{10.81}$$

thus the optimal criterion does not depend on the control u.

In an interval $[t, t + \Delta t]$, the Bellman optimality principle as given in the recurrent Eq. (10.79) can be formulated as

$$J^*(\boldsymbol{x}(t), t) = \min_{u(t)}\left\{\int_t^{t+\Delta t} r(\boldsymbol{x}, u, \tau)d\tau + J^*(\boldsymbol{x}(t+\Delta t), t+\Delta t)\right\} \tag{10.82}$$

This can be expressed in continuous form as a Taylor series expansion in the neighborhood of the state $\boldsymbol{x}(t)$ and time t

$$J^*(\boldsymbol{x}(t), t) = \min_{u(t)}\left\{\begin{array}{l} r(\boldsymbol{x}, u, t)\,\Delta t + J^*(\boldsymbol{x}(t), t) + \dfrac{\partial J^*}{\partial t}\Delta t \\ + \left(\dfrac{\partial J^*}{\partial \boldsymbol{x}}\right)^T f(\boldsymbol{x}, u, t)\,\Delta t + 0(\Delta t) \end{array}\right\} \tag{10.83}$$

Taking the limit when $\Delta t \rightarrow 0$ results in the Hamilton-Jacobi-Bellman equation

$$-\frac{\partial J^*}{\partial t} = \min_{u(t)} \left\{ r(\boldsymbol{x},u,t) + \left(\frac{\partial J^*}{\partial \boldsymbol{x}}\right)^T f(\boldsymbol{x},u,t) \right\} \tag{10.84}$$

As the optimal criterion does not depend on control u, it yields $J^*(\boldsymbol{x}(t_f), t_f) = W(\boldsymbol{x}(t_f))$, which gives the boundary condition for the Hamilton-Jacobi-Bellman equation (10.84)

$$J^*(\boldsymbol{x}, t_f) = W(\boldsymbol{x}), \quad \forall \boldsymbol{x} \tag{10.85}$$

The solution of Eq. (10.84) is the optimal control law

$$u^* = g\left(\frac{\partial J^*}{\partial \boldsymbol{x}}, \boldsymbol{x}, t\right) \tag{10.86}$$

which, when introduced into Eq. (10.84), gives

$$-\frac{\partial J^*}{\partial t} = r(\boldsymbol{x},g,t) + \left(\frac{\partial J^*}{\partial \boldsymbol{x}}\right)^T f(\boldsymbol{x},g,t) \tag{10.87}$$

whose solution is $J^*(\boldsymbol{x}, t)$ subject to the boundary condition (10.85). Eq. (10.87) should be compared to the Hamilton-Jacobi equation (10.43). Then, the gradient $\partial J^*/\partial \boldsymbol{x}$ should be calculated and returned in Eq. (10.86), which gives the optimal state-feedback control law

$$u^* = g\left(\frac{\partial J^*}{\partial \boldsymbol{x}}, \boldsymbol{x}, t\right) = h(\boldsymbol{x}, t) \tag{10.88}$$

This corresponds to a closed-loop optimal control law given as a state feedback.

10.5 Linear Quadratic Control

Numerous publications deal with linear optimal control, are, in particular, the books by Refs. 1,6,35–41, and more recently, in robust control.[42] Furthermore, among reference papers, cite Refs. 43–45. Even, Pannocchia et al.[46] proposed constrained LQ control to replace the classical proportional-integral-derivative (PID) control for which they see no advantage. LQ control is presented here in the previously discussed general framework of optimal control (see Ref. 11 for more description with detailed examples).

10.5.1 Continuous-Time Linear Quadratic Control

In continuous time, the system is represented in the state space by the deterministic linear model

$$\begin{cases} \dot{\boldsymbol{x}}(t) = \boldsymbol{A}\,\boldsymbol{x}(t) + \boldsymbol{B}\,\boldsymbol{u}(t) \\ \boldsymbol{y}(t) = \boldsymbol{C}\,\boldsymbol{x}(t) \end{cases} \tag{10.89}$$

where $\boldsymbol{u}$ is the control vector of dimension n_u, $\boldsymbol{x}$ the state vector of dimension n, and $\boldsymbol{y}$ the output vector of dimension n_y. $\boldsymbol{A}$, $\boldsymbol{B}$, $\boldsymbol{C}$ are matrices of respective sizes $n \times n$, $n \times n_u$, $n_y \times n$.

The control $\boldsymbol{u}$ must minimize the classical quadratic criterion

$$J = 0.5\,\boldsymbol{x}^T(t_f)\,\boldsymbol{Q}_f\,\boldsymbol{x}(t_f) + 0.5\int_{t_0}^{t_f}[\boldsymbol{x}^T(t)\,\boldsymbol{Q}\,\boldsymbol{x}(t) + \boldsymbol{u}^T(t)\,\boldsymbol{R}\,\boldsymbol{u}(t)]dt \tag{10.90}$$

where matrices $\boldsymbol{Q}_f$, $\boldsymbol{Q}$ are symmetrical semipositive definite, whereas $\boldsymbol{R}$ is symmetrical positive definite. This criterion tends to bring the state $\boldsymbol{x}$ toward 0. The first part of the criterion represents the performance whereas the second part is the energy spent to bring the state toward zero. Several cases can be distinguished with respect to the criterion according to whether the final time t_f is fixed or free and the final state is fixed or free.[1,6]

Other criteria have been derived from the original criterion by replacing the state $\boldsymbol{x}$ by $\boldsymbol{z}$ with $\boldsymbol{z} = \boldsymbol{M}\boldsymbol{x}$, where $\boldsymbol{z}$ represents a linear combination of the states, for example, a measurement, or the output if $\boldsymbol{M} = \boldsymbol{C}$. The problem is then the regulation of $\boldsymbol{z}$. It is possible to incorporate the tracking of a reference trajectory $\boldsymbol{z}^r = \boldsymbol{M}\boldsymbol{x}^r$ by replacing the state $\boldsymbol{x}$ with the tracking error $(\boldsymbol{z}^r - \boldsymbol{z})$.

Thus the most general criterion can be considered, which takes into account the different previous cases

$$\begin{aligned} J &= 0.5(\boldsymbol{z}^r - \boldsymbol{z})^T(t_f)\,\boldsymbol{Q}_f(\boldsymbol{z}^r - \boldsymbol{z})(t_f) \\ &\quad + 0.5\int_{t_0}^{t_f}[(\boldsymbol{z}^r - \boldsymbol{z})^T(t)\,\boldsymbol{Q}(\boldsymbol{z}^r - \boldsymbol{z})(t) + \boldsymbol{u}^T(t)\,\boldsymbol{R}\,\boldsymbol{u}(t)]dt \end{aligned} \tag{10.91}$$

or

$$\begin{aligned} J &= 0.5(\boldsymbol{x}^r - \boldsymbol{x})^T(t_f)\,\boldsymbol{M}^T\,\boldsymbol{Q}_f\,\boldsymbol{M}(\boldsymbol{x}^r - \boldsymbol{x})(t_f) \\ &\quad + 0.5\int_{t_0}^{t_f}[(\boldsymbol{x}^r - \boldsymbol{x})^T(t)\,\boldsymbol{M}^T\,\boldsymbol{Q}\,\boldsymbol{M}(\boldsymbol{x}^r - \boldsymbol{x})(t) + \boldsymbol{u}^T(t)\,\boldsymbol{R}\,\boldsymbol{u}(t)]dt \end{aligned} \tag{10.92}$$

The matrices $\boldsymbol{Q}_f$, $\boldsymbol{Q}$, $\boldsymbol{R}$ must have dimensions adapted to the retained criterion.

According to the techniques of variational calculation applied to optimal control, in particular Hamilton-Jacobi method, let us introduce the Hamiltonian as in Eq. (10.29)

$$H = -0.5[(\boldsymbol{x}^r - \boldsymbol{x})^T(t)\,\boldsymbol{M}^T\,\boldsymbol{Q}\,\boldsymbol{M}(\boldsymbol{x}^r - \boldsymbol{x})(t) + \boldsymbol{u}^T(t)\,\boldsymbol{R}\,\boldsymbol{u}(t)] + \boldsymbol{\psi}(t)^T[\boldsymbol{A}\,\boldsymbol{x} + \boldsymbol{B}\,\boldsymbol{u}] \tag{10.93}$$

The Hamilton canonical equations (10.30) provide, besides the state Eq. (10.89), the derivative of the costate vector

$$-\frac{\partial H}{\partial x} = \dot{\boldsymbol{\psi}}(t) = -\boldsymbol{M}^T\,\boldsymbol{Q}\,\boldsymbol{M}(\boldsymbol{x}^r - \boldsymbol{x}) - \boldsymbol{A}^T\,\boldsymbol{\psi} \tag{10.94}$$

with the final transversality condition (assuming that the state $\boldsymbol{x}_f$ is free, i.e., not constrained)

$$\boldsymbol{\psi}(t_f) = \boldsymbol{M}^T\,\boldsymbol{Q}_f\,\boldsymbol{M}(\boldsymbol{x}^r - \boldsymbol{x})_f \tag{10.95}$$

Moreover, in the absence of constraints, the condition of maximization of the Hamiltonian with respect to $\boldsymbol{u}$ gives

$$\frac{\partial H}{\partial \boldsymbol{u}} = 0 = -\boldsymbol{R}\,\boldsymbol{u} + \boldsymbol{B}^T\,\boldsymbol{\psi} \tag{10.96}$$

hence the optimal control

$$\boldsymbol{u}(t) = \boldsymbol{R}^{-1}\,\boldsymbol{B}^T\,\boldsymbol{\psi}(t) \tag{10.97}$$

Moreover, it can be noticed that $\boldsymbol{H}_{uu} = -\boldsymbol{R}$, which is thus symmetric negative, so that the Hamiltonian is maximum at the optimal control.

Gathering all these results, the system to be solved becomes a two-point boundary-value problem

$$\begin{aligned}
&\begin{bmatrix} \dot{\boldsymbol{x}}(t) \\ \dot{\boldsymbol{\psi}}(t) \end{bmatrix} = \begin{bmatrix} \boldsymbol{A} & \boldsymbol{B}\,\boldsymbol{R}^{-1}\,\boldsymbol{B}^T \\ \boldsymbol{M}^T\,\boldsymbol{Q}\,\boldsymbol{M} & -\boldsymbol{A}^T \end{bmatrix} \begin{bmatrix} \boldsymbol{x}(t) \\ \boldsymbol{\psi}(t) \end{bmatrix} - \begin{bmatrix} 0 & 0 \\ \boldsymbol{M}^T\,\boldsymbol{Q}\,\boldsymbol{M} & 0 \end{bmatrix} \begin{bmatrix} \boldsymbol{x}^r(t) \\ 0 \end{bmatrix} \\
&\boldsymbol{x}(t_0) = \boldsymbol{x}_0 \\
&\boldsymbol{\psi}(t_f) = \boldsymbol{M}^T\,\boldsymbol{Q}_f\,\boldsymbol{M}(\boldsymbol{x}^r - \boldsymbol{x})_f
\end{aligned} \tag{10.98}$$

10.5.1.1 *Regulation case:* $\mathbf{x}^r = 0$

A classical approach consists of introducing the transition matrix corresponding to the previous differential system, which can be partitioned such that

$$\begin{bmatrix} \boldsymbol{x}(\tau) \\ \boldsymbol{\psi}(\tau) \end{bmatrix} = \begin{bmatrix} \boldsymbol{\Phi}_{xx}(\tau,t) & \boldsymbol{\Phi}_{x\psi}(\tau,t) \\ \boldsymbol{\Phi}_{\psi x}(\tau,t) & \boldsymbol{\Phi}_{\psi\psi}(\tau,t) \end{bmatrix} \begin{bmatrix} \boldsymbol{x}(t) \\ \boldsymbol{\psi}(t) \end{bmatrix}, \quad \tau \in [t, t_f] \tag{10.99}$$

This equation can be used at the final instant $\tau = t_f$ (where $\boldsymbol{\psi}(t_f)$ is known) thus

$$\begin{aligned}
\boldsymbol{x}(t_f) &= \boldsymbol{\Phi}_{xx}(t_f,t)\,\boldsymbol{x}(t) + \boldsymbol{\Phi}_{x\psi}(t_f,t)\,\boldsymbol{\psi}(t) \\
\boldsymbol{\psi}(t_f) &= \boldsymbol{\Phi}_{\psi x}(t_f,t)\,\boldsymbol{x}(t) + \boldsymbol{\Phi}_{\psi\psi}(t_f,t)\,\boldsymbol{\psi}(t)
\end{aligned} \tag{10.100}$$

hence

$$\begin{aligned}
\boldsymbol{\psi}(t) = &-[\boldsymbol{\Phi}_{\psi\psi}(t_f,t) + \boldsymbol{M}^T\,\boldsymbol{Q}_f\,\boldsymbol{M}\boldsymbol{\Phi}_{x\psi}(t_f,t)]^{-1} \\
&\times [\boldsymbol{\Phi}_{\psi x}(t_f,t) + \boldsymbol{M}^T\,\boldsymbol{Q}_f\,\boldsymbol{M}\boldsymbol{\Phi}_{xx}(t_f,t)]\,\boldsymbol{x}(t)
\end{aligned} \tag{10.101}$$

a relation which can be denoted by

$$\boldsymbol{\psi}(t) = -\boldsymbol{M}^T\,\boldsymbol{S}(t)\,\boldsymbol{M}\,\boldsymbol{x}(t) = -\boldsymbol{P}_c(t)\,\boldsymbol{x}(t) \tag{10.102}$$

both to express the proportionality and to verify the terminal condition (at t_f), $\boldsymbol{M}$ being any constant matrix. The subscript c of $\boldsymbol{P}_c$ means that we are treating the control problem (to be compared with $\boldsymbol{P}_f$ later used for Kalman filtering). The relation (10.102) is also called a backward sweep solution.[7]

This relation joined to the optimal control expression gives

$$u^*(t) = -R^{-1} B^T M^T S(t) M \; x(t) = -R^{-1} B^T P_c(t) x(t) \tag{10.103}$$

which shows that this is a state feedback control. By setting $\boldsymbol{M} = \boldsymbol{I}$, we find the classical formula which equals $\boldsymbol{P}_c(t)$ and $\boldsymbol{S}(t)$

$$u^*(t) = -R^{-1} B^T S(t) x(t) \tag{10.104}$$

In fact, the matrix $\boldsymbol{P}_c(t)$ can be calculated directly. Use the relation

$$\psi(t) = -P_c(t) x(t) \tag{10.105}$$

inside the system (10.98). The continuous differential Riccati equation results

$$\begin{aligned} \dot{P}_c(t) &= -P_c(t) A - A^T P_c(t) + P_c(t) B R^{-1} B^T P_c(t) - M^T Q M \\ \text{with} \quad P_c(t_f) &= M^T Q_f M \end{aligned} \tag{10.106}$$

where the matrix $\boldsymbol{P}_c(t)$ is symmetrical semipositive definite. Knowing the solution of this differential equation, the optimal control law can be calculated

$$u^*(t) = -R^{-1} B^T P_c(t) x(t) = -K_c(t) x(t) \tag{10.107}$$

Notice that the differential Riccati equation (10.106), being known by its final condition, can be integrated backwards to deduce $\boldsymbol{P}_c(t_0)$, which will allow us to exploit the optimal control law in relation to the differential system (10.89).

If the horizon t_f is infinite, the control law is

$$u^*(t) = -R^{-1} B^T P_c x(t) \tag{10.108}$$

where the matrix $\boldsymbol{P}_c$ is the solution of the algebraic Riccati equation

$$P_c A + A^T P_c - P_c B R^{-1} B^T P_c + M^T Q M = 0 \tag{10.109}$$

which is the steady-state form of the differential Riccati equation (10.106). In this case, the condition $\boldsymbol{P}_c(t_f) = \boldsymbol{M}^T \boldsymbol{Q}_f \boldsymbol{M}$ disappears. Noting $\boldsymbol{P}_{c,\infty}$ the solution of the algebraic Riccati equation, the constant gain results

$$K_{c,\infty} = R^{-1} B^T P_{c,\infty} \tag{10.110}$$

hence the constant state-variable feedback

$$u(t) = -K_{c,\infty} x(t) \tag{10.111}$$

so that the plant dynamics is

$$\dot{x}(t) = (A - BK_{c,\infty})x(t) \quad (10.112)$$

Noting $\sqrt{Q}$ ("square root" of Q) the matrix such that $Q = \sqrt{Q}^T \sqrt{Q}$, the stabilization of the system is guaranteed if the pair $(\sqrt{Q}, A)$ is observable and the pair (Q, A) is stabilizable, Riccati equation (10.109) possesses a unique solution and the closed-loop plant $(A - BK_{c,\infty})$ is asymptotically stable. Compared to the variable gain $K_c(t)$, the constant gain $K_{c,\infty}$ is suboptimal, but when t_f becomes large the gain $K_c(t)$ tends toward $K_{c,\infty}$.

Different methods have been published to solve Eq. (10.109), which poses serious numerical problems. A solution can be to integrate backward the differential Riccati equation (10.106) until a stationary solution is obtained. Another solution, which is numerically robust and is based on a Schur decomposition method, is proposed by Arnold and Laub[47] and Laub.[48]

A matrix A of dimension $(2n \times 2n)$ is called Hamiltonian if $J^{-1} A^T J = -A$ or $J = -A^{-T} J A$, where J is equal to: $\begin{bmatrix} 0 & I \\ I & 0 \end{bmatrix}$. An important property[48] of Hamiltonian matrices is that if λ is an eigenvalue of a Hamiltonian matrix, $-\lambda$ is also an eigenvalue with the same multiplicity.

Consider the Hamiltonian matrix H

$$H = \begin{bmatrix} A & -B R^{-1} B^T \\ -M^T Q M & -A^T \end{bmatrix} \quad (10.113)$$

whose eigenvalues and eigenvectors are sought. Order the matrix U of the eigenvectors of the Hamiltonian matrix of dimension $2n \times 2n$ in such a way that the first n columns are the eigenvectors corresponding to the stable eigenvalues (negative real or complex with a negative real part), in the form

$$U = \begin{bmatrix} U_{11} & U_{12} \\ U_{21} & U_{22} \end{bmatrix} \quad (10.114)$$

where the blocks U_{ij} have dimension $n \times n$. The solution to the algebraic Riccati equation (10.109) is then

$$P_c = U_{21} U_{11}^{-1} \quad (10.115)$$

Note that the stationary solution to this problem can also be obtained by iterative methods.[47]

The stable eigenvalues of the Hamiltonian matrix H are the poles of the optimal closed-loop system

$$\dot{x}(t) = (A - BK_c)x(t) \quad (10.116)$$

10.5.1.2 Tracking case: $\mathbf{x}^r \neq 0$

In the presence of the tracking term $\boldsymbol{x}^r$, the differential system (10.98) is not homogeneous anymore, and it is necessary to add a term to Eq. (10.99), as

$$\begin{bmatrix} \boldsymbol{x}(\tau) \\ \boldsymbol{\psi}(\tau) \end{bmatrix} = \begin{bmatrix} \boldsymbol{\Phi}_{xx}(\tau,t) & \boldsymbol{\Phi}_{x\psi}(\tau,t) \\ \boldsymbol{\Phi}_{\psi x}(\tau,t) & \boldsymbol{\Phi}_{\psi\psi}(\tau,t) \end{bmatrix} \begin{bmatrix} \boldsymbol{x}(t) \\ \boldsymbol{\psi}(t) \end{bmatrix} + \begin{bmatrix} \boldsymbol{g}_x(\tau,t) \\ \boldsymbol{g}_\psi(\tau,t) \end{bmatrix} \tag{10.117}$$

In the same manner as previously, at the final instant $\tau = t_{\mathrm{f}}$, we obtain

$$\begin{aligned} \boldsymbol{x}(t_{\mathrm{f}}) &= \boldsymbol{\Phi}_{xx}(t_{\mathrm{f}},t)\,\boldsymbol{x}(t) + \boldsymbol{\Phi}_{x\psi}(t_{\mathrm{f}},t)\boldsymbol{\psi}(t) + \boldsymbol{g}_x(t_{\mathrm{f}},t) \\ \boldsymbol{\psi}(t_{\mathrm{f}}) &= \boldsymbol{\Phi}_{\psi x}(t_{\mathrm{f}},t)\,\boldsymbol{x}(t) + \boldsymbol{\Phi}_{\psi\psi}(t_{\mathrm{f}},t)\boldsymbol{\psi}(t) + \boldsymbol{g}_\psi(t_{\mathrm{f}},t) \end{aligned} \tag{10.118}$$

hence

$$\begin{aligned} \boldsymbol{\psi}(t) \; &= -\left[\boldsymbol{\Phi}_{\psi\psi}(t_{\mathrm{f}},t) + \boldsymbol{M}^T\,\boldsymbol{Q}_{\mathrm{f}}\,\boldsymbol{M}\,\boldsymbol{\Phi}_{x\psi}(t_{\mathrm{f}},t)\right]^{-1} \\ &\quad \times \left\{\left[\boldsymbol{\Phi}_{\psi x}(t_{\mathrm{f}},t) + \boldsymbol{M}^T\,\boldsymbol{Q}_{\mathrm{f}}\,\boldsymbol{M}\boldsymbol{\Phi}_{xx}(t_{\mathrm{f}},t)\right]\boldsymbol{x}(t) \right. \\ &\quad \left. + \boldsymbol{M}^T\,\boldsymbol{Q}_{\mathrm{f}}\,\boldsymbol{M}(\boldsymbol{g}_x(t_{\mathrm{f}},t) - x_{r,f}) + \boldsymbol{g}_\psi(t_{\mathrm{f}},t)\right\} \end{aligned} \tag{10.119}$$

giving an expression in the form

$$\boldsymbol{\psi}(t) = -\boldsymbol{P}_{\mathrm{c}}(t)\,\boldsymbol{x}(t) + \boldsymbol{s}(t) \tag{10.120}$$

By introducing this expression in Eq. (10.94), we again get the differential Riccati equation (10.106) whose matrix $\boldsymbol{P}_{\mathrm{c}}$ is a solution, with the same terminal condition. Moreover, we obtain the differential equation giving the vector $\boldsymbol{s}$

$$\begin{aligned} \dot{\boldsymbol{s}}(t) &= \left[\boldsymbol{P}_{\mathrm{c}}(t)\,\boldsymbol{B}\,\boldsymbol{R}^{-1}\,\boldsymbol{B}^T - \boldsymbol{A}^T\right]\boldsymbol{s}(t) - \boldsymbol{M}^T\,\boldsymbol{Q}\,\boldsymbol{M}\,\boldsymbol{x}^r \\ &\text{with } \; \boldsymbol{s}(t_{\mathrm{f}}) = \boldsymbol{M}^T\,\boldsymbol{Q}_{\mathrm{f}}\,\boldsymbol{M}\,\boldsymbol{x}^r_{\mathrm{f}} \end{aligned} \tag{10.121}$$

This equation is often termed feedforward. Like the differential Riccati equation (10.106), it must be integrated backward in time, so that both equations must be integrated offline before implementing the control and require the knowledge of the future reference trajectory, thus posing a problem for actual online implementation, which will lead to the suboptimal solutions to avoid this difficulty.[41] Knowing the solutions of this differential equation and of the differential Riccati equation (10.106), the optimal control law can be calculated and applied

$$\boldsymbol{u}^*(t) = -\boldsymbol{R}^{-1}\,\boldsymbol{B}^T\,\boldsymbol{P}_{\mathrm{c}}(t)\,\boldsymbol{x}(t) + \boldsymbol{R}^{-1}\,\boldsymbol{B}^T\;\;\boldsymbol{s}(t) = \boldsymbol{u}_{\mathrm{fb}}(t) + \boldsymbol{u}_{\mathrm{ff}}(t) \tag{10.122}$$

In this form, $\boldsymbol{u}_{\mathrm{fb}}(t)$ represents a state feedback control (term in $x(t)$), as the gain $\boldsymbol{K}_{\mathrm{c}}$ depends at each instant on the solution of the Riccati equation, and $\boldsymbol{u}_{\mathrm{ff}}(t)$ represents a feedforward control (term in $s(t)$). This structure is visible in Fig. 10.3. The practical use of the LQ

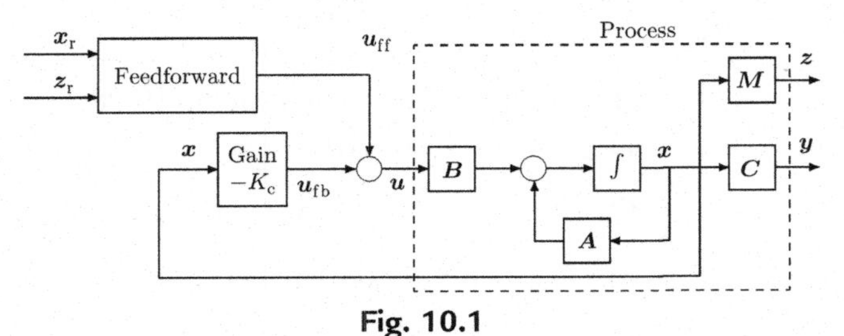

Fig. 10.1
Structure of linear quadratic control.

regulator is thus decomposed into two parts, according to a hierarchical manner, first offline calculation of the optimal gain, then actual control using feedback (Fig. 10.1).

In the same manner as in regulation, when the horizon is infinite, $\boldsymbol{P}_c$ is solution of the algebraic Riccati equation (10.109) and $\boldsymbol{s}$ is solution of the algebraic equation

$$\boldsymbol{s} = [\boldsymbol{P}_c \boldsymbol{B} \boldsymbol{R}^{-1} \boldsymbol{B}^T - \boldsymbol{A}^T]^{-1} \boldsymbol{M}^T \boldsymbol{Q} \boldsymbol{M} \boldsymbol{x}^r \tag{10.123}$$

This solution is frequently adopted in actual practice.

Lin[49] recommends choosing, as a first approach, diagonal criterion weighting matrices with their diagonal terms equal to

$$q_i = 1/(z_i)^2_{\max}, \quad r_i = 1/(u_i)^2_{\max} \tag{10.124}$$

in order to realize a compromise between the input variations and the performance with respect to the output, while Anderson and Moore[37] propose taking

$$q_i = 1 \Big/ \int_0^\infty z_i^2 dt, \quad r_i = 1 \Big/ \int_0^\infty u_i^2 dt \tag{10.125}$$

It frequently happens, as in the simple linear example previously treated, that some components of the control vector are bounded

$$|u_i| \leq u_{i,\max} \tag{10.126}$$

In this case, the optimal control is equal to

$$\boldsymbol{u}^* = \operatorname{sat}(\boldsymbol{R}^{-1} \boldsymbol{B}^T \boldsymbol{\psi}) \tag{10.127}$$

defining the saturation function by

$$\operatorname{sat}(u_i) = \begin{cases} u_i & \text{if } |u_i| \leq u_{i,\max} \\ u_{i,\max} & \text{if } |u_i| \geq u_{i,\max} \end{cases} \tag{10.128}$$

Linear quadratic control of a chemical reactor

A continuous perfectly stirred chemical reactor[11] is modeled with four states (x_1, concentration of A; x_2, reactor temperature; x_3, jacket temperature; and x_4, liquid volume), two manipulated inputs corresponding to valve positions that allow to control the inlet temperature in the jacket $T_{j,\,in}$ and the feed concentration $C_{A,\,f}$, respectively, two controlled outputs, the concentration of A and the reactor temperature, respectively. The nonlinear model of the reactor is

$$\begin{aligned}
\dot{x}_1 &= \frac{F_f}{x_4}(C_{A,f} - x_1) - k\,x_1 \\
\dot{x}_2 &= \frac{F_f}{x_4}(T_f - x_2) - \frac{\Delta H\,k\,x_1}{\rho\,C_p} - \frac{UA(x_2 - x_3)}{\rho\,C_p\,x_4} \\
\dot{x}_3 &= \frac{F_j}{V_j}(T_{j,in} - x_2) + \frac{UA(x_2 - x_3)}{\rho_j\,C_{pj}\,V_j} \\
\dot{x}_4 &= F_f - F_{sp} + K_r\,(V_{sp} - x_4) y_1 = x_1 \\
y_2 &= x_2
\end{aligned} \tag{10.129}$$

with

$$k = k_0 \exp\left(-\frac{E_a}{R\,x_2}\right);\quad C_{A,f} = u_2\,C_{A,f};\quad T_{j,in} = u_1\,T_{hot} + (1 - u_1)\,T_{cold} \tag{10.130}$$

where T_{cold} and T_{hot} are the temperatures of two cold and hot heat exchangers assumed constant. The volume x_4 is controlled independently by a proportional controller. Both manipulated inputs are bounded in the interval [0, 1].

The physical parameters of the chemical reactor are given in Table 10.1.

The linearized model of the chemical reactor calculated at initial time and maintained constant during the study is given in the state space by

$$\boldsymbol{A} = \begin{bmatrix} -0.0010 & -0.0254 & 0 & -0.2052 \\ 0 & -0.0036 & 0.0040 & 0.0048 \\ 0 & -0.4571 & -0.0429 & 0 \\ 0 & 0 & 0 & -0.05 \end{bmatrix} \quad \boldsymbol{B} = \begin{bmatrix} 0 & 0.78 \\ 0 & 0 \\ 40 & 0 \\ 0 & 0 \end{bmatrix} \tag{10.131}$$

$$\boldsymbol{C} = \begin{bmatrix} 1 & 0 & 0 & 0 \\ 0 & 1 & 0 & 0 \end{bmatrix} \quad \boldsymbol{D} = \begin{bmatrix} 0 & 0 \\ 0 & 0 \end{bmatrix}$$

Table 10.1 Initial variables and main parameters of the CSTR

Parameter	Value
Flow rate of the feed	$F_f = F_{sp} = 3 \times 10^{-4}$ m^3 s^{-1}
Concentration of reactant A in the feed	$C_{A,f} = 3900$ mol m^{-3}
Temperature of the feed	$T_f = 295$ K
Volume of reactor	$V(t=0) = V_{sp} = 1.5$ m^3
Kinetic constant	$k_0 = 2 \times 10^8$ s^{-1}
Activation energy	$E = 7 \times 10^4$ J mol^{-1}
Heat of reaction	$\Delta H = -7 \times 10^4$ J mol^{-1}
Density of reactor contents	$\rho = 1000$ kg m^{-3}
Heat capacity of reactor contents	$C_p = 3000$ J kg^{-1} K^{-1}
Temperature of the cold heat exchanger	$T_c = 280$ K
Temperature of hot heat exchanger	$T_h = 360$ K
Flow rate of the heat-conducting fluid	$F_j = 5 \times 10^{-2}$ m^3 s^{-1}
Volume of jacket	$V_j = 0.1$ m^3
Heat-transfer coefficient between the jacket and the reactor contents	$U = 900$ W m^{-2} K^{-1}
Heat-exchange area	$A = 20$ m^2
Density of the heat-conducting fluid	$\rho_j = 1000$ kg m^{-3}
Heat capacity of the heat-conducting fluid	$C_{pj} = 4200$ J kg^{-1} K^{-1}
Proportional gain of the level controller	$K_r = 0.05$ s^{-1}

The form (10.92) of the quadratic criterion is here used with $\boldsymbol{M} = \boldsymbol{C}$. The linearized model is used to calculate the control law in terms of deviation variables, which is then applied to the plant model which provides the values of the theoretical states. Due to the exothermic chemical reaction, this reactor exhibits a strongly nonlinear behavior. The matrix $\boldsymbol{Q}$ is identity $\boldsymbol{I}$ and matrix $\boldsymbol{R}$ is $10\boldsymbol{I}$ giving more importance to robustness. This avoids too frequent variations of the manipulated inputs. The set point changes are decoupled (Fig. 10.2). At $t = 1000$ s when the concentration set point variation occurs, the input u_2 used to control the inlet temperature in the jacket saturates due to the large exothermicity whereas at $t = 2000$ s, when the temperature set point variation occurs, the input u_1 controlling the inlet concentration in the reactor saturates. This demonstrates the large coupling between concentration and temperature due to the reaction exothermicity. In spite of these limitations, the LQ control law allows to perfectly control this nonlinear plant, which is very representative of the behavior of an actual perfectly stirred chemical reactor.

10.5.2 Linear Quadratic Gaussian Control

In LQ control, such as was previously discussed, the states are assumed to be perfectly known. In fact, this is seldom the case. Indeed, often the states have no physical reality and, if they have one, frequently they are not measurable or unmeasured. Thus it is necessary to estimate the states in order to use their estimation in the control model.

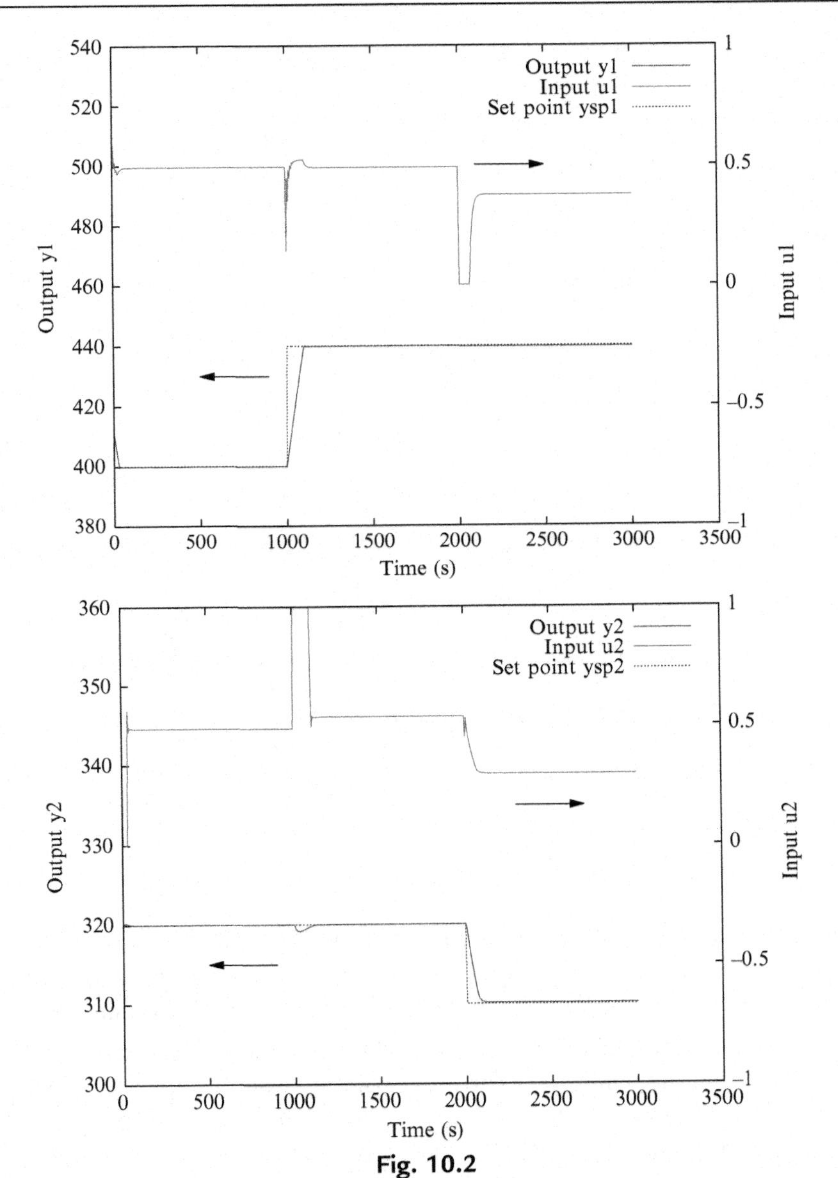

Fig. 10.2
Linear quadratic control of chemical reactor (criterion weighting: $\boldsymbol{Q} = 0.1\ \boldsymbol{I}; \boldsymbol{R} = \boldsymbol{I}$). *Top*: Input u_1 and output y_1 (concentration of A). *Bottom*: Input u_2 and output y_2 (reactor temperature).

In continuous time, the system is represented in the state space by the stochastic linear model

$$\begin{cases} \dot{\boldsymbol{x}}(t) = \boldsymbol{A}\,\boldsymbol{x}(t) + \boldsymbol{B}\,\boldsymbol{u}(t) + \boldsymbol{w}(t) \\ \boldsymbol{y}(t) = \boldsymbol{C}\,\boldsymbol{x}(t) + \boldsymbol{v}(t) \end{cases} \tag{10.132}$$

where $w(t)$ and $v(t)$ are uncorrelated Gaussian white noises, respectively, of state and measurement (or output), of the respective covariance matrices

$$\mathrm{E}\{ww^T\} = W \geq 0, \ \mathrm{E}\{vv^T\} = V > 0 \tag{10.133}$$

Denote by $\hat{x}$ the state estimation, so that the state reconstruction error is $e(t) = x - \hat{x}$. An optimal complete observer such as

$$\dot{\hat{x}} = A\,\hat{x}(t) + B\,u(t) + K_f(t)[y(t) - C\,\hat{x}(t)] \tag{10.134}$$

minimizes the covariance matrix of the state reconstruction error, thus

$$\mathrm{E}\left\{(x - \hat{x})P_w(x - \hat{x})^T\right\} \tag{10.135}$$

where P_w is a weighting matrix (possibly, the identity matrix).

Kalman and Bucy[44] solved this problem and showed that the estimator gain matrix K_f is equal to

$$K_f(t) = P_f(t)\,C^T\,V^{-1} \tag{10.136}$$

where $P_f(t)$ is the solution of the continuous differential Riccati equation

$$\begin{aligned} &\dot{P}_f(t) = A\,P_f(t) + P_f(t)\,A^T - P_f(t)\,C^T\,V^{-1}\,C\,P_f(t) + W \\ &\text{with } \ P_f(t_0) = P_0 \end{aligned} \tag{10.137}$$

Moreover, the initial estimator condition is

$$\hat{x}(t_0) = \hat{x}_0 \tag{10.138}$$

The Kalman-Bucy filter thus calculated is the best state estimator or observer in the sense of linear least squares. It must be noticed that the determination of the Kalman filter is a dual problem of the LQ optimal control problem: to go from the control problem to the estimation one, it suffices to make the following correspondences: $A \to A^T$, $B \to C^T$, $M^TQM \to W$, $R \to V$, $P_c \to P_f$; on the one hand, the control Riccati equation progresses backwards with respect to time, on the other hand, the estimation Riccati equation progresses forwards with respect to time. This latter remark obliges us to carefully manipulate all time-depending functions of the solutions of the Riccati equations[40]: $P_c(t)$ (control problem) is equal to $P_f(t_0 + t_f - t)$ (estimation problem), where t_0 is the initial time of the estimation problem and t_f the final time of the control problem.

When the estimation horizon becomes very large, in general the solution of the Riccati equation (10.137) tends toward a steady-state value, corresponding to the solution of the following algebraic Riccati equation

$$A\,P_f + P_f\,A^T - P_f\,C^T\,V^{-1}\,C\,P_f + W = 0 \tag{10.139}$$

giving the steady-state gain matrix of the estimator

$$K_f = P_f \, C^T \, V^{-1} \tag{10.140}$$

Kwakernaak and Sivan[40] detail the conditions of convergence. For reasons of duality, the solving of the algebraic Riccati equation (10.139) is completely similar to that of Eq. (10.109).

Consider the general case of tracking, the control law similar to Eq. (10.122) is now based on the state estimation

$$u^*(t) = -R^{-1} B^T P_c(t) \hat{x}(t) + R^{-1} B^T \; s(t) = -K_c \hat{x}(t) + u_{ff}(t) \tag{10.141}$$

The system state equation can be written as

$$\dot{x}(t) = A\,x(t) - B\,K_c(t)\,\hat{x}(t) + B\,u_{ff}(t) + w(t) \tag{10.142}$$

so that the complete scheme of the Kalman filter and state feedback optimal control (Fig. 10.3) is written as

$$\begin{bmatrix} \dot{x} \\ \dot{\hat{x}} \end{bmatrix} = \begin{bmatrix} A & -B\,K_c \\ K_f\,C & A - K_f\,C - B\,K_c \end{bmatrix} \begin{bmatrix} x \\ \hat{x} \end{bmatrix} + \begin{bmatrix} B\,u_{ff}(t) + w \\ B\,u_{ff}(t) + K_f\,v \end{bmatrix} \tag{10.143}$$

which can be transformed by use of the estimation error $e(t) = x(t) - \hat{x}(t)$

$$\begin{bmatrix} \dot{x} \\ \dot{e} \end{bmatrix} = \begin{bmatrix} A - B\,K_c & B\,K_c \\ 0 & A - K_f\,C \end{bmatrix} \begin{bmatrix} x \\ e \end{bmatrix} + \begin{bmatrix} B\,u_{ff}(t) + w \\ w - K_f\,v \end{bmatrix} \tag{10.144}$$

The closed-loop eigenvalues are the union of the eigenvalues of the state feedback optimal control scheme and the eigenvalues of Kalman filter. Thus it is possible to separately determine the observer and the state feedback optimal control law, which constitutes the separation

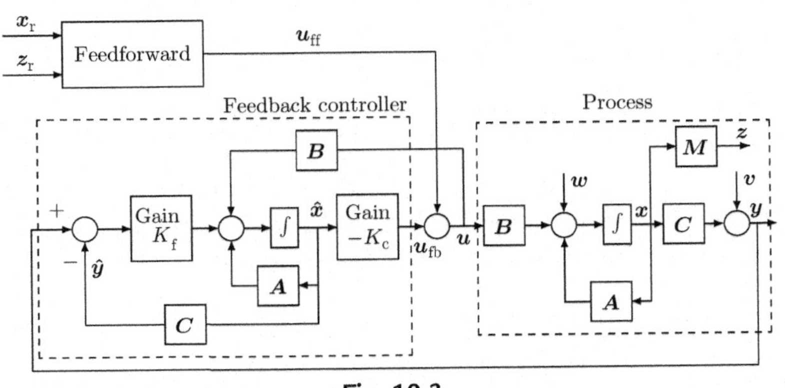

Fig. 10.3
Structure of linear quadratic Gaussian control.

principle of linear quadratic Gaussian (LQG) control. This property that we have just verified for a complete observer is also verified for a reduced observer.

It is useful to notice that the Kalman filter gain is proportional to $\boldsymbol{P}$ (which will vary, but must be initialized) and inversely proportional to the measurement covariance matrix $\boldsymbol{V}$. Thus, if $\boldsymbol{V}$ is low, the filter gain will be very large, as the confidence in the measurement will be large; the risk of low robustness is then high. The Kalman filter can strongly deteriorate the stability margins.[50] The characteristic matrices of the Kalman filter can also be considered as tuning parameters. It is also possible to introduce an integrator per input-output channel in order to effectively realize the set point tracking; the modeled system represents, in this case, the group of the process plus the integrators. It is possible to add an output feedback[49] according to Fig. 10.4, which improves the robustness of regulation and tracking.

Lewis[51] discusses in detail the problem of the LQ regulator with output feedback set as

$$u = -K\, y \tag{10.145}$$

which is more difficult to solve than the classical state feedback, but presents much practical interest. The problem results in a Lyapunov equation that can be solved.

In several applications, LQG revealed that it was sensitive to system uncertainty, so that new approaches were necessary. An important development in LQG control is the consideration of robustness so as to satisfy frequency criteria concerning the sensitivity and complementary sensitivity functions, or gain and phase margins. Actually, the stability margins of LQG control may reveal themselves to be insufficient. LQG/loop transfer recovery (LTR) means use of the LQG design together with a robust control design,[42, 51–54] thus it guarantees closed-loop robustness, both as performance and stability robustness. However, the increase of robustness by use of LQG/LTR results unavoidably in a decrease of performance which can be handled by special optimization techniques.[55]

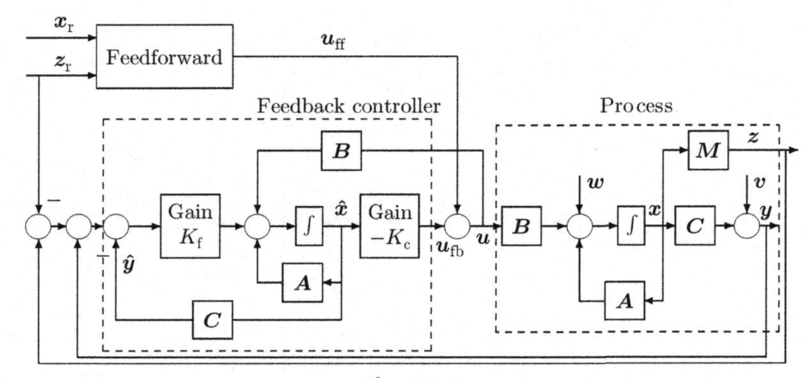

Fig. 10.4
Structure of linear quadratic Gaussian control with added output feedback.

Linear quadratic Gaussian control of a chemical reactor

The same chemical reactor as for LQ control is used to evaluate LQG control. The same set points are imposed. In this case the theoretical states describe the behavior of the plant but noisy measurements are used as process outputs. The standard deviations of concentration and temperature are equal to 2 and 0.5, respectively.

The same control law as for LQ control is used, but it here takes into account the estimated states which themselves depend on the noisy measurements. Two types of Kalman estimators have been used to estimate the states. The linear Kalman filter based on the constant linear model of the plant, calculated at initial time, resulted in deviations of the outputs with respect to the set points. To avoid these deviations and better estimate the states, an extended Kalman filter was used. It is based on the linearized model of the process which is no more constant, being updated at each sampling period equal to 10 s, and takes better into account the strong nonlinearity of the process due to a large exothermicity. The matrix $\boldsymbol{Q}$ is identity $\boldsymbol{I}$ and matrix $\boldsymbol{R}$ is augmented as $50\boldsymbol{I}$ to avoid too large moves of the manipulated inputs. The Kalman matrices are $\boldsymbol{W} = 0.001\boldsymbol{I}$, $\boldsymbol{V} = 0.05\boldsymbol{I}$, and $\boldsymbol{P} = \boldsymbol{I}$, except the last diagonal term equal to 0.1 for this latter. In Fig. 10.5, it is clear that the inputs still show relatively large variations, in particular u_2, but if $\boldsymbol{R}$ is still augmented, the performance is degraded and larger deviations between the controlled outputs and the set points, especially temperature, are observed. An output feedback such as Eq. (10.145) could be used, but it was not implemented here to demonstrate the potential of LQG control. Globally, in spite of a strongly nonlinear process, LQG control is very satisfactory.

10.5.3 Discrete-Time Linear Quadratic Control

In discrete time, the system is represented in the state space by the deterministic linear model

$$\begin{cases} \boldsymbol{x}_{k+1} = \boldsymbol{F}\,\boldsymbol{x}_k + \boldsymbol{G}\,\boldsymbol{u}_k \\ \boldsymbol{y}_k = \boldsymbol{H}\,\boldsymbol{x}_k \end{cases} \tag{10.146}$$

The sought control must minimize a quadratic criterion similar to Eq. (10.91) thus

$$\begin{aligned} J &= 0.5\,[z_N^r - z_N]^T\,\boldsymbol{Q}_N\,[z_N^r - z_N] \\ &\quad + 0.5\sum_{k=0}^{N-1}\left\{[z_k^r - z_k]^T\,\boldsymbol{Q}\,[z_k^r - z_k] + \boldsymbol{u}_k^T\,\boldsymbol{R}\,\boldsymbol{u}_k\right\} \\ &= 0.5\,[\boldsymbol{x}_N^r - \boldsymbol{x}_N]^T\,\boldsymbol{M}^T\,\boldsymbol{Q}_N\,\boldsymbol{M}\,[\boldsymbol{x}_N^r - \boldsymbol{x}_N] \\ &\quad + 0.5\sum_{k=0}^{N-1}\{[\boldsymbol{x}_k^r - \boldsymbol{x}_k]^T\,\boldsymbol{M}^T\,\boldsymbol{Q}\,\boldsymbol{M}\,[\boldsymbol{x}_k^r - \boldsymbol{x}_k] + \boldsymbol{u}_k^T\,\boldsymbol{R}\,\boldsymbol{u}_k\} \end{aligned} \tag{10.147}$$

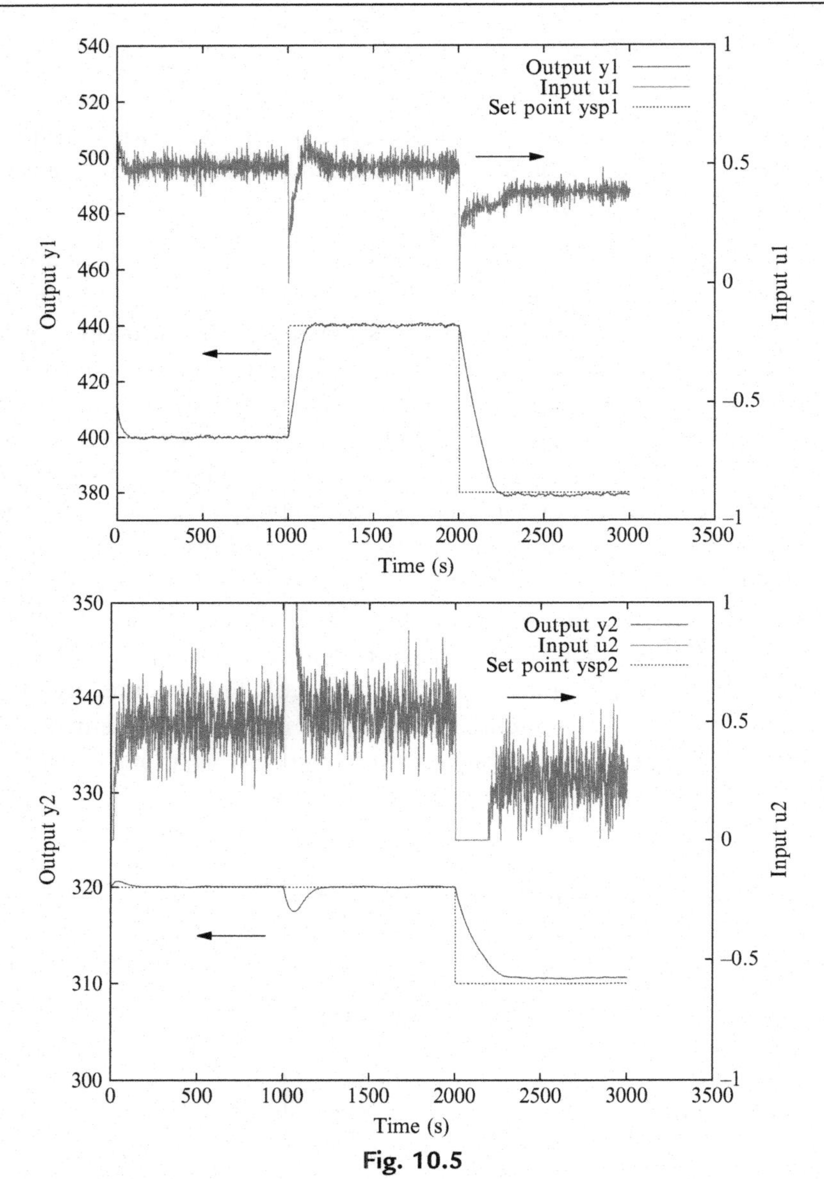

Fig. 10.5

Linear quadratic Gaussian control of chemical reactor (criterion weighting: $\boldsymbol{Q} = 0.02\ \boldsymbol{I}; \boldsymbol{R} = \boldsymbol{I}$). *Top*: Input u_1 and output y_1 (concentration of A). *Bottom*: Input u_2 and output y_2 (reactor temperature).

where the matrices $\boldsymbol{Q}_N$, $\boldsymbol{Q}$ are semipositive definite and $\boldsymbol{R}$ is positive definite. Furthermore $z = \boldsymbol{M}\boldsymbol{x}$ represents measurements or outputs. It would be possible to use variational methods to deduce from them the optimal control law,[4,41] which would provide a system perfectly similar to Eq. (10.98). However, variational methods are a priori designed in the framework of continuous variables, thus for continuous time; on the other hand, dynamic programming is perfectly adapted to the discrete case. Thus we will sketch out the reasoning in this framework. For more details, it is possible, for example, to refer to Refs. 56,57.

The system is considered at any instant i included in the interval $[0, N]$, assuming that the policy preceding that instant, thus the sequence of the $\{\boldsymbol{u}_k, k \in [i+1, N]\}$, is optimal (the final instant N is the starting point for performing the procedure of dynamic programming). In these conditions, the criterion of interest is in the form

$$\begin{aligned} J_i &= 0.5\, [z_N^r - z_N]^T \boldsymbol{Q}_N\, [z_N^r - z_N] \\ &+ 0.5 \sum_{k=i}^{N-1} \left\{ [z_k^r - z_k]^T \boldsymbol{Q}\, [z_k^r - z_k] + \boldsymbol{u}_k^T \boldsymbol{R}\, \boldsymbol{u}_k \right\} \\ &= 0.5\, [\boldsymbol{x}_N^r - \boldsymbol{x}_N]^T \boldsymbol{M}^T \boldsymbol{Q}_N \boldsymbol{M}\, [\boldsymbol{x}_N^r - \boldsymbol{x}_N] + \sum_{k=i}^{N-1} L_k(\boldsymbol{x}_k, \boldsymbol{u}_k) \end{aligned} \tag{10.148}$$

with the revenue function L_k defined by

$$L_k(\boldsymbol{x}_k, \boldsymbol{u}_k) = 0.5\, \{[\boldsymbol{x}_k^r - \boldsymbol{x}_k]^T \boldsymbol{M}^T \boldsymbol{Q} \boldsymbol{M}\, [\boldsymbol{x}_k^r - \boldsymbol{x}_k] + \boldsymbol{u}_k^T \boldsymbol{R}\, \boldsymbol{u}_k\} \tag{10.149}$$

According to Bellman optimality principle, the optimal value J_i^* of the criterion can be expressed in a recurrent form

$$J_i^* = \min_{\boldsymbol{u}} \{0.5 [\boldsymbol{x}_i^r - \boldsymbol{x}_i]^T \boldsymbol{M}^T \boldsymbol{Q} \boldsymbol{M}\, [\boldsymbol{x}_i^r - \boldsymbol{x}_i] + \boldsymbol{u}_i^T \boldsymbol{R}\, \boldsymbol{u}_i + J_{i+1}^*\} \tag{10.150}$$

The final state is assumed to be free. It is necessary to know the expression of J_{i+1}^* with respect to the state to be able to perform the minimization. Reasoning by recurrence, suppose that it is in quadratic form

$$J_{i+1}^* = 0.5 \left\{ \boldsymbol{x}_{i+1}^T \boldsymbol{S}_{i+1} \boldsymbol{x}_{i+1} + 2\, \boldsymbol{g}_{i+1} \boldsymbol{x}_{i+1} + h_{i+1} \right\} \tag{10.151}$$

hence

$$\begin{aligned} J_{i+1}^* &= 0.5 \left\{ [\boldsymbol{F} \boldsymbol{x}_i + \boldsymbol{G} \boldsymbol{u}_i]^T \boldsymbol{S}_{i+1}\, [\boldsymbol{F} \boldsymbol{x}_i + \boldsymbol{G} \boldsymbol{u}_i] \right. \\ &\left. + 2\, \boldsymbol{g}_{i+1}\, [\boldsymbol{F} \boldsymbol{x}_i + \boldsymbol{G} \boldsymbol{u}_i] + h_{i+1} \right\} \end{aligned} \tag{10.152}$$

We deduce

$$\begin{aligned} J_i^* &= 0.5 \min_{\boldsymbol{u}_i} \{ [\boldsymbol{x}_i^r - \boldsymbol{x}_i]^T \boldsymbol{M}^T \boldsymbol{Q} \boldsymbol{M}\, [\boldsymbol{x}_i^r - \boldsymbol{x}_i] + \boldsymbol{u}_i^T \boldsymbol{R}\, \boldsymbol{u}_i \\ &+ [\boldsymbol{F} \boldsymbol{x}_i + \boldsymbol{G} \boldsymbol{u}_i]^T \boldsymbol{S}_{i+1}\, [\boldsymbol{F} \boldsymbol{x}_i + \boldsymbol{G} \boldsymbol{u}_i] \\ &+ 2\, \boldsymbol{g}_{i+1}\, [\boldsymbol{F} \boldsymbol{x}_i + \boldsymbol{G} \boldsymbol{u}_i] + h_{i+1} \} \end{aligned} \tag{10.153}$$

We search the minimum with respect to $\boldsymbol{u}_i$ thus

$$\boldsymbol{R}\,\boldsymbol{u}_i^* + \boldsymbol{G}^T \boldsymbol{S}_{i+1}[\boldsymbol{F}\,\boldsymbol{x}_i + \boldsymbol{G}\,\boldsymbol{u}_i^*] + \boldsymbol{G}^T \boldsymbol{g}_{i+1} = 0 \tag{10.154}$$

or

$$\boldsymbol{u}_i^* = -[\boldsymbol{R} + \boldsymbol{G}^T \boldsymbol{S}_{i+1}\,\boldsymbol{G}]^{-1}\,[\boldsymbol{G}^T \boldsymbol{S}_{i+1}\,\boldsymbol{F}\,\boldsymbol{x}_i + \boldsymbol{G}^T \boldsymbol{g}_{i+1}] \tag{10.155}$$

provided that the matrix $[\boldsymbol{R} + \boldsymbol{G}^T \boldsymbol{S}_{i+1}\,\boldsymbol{G}]$ is invertible. Notice that the optimal control is in the form

$$\boldsymbol{u}_i^* = -\boldsymbol{K}_i\,\boldsymbol{x}_i + \boldsymbol{k}_i\,\boldsymbol{g}_{i+1} \tag{10.156}$$

revealing the state feedback with the gain matrix $\boldsymbol{K}_i$ and feedforward with the gain $\boldsymbol{k}_i$. Thus we set

$$\begin{aligned} \boldsymbol{K}_i &= [\boldsymbol{R} + \boldsymbol{G}^T \boldsymbol{S}_{i+1}\,\boldsymbol{G}]^{-1}\,\boldsymbol{G}^T \boldsymbol{S}_{i+1}\,\boldsymbol{F};\quad \boldsymbol{S}_N = \boldsymbol{M}^T \boldsymbol{Q}_N\,\boldsymbol{M} \\ \boldsymbol{k}_i &= -[\boldsymbol{R} + \boldsymbol{G}^T \boldsymbol{S}_{i+1}\,\boldsymbol{G}]^{-1}\,\boldsymbol{G}^T \end{aligned} \tag{10.157}$$

It is then possible to verify that J_i^* is effectively in quadratic form; thus we find

$$\begin{aligned} \boldsymbol{S}_i &= \boldsymbol{M}^T \boldsymbol{Q}\boldsymbol{M} + \boldsymbol{F}^T \boldsymbol{S}_{i+1}\,(\boldsymbol{F} - \boldsymbol{G}\boldsymbol{K}_i) \\ \boldsymbol{g}_i &= -\boldsymbol{M}^T \boldsymbol{Q}\,\boldsymbol{z}_i^r + (\boldsymbol{F}^T - \boldsymbol{G}\,\boldsymbol{K}_i)^T \boldsymbol{g}_{i+1};\quad \boldsymbol{g}_N = \boldsymbol{M}\,\boldsymbol{Q}_N\,\boldsymbol{z}_N^r \end{aligned} \tag{10.158}$$

The group of Eqs. (10.156)–(10.158) allows us to determine the inputs $\boldsymbol{u}$. When not all states are known, of course it is necessary to use a discrete Kalman filter which will work with the optimal control law according to the same separation principle as in the continuous case.

It can be shown that Eq. (10.158) is equivalent to the discrete differential Riccati equation

$$\boldsymbol{S}_i = (\boldsymbol{F} - \boldsymbol{G}\boldsymbol{K}_i)^T \boldsymbol{S}_{i+1}\,(\boldsymbol{F} - \boldsymbol{G}\boldsymbol{K}_i) + \boldsymbol{K}_i^T \boldsymbol{R}\,\boldsymbol{K}_i + \boldsymbol{M}^T \boldsymbol{Q}\boldsymbol{M} \tag{10.159}$$

here presented in Joseph form and better adapted to numerical calculation.

Let us apply the Hamilton-Jacobi principle to the discrete-time optimal regulator. If the control law $\boldsymbol{u}_i^*$ and the corresponding states $\boldsymbol{x}_i^*$ are optimal, according to the Hamilton-Jacobi principle, there exists a costate vector $\boldsymbol{\psi}_i^*$ such that $\boldsymbol{u}_i^*$ is the value of the control $\boldsymbol{u}_i$ which maximizes the Hamiltonian function H_a

$$\begin{aligned} H_a &= -L_i(\boldsymbol{x}_i^*, \boldsymbol{u}_i) + \boldsymbol{\psi}_{i+1}^{*T}\,\boldsymbol{x}_{i+1} \\ &= -L_i(\boldsymbol{x}_i^*, \boldsymbol{u}_i) + \boldsymbol{\psi}_{i+1}^{*T}\,[\boldsymbol{F}\,\boldsymbol{x}_i + \boldsymbol{G}\,\boldsymbol{u}_i] \\ &= -0.5\,\{[\boldsymbol{x}_i^r - \boldsymbol{x}_i]^T \boldsymbol{M}^T \boldsymbol{Q}\boldsymbol{M}\,[\boldsymbol{x}_i^r - \boldsymbol{x}_i] + \boldsymbol{u}_i^T \boldsymbol{R}\,\boldsymbol{u}_i\} + \boldsymbol{\psi}_{i+1}^{*T}\,[\boldsymbol{F}\,\boldsymbol{x}_i + \boldsymbol{G}\,\boldsymbol{u}_i] \end{aligned} \tag{10.160}$$

The Hamilton-Jacobi conditions give

$$\psi_i^* = -\frac{\partial H_a}{\partial x_i^*} = M^T Q M [x_i^* - x_i^r] - F^T \psi_{i+1} \quad \text{with } \psi_N = M^T Q M [x_N^* - x_N^r] \tag{10.161}$$

and the control which maximizes the Hamiltonian function H_a is such that

$$\frac{dH_a}{du_i} = 0 \Longrightarrow -R u_i + G^T \psi_{i+1} = 0 \Longrightarrow u_i^* = R^{-1} G^T \psi_{i+1} \tag{10.162}$$

hence

$$x_{i+1} = F x_i + G u_i = F x_i + G R^{-1} G^T \psi_{i+1} \tag{10.163}$$

Introduce the matrix $\mathcal{H}$ such that

$$\begin{bmatrix} x_i \\ \psi_i \end{bmatrix} = \mathcal{H} \begin{bmatrix} x_{i+1} \\ \psi_{i+1} \end{bmatrix} \quad \text{or} \quad \begin{bmatrix} x_{i+1} \\ \psi_{i+1} \end{bmatrix} = \mathcal{H}^{-1} \begin{bmatrix} x_i \\ \psi_i \end{bmatrix} \tag{10.164}$$

Assume $x_i^r = 0$ for the regulation case. The two conditions (10.161), (10.162) can be grouped as

$$\begin{aligned} \psi_i^* &= M^T Q M x_i^* - F^T \psi_{i+1} \\ x_{i+1} &= F x_i + G R^{-1} G^T \psi_{i+1} \end{aligned} \tag{10.165}$$

from which we deduce the matrix $\mathcal{H}$

$$\mathcal{H} = \begin{bmatrix} F^{-1} & -F^{-1} G R^{-1} G^T \\ M^T Q M F^{-1} & -F^T - M^T Q M F^{-1} G R^{-1} G^T \end{bmatrix} \tag{10.166}$$

In the case where a steady-state gain K_∞ is satisfactory, which can be realized when the horizon N is large, the gain matrix can be obtained after solving the algebraic Riccati equation

$$S = F^T [S - S G (G^T S G + R)^{-1} G^T S] F + M Q M \tag{10.167}$$

whose solution (corresponding to discrete time) is obtained in a parallel manner to the continuous case, by first considering the matrix H. Its inverse is the symplectic matrix H^{-1} equal to

$$H^{-1} = \begin{bmatrix} F + G R^{-1} G^T F^{-T} M^T Q M & -G R^{-1} G^T F^{-T} \\ F^{-T} M^T Q M & -F^{-T} \end{bmatrix} \tag{10.168}$$

A matrix A is symplectic, when, given the matrix $J = \begin{bmatrix} 0 & I \\ -I & 0 \end{bmatrix}$, the matrix A verifies $A^T J A = J$. If λ is an eigenvalue of a symplectic matrix A, $1/\lambda$ is also an eigenvalue of A; λ is thus also an eigenvalue of A^{-1}.[48]

We seek the eigenvalues and associated eigenvectors of $\boldsymbol{H}^{-1}$. Then, we form the matrix $\boldsymbol{U}$ of the eigenvectors so that the first n columns correspond to the stable eigenvalues (inside the unit circle), in the form

$$\boldsymbol{U} = \begin{bmatrix} \boldsymbol{U}_{11} & \boldsymbol{U}_{12} \\ \boldsymbol{U}_{21} & \boldsymbol{U}_{22} \end{bmatrix} \tag{10.169}$$

where the blocks $\boldsymbol{U}_{ij}$ have dimension $n \times n$. The solution of the discrete Riccati algebraic equation (10.167) is then

$$\boldsymbol{S}_\infty = \boldsymbol{U}_{21}\,\boldsymbol{U}_{11}^{-1} \tag{10.170}$$

giving the steady-state matrix.

For the tracking problem, in parallel to the stationary solution for the gain matrix $\boldsymbol{K}$, the stationary solution for the feedforward gain is deduced from Eq. (10.158) and is given by

$$\boldsymbol{g}_i = [\boldsymbol{I} - (\boldsymbol{F}^T - \boldsymbol{G}\,\boldsymbol{K}_\infty)^T]^{-1}\,[-\boldsymbol{M}^T\,\boldsymbol{Q}\,\boldsymbol{z}_i^r] \tag{10.171}$$

The use of stationary gains provides a suboptimal solution but is more robust than using the optimal gains coming from Eqs. (10.157), (10.158).

The discrete LQ Gaussian control is derived from the previously described discrete LQ control by coupling a discrete linear Kalman filter in order to estimate the states.

Remark Recall the operating conditions of quadratic control. In general, it is assumed that the pair $(\boldsymbol{A}, \boldsymbol{B})$ in continuous time, or $(\boldsymbol{F}, \boldsymbol{G})$ in discrete time, is controllable. Moreover, when the horizon is infinite and when we are looking for steady-state solutions of the Riccati equation, the condition that the pair $(\boldsymbol{A}, \sqrt{\boldsymbol{Q}})$ in continuous time, or $(\boldsymbol{F}, \sqrt{\boldsymbol{Q}})$ in discrete time, is observable (the notation $\boldsymbol{H} = \sqrt{\boldsymbol{Q}}$ means that $\boldsymbol{Q} = \boldsymbol{H}^T\,\boldsymbol{H}$) must be added.

References

1. Kirk D. *Optimal control theory. An introduction.* Englewood Cliffs (NJ): Prentice Hall; 1970.
2. Pontryaguine L, Boltianski V, Gamkrelidze R, Michtchenko E. *Théorie mathématique des processus optimaux.* Moscou: Mir; 1974. Edition Française.
3. Bellman R. *Dynamic programming.* Princeton (NJ): Princeton University Press; 1957.
4. Borne P, Dauphin-Tanguy G, Richard J, Rotella F, Zambettakis I. *Commande et optimisation des processus.* Paris: Technip; 1990.
5. Boudarel R, Delmas J, Guichet P. *Commande optimale des processus.* Paris: Dunod; 1969.
6. Bryson A, Ho Y. *Applied optimal control.* Washington (DC): Hemisphere; 1975.
7. Bryson A. *Dynamic optimization.* Menlo Park (CA): Addison Wesley; 1999.
8. Feldbaum A. *Principes théoriques des systèmes asservis optimaux.* Moscou: Mir; 1973 Edition Française.
9. Pun L. *Introduction à la pratique de l'optimisation.* Paris: Dunod; 1972.
10. Ray W, Szekely J. *Process optimization with applications in metallurgy and chemical engineering.* New York: Wiley; 1973.

11. Corriou J. *Process control—theory and applications*. London: Springer; 2004.
12. Corriou J. *Commande des Procédés*. 3rd ed. Paris: Lavoisier, Tec. & Doc; 2012.
13. Lamnabhi-Lagarrigue F. Singular optimal control problems: on the order of a singular arc. *Syst Control Lett* 1987;**9**:173–82.
14. Stengel R. *Optimal control and estimation*. New York: Courier Dover Publications; 1994.
15. Goh C, Teo K. Control parametrization: a unified approach to optimal control problems with general constraints. *Automatica* 1988;**24**:3–18.
16. Teo K, Goh C, Wong K. *A unified computational approach to optimal control problems*. Harlow (Essex, England): Longman Scientific & Technical; 1991.
17. Biegler L. Solution of dynamic optimization problems by successive quadratic programming and orthogonal collocation. *Comp Chem Eng* 1984;**8**:243–8.
18. Luus R. Numerical convergence properties of iterative dynamic programming when applied to high dimensional systems. *Trans IChemE A* 1996;**74**:55–62.
19. Banga J, Carrasco E. Rebuttal to the comments of Rein Luus on "dynamic optimization of batch reactors using adaptive stochastic algorithms" *Ind Eng Chem Res* 1998;**37**:306–7.
20. Bojkov B, Luus R. Optimal control of nonlinear systems with unspecified final times. *Chem Eng Sci* 1996;**51** (6):905–19.
21. Carrasco E, Banga J. Dynamic optimization of batch reactors using adaptive stochastic algorithms. *Ind Eng Chem Res* 1997;**36**:2252–61.
22. Luus R, Hennessy D. Optimization of fed-batch reactors by the Luus-Jaakola optimization procedure. *Ind Eng Chem Res* 1999;**38**:1948–55.
23. Mekarapiruk W, Luus R. Optimal control of inequality state constrained systems. *Ind Eng Chem Res* 1997;**36**:1686–94.
24. Bellman R, Dreyfus S. *Applied dynamic programming*. Princeton (NJ): Princeton University Press; 1962.
25. Aris R. *The optimal design of chemical reactors: a study in dynamic programming*. New York: Academic Press; 1961.
26. Roberts S. *Dynamic programming in chemical engineering and process control*. New York: Academic Press; 1964.
27. Aris R, Rudd D, Amundson N. On optimum cross current extraction. *Chem Eng Sci* 1960;**12**:88–97.
28. Aris R. Studies in optimization. II. Optimal temperature gradients in tubular reactors. *Chem Eng Sci* 1960;**13** (1):18–29.
29. Roberts S, Laspe C. Computer control of a thermal cracking reaction. *Ind Eng Chem* 1961;**53**(5):343–8.
30. Kaufmann A, Cruon R. *La programmation dynamique. Gestion scientifique séquentielle*. Paris: Dunod; 1965.
31. Luus R. Application of dynamic programming to high-dimensional nonlinear optimal control systems. *Int J Control* 1990;**52**(1):239–50.
32. Luus R. Application of iterative dynamic programming to very high-dimensional systems. *Hung J Ind Chem* 1993;**21**:243–50.
33. Luus R, Bojkov B. Application of iterative dynamic programming to time-optimal control. *Chem Eng Res Des* 1994;**72**:72–80.
34. Luus R. Optimal control of bath reactors by iterative dynamic programming. *J Proc Control* 1994;**4**(4): 218–26.
35. Athans M, Falb P. *Optimal control: an introduction to the theory and its applications*. New York: MacGraw-Hill; 1966.
36. Anderson B, Moore J. *Linear optimal control*. Englewood Cliffs (NJ): Prentice Hall; 1971.
37. Anderson B, Moore J. *Optimal control, linear quadratic methods*. Englewood Cliffs (NJ): Prentice Hall; 1990.
38. Grimble M, Johnson M. *Optimal control and stochastic estimation: deterministic systems*. vol. 1. Chichester: Wiley; 1988.
39. Grimble M, Johnson M. *Optimal control and stochastic estimation: stochastic systems*. vol. 2. Chichester: Wiley; 1988.
40. Kwakernaak H, Sivan R. *Linear optimal control systems*. New York: Wiley-Interscience; 1972.

41. Lewis F. *Optimal control*. New York: Wiley; 1986.
42. Maciejowski J. *Multivariable feedback design*. Wokingham (England): Addison-Wesley; 1989.
43. Kalman R. A new approach to linear filtering and prediction problems. *Trans ASME Ser D J Basic Eng* 1960;**82**:35–45.
44. Kalman R, Bucy R. New results in linear filtering and prediction theory. *Trans ASME Ser D J Basic Eng* 1961;**83**:95–108.
45. Kalman R. Mathematical description of linear dynamical systems. *J SIAM Control Ser A* 1963;**1**:152–92.
46. Pannocchia G, Laachi N, Rawlings J. A candidate to replace PID control: SISO-constrained LQ control. *AIChE J* 2005;**51**(4):1178–89.
47. Arnold W, Laub A. Generalized eigenproblem algorithms and software for algebraic Riccati equations. *IEEE Proc* 1984;**72**(12):1746–54.
48. Laub A. A Schur method for solving algebraic Riccati equations. *IEEE Trans Autom Control* 1979;**24**(6): 913–21.
49. Lin C. *Advanced control systems design*. Englewood Cliffs (NJ): Prentice Hall; 1994.
50. Doyle J. Guaranteed margins for LQG regulators. *IEEE Trans Autom Control* 1978;**23**:756–7.
51. Lewis F. *Applied optimal control and estimation*. Englewood Cliffs (NJ): Prentice Hall; 1992.
52. Athans M. *A tutorial on the LQG/LTR method*. Laboratory for Information and Decision Systems, Massachusetts Institute of Technology; 1986.
53. Kwakernaak H. Robust control and $\mathcal{H}\infty$-optimization—tutorial paper. *Automatica* 1993;**29**(2):255–73.
54. Stein G, Athans M. The LQG/LTR procedure for multivariable feedback control design. *IEEE Trans Autom Control* 1987;**32**(2):105–14.
55. Apkarian P, Noll D. Nonsmooth H_∞ synthesis. *IEEE Trans Autom Control* 2006;**51**:71–86.
56. Dorato J, Levis A. Optimal linear regulator: the discrete-time case. *IEEE Trans Autom Control* 1971;**16** (6):613–20.
57. Foulard C, Gentil S, Sandraz J. *Commande et régulation par calculateur numérique*. Paris: Eyrolles; 1987.

Further Reading

Lee E, Markus L. *Foundations of optimal control theory*. Malabar (FL): Krieger; 1967.
Kailath T. *Linear systems theory*. Englewood Cliffs (NJ): Prentice Hall; 1980.

CHAPTER 11

Control and Optimization of Batch Chemical Processes

Dominique Bonvin*, Grégory François†

*Ecole Polytechnique Fédérale de Lausanne (EPFL), Lausanne, Switzerland, †The University of Edinburgh, Edinburgh, United Kingdom

11.1 Introduction

Batch processing is widely used in the manufacturing of goods and commodity products, in particular in the chemical, pharmaceutical, and food industries.[1] Batch operation differs significantly from continuous operation. While in continuous operation the process is maintained at an economically desirable operating point, in batch operation the process evolves from an initial to a final state. In the chemical industry for example, since the design of a continuous plant requires substantial engineering effort, continuous operation is rarely used for low-volume production. Discontinuous operations can be of the batch or semibatch type. In batch operations, the products to be processed are loaded in a vessel and processed without material addition or removal. This operation permits more flexibility than continuous operation by allowing adjustment of the operating conditions and the final time. Additional flexibility is available in semibatch operations, where reactants are continuously added by adjusting the feedrate profiles, or products are removed via some outflow. We use the term batch process to include semibatch processes.

In the chemical industry, many batch processes deal with reaction and separation operations. Reactions are central to chemical processing and can be performed in an homogeneous (single-phase) or heterogeneous (multiphase) environment. Separation processes can be of very different types, such as distillation, absorption, extraction, adsorption, chromatography, crystallization, drying, filtration, and centrifugation. The operation of batch processes follows recipes developed in the laboratory. A sequence of operations is performed in a prespecified order in specialized process equipment, yielding a certain amount of product. The sequence of tasks to be carried out on each piece of equipment and drying, is predefined. The desired production volume is then achieved by repeating the processing steps on a predetermined schedule.

Coulson and Richardson's Chemical Engineering. http://dx.doi.org/10.1016/B978-0-08-101095-2.00011-4

This chapter describes the various control and optimization strategies available for the operation of batch processes. Section 11.2 discusses the main features of batch processes, while Section 11.3 presents the types of models that are available for either online or run-to-run operation, respectively. The rest of the chapter contains two parts, namely Part A comprising Sections 11.4–11.7 and concerned with control (online control, run-to-run control, batch automation, and control applications), and Part B comprising Sections 11.8–11.10 and dealing with optimization (numerical optimization, real-time optimization, and optimization applications). Finally, a summary and outlook are provided in Section 11.11.

11.2 Features of Batch Processes

Process engineers have developed considerable expertise in designing and operating continuous processes. On the other hand, chemists have been trained to develop new routes for synthesizing chemicals, often using a batch or semibatch mode of operation. The operation of batch processes requires considerable attention, in particular regarding the coordination in time of various processing tasks such as charging, heating, reaction, separation, cooling, and the determination of optimal temperature and feeding profiles. Process control engineers have been eager to use their expertise in controlling and optimizing continuous processes to achieve comparable success with batch processes. However, this is rarely possible, the reason being significant differences between continuous and batch processes. The two main distinguishing features are discussed first[2]:

- *Distinguishing feature 1*: No steady-state operating point. In batch processes, chemical and physical transformations proceed from an initial state to a very different final state. In a batch reactor, for instance, even if the reactor temperature is kept constant, the concentrations, and thus also the reaction rates, change significantly over the duration of the batch. Consequently, there does not exist a steady-state operating point around which the control system can be designed. The decision variables are infinite-dimensional time profiles. Furthermore, important process characteristics—such as static gains, time constants, and time delays—are time varying.
- *Distinguishing feature 2*: Repetitive nature. Batch processing is characterized by the frequent repetition of batch runs. Hence, it is appealing to use the results from previous runs to optimize the operation of subsequent ones. This has generated the industrially relevant topics of run-to-run control and run-to-run optimization.

In addition to these two features, a number of issues tend to complicate the operation of batch processes:

- *Nonlinear behavior*. Constitutive equations, such as reaction rates and thermodynamic relationships, are typically nonlinear. Since a batch process operates over a wide range of conditions, it is not possible to use, for the purpose of control design and optimization, models that have been linearized around a steady-state operating point, as this is typically done for continuous processes.

- *Poor models.* There is little time in batch processing for thorough investigations of the reaction, mixing, heat- and mass-transfer issues. Consequently, the models are often poor. For example, it may well happen that, in the production of specialty chemicals, the number of significant reactions is unknown, not to mention their stoichiometry or kinetics.
- *Few specific measurements.* The sensors that allow measuring concentrations online are rare. Chemical composition is usually determined by drawing a sample and analyzing it offline, that is, by using invasive and destructive methods. Furthermore, the available measurements—often physical quantities such as temperature, pressure, torque, turbidity, reflective index, and electric conductivity—might exhibit low accuracy due to the wide range of operation that the measuring instrument has to cover.
- *Constrained operation.* A process is typically designed to operate in a limited region of the state space. In addition, the values that the inputs can take are upper and lower bounded. The presence of these limitations (labeled constraints) complicates the design of operational strategies for two main reasons: (i) even if the process is linear or has been linearized along a reference trajectory, constraints make it nonlinear, and (ii) controllability might be lost when a manipulated variable (MV) hits a constraint. Due to the wide operating range of batch processes, it is rarely possible to design the process so as to enforce feasible operation close to constraints, as this is typically done for continuous processes. In fact, it has been our experience that safety and operational constraints dominate the operation of batch processes.
- *Presence of disturbances.* Operator errors (e.g., wrong stirrer or solvent choice, incorrect material loading) and processing problems (e.g., fouling of sensors and reactor walls, insufficient mixing, incorrect feeding profiles, sensor failures) represent major disturbances that, unfortunately, cannot be totally ruled out. There are other unmeasured disturbances entering the process as the result of upstream process variability such as impurities in the raw materials.
- *Irreversible behavior.* In processes with history-dependent product properties, such as polymerization and crystallization, it is often impossible to introduce remedial corrections once off-specification material has been produced. This contrasts with continuous processes, where appropriate control action can bring the process back to the desired steady state following an upset in operating conditions.
- *Limited corrective action.* The ability to influence the process typically decreases with time. This, together with the finite duration of a batch run, limits the impact of corrective actions. Often, if a batch run shows a deviation in product quality, the charge has to be discarded.

11.3 Models of Batch Processes

Model-based control and optimization techniques rely on appropriate mathematical representations. The meaning of the qualifier "appropriate" depends on the system at hand and the objectives of the study. Some modeling aspects of batch processes are addressed next.[3]

11.3.1 What to Model?

Consider the example of a batch chemical reactor. The reactor system comprises the reactor vessel and one or several reactions. For reasons mentioned earlier, the reactions are often poorly known in the batch environment. Even when the desired reaction and the main side reactions are well documented, there might exist additional poorly known or totally unknown reactions. For example, it is frequently observed that the desired products and the known side products do not account for all the transformed reactants. Hence, since the stoichiometry of the reaction system is not completely known, the kinetic models are often of aggregate nature, that is, they encompass real and pseudo (lumped) reactions in order to describe the observed concentrations. This way, the number of modeled reactions, and thereby the number of kinetic parameters to be estimated, can be kept low. Furthermore, to determine the operational strategy for a reactor, it is necessary to consider the reaction kinetics, reactor dynamics, and operational constraints. In industrial situations, the dynamics associated with the thermal exchange between the reactor and the jacket are often dominant and must necessarily be included in the model. The reasoning developed here for batch reactors[4,5] extends similarly to other batch units such as batch distillation columns[6,7] and batch crystallizers.[8,9]

11.3.2 Model Types

The models used can be of several types, the characteristics of which are detailed next.

- *Data-driven black-box models.* Despite their simplicity, empirical input-output models are often able to satisfactorily represent the relationship between manipulated and observed variables. Linear and nonlinear ARMAX-type models are readily used to represent dynamic systems.[10,11] Neural network methods have also been proposed to model the dynamics of batch processes.[12] More recently, researchers have investigated the use of multivariate statistical partial least-squares (PLS) models as a means to predict product quality in batch reactors.[13] Experimental design techniques can help assess and increase the validity of data-driven models.[14] Although simple and relatively easy to obtain, data-driven input-output models have certain drawbacks:
  1. They often exhibit good interpolative capabilities, yet they are inadequate for predicting the process behavior outside the experimental domain in which the data were collected for model building. Such a feature significantly limits the applicability

of these models for optimization, that is, the determination of better profiles that have not been seen before.

2. Input-output models represent a dynamic relationship only between variables that are manipulated or measured. Unfortunately, some key variables, such as concentrations, often remain unmeasured in batch processes and thus cannot be modeled via data-driven models.

- *Knowledge-driven white-box models.* A mechanistic state-space representation based on energy and material balances is the preferred approach for modeling batch processes. The rate expressions describe the effects that the temperature and the concentrations have on the various rates. The model of a given unit relates the independent variables (inputs and disturbances) to the states (concentrations, temperature, and volume) and the corresponding outputs.

  The dynamic model of a batch process can be written as follows[15]:

$$\dot{x}_k(t) = F(x_k(t), u_k(t)), \quad x_k(0) = x_{0,k}, \tag{11.1}$$

$$y_k(t) = H(x_k(t), u_k(t)), \tag{11.2}$$

$$z_k = Z(x_k[0, t_f], u_k[0, t_f]), \tag{11.3}$$

  where t denotes the run time, k the batch or run index, x the n-dimensional state vector, and u the m-dimensional input vector. There are two types of measured outputs, namely, the p-dimensional vector of run-time outputs $y(t)$ that are available online and the q-dimensional vector of run-end outputs z that are available at the final time t_f. Note that z is not limited to quantities measured at the final time but can include quantities inferred from the profiles $x_k[0, t_f]$ and $u_k[0, t_f]$, such as the maximal temperature reached during the batch. In addition to the run-time dynamics specific to a given run, one has the possibility of updating the initial conditions and the inputs on a run-to-run basis as follows[1]:

$$x_{0,k+1} = I(x_k[0, t_f], u_k[0, t_f]), \quad x_{0,0} = x_{\text{init}}, \tag{11.4}$$

$$u_{k+1}[0, t_f] = K(x_k[0, t_f], u_k[0, t_f]), \quad u_0[0, t_f] = u_{\text{init}}[0, t_f]. \tag{11.5}$$

  Note that, in this formulation, the final time t_f is the same for all batches.

  Mechanistic models are typically derived from physico-chemical laws. They are well suited for a wide range of process operations. However, they are difficult and time-consuming to build for industrially relevant processes. A sensitivity analysis can help evaluate the dominant terms in a model and retain those that are most relevant to the processing objectives.[16] No realistic model is purely mechanistic, as a few physical parameters typically need to be estimated from process data. As with data-driven models, experimental

[1] The initial conditions and inputs can be updated on the basis of several prior batches; the relationships proposed here consider only the previous batch for simplicity of notation.

design techniques[17] are useful tools for building sound models from a limited amount of data.

- *Hybrid gray-box models.* As a combination of the two extreme cases listed above, hybrid models can be very helpful in certain situations. They typically possess a simple structure that is based on some qualitative knowledge of the process.[18] The model parameters are particularly easy to identify if the model structure can be reduced to an ARMAX-type form.[19] For example, reaction lumping is often exercised. Because of the aggregate structure of the model, it is necessary to adjust certain model parameters online as, for example, the rate parameters might represent a different aggregation initially than toward the end of the batch. This can be done continuously or in some ad hoc fashion, but clearly at a rate slower than that of the main dynamics. Furthermore, tendency models have been proposed.[20] These models retain the physical understanding of the system but are expressed in a form that is suitable to be fitted to local behavior. Although tendency modeling appears to be of high industrial relevance, it has received little attention in the academic research community.

11.3.3 Static View of a Batch Process

Due to the absence of a steady state, most signals in a batch process evolve with time. Consequently, the decision variables (here the input profiles $u[0, t_f]$ for a given batch run) are of infinite dimension. It is, however, possible to represent a batch process as a *static map* between a finite number of input parameters (used to define the input profiles before batch start) and the run-end outputs (representing the outcome at batch end). The key element is the parameterization of the input profiles as

$$u[0, t_f] = \mathcal{U}(\pi, t), \tag{11.6}$$

with the input parameters π. This parameterization is achieved by dividing the input profiles into time intervals and using a polynomial approximation (of which the simplest form is a constant value) within each interval. The input parameter vector π can also include switching times between intervals.[21]

With such a parameterization, the batch process can be seen as a static map between the finite set of input parameters π and the run-end outputs z:

$$z = M(\pi). \tag{11.7}$$

This static model indicates that, once the input parameters π have been specified, it is possible to compute the run-end outputs z. For this, one needs to generate the input profiles $u[0, t_f]$ using Eq. (11.6), integrate Eq. (11.1) and generate z via Eq. (11.3). Alternatively, with an experimental set-up, one can generate the input profiles $u[0, t_f]$, apply them to the batch process, and collect information regarding the run-end outputs z. Although the static model (11.7)

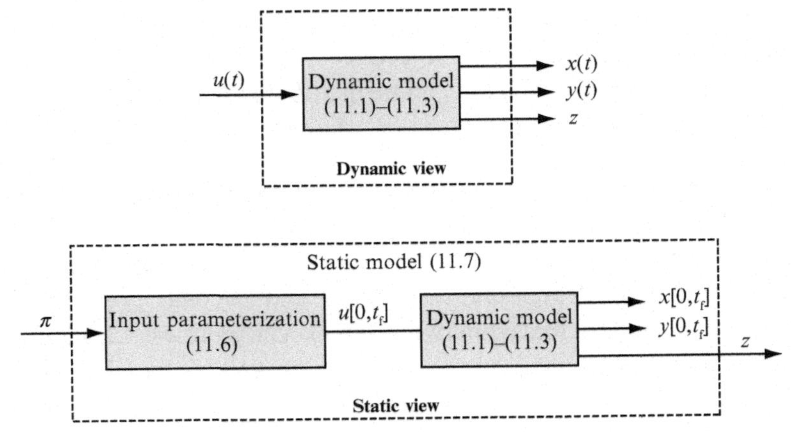

Fig. 11.1

Two different views of the same batch run. *Dynamic view*: The dynamic model describes the behavior of the states $x(t)$, the run-time outputs $y(t)$, and the run-end outputs z as functions of the inputs $u(t)$. *Static view*: The static model $z = M(\pi)$ explicits the relationship between the input parameters π and the run-end outputs z.

looks simple, one has to keep in mind that the batch dynamics are hidden by the fact that only the relationship between the input parameters (before the run) and the run-end outputs (after the run) is considered. The dynamics are represented implicitly in the static map between π and z. These two different ways of looking at the operation of a given batch are shown in Fig. 11.1. Naturally, the "Dynamic view" will be relevant for online operations, while the "Static view" will be more suited for run-to-run operations.

Part A Control

Batch process control has gained popularity over the past 40 years. This is due to the development of novel algorithms with proven ability to improve the performance of industrial processes despite uncertainty and perturbations. But the development of new sensors and of automation tools also contributed to this success. At this stage, it is important to distinguish between control and automation. In practice, industry often tends to consider that control and automation are two identical topics and use the two words interchangeably. Although arguable, we will consider hereafter that *control* corresponds to the theory, that is, the research field, whereby new control methods, algorithms, and controller structures are sought. On the other hand, *automation* represents the technology (the set of devices and their organization) that is needed to program and implement control solutions.

The control of batch processes differs from the control of continuous processes because of the two main distinguishing features presented in Section 11.2. First, since batch processes have no steady-state operating point, the setpoints (SPs) to track are *time-varying trajectories*. Second, batch processes are repeated over time and are characterized by *two independent variables*: the run time t and the run index k. The independent variable k provides additional degrees of freedom for meeting the control objectives when these objectives do not necessarily have to be completed in a single batch but can be distributed over several successive batches. Batch process control aims to track two types of references:

(a) *The run-time references $y_{ref}(t)$ or $y_{ref}[0, t_f]$* are determined offline either from accumulated experience and knowledge of the process, or by solving a dynamic optimization problem (see Section 11.8.1). Examples of trajectories to be tracked include concentration, temperature, pressure, conversion, and viscosity profiles. In general, trajectory tracking can be achieved in two ways: (i) via "online control" in a single batch, whereby input adjustments are made as the batch progresses based on online measurements, or (ii) via "run-to-run control" over several batches, where the input trajectories are computed between batches using information gathered from previous batches. It is also possible to combine approaches (i) and (ii) as will be detailed later.

(b) *The run-end references z_{ref}* are associated with the run-end outputs that are only available at the end of the batch. The most common run-end outputs are product quality, productivity, and selectivity. As with run-time references, the run-end references can be tracked online or on a run-to-run basis.

With these two types of control objectives and the two different ways of reaching them (online and over several runs), there are four different control strategies as shown in Fig. 11.2. It is

Implementation aspect	Control objective: Run-time references $y_{ref}(t)$ or $y_{ref}[0,t_f]$	Control objective: Run-end references z_{ref}
Online --> $u(t)$, $u[t,t_f]$	(1) Feedback control: $u(t)$ → P → $y(t)$; FbC	(2) Predictive control: $u[t,t_f]$ → P → $y(t)$ → E → $\hat{x}(t)$; $z_{pred}(t)$; MPC
Run-to-run --> $u[0,t_f]$	(3) Iterative learning control: $u[0,t_f]$ → P → $y[0,t_f]$; ILC with run update	(4) Run-to-run control: $\pi \xrightarrow{\mathcal{U}(\pi)} u[0,t_f]$ → P → z; RtR with run update

Fig. 11.2
Control strategies for batch processes. The strategies are classified according to the control objective (*horizontal division*) and the implementation aspect (*vertical division*). Each objective can be met either online or on a run-to-run basis depending on the type of measurements available. The signal $u(t)$ represents the input vector at time t, $u[0, t_f]$ the corresponding input trajectories, $y(t)$ the run-time outputs measured online at time t, and z the run-end outputs available at the end of the batch. When online and run-to-run control are combined, the run-to-run part $u[0, t_f]$ can be considered as the feedforward contribution, while the online part $u(t)$ is the feedback part. *P*, plant; *E*, state estimator; *FbC*, feedback control; *ILC*, iterative learning control; *MPC*, model predictive control; *RtR*, run-to-run control.

interesting to see what view (dynamic or static) each control strategy uses. It turns out that Strategies 1–3 use a dynamic view of the batch process, with only Strategy 4 is based on the static view. In the following, Section 11.4 addresses online control (the first row in Fig. 11.1, Strategies 1 and 2), while Section 11.5 deals with run-to-run control (the second row in Fig. 11.1, Strategies 3 and 4).

11.4 Online Control

The idea of "online control" is to take corrective action during the batch, that is, in run-time t. It is possible to do so with two objectives in mind, namely, the run-time reference trajectories $y_{ref}(t)$ and the run-end references z_{ref}. This will be discussed in the next two sections.

11.4.1 Feedback Control of Run-Time Outputs (Strategy 1)

We will address successively the application of conventional feedback control to batch processes, the thermal control of batch reactors, and the stability issue for batch processes.

11.4.1.1 Conventional feedback control

The application of online feedback control to batch processes has the peculiarity that the SPs are often time-varying trajectories. Even if some of the controlled variables (CVs), such as the temperature in isothermal operation, remain constant, the key process parameters (static gains, time constants, and time delays) can vary considerably during the duration of the batch.

Feedback control is implemented using proportional-integral-derivative (PID) techniques or more sophisticated alternatives such as cascade control, predictive control, disturbance compensation (feedforward control), and time-delay compensation.[22,23] The online feedback controller can be written formally as

$$u(t) = \mathcal{K}(y(t), y_{ref}(t)), \tag{11.8}$$

where $\mathcal{K}$ is the online control law for run-time outputs as shown in Fig. 11.3.

One way to adapt for variations in process characteristics is via adaptive control.[24] However, adaptive control requires continuous process identification, which is often impractical in batch processes due to the finite batch duration. Instead, practitioners use conventional feedback control and schedule the values of some of the controller parameters (mostly the gains, thus

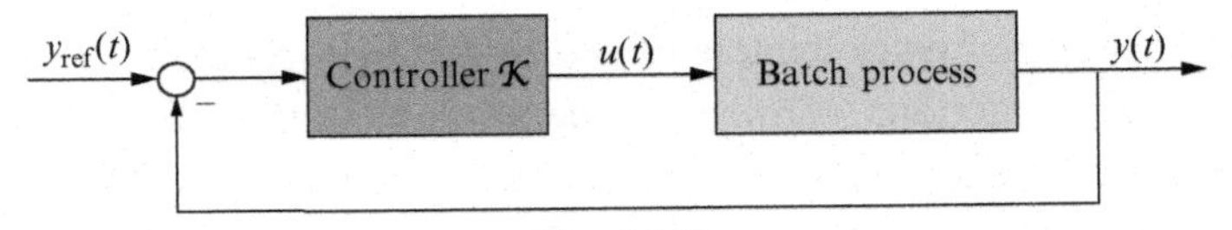

Fig. 11.3
Online feedback control of the run-time outputs y in a batch process.

leading to gain scheduling[25]). Appropriate values of the control parameters can be computed offline ahead of time for the various phases of the batch run and stored for later use.

A few industrial applications of advanced feedback control are available in the literature. For example, a successful implementation of temperature control in an industrial 35-m^3 semibatch polymerization reactor using a flatness-based two-degree-of-freedom controller has been reported.[26] Also, the performance of four different controllers (standard proportional-integral (PI) control, self-tuning PID control, and two nonlinear controllers) for regulating the reactor temperature in a 5-L jacketed batch suspension methyl methacrylate polymerization reactor has been compared.[27] As expected, the performance of the standard PI controller was the poorest since the controller parameters were fixed and not adapted to changing process characteristics. The self-tuning PID control performed better because measurements were used to adapt the controller to the varying process characteristics. The two nonlinear controllers, which were based on differential-geometric techniques requiring full-state measurement,[28] showed excellent performance despite significant uncertainty in the heat-transfer coefficient.

11.4.1.2 Control of heat generation

The control of exothermic reactions in batch reactors is a challenging problem that is tackled using one of various operation modes[29]:

(1) *Isothermal batch operation.* Most reaction systems are investigated isothermally in the laboratory. For safety and selectivity reasons, pilot-plant and industrial reactors are also run isothermally at the same temperature specifications. However, the transfer function between the jacket inlet temperature and the reactor temperature depends on the amount of heat evolved, and it can become unstable for certain combinations of the reaction parameters.[29] Adaptation of the controller parameters is therefore necessary. Most often, this is done by using precomputed values, that is, via gain scheduling. An energy balance around the reactor is sometimes used to estimate the heat generated by the chemical reactions, which represents the major disturbance for the control system. This estimate can then be used very effectively in a feedforward scheme. It is important to mention that isothermal batch operation often exhibits low productivity because it is limited by the maximal heat-generation rate (corresponding to the maximal heat-removal capacity of the reactor vessel), which typically occurs initially for only a short period of time.
(2) *Isothermal semibatch operation.* The productivity of isothermal batch reactors can be increased through semibatch operation. This way, the reactant concentrations can be increased once the initial phase characterized by the heat-removal limitation is over. For strongly exothermic reactions, semibatch operation has the additional advantage of improving the effective heat-removal capacity through feeding of cold reactants. This, in turn, increases the productivity of the reactor by allowing higher temperatures. Hence, isothermal semibatch operation is often the preferred way of running discontinuous

reactors in industry. The temperature is controlled by manipulating the feedrate of the limiting reactant. However, the system exhibits nonminimum phase behavior, that is, a flowrate increase of the cold feed first reduces the temperature before the temperature raises due to higher reaction rates, which significantly complicates the control.[30] Furthermore, in most semibatch operations, it is useful to finish off with a batch phase so as to fully consume the limiting reactant. Without a kinetic model, it is not obvious when to switch from the semibatch to the batch mode. Finally, although isothermal, the operation of a semibatch reactor should be such that the heat-removal constraint is never violated. This has forced industrial practice to be rather conservative, with a temperature SP often chosen from adiabatic considerations.

(3) *Constant-rate batch operation.* Since a key objective in the operation of batch reactors is to master the amount of heat produced by the chemical transformations, it is meaningful to maintain a constant reaction rate. The temperature is kept low initially when the concentrations are high, to be increased with time as the reactants deplete. This way, the reactor productivity is considerably higher than through isothermal operation. However, large temperature excursions can have a detrimental effect on product selectivity. If reaction kinetics are known, it is a simple matter to compute the temperature profile that results in the desired heat generation. If the heat evolution can be measured or estimated online, feedback control can maintain it at the (constant) SP. Laboratory-scale reaction calorimeters have been devised to measure or estimate the heat-evolution profiles of (possibly unknown) reaction systems.[31] The temperature profiles obtained in the laboratory can then be used in production environments, at least in qualitative terms, provided that sufficient heat-removal capacity is available.

11.4.1.3 Run-time stability of batch processes

Stability is of uppermost importance to prevent runaway-type behavior. However, run-time *asymptotic* stability is of little value since t_f is finite. Hence, one can say that "instability" can be tolerated for a batch process because of its finite duration.

Yet, an important issue in run time is reproducibility, which addresses the question of whether the trajectories of various runs with sufficiently close initial conditions and identical input profiles will remain close during the run. The problem of stability is in fact related to the sensitivity to perturbations. The basic framework for assessing this sensitivity is as follows:

- The system is perturbed by either variations of the initial conditions or external disturbances that affect the system states.
- Sensitivity is assessed as a norm indicating the relative effect of the perturbations.

For a system to be stable, this norm must remain bounded for a bounded perturbation, and it must go to zero with time for a vanishing perturbation. However, when dealing with finite-time systems, it is difficult to infer stability from this norm since, except for some special cases such as finite escape time, boundedness is guaranteed. Also, the vanishing behavior with time cannot be analyzed since t_f is finite. The element of interest is in fact the numerical value of the norm, and the issue becomes quantitative rather than binary (yes-no). Details can be found elsewhere.[15]

11.4.2 Predictive Control of Run-End Outputs (Strategy 2)

With a sufficiently accurate process model, and in the absence of disturbances, tracking the profiles determined offline is often sufficient to meet the batch-end product quality requirements.[32] However, in the presence of disturbances, following prespecified profiles is unlikely to lead to the desired product quality. Hence, the following question arises: Is it possible to design an *online* control scheme for effective control of *run-end* outputs using run-time measurements? Since such an approach amounts to controlling a quantity that has not yet been measured, it is necessary to *predict* run-end outputs to compute the required corrective control action. Model predictive control (MPC) is well suited for the task of controlling future quantities that need to be predicted.[33] This approach to batch control can, therefore, be formulated as an MPC problem with a shrinking prediction horizon (equal to the remaining duration of the batch) and an objective function that penalizes deviations from the desired product quality at batch end.[34]

11.4.2.1 Model predictive control

With the MPC strategy, two steps need to be implemented online at each sampling time, namely, the estimation of the current states and the computation of piecewise-constant profiles of future control moves.

First step: Online estimation of $x(t)$

This is done via state estimation using models (11.1), (11.2) and measurements of the run-time outputs $y(t)$.[35,36] The resulting state estimate is denoted $\hat{x}(t)$. However, this task is difficult to implement with sufficient accuracy because of some of the limiting characteristics of batch processes, in particular their nonlinear behavior, the presence of large model inaccuracies, and the lack of specific measurements.

Second step: Online prediction of z and computation of $u[t, t_f]$

In the context of batch operation, predictive control consists in determining the inputs $u[t, t_f]$ to meet z_{ref} by solving at time t a finite-horizon open-loop optimal control problem. Concretely, we consider the dynamic model given by Eqs. (11.1), (11.3) and solve the following optimization problem at the discrete time instant t_i:

$$\min_{u[t_i,\, t_f]} \quad J := \frac{1}{2}\Delta z^T(t_i) P\, \Delta z(t_i) + \frac{1}{2}\int_{t_i}^{t_f} u^T(t) R u(t)\, dt \tag{11.9}$$

$$\text{s.t.} \quad \dot{x}(t) = F(x(t), u(t)), \quad x(t_i) = \hat{x}(t_i), \tag{11.10}$$

$$z_{\text{pred}}(t_i) = Z(x[0, t_f], u[0, t_f]), \tag{11.11}$$

$$x(t_f) \in \mathcal{X}, \tag{11.12}$$

where $z_{\text{pred}}(t_i)$ is the run-end outputs predicted at time t_i, $\Delta z(t_i) := z_{\text{pred}}(t_i) - z_{\text{ref}}$ the predicted run-end errors, P and R are positive-definite weighting matrices of appropriate dimensions, $\mathcal{X}$ is the bounded region of state space where the final state should lie, t_i is the current discrete time instant at which the optimization is performed, and t_f the final time. Note that the profiles $u[0, t_f]$ and $x[0, t_f]$ include (i) the parts $u[0, t_i]$ and $x[0, t_i]$ that were determined prior to the time t_i and (ii) the parts $u[t_i, t_f]$ and $x[t_i, t_f]$ that are computed at t_i. The integral term in the objective function (11.9) is optional and serves to penalize large control efforts. Numerical optimization yields the control sequence, $u^*[t_i, t_f]$, of which only the first part, $u^*[t_i, t_i + h]$, is applied to the plant in an open-loop fashion, where h is the sampling period. Numerical optimization is then repeated at every sampling instant. Note that it is important to include a bounded region for the final states for the sake of stability.[33]

The control law can be written formally as

$$u[t, t_f] = \mathcal{P}(z_{\text{pred}}(t), z_{\text{ref}}), \tag{11.13}$$

where $\mathcal{P}$ is the predictive control law for run-end outputs and $z_{\text{pred}}(t)$ is the prediction of z available at time t as shown in Fig. 11.4.

Linear MPC is widely used in industry, more so than nonlinear MPC. However, one may ask whether linear MPC is sufficient to deal with highly nonlinear batch processes? Note that nonlinear MPC is an active area of research. For example, a nonlinear MPC scheme has been proposed to regulate run-end molecular properties within bounds in the context of batch polymerization reactors.[37]

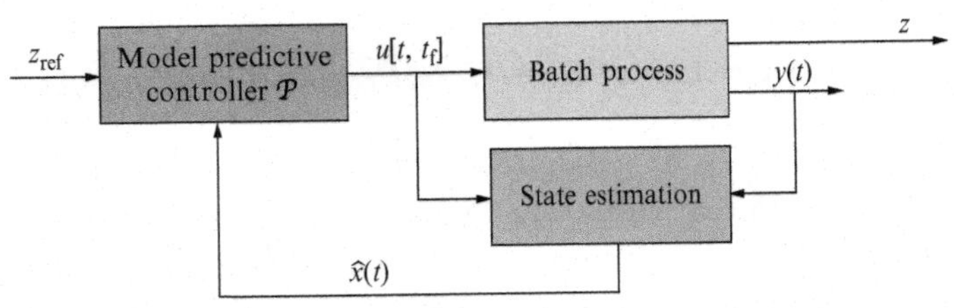

Fig. 11.4
Online predictive control of the run-end outputs z in a batch process. The model in the "Model Predictive Controller" block is used to compute $z_{\text{pred}}(t)$ from $\hat{x}(t)$.

11.4.2.2 Alternative ways of meeting z_{ref} online

In MPC, z is predicted online as $z_{\mathrm{pred}}(t)$ from the state estimates $\hat{x}(t)$, and then the control profile $u[t, t_f]$ is computed via numerical optimization. This is rather demanding, in particular regarding the availability of an accurate process model for both state estimation and input computation. This requirement can be reduced significantly with the approaches proposed next.

Use of data-driven PLS models

When the detailed dynamic model (11.1) is not available, it is possible to use PLS regression to identify models from historical data.[38] Then, in real time during the progress of the batch, the PLS model is used to estimate the run-end outputs and track z_{ref} by adjusting $u(t)$ via online control.

Another less ambitious but very efficient control approach has also been proposed, whereby, if the prediction of z falls outside a predefined no-control region, a mid-course correction is implemented.[39] Note that the approach can easily be modified to accommodate more than one mid-course correction.

Still another variant of the use of PLS models involves controlling the process in the reduced space (scores) of a latent variable model rather than in the space of the discretized inputs.[40]

Use of online tracking

An alternative for meeting run-end references using online measurements consists in tracking some feasible trajectory, $z_{\mathrm{ref}}(t)$, whose main purpose is to enforce z_{ref} at final time, that is, $z_{\mathrm{ref}}(t_f) = z_{\mathrm{ref}}$.[41] This necessitates to be able to measure or estimate $z(t)$ during the batch.

11.5 Run-to-Run Control

Run-to-run control takes advantage of the repetition of batches that is characteristic of batch processing. One uses information from previous batches to improve the performance of the next batch. This can be achieved with respect to both run-time and run-end objectives, as discussed in the next two sections.

11.5.1 Iterative Learning Control of Run-Time Profiles (Strategy 3)

The time profiles of the MVs can be generated using *iterative learning control* (ILC), which exploits information from previous runs to improve the performance of the current run.[42] This strategy exhibits the limitations of open-loop control with respect to the current run, in particular the fact that there is no feedback correction for run-time disturbances. Nevertheless, this scheme is useful for generating time-varying feedforward input terms. The ILC controller has the formal structure

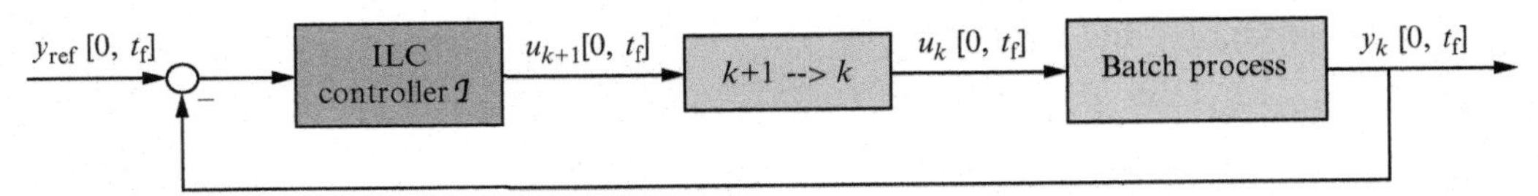

Fig. 11.5
ILC of the run-time profiles $y_k[0, t_f]$ in a batch process. The run update is indicated by $k + 1 \to k$.

$$u_{k+1}[0,t_f] = \mathcal{I}(y_k[0,t_f], y_{\mathrm{ref}}[0,t_f]), \tag{11.14}$$

where $\mathcal{I}$ is the ILC law for run-time outputs. ILC uses the entire profiles of the previous run to generate the input profiles for the next run as shown in Fig. 11.5. ILC has been successfully applied in robotics[43,44] and in batch chemical processing.[45,46]

11.5.1.1 ILC problem formulation

A repetitive batch process is considered in operator notation:

$$y_k[0,t_f] = \mathcal{G}(u_k[0,t_f], x_k(0)), \tag{11.15}$$

where $u_k[0, t_f]$ and $x_k(0)$ represent the input profiles and the initial conditions in run k, $\mathcal{G}$ is the operator representing the system. This relationship can be obtained from models (11.1), (11.2) as follows:

- With the inputs $u_k[0, t_f]$ and the initial conditions $x_k(0)$ specified, integrate equation (11.1) to obtain $x_k[0, t_f]$.
- Eq. (11.2) allows computing $y_k[0, t_f]$ from $x_k[0, t_f]$ and $u_k[0, t_f]$.

For the case where the system operator is linear and the signals are in discrete time, Eq. (11.15) can be expressed in matrix form as follows:

$$y_k = Gu_k + y_{0,k}, \tag{11.16}$$

with $y_k \in \Re^{Np}$, $u_k \in \Re^{Nm}$, and $G \in \Re^{Np \times Nm}$, where N is the finite number of time samples, p the number of outputs, and m the number of inputs. $y_{0,k} \in \Re^{Np}$ represents the response to the initial conditions $x_k(0)$.

ILC tries to improve trajectory following by utilizing the previous-cycle tracking errors. The ILC update law for the inputs u_{k+1} is given as

$$u_{k+1} = Au_k + Be_k, \quad e_k = y_{\mathrm{ref}} - y_k, \tag{11.17}$$

where $y_{\mathrm{ref}} \in \Re^{Np}$ is the references to be tracked, and $A \in \Re^{Nm \times Nm}$ and $B \in \Re^{Nm \times Np}$ are operators applied to the previous-cycle inputs and tracking errors, respectively. In the remainder of this section dealing with ILC, the signals without an explicit time dependency are expressed as vectors containing the N finite time samples, that is, $u_k \in \Re^{Nm}$, $y_k \in \Re^{Np}$, and $e_k \in \Re^{Np}$.

11.5.1.2 ILC convergence and residual errors

An ILC law is convergent if the following limits exist:

$$\lim_{k\to\infty} y_k = y_\infty \quad \text{and} \quad \lim_{k\to\infty} u_k = u_\infty. \tag{11.18}$$

One way to ensure convergence is that the relation between u_k and u_{k+1} be a contraction mapping in some appropriate norm:

$$\| u_{k+1} \| \le \rho \| u_k \|, \quad 0 \le \rho < 1. \tag{11.19}$$

Using Eq. (11.16) in Eq. (11.17) gives:

$$u_{k+1} = (A - BG)u_k + B(y_{\text{ref}} - y_{0,k}). \tag{11.20}$$

Note that $(y_{\text{ref}} - y_{0,k})$ does not represent errors as in Eq. (11.17), but the differences between the reference trajectories and the responses to the initial conditions. Also, in contrast to Eq. (11.17), where e_k depends on u_k through y_k, $(y_{\text{ref}} - y_{0,k})$ is independent of u_k. Hence, the convergence of the algorithm depends on the homogenous part of Eq. (11.20) and the condition for convergence is

$$\| A - BG \| < 1. \tag{11.21}$$

If the iterative scheme converges, with $u_k = u_{k+1} = u_\infty$ and $y_{0,k} = y_{0,k+1} = y_0$, Eq. (11.20) gives:

$$u_\infty = (I - A + BG)^{-1} B(y_{\text{ref}} - y_0). \tag{11.22}$$

Since the converged outputs y_∞ are not necessarily equal to the desired outputs y_{ref}, the final tracking errors can be different from zero. The final tracking errors are given by

$$e_\infty = y_{\text{ref}} - (Gu_\infty + y_0) = (I - G(I - A + BG)^{-1} B)(y_{\text{ref}} - y_0). \tag{11.23}$$

If BG is invertible, this equation can be rewritten as

$$e_\infty = G(BG)^{-1}(I - A)(I - A + BG)^{-1} B(y_{\text{ref}} - y_0). \tag{11.24}$$

Note that the residual errors are zero when $A = I$ and nonzero otherwise. Hence, the choice $A = I$ provides integral action along the run index k. Zero errors imply that perfect system inversion has been achieved.

11.5.1.3 ILC with current-cycle feedback

In conventional ILC schemes, only the tracking errors of the previous cycle are used to adjust the inputs in the current cycle. To be able to reject within-run perturbations, the errors occurring during the current cycle can be used as well, which leads to modified update laws. However, from the points of view of convergence and error analysis, these modified schemes simply correspond to different choices of the operators A and B in Eq. (11.17) as shown next. Let

$$u_{k+1} = u_{k+1}^{\mathrm{ff}} + u_{k+1}^{\mathrm{fb}}, \tag{11.25}$$

$$\text{with } u_{k+1}^{\mathrm{ff}} = \bar{A}u_k^{\mathrm{ff}} + \bar{B}e_k, \quad u_{k+1}^{\mathrm{fb}} = Ce_{k+1}, \tag{11.26}$$

where the superscripts $(\cdot)^{\mathrm{ff}}$ and $(\cdot)^{\mathrm{fb}}$ are used to represent the feedforward and feedback parts of the inputs, respectively, and $\bar{A}$, $\bar{B}$, and C are operators.

Using Eq. (11.26) in Eq. (11.25) and combining it with Eqs. (11.16), (11.17) gives:

$$\begin{aligned} u_{k+1} &= (I+CG)^{-1}\left(\bar{A}(I+CG)-\bar{B}G\right)u_k + (I+CG)^{-1}\left(\bar{B}+C-\bar{A}C\right)(y_{\mathrm{ref}}-y_0) \\ &= (A-BG)u_k + B(y_{\mathrm{ref}}-y_0), \end{aligned} \tag{11.27}$$

where $A=(I+CG)^{-1}(\bar{A}+CG)$ and $B=(I+CG)^{-1}(\bar{B}+(I-\bar{A})C)$. The convergence condition and residual errors can be analyzed using the operators A and B, similarly to what was done in Section 11.5.1.2.

11.5.1.4 ILC with improved performance

The standard technique for approximate inversion is to introduce a forgetting factor in the input update.[47] This causes the residual tracking errors to be nonzero over the entire interval. Note, however, that the main difficulty with the feasibility of inversion arises during the first part of the trajectory due to unmatched initial conditions.[48] After a certain catch-up time, trajectory following is relatively easy. Hence, the idea is to allow nonzero tracking errors early in the run and have the errors decrease with run time t, which can be achieved with an input shift. This is documented next as part of an ILC implementation that includes (i) input shift, (ii) shift of the previous-cycle errors, and (iii) current-cycle feedback.[49]

The iterative update law is written as follows:

$$u_{k+1}(t) = u_{k+1}^{\mathrm{ff}}(t) + K^{\mathrm{fb}}e_{k+1}(t), \tag{11.28}$$

where $K^{\mathrm{fb}} \in \Re^{Nm\times Np}$ is the proportional gains of the online feedback controller. The feedforward part of the current inputs, $u_{k+1}^{\mathrm{ff}}(t)$, consists of shifted versions of the feedforward part of the previous inputs and the previous-cycle tracking errors:

$$u_{k+1}^{\mathrm{ff}}(t) = u_k^{\mathrm{ff}}(t+\delta_u) + K^{\mathrm{ff}}e_k(t+\delta_e), \tag{11.29}$$

where $K^{\mathrm{ff}} \in \Re^{Nm\times Np}$ is the proportional gains of the feedforward controller, δ_u the time shift of the feedforward input trajectories, and δ_e the time shift of the previous-run error trajectory. The remaining parts of the inputs and errors are kept constant, that is, $u_k^{\mathrm{ff}}[t_f-\delta_u, t_f] = u_k^{\mathrm{ff}}(t_f-\delta_u)$ and $e_k[t_f-\delta_e, t_f] = e_k(t_f-\delta_e)$, respectively.

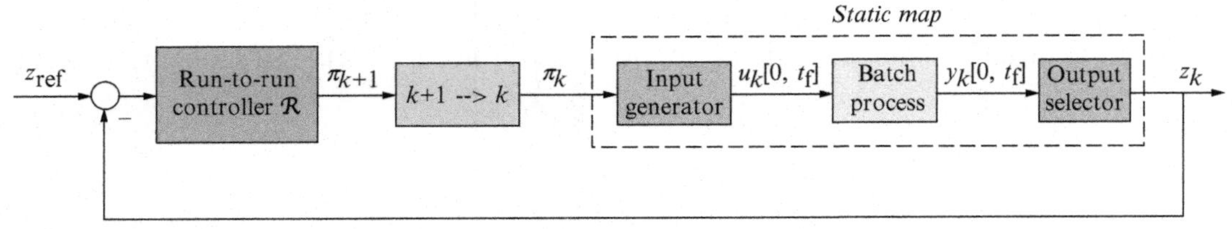

Fig. 11.6

Run-to-run control of the run-end outputs z in a batch process. The input generator constructs $u_k[0, t_f]$ from π_k according to Eq. (11.31), while the output selector selects the run-end outputs.

11.5.2 Run-to-Run Control of Run-End Outputs (Strategy 4)

In this case, the interest is to steer the run-end outputs z toward z_{ref} using the input parameters π. For this, one uses the static model (11.7) that results from the input parameterization (Eq. 11.6). The run-to-run feedback control law can be written as

$$\pi_{k+1} = \mathcal{R}(z_k, z_{ref}), \tag{11.30}$$

$$u_k[0, t_f] = \mathcal{U}(\pi_k, t), \tag{11.31}$$

where $\mathcal{U}$ represents the input parameterization and $\mathcal{R}$ is the run-to-run feedback control law for run-end outputs as shown in Fig. 11.6. For example, one can use the discrete integral control law

$$\pi_{k+1} = \pi_k + K(z_{ref} - z_k), \tag{11.32}$$

where K is an appropriate gain matrix. To be able to meet exactly all references, it is necessary to have at least as many inputs parameters as there are run-end outputs.

11.5.2.1 Control algorithm

Run-to-run control proceeds as follows:

1. *Initialization*. Parameterize the input profiles $u(t)$ as $\mathcal{U}(\pi, t)$ in such a way that the input parameter vector π has the same dimension as the vector of run-end outputs z in order to generate a square control problem. Set $k = 1$ and select an initial guess for π_k.
2. *Iteration*. Construct the inputs $u_k[0, t_f]$ using Eq. (11.31) and apply them to the process. Complete the run and determine z_k from the measurements.
3. Update the input parameters using the control law (11.30). Set $k := k + 1$ and repeat Steps 2–3 until convergence.

11.5.2.2 Convergence analysis

The convergence of the run-to-run algorithm can be determined by analyzing the closed-loop error dynamics, where the errors are $e_k := z_{ref} - z_k$.

We discuss next a convergence analysis that considers a linearized version of system (11.7) and the linear integral control law (11.32). With the linear model,

$$z_k = S\pi_k, \tag{11.33}$$

where $S = \left.\frac{\partial M}{\partial \pi}\right|_{\overline{\pi}}$ is the sensitivity matrix computed at the nominal operating point corresponding to $\overline{\pi}$, the linearized error dynamics become

$$e_{k+1} = e_k - SKe_k = (I - SK)e_k. \tag{11.34}$$

The eigenvalues of the matrix $(I - S\,K)$ should be within the unit circle for the algorithm to converge. Convergence is determined by both the sensitivity matrix S and the controller gains K. Convergence analysis of run-to-run control algorithms is also possible when the relationship between π_k and z_k is nonlinear, but it typically requires additional assumptions regarding the nature of the nonlinearities.[50]

11.5.2.3 Run-to-run stability

The interest in studying stability in the run index k arises from the necessity to guarantee convergence of run-to-run control schemes. Here, the standard notion of stability applies as the independent variable k goes to infinity. The main conceptual difference with the stability of continuous processes is that "equilibrium" refers to entire trajectories. Hence, the norms have to be defined in the space of functions $\mathcal{L}$ such as the integral squared norm l_2 of the signal $x(t)$:

$$\| x[0, t_f] \|_{l_2} = \int_0^{t_f} \left(x^2(t)dt\right)^{1/2}. \tag{11.35}$$

For studying stability with respect to the run index k, system (11.1) is considered under closed-loop operation, that is, with all possible online and run-to-run feedback loops. Hence, we no longer restrict our attention to the input-output run-to-run π-z behavior, but we consider the entire state vector. For simplicity, let us assume that t_f is constant for all runs. When dealing with the kth run, the trajectories of the $(k-1)$st run are known, which fixes $u_k[0, t_f]$ according to the ILC control law (11.14). These input profiles, along with the online feedback law (11.8) or (11.13), are applied to system (11.1) to obtain $x_k(t)$ for all t and thus $x_k[0, t_f]$. All these operations can be represented formally as:

$$x_k[0, t_f] = \mathcal{F}(x_{k-1}[0, t_f]), \quad x_0[0, t_f] = x_{\text{init}}[0, t_f], \tag{11.36}$$

where $x_{\text{init}}[0, t_f]$ is the initial state trajectories obtained by integration of Eq. (11.1) for $k = 0$, $x_{0,0} = x_{\text{init}}$, and $u_0[0, t_f] = u_{\text{init}}[0, t_f]$. Eq. (11.36) describes the run-to-run dynamics associated with all control activities. Run-to-run stability is considered around the equilibrium trajectory computed from Eq. (11.36), that is $\overline{x}[0, t_f] = \mathcal{F}(\overline{x}[0, t_f])$, and is investigated using a Lyapunov-function method.[51]

11.6 Batch Automation

Although batch processing was dominant until the 1930s, batch process control received significant interest only 50 years later.[52] This is mainly due to the fact that, when control theory started to emerge in the 1940s, there was a shift from the dominance of batch processing to continuous operations. Hence, the dynamics and control of batch processes became a subject of investigation in the 1980s. Engineers then tried to carry over to batch processes the experience gained over the years with the control of continuous processes. However, the specificities of batch processes make their control quite challenging.

This is also true at the implementation level. If continuous processes typically exhibit four main regimes, namely, start-up, continuous operation, possibly grade transition, and shut down, batch processes are much more versatile since, in the absence of a steady state, the system evolves freely from a set of initial conditions to a final state. Hence, the recipe can be rather varied, ranging from basic isothermal operation to a succession of complex operations that can involve discrete decisions. Also, batch processes are often repeated over time, with or without changes between batches, which makes the task of designing a control strategy much more involved than for continuous operations.

The control of batch operations relies on various types of controllers that include stand-alone controllers, programmable logic controllers (PLCs), distributed control systems (DCS), and personal computers (PCs). These four platforms are discussed next.

11.6.1 Stand-Alone Controllers

The basic stand-alone controllers are single-loop controllers (SLCs).[53] These devices generally embed a microprocessor with fixed functionalities such as PID control. The popularity of SLCs arose from their simplicity, low cost, and small size. Originally, stand-alone controllers were capable of integrating dual-loop control, that is, they can handle two control loops with basic on/off and PID control. More recently, these devices have incorporated self-tuning algorithms and, more importantly for batch processes, time scheduling and sequencing functionalities. Still, the main advantage of SLCs is their low cost per control loop. Furthermore, they are often used in parallel or in combination with more advanced control structures such as DCS, mainly because they can achieve very acceptable control performance for minimal investment regarding cost, maintenance, and required knowledge.

11.6.2 Programmable Logic Controllers

PLCs were introduced in the process industries in the early 1970s, following the development in the automotive industry, as computing systems "that had the flexibility of a computer, yet could be programmed and maintained by plant engineers and technicians."[53] Before PLCs, relays, counters, and timers were mainly used, but they offered much less flexibility. PLCs can perform complex computations as they have a computational power that is comparable to a small PC.[54]

Considerable progress was made with the introduction of micro-PLCs, which can be installed close to the process at a much lower cost. The development of Supervisory Control and Data Acquisition (SCADA) constitutes another breakthrough that has increased the scope of applications. It is possible to handle numerous operations distributed over a large distance, thereby opening up the applicability of PLCs to cases where batch process operations are fully integrated in the plant-wide operation of a production site.

11.6.3 Distributed Control Systems

DCS are control systems with control elements spread around the plant. They generally propose a hierarchical approach to control, with several interconnected layers operating at different time scales. At the top of the hierarchy is the production/scheduling layer, where the major decisions are taken on the basis of market considerations and measurements collected from the plant. These decisions are sent to an intermediate layer, where they are used, together with plant measurements and estimated fluctuations on price and raw-material quality, to optimize the plant performance. These decisions can take the form of SP trajectories for low-level controllers. This intermediate layer, which is often referred to as the real-time optimization (RTO) layer, is implemented via DCS.[55] Because of their distributed nature, DCS can help integrate batch operations into the continuous operation of a plant, which has considerable impact on the planning, scheduling, and RTO tasks.[56] The control and optimization hardware has improved continuously over the past decades, thereby keeping pace with the development of advanced control and optimization methods.

Although the differences between DCS and SCADA can appear to be subtle, one could state the following[56]: DCS (i) are mainly process driven, (ii) are focused on small geographic areas, (iii) are suited to large integrated chemical plants, (iv) rely on good data quality, and (v) incorporate powerful closed-loop control hardware. On the other hand, SCADA is generally more "event driven," which makes it more suited to the supervision of multiple independent systems.

11.6.4 Personal Computers

PCs can also be used for the control and optimization of batch processes. It has been reported that 25% of batch process control involves personal computers. Of course, this figure does not mean that 25% of batch process control involves only PCs, as PCs are often used in combination with other controllers such as SLCs or PLCs. In industry, PCs are not only used to program and run control algorithms, but they include many other elements such as interfaces, communication protocols, and networking.[53] The development of user-friendly software interfaces, of simulation tools, and of fast and reliable optimization software has been key to the development of batch processing. With the relatively low cost of PCs, it has become economically viable to use them routinely for control. While a PC gives a lot of flexibility, especially for taking educated decisions in the context of a multilayered control structure, it can also be used as a stand-alone advanced controller or even as an online optimization tool. In practice, however, PC-based control is often limited to supervisory control, where the PC sends recommendations to low-level controllers. This can take the form of input profiles to implement or of output trajectories to track during batch operation. These recommendations can then be implemented directly by means of low-level controllers, or manually by an operator who can take the final decision of implementing or not the recommendations.[53]

11.7 Control Applications

11.7.1 Control of Temperature and Final Concentrations in a Semibatch Reactor

11.7.1.1 Reaction system

Consider the reaction of pyrrole A with diketene B to produce 2-acetoacetyl pyrrole C. Diketene is a very aggressive compound that reacts with itself and other species to produce the undesired products D, E, and F.[57] The desired and side reactions are

$$A + B \rightarrow C,$$

$$2B \rightarrow D,$$

$$C + B \rightarrow E,$$

$$B \rightarrow F.$$

The reactions are highly exothermic, that is, they produce heat. The reactor is operated in semibatch mode with A present initially in the reactor and B added with the feedrate profile $u(t)$. Moreover, the reaction system is kept isothermal by removing the heat produced by the chemical reactions through a cooling jacket surrounding the reactor. The flowrate of cooling fluid is adjusted to keep the reactor temperature at its desired SP.

11.7.1.2 Model of the reactor

The nonlinear dynamic model of the reaction system is obtained from a material balance for each species and a total mass balance[58]:

$$\frac{dc_A}{dt} = -\frac{u}{V}c_A - k_1 c_A c_B, \qquad (11.37)$$

$$\frac{dc_B}{dt} = \frac{u}{V}(c_B^{in} - c_B) - k_1 c_A c_B - 2k_2 c_B^2 - k_3 c_C c_B - k_4 c_B, \tag{11.38}$$

$$\frac{dc_C}{dt} = -\frac{u}{V} c_C + k_1 c_A c_B - k_3 c_C c_B, \tag{11.39}$$

$$\frac{dc_D}{dt} = -\frac{u}{V} c_D + k_2 c_B^2, \tag{11.40}$$

$$\frac{dc_E}{dt} = -\frac{u}{V} c_E + k_3 c_C c_B, \tag{11.41}$$

$$\frac{dc_F}{dt} = -\frac{u}{V} c_F + k_4 c_B, \tag{11.42}$$

$$\frac{dV}{dt} = u, \tag{11.43}$$

where c_i is the concentration of the ith species, V is the reactor volume, k_j is the rate constant of the jth reaction, c_B^{in} is the inlet concentration of B, and u is the volumetric feedrate of B.

11.7.1.3 Control objective

One would like to operate the reactor safely and efficiently. If the heat produced cannot be removed by the cooling jacket, the reactions accelerate and produce even more heat. This positive feedback effect, known as thermal runaway, can result in high temperature and pressure and thus lead to an explosion; this is precisely what caused the Bhopal disaster.[59] The performance of the reactor is evaluated by its productivity, namely, the amount of desired product available at final time, as well as its selectivity, namely, the portion of converted reactant B that forms the desired product C. Note that the performance criteria (productivity and selectivity) are evaluated at the final time, whereas the MV (feedrate of B) is a run-time profile. The control objective is to operate isothermally at 50°C and match the final concentrations $c_{B,ref}$ and $c_{D,ref}$ obtained in the laboratory.

11.7.1.4 Online control, ILC, and run-to-run control

The control strategies, shown in Fig. 11.7, include (i) a cascade scheme for online feedback control of $T_r(t)$ with ILC update of $T_{j,ref}^{ff}[0, t_f]$ and (ii) run-to-run control of the final concentrations through update of the input parameters u_1 and u_2.

Control of the reactor temperature T_r is implemented by means of a PI/P-cascade scheme that adjusts the flowrate of cooling fluid. The master loop contributes a feedback term to the reference value for the jacket temperature T_j. This reference value, which also includes a feedforward term, reads

$$T_{j,ref,k}(t) = T_{j,ref,k}^{ff}(t) + T_{j,ref,k}^{fb}(t), \tag{11.44}$$

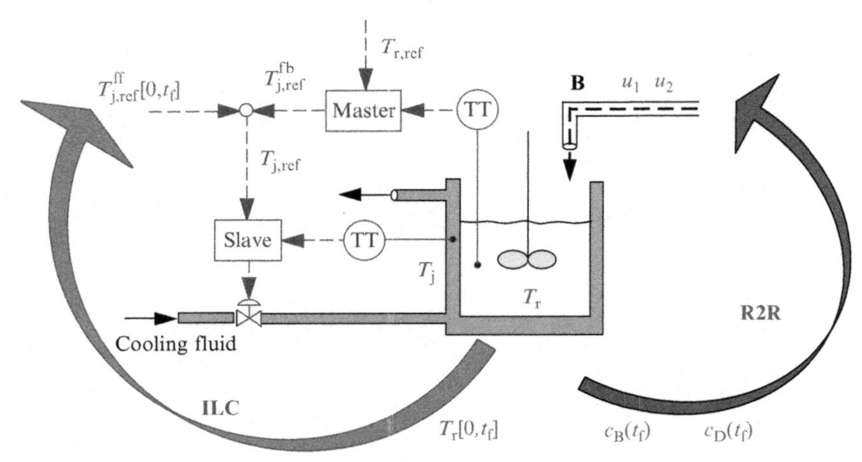

Fig. 11.7
Online and run-to-run strategies for controlling the reactor temperature $T_r(t)$ and the final concentrations $c_B(t_f)$ and $c_D(t_f)$.

$$T^{fb}_{j,ref,k}(t) = K_R\left(e_k(t) + \frac{1}{\tau_I}\int_0^t e_k(\tau)d\tau\right), \tag{11.45}$$

$$T^{ff}_{j,ref,k+1}[0,t_f] = T^{ff}_{j,ref,k}[0,t_f] + K_{ILC}e_k[0,t_f], \tag{11.46}$$

where $e_k(t) := T_{r,ref}(t) - T_{r,k}(t)$, K_R is the proportional gain, and τ_I is the integral time constant of the PI master controller. K_{ILC} is the gain of the ILC controller. Eq. (11.45) computes the feedback term $T^{fb}_{j,ref,k}(t)$ using a PI feedback law, while Eq. (11.46) implements run-to-run adaptation of the feedforward term $T^{ff}_{j,ref,k}(t)$ based on ILC.

The second MV is u, the feedrate of reactant B, through which the two reactions can be steered and brought to the desired final concentrations. Since these final concentrations are measured only at the end of the batch, the feedrate of B is adjusted on a run-to-run basis. Two input parameters are needed to control these two concentrations. Hence, the feedrate profile $u[0, t_f]$ is parameterized using the two feedrate levels u_1 and u_2, each valid over half the batch time. The sensitivities of the final concentrations $z = (c_B(t_f), c_D(t_f))^T$ with respect to $\pi = (u_1, u_2)^T$ are evaluated experimentally, which gives $S = \left.\frac{\partial z}{\partial \pi}\right|_k$. With this notation, the discrete integral control law has the form

$$\pi_{k+1} = \pi_k + K_{RtR}S^{-1}[z_{ref} - z_k], \tag{11.47}$$

where K_{RtR} is the 2 × 2 diagonal gain matrix of the run-to-run controller.

The contributions of ILC and run-to-run control to the control of reactor temperature and final concentrations are shown in Figs. 11.8 and 11.9, respectively. One sees that adjustment of $T^{ff}_{j,ref}$ using ILC reduces the maximal temperature excursion from 57.1 to 52.7°C.

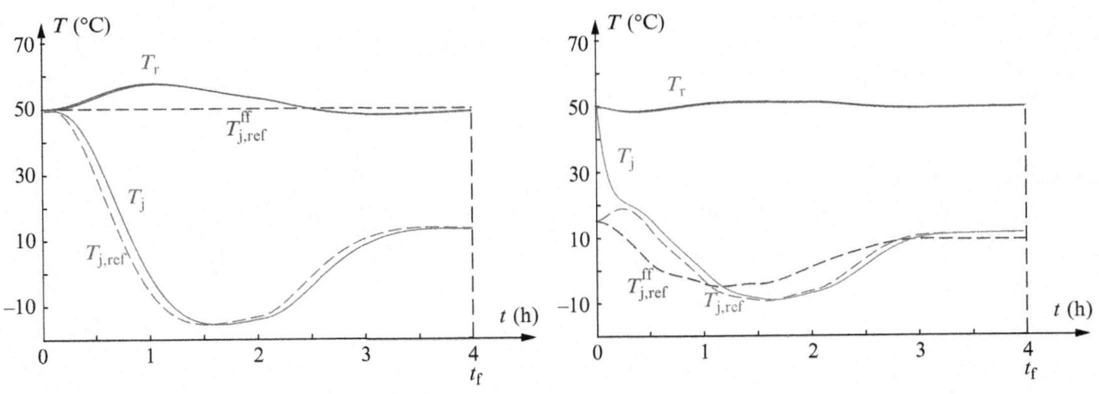

Fig. 11.8
Temperature control around the constant setpoint $T_{r,ref} = 50°C$, without ILC (*left*), and after three ILC iterations (*right*). The *solid lines* represent the reactor temperature T_r and the jacket temperature T_j. The *dashed lines* represent the feedforward term of the jacket temperature setpoint $T^{ff}_{j,ref}$ and the resulting jacket temperature setpoint $T_{j,ref}$.

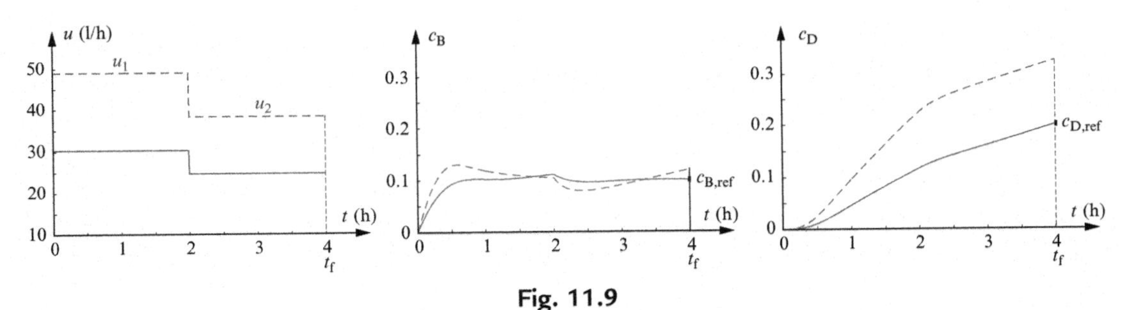

Fig. 11.9
Run-to-run control for meeting the run-end concentrations $c_{B,ref}$ and $c_{D,ref}$. The inlet feedrate is parameterized using the two levels u_1 and u_2 (*left*). The concentration profiles for $c_B(t)$ and $c_D(t)$ are shown in the *center* and at *right*, respectively. The *dashed lines* show the initial profiles, while the *solid lines* show the corresponding profiles after two iterations. The references $c_{B,ref}$ and $c_{D,ref}$ are reached via run-to-run adjustment of u_1 and u_2.

11.7.2 Scale-Up via Feedback Control

Short times to market are required in the specialty chemicals industry. One way to reduce this time to market is by skipping the pilot-plant investigations. However, due to scale-related differences in operating conditions, direct extrapolation of conditions obtained in the laboratory is often impossible, especially when terminal objectives must be met and path constraints respected. In fact, ensuring feasibility at the industrial scale is of paramount importance. This section presents an example for which the combination of online and run-to-run control allows meeting production requirements over a few batches.

11.7.2.1 Problem formulation

Consider the following parallel reaction scheme[60]:

$$A + B \rightarrow C, \quad 2B \rightarrow D.$$

The desired product is C, while D is undesired. The reactions are exothermic. A 1-L reactor is used in the laboratory, while a jacketed reactor of 5 m^3 is used in production. The manipulated inputs are the feedrate $F(t)$ and the coolant flowrate through the jacket $F_j(t)$. The operational requirements can be formulated as:

$$T_j(t) \geq 10°C, \tag{11.48}$$

$$y_D(t_f) = \frac{2n_D(t_f)}{n_C(t_f) + 2n_D(t_f)} \leq 0.18, \tag{11.49}$$

where $T_j(t)$ is the jacket temperature, and $n_C(t_f)$ and $n_D(t_f)$ denote the numbers of moles of C and D at final time, respectively.

11.7.2.2 Laboratory recipe

The recipe obtained in the laboratory proposes to initially fill the reactor with A, and then feed B at some constant feedrate $\overline{F}$, while maintaining the reactor isothermal at $T_r = 40°C$. As cooling is not an issue for the laboratory reactor equipped with an efficient jacket, experiments were carried out using a scale-down approach, that is, the cooling rate was artificially limited so as to anticipate the limited cooling capacity of the industrial reactor. Scaling down is performed by the introduction of an operating constraint that limits the cooling capacity; in this case, the maximal cooling capacity of the industrial reactor is simply divided by the scale-up factor r:

$$[q_{c,max}]_{lab} = \frac{\left[(T_r - T_{j,min})UA\right]_{prod}}{r}, \tag{11.50}$$

with $r = 5000$ and where $UA = 3.7 \times 10^4$ J/mol°C is the estimated heat-transfer capacity of the production reactor. With $T_r - T_{j,min} = 30°C$, the maximal cooling rate is 222 J/min. Table 11.1 summarizes the key parameters of the laboratory recipe and the (simulated) laboratory results.

Table 11.1 Laboratory recipe and results for the scale-up problem

Recipe Parameters		Laboratory Results
$T_r = 40°C$ $c_{A,o} = 0.5$ mol/L $V_0 = 1$ L $\overline{F} = 4 \times 10^{-4}$ L/min	$c_{B,in} = 5$ mol/L $c_{B,o} = 0$ mol/L $t_f = 240$ min	$n_C(t_f) = 0.346$ mol $y_D(t_f) = 0.170$ $\max_t q_c(t) = 182.6$ J/min

11.7.2.3 Scale-up via online control, ILC, and run-to-run control

The goal of scale-up is to reproduce in production the productivity and selectivity that are obtained in the laboratory, while enforcing the desired reactor temperature. With selectivity and productivity as run-end outputs, the feedrate profile $F[0, t_f]$ is parameterized using two adjustable parameters, namely, the constant feedrate levels F_1 and F_2, each one valid over half the batch time. Temperature control is done via a combined feedforward/feedback scheme as shown in Section 11.7.1.

A control problem can be formulated with the following MVs, CVs, and SPs:

- MVs: $u(t) = T_{j,ref}(t)$, $\pi = [F_1,\ F_2]^T$;
- CVs: $y(t) = T_r(t)$, $z = [n_C(t_f),\ y_D(t_f)]^T$;
- SPs: $y_{ref} = 40°C$, $z_{ref} = [1630\text{ mol},\ 0.17]^T$.

The SPs are chosen as follows:

- $T_{r,ref} = 40°C$ is the value used in the laboratory.
- $n_{C,ref}$ is the value obtained in the laboratory multiplied by the scale-up factor r minus a 100-mol backoff to account for scale-related uncertainties and run-time disturbances. The introduction of backoff makes the control problem more flexible.
- $y_{D,ref}$ is the value obtained in the lab, which respects the upper bound of 0.18.

The control scheme is shown in Fig. 11.10. The input profiles are updated using (i) the cascade feedback controller $\mathcal{K}$ to control the reactor temperature $T_r(t)$ online, (ii) the ILC controller $\mathcal{I}$ to improve the reactor temperature by adjusting $T^{ff}_{j,ref}(t)$, and (iii) the run-to-run controller $\mathcal{R}$ to control $z = [n_C(t_f), y_D(t_f)]^T$ by adjusting $\pi = [F_1, F_2]^T$. Details regarding the implementation of the different control elements can be found elsewhere.[60]

11.7.2.4 Simulation results

The recipe presented in Table 11.1 is applied to the 5-m^3 industrial reactor, equipped with a 2.5-m^3 jacket. In this simulated case study, uncertainty is introduced artificially by modifying the two kinetic parameters, which are reduced by 25% and 20%, respectively. Also, Gaussian noise with standard deviations of 0.001 mol/L and 0.1°C is considered for the measurements of the final concentrations and the reactor temperature, respectively. It turns out that, for the first run, application of the laboratory recipe with $\pi = [r\overline{F},\ r\overline{F}]^T$ violates the final selectivity of D. Upon adapting the MVs with the proposed scale-up algorithm, the free parts of the recipe are modified iteratively to achieve the production targets for the industrial reactor, as shown in Fig. 11.11.

11.7.3 Control of a Batch Distillation Column

A binary batch distillation column is used to illustrate in simulation the application of ILC. This example is described in detail elsewhere.[61]

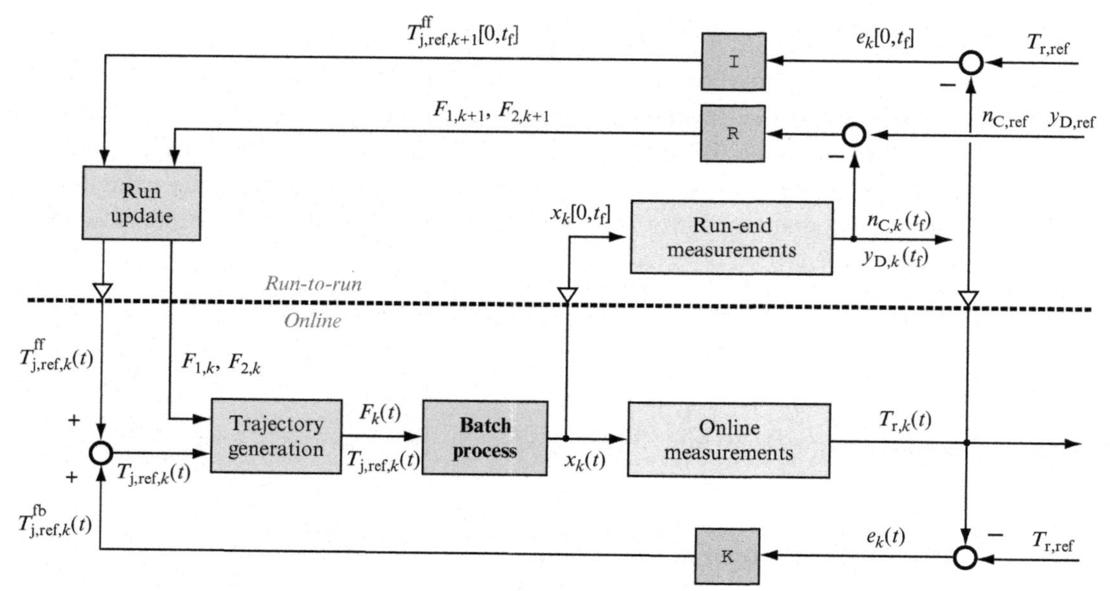

Fig. 11.10

Control scheme for scale-up implementation. Notice the distinction between online and run-to-run activities. The symbol ∇ represents the concentration/expansion of information between a profile (e.g., $x_k[0, t_f]$) and an instantaneous value (e.g., $x_k(t)$).

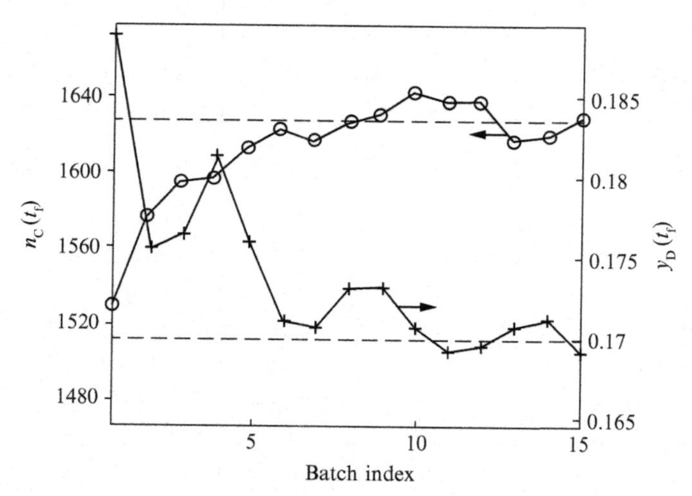

Fig. 11.11

Evolution of the production of C, $n_C(t_f)$ and the yield of D, $y_D(t_f)$ for the large-scale industrial reactor. Most of the improvement is achieved in the first seven runs (the *dashed lines* represent the target values).

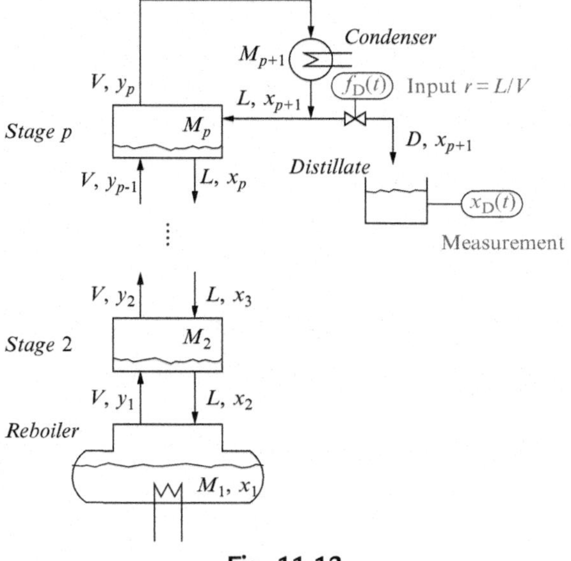

Fig. 11.12
Binary batch distillation column.

11.7.3.1 Model of the column

A binary batch distillation column with p equilibrium stages is considered (see Fig. 11.12). Using standard assumptions[62] and writing molar balance equations for the holdup in the reboiler and for the liquid on the various stages and in the condenser, the following model of order (p + 2) is obtained:

$$\dot{M}_1 = (r-1)V, \tag{11.51}$$

$$\dot{x}_1 = \frac{V}{M_1}(x_1 - y_1 + r x_2), \tag{11.52}$$

$$\dot{x}_i = \frac{V}{M_i}(y_{i-1} - y_i + r(x_{i+1} - x_i)), \quad i = 2, \ldots, p, \tag{11.53}$$

$$\dot{x}_{p+1} = \frac{V}{M_{p+1}}(y_p - x_{p+1}), \tag{11.54}$$

where x_i is the molar liquid fraction, y_i the molar vapor fraction, M_i the molar holdup on Stage i, V the vapor flowrate, and D the distillate flowrate. Stage 1 refers to the reboiler, Stage p to the top of the column, and Stage p + 1 to the condenser. The internal reflux ratio $r = \frac{L}{V} = \frac{V-D}{V}$ is considered as the MV. The vapor-liquid equilibrium relationship is

$$y_i = \frac{\alpha x_i}{1 + (\alpha - 1)x_i}, \quad i = 1, \ldots, p, \tag{11.55}$$

Table 11.2 Model parameters and initial conditions, $i = 2, \ldots, p$

p	10		$M_1(0)$	100	kmol
α	1.6		$x_1(0)$	0.5	
M_i	0.2	kmol	$x_i(0)$	0.5	
M_{p+1}	2	kmol	$x_{p+1}(0)$	0.5	
V	15	kmol/h	$x_{D,ref}$	0.9	
t_f	10	h	h	0.1	h

where α is the relative volatility. The composition of the accumulated distillate, x_D, which is measured with the sampling time $h = 0.1$ h, is given by:

$$x_D(t) = \frac{\sum_{i=1}^{p} (x_i(t)M_i(t) - x_i(0)M_i(0))}{M_1(t) - M_1(0)}. \tag{11.56}$$

The model parameters, the initial conditions, and the control objective $x_{D,ref}$ are given in Table 11.2.

11.7.3.2 Operational objective

A batch is divided into two operational phases:

1. Start-up phase with full reflux up to time t_s: $r = 1$, $t = [0, t_s]$, $t_s = 1.415$ h.
2. Distillation phase: $r \in [0, 1]$, $t = (t_s, t_f]$, $t_f = 10$ h.

The objective is to adjust the reflux ratio to get the distillate purity $x_D(t_f) = 0.9$ at the end of the batch. This will be obtained by tracking an appropriate reference trajectory for the distillate purity $x_D(t)$ in the second phase. This trajectory, which has the property of ending at the desired distillate purity at final time, is chosen to be linear, with $x_{D,ref}(t_s) = 0.925$ and $x_{D,ref}(t_f) = 0.9$. The accumulated distillate purity $x_D(t)$ is measured in Phase 2.

In order to obtain a realistic test scenario, the following uncertainty is considered:

- *Perturbation*: The vapor rate fluctuates every 0.5 h following an uniform distribution in the range $V = [13, 17]$ kmol/h.
- *Measurement noise*: 5% multiplicative Gaussian noise is added to the product composition $x_D(t)$.

The values of the squared tracking error $\sum_{t_s}^{t_f} e^2(t)$ and the final tracking error $e(t_f)$ upon convergence are averaged over 20 realizations of the perturbation and measurement noise. Also, the variance $v_e(t_f)$ of the final tracking error is calculated from 20 realizations.

11.7.3.3 Trajectory tracking via ILC

Trajectory tracking involving a single output, $x_D(t)$, and a single input, $r(t)$, is implemented on a run-to-run basis via ILC. The initial input trajectory used for ILC is linear with $r(t_s) = 0.898$ and $r(t_f) = 0.877$. The residual tracking error cannot be reduced to zero for all times because of the nonzero tracking error at time t_s arising from uncertainties in the start-up phase, that is, $x_D(t_s) \neq x_{D,ref}(t_s)$. As a consequence, the ILC schemes that try to enforce zero residual tracking error do not converge in this case. Instead, nonzero tracking error has to be tolerated to enforce convergence. This can be accomplished by applying a forgetting factor or a time shift to the feedforward trajectory. Three ILC schemes without current-cycle feedback are considered, with $N = 100$ time samples:

- $\beta = 0.999$: A forgetting factor β is applied to the feedforward input trajectory in Eq. (11.26) with $\bar{A} = \beta I$ and $\overline{B} = K$.
- $\delta_u = 0.25$ h: A small shift is applied to the feedforward input trajectory in Eq. (11.29).
- $\delta_u = 1$ h, $\delta_e = 1$ h: The same large shift is applied to the feedforward input and the error trajectory in Eq. (11.29).

The proportional gain $K = 0.1\ I$ is determined as a compromise between robustness and performance. ILC with forgetting factor converges after 30 runs, while the schemes with time shift of the trajectories converge after 25 runs as shown in Fig. 11.13. The final tracking error is slightly smaller with the latter methods, especially when the time shift of the feedforward

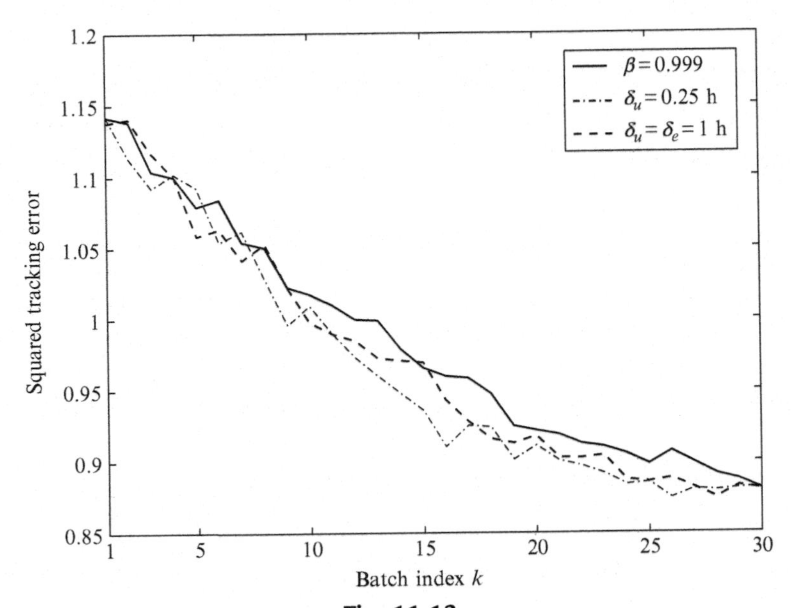

Fig. 11.13
Evolution of the squared tracking error for three ILC schemes.

Table 11.3 Comparison of the three tracking schemes after 30 runs in terms of the squared tracking error $\sum_{t_s}^{t_f} e^2(t)$, the tracking error at final time $e(t_f)$, and its variance $v_e(t_f)$. In bold, the lowest value of $e(t_f)$.

Strategy	$\sum_{t_s}^{t_f} e^2(t)$	$\lvert e(t_f)\rvert \times 10^3$	$v_e(t_f) \times 10^5$
$\beta = 0.999$	0.8738	6.5	4.1
$\delta_u = 0.25$ h	0.8745	**0.5**	3.0
$\delta_u = 1$ h, $\delta_e = 1$ h	0.8833	2.2	3.5

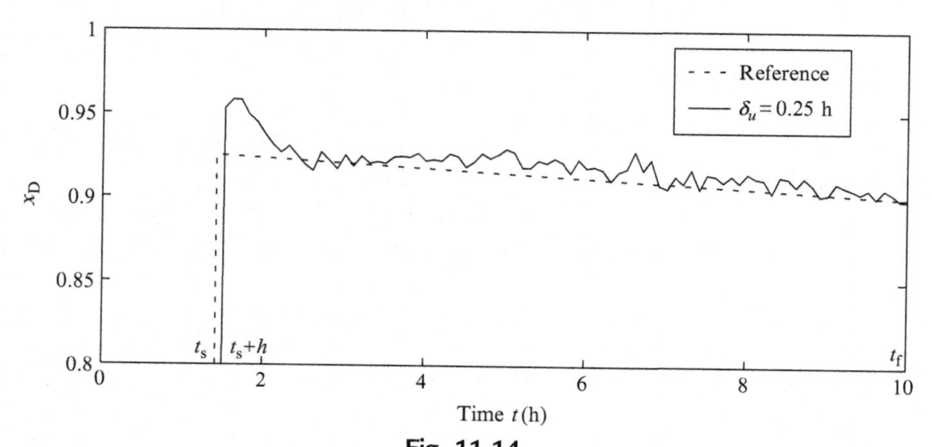

Fig. 11.14
Tracking performance using ILC with input shift.

trajectory is reduced to $\delta_u = 0.25$ h as shown in Table 11.3. Fig. 11.14 shows the tracking performance of ILC with input shift after 20 runs. Note that, since it is necessary to have distillate in the product tank to be able to take measurements, tracking starts one sampling time after the start of Phase 2, that is, at $t_s + h$.

Part B Optimization

Process optimization is the method of choice for improving the performance of chemical processes while enforcing operational constraints.[63] Long considered as an appealing tool but only applicable to academic problems, optimization has now become a viable technology.[64,65] Still, one of the strengths of optimization, namely, its inherent mathematical rigor, can also be perceived as a weakness, since engineers might sometimes find it difficult to obtain an appropriate mathematical formulation to solve their practical problems. Furthermore, even when process models are available, the presence of plant-model mismatch and process disturbances makes the direct use of model-based optimal inputs hazardous.

In the last 30 years, the field of RTO has emerged to help overcome the aforementioned modeling difficulties. RTO integrates process measurements into the optimization framework. This way, process optimization does not rely exclusively on a (possibly inaccurate) process model but also on process information stemming from measurements. The first widely available RTO approach was the two-step approach that adapts the model parameters on the basis of differences between predicted and measured outputs and uses the updated process model to recompute the optimal inputs.[66,67] However, in the presence of *structural* plant-model mismatch, this method is very unlikely to drive the plant to optimality.[68,69] Hence, alternatives to the two-step approach have recently been developed. For example, the modifier-adaptation scheme also proposes to solve a model-based optimization problem, but using a plant model with fixed model parameters.[69] Correction for uncertainty is made via modifier terms that are added to the cost and constraint functions of the optimization problem. As the modifiers include information on the differences between the predicted and the plant necessary conditions of optimality (NCO), this approach is prone to reach the plant optimum upon convergence. Yet another field has emerged, for which numerical optimization is *not* used online. With the so-called self-optimizing approaches,[70–72] the optimization problem is recast as a control problem that uses measurements to enforce certain optimality features for the plant. However, note that self-optimizing control[70] and extremum-seeking control,[71] which have been developed to optimize continuous plants, are not suited to the optimization of batch processes.

The optimization of a batch process falls naturally in the category of *dynamic optimization*, that is, with the infinite-dimensional profiles $u(t)$ as decision variables. However, the introduction of input parameterization can transform the dynamic optimization problem into a static optimization problem with a countable number of decision variables, a so-called nonlinear program (NLP).

This optimization part is organized as follows. Various aspects of numerical optimization for both static and dynamic optimizations are presented first, followed by a discussion of two RTO classes, namely, repeated numerical optimization and optimizing control. The theoretical developments will then be illustrated by three case studies.

11.8 Numerical Optimization

Apart from very specific cases, the standard way of solving an optimization problem is via numerical optimization, for which a model of the process is required. As already mentioned, the optimization of a batch process can be formulated as a dynamic optimization problem. In this chapter, we will consider the following constrained dynamic optimization problem:

$$\min_{u[0,\, t_f]} \quad J := \phi(x(t_f)) \tag{11.57}$$

$$\text{s.t.} \quad \dot{x}(t) = F(x(t), u(t)), \quad x(0) = x_0, \tag{11.58}$$

$$S(x(t), u(t)) \leq 0, \tag{11.59}$$

$$T(x(t_f)) \leq 0, \quad (11.60)$$

where ϕ is the terminal-time cost functional to be minimized, $x(t)$ the n-dimensional vector of state profiles with the known initial conditions x_0, $u(t)$ the m-dimensional vector of input profiles, S the n_S-dimensional vector of path constraints that hold during the interval $[0, t_f]$, T the n_T-dimensional vector of terminal constraints that hold at t_f, and t_f the final time, which can be either free or fixed.[2] The optimization problem (11.57)–(11.60) is said to be in the Mayer form, that is, J is a terminal-time cost. When an integral cost is added to ϕ, the corresponding problem is said to be in the Bolza form, while when it only incorporates the integral cost, it is referred to as being in the Lagrange form. Note that these three formulations can be made equivalent by the introduction of additional states.[73]

The nonlinear dynamic optimization problem (11.57)–(11.60) is difficult to solve due to the time dependency of the various variables. In particular, the inputs $u(t)$ represent an infinite-dimensional decision vector. This time dependency can be removed via time discretization using, for example, the method of orthogonal collocation,[74] which, however, results in a large vector of decision variables. Another approach consists in viewing the batch process as a static map between a small number of input parameters π and a small number of run-end outputs z as discussed in Section 11.3.3. These approximations reduce the dynamic optimization problem into a static problem. Next, the following problems will be addressed successively: dynamic optimization, reformulation of a dynamic optimization problem as a static optimization problem, static optimization, and effect of uncertainty.

11.8.1 Dynamic Optimization

Consider the optimization problem (11.57)–(11.60), for which the NCO are given next.

11.8.1.1 Pontryagin's minimum principle

The NCO for a dynamic optimization problem are given by Pontryagin's minimum principle (PMP). The application of PMP provides insight in the optimal solution and generates explicit conditions for meeting active path and terminal constraints and forcing certain sensitivities to zero.

Let us define the Hamiltonian function $H(t)$,

$$H(t) = \lambda^T(t)F(x(t), u(t)) + \mu^T(t)S(x(t), u(t)), \quad (11.61)$$

and the augmented terminal cost

[2] If t_f is free, it becomes a decision variable. Note that it is also possible to have additional *constant* decision variables. In this case, the decision variables would include the vector of input profiles $u[0, t_f]$ and the vector of constant quantities, ρ. This general case will not be considered here for simplicity of exposition, see Ref. 72 for details.

Table 11.4 NCO for a dynamic optimization problem[73]

	Path	Terminal
Constraints	$\mu^T(t)S(x(t),u(t))=0,\ \mu(t)\geq 0$	$\nu^T T(x(t_f))=0,\ \nu\geq 0$
Sensitivities	$\frac{\partial H}{\partial u}(t)=0$	–[a]

[a]If the decision variables include the *constant* vector ρ, there will be terminal sensitivity elements for the determination of ρ, see Ref. 72 for details.

$$\Phi(t_f)=\phi(x(t_f))+\nu^T T(x(t_f)), \tag{11.62}$$

where $\lambda^T(t)$ are the adjoint variables such that

$$\dot{\lambda}^T(t)=-\frac{\partial H}{\partial x}(t),\ \ \lambda^T(t_f)=\frac{\partial \Phi}{\partial x(t_f)}, \tag{11.63}$$

$\mu(t)\geq 0$ is the Lagrange multipliers associated with the path constraints, and $\nu \geq 0$ is the Lagrange multipliers associated with the terminal constraints. In dynamic optimization problems, there are both path and terminal objectives to satisfy, with these objectives being either constraints or sensitivities. PMP and the introduction of the Hamiltonian function $H(t)$ allow (i) assigning the NCO separately to path and terminal objectives, and (ii) expressing them in terms of meeting constraints and sensitivities, as shown in Table 11.4. The solution is generally discontinuous and consists of several intervals or arcs.[73,75] Each arc is characterized by a different set of active path constraints; that is, this set changes between successive arcs.

11.8.1.2 Solution methods

Solving the dynamic optimization problem (11.57)–(11.60) corresponds to finding the best input profiles $u[0, t_f]$ such that the cost functional is minimized, while meeting both the path and terminal constraints. Since the decision variables $u[0, t_f]$ are infinite dimensional, the inputs need to be parameterized using a finite set of parameters to be able to use numerical techniques. These techniques can be classified in two main categories according to the underlying formulation,[73] namely, direct optimization methods that solve the optimization problem (11.57)–(11.60) directly, and PMP-based methods that attempt to satisfy the NCO given in Table 11.4.

Direct optimization methods are distinguished further depending on whether the system equations are integrated explicitly or not.[73] In the *sequential approach*, the system equations are integrated explicitly, and the optimization is carried out in the space of the decision variables only. This corresponds to a "feasible-path" approach because the differential equations are satisfied at each step of the optimization. A piecewise-constant or piecewise-polynomial approximation of the inputs is often used. The computationally expensive part of the sequential approach is the accurate integration of the system equations, which needs to be performed even when the decision variables are far from the optimal solution! In the *simultaneous approach*, an approximation of the system equations is introduced to avoid explicit integration for each candidate set of input profiles, thereby reducing the computational burden. Because the optimization is carried out in the full space of discretized inputs and states, the differential equations are satisfied only at the solution.[74,76] This is therefore called an "infeasible-path" approach. The direct approaches are by far the most commonly used. However, note that the input parameterization is often chosen arbitrarily by the user, which can affect both the efficiency and the accuracy of the approach.[77]

On the other hand, *PMP-based methods* try to satisfy the first-order NCO given in Table 11.4. The NCO involve the state and adjoint variables, which need to be computed via integration. The differential equation system is a two-point boundary-value problem because initial conditions are available for the states and terminal conditions for the adjoints. The optimal inputs can be expressed analytically from the NCO in terms of the states and the adjoints, that is, $u^*[0, t_f] = \mathbf{U}(x[0, t_f], \lambda[0, t_f])$. The resulting differential-algebraic system of equations can be solved using a shooting approach,[78] that is, the decision variables include the initial conditions $\lambda(0)$ that are chosen in order to satisfy $\lambda(t_f)$.

11.8.2 Reformulation of a Dynamic Optimization Problem as a Static Optimization Problem

Consider the optimization problem (11.57)–(11.60). Numerically, this problem can be solved by control vector parameterization (CVP),[76] whereby the infinite-dimensional inputs $u(t)$ are parameterized using a finite number of parameters π. Typically, the inputs are divided into small intervals, and a polynomial approximation of the inputs is used within each interval. The input parameter vector π can also include the switching times between the intervals. It follows that the inputs can be expressed as $u(t) = \mathcal{U}(\pi, t)$ and the optimization performed with respect to π, where π is the n_π-dimensional vector of *constant* decision variables.

The difficulty with such a parameterization is that the path constraints $S(x, u)$ have to be discretized as well, resulting in as many constraints as there are discretization points (typically a large number). However, if structural information regarding the shape of the optimal solution is available, the path constraints can be handled differently using an alternative parameterization as explained next.

The solution to a constrained terminal-time dynamic optimization problem is typically discontinuous and consists of various arcs. For a given input, the ith arc $\eta_i(t)$ can be of two types:

1. $\eta_{i,\text{path}}$: Arc $\eta_i(t)$ is determined by an active path constraint (constraint-seeking arc).
2. $\eta_{i,\text{sens}}$: Arc $\eta_i(t)$ is not governed by an active path constraint, it is sensitivity seeking.

Let us assume that the path constraints are met using either (i) assignment of input values in the case of active input bounds, or (ii) online feedback control via the measurement or estimation of the constrained quantities and their regulation to track the constraint values.

The other parts of the inputs are parameterized as in CVP as $u(t) = \mathcal{U}(\pi, t)$. Since the path constraints are handled directly, they do not involve the decision variables π. Note that the knowledge of the active path constraints in various intervals is crucial for this parameterization. The final states, which affect the objective function and the terminal constraints, can be expressed in terms of the decision variables π as follows:

$$x(t_f) = x_0 + \int_0^{t_f} F(x(t), \mathcal{U}(\pi, t))dt := \mathcal{F}(\pi), \tag{11.64}$$

as t_f is an element of π. Hence, the dynamic optimization problem (11.57)–(11.60) can be reformulated as:

$$\min_{\pi} \quad \phi(\mathcal{F}(\pi)) \tag{11.65}$$

$$\text{s.t.} \quad T(\mathcal{F}(\pi)) \leq 0, \tag{11.66}$$

because the system dynamics are included in $\mathcal{F}$ and the path constraints are taken care of as mentioned earlier. It follows that the dynamic optimization problem can be reformulated as the following static NLP:

$$\min_{\pi} \quad J := \Phi(\pi) \tag{11.67}$$

$$\text{s.t.} \quad G(\pi) \leq 0. \tag{11.68}$$

The scalar cost function to be minimized is written $\Phi(\pi) := \phi(\mathcal{F}(\pi))$, and the n_g-dimensional vector of constraints is written generically $G(\pi) := T(\mathcal{F}(\pi))$.

Some remarks are necessary at this point:

- This reformulation is a mathematical way of viewing the dynamic process as a static input-output "π-(Φ, G)" map. However, the process remains dynamic, with the terminal cost and the terminal constraints being affected by the process dynamics.
- This reformulation is clearly an approximation because it relies on (i) a parameterization of the inputs, and (ii) approximate tracking of the state constraints.
- NLP (11.67, 11.68) is a complex problem if one wants to predict Φ and G from π because the system dynamics need to be integrated according to Eq. (11.64). However, the situation is much simpler in the experimental context, as one simply apply the inputs π to the plant and measure both Φ and G. In this case, the model used to predict Φ and G is replaced by the plant itself. This is a perfect example of the power of measurement-based optimization compared with numerical optimization, as will be detailed later.

11.8.3 Static Optimization

The optimization problem is of algebraic nature with a finite number of constant decision variables.

11.8.3.1 KKT necessary conditions of optimality

With the formulation (11.67, 11.68) and the assumption that the cost function Φ and the constraint functions G are differentiable, the Karush-Kuhn-Tucker (KKT) conditions read[79]:

$$G(\pi^*) \leq 0, \tag{11.69}$$

$$\nabla\Phi(\pi^*) + (\nu^*)^T \nabla G(\pi^*) = 0, \tag{11.70}$$

$$\nu^* \geq 0, \tag{11.71}$$

$$(\nu^*)^T G(\pi^*) = 0, \tag{11.72}$$

where π^* denotes the solution, ν^* the n_g-dimensional vector of Lagrange multipliers associated with the constraints, $\nabla\Phi(\pi^*)$ the n_π-dimensional row vector denoting the cost gradient evaluated at π^*, and $\nabla G(\pi^*)$ the $(n_g \times n_\pi)$-dimensional Jacobian matrix computed at π^*. For these equations to be necessary conditions, π^* needs to be a regular point for the constraints, which calls for linear independence of the active constraints, that is, rank $\{\nabla G_a(\pi^*)\} = n_{g,a}$, where G_a represents the set of active constraints, whose cardinality is $n_{g,a}$.

Condition (11.69) is the primal feasibility condition, Condition (11.71) is the dual feasibility condition, and Condition (11.72) is the complementarity slackness condition. The stationarity condition (11.70) indicates that, at the solution, collinearity between the cost gradient and a linear combination of the gradients of the active constraints prevents from finding a search direction that would result in cost reduction while still keeping the constraints satisfied.

11.8.3.2 Solution methods

Static optimization can be solved by state-of-the-art nonlinear programming techniques. In the presence of constraints, the three most popular approaches are[80]: (i) penalty-function methods, (ii) interior-point methods, and (iii) sequential quadratic programming (SQP). The main idea in penalty-function methods is to replace the solution to a constrained optimization problem by the solution to a sequence of unconstrained optimization problems. This is made possible by incorporating the constraints in the objective function via a penalty term, which penalizes any violation of the constraints, while guaranteeing that the two problems share the same solution (by selecting sufficiently large weights). Interior-point methods also incorporate the constraints in the objective function.[81] However, the constraints are approached from the feasible region, and the additive terms increase to become infinitely large at the value of the constraints, thereby acting more like a barrier than a penalty term. A clear advantage of interior-point methods is that feasible iterates are generated, while for penalty function methods, feasibility is only guaranteed upon convergence. Note that a barrier-penalty function that combines the advantages of both approaches has also been proposed.[82] Another way of computing the solution to a static optimization problem is to solve the set of NCO, for example, iteratively using SQP. SQP methods solve a sequence of optimization subproblems, each one minimizing a quadratic approximation to the Lagrangian function $L := \Phi + \nu^T G$ subject to a linear approximation of the constraints. SQP typically uses Newton or quasi-Newton methods to solve the KKT conditions.

11.8.4 Effect of Uncertainty

11.8.4.1 Plant-model mismatch

The model used for optimization consists of a set of equations that represent an abstract view, yet always a simplification, of the real process. Such a model is built based on conservation laws (for mass, numbers of moles, energy) and constitutive relationships that express kinetics, equilibria, and transport phenomena. The simplifications that are introduced at the modeling stage to obtain a tractable model affect the quality of the process model in two ways:

1. some physical or chemical phenomena are assumed to be negligible and are discarded, and
2. some dynamic equations are assumed to be at quasi-steady state.

Hence, the structure of the model that is used differs from the "true" model structure. This dichotomy gives rise to the so-called structural plant-model mismatch. Furthermore, the model involves a number of physical parameters, whose values are not known accurately. These parameters are identified using process measurements and, consequently, are only known to belong to some interval with a certain probability. For the sake of simplicity, we will consider thereafter that all modeling uncertainties, although unknown, can be incorporated in the vector of uncertain parameters θ.

11.8.4.2 Model adequacy

Uncertainty is detrimental to the quality of both model predictions and optimal solutions. If the model is not able to predict the process outputs accurately, it will most likely not be able to predict the NCO correctly. On the other hand, even if the model is able to predict the process outputs accurately, it is often unable to predict the NCO correctly because it has been trained to predict the outputs and not, for instance, the gradients that are key constituents of the NCO. Hence, numerical optimization is capable of computing optimal inputs for the *model*, but it often fails to reach *plant* optimality.

The property that ensures that a model-based optimization problem will be able to determine the optimal inputs for the plant is referred to in the literature as "model adequacy." For a given model-based RTO scheme, a model is adequate if the RTO scheme is able to predict the correct set of active *plant* constraints and the correct alignment of *plant* gradients. Model adequacy represents a major challenge in process optimization because, as discussed earlier, models are trained to predict the plant outputs rather than the plant NCO. In practice, application of model-based optimal inputs leads to suboptimal, and often infeasible, operation.

11.9 Real-Time Optimization

In the presence of modeling errors and process disturbances, the control trajectories computed offline lose their optimal character. One way to reject the effect of uncertainty on the overall performance (with respect to both optimality and feasibility) is by adequately incorporating process measurements in the optimization framework. This is the field of RTO. Measurements can be incorporated in two different ways as shown in Fig. 11.15.

(a) *Adapt the process model and repeat the optimization.* At each iteration, the model parameters are updated and the optimization problem solved numerically to generate $u^*[0, t_f]$. This adaptation and optimization can also be repeated online, in a single run, to estimate at time t the current states $\hat{x}(t)$ and compute the inputs $u^*[t, t_f]$ for the remaining part of the batch.

(b) *Adapt the inputs through optimizing feedback control.* Here, the optimization is implicit because optimality and feasibility are enforced via feedback control to satisfy the NCO. The control scheme often involves (i) online elements that generate parts of the inputs, $u_a{}^*(t)$, and (ii) run-to-run elements that generate the other part of the inputs, $u_b{}^*[0, t_f]$, via the input parameters π^*.

These RTO schemes will be discussed in the following sections.

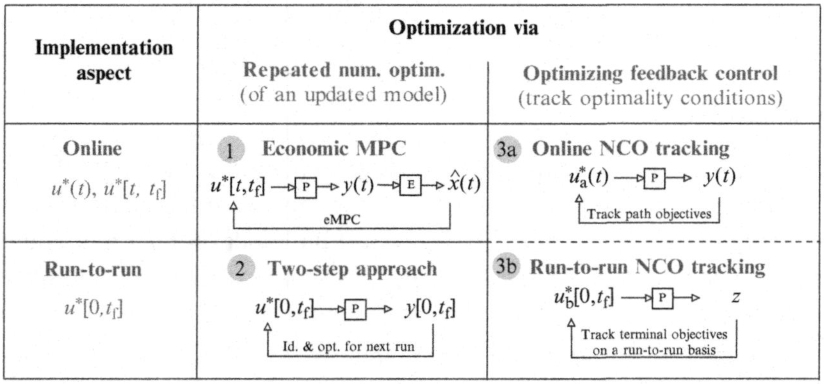

Implementation aspect	Optimization via	
	Repeated num. optim. (of an updated model)	Optimizing feedback control (track optimality conditions)
Online $u^*(t)$, $u^*[t, t_f]$	1 Economic MPC $u^*[t,t_f]$ → P → $y(t)$ → E → $\hat{x}(t)$ eMPC	3a Online NCO tracking $u_a^*(t)$ → P → $y(t)$ Track path objectives
Run-to-run $u^*[0,t_f]$	2 Two-step approach $u^*[0,t_f]$ → P → $y[0,t_f]$ Id. & opt. for next run	3b Run-to-run NCO tracking $u_b^*[0,t_f]$ → P → z Track terminal objectives on a run-to-run basis

Fig. 11.15

RTO schemes for batch processes. The schemes are classified according to whether the optimization is implemented via repeated numerical optimization or feedback control (*horizontal division*) and whether it is implemented online or on a run-to-run basis (*vertical division*). Note that NCO tracking can be implemented using the four control approaches given in Fig. 11.2 to track the path and terminal objectives either online or on a run-to-run basis. Most often, the path constraints are implemented online, while the terminal objectives are implemented on a run-to-run manner (see the similarities with Table 11.4). *P*, plant; *E*, state estimator; *eMPC*, economic model predictive control.

11.9.1 Repeated Numerical Optimization

As additional information about the process becomes available, either during or at the end of the run, this information can be used to improve future operations. This will be documented next for both online implementation using economic MPC and run-to-run implementation via the two-step approach.

11.9.1.1 Economic MPC (Strategy 1)

The approach is similar to MPC discussed in Section 11.4.2.1, but for the fact that the cost function is no longer tailored to tracking run-end references, but rather to minimize an economic cost function. The problem can be formulated as follows:

$$\min_{u[t_i,\, t_f]} \quad J := \phi(x(t_f)) \tag{11.73}$$

$$\text{s.t.} \quad \dot{x}(t) = F(x(t), u(t)), \quad x(t_i) = \hat{x}(t_i), \tag{11.74}$$

$$S(x(t), u(t)) \le 0, \tag{11.75}$$

$$T(x(t_f)) \le 0, \tag{11.76}$$

$$x(t_f) \in \mathcal{X}, \tag{11.77}$$

where t_i is the discrete time instant at which the optimization is performed, $\hat{x}(t_i)$ is the state estimate at that time, and $\mathcal{X}$ is the bounded region of state space where the final state should lie. Numerical optimization yields the control sequence, $u^*[t_i, t_f]$, of which only the first part, $u^*[t_i, t_i + h]$, is applied in an open-loop fashion to the plant. Numerical optimization is then repeated at every sampling instant. The scheme is shown in Fig. 11.16.

Note that reoptimization performed during a run requires a valid process model, including estimates of the current states. Hence, the main engineering challenge in the context of batch processes lies in estimating the states and the parameters from a few noisy measurements. Although the use of extended Kalman filters[83] or moving-horizon estimation[84] has become

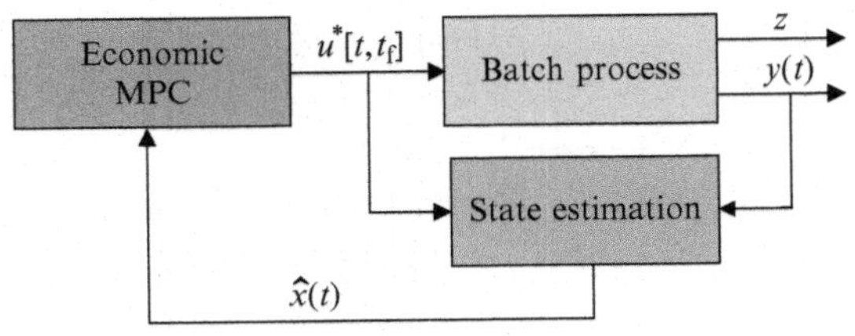

Fig. 11.16
Economic MPC to optimize a batch process repeatedly online.

increasingly common and successful for continuous processes, these methods are difficult to use for batch processes due to the large variations that the states typically go through and the limited time of operation. Good accounts of these difficulties are available.[85,86]

The weakness of this method is clearly its reliance on the model; if the model parameters are not updated, model accuracy plays a crucial role. However, if the model is updated, there is a conflict between parameter estimation and optimization, the so-called dual control problem,[87] since parameter estimation requires persistency of excitation, that is, the inputs must be varied sufficiently to uncover the unknown parameters, a condition that is usually not satisfied when near-optimal inputs are applied. Finally, note that the scheme is called "economic MPC" because MPC typically addresses the tracking of given trajectories. Here, there are no optimal trajectories to track, but rather some economic cost function to minimize. This topic has gained a lot of interest in recent years.[88]

11.9.1.2 Two-step approach (Strategy 2)

In the two-step approach, measurements are used to refine the model, which is then used to optimize the process.[67] The two-step approach has gained popularity over the past 30 years mainly because of its conceptual simplicity. Yet, the two-step approach is characterized by certain intrinsic difficulties that are often overlooked. In its iterative version, the two-step approach involves two optimization problems, namely, one each for parameter identification and process optimization as shown in Fig. 11.17 and described next:

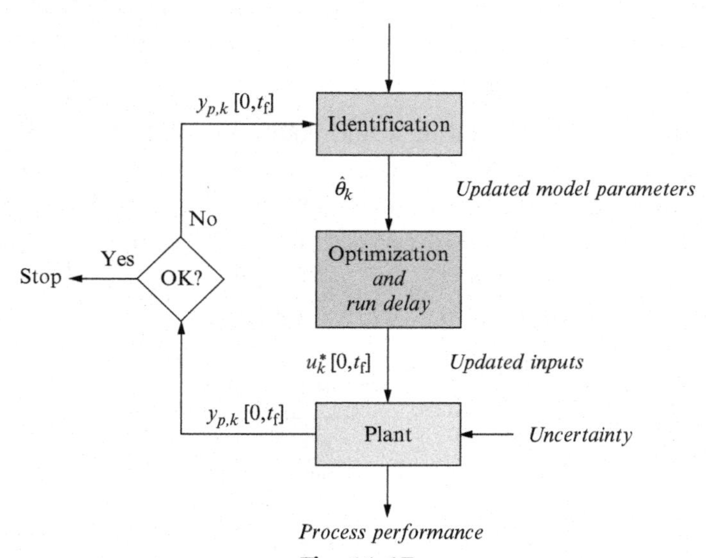

Fig. 11.17
Basic idea of the two-step approach with its two optimization problems.

$$\text{Identification} \quad \hat{\theta}_k := \arg\min_{\theta} \left(\| y_{p,k}[0,t_f] - y_k[0,t_f] \|_{l_2} \right)$$
$$\text{s.t.} \quad \dot{x}_k(t) = F\left(x_k(t), u_k^*(t), \theta\right), \quad x_k(0) = x_{0,k}, \tag{P1}$$
$$y_k(t) = H\left(x_k(t), u_k^*(t), \theta\right),$$
$$\theta \in \Theta$$

$$\text{Optimization} \quad u_{k+1}^*[0,t_f] := \arg\min_{u[0,t_f]} \phi(x_k(t_f))$$
$$\text{s.t.} \quad \dot{x}_k(t) = F\left(x_k(t), u(t), \hat{\theta}_k\right), \quad x(0) = x_{0,k}, \tag{P2}$$
$$S(x_k(t), u(t), \hat{\theta}_k) \leq 0,$$
$$T\left(x_k(t_f), \hat{\theta}_k\right) \leq 0,$$

where Θ indicates the set in which the uncertain parameters θ are assumed to lie. The subscript $(\cdot)_p$ is used to indicate that the outputs are the plant measurements $y_{p,k}[0, t_f]$, by opposition to the values predicted by the model, $y_k[0, t_f]$.

The first step identifies best values for the uncertain parameters by minimizing the l_2-norm of the output prediction errors. The updated model is then used for computing the optimal inputs for the next iteration. Algorithmically, the optimization of the performance of a batch process proceeds as follows:

1. *Initialization.* Set $k = 1$ and select an initial guess for the input profiles $u_k^*[0, t_f]$.
2. *Iteration.* Apply the inputs $u_k^*[0, t_f]$ to the plant. Complete the batch while measuring the run-time outputs $y_{p,k}[0, t_f]$.
3. Compute the distance between the predicted and measured outputs and continue if this distance exceeds the tolerance, otherwise stop.
4. Solve the identification problem (P1) and compute $\hat{\theta}_k$.
5. Solve the optimization problem (P2) and compute $u_{k+1}^*[0, t_f]$. Set $k := k + 1$ and repeat Steps 2–5.

The two-step approach suffers from two main limitations. First, the identification problem requires sufficient excitation, which is, however, rarely the case since the inputs are computed for optimality rather than for the sake of identification. Hence, one has to make sure that there is sufficient excitation, for example, via a dual optimization approach that adds a constraint to the optimization problem regarding the accuracy of the estimated parameters.[89] The second limitation is inherent to the philosophy of the method. Since the adjustable handles are the model parameters, the method assumes that (i) all the uncertainty (including process disturbances) can be represented by the set of uncertain parameters, which is rarely the case.

11.9.2 Optimizing Feedback Control

The second class of RTO methods proposes to adapt the process inputs via feedback control, that is, without repeating the numerical optimization. Since feedback control is used to enforce the NCO of the dynamic optimization problem, the resulting scheme is called "NCO tracking."

11.9.2.1 NCO tracking (Strategy 3)

NCO tracking is a feedback control scheme, where the CVs correspond to measurements or estimates of the plant NCO, and the MVs are appropriate elements of the input profiles. Enforcing the plant NCO is indeed an indirect way of solving the optimization problem for the plant, along the lines of the PMP-based methods discussed in Section 11.8.1.2.

Necessary conditions of optimality

The NCO of the dynamic optimization problem (11.57)–(11.60) encompass four parts as shown in Table 11.4: (i) the path constraints, (ii) the path sensitivities, (iii) the terminal constraints, and (iv) the terminal sensitivities. There are as many NCO as there are degrees of freedom in the optimization problem, thus making the system of equations perfectly determined. NCO tracking proposes to enforce the four NCO parts via tailored feedback control. In particular, some objectives are met using online control, while others are met via run-to-run control, thereby fully exploiting the versatility of control approaches for batch processes, as discussed in Sections 11.4 and 11.5. The NCO-tracking scheme is therefore a two-level (online and run-to-run) multivariable feedback control problem, as shown in Fig. 11.18. The design of the NCO-tracking controller is supported by the concept of "solution model."[72]

Solution model

The solution model is a qualitative model of the optimal solution, which includes (i) the types and sequence of arcs, thus assuming no change in active constraints, (ii) the degrees of freedom (MVs) that one would like to use for control, and (iii) the corresponding

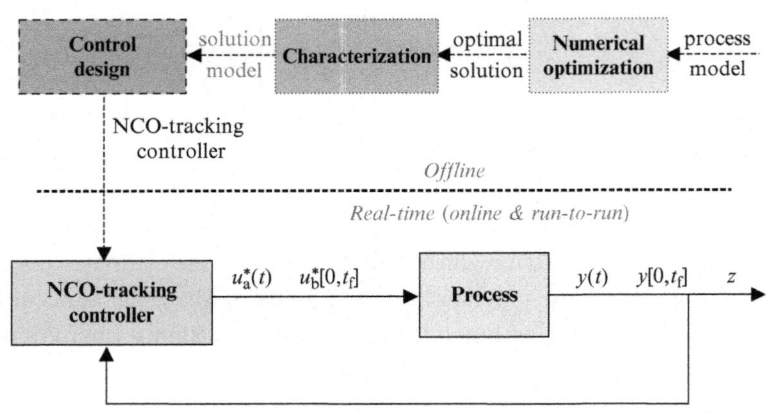

Fig. 11.18
NCO-tracking scheme. The offline activities include the generation of a solution model and the design of an NCO-tracking controller. The real-time activities involve online and run-to-run control to enforce the plant NCO; $u_a^*(t)$ and $u_b^*[0, t_f]$ are the input parts that are generated online and on a run-to-run basis, respectively.

NCO (CVs). It is important to understand that the solution model is simply a tool that helps solve the problem at hand in an efficient way. Although there may be an exact solution model for the process model at hand, this is certainly not the case for the unknown plant. Hence, the designer will be able to propose alternative (from simple to more complex) solution models to tackle the design of the multivariable control problem. This aspect, which represents one of the strengths of NCO tracking, will be detailed in the first case study.

The development of a solution model involves three main steps:

1. Characterize the optimal solution in terms of the types and sequence of arcs by performing numerical optimization using the best available plant model. This plant model need not be very accurate, but it ought to provide the correct types and sequence of arcs. One typically performs a robustness analysis to ensure that the qualitative solution remains structurally valid in presence of uncertainty.
2. Select a finite set of input arcs and parameters to represent (or approximate) the input profiles, and formulate the NCO for this choice of degrees of freedom. Note that the NCO will change with the choice of the degrees of freedom.
3. Pair the MVs and the NCO to form a multivariable control problem.

Steps 2 and 3 might require iterations because they greatly affect the implementation that is discussed next.

Implementation aspects

In its general form, NCO tracking involves both online and run-to-run control, as determined by the solution model and the resulting MV-CV pairing. Some NCO elements are typically implemented online, while others are easier on a run-to-run basis. For example, a path constraint is easily enforced online via constraint control, while both terminal constraints and terminal sensitivities are easier to meet iteratively over several runs. The decision regarding which NCO elements to implement online and which on a run-to-run basis depends on the nature of the various arcs. For a given input, the various arcs can be of two types:

(i) A *constraint-seeking input arc* is associated with a path constraint being active in a given time interval. If the path constraint is an input bound, the input is simply set at the bound, while in the case of a state constraint, feedback control can be used to track the value of the constrained quantity.

(ii) A *sensitivity-seeking input arc* requires $\frac{\partial H}{\partial u}(t) = 0$, which is difficult to implement as such because $H(t)$ is a function of the adjoint variables $\lambda(t)$. Hence, one tries to approximate the sensitivity-seeking arc using a parsimonious input parameterization with only a few constant parameters. This way, the number of degrees of freedom is no longer the infinite-dimensional $u(t)$, but rather the few constant input parameters.

Furthermore, with the input parameterization, the batch process is viewed as the static map $z = M(\pi)$ given by Eq. (11.7). Consequently, the path condition $\frac{\partial H}{\partial u}(t) = 0$ is replaced by terminal conditions of the type $\frac{\partial \phi(x(t_f))}{\partial \pi} = 0$, which can be enforced on a run-to-run basis. Note that the effect of the approximations introduced at the solution level can be assessed in terms of optimality loss.

In summary, the ease of implementation and therefore the success of NCO tracking depends on the quality of the approximations that are introduced to generate the solution model (choice of MVs, corresponding NCO, and pairing MV-NCO). These are true engineering decisions that are made with *plant optimality* in mind!

11.10 Optimization Applications

11.10.1 Semibatch Reactor With Safety and Selectivity Constraints

A simple semibatch reactor with jacket cooling is considered to illustrate the NCO-tracking approach and, in particular, the generation of alternative solution models.[72]

11.10.1.1 Problem formulation

- *Reaction system*: $A + B \rightarrow C$, $2B \rightarrow D$, isothermal, exothermic reactions.
- *Objective*: Maximize the amount of C at a given final time.
- *Manipulated input*: Feedrate of B.
- *Path constraints*: Input bounds; heat-removal constraint expressed as a lower bound on the cooling jacket temperature.
- *Terminal constraint*: Upper bound on the amount of D at final time.

Model equations

Assuming perfect control of the reactor temperature through adjustment of the cooling jacket temperature, the model equations read:

$$\dot{c_A} = -k_1 c_A c_B - \frac{u}{V} c_A, \quad c_A(0) = c_{A,0}, \tag{11.78}$$

$$\dot{c_B} = -k_1 c_A c_B - 2k_2 c_B^2 + \frac{u}{V}(c_{B_{in}} - c_B), \quad c_B(0) = c_{B,0}, \tag{11.79}$$

$$\dot{V} = u, \quad V(0) = V_0, \tag{11.80}$$

$$T_j = T_r - \frac{V}{UA}[(-\Delta H_1) k_1 c_A c_B + (-\Delta H_2) k_2 c_B^2], \tag{11.81}$$

$$n_C = n_{A,0} - n_A, \tag{11.82}$$

Table 11.5 Model parameters, operating bounds, and initial conditions

k_1	0.11	L/(mol min)	k_2	0.13	L/(mol min)
ΔH_1	-8×10^4	J/mol	ΔH_2	-10^5	J/mol
UA	1.25×10^4	J/(min °C)	$c_{B_{in}}$	5	mol/L
T_r	30	°C	$T_{j,min}$	10	°C
u_{max}	1	L/min	$n_{D,max}$	100	mol
$c_{A,0}$	0.5	mol/L	$c_{B,0}$	0	mol/L
V_0	1000	L	t_f	180	min

$$n_D = \frac{1}{2}[(n_A - n_{A,0}) - (n_B - n_{B,0}) + c_{B_{in}}(V - V_0)], \tag{11.83}$$

where the number of moles of Species X is defined as $n_X(t) = V(t)c_X(t)$.

Variables and parameters

c_X: concentrations of Species X, n_X: number of moles of species X, V: reactor volume, k_i: kinetic coefficient of reaction i, u: feedrate of B, $c_{B_{in}}$: inlet concentration of B, ΔH_i: enthalpy of reaction i, T_r: reactor temperature, T_j: cooling jacket temperature, U: heat-transfer coefficient, and A: reactor heat-exchange area. The model, operating, and optimization parameters are given in Table 11.5.

Optimization problem

The objective of maximizing the amount of C at the given final time t_f can be written mathematically as follows:

$$\max_{u[0,t_f]} \quad n_C(t_f) \tag{11.84}$$

$$\text{s.t.} \quad \text{dynamic model}(11.78)-(11.83), \quad 0 \le u(t) \le u_{max}, \tag{11.85}$$

$$T_j(t) \ge T_{j,min}, \tag{11.86}$$

$$n_D(t_f) \le n_{D,max}. \tag{11.87}$$

11.10.1.2 Characterization of the optimal solution

The optimal feedrate profile shown in Fig. 11.19 exhibits three intervals:

- The feedrate is initially at its upper bound, $u_{bound}(t) = u_{max}$, in order to attain the heat-removal constraint as quickly as possible.
- Then, $u_{path}(t)$ keeps the path constraint $T_j(t) = T_{j,min}$ active.

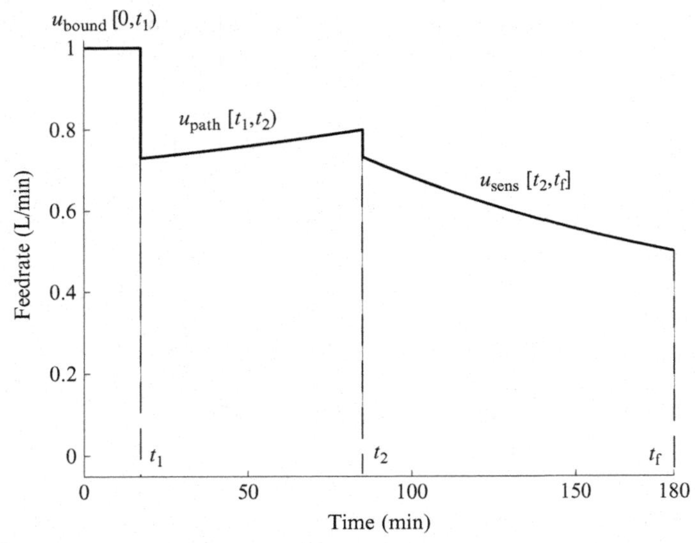

Fig. 11.19
Optimal feedrate consisting of the three arcs $u_{bound}[0, t_1)$, $u_{path}[t_1, t_2)$, and $u_{sens}[t_2, t_f]$.

- Finally, the input switches to $u_{sens}(t)$ to take advantage of the best compromise between producing the desired C and the undesired D. The switching time t_2 is determined in such a way that the terminal constraint $n_D(t_f) = n_{D,max}$ is met at final time.

Using PMP, it can be shown that the competition between the two reactions results in a sensitivity-seeking feedrate reflecting the optimal compromise between producing C and D. This sensitivity-seeking input can be calculated from the second time derivative of $H_u = 0$ as:

$$u_{sens}(t) = \frac{n_B(-k_1 c_A c_B + 2k_1 c_A c_{Bin} + 4k_2 c_B c_{Bin})}{2c_{Bin}(c_{Bin} - c_B)}. \tag{11.88}$$

On the other hand, the constraint-seeking arc u_{path} corresponds to riding along the path constraint $T_j(t) = T_{j,min}$. The input is obtained by differentiating the path constraint once with respect to time, that is, from $\dot{T}_j = 0$:

$$u_{path}(t) = \frac{c_B V(\Delta H_1 k_1 c_A(k_1 c_B + k_1 c_A + 2k_2 c_B) + 2\Delta H_2 k_2 c_B(k_1 c_A + 2k_2 c_B))}{\Delta H_1 k_1 c_A(c_{Bin} - c_B) + \Delta H_2 k_2 c_B(2c_{Bin} - c_B)}. \tag{11.89}$$

Since analytical expressions exist for all three input arcs, and t_1 can be determined upon $T_j(t)$ reaching $T_{j,min}$, the optimal solution can be parameterized using *a single parameter*, namely, the switching time t_2 between $u_{path}(t)$ and $u_{sens}(t)$. Note that the input arc $u_{path}(t)$ given by Eq. (11.89) will keep the reactor on the path constraint only for the case of a perfect model and perfect measurements. Fortunately, in the presence of uncertainty, one can implement $u_{path}(t)$ by forcing $T_j(t)$ to track the reference $T_{j,min}$, which is straightforward to implement using, for example, PI control. Similarly, the arc $u_{sens}(t)$ is only optimal for the case of a perfect model and perfect measurements. The way around this difficulty is best handled with the selection of an appropriate solution model.

11.10.1.3 Alternative solution models

A key feature of NCO tracking is the possibility to choose appropriate MVs and CVs and, if necessary, to introduce approximations. Hence, the choice of MVs and CVs is not made once for all, but it may require iterations. In fact, the appropriateness of these choices is often key to the success of NCO tracking in practice. This is illustrated next for this example, whereby three alternative solution models are proposed, all with the same objective but with very different implementation aspects.

Solution model A corresponding to the optimal input for the plant model

This solution model has the structure of the optimal solution given in Fig. 11.19. The decision variables (MVs) associated with the feedrate input $u[0, t_f]$ are the three arcs $u[0, t_1)$, $u[t_1, t_2)$, and $u[t_2, t_f]$ and the two switching times t_1 and t_2. The corresponding NCO (CVs) include three path constraints and two pointwise constraints at times t_1 and t_f (Table 11.6).

The pairing between MVs and CVs is rather straightforward:

1. $u[0, t_1)$ is implemented open loop as $u[0,t_1) = u_{\max}$.
2. t_1 is determined when $T_j(t)$ reaches $T_{j,\min}$.
3. $u[t_1, t_2)$ is determined by tracking the path constraint $T_j(t) = T_{j,\min}$.
4. t_2 is determined from the terminal condition $n_D(t_f) = n_{D,\max}$.
5. And finally, $u[t_2, t_f]$ is determined from the path sensitivity $\frac{\partial H}{\partial u}[t_2, t_f] = 0$.

If the pairing is easily formulated, it is rather difficult to implement this control strategy in practice. Steps 1–3 are straightforward. Step 4 requires prediction, which requires a model to be done online or, otherwise, could be implemented on a run-to-run basis. Step 5 requires a model and the computation of the Hamiltonian function $H(t)$ that includes the adjoint variables $\lambda(t)$. Hence, Solution model A is not very useful for implementation.

Table 11.6 NCO for Solution model A

	Path	Terminal
Constraints	$u[0,t_1) = u_{\max}$ $\underline{T_j(t_1) = T_{j,\min}}$ $T_j[t_1,t_2) = T_{j,\min}$	$\underline{n_D(t_f) = n_{D,\max}}$
Sensitivities	$\frac{\partial H}{\partial u}[t_2, t_f] = 0$	–

Note: Pointwise constraints are underlined.

Table 11.7 NCO for Solution model B

	Path	Terminal
Constraints	$u[0,t_1) = u_{max}$ $\underline{T_j(t_1) = T_{j,min}}$ $T_j[t_1,t_2) = T_{j,min}$	$\underline{n_D(t_f) = n_{D,max}}$
Sensitivities	–	$\underline{\frac{\partial n_C(t_f)}{\partial \pi} = 0}$

Note: Pointwise constraints are underlined.

Solution model B introducing approximation to the sensitivity-seeking arc

We introduce the parameterization $u_{sens}[t_2, t_f] = \pi$, which allows approximating the last arc with a single parameter. The corresponding NCO include two path constraints and three pointwise constraints at times t_1 and t_f (Table 11.7).

The pairing between MVs and CVs is now as follows:

1. $u[0, t_1)$ is implemented open loop as $u[0,t_1) = u_{max}$.
2. t_1 is determined when $T_j(t)$ reaches $T_{j,min}$.
3. $u[t_1, t_2)$ is determined by tracking the path constraint $T_j(t) = T_{j,min}$.
4. t_2 is determined from the terminal condition $n_D(t_f) = n_{D,max}$.
5. π is determined from the terminal condition $\frac{\partial n_C(t_f)}{\partial \pi} = 0$.

Steps 1–3 are the same as with Solution model A and are therefore straightforward to implement. Now, both Steps 4 and 5 require prediction, which requires a model to be done online or, otherwise, could be implemented on a run-to-run basis. Steps 4 and 5 represent a 2×2 control problem. Note that it is no longer necessary to compute the Hamiltonian function $H(t)$.

Solution model C using a reference trajectory to meet the terminal constraint

This solution model attempts to meet the terminal constraint $n_D(t_f) = n_{D,max}$ online within a single run. For this, we define the profile $n_{D,term}(t)$ in the time interval $[t_2, t_f]$ that has the property to end up at $n_{D,max}$ at final time. One could, for example, define a profile that increases linearly between $n_{D,2}$ at t_2 and $n_{D,max}$ at t_f:

$$n_{D,term}(t) = n_{D,2} + \frac{n_{D,max} - n_{D,2}}{t_f - t_2}(t - t_2). \tag{11.90}$$

This way, the corresponding NCO include three path constraints and three pointwise constraints at times t_1 and t_2 and t_f (Table 11.8).

Table 11.8 NCO for Solution model C

	Path	Terminal
Constraints	$u[0,t_1) = u_{max}$ $\underline{T_j(t_1) = T_{j,min}}$ $T_j[t_1, t_2) = T_{j,min}$ $\underline{n_D(t_2) = n_{D,2}}$ $n_D[t_2, t_f] = n_{D,term}[t_2, t_f]$	–
Sensitivities	–	$\underline{\frac{\partial n_C(t_f)}{\partial n_{D,2}} = 0}$

Note: Pointwise constraints are underlined.

The pairing between MVs and CVs is now as follows:

1. $u[0, t_1)$ is implemented open loop as $u[0,t_1) = u_{max}$.
2. t_1 is determined when $T_j(t)$ reaches $T_{j,min}$.
3. $u[t_1, t_2)$ is determined by tracking the path constraint $T_j(t) = T_{j,min}$.
4. t_2 is determined when $n_D(t)$ reaches the predefined value $n_{D,2}$ for the current batch.
5. $u[t_2, t_f]$ is determined by tracking the path constraint $n_D(t) = n_{D,term}(t)$.

Again, Steps 1–3 are the same as with Solution models A and B. What is new is that Steps 4 and 5 are now also straightforward to implement online, provided the concentration of the Species D can be measured or estimated online.

A comparison of Tables 11.6–11.8 calls for the following comments:

- Solution model A, which is able to generate the numerical solution obtained from the plant model, is very impractical for implementation using feedback control. However, note that this parameterization might be rather handy for numerical optimization, for example, as an alternative to CVP.
- Solution model B can be implemented via a combination of online and run-to-run control because it does not contain path sensitivities that typically require a model for computation.
- Solution model C contains mostly path constraints that can be handled via online control. The terminal sensitivity $\frac{\partial n_C(t_f)}{\partial n_{D,2}} = 0$ is a way of determining the optimal value of $n_{D,2}$. However, for any practical purpose, the value of $n_{D,2}$ can be chosen conservatively (slightly smaller than $n_{D,max}$) with negligible effect on the cost.
- All the corresponding tracking schemes attempt to meet the active constraints and push certain gradients to zero. Like with any control scheme, one cannot guarantee, in the presence of disturbances, that the SPs are met at all times. However, the presence of feedback helps reduce the sensitivity to these disturbances. One finds here all the pros and cons of multivariable feedback control!

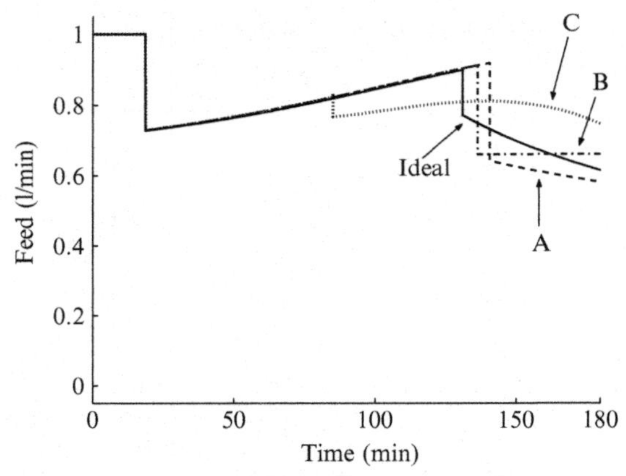

Fig. 11.20
Feedrate profile: Ideal profile computed with complete knowledge of the uncertainty (*solid*) and after the fifth run with NCO-tracking control based on the three solution models A, B, and C.

11.10.1.4 Simulation results

Optimization results using NCO tracking with the there aforementioned solution models are presented next. It is assumed that uncertainty is present in the form of time-varying kinetic coefficients for the plant:

$$k_1(t) = k_{1,0}\frac{\alpha}{\alpha + t}, \quad k_2(t) = k_{2,0}\frac{\beta}{\alpha - t}, \tag{11.91}$$

with the nominal model values $k_{1,0} = 0.11$ L/(mol min), $k_{2,0} = 0.13$ L/(mol min), $\alpha = 1800$ min, and $\beta = 900$ min, and where t is the time in minutes. These variations might correspond to a change in catalyst activity with time. The information regarding these variations is, of course, not revealed to the optimization algorithms. If this information were available, the ideal cost value would be $J^* = 392.3$ mol of the desired product C.

The input profiles with NCO tracking using Solution models A, B, and C are shown in Fig. 11.20.[3] Though the third arc is quite different in the three cases, the optimal costs are nearly the same, as will be documented later.

Optimality measure

With several solution model candidates, it is important to be able to assess the quality of the approximation. For this, the measure Θ that expresses the ratio of two cost differences is introduced[90]:

[3] Solution model A requires a process model to enforce the path condition $H_u[t_2, t_f] = 0$. In this simulation study, since the process model is available, it is possible to "implement" NCO tracking based on Solution model A. In practice, however, NCO tracking works exclusively on the basis of online and run-to-run feedback control, that is, without the need of a process model.

$$\Theta = \frac{J(u^{\text{cons}}) - J(u^{\text{NCO}})}{J(u^{\text{cons}}) - J(u^*)}, \tag{11.92}$$

where u^{NCO} represents the inputs obtained via NCO tracking, u^* the true optimal solution, and u^{cons} a conservative solution used in practice. The true optimal solution u^* is typically unknown. However, note that the measure Θ does not require the optimal inputs u^* themselves, but rather the optimal cost $J(u^*)$, which could possibly be obtained by extrapolation of the costs obtained with models of increasing complexity. The optimality measure defined earlier runs between 0 and 1 and provides the fraction of potential improvement that has been realized. The closer Θ is to 1, the better the optimality.

Quantitative comparison

The results are compared in Table 11.9, where, for all cases, convergence is achieved in less than 10 runs. For the sake of comparison, application of the optimal input calculated offline using the nominal plant model, u^{cons}, gives the cost $J(u^{\text{cons}}) = 374.6$ mol. The main observations are as follows:

- The constraint on the jacket temperature is active in all cases because it is handled online. The first run with Controllers A and B indicates how much can be gained by meeting that path constraint alone.
- Enforcing the path sensitivity in Controller A using a model and neighboring-extremal control[75] is not really better (here even slightly worse) than a crude approximation by a constant value as implemented by Controller B.

Table 11.9 Comparison of NCO-tracking scenarios for three different controllers and various numbers of runs

Distance to the two constraints, cost J, and optimality measure Θ

Controller	Run Index	Min. Jacket Temperature ($T_{j,min} = 10°C$)	Final Amount of D (mol) ($n_{D,max} = 100$ mol)	Final Amount of C (mol) Cost J	Optimality Measure Θ
Optimal		10	100	392.3	1
Open loop		10.1	72.5	374.6	0
	1	10.0	79.6	376.8	0.12
A	5	10.0	96.2	389.8	0.86
	10	10.0	99.6	392.1	0.99
	1	10.0	83.6	383.3	0.49
B	5	10.0	98.1	391.6	0.96
	10	10.0	99.9	392.2	0.99
	1	10.0	100	391.8	0.97
C	5	10.0	100	391.8	0.97
	10	10.0	100	391.8	0.97

Notes: "Optimal" is with respect to the perturbed system. "Open loop" means open-loop application of the optimal input computed for the nominal (unperturbed) system. Note that open-loop application of the nominal optimal input does not meet the two constraints and is therefore widely suboptimal.

- Meeting the terminal constraint improves the cost significantly. This is done via run-to-run control for Controllers A and B and online via tracking of $n_{D,term}(t)$ in Controller C. Hence, with Controller C, all the constraints are met online and thus the cost is nearly optimal from the first run on. The parameter $n_{D,2}$, which indirectly selects the switching time t_2, has a negligible effect on the cost and therefore is not adapted in this study.[4] Hence, all the adjustments with Controller C are made online.

11.10.2 Industrial Batch Polymerization

The last case study illustrates the use of NCO tracking for the optimization of an industrial reactor for the copolymerization of acrylamide.[91] As the polymer is repeatedly produced in a batch reactor, run-to-run NCO tracking using run-end measurements is applied.

11.10.2.1 A brief description of the process

The 1-ton industrial reactor investigated in this section is dedicated to the inverse-emulsion copolymerization of acrylamide and quaternary ammonium cationic monomers, a heterogeneous water-in-oil polymerization process. Nucleation and polymerization are confined to the aqueous monomer droplets, while the polymerization follows a free-radical mechanism. Table 11.10 summarizes the reactions that are known to occur.

A tendency model capable of predicting the conversion and the average molecular weight was developed. The seventh-order model includes balance equations for the two monomers and

Table 11.10 Main reactions in the inverse-emulsion process

Oil-phase reactions
Initiation by initiator decomposition
Reactions of primary radicals
Propagation reactions
Transfer between phases
Initiator
Comonomers
Primary radicals
Aqueous-phase reactions
Reactions of primary radicals
Propagation reactions
Unimacromolecular termination with emulsifier
Reactions of emulsifier radicals
Transfer to monomer
Addition to terminal double bond
Termination by disproportionation

[4] In this example, the terminal sensitivities are so low that there is no need for adaptation. This is valid for π in Controller B and $n_{D,2}$ in Controller C.

for the chain-transfer agent, dynamic equations for the zeroth-, first-, and second-order moments of the molecular weight distribution and for the efficiency of the initiator. The model parameters have been fitted to match observed data. Note that certain effect are nearly impossible to model. For instance, the efficiency of the initiator can vary significantly between batches because of residual oxygen concentration at the outset of the reaction. Chain-transfer agents and reticulants are also added to help control the molecular weight distribution. These small variations in the recipe are not incorporated in the tendency model. Hence, optimization of this process clearly calls for the use of measurement-based techniques.

11.10.2.2 Nominal optimization of the tendency model

The objective is to minimize the reaction time, while meeting four constraints, namely, (i) the terminal molecular weight $X(t_f)$ has to exceed the target value X_{min} to ensure total conversion of acrylamide, (ii) the terminal conversion $\overline{M}_w(t_f)$ has to exceed a target value to ensure total conversion of acrylamide, (iii) heat removal is limited, which is incorporated in the optimization problem by the lower bound $T_{j,in,min}$ on the jacket inlet temperature $T_{j,in}(t)$, and (iv) the reactor temperature $T_r(t)$ is upper bounded. The MVs are the reactor temperature profile $T_r[0, t_f]$ and the reaction time t_f. The dynamic optimization problem can be formulated as follows:

$$\min_{T_r[0,\,t_f],\,t_f} \quad J := t_f \tag{11.93}$$

$$\text{s.t.} \quad \text{dynamic reactor model with initial conditions} \quad X(t_f) \geq X_{min}, \tag{11.94}$$

$$\overline{M}_w(t_f) \geq \overline{M}_{w,min}, \tag{11.95}$$

$$T_{j,in}(t) \geq T_{j,in,min}, \tag{11.96}$$

$$T_r(t) \leq T_{r,max}. \tag{11.97}$$

11.10.2.3 Solution model

The results of nominal optimization are shown in Fig. 11.21, with normalized values of the reactor temperature $T_r(t)$ and of the time t.

The nominal optimal solution consists of two arcs with the following interpretation:

- *Heat-removal limitation.* Up to a certain level of conversion, the temperature is limited by heat removal. Initially, the operation is isothermal and corresponds closely to what is used in industrial practice. Also, this first isothermal arc ensures that the terminal constraint on molecular weight will be satisfied as it is mostly determined by the concentration of chain-transfer agent.

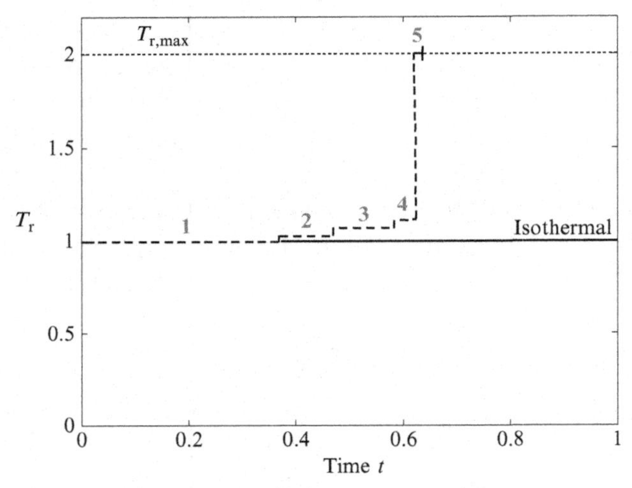

Fig. 11.21
Optimal reactor temperature calculated from the nominal model using five piecewise-constant temperature elements of variable duration.

- *Intrinsic compromise*. The second arc represents a compromise between reaction speed and quality. The decrease in reaction rate due to lower monomer concentration is compensated by an increase in temperature, which accelerates the reaction but decreases molecular weight.

This interpretation of the nominal solution is the basis for the solution model. Since operators are reluctant to change the temperature policy during the first part of the batch and the reaction is highly exothermic, it has been decided to

- implement the first arc isothermally, with the temperature kept at the value used in industrial practice; and
- implement the second arc adiabatically, that is, without jacket cooling. The reaction mixture is heated up by the reaction, which allows linking the maximal reachable temperature to the amount of reactants (and thus the conversion) at the time of switching.

With this so-called "semiadiabatic" temperature profile, there are only two degrees of freedom, namely, the switching time between the two arcs t_{sw} and the final time t_f. The dynamic optimization problem can be rewritten as the following *static* problem (see Section 11.8.2):

$$\min_{t_{sw},\, t_f} \quad t_f \tag{11.98}$$

$$\text{s.t.} \quad \begin{aligned} &\text{static model (in this study, the plant)} \\ &X(t_f) \geq X_{\min}, \end{aligned} \tag{11.99}$$

$$\overline{M}_w(t_f) \geq \overline{M}_{w,\min} \tag{11.100}$$

$$T_r(t_f) \leq T_{r,\max}. \tag{11.101}$$

Such a reformulation is made possible since:

1. The switching time t_{sw} and the final time t_f are fixed at the beginning of the batch, while performance and constraints are evaluated at batch end. This way, the dynamics are lumped into the static map (Eq. 11.7) $\{\pi: t_{sw}, t_f\} \rightarrow \{z: t_f, X(t_f), \overline{M}_w(t_f), T_r(t_f)\}$.
2. Maintaining the temperature constant initially at its current practice value ensures that the heat-removal limitation is satisfied. This constraint can thus be removed from the problem formulation.

Because (i) the constraint on the molecular weight is less restrictive than that on the reactor temperature, (ii) the final time is defined upon meeting the desired conversion, and (iii) the terminal constraint on reactor temperature is active at the optimum, the NCO reduce to the following two conditions:

$$T_r(t_f) - T_{r,\max} = 0, \tag{11.102}$$

$$\frac{\partial t_f}{\partial t_{sw}} + \nu \frac{\partial [T_r(t_f) - T_{r,\max}]}{\partial t_{sw}} = 0, \tag{11.103}$$

where ν is the Lagrange multiplier associated with the constraint on final temperature. The first equation determines the switching time, while the second equation can be used for computing ν, which, however, is not used in this study.

11.10.2.4 Industrial results

The solution to the original dynamic optimization problem can be approximated by adjusting the switching time so as to meet the terminal constraint on reactor temperature. This can be implemented using a simple integral run-to-run controller as shown in Fig. 11.22.

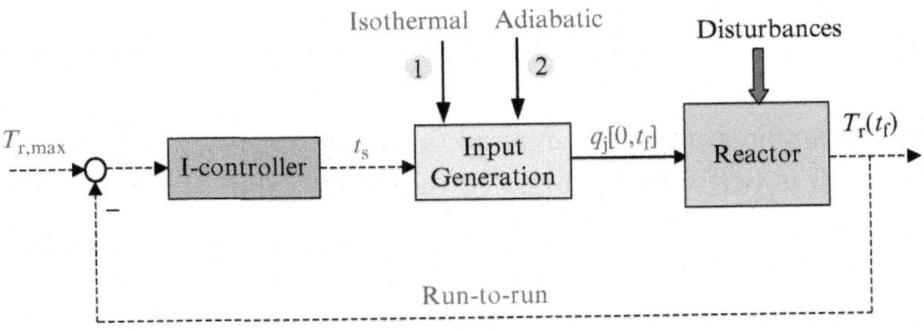

Fig. 11.22

Run-to-run control of the final reactor temperature by adjusting the switching time t_s. The reactor is initially operated isothermally at the normalized temperature $T_{r,ref} = 1$. Adiabatic operation starts at t_s, which is adjusted on a run-to-run basis to achieve $T_r(t_f) = T_{r,max}$. The manipulated input is the flowrate of cooling medium in the jacket $q_j(t)$, which is generated by a feedback control law to regulate $T_r(t)$ around $T_{r,ref} = 1$ in the isothermal phase for $t < t_s$; then, $q_j[t_s, t_f] = 0$ is set in the adiabatic phase.

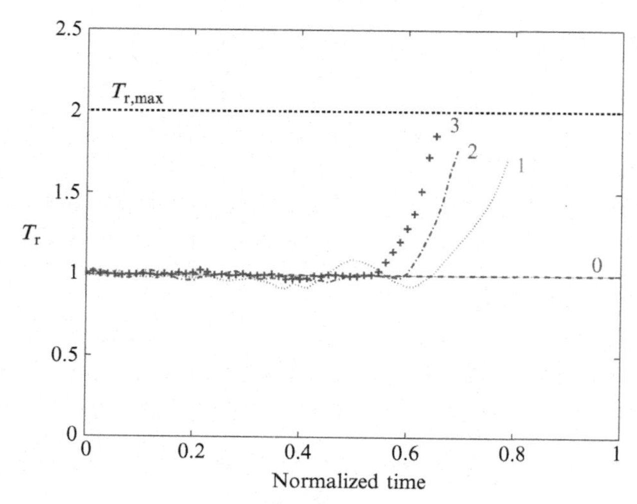

Fig. 11.23
Normalized temperature profiles measured in an industrial reactor as a result of reducing the switching time between isothermal and adiabatic operations. Compared with the normalized reaction time of 1 for the current-practice isothermal operation, the reaction time is successively reduced to 0.78, 0.72, and 0.65 in three batches. Note that a backoff from $T_{r,max}$ is implemented for safety purposes, mainly to account for run-time disturbances that cannot be handled by this technique.

Table 11.11 Run-to-run optimization results for a 1-ton copolymerization reactor

Batch	Strategy	t_{sw}	$T_r(t_f)$	t_f
0	Isothermal		1.00	1.00
1	Semiadiabatic	0.65	1.70	0.78
2	Semiadiabatic	0.58	1.78	0.72
3	Semiadiabatic	0.53	1.85	0.65

Fig. 11.23 shows the application of the method to the optimization of the 1-ton industrial reactor. The first batch is performed using a conservative value of the switching time. The reaction time is significantly reduced after only two batches, without any off-spec product. Table 11.11 summarizes the adaptation results, highlighting the 35% reduction in reaction time compared to the isothermal policy used in industrial practice. Results could have been even more impressive, but a backoff from the constraint on the final temperature was added for safety purposes. This semiadiabatic policy has become standard practice for our industrial partner. The same policy has also been implemented, together with the adaptation scheme, to other polymer grades and to larger reactors.

11.11 Conclusions

11.11.1 Summary

This chapter has analyzed the operation of batch and semibatch processes with respect to both control and optimization. There often is a great potential for process improvement in batch processes. Although the major gains are concerned with chemical choices regarding the reaction and separation steps, process operation also carries a significant potential for cost reduction. However, cost savings in process operations are often associated with automation rather than advanced control. The former term encompasses the hardware and software elements needed to operate an industrial process as automatically as possible. These include automatic start-up and shut-down operations, process monitoring, fault diagnosis, alarm handling, automatic sampling, and analysis. The latter term is concerned mainly with algorithmic solutions for solving well-defined problems and, as such, might only represent 10% of the control and automation activities in the batch process industry.

This chapter has also shown that incorporating measurements in the optimization framework helps improve the performances of chemical processes despite the use of models of limited accuracy. The various RTO methods differ in the way measurements are used and the inputs adjusted to reject the effect of uncertainty. Measurements can be utilized to iteratively (i) update the states or the parameters of the model that is used for optimization, or (ii) adjust the inputs via feedback control to enforce the NCO. It has been argued that the latter technique has the ability of rejecting the effect of uncertainty in the form of plant-model mismatch and process disturbances.

11.11.2 Future Challenges

While several of the techniques presented here have been implemented successfully in industrial practice, many challenging problems still remain. Here are a few of them:

- Advances in optimization techniques in general, and in MPC in particular, have undoubtedly influenced industrial implementation. However, these techniques depend on the availability of process models of reasonably high fidelity and modest complexity, which is difficult to achieve in practice. Methods of nonlinear model reduction capable of representing complex process dynamics effectively with reduced-order, control-relevant models will provide a significant boost to the industrial application of advanced control and optimization techniques.
- While advances in state estimation have enabled the estimation of infrequently measured product characteristics, these estimates can never completely replace the actual measurements themselves. And as the manufacturing chain becomes inexorably more

tightly integrated, each successive downstream customer will place increasingly stringent performance demands on the products they receive from each of their suppliers. Meeting these demands will ultimately require process control performance levels that cannot be attained without better measurement or estimation of product properties. Advances in sensors, analyzers, and ancillary measurement technology will be required in order to make measurements of product properties more frequently available than is currently possible.

- As the structure of process control systems gets more complex, analyses of model dynamics, overall system stability, and achievable performance will be essential, especially for providing guidance in selecting the best alternative for each problem.

Acknowledgments

The authors would like to thank the former and present group members at EPFL's Laboratoire d'Automatique, who contributed many of the insights and results presented here. A particular thank goes to Dr. Alejandro Marchetti and Dr. René Schneider who provided useful feedback on the manuscript.

References

1. Bonvin D. Control and optimization of batch processes. In: Baillieul J, Samad T, editors. *Encyclopedia of systems and control*. Berlin: Springer; 2014. p. 1–6.
2. Bonvin D. Optimal operation of batch reactors—a personal view. *J Process Control* 1998;**8**(5–6):355–68.
3. Diwekar U. *Batch processing: modeling and design*. Hoboken (NJ): CRC Press; 2014.
4. Rani KY. *Optimization and control of semi-batch reactors: data-driven model-based approaches*. Saarbrücken, Germany: LAP Lambert Academic Publishing; 2011.
5. Caccavale F, Iamarino M, Pierri F, Tufano V. *Control and monitoring of chemical batch reactors. Advances in industrial control*, New York: Springer-Verlag; 2011.
6. Mujtaba IM. *Batch distillation: design and operation. Series on Chemical Engineering*, vol. 3. London: Imperial College Press; 2004.
7. Diwekar U. *Batch distillation: simulation, optimal design, and control. Series in chemical and mechanical engineering*, Hoboken (NJ): CRC Press; 1995.
8. Seki H, Furuya N, Hoshino S. Evaluation of controlled cooling for seeded batch crystallization incorporating dissolution. *Chem Eng Sci* 2012;**77**(10):10–7.
9. Wieckhusen D. Development of batch crystallizations. In: Beckmann W, editor. *Crystallization: basic concepts and industrial applications*. Chichester: Wiley-VCH; 2013. p. 187–202.
10. Ljung L. *System identification: theory for the user*. Englewood Cliffs (NJ): Prentice-Hall; 1999.
11. Pearson RK, Ogunnaike BA. *Nonlinear process identification*. Upper Saddle River (NJ): Prentice-Hall; 1996.
12. Himmelblau DM. Applications of artificial neural networks in chemical engineering. *Korean J Chem Eng* 2000;**17**(4):373–92.
13. Russel SA, Kesavan P, Lee JH, Ogunnaike BA. Recursive data-based prediction and control of batch product quality. *AIChE J* 1998;**44**(11):2442–58.
14. Cochran WG, Cox GM. *Experimental designs*. New York: John Wiley; 1992.
15. Srinivasan B, Bonvin D. Controllability and stability of repetitive batch processes. *J Process Control* 2007;**17**:285–95.

16. Pertev C, Turker M, Berber R. Dynamic modelling, sensitivity analysis and parameter estimation of industrial yeast fermenters. *Comp Chem Eng* 1997;**21**:S739–44.
17. Montgomery DC. *Design and analysis of experiments.* 8th ed. New York: John Wiley & Sons; 2013.
18. Jorgensen SB, Hangos KM. Grey-box modelling for control: qualitative models as unifying framework. *Int J Adaptive Control Signal Proc* 1995;**9**(6):547–62.
19. Tulleken HJAF. Application of the grey-box approach to parameter estimation in physico-chemical models. In: *IEEE CDC*; 1991. p. 1177–83. Brighton.
20. Filippi C, Greffe JL, Bordet J, Villermaux J, Barnay JL, Ponte B, et al. Tendency modeling of semi-batch reactors for optimization and control. *Comp Chem Eng* 1986;**41**:913–20.
21. François G, Srinivasan B, Bonvin D. Use of measurements for enforcing the necessary conditions of optimality in the presence of constraints and uncertainty. *J Process Control* 2005;**15**(6):701–12.
22. Ogunnaike BA, Ray WH. *Process dynamics, modeling and control.* New York: Oxford University Press; 1994.
23. Seborg DE, Edgar TF, Mellichamp DA, Doyle FJ. *Process dynamics and control.* New York: John Wiley; 2004.
24. Aström KJ, Wittenmark B. *Adaptive control.* 2nd ed. Reading (MA): Addison-Wesley; 1995.
25. Apkarian P, Adams R. Advanced gain-scheduling techniques for uncertain systems. *IEEE Trans Control Syst Technol* 1998;**6**:21–32.
26. Hagenmeyer V, Nohr M. Flatness-based two-degree-of-freedom control of industrial semi-batch reactors using a new observation model for an extended Kalman filter approach. *Int J Control* 2008;**81**(3):428–38.
27. Shahrokhi M, Ali Fanaei M. Nonlinear temperature control of a batch suspension polymerization reactor. *Poly Eng Sci* 2002;**42**(6):1296–308.
28. Soroush M, Kravaris C. Nonlinear control of a batch polymerization reactor: an experimental study. *AIChE J* 1992;**38**(9):1429–48.
29. Juba MR, Hamer JW. Progress and challenges in batch process control. In: *Chemical process control—CPC-III*; 1986. p. 139–83. Asilomar (CA).
30. Wright RA, Kravaris C. Nonminimum-phase compensation for nonlinear processes. *AIChE J* 1992;**38** (1):26–40.
31. *Mettler-Toledo. Optimax heat-flow calorimetry*; http://www.mt.com/ch/en/home/products/L1_AutochemProducts/Reaction-Calorimeters-RC1-HFCal/OptiMax-HFCal-Heat-Flow-Calorimeter.html Accessed on: May 17, 2017.
32. Ogunnaike B, François G, Soroush M, Bonvin D. Control of polymerization processes. In: Levine WS, editor. *The control handbook—control system applications.* Boca Raton (FL): CRC Press; 2011. 12.1–23.
33. Rawlings JB, Mayne DQ. *Model predictive control: theory and design.* Madison (WI): Nob Hill Pub; 2009.
34. Nagy ZK, Braatz RD. Robust nonlinear model predictive control of batch processes. *AIChE J* 2003;**49** (7):1776–86.
35. Brown RG. *Introduction to random signal analysis and Kalman filtering.* New York: John Wiley & Sons; 1983.
36. Kamen EW, Su JK. *Introduction to optimal estimation. Advanced Textbooks in Control and Signal Processing,* Dordrecht: Springer-Verlag; 1999.
37. Valappil J, Georgakis C. Nonlinear model predictive control of end-use properties in batch reactors. *AIChE J* 2002;**48**(9):2006–21.
38. Flores-Cerrillo J, MacGregor JF. Within-batch and batch-to-batch inferential-adaptive control of semibatch reactors: a partial least squares approach. *Ind Eng Chem Res* 2003;**42**:3334–5.
39. Yabuki Y, MacGregor JF. Product quality control in semi-batch reactors using mid-course correction policies. *Ind Eng Chem Res* 1997;**36**:1268–75.
40. Flores-Cerrillo J, MacGregor JF. Control of batch product quality by trajectory manipulation using latent variable models. *J Process Control* 2004;**14**:539–53.
41. Welz C, Srinivasan B, Bonvin D. Measurement-based optimization of batch processes: meeting terminal constraints on-line via trajectory following. *J Process Control* 2008;**18**(3–4):375–82.
42. Moore KL. *Iterative learning control for deterministic systems. Advances in industrial control,* London: Springer-Verlag; 1993.
43. Arimoto S, Kawamura S, Miyazaki F. Bettering operation of robots by learning. *J Robot Syst* 1984;**1**(2): 123–40.

44. Wallen J, Gunnarsson S, Henriksson R, Moberg S, Norrlof M. ILC applied to a flexible two-link robot model using sensor-fusion-based estimates. In: *IEEE CDC*; 2009. p. 458–63. Shanghai, China.
45. Lee JH, Lee KS. Iterative learning control applied to batch processes: an overview. *Control Eng Practice* 2007; **15**(10):1306–18.
46. Wang Y, Gao F, Doyle FJ. Survey on iterative learning control, repetitive control, and run-to-run control. *J Process Control* 2009;**19**(10):1589–600.
47. Arimoto S, Naniwa T, Suzuki H. Robustness of P-type learning control with a forgetting factor for robotic motion. In: *29th conference on decision and control*; 1990. p. 2640–5. Honolulu, HI.
48. Heinzinger G, Fenwick D, Paden B, Miyazaki F. Stability of learning control with disturbances and uncertain initial conditions. *IEEE Trans Autom Control* 1992;**37**:110–4.
49. Welz C, Srinivasan B, Bonvin D. Iterative learning control with input shift. In: *IFAC DYCOPS'7*; 2004. Cambridge, MA.
50. François G, Srinivasan B, Bonvin D. A globally convergent algorithm for the run-to-run control of systems with sector nonlinearities. *Ind Eng Chem Res* 2011;**50**(3):1410–8.
51. Slotine J-JE, Li W. *Applied nonlinear control.* Englewood Cliffs (NJ): Prentice-Hall; 1991.
52. Thompson WC. Batch process control. In: McMillan GK, Considine DM, editors. *Process/industrial instruments and control handbook.* New York: MacGraw-Hill; 1999. 3.70–84.
53. Hudak T. Programmable controllers. In: McMillan GK, Considine DM, editors. *Process/industrial instruments and control handbook.* New York: MacGraw-Hill; 1999. 3.32–50.
54. Netto R, Bagri A. Programmable logic controllers. *Int J Comput Appl* 2013;**77**(11):27–31.
55. Marchetti AG. Modifier-adaptation methodology for real-time optimization. Doctoral thesis No. 4449; EPFL Lausanne, Switzerland; 2009.
56. Galloway B, Hancke GP. Introduction to industrial control networks. *IEEE Commun Surv Tut* 2013;**15** (2):860–80.
57. Ruppen D, Bonvin D, Rippin DWT. Implementation of adaptive optimal operation for a semi-batch reaction system. *Comp Chem Eng* 1998;**22**:185–9.
58. Bequette W. *Process dynamics: modeling, analysis and simulation.* Englewood Cliffs (NJ): Prentice-Hall; 1998.
59. Eckerman I. *THE BHOPAL SAGA—causes and consequences of the World's largest industrial disaster.* Hyderabad: University Press (India) Private Limited; 2005.
60. Marchetti A, Amrhein M, Chachuat B, Bonvin D. Scale-up of batch processes via decentralized control. In: *IFAC ADCHEM'06*; 2006. Gramado, Brazil.
61. Welz C, Srinivasan B, Bonvin D. Combined online and run-to-run optimization of batch processes with terminal constraints. In: *IFAC ADCHEM'04*; 2004. Hong Kong.
62. Stephanopoulos G. *Chemical process control: an introduction to theory and practice.* Englewood Cliffs, NJ: Prentice-Hall; 1984.
63. François G, Bonvin D. Measurement-based real-time optimization of chemical processes. In: Pushpavanam S, editor. *Control and optimisation of process systems. Advances in chemical engineering,* Amsterdam: Elsevier; 2013.
64. Rotava O, Zanin AC. Multivariable control and real-time optimization—an industrial practical view. *Hydrocarb Process* 2005;**84**(6):61–71.
65. Boyd S, Vandenberghe L. *Convex optimization.* Cambridge: Cambridge University Press; 2004.
66. Marlin TE, Hrymak AN. Real-time operations optimization of continuous processes. In: *AIChE Symposium Series—CPC-V.* vol. 93; 1997. p. 156–64.
67. Zhang Y, Monder D, Forbes JF. Real-time optimization under parametric uncertainty: a probability constrained approach. *J Process Control* 2002;**12**:373–89.
68. Forbes JF, Marlin TE, MacGregor JF. Model adequacy requirements for optimizing plant operations. *Comp Chem Eng* 1994;**18**(6):497–510.
69. Marchetti A, Chachuat B, Bonvin D. Modifier-adaptation methodology for real-time optimization. *Ind Eng Chem Res* 2009;**48**(13):6022–33.
70. Skogestad S. Self-optimizing control: the missing link between steady-state optimization and control. *Comp Chem Eng* 2000;**24**:569–75.
71. Ariyur KB, Krstic M. *Real-time optimization by extremum-seeking control.* New York: John Wiley; 2003.

72. Srinivasan B, Bonvin D. Real-time optimization of batch processes via tracking the necessary conditions of optimality. *Ind Eng Chem Res* 2007;**46**(2):492–504.
73. Srinivasan B, Palanki S, Bonvin D. Dynamic optimization of batch processes: I. Characterization of the nominal solution. *Comp Chem Eng* 2003;**44**:1–26.
74. Biegler LT. *Nonlinear programming: concepts, algorithms, and applications to chemical processes. MOS-SIAM Series on Optimization.* Philadelphia: MO10, SIAM; 2010.
75. Bryson AE, Ho YC. *Applied optimal control.* Washington (DC): Hemisphere; 1975.
76. Vassiliadis VS, Sargent RWH, Pantelides CC. Solution of a class of multistage dynamic optimization problems. 1. Problems without path constraints. *Ind Eng Chem Res* 1994;**33**(9):2111–22.
77. Logsdon JS, Biegler LT. Accurate solution of differential-algebraic optimization problems. *Ind Eng Chem Res* 1989;**28**(11):1628–39.
78. Bryson AE. *Dynamic optimization.* Menlo Park (CA): Addison-Wesley; 1999.
79. Bazarra MS, Sherali HD, Shetty CM. *Nonlinear programming: theory and algorithms.* 2nd ed. New York: John Wiley & Sons; 1993.
80. Gill PE, Murray W, Wright MH. *Practical optimization.* London: Academic Press; 1981.
81. Forsgren A, Gill PE, Wright MH. Interior-point methods for nonlinear optimization. *SIAM Rev* 2002;**44**(4): 525–97.
82. Srinivasan B, Biegler LT, Bonvin D. Tracking the necessary conditions of optimality with changing set of active constraints using a barrier-penalty function. *Comp Chem Eng* 2008;**32**(3):572–9.
83. Chui CK, Chen G. *Kalman filtering with real-time applications.* 4th ed. Dordrecht: Springer-Verlag; 2009.
84. Rawlings JB. Moving horizon estimation. In: Baillieul J, Samad T, editors. *Encyclopedia of systems and control.* Berlin: Springer; 2014. p. 1–7.
85. Kozub DJ, MacGregor JF. State estimation for semi-batch polymerization reactors. *Chem Eng Sci* 1992;**47**: 1047–62.
86. Dochain D. State and parameter estimation in chemical and biochemical processes: a tutorial. *J Process Control* 2003;**13**(8):801–18.
87. Wittenmark B. Adaptive dual control methods: an overview. In: *IFAC symp on adaptive syst in control and signal proc*; 1995. p. 67–72. Budapest.
88. Angeli D. Economic model predictive control. In: Baillieul J, Samad T, editors. *Encyclopedia of systems and control.* Berlin: Springer; 2014. p. 1–9.
89. Marchetti AG, Chachuat B, Bonvin D. A dual modifier-adaptation approach for real-time optimization. *J Process Control* 2010;**20**(9):1027–37.
90. Welz C, Marchetti AG, Srinivasan B, Bonvin D, Ricker NL. Validation of a solution model for the optimization of a batch distillation column. In: *American control conference*; 2005. Portland, OR.
91. François G, Srinivasan B, Bonvin D, Hernandez Barajas J, Hunkeler D. Run-to-run adaptation of a semi-adiabatic policy for the optimization of an industrial batch polymerization process. *Ind Eng Chem Res* 2004;**43** (23):7238–42.

CHAPTER 12

Nonlinear Control

Jean-Pierre Corriou
Lorraine University, Nancy Cedex, France

12.1 Introduction

Nonlinear behavior is the general rule in physics and nature. Linear models, obtained by linearization or identification, are in general crude approximations of nonlinear behaviors of plants in the neighborhood of an operating point. However, in many cases such as startup, shutdown, or important transient regimes, study of batch and fed-batch processes, a linear model is insufficient to correctly reproduce the reality, and the resulting linear controller cannot guarantee stability and performance. Yet, because of the difficulty to cope with nonlinear control, linear models and linear controllers are by far dominant.

Nevertheless, efficient methods exist that can be used with nonlinear models provided the end-users are willing to carry out some effort.

Among existing theories, one can find backstepping, sliding mode control,[1–3], flatness-based control,[4,5], and methods based on Lyapunov stability, nonlinear model predictive control.[6–8] These methods are powerful and would deserve a long development.

In this chapter, a particular method of nonlinear control, often called nonlinear geometric control[2,9–11] will be presented and discussed. It is based on differential geometry but can be understood in simpler words. Differential geometry is devoted in particular to the theory of differential equations in relation with geometry, surfaces, and manifolds.

12.2 Some Mathematical Notions Useful in Nonlinear Control

The theory of linear control was developed long before nonlinear control and some tools available in linear control can be adapted to nonlinear control without performing the usual linear approximation of dynamics by calculation of the Jacobian. Several textbooks are devoted to nonlinear systems, analysis, and control, among which Refs. 3,9,10,12–14.

Coulson and Richardson's Chemical Engineering. http://dx.doi.org/10.1016/B978-0-08-101095-2.00012-6

The important first point about nonlinear geometric control is that it can be used for systems that are affine with respect to the manipulated input, that is, that can be described like the following single-input single-output (SISO) plant as

$$\begin{cases} \dot{\boldsymbol{x}} = \boldsymbol{f}(\boldsymbol{x}) + \boldsymbol{g}(\boldsymbol{x})\, u \\ y = h(\boldsymbol{x}) \end{cases} \tag{12.1}$$

where $\boldsymbol{x}$ is the state vector of dimension n, u the control input, and y the controlled output. This might seem a severe restriction, but in chemical engineering, most systems are of the form (12.1). This is because the manipulated input is in general a flow rate, a position of valve, which appears linearly in nonlinear models. $\boldsymbol{f}(\boldsymbol{x})$ and $\boldsymbol{g}(\boldsymbol{x})$ are, respectively, called vector fields of the dynamics and the control. They are assumed smooth mappings and $h(\boldsymbol{x})$ is a smooth function.

The parallel between the system (12.1) and the linear state-space model

$$\begin{cases} \dot{\boldsymbol{x}} = \boldsymbol{A}\,\boldsymbol{x} + \boldsymbol{B}\, u \\ y = \boldsymbol{C}\boldsymbol{x} \end{cases} \tag{12.2}$$

is obvious. The notions of linear control that will be developed and applied to nonlinear control can be found in Refs. 10,15. The presentation will here deal only with nonlinear systems and control-related notions.

12.2.1 Notions of Differential Geometry

The derivative of a function $\lambda(\boldsymbol{x})$ in the direction of the field $\boldsymbol{f}$ (directional derivative) is called the Lie derivative and is defined by

$$L_f\lambda(\boldsymbol{x}) = \sum_{i=1}^{n} \frac{\partial \lambda}{\partial x_i} f_i(\boldsymbol{x}) = \left\langle \frac{\partial \lambda}{\partial \boldsymbol{x}}, \boldsymbol{f}(\boldsymbol{x}) \right\rangle \tag{12.3}$$

It plays a very important role in nonlinear control. Indeed, for the system (12.1)

$$\begin{aligned} \frac{dy}{dt} &= \sum_{i=1}^{n} \frac{\partial h}{\partial x_i} \frac{dx_i}{dt} \\ &= \sum_{i=1}^{n} \frac{\partial h}{\partial x_i} (f_i(\boldsymbol{x}) + g_i(\boldsymbol{x})\, u) \\ &= L_f h(\boldsymbol{x}) + L_g h(\boldsymbol{x})\, u \end{aligned} \tag{12.4}$$

thus the time derivative of the output is simply expressed with respect to the Lie derivatives. The Lie derivative $L_f\lambda(\boldsymbol{x})$ is the derivative of λ along the integral curves of the vector field $\boldsymbol{f}$. The integral curves are the curves of the solution $\boldsymbol{x}(t)$ passing by $\boldsymbol{x}^\circ$ for the state-space system

$$\dot{\boldsymbol{x}}(t) = \boldsymbol{f}(\boldsymbol{x}(t)); \quad \boldsymbol{x}(0) = \boldsymbol{x}^\circ$$

Consider successive differentiations, such as the differentiation of λ in the direction of $\boldsymbol{f}$, then in the direction of $\boldsymbol{g}$, that is

$$L_g L_f \lambda(\boldsymbol{x}) = \frac{\partial L_f \lambda}{\partial \boldsymbol{x}} \boldsymbol{g}(\boldsymbol{x}) \tag{12.5}$$

or further, to differentiate λ, k times in the direction of $\boldsymbol{f}$

$$L_f^k \lambda(\boldsymbol{x}) = \frac{\partial L_f^{k-1} \lambda}{\partial \boldsymbol{x}} \boldsymbol{f}(\boldsymbol{x}) \quad \text{with } L_f^0 \lambda(\boldsymbol{x}) = \lambda(\boldsymbol{x}) \tag{12.6}$$

The Lie bracket is defined by

$$[\boldsymbol{f}, \boldsymbol{g}](\boldsymbol{x}) = \frac{\partial \boldsymbol{g}}{\partial \boldsymbol{x}} \boldsymbol{f}(\boldsymbol{x}) - \frac{\partial \boldsymbol{f}}{\partial \boldsymbol{x}} \boldsymbol{g}(\boldsymbol{x}) \tag{12.7}$$

where $\partial \boldsymbol{f}/\partial \boldsymbol{x}$ is the Jacobian matrix of $\boldsymbol{f}$ equal to (same for $\partial \boldsymbol{g}/\partial \boldsymbol{x}$, the Jacobian matrix of $\boldsymbol{g}$)

$$D\boldsymbol{f}(\boldsymbol{x}) = \frac{\partial \boldsymbol{f}}{\partial \boldsymbol{x}} = \begin{bmatrix} \frac{\partial f_1}{\partial x_1} & \cdots & \frac{\partial f_1}{\partial x_n} \\ \vdots & \vdots & \\ \frac{\partial f_n}{\partial x_1} & \cdots & \frac{\partial f_n}{\partial x_n} \end{bmatrix} \tag{12.8}$$

The operation on the Lie bracket of $\boldsymbol{g}$ by iterating on $\boldsymbol{f}$ can be repeated and the following notation is adopted

$$\mathrm{ad}_f^k \boldsymbol{g}(\boldsymbol{x}) = [\boldsymbol{f}, \mathrm{ad}_f^{k-1} \boldsymbol{g}](\boldsymbol{x}) \quad \text{for } k > 1 \quad \text{with } \mathrm{ad}_f^0 \boldsymbol{g}(\boldsymbol{x}) = \boldsymbol{g}(\boldsymbol{x}) \tag{12.9}$$

The Lie bracket is a bilinear, skew-symmetric mapping and satisfies the Jacobi identity

$$[\boldsymbol{f}, [\boldsymbol{g}, \boldsymbol{p}]] + [\boldsymbol{g}, [\boldsymbol{p}, \boldsymbol{f}]] + [\boldsymbol{p}, [\boldsymbol{f}, \boldsymbol{g}]] = 0 \tag{12.10}$$

where $\boldsymbol{f}$, $\boldsymbol{g}$, $\boldsymbol{p}$ are vector fields.

12.2.2 Relative Degree of a Monovariable Nonlinear System

For a linear transfer function

$$G(s) = \frac{N(s)}{D(s)} = \frac{b_0 + b_1 s + \cdots + b_m s^m}{a_0 + a_1 s + \cdots + a_n s^n} \tag{12.11}$$

that has no common poles and zeros, the roots of the denominator are called the poles, the roots of the numerator are the transmission zeros, and the difference $n - m$ is the relative degree of the transfer function.

Considering the SISO nonlinear system (12.1), the relative degree, or relative order, or characteristic index, stems from the following definition[16]: the relative degree of the nonlinear system (12.1) over a domain U is the smallest integer r for which

$$L_g L_f^{r-1} h(\boldsymbol{x}) \neq 0 \quad \text{for all } \boldsymbol{x} \text{ in } U \tag{12.12}$$

For the linear system (12.2), it would yield

$$L_g L_f^{r-1} h(\boldsymbol{x}) = CA^{r-1}B \neq 0 \tag{12.13}$$

and is consistent with the definition of the relative degree for linear systems.

Thus the nonlinear system (12.1) possesses a relative degree r equal to

$$\begin{aligned} &r=1 \quad \text{if } L_g h(\boldsymbol{x}) \neq 0 \\ &r=2 \quad \text{if } L_g h(\boldsymbol{x}) = 0 \text{ and } L_g L_f h(\boldsymbol{x}) \neq 0 \\ &r=3 \quad \text{if } L_g h(\boldsymbol{x}) = L_g L_f h(\boldsymbol{x}) = 0 \text{ and } L_g L_f^2 h(\boldsymbol{x}) \neq 0 \\ &\vdots \end{aligned} \tag{12.14}$$

Using this definition, the relative degree r can be obtained from the successive time derivatives of the output y as

$$\begin{aligned} \frac{dy}{dt} &= L_f h(\boldsymbol{x}) + L_g h(\boldsymbol{x})\, u \\ &= L_f h(\boldsymbol{x}) \quad \text{if } 1 < r \\ &\vdots \\ \frac{d^k y}{dt^k} &= L_f^k h(\boldsymbol{x}) + L_g L_f^{k-1} h(\boldsymbol{x})\, u \\ &= L_f^k h(\boldsymbol{x}) \quad \text{if } k < r \\ \frac{d^r y}{dt^r} &= L_f^r h(\boldsymbol{x}) + L_g L_f^{r-1} h(\boldsymbol{x})\, u \quad \text{as } L_g L_f^{r-1} h(\boldsymbol{x}) \neq 0 \end{aligned} \tag{12.15}$$

Thus the relative degree is the smallest degree of differentiation of the output y, which depends explicitly on the input u. This can be a convenient way to find the relative degree of an SISO system.

It may happen that, for example, the first Lie derivative $L_g h(\boldsymbol{x})$ of the sequence $L_g L_f^{k-1} h(\boldsymbol{x})$ is zero at a given point. In that case, according to Isidori[10], the relative degree cannot be defined strictly at $\boldsymbol{x}^\circ$, but will be defined in the neighborhood U (notion of dense open subset). This will be accepted in the following.

If it happens that

$$L_g L_f^k h(\boldsymbol{x}) = 0 \quad \text{for all } k, \quad \text{for all } \boldsymbol{x} \text{ in } U \tag{12.16}$$

the relative degree cannot be defined in the neighborhood of $\boldsymbol{x}^\circ$ and the output is not affected by the input u.

It can be shown[10] that the matrix

$$\begin{bmatrix} Dh(\boldsymbol{x}) \\ DL_f h(\boldsymbol{x}) \\ \vdots \\ DL_f^{r-1} h(\boldsymbol{x}) \end{bmatrix} \left[\boldsymbol{g}(\boldsymbol{x}) \;\; \mathrm{ad}_f \boldsymbol{g}(\boldsymbol{x}) \;\; \ldots \;\; \mathrm{ad}_f^{r-1} \boldsymbol{g}(\boldsymbol{x}) \right] \tag{12.17}$$

has rank r. This implies that the row vectors $Dh(\boldsymbol{x}), DL_f h(\boldsymbol{x}), \ldots, DL_f^{r-1} h(\boldsymbol{x})$ are linearly independent. Thus the r functions $h(\boldsymbol{x}), L_f h(\boldsymbol{x}), \ldots, L_f^{r-1} h(\boldsymbol{x})$ can form a new set of coordinates in the neighborhood of point $\boldsymbol{x}^\circ$.

12.2.3 Frobenius Theorem

The Frobenius theorem gives a necessary and sufficient condition of integrability of a system of first-order partial differential equations whose right member depends only on variables or unknowns but not on partial derivatives of the unknowns. It is also called a Pfaff system.

It will be presented according to Isidori[10]

(a) First, let us consider d smooth vector fields $\boldsymbol{f}_i(\boldsymbol{x})$, defined on Ω°, which span a distribution Δ, denoted by

$$\Delta = \mathrm{span}\{\boldsymbol{f}_1(\boldsymbol{x}), \ldots, \boldsymbol{f}_d(\boldsymbol{x})\} \tag{12.18}$$

To define a distribution, consider smooth vector fields $\boldsymbol{f}_1(\boldsymbol{x}), \ldots, \boldsymbol{f}_d(\boldsymbol{x})$ that span at a point $\boldsymbol{x}$ of U a vector space dependent on $\boldsymbol{x}$ that can be denoted by $\Delta(\boldsymbol{x})$. The mapping assigning this vector space to any point $\boldsymbol{x}$ is called a smooth distribution.

In the same neighborhood Ω°, the codistribution W of dimension $n-d$ is spanned by $n-d$ covector fields, $\boldsymbol{w}_1, \ldots, \boldsymbol{w}_{n-d}$, such that

$$\langle \boldsymbol{w}_j(\boldsymbol{x}), \boldsymbol{f}_i(\boldsymbol{x}) \rangle = 0 \;\; \forall\, 1 \le i \le d, \;\; 1 \le j \le n-d \tag{12.19}$$

Due to that property, the codistribution is denoted as: $W = \Delta^\perp$, and $\boldsymbol{w}_j$ is the solution of the equation

$$\boldsymbol{w}_j(\boldsymbol{x}) \boldsymbol{F}(\boldsymbol{x}) = 0 \tag{12.20}$$

where $\boldsymbol{F}(\boldsymbol{x})$ is the matrix of dimension $n \times d$, of rank d, equal to

$$\boldsymbol{F}(\boldsymbol{x}) = [\boldsymbol{f}_1(\boldsymbol{x}) \ \ \dots \ \ \boldsymbol{f}_d(\boldsymbol{x})] \tag{12.21}$$

The row vectors $\boldsymbol{w}_j$ form a basis of the space of the solutions of Eq. (12.20).

(b) We look for solutions such that

$$\boldsymbol{w}_j = \frac{\partial \lambda_j}{\partial \boldsymbol{x}} \tag{12.22}$$

correspond to smooth functions λ_j, that is, we look for $n - d$ independent solutions (the row vectors $\partial\lambda_1/\partial\boldsymbol{x}, \dots, \partial\lambda_{n-d}/\partial\boldsymbol{x}$ are independent) of the following differential equation

$$\frac{\partial \lambda_j}{\partial \boldsymbol{x}} \boldsymbol{F}(\boldsymbol{x}) = \frac{\partial \lambda_j}{\partial \boldsymbol{x}} [\boldsymbol{f}_1(\boldsymbol{x}) \ \ \dots \ \ \boldsymbol{f}_d(\boldsymbol{x})] = 0 \tag{12.23}$$

(c) We search the condition of existence of $n - d$ independent solutions of differential equation (12.23), which amounts to seeking the integrability of the distribution Δ: a distribution of dimension d, defined on an open domain U of $\mathbb{R}^n$, is completely integrable if, for any point $\boldsymbol{x}^\circ$ of U, there exist $n - d$ smooth functions, taking real values, defined on a neighborhood of $\boldsymbol{x}^\circ$, such that

$$\text{span}\left\{\frac{\partial \lambda_1}{\partial \boldsymbol{x}}, \dots, \frac{\partial \lambda_{n-d}}{\partial \boldsymbol{x}}\right\} = \Delta^\perp \tag{12.24}$$

The condition of existence is produced by the Frobenius theorem.

(d) Frobenius theorem: A distribution is nonsingular if and only if it is involutive.
A distribution Δ is defined as involutive if the Lie bracket of any couple of vector fields belonging to Δ belongs to Δ

$$\boldsymbol{f}_1 \text{ and } \boldsymbol{f}_2 \in \Delta \Longrightarrow [\boldsymbol{f}_1, \boldsymbol{f}_2] \in \Delta \Leftrightarrow$$
$$[\boldsymbol{f}_i, \boldsymbol{f}_j](\boldsymbol{x}) = \sum_{k=1}^{m} \alpha_{ijk} \boldsymbol{f}_k(\boldsymbol{x}) \quad \forall i, j$$

In the case where $\boldsymbol{F}$ is reduced to only a vector field $\boldsymbol{f}_1$ ($d = 1$), Eq. (12.23) can be geometrically interpreted as:

- The gradient of λ is orthogonal to $\boldsymbol{f}_1$.
- The vector $\boldsymbol{f}_1$ is tangent to the surface λ = constant passing by this point.
- The integral curve of $\boldsymbol{f}_1$ passing by this point is entirely on the surface λ = constant.

12.2.4 Coordinates Transformation

The objective of the change of coordinates is to present the system in a simpler form in the new coordinates.

A function Φ of $\mathbb{R}^n$ in $\mathbb{R}^n$, defined in a domain U, is called a diffeomorphism if it is smooth and if its inverse Φ^{-1} exists and is smooth. If the domain U is the whole space, the diffeomorphism is global; otherwise, it is local. The diffeomorphism is thus a nonlinear coordinate change possessing the previous properties.

Consider a function Φ defined in a domain U of $\mathbb{R}^n$. $\Phi(\boldsymbol{x})$ defines a local diffeomorphism on a subdomain Ω° of Ω, if and only if the Jacobian matrix $\partial\Phi/\partial\boldsymbol{x}$ is nonsingular at $\boldsymbol{x}^\circ$ belonging to Ω,

A diffeomorphism allows us to transform a nonlinear system into another nonlinear system defined with regard to new states.

Given the SISO nonlinear system (12.1) of relative degree r at $\boldsymbol{x}^\circ$, set the r first functions

$$\begin{aligned} \phi_1(\boldsymbol{x}) &= h(\boldsymbol{x}) \\ \phi_2(\boldsymbol{x}) &= L_f h(\boldsymbol{x}) \\ &\vdots \\ \phi_r(\boldsymbol{x}) &= L_f^{r-1} h(\boldsymbol{x}) \end{aligned} \tag{12.25}$$

If $r < n$, it is possible to find $n - r$ functions $\phi_{r+1}(\boldsymbol{x}), \ldots, \phi_n(\boldsymbol{x})$ such that the mapping

$$\Phi(\boldsymbol{x}) = \begin{bmatrix} \phi_1(\boldsymbol{x}) \\ \vdots \\ \phi_n(\boldsymbol{x}) \end{bmatrix} \tag{12.26}$$

has its Jacobian matrix nonsingular and thus constitutes a possible coordinate change at $\boldsymbol{x}^\circ$.

The value taken by the additional functions $\phi_{r+1}(\boldsymbol{x}), \ldots, \phi_n(\boldsymbol{x})$ at $\boldsymbol{x}^\circ$ is not important and these functions can be chosen such that

$$< D\phi_i(\boldsymbol{x}), \boldsymbol{g}(\boldsymbol{x}) > = L_g \phi_i(\boldsymbol{x}) = 0 \quad \text{for all } r+1 \le i \le n \quad \text{for all } \boldsymbol{x} \text{ in } \Omega \tag{12.27}$$

The demonstration makes use of the Frobenius theorem.[10]

12.2.5 Normal Form

Given the vector $z = \Phi(\boldsymbol{x})$, making use of the r first new coordinates z_i defined by $z_i = y^{(i-1)} = \phi_i(\boldsymbol{x})$, $(i = 1, \ldots, r)$, defined according to the relations (12.25), the nonlinear system (12.1) can be described as

$$
\begin{aligned}
\frac{dz_1}{dt} &= \frac{\partial \phi_1}{\partial \boldsymbol{x}} \frac{d\boldsymbol{x}}{dt} = \frac{\partial h}{\partial \boldsymbol{x}} \frac{d\boldsymbol{x}}{dt} = L_f h(\boldsymbol{x}(t)) = \phi_2(\boldsymbol{x}(t)) = z_2(t) \\
&\vdots \\
\frac{dz_{r-1}}{dt} &= \frac{\partial \phi_{r-1}}{\partial \boldsymbol{x}} \frac{d\boldsymbol{x}}{dt} = \frac{\partial L_f^{r-2} h}{\partial \boldsymbol{x}} \frac{d\boldsymbol{x}}{dt} = L_f^{r-1} h(\boldsymbol{x}(t)) = \phi_r(\boldsymbol{x}(t)) = z_r(t) \\
\frac{dz_r}{dt} &= \frac{\partial \phi_r}{\partial \boldsymbol{x}} \frac{d\boldsymbol{x}}{dt} = \frac{\partial L_f^{r-1} h}{\partial \boldsymbol{x}} \frac{d\boldsymbol{x}}{dt} = L_f^r h(\boldsymbol{x}(t)) + L_g L_f^{r-1} h(\boldsymbol{x}(t))\, u(t)
\end{aligned}
\tag{12.28}
$$

The expression of $\dot{z}_r(t)$ must be transformed with respect to $z(t)$ by using the inverse relation $\boldsymbol{x}(t) = \Phi^{-1}(z(t))$, yielding

$$
\begin{aligned}
\frac{dz_r}{dt} &= L_f^r h(\Phi^{-1}(z(t))) + L_g L_f^{r-1} h(\Phi^{-1}(z(t)))\, u(t) \\
&= b(z(t)) + a(z(t))\, u(t)
\end{aligned}
\tag{12.29}
$$

by setting

$$
a(z(t)) = L_g L_f^{r-1} h(\Phi^{-1}(z(t))); \quad b(z(t)) = L_f^r h(\Phi^{-1}(z(t)))
\tag{12.30}
$$

and by noticing that, by definition of the relative degree, $a(z^\circ) \neq 0$ at $z^\circ = \Phi(\boldsymbol{x}^\circ)$.

The coordinates z_i, $r < i \leq n$, can be chosen according to Eq. (12.27) so that $L_g \phi_i(\boldsymbol{x}) = 0$, which gives

$$
\begin{aligned}
\frac{dz_i}{dt} &= \frac{\partial \phi_i}{\partial \boldsymbol{x}} \frac{d\boldsymbol{x}}{dt} = \frac{\partial \phi_i}{\partial \boldsymbol{x}} (\boldsymbol{f}(\boldsymbol{x}(t)) + \boldsymbol{g}(\boldsymbol{x}(t))\, u(t)); \quad r < i \leq n \\
&= L_f \phi_i(\boldsymbol{x}(t)) + L_g \phi_i(\boldsymbol{x}(t))\, u(t) = L_f \phi_i(\boldsymbol{x}(t)) \\
&= L_f \phi_i(\Phi^{-1}(z(t)))
\end{aligned}
\tag{12.31}
$$

Set

$$
q_i(z(t)) = L_f \phi_i(\Phi^{-1}(z(t))); \quad r < i \leq n
\tag{12.32}
$$

Taking into account the previous equations, the normal form[10,17] results

$$
\begin{aligned}
\dot{z}_1 &= z_2 \\
&\vdots \\
\dot{z}_{r-1} &= z_r \\
\dot{z}_r &= b(z) + a(z)\, u(t) \\
\dot{z}_{r+1} &= q_{r+1}(z) \\
\dot{z}_n &= q_n(z)
\end{aligned}
\tag{12.33}
$$

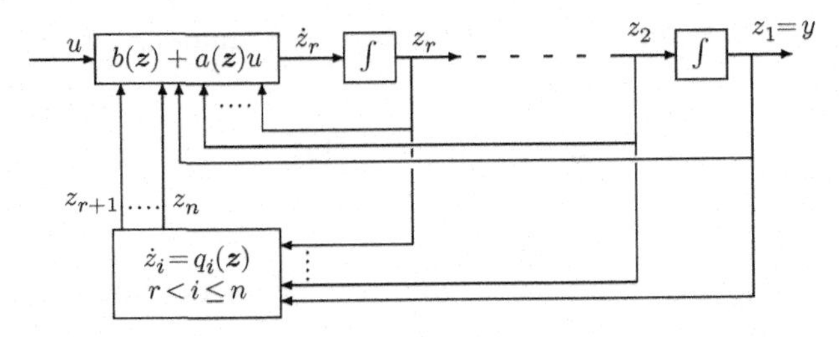

Fig. 12.1
Description of the normal form.

to which the equation of the output must be added

$$y = h(\boldsymbol{x}) = z_1 \tag{12.34}$$

This result can be symbolized in a block diagram (Fig. 12.1) using the chain of r integrators necessary to go from the control input to the output.

The condition $L_g\phi_i(\boldsymbol{x}) = 0$, fundamental for seeking the functions ϕ_i, $r < i \leq n$, can be difficult to fulfill, as it corresponds to the solving of a system of $n - r$ partial differential equations. To define a coordinate change, it can be sufficient to find these functions so that the matrix $\boldsymbol{\Phi}$ is simply nonsingular.

12.2.6 Controllability and Observability

Consider the nonlinear system

$$\dot{\boldsymbol{x}} = \boldsymbol{f}(\boldsymbol{x}) + \boldsymbol{g}(\boldsymbol{x})\, u \tag{12.35}$$

defined in a domain U.

The system (12.35) is controllable if, given two arbitrary states $\boldsymbol{x}_0$ and $\boldsymbol{x}_1$, there exists an admissible input $u(t)$ such that the system can be steered from the state $\boldsymbol{x}_0$ to the desired state $\boldsymbol{x}_1$ in finite time T.

The controllability of this nonlinear system can be studied by proceeding to a linearization of the system

$$\dot{z} = \frac{\partial f}{\partial \boldsymbol{x}} z + \boldsymbol{g}(\boldsymbol{x})\, v \tag{12.36}$$

and by studying the controllability matrix, in a way similar (except that it deals with distributions, see Eq. 12.54) to the controllability of linear systems where it is defined as the rank of the controllability matrix

$$C = [\boldsymbol{B} \;\; \boldsymbol{A}\boldsymbol{B} \;\; \dots \;\; \boldsymbol{A}^{n-2}\boldsymbol{B} \;\; \boldsymbol{A}^{n-1}\boldsymbol{B}] \tag{12.37}$$

which should be of rank n, given the model of the linear state-space system

$$\begin{aligned}\dot{\boldsymbol{x}}(t) &= \boldsymbol{A}\boldsymbol{x}(t) + \boldsymbol{B}\boldsymbol{u}(t)\\ \boldsymbol{y}(t) &= \boldsymbol{C}\boldsymbol{x}(t) + \boldsymbol{D}\boldsymbol{u}(t)\end{aligned} \tag{12.38}$$

with the state $\boldsymbol{x}$ of dimension n.

However, this approach is not always satisfying; actually, a nonlinear system can be controllable whereas its linear approximation is not. It is necessary to introduce the notion of reachability,[10,13] a weaker form of controllability.

For observability that also requires complex topological notions, both previous books are recommended. Observability can be defined in an approximate manner by the following property:

In parallel to the definition of controllability, if, given two different initial conditions $\boldsymbol{x}_1(0)$ and $\boldsymbol{x}_2(0)$, there exists a control input $u(t)$ defined in $[0, T]$ such that the corresponding outputs $y_1(\boldsymbol{x}, u, t)$ and $y_2(\boldsymbol{x}, u, t)$ are not totally similar in $[0, T]$, the system is observable. It means that, given the measurable input and output, it is possible to determine the state. The input $u(t)$ distinguishes the initial conditions $\boldsymbol{x}_1(0)$ and $\boldsymbol{x}_2(0)$ in $[0, T]$. If $u(t)$ distinguishes any pair $(\boldsymbol{x}_1, \boldsymbol{x}_2)$ in $[0, T]$, the input $u(t)$ is universal.

12.2.7 Principle of Feedback Linearization

The objective is to design a control law that is a function of the states so that, to the resulting linearized system, efficient methods of linear control can be then applied. To perform this linearization, two types of feedback can be used, state feedback or output feedback, corresponding to input-state or input-output linearization, respectively. The states are assumed to be known. In the case where all the states are not known, it is necessary to couple a state estimator, called an observer (for linear systems, the linear Kalman filter is an optimal observer) to the control system. When the state feedback control law depends only on the values of the states $\boldsymbol{x}$ and the external input v, it is a static-state feedback. If the control law corresponds to the output of a dynamic system, itself depending on the states $\boldsymbol{x}$ and on the external input v, it is a dynamic-state feedback.

Consider the SISO nonlinear system affine with respect to the input

$$\dot{\boldsymbol{x}} = \boldsymbol{f}(\boldsymbol{x}) + \boldsymbol{g}(\boldsymbol{x})\, u \tag{12.39}$$

defined in a neighborhood U of $\boldsymbol{x}^\circ$ and such that: $f(\boldsymbol{x}^\circ) = 0$. The problem of feedback linearization is to find smooth functions p and q with $q(\boldsymbol{x}^\circ) \neq 0$, and a diffeomorphism Φ with $\Phi(\boldsymbol{x}^\circ) = 0$ such that by defining:

- an external input $v = p(\boldsymbol{x}) + q(\boldsymbol{x})\, u$,
- the transformed variables $z = \Phi(\boldsymbol{x})$,

the resulting system is linear under the form

$$\dot{z} = \boldsymbol{A}\, z + \boldsymbol{B}\, v \tag{12.40}$$

where the pair $(\boldsymbol{A}, \boldsymbol{B})$ is controllable. The new state z is called a linearizing state and the control law is a linearizing control law.

Given the state $\boldsymbol{x}$, the control law (Fig. 12.2) is

$$u(t) = \frac{-p(\boldsymbol{x})}{q(\boldsymbol{x})} + \frac{v}{q(\boldsymbol{x})} = \alpha(\boldsymbol{x}) + \beta(\boldsymbol{x})\, v \tag{12.41}$$

12.2.8 Exact Input-State Linearization for a System of Relative Degree Equal to n

The input-state linearization is often called exact linearization.[10] First, consider the system (12.39) possessing a relative degree $r = n$, thus equal to the dimension of the state vector, at a point $\boldsymbol{x}^\circ$.

In this case, the coordinate change necessary to obtain the normal form (Section 12.2.5) is

$$\Phi(\boldsymbol{x}) = \begin{bmatrix} \phi_1(\boldsymbol{x}) \\ \vdots \\ \phi_n(\boldsymbol{x}) \end{bmatrix} = \begin{bmatrix} h(\boldsymbol{x}) \\ \vdots \\ L_f^{n-1} h(\boldsymbol{x}) \end{bmatrix} \tag{12.42}$$

and the resulting normal form is

$$\begin{aligned} \dot{z}_1 &= z_2 \\ &\vdots \\ \dot{z}_{n-1} &= z_n \\ \dot{z}_n &= b(z) + a(z)\, u(t) \end{aligned} \tag{12.43}$$

with $a(z^\circ) \neq 0$ because of the definition of the relative degree (Fig. 12.3).

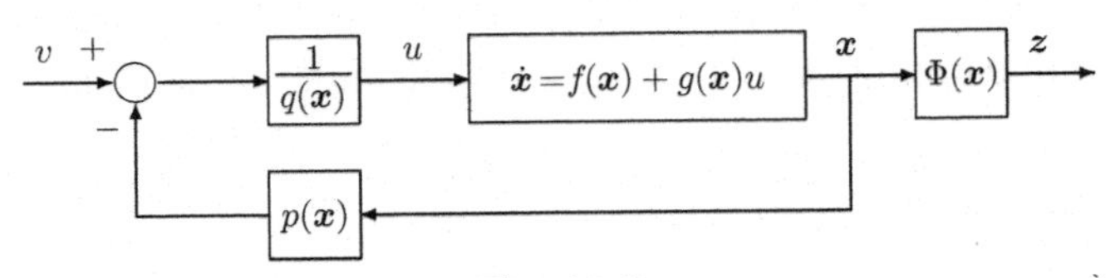

Fig. 12.2
Linearizing feedback.

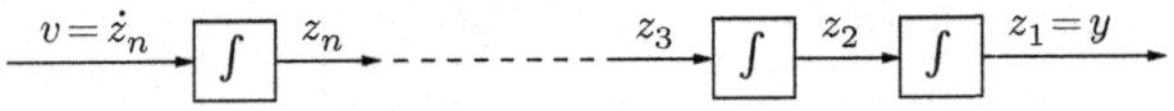

Fig. 12.3
Exactly linearized system by static-state feedback.

The state feedback control law can be chosen as

$$u(t) = -\frac{b(z)}{a(z)} + \frac{v}{a(z)} \tag{12.44}$$

hence the resulting closed-loop system (Fig. 12.2 with its chain of integrators)

$$\begin{aligned} \dot{z}_1 &= z_2 \\ &\vdots \\ \dot{z}_{n-1} &= z_n \\ \dot{z}_n &= v \end{aligned} \tag{12.45}$$

that is a linear and controllable system expressed in the Brunovsky canonical form[18]

$$\dot{z} = \begin{bmatrix} 0 & 1 & 0 & \dots & 0 \\ 0 & 0 & 1 & \ddots & \vdots \\ \vdots & & \ddots & \ddots & 0 \\ \vdots & & & \ddots & 1 \\ 0 & \dots & \dots & 0 & \end{bmatrix} z + \begin{bmatrix} 0 \\ \vdots \\ \vdots \\ 0 \\ 1 \end{bmatrix} v \tag{12.46}$$

To obtain the control law (12.44), a coordinate change and a state feedback taken into any order, thus exchangeable, are used. If the state feedback is first used and then the coordinate change, the following control law results

$$\begin{aligned} u(t) &= -\frac{b(\Phi(\boldsymbol{x}))}{a(\Phi(\boldsymbol{x}))} + \frac{v}{a(\Phi(\boldsymbol{x}))} \\ &= \frac{-L_f^n h(\boldsymbol{x}) + v}{L_g L_f^{n-1} h(\boldsymbol{x})} \end{aligned} \tag{12.47}$$

often used under this form that corresponds to the same controllable linear system (12.45). This control law is called linearizing state feedback and the coordinates $\boldsymbol{\Phi}$ are the linearizing coordinates.

Two remarks[10] are particularly important:

- It was assumed that $\boldsymbol{x}^\circ$ is a stationary point for the system (12.1), meaning that $f(\boldsymbol{x}^\circ) = 0$ and $h(\boldsymbol{x}^\circ) = 0$, hence

$$\phi_1(\boldsymbol{x}^\circ) = h(\boldsymbol{x}^\circ) = 0$$

$$\phi_i(\boldsymbol{x}^\circ) = \frac{\partial L_f^{i-2} h}{\partial \boldsymbol{x}} \boldsymbol{f}(\boldsymbol{x}^\circ) = 0, \quad 1 < i \leq n \tag{12.48}$$

or $\boldsymbol{z}^\circ = 0$. It is always possible to come back to $h(\boldsymbol{x}^\circ) = 0$ by an appropriate translation.

– It is possible to perform a pole placement or to satisfy an optimality criterion, by imposing a feedback (Fig. 12.4) as

$$v_2 = \boldsymbol{K}\,\boldsymbol{z}, \quad \text{with the gain vector}: \boldsymbol{K} = [c_0 \ldots c_{n-1}] \tag{12.49}$$

equivalent to

$$v_2 = c_0 h(\boldsymbol{x}) + c_1 L_f h(\boldsymbol{x}) + \cdots + c_{n-1} L_f^{n-1} h(\boldsymbol{x}) \tag{12.50}$$

which is a nonlinear state feedback with respect to $\boldsymbol{x}$. The state feedback control law becomes, in this case,

$$u(t) = \frac{-L_f^n h(\boldsymbol{x}) + \dot{z}_n}{L_g L_f^{n-1} h(\boldsymbol{x})} = \frac{-L_f^n h(\boldsymbol{x}) - \sum_{i=0}^{n-1} c_i L_f^i h(\boldsymbol{x}) + v}{L_g L_f^{n-1} h(\boldsymbol{x})} \tag{12.51}$$

and appears as an extension of Eq. (12.47). The external input v can be chosen in many different ways. For example, it could be equal to the set point y_{ref}. When $v = 0$, this corresponds to the local asymptotic equilibrium $\boldsymbol{z} = 0$ which is preserved.
By expressing the transfer function $Y(s)/V(s)$, the characteristic polynomial associated with Eq. (12.51) then equal to

$$c_0 + c_1\, s + \cdots + c_{n-1}\, s^{n-1} + s^n \tag{12.52}$$

whose coefficients can be chosen so as to realize the adequate pole placement.
By simply considering the system (12.39) a priori defined without output, Isidori[10] shows that the exact input-state linearization is possible in a neighborhood U of $\boldsymbol{x}^\circ$ if and only if there exists a scalar function $\lambda(\boldsymbol{x})$ such that the system with the "output" redefined

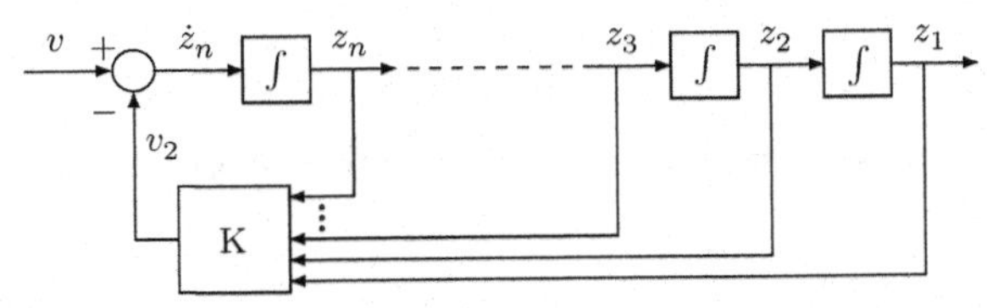

Fig. 12.4
Nonlinear control with pole placement for a system of relative degree equal to n (exact input-state linearization).

$$\dot{\boldsymbol{x}} = \boldsymbol{f}(\boldsymbol{x}) + \boldsymbol{g}(\boldsymbol{x})\, u$$
$$y = \lambda(\boldsymbol{x}) \tag{12.53}$$

has a relative degree equal to n at $\boldsymbol{x}^\circ$. Referring to Eq. (12.46) and Fig. 12.2, the function $\lambda(\boldsymbol{x})$ is equal to $z_1(\boldsymbol{x})$.

The following general theorem[19,20] is particularly important.

Theorem *The system* (12.39) *is exactly linearizable in state space* (*input-state linearization*) *in a neighborhood U of* $\boldsymbol{x}^\circ$ *if and only if the following conditions are satisfied:*

1. *The vector fields* $\{\boldsymbol{g}(\boldsymbol{x}^\circ), \mathrm{ad}_f\boldsymbol{g}(\boldsymbol{x}^\circ), \ldots, \mathrm{ad}_f^{n-1}\boldsymbol{g}(\boldsymbol{x}^\circ)\}$ are linearly independent.
2. *The distribution* $\mathrm{span}\{g, \mathrm{ad}_f g, \ldots, \mathrm{ad}_f^{n-2} g\}$ is involutive in U.

Condition 1 can be written as "the following matrix

$$[\boldsymbol{g}(\boldsymbol{x}^\circ), \mathrm{ad}_f\boldsymbol{g}(\boldsymbol{x}^\circ), \ldots, \mathrm{ad}_f^{n-1}\boldsymbol{g}(\boldsymbol{x}^\circ)] \tag{12.54}$$

has rank n." It is a controllability condition of the nonlinear system. This matrix must be invertible. For linear state-space systems, this matrix is the controllability matrix

$$[\boldsymbol{B}, \boldsymbol{AB}, \ldots, \boldsymbol{A}^{n-1}\boldsymbol{B}]$$

To realize the exact input-state linearization, it must be proceeded to the following stages:

- *Build the vector fields* $\boldsymbol{g}(\boldsymbol{x}^\circ), \mathrm{ad}_f\boldsymbol{g}(\boldsymbol{x}^\circ), \ldots, \mathrm{ad}_f^{n-1}\boldsymbol{g}(\boldsymbol{x}^\circ)$.
- *Check if the conditions of controllability and involutivity are verified.*
- *If these conditions are verified, find the function* $\lambda(\boldsymbol{x})$ *from the equations*

$$L_g\lambda(\boldsymbol{x}^\circ) = L_g L_f \lambda(\boldsymbol{x}^\circ) = \cdots = L_g L_f^{n-2}\lambda(\boldsymbol{x}^\circ) = 0$$
$$L_g L_f^{n-1}\lambda(\boldsymbol{x}^\circ) \neq 0$$

It is often mentioned that finding this function $\lambda(\boldsymbol{x})$ *is a difficult task.*[3]

- *Calculate the coordinate change*

$$\Phi(\boldsymbol{x}) = [\lambda(\boldsymbol{x}), L_f\lambda(\boldsymbol{x}), \ldots, L_f^{n-1}\lambda(\boldsymbol{x})] \tag{12.55}$$

12.2.9 *Input-Output Linearization of a System With Relative Degree* r *Less than or Equal to* n

Two cases are distinguished:

- Isidori[10] notes that the nonlinear system affine with respect to the input

$$\begin{cases} \dot{x} = f(x) + g(x)\, u \\ y = h(x) \end{cases} \tag{12.56}$$

having a relative degree $r < n$ can satisfy the conditions of Hunt-Su-Meyer theorem. However, in this case, it was shown that there exists a different "output" λ such that the system has a relative degree equal to n. Thus the newly defined system satisfies the previous theorem; by using a feedback $u = \alpha(x) + \beta(x)\, v$ and a coordinate change $\Phi(x)$, it is transformed into a controllable linear system, but the real output, in general, is not linear with respect to the new one

$$y = h(\Phi^{-1}(z)) \tag{12.57}$$

- If the output y is fixed by $y = h(x)$ and if the system possesses a relative degree r less than or equal to n, by using the coordinate change $\Phi(x)$ as in Eq. (12.28), it is possible to transform the system into the normal form (12.33) and to set $v = \dot{z}_r$ so that the system is simply expressed in the transformed coordinates in Byrnes-Isidori canonical form

$$\begin{aligned} \dot{z}_1 &= z_2 \\ &\vdots \\ \dot{z}_{r-1} &= z_r \\ \dot{z}_r &= v = b(z) + a(z)\, u(t) \\ \dot{z}_{r+1} &= q_{r+1}(z) \\ \dot{z}_n &= q_n(z) \\ y &= z_1 \end{aligned} \tag{12.58}$$

with

$$b(z) = L_f^r h(x), \quad a(z) = L_g L_f^{r-1} h(x) \tag{12.59}$$

The control law is deduced

$$\begin{aligned} u(t) &= -\frac{b(z)}{a(z)} + \frac{v}{a(z)} \\ &= \frac{-L_f^r h(x) + v}{L_g L_f^{r-1} h(x)} \end{aligned} \tag{12.60}$$

The resulting system is only partially linear, but the output is influenced by the external input v only through a chain of r integrators (Fig. 12.1) related to the new states $z_1, \ldots, z_r$

$$y^{(r)} = L_f^r h(x) + L_g L_f^{r-1} h(x)\, u = v \tag{12.61}$$

The new states $z_{r+1}, \ldots, z_n$ which constitute the nonlinear part of the system do not influence the output y.

Following the second case of this description of input-output linearization, the control law (12.60) can be used for many SISO nonlinear plants affine with respect to the input, for which the relative degree is lower than n.

12.2.10 Zero Dynamics

For a linear system having a strictly proper transfer function (12.11), that is, the degree of the numerator is strictly lower than that of the denominator, when positive zeros are present, the system is called nonminimum phase. It must be recalled that positive zeros become unstable poles for the inverse of the transfer function that can constitute the ideal controller. Note that, for this reason, positive zeros are considered apart in internal model control.[21]

For a SISO system, the zero dynamics amounts to find an input u and initial conditions $\boldsymbol{x}_0$ such that $y(t) = 0, \forall t$. This implies that not only $y = 0$, but also its derivatives $y^{(i)} = 0, i = 0, \ldots, r$. The dynamics of the system corresponding to these conditions is called zero dynamics.

In the case of a linear time-invariant system of relative degree $r = n$, the numerator of the system transfer function is reduced to a constant, the transfer function has no zeros and the system does not have zero dynamics. Consequently, when $r < n$, the study of zero dynamics is important.

Similarly, for a nonlinear system, to study the zero dynamics, only the case where the relative degree r is lower than n is considered. The vector is represented in normal form (12.33) by separating the linear part of dimension r and the nonlinear part of dimension $n - r$ as

$$\boldsymbol{\xi} = \begin{bmatrix} z_1 \\ z_2 \\ \vdots \\ z_r \end{bmatrix} = \begin{bmatrix} y \\ \dot{y} \\ \vdots \\ y^{(r-1)} \end{bmatrix} ; \; \boldsymbol{\eta} = \begin{bmatrix} z_{r+1} \\ \vdots \\ z_n \end{bmatrix} \tag{12.62}$$

so that the system can be rewritten as

$$\begin{aligned} \dot{z}_1 &= z_2 \\ &\vdots \\ \dot{z}_{r-1} &= z_r \\ \dot{z}_r &= b(\boldsymbol{\xi},\boldsymbol{\eta}) + a(\boldsymbol{\xi},\boldsymbol{\eta})\, u(t) \\ \dot{\boldsymbol{\eta}} &= q(\boldsymbol{\xi},\boldsymbol{\eta}) \end{aligned} \tag{12.63}$$

where $\boldsymbol{\xi}$ and $\boldsymbol{\eta}$ constitute the normal coordinates or normal states.

The dynamics of the nonlinear system is thus decomposed into an external input-output part and an internal unobservable part. Whereas the external part is simple to design, there remains the problem of the internal stability corresponding to the last $(n - r)$ equations: $\dot{\boldsymbol{\eta}} = q(\boldsymbol{\xi},\boldsymbol{\eta})$.

Considering $\boldsymbol{x}^\circ$ as an equilibrium point of the system, it results $f(\boldsymbol{x}^\circ) = 0$ and it is possible to choose $h(\boldsymbol{x}^\circ) = 0$. In the normal coordinates $(\boldsymbol{\xi}, \boldsymbol{\eta})$, it can be assumed that the point (0, 0) is the equilibrium point, hence $b(0, 0) = 0$ and $q(0, 0) = 0$.

The aim is to make the output zero for all t in the neighborhood of $t = 0$. In the normal form, this would amount to imposing

$$z_1 = \cdots = z_r = 0 \Leftrightarrow \dot{z}_1 = \cdots = \dot{z}_r = 0 \Leftrightarrow \boldsymbol{\xi} = 0 \quad \text{for all } t \tag{12.64}$$

as, moreover, the output is imposed $y = z_1 = 0$. The input u results such that

$$0 = b(0,\boldsymbol{\eta}) + a(0,\boldsymbol{\eta})\, u(t) \tag{12.65}$$

with $a(0, \boldsymbol{\eta}) \neq 0$ still in the neighborhood of $t = 0$. Moreover, the variable $\boldsymbol{\eta}$ is such that

$$\dot{\boldsymbol{\eta}} = q(0,\boldsymbol{\eta}), \quad \text{with } \boldsymbol{\eta}(0) = \boldsymbol{\eta}^\circ \tag{12.66}$$

which is an autonomous system of differential equations whose solution is the variable $\boldsymbol{\eta}(t)$. It yields the unique input that imposes a zero output in the neighborhood of $t = 0$

$$u(t) = -\frac{b(0,\boldsymbol{\eta}(t))}{a(0,\boldsymbol{\eta}(t))} \tag{12.67}$$

The dynamics of Eq. (12.66), which results from the condition of zero output, is called zero dynamics or unforced zero dynamics, it describes the internal behavior of the system. The zero dynamics is the dynamics of the inverse of the system.

The search of the zero output could have been realized in the original state space, by setting

$$y(t) = \dot{y}(t) = \cdots = y^{(r-1)}(t) = y^{(r)}(t) = 0 \quad \text{for all } t \tag{12.68}$$

or further, in the neighborhood of $\boldsymbol{x}^\circ$

$$\begin{aligned} & h(\boldsymbol{x}) = L_f h(\boldsymbol{x}) = \cdots = L_f^{r-1} h(\boldsymbol{x}) = 0 \\ & L_f^r h(\boldsymbol{x}) + L_g L_f^{r-1} h(\boldsymbol{x})\, u(t) = 0 \end{aligned} \tag{12.69}$$

The case of tracking a reference output y_{ref} is deducted by translation from the previous case of a zero output. In the neighborhood of $t = 0$, the output is imposed

$$y(t) = y_{\text{ref}}(t) \tag{12.70}$$

giving in the new coordinates

$$z_i(t) = y_{\text{ref}}^{(i-1)}(t), \quad 1 \leq i \leq r \tag{12.71}$$

By analogy with the previous case, we set

$$\boldsymbol{\xi}_{\text{ref}} = \begin{bmatrix} z_1 \\ z_2 \\ \vdots \\ z_r \end{bmatrix} = \begin{bmatrix} y_{\text{ref}} \\ y_{\text{ref}}^{(1)} \\ \vdots \\ y_{\text{ref}}^{(r-1)} \end{bmatrix} \tag{12.72}$$

The equation that imposes the control results

$$y_{\text{ref}}^{(r)}(t) = b(\boldsymbol{\xi}_{\text{ref}}(t), \boldsymbol{\eta}(t)) + a(\boldsymbol{\xi}_{\text{ref}}(t), \boldsymbol{\eta}(t))\, u(t) \tag{12.73}$$

where $\boldsymbol{\eta}$ is the solution of the following autonomous differential system

$$\dot{\boldsymbol{\eta}}(t) = q(\boldsymbol{\xi}_{\text{ref}}(t), \boldsymbol{\eta}(t)), \quad \text{with } \boldsymbol{\eta}(0) = \boldsymbol{\eta}^{\circ} \tag{12.74}$$

From Eq. (12.73), results Eq. (1.75) of the unique control imposing on the output to exactly track the reference

$$u(t) = \frac{y_{\text{ref}}^{(r)}(t) - b(\boldsymbol{\xi}_{\text{ref}}(t), \boldsymbol{\eta}(t))}{a(\boldsymbol{\xi}_{\text{ref}}(t), \boldsymbol{\eta}(t))} \tag{12.75}$$

The system of differential equations (12.74) coupled with Eq. (12.75) yields the forced zero dynamics or dynamics of the inverse of the system (12.56), corresponding to a control such that the output exactly tracks the reference. $\boldsymbol{\eta}$ is the state of the dynamics of the inverse, $\boldsymbol{\xi}_{\text{ref}}$ its control input and u its output.

12.2.11 Asymptotic Stability

Consider the system under its normal form

$$\begin{aligned} \dot{z}_1 &= z_2 \\ &\vdots \\ \dot{z}_{r-1} &= z_r \\ \dot{z}_r &= b(\boldsymbol{\xi}, \boldsymbol{\eta}) + a(\boldsymbol{\xi}, \boldsymbol{\eta})\, u(t) \\ \dot{\boldsymbol{\eta}} &= q(\boldsymbol{\xi}, \boldsymbol{\eta}) \end{aligned} \tag{12.76}$$

assuming as previously that $(\boldsymbol{\xi}, \boldsymbol{\eta}) = (0, 0)$ is an equilibrium point. In parallel to the state feedback (Eq. 12.51), consider the external input

$$v = -\boldsymbol{K}\boldsymbol{z}, \quad \text{with the gain vector} : \boldsymbol{K} = [c_0 \ldots c_{r-1}] \tag{12.77}$$

so that the state feedback becomes

$$u(t) = \frac{-b(\boldsymbol{\xi},\boldsymbol{\eta}) - \sum_{i=0}^{r-1} c_i z_{i+1}}{a(\boldsymbol{\xi},\boldsymbol{\eta})} = \frac{-L_f^r h(\boldsymbol{x}) - \sum_{i=0}^{r-1} c_i L_f^i h(\boldsymbol{x})}{L_g L_f^{r-1} h(\boldsymbol{x})} \tag{12.78}$$

giving the closed-loop system

$$\begin{aligned} \dot{\boldsymbol{\xi}} &= A\boldsymbol{\xi} \\ \dot{\boldsymbol{\eta}} &= q(\boldsymbol{\xi},\boldsymbol{\eta}) \end{aligned} \tag{12.79}$$

where $\boldsymbol{A}$ is a companion controllability matrix equal to

$$\begin{bmatrix} 0 & 1 & 0 & \dots & 0 \\ \vdots & \ddots & 1 & \vdots & \\ \vdots & \ddots & 0 & & \\ 0 & \dots & 0 & 1 & \\ -c_0 & -c_1 & \dots & \dots & -c_{r-1} \end{bmatrix} \tag{12.80}$$

which has the characteristic polynomial

$$c_0 + c_1\, s + \cdots + c_{r-1}\, s^{r-1} + s^r \tag{12.81}$$

If, first the coefficients are chosen so that the roots of this polynomial have a negative real part and, secondly the zero dynamics corresponding to $\dot{\boldsymbol{\eta}} = q(0,\boldsymbol{\eta})$ is asymptotically locally stable, then the state feedback (Eq. 12.78) stabilizes asymptotically locally the system (12.79) in the neighborhood of the equilibrium $(\boldsymbol{\xi}, \boldsymbol{\eta}) = (0, 0)$.

The role and the importance of zero dynamics thus clearly appear at this level. If the linear approximation of the system possesses uncontrollable modes, the latter necessarily correspond to eigenvalues of the linear approximation Q of the zero dynamics. The linear approximation of the system is given by

$$\begin{aligned} \dot{z}_1 &= z_2 \\ &\vdots \\ \dot{z}_{r-1} &= z_r \\ \dot{z}_r &= b(\boldsymbol{\xi},\boldsymbol{\eta}) + a(\boldsymbol{\xi},\boldsymbol{\eta})\, u(t) \approx \boldsymbol{R}\,\boldsymbol{\xi} + \boldsymbol{S}\,\boldsymbol{\eta} + K\,u \\ \dot{\boldsymbol{\eta}} &= q(\boldsymbol{\xi},\boldsymbol{\eta}) \approx \boldsymbol{P}\,\boldsymbol{\xi} + \boldsymbol{Q}\,\boldsymbol{\eta} \end{aligned} \tag{12.82}$$

with the partial derivative matrices considered at $(\boldsymbol{\xi}, \boldsymbol{\eta}) = (0, 0)$

$$\boldsymbol{R} = \left[\frac{\partial b}{\partial \xi}\right], \quad \boldsymbol{S} = \left[\frac{\partial b}{\partial \eta}\right], \quad \boldsymbol{P} = \left[\frac{\partial q}{\partial \xi}\right], \quad \boldsymbol{Q} = \left[\frac{\partial q}{\partial \eta}\right] \tag{12.83}$$

Note that it is not necessary that the linear approximation be asymptotically stable for the nonlinear system to be stable.

As already realized for a system of relative degree n with the state feedback (Eq. 12.51), an external input v (Fig. 12.5) can be taken into account as

$$u(t) = \frac{-L_f^r h(\boldsymbol{x}) - \sum_{i=0}^{r-1} c_i L_f^i h(\boldsymbol{x}) + v}{L_g L_f^{r-1} h(\boldsymbol{x})} \tag{12.84}$$

so that the system (12.76) is transformed into

$$\begin{aligned} \dot{\boldsymbol{\xi}} &= \boldsymbol{A}\boldsymbol{\xi} + \boldsymbol{B}\, v \\ \dot{\boldsymbol{\eta}} &= q(\boldsymbol{\xi}, \boldsymbol{\eta}) \end{aligned} \tag{12.85}$$

with $\boldsymbol{B} = [0 \ldots 0\ 1]^T$. Provided that the zero dynamics is stable, the stability will depend on the characteristic polynomial

$$c_0 + c_1\, s + \cdots + c_{r-1}\, s^{r-1} + s^r \tag{12.86}$$

whose coefficients can be imposed so as to achieve the desired pole placement.

By analogy with linear systems, a nonlinear system is called minimum phase if its unforced zero dynamics is asymptotically locally stable at (0, 0).

12.2.12 Tracking of a Reference Trajectory

To make the output $y = z_1$ converge asymptotically toward a reference trajectory y_{ref}, it suffices to gather the elements of both previous sections, the forced zero dynamics with reference trajectory and the asymptotic stability.

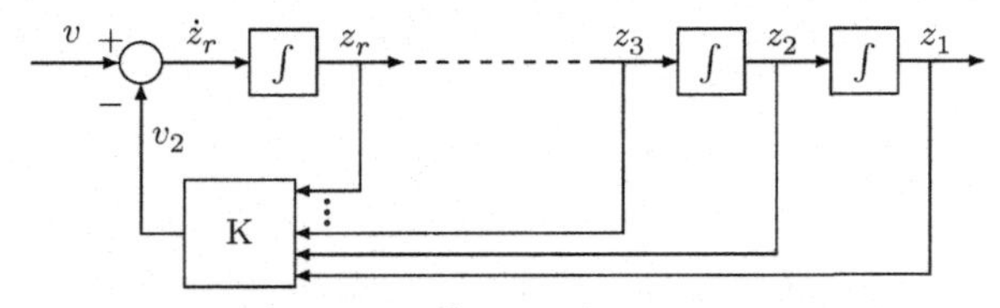

Fig. 12.5
Nonlinear control with pole placement for a system of relative degree $r \leq n$.

For the system in its normal form (12.76), consider the state feedback

$$u(t) = \frac{-b(\boldsymbol{\xi},\boldsymbol{\eta}) + y_{\text{ref}}^{(r)} - \sum_{i=0}^{r-1} c_i (z_{i+1} - y_{\text{ref}}^{(i)})}{a(\boldsymbol{\xi},\boldsymbol{\eta})} = \frac{-L_f^r h(\boldsymbol{x}) + y_{\text{ref}}^{(r)} - \sum_{i=0}^{r-1} c_i (L_f^i h(\boldsymbol{x}) - y_{\text{ref}}^{(i)})}{L_g L_f^{r-1} h(\boldsymbol{x})} \tag{12.87}$$

The error defined by

$$e(t) = y(t) - y_{\text{ref}}(t) \tag{12.88}$$

is the solution of the following differential equation

$$e^{(r)} + c_{r-1}\, e^{(r)} + \cdots + c_1\, e^{(1)} + c_0\, e = 0 \tag{12.89}$$

whose parallel with the characteristic polynomial Eq. (12.86) is obvious. By choosing adequately the coefficients c_i according to a pole placement strategy, the exponential convergence of the error toward 0 when $t \to \infty$ can be guaranteed. In the same manner as in the previous section, it is necessary that the zero dynamics (here forced) corresponding to

$$\dot{\boldsymbol{\eta}} = q(\boldsymbol{\xi}_{\text{ref}}, \boldsymbol{\eta}), \quad \text{with } \boldsymbol{\eta}_{\text{ref}}(0) = 0 \tag{12.90}$$

is stable (where $\boldsymbol{\eta}_{\text{ref}}(t)$ is the solution of this differential system) with the vector

$$\boldsymbol{\xi}_{\text{ref}} = \left[y_{\text{ref}}(t), y_{\text{ref}}^{(1)}(t), \ldots, y_{\text{ref}}^{(r-1)}(t) \right]^T \tag{12.91}$$

Isidori[10] studied the particular case where the reference y_{ref} is defined as a linear reference model. In the case where an external input is taken into account, the state feedback (Eq. 12.87) becomes

$$u(t) = \frac{v - L_f^r h(\boldsymbol{x}) + y_{\text{ref}}^{(r)} - \sum_{i=0}^{r-1} c_i \left(L_f^i h(\boldsymbol{x}) - y_{\text{ref}}^{(i)} \right)}{L_g L_f^{r-1} h(\boldsymbol{x})} \tag{12.92}$$

Two SISO applications of nonlinear geometric control performed in simulation with reference trajectories for realistic models are discussed in Corriou.[15] The first one concerns a realistic continuous stirred tank reactor where a chemical reaction takes place. The second one concerns a fed-batch biological reactor. These reactors present different relative degrees. For each of them, an extended Kalman filter is used to estimate the states. An example of pilot application to a copolymerization reactor preceded by a study of dynamic optimization in order to obtain the optimal profile temperature is given by Gentric et al.[22]

12.2.13 Decoupling With Respect to a Disturbance

We assume that a modelable disturbance d acts in an affine manner on the system thus reformulated

$$\begin{cases} \dot{\boldsymbol{x}} = \boldsymbol{f}(\boldsymbol{x}) + \boldsymbol{g}(\boldsymbol{x})\, u + \boldsymbol{w}(\boldsymbol{x})\, d \\ y = h(\boldsymbol{x}) \end{cases} \tag{12.93}$$

The aim is to define an input u by static-state feedback such that the output y does not depend (is decoupled) on the disturbance d, at least theoretically. Considering $z_1 = y$, the derivative of the output can be expressed with the normal coordinates following (Eq. 12.28)

$$\dot{z}_1 = L_f h(\boldsymbol{x}(t)) + L_g h(\boldsymbol{x}(t))\, u(t) + L_w h(\boldsymbol{x}(t))\, d(t) \tag{12.94}$$

Using the relative degree r of the system gives $L_g h = 0$, if $r > 1$, and it can be imposed that $L_w h = 0$ so that the disturbance has no influence. For the following coordinates, a similar condition is used

$$L_w L_f^{i-1} h = 0, \quad 1 \leq i \leq r \tag{12.95}$$

hence the system in the normal form including the influence of the disturbance from rank $r + 1$

$$\begin{aligned} \dot{z}_1 &= z_2 \\ &\vdots \\ \dot{z}_{r-1} &= z_r \\ \dot{z}_r &= L_f^r h(\boldsymbol{x}(t)) + L_g L_f^{r-1} h(\boldsymbol{x}(t))\, u(t) = b(\boldsymbol{\xi},\boldsymbol{\eta}) + a(\boldsymbol{\xi},\boldsymbol{\eta})\, u \\ \dot{\boldsymbol{\eta}} &= q(\boldsymbol{\xi},\boldsymbol{\eta}) + r(\boldsymbol{\xi},\boldsymbol{\eta})\, d \end{aligned} \tag{12.96}$$

with, moreover $y = z_1$. Clearly from the normal form (12.96), it follows that the input defined by the static-state feedback

$$u = \frac{-b(\boldsymbol{\xi},\boldsymbol{\eta}) + v}{a(\boldsymbol{\xi},\boldsymbol{\eta})} \tag{12.97}$$

which gives

$$\dot{z}_r = v \tag{12.98}$$

perfectly decouples the output $y = z_1$ from the disturbance d.

But the decoupling condition is contained in the system of Eq. (12.95) which can be expressed as

$$\begin{aligned} &< L_f^{i-1} h, \boldsymbol{w} > = 0, \quad 1 \leq i \leq r \Leftrightarrow \\ &\boldsymbol{w}(\boldsymbol{x}) \text{ belongs to the codistribution } \Delta^{\perp} \text{ in the neighborhood of } \boldsymbol{x}^{\circ} \end{aligned} \tag{12.99}$$

where the distribution Δ is equal to

$$\Delta = \text{span}\{Dh, DL_f h, \ldots, DL_f^{r-1} h\} \tag{12.100}$$

The previous decoupling has been realized by state feedback. In some cases, the disturbance can be measured, therefore it is attractive to consider a feedforward term in the control law, now formulated as

$$u(t) = \alpha(\boldsymbol{x}) + \beta(\boldsymbol{x})\, v + \gamma(\boldsymbol{x})\, d \tag{12.101}$$

hence the closed-loop system equations

$$\begin{aligned} \dot{\boldsymbol{x}} &= \boldsymbol{f}(\boldsymbol{x}) + \boldsymbol{g}(\boldsymbol{x})[\alpha(\boldsymbol{x}) + \beta(\boldsymbol{x})\, v] + [\boldsymbol{w}(\boldsymbol{x}) + \boldsymbol{g}(\boldsymbol{x})\gamma(\boldsymbol{x})]\, d \\ y &= h(\boldsymbol{x}) \end{aligned} \tag{12.102}$$

The parallel to the decoupling by state feedback can be noticed. It suffices that

$$\begin{aligned} &< L_f^{i-1} h, \boldsymbol{w} + \boldsymbol{g}\gamma > = 0, \quad 1 \le i \le r \quad \Leftrightarrow \\ &[\boldsymbol{w}(\boldsymbol{x}) + \boldsymbol{g}(\boldsymbol{x})\gamma(\boldsymbol{x})] \text{ belongs to the codistribution } \Delta^{\perp} \text{ in the neighborhood of } \boldsymbol{x}^{\circ} \end{aligned} \tag{12.103}$$

This condition can be simplified by expressing the Lie bracket

$$\begin{aligned} < L_f^{i-1} h, \boldsymbol{w} + \boldsymbol{g}\gamma > &= L_{w+g\gamma} L_f^{i-1} h(\boldsymbol{x}) \\ &= L_w L_f^{i-1} h(\boldsymbol{x}) + L_g L_f^{i-1} h(\boldsymbol{x})\gamma(\boldsymbol{x}) = 0, \quad 1 \le i \le r \end{aligned} \tag{12.104}$$

yielding the expression of the function γ

$$\gamma(\boldsymbol{x}) = -\frac{L_w L_f^{r-1} h(\boldsymbol{x})}{L_g L_f^{r-1} h(\boldsymbol{x})} \tag{12.105}$$

which must be reinjected in control law (12.101).

12.2.14 Case of Nonminimum-Phase Systems

As previously described, the nonlinear control law (12.60) consisted essentially in inverting the nonlinear model of the system in a close way to the ideal controller of a linear system. However, for a linear system, the positive zeros become unstable poles and must be discarded from the effective controller like in internal model control. Rigorously and similarly, the nonlinear control law (12.60) cannot be applied to nonminimum-phase systems, as they are not invertible. Nevertheless, it is possible[3] to apply to them a control law that gives a small tracking error or to redefine the output so that the modified zero dynamics is stable. It requires that the perfect tracking of the newly defined output also leads to a good tracking of the actual controlled output. Some rare examples of control laws for unstable systems are available in the literature where Kravaris and Daoutidis[23] study the case of second-order nonminimum-phase systems or

Engell and Klatt[24] study an unstable CSTR. Clearly, nonlinear geometric control is not well adapted for these cases.

12.2.15 Globally Linearizing Control

This linearizing control of an input-output type proposed[25–29] is based on the same concepts of differential geometry as those previously developed and can be considered as an extremely close variant. The control law applied to a minimum-phase system of relative degree r is

$$u = \frac{v - L_f^r h(\boldsymbol{x}) - \beta_1 L_f^{r-1} h(\boldsymbol{x}) - \cdots - \beta_{r-1} L_f h(\boldsymbol{x}) - \beta_r h(\boldsymbol{x})}{L_g L_f^{r-1} h(\boldsymbol{x})} \tag{12.106}$$

This is exactly Eq. (12.78) where the c_i's are replaced by β_i's. It gives the following input-output linear dynamics

$$y^{(r)} + \beta_1 \, y^{(r-1)} + \cdots + \beta_{r-1} \, y^{(1)} + \beta_r \, y = v \tag{12.107}$$

To guarantee a zero asymptotic error and improve robustness, that is, in the presence of modeling errors and step disturbances (for a PI), the external input can be supplied by the following controller

$$v(t) = \int_0^t c(t-\tau) \, [y_{\text{ref}}(\tau) - y(\tau)] \, d\tau \tag{12.108}$$

where the function $c(t)$, for example, can be chosen as the inverse of a given transfer function. Frequently, v will be a proportional-integral (PI) controller (Fig. 12.6), for example,

$$v(t) = K_c \left[y_{\text{ref}}(t) - y(t) + \frac{1}{\tau_i} \int_0^t (y_{\text{ref}}(\tau) - y(\tau)) \, d\tau \right] \tag{12.109}$$

In this case, the system stability is conditioned by the roots of the characteristic polynomial

$$s^{r+1} + \beta_1 \, s^r + \cdots + \beta_{r-1} \, s^2 + (\beta_r + K_c) \, s + \frac{K_c}{\tau_I} = 0 \tag{12.110}$$

where the roles of the nonlinear part and linear PI can be weighted in the choice of the parameters in the term ($\beta_r + K_c$). This control has been tried out and extended under different

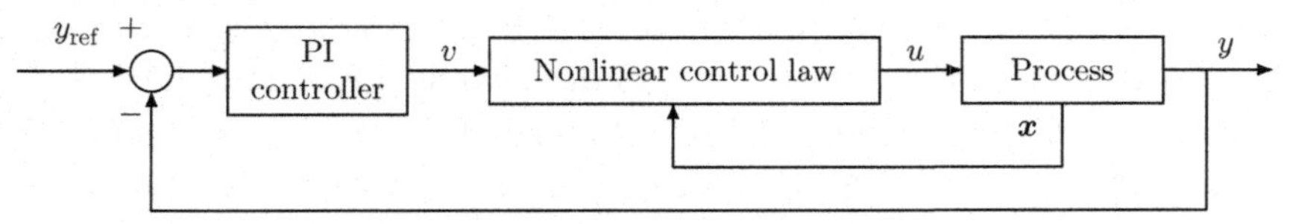

Fig. 12.6
Globally linearizing control with PI controller.

forms such as combination with a feedforward controller[30] to take into account disturbances, combination of a Smith predictor and a state observer to take into account time delays,[31] with modified application to nonminimum-phase second-order systems,[23] application to a polymerization reactor,[32] with a robustness study.[33]

12.2.16 Generic Model Control

Generic model control[34] (GMC) proposed by Lee and Sullivan[35] has not been formulated by using the concepts of differential geometry. For a multiinput multioutput plant, it uses a different and general nonlinear model

$$\begin{cases} \dot{\boldsymbol{x}} = \boldsymbol{f}(\boldsymbol{x},\boldsymbol{u},\boldsymbol{d},t) \\ \boldsymbol{y} = \boldsymbol{h}(\boldsymbol{x}) \end{cases} \tag{12.111}$$

where the input does not appear any more in a linear way. It uses the expression of the derivative of the output as

$$\dot{\boldsymbol{y}} = \frac{\partial \boldsymbol{h}}{\partial \boldsymbol{x}} \boldsymbol{f}(\boldsymbol{x},\boldsymbol{u},\boldsymbol{d},t) = \boldsymbol{H}_x \boldsymbol{f}(\boldsymbol{x},\boldsymbol{u},\boldsymbol{d},t) \tag{12.112}$$

This derivative is compared to a reference that is an arbitrary function $\boldsymbol{r}^*$ so that $\dot{\boldsymbol{y}}^* = \boldsymbol{r}^*$ and

$$\dot{\boldsymbol{y}}^* = \boldsymbol{K}_1(\boldsymbol{y}^* - \boldsymbol{y}) + \boldsymbol{K}_2 \int (\boldsymbol{y}^* - \boldsymbol{y}) dt \tag{12.113}$$

where $\boldsymbol{K}_1$ and $\boldsymbol{K}_2$ are constant matrices. After posing the control problem in an optimization setting, finally, using an approximate model $\tilde{\boldsymbol{f}}$ and $\tilde{\boldsymbol{h}}$, the problem is set as a system of equations

$$\tilde{\boldsymbol{H}}_x \tilde{\boldsymbol{f}}(\boldsymbol{x},\boldsymbol{u},\boldsymbol{d},t) - \boldsymbol{K}_1(\boldsymbol{y}^* - \boldsymbol{y}) - \boldsymbol{K}_2 \int (\boldsymbol{y}^* - \boldsymbol{y}) dt = 0 \tag{12.114}$$

that Lee and Sullivan[35] solve for a few cases.

Even if the system is not affine with respect to the input, it can be considered in the framework of nonlinear geometric control. Lee and Sullivan[35] mention that when the relative degree is 1, an explicit solution is found. Indeed it is based on a realization of the inverse of the model. For this reason, its applicability is limited,[36] as it is reserved to systems of relative degree equal to 1.

Some authors[37–39] compare globally linearizing control (GLC) and GMC to nonlinear control developed by differential geometry, as has been discussed by Isidori[10] GLC differs by nonlinear control only by the expression of the external input. GMC is much more different but presents some similarities.

12.3 Multivariable Nonlinear Control

Geometric nonlinear control up to now was presented for SISO systems but it can be extended to multiinput multioutput systems. It will be presented only for square systems, that is, with same number of inputs and outputs. Detailed demonstrations can be found, in particular, in the textbook by Isidori[10] and only the main points will be cited here.

Multivariable nonlinear systems, affine with respect to the inputs, are modeled as

$$\begin{aligned} \dot{\boldsymbol{x}} &= \boldsymbol{f}(\boldsymbol{x}) + \sum_{i=1}^{m} g_i(\boldsymbol{x})\, u_i \\ y_i &= h_i(\boldsymbol{x}), \quad i = 1, \ldots, m \end{aligned} \tag{12.115}$$

with

$$\boldsymbol{u} = \begin{bmatrix} u_1 \\ \vdots \\ u_m \end{bmatrix}; \; \boldsymbol{y} = \begin{bmatrix} y_1 \\ \vdots \\ y_m \end{bmatrix} \tag{12.116}$$

where $\boldsymbol{f}(\boldsymbol{x})$, $g_i(\boldsymbol{x})$, $h_i(\boldsymbol{x})$ are smooth vector fields. In a more compact form, the system is modeled by

$$\begin{aligned} \dot{\boldsymbol{x}} &= \boldsymbol{f}(\boldsymbol{x}) + \boldsymbol{g}(\boldsymbol{x})\, \boldsymbol{u} \\ \boldsymbol{y} &= \boldsymbol{h}(\boldsymbol{x}) \end{aligned} \tag{12.117}$$

12.3.1 Relative Degree

The concepts of nonlinear geometric control for multivariable systems are an extension of the SISO case. However, some of them are specific. In the multivariable case, the notion of relative degree is extended as

- The vector of relative degrees $[r_1, \ldots, r_m]^T$ defined in a neighborhood of $\boldsymbol{x}^\circ$ by

$$L_{g_j} L_f^k h_i(\boldsymbol{x}) = 0, \quad \forall j = 1, \ldots, m, \quad \forall k < r_i - 1, \quad \forall i = 1, \ldots, m \tag{12.118}$$

- The matrix $\boldsymbol{A}(\boldsymbol{x})$ defined by

$$\boldsymbol{A}(\boldsymbol{x}) = \begin{bmatrix} L_{g_1} L_f^{r_1-1} h_1(\boldsymbol{x}) & \ldots & L_{g_m} L_f^{r_1-1} h_1(\boldsymbol{x}) \\ \vdots & & \vdots \\ L_{g_1} L_f^{r_m-1} h_m(\boldsymbol{x}) & \ldots & L_{g_m} L_f^{r_m-1} h_m(\boldsymbol{x}) \end{bmatrix} \tag{12.119}$$

must be nonsingular at $\boldsymbol{x}^\circ$.

Given an output of subscript i, the vector of the Lie derivatives verifies

$$\begin{aligned} &\left[L_{g_1}L_f^k h_i(\boldsymbol{x}) \;\ldots\; L_{g_m}L_f^k h_i(\boldsymbol{x})\right] = 0 \quad \forall k < r_i - 1 \\ &\left[L_{g_1}L_f^{r_i-1} h_i(\boldsymbol{x}) \;\ldots\; L_{g_m}L_f^{r_i-1} h_i(\boldsymbol{x})\right] \neq 0 \end{aligned} \tag{12.120}$$

where r_i is the number of times that the output $y_i(t)$ must be differentiated to make at least one component of the vector $\boldsymbol{u}(t)$ appear. There exists at least one couple (u_j, y_i) having r_i as the relative degree.

Consider a system having $[r_1, \ldots, r_m]^T$ as the vector of relative degrees. The vectors

$$\begin{gathered} \left[Dh_1(\boldsymbol{x}^\circ),\; DL_f h_1(\boldsymbol{x}^\circ),\; \ldots,\; DL_f^{r_1-1} h_1(\boldsymbol{x}^\circ)\right] \\ \vdots \\ \left[Dh_m(\boldsymbol{x}^\circ),\; DL_f h_m(\boldsymbol{x}^\circ),\; \ldots,\; DL_f^{r_m-1} h_m(\boldsymbol{x}^\circ)\right] \end{gathered} \tag{12.121}$$

are linearly independent.

12.3.2 Coordinate Change

The proposed coordinate change for a multivariable system is exactly analogous to the SISO case. Consider a system of relative degree vector $[r_1, \ldots, r_m]^T$. Let the coordinate change

$$\forall\, 1 \leq i \leq m \qquad \begin{aligned} \phi_1^i(\boldsymbol{x}) &= h_i(\boldsymbol{x}) \\ \phi_2^i(\boldsymbol{x}) &= L_f h_i(\boldsymbol{x}) \\ &\vdots \\ \phi_{r_i}^i(\boldsymbol{x}) &= L_f^{r_i-1} h_i(\boldsymbol{x}) \end{aligned} \tag{12.122}$$

The total relative degree is denoted by $r = r_1 + \cdots + r_m$.

If $r < n$, it is possible to find $(n - r)$ additional functions $\phi_{r+1}(\boldsymbol{x}), \ldots, \phi_n(\boldsymbol{x})$ such that the mapping

$$\Phi(\boldsymbol{x}) = \left[\phi_1^1(\boldsymbol{x}), \ldots, \phi_{r_1}^1(\boldsymbol{x}), \ldots, \phi_1^m(\boldsymbol{x}), \ldots, \phi_{r_m}^m(\boldsymbol{x}), \phi_{r+1}(\boldsymbol{x}), \ldots, \phi_n(\boldsymbol{x})\right]^T \tag{12.123}$$

has its Jacobian matrix nonsingular at $\boldsymbol{x}^\circ$ and is a potential coordinate change in a neighborhood U of $\boldsymbol{x}^\circ$.

Moreover, if the distribution

$$G = \text{span}\{\boldsymbol{g}_1, \ldots, \boldsymbol{g}_m\} \tag{12.124}$$

is involutive in U, the additional functions: $\phi_{r+1}(\boldsymbol{x})$, ..., $\phi_m(\boldsymbol{x})$ can be chosen such that

$$L_{g_j}\phi_i(\boldsymbol{x})=0, \quad r+1 \leq i \leq n, \quad 1 \leq j \leq m \tag{12.125}$$

The condition of nonsingularity of matrix $\boldsymbol{A}(\boldsymbol{x})$ can be extended to a system having more inputs than outputs. It becomes a rank condition: the rank of the matrix must be equal to the number of its rows, and, furthermore, the system must have more inputs than outputs, which is usual.

12.3.3 Normal Form

The coordinate change gives for ϕ_i^1 (similarly for others ϕ_i^j)

$$\begin{aligned}
\dot{\phi}_1^1 &= \phi_2^1(\boldsymbol{x}) \\
&\vdots \\
\dot{\phi}_{r_1-1}^1 &= \phi_{r_1}^1(\boldsymbol{x}) \\
\dot{\phi}_{r_1-1}^1 &= L_f^{r_1} h_1(\boldsymbol{x}) + \sum_{j=1}^{m} L_{g_j} L_f^{r_1-1} h_1(\boldsymbol{x})\, u_j(t)
\end{aligned} \tag{12.126}$$

The normal coordinates are given by

$$\begin{aligned}
&\boldsymbol{\xi} = [\boldsymbol{\xi}^1, \ldots, \boldsymbol{\xi}^m] \text{ with } \boldsymbol{\xi}^i = \begin{bmatrix} \xi_1^i \\ \vdots \\ \xi_{r_i}^i \end{bmatrix} = \begin{bmatrix} \phi_1^i(\boldsymbol{x}) \\ \vdots \\ \phi_{r_i}^i(\boldsymbol{x}) \end{bmatrix}, \quad i = 1, \ldots, m \\
&\boldsymbol{\eta} = \begin{bmatrix} \eta_1 \\ \vdots \\ \eta_{n-r} \end{bmatrix} = \begin{bmatrix} \phi_{r+1}(\boldsymbol{x}) \\ \vdots \\ \phi_n(\boldsymbol{x}) \end{bmatrix}
\end{aligned} \tag{12.127}$$

which gives the normal form

$$\begin{aligned}
\dot{\xi}_1^i &= \xi_2^i \\
&\vdots \\
\dot{\xi}_{r_i-1}^i &= \xi_{r_i}^i \\
\dot{\xi}_{r_i}^i &= b_i(\boldsymbol{\xi}, \boldsymbol{\eta}) + \sum_{j=1}^{m} a_{ij}(\boldsymbol{\xi}, \boldsymbol{\eta})\, u_j \\
\dot{\boldsymbol{\eta}} &= q(\boldsymbol{\xi}, \boldsymbol{\eta}) + p(\boldsymbol{\xi}, \boldsymbol{\eta})\, u \\
y &= \xi_1^i
\end{aligned} \tag{12.128}$$

with

$$a_{ij}(\boldsymbol{\xi},\boldsymbol{\eta}) = L_{g_j} L_f^{r_i-1} h_i(\Phi^{-1}(\boldsymbol{\xi},\boldsymbol{\eta})), \quad i,j = 1,\ldots,m$$
$$b_i(\boldsymbol{\xi},\boldsymbol{\eta}) = L_f^{r_i} h_i(\Phi^{-1}(\boldsymbol{\xi},\boldsymbol{\eta})), \quad i = 1,\ldots,m \tag{12.129}$$

where a_{ij} are the coefficients of matrix $\boldsymbol{A}(\boldsymbol{x})$.

12.3.4 Zero Dynamics

The zero dynamics is defined in the same way as for SISO systems (Section 12.2.10). The inputs and the initial conditions are sought so that the outputs are identically zero in a neighborhood U of $\boldsymbol{x}^\circ$, implying $\boldsymbol{\xi} = 0$. The unique control vector results

$$\boldsymbol{u}(t) = -\boldsymbol{A}^{-1}(0,\boldsymbol{\eta}(t))\,\boldsymbol{b}(0,\boldsymbol{\eta}(t)) \tag{12.130}$$

and the zero dynamics or unforced dynamics is solution of the differential equations

$$\dot{\boldsymbol{\eta}}(t) = \boldsymbol{q}(\boldsymbol{\xi},\boldsymbol{\eta}) - \boldsymbol{p}(\boldsymbol{\xi},\boldsymbol{\eta})\,\boldsymbol{A}^{-1}(\boldsymbol{\xi},\boldsymbol{\eta})\,\boldsymbol{b}(\boldsymbol{\xi},\boldsymbol{\eta}), \quad \boldsymbol{\eta}(0) = \boldsymbol{\eta}^\circ \tag{12.131}$$

The control (Eq. 12.130) can also be expressed in the original state space as

$$\boldsymbol{u}^*(\boldsymbol{x}) = -\boldsymbol{A}^{-1}(\boldsymbol{x})\,\boldsymbol{b}(\boldsymbol{x}) \tag{12.132}$$

The change from the unforced dynamics to the forced dynamics would be realized in the same manner as for SISO systems.

12.3.5 Exact Linearization by State Feedback and Diffeomorphism

Let the system

$$\dot{\boldsymbol{x}} = \boldsymbol{f}(\boldsymbol{x}) + \boldsymbol{g}(\boldsymbol{x})\,\boldsymbol{u} \tag{12.133}$$

without considering the outputs.

The distributions are defined as

$$\begin{aligned}
G_0 &= \text{span}\{g_1,\ldots,g_m\} \\
G_1 &= \text{span}\{g_1,\ldots,g_m,\, \text{ad}_f g_1,\ldots,\, \text{ad}_f g_m\} \\
&\vdots \\
G_i &= \text{span}\left\{\text{ad}_f^k g_1,\ldots,\, \text{ad}_f^k g_m\,; 0 \le k \le i\right\}; \quad i = 0,\ldots,n-1
\end{aligned} \tag{12.134}$$

The matrix $\boldsymbol{g}(\boldsymbol{x}^\circ)$ is assumed of rank m. The exact linearization is possible if and only if:

- the distribution $G_i (i = 0, \ldots, n - 1)$ has a constant dimension in the neighborhood of $\boldsymbol{x}^\circ$;

- the distribution G_{n-1} has a dimension equal to n; and
- the distribution $G_i (i = 0, \ldots, n-2)$ is involutive.

It is necessary to check:

- the nonsingularity of matrix $A(x)$ (Eq. 12.119);
- the total relative degree r must be equal to n; and
- as for SISO systems, the outputs y_i are given by the solutions $\lambda_i(x)$, $(j=1,\ldots,m)$ of the equations

$$L_{g_j} L_f^k \lambda_i(x) = 0, \quad 0 \leq k \leq r_i - 2, \; j = 1, \ldots, m \tag{12.135}$$

Referring to Eqs. (12.119), (12.129), the linearizing state feedback is

$$\boldsymbol{u} = \boldsymbol{A}^{-1}(\boldsymbol{x})\,(-\boldsymbol{b}(\boldsymbol{x}) + \boldsymbol{v}) \tag{12.136}$$

and the linearizing normal coordinates are

$$\xi_k^i(\boldsymbol{x}) = L_f^{k-1} h_i(\boldsymbol{x}), \quad 1 \leq k \leq r_i, \; i = 1, \ldots, m \tag{12.137}$$

12.3.6 Nonlinear Control Perfectly Decoupled by Static-State Feedback

The decoupling for the system (12.117) will be perfectly realized when any output y_i $(1 \leq i \leq m)$ is influenced only by the corresponding input v_i. This problem has a solution only if the decoupling matrix $\boldsymbol{A}(\boldsymbol{x})$ is nonsingular at $\boldsymbol{x}^\circ$, that is, if the system possesses a vector of relative degrees.

The static-state feedback is given by Eq. (12.136). To perform the decoupling, it suffices to consider the system in its normal form (12.128) and to propose the following control law

$$\boldsymbol{u} = -\boldsymbol{A}^{-1}(\boldsymbol{\xi},\boldsymbol{\eta})\,\boldsymbol{b}(\boldsymbol{\xi},\boldsymbol{\eta}) + \boldsymbol{A}^{-1}(\boldsymbol{\xi},\boldsymbol{\eta})\,\boldsymbol{v} \tag{12.138}$$

which transforms the system into

$$\begin{aligned}
\dot{\xi}_1^i &= \xi_2^i \\
&\vdots \\
\dot{\xi}_{r_i-1}^i &= \xi_{r_i}^i \\
\dot{\xi}_{r_i}^i &= b_i(\boldsymbol{\xi},\boldsymbol{\eta}) + \sum_{j=1}^{m} a_{ij}(\boldsymbol{\xi},\boldsymbol{\eta})\, u_j \\
&= L_f^{r_i} h_i(\Phi^{-1}(\boldsymbol{\xi},\boldsymbol{\eta})) + \sum_{j=1}^{m} L_{g_j} L_f^{r_i-1} h_i(\Phi^{-1}(\boldsymbol{\xi},\boldsymbol{\eta}))\, u_j \\
&= v_i \\
\dot{\boldsymbol{\eta}} &= \boldsymbol{q}(\boldsymbol{\xi},\boldsymbol{\eta}) + \boldsymbol{p}(\boldsymbol{\xi},\boldsymbol{\eta})\,\boldsymbol{u} \\
y &= \xi_1^i
\end{aligned} \tag{12.139}$$

as we could write

$$\begin{bmatrix} \dot{\xi}_{r_1}^1 \\ \vdots \\ \dot{\xi}_{r_m}^m \end{bmatrix} = \boldsymbol{b}(\boldsymbol{\xi},\boldsymbol{\eta}) + \boldsymbol{A}(\boldsymbol{\xi},\boldsymbol{\eta})\,\boldsymbol{u} = \boldsymbol{v} \tag{12.140}$$

from Eq. (12.129), giving the coefficients a_{ij}, the control law (12.138), and the normal form.

Two cases can take place:

- the total relative degree r is lower than n. In that case, there exists an unobservable part in the system, influenced by the inputs and the states, but not affecting the outputs; and
- the total relative degree r is equal to n. Then, the system can be decoupled into m chains, each one composed of r_i integrators. The system is thus transformed into a completely linear and controllable system.

Evidently, it must be verified that the decoupling matrix $\boldsymbol{A}(\boldsymbol{x})$ is nonsingular at $\boldsymbol{x}^\circ$.

In the same way as for SISO systems, pole placements can be performed by adding state feedbacks as

$$v_i = -c_0^i \xi_1^i - \cdots - c_{r_i-1}^i \xi_{r_i}^i \tag{12.141}$$

The different items treated for SISO systems, that is, asymptotic stability of the zero dynamics, disturbance rejection, reference model tracking, are studied in a very similar manner.[10]

Examples of application of multivariable nonlinear geometric control to plants are not many in the literature.[27,32,40] Among them, To et al.[41] describe nonlinear control of a simulated industrial evaporator by means of different techniques, including input-output linearization, Su-Hunt-Meyer transformation and generic model control. The process model possesses three states, two manipulated inputs, and two controlled outputs.

12.3.7 Obtaining a Relative Degree by Dynamic Extension

It happens that some multivariable systems possess no relative degree vector. In this case, no static-state feedback can change this result, as the relative degree property is independent of it.

Suppose that the nonlinear system (12.117) possesses no total relative degree, as the rank of matrix $\boldsymbol{A}(\boldsymbol{x})$ is smaller than m, number of inputs and outputs, for this square system. To obtain the relative degree, a dynamic part is added[10] between the old inputs $\boldsymbol{u}$ and the new inputs $\boldsymbol{v}$, modeled by

$$\begin{aligned} \boldsymbol{u} &= \alpha(\boldsymbol{x},\boldsymbol{\zeta}) + \beta(\boldsymbol{x},\boldsymbol{\zeta})\,\boldsymbol{v} \\ \dot{\boldsymbol{\zeta}} &= \gamma(\boldsymbol{x},\boldsymbol{\zeta}) + \delta(\boldsymbol{x},\boldsymbol{\zeta})\,\boldsymbol{v} \end{aligned} \tag{12.142}$$

$v_i=\dot{\zeta}_k$ ∫ ζ_k -------- ζ_2 ∫ $u_i=\zeta_1$

Fig. 12.7
Dynamic extension.

A simple type of dynamics used is the interposition of integrators between an input v_i and the input u_i (Fig. 12.7). In the case of two integrators, it is modeled by

$$\begin{aligned} u_i &= \zeta_1 \\ \dot{\zeta}_1 &= \zeta_2 \\ \dot{\zeta}_2 &= v_i \end{aligned} \tag{12.143}$$

The modified system will be

$$\begin{aligned} \dot{\boldsymbol{x}} &= \boldsymbol{f}(\boldsymbol{x}) + \boldsymbol{g}(\boldsymbol{x})\,\alpha(\boldsymbol{x},\boldsymbol{\zeta}) + \boldsymbol{g}(\boldsymbol{x})\,\beta(\boldsymbol{x},\boldsymbol{\zeta})\,\boldsymbol{v} \\ \dot{\boldsymbol{\zeta}} &= \gamma(\boldsymbol{x},\boldsymbol{\zeta}) + \delta(\boldsymbol{x},\boldsymbol{\zeta})\,\boldsymbol{v} \\ \boldsymbol{y} &= \boldsymbol{h}(\boldsymbol{x}) \end{aligned} \tag{12.144}$$

The dynamic extension[10] is performed according to an iterative procedure increasing the rank of matrix $\boldsymbol{A}(\boldsymbol{x}, \boldsymbol{\zeta})$ corresponding to the modified system until it is equal to m. This specific algorithm contains a procedure of identification of the inputs on which it is necessary to act. Thus, if the relative degree is obtained by dynamic extension and if the total relative degree of the extended system is equal to n, it is possible to assure that the original system can be transformed into a completely linear and controllable system by a dynamic-state feedback and a coordinate change.

Such cases are rare in the literature. Soroush and Kravaris[42] describe a continuous polymerization reactor where conversion and temperature are controlled by manipulating two coordinated flow rates and two coordinated heat inputs. According to this model, the characteristic matrix of the system is singular. The authors use a dynamic input-output linearizing state feedback by redefining the second input simply using the rate of change, that is, the derivative, of monomer flow rate instead of the original monomer flow rate. This amounts to adding an integrator.

12.3.8 Nonlinear Adaptive Control

Frequently, the process model is uncertain or time-varying. Adaptive control was first developed for linear systems and is the subject of many books,[43–46] in particular for generalized predictive control.[47] In most cases, the identification deals with parameters of the discrete-time models, often transfer functions, without physical signification,[48–51] posing serious problems of robustness.[52] However, it may occur that online identification of physical parameters influencing the model is necessary.[53–55]

It has been extended to the nonlinear systems and nonlinear adaptive control is described in dedicated books[1,45] and papers such as Refs. 56–58.

A short description of a chemical process example is given in the following. The used technique is inspired from procedures of recursive identification presented by Sastry and Bodson.[45] Wang et al.[57] studied a batch styrene polymerization reactor. At the beginning of reaction, the extent of polymerization is still low and the viscosity of the reactor contents is close to that of the solvent. When the monomer conversion increases, the viscosity strongly increases because of the gel effect, and the heat-transfer coefficient decreases significantly because the stirring progressively moves from a turbulent regime to a laminar one, creating reactor runaway hazard. The nonlinear model[57] describes these phenomena, which have been taken into account in the identification and nonlinear geometric control by using an augmented state vector. Besides the traditional states, which are concentrations and temperatures, the gel effect coefficient and the heat-transfer coefficient are estimated during the reaction.

12.4 Nonlinear Multivariable Control of a Chemical Reactor

A continuous perfectly stirred chemical reactor[15] is modeled with four states (x_1, concentration of A; x_2, reactor temperature; x_3, jacket temperature; x_4, liquid volume), two manipulated inputs corresponding to valve positions that allow to control the inlet temperature in the jacket $T_{j,\,in}$ and the feed concentration $C_{A,\,f}$, respectively, two controlled outputs, the concentration of A and the reactor temperature. The nonlinear model of the reactor is

$$\begin{aligned}
\dot{x}_1 &= \frac{F_f}{x_4}\,(C_{A,f} - x_1) - k\,x_1 \\
\dot{x}_2 &= \frac{F_f}{x_4}\,(T_f - x_2) - \frac{\Delta H\,k\,x_1}{\rho\,C_p} - \frac{UA(x_2 - x_3)}{\rho\,C_p\,x_4} \\
\dot{x}_3 &= \frac{F_j}{V_j}(T_{j,in} - x_2) + \frac{UA(x_2 - x_3)}{\rho_j\,C_{pj}\,V_j} \\
\dot{x}_4 &= F_f - F_{sp} + K_r\,(V_{sp} - x_4) \\
y_1 &= x_1 \\
y_2 &= x_2
\end{aligned} \tag{12.145}$$

with

$$k = k_0 \exp\left(-\frac{E_a}{R\, x_2}\right); \quad C_{A,f} = u_1\, C_{A,0} \quad T_{j,in} = u_2\, T_{hot} + (1 - u_2)\, T_{cold} \tag{12.146}$$

where T_{cold} and T_{hot} are the temperatures of two cold and hot heat exchangers assumed constant, respectively. The volume x_4 is controlled independently by a proportional controller. Both manipulated inputs are bounded in the interval [0, 1].

The physical parameters of the chemical reactor are given in Table 12.1.

The nonlinear state-space model (12.145) is affine with respect to the inputs. It is a square system that can be written as (Eq. 12.115) exactly as

$$\begin{aligned} \dot{\boldsymbol{x}} &= \boldsymbol{f}(\boldsymbol{x}) + \sum_{i=1}^{2} \boldsymbol{g}_i(\boldsymbol{x})\, u_i \\ y_i &= h_i(\boldsymbol{x}), \quad i = 1, \ldots, 2 \end{aligned} \tag{12.147}$$

Table 12.1 Initial variables and main parameters of the CSTR

Flow rate of the feed	$F_f = F_{sp} = 3 \times 10^{-3}\ \text{m}^3\ \text{s}^{-1}$
Concentration of reactant A before the feed	$C_{A,\,0} = 3900\ \text{mol m}^{-3}$
Temperature of the feed	$T_f = 295$ K
Volume of reactor	$V(t = 0) = V_{sp} = 1.5\ \text{m}^3$
Kinetic constant	$k_0 = 2 \times 10^7\ \text{s}^{-1}$
Activation energy	$E = 7 \times 10^4\ \text{J mol}^{-1}$
Heat of reaction	$\Delta H = -7 \times 10^4\ \text{J mol}^{-1}$
Density of reactor contents	$\rho = 1000\ \text{kg m}^{-3}$
Heat capacity of reactor contents	$C_p = 3000\ \text{J kg}^{-1}\ \text{K}^{-1}$
Temperature of the cold heat exchanger	$T_c = 280$ K
Temperature of hot heat exchanger	$T_h = 360$ K
Flow rate of the heat-conducting fluid	$F_j = 5 \times 10^{-2}\ \text{m}^3\ \text{s}^{-1}$
Volume of jacket	$V_j = 0.1\ \text{m}^3$
Heat-transfer coefficient between the jacket and the reactor contents	$U = 900\ \text{W m}^{-2}\ \text{K}^{-1}$
Heat-exchange area	$A = 20\ \text{m}^2$
Density of the heat-conducting fluid	$\rho_j = 1000\ \text{kg m}^{-3}$
Heat capacity of the heat-conducting fluid	$C_{pj} = 4200\ \text{J kg}^{-1}\ \text{K}^{-1}$
Proportional gain of the level controller	$K_r = 0.05\ \text{s}^{-1}$

with the vector fields

$$
f(\boldsymbol{x}) = \begin{bmatrix} -\dfrac{F_\mathrm{f}}{x_4} x_1 - k\, x_1 \\ \dfrac{F_\mathrm{f}}{x_4}(T_\mathrm{f} - x_2) - \dfrac{\Delta H\, k\, x_1}{\rho\, C_p} - \dfrac{UA(x_2 - x_3)}{\rho\, C_p\, x_4} \\ \dfrac{F_\mathrm{j}}{V_\mathrm{j}}(T_\mathrm{cold} - x_2) + \dfrac{UA(x_2 - x_3)}{\rho_j\, C_{pj}\, V_\mathrm{j}} \\ F_\mathrm{f} - F_\mathrm{sp} + K_r\,(V_\mathrm{sp} - x_4) \end{bmatrix}
$$

$$
g_1(\boldsymbol{x}) = \begin{bmatrix} \dfrac{F_\mathrm{f}}{x_4} C_{\mathrm{A,f}} \\ 0 \\ 0 \\ 0 \end{bmatrix}; \quad g_2(\boldsymbol{x}) = \begin{bmatrix} 0 \\ 0 \\ \dfrac{F_\mathrm{j}}{V_\mathrm{j}}(T_\mathrm{hot} - T_\mathrm{cold}) \\ 0 \end{bmatrix} \tag{12.148}
$$

First, the relative degree of this system is to be determined. For that, the following Lie derivatives are calculated, beginning by considering the first output.

- The Lie derivative of $h_1(\boldsymbol{x})$ in the direction of the vector field f

$$
L_f h_1(\boldsymbol{x}) = \sum_i \frac{\partial h_1}{\partial x_i} f_i(\boldsymbol{x}) = f_1(\boldsymbol{x}) \tag{12.149}
$$

- Then $L_{g_1} L_f h_1$ is calculated

$$
L_{g_1} L_f h_1 = L_{g_1} f_1 = \sum_i \frac{\partial f_1}{\partial x_i} g_{1,i} = \frac{\partial f_1}{\partial x_1} g_{1,1} \neq 0 \tag{12.150}
$$

- Then $L_{g_2} L_f h_1$ is calculated

$$
L_{g_2} L_f h_1 = L_{g_2} f_1 = \sum_i \frac{\partial f_1}{\partial x_i} g_{2,i} = \frac{\partial f_1}{\partial x_3} g_{2,3} = 0 \tag{12.151}
$$

thus the vector

$$
\left[L_{g_1} L_f h_1 \,,\; L_{g_2} L_f h_1 \right] \neq 0 \tag{12.152}
$$

hence $r_1 = 2$.
Now, the second output is concerned.

– The Lie derivative of $h_2(\boldsymbol{x})$ in the direction of the vector field f

$$L_f h_2(\boldsymbol{x}) = \sum_i \frac{\partial h_2}{\partial x_i} f_i(\boldsymbol{x}) = f_2(\boldsymbol{x}) \tag{12.153}$$

– Then $L_{g_1} L_f h_2$ is calculated

$$L_{g_1} L_f h_2 = L_{g_1} f_2 = \sum_i \frac{\partial f_2}{\partial x_i} g_{1,i} = \frac{\partial f_2}{\partial x_1} g_{1,1} \neq 0 \tag{12.154}$$

– Then $L_{g_2} L_f h_2$ is calculated

$$L_{g_2} L_f h_2 = L_{g_2} f_2 = \sum_i \frac{\partial f_2}{\partial x_i} g_{2,i} = \frac{\partial f_2}{\partial x_3} g_{2,3} \neq 0 \tag{12.155}$$

This implies that

$$\left[L_{g_1} L_f h_2 \, , \, L_{g_2} L_f h_2 \right] \neq 0 \tag{12.156}$$

hence $r_2 = 2$.

The matrix $\boldsymbol{A}(\boldsymbol{\Phi}(x))$ is then equal to

$$\boldsymbol{A}(\boldsymbol{\Phi}(x)) = \begin{bmatrix} L_{g_1} L_f h_1 & L_{g_2} L_f h_1 \\ L_{g_1} L_f h_2 & L_{g_2} L_f h_2 \end{bmatrix} = \begin{bmatrix} \frac{\partial f_1}{\partial x_1} g_{1,1} & 0 \\ \frac{\partial f_2}{\partial x_1} g_{1,1} & \frac{\partial f_2}{\partial x_3} g_{2,3} \end{bmatrix} \tag{12.157}$$

which is nonsingular for most $\boldsymbol{x}$ in the physical domain. The total relative degree is $r = r_1 + r_2 = 4$. Note that $r = n$, dimension of the system.

The proposed change of coordinates is

$$\begin{bmatrix} \phi_1^1(\boldsymbol{x}) = h_1(\boldsymbol{x}) = x_1 & \phi_1^2(\boldsymbol{x}) = h_2(\boldsymbol{x}) = x_2 \\ \phi_2^1(\boldsymbol{x}) = L_f h_1(\boldsymbol{x}) = f_1 & \phi_2^2(\boldsymbol{x}) = L_f h_2(\boldsymbol{x}) = f_2 \end{bmatrix} \tag{12.158}$$

giving

$$\boldsymbol{\Phi}(\boldsymbol{x}) = [\phi_1^1 \, , \, \phi_2^1 \, , \, \phi_1^2 \, , \, \phi_2^2] \tag{12.159}$$

must have its Jacobian matrix nonsingular. It can be verified as this Jacobian matrix is equal to

$$\begin{bmatrix} 1 & \frac{\partial f_1}{\partial x_1} & 0 & \frac{\partial f_2}{\partial x_1} \\ 0 & \frac{\partial f_1}{\partial x_2} & 1 & \frac{\partial f_2}{\partial x_2} \\ 0 & 0 & 0 & \frac{\partial f_2}{\partial x_3} \\ 0 & \frac{\partial f_1}{\partial x_4} & 0 & \frac{\partial f_2}{\partial x_4} \end{bmatrix} \tag{12.160}$$

where the present partial derivatives are not zero.

The normal coordinates result

$$\boldsymbol{\xi}^1 = \begin{bmatrix} x_1 \\ L_f h_1 = f_1 \end{bmatrix}; \; \boldsymbol{\xi}^2 = \begin{bmatrix} x_2 \\ L_f h_2 = f_2 \end{bmatrix} \tag{12.161}$$

and $\boldsymbol{\eta}$ has a null dimension. The normal form is then

$$\begin{aligned}
&\dot{\phi}_1^1(\boldsymbol{x}) = \phi_2^1(\boldsymbol{x}) \\
&\dot{\phi}_2^1(\boldsymbol{x}) = \sum_i \frac{\partial f_1}{\partial x_i}\dot{x}_i = \frac{\partial f_1}{\partial x_1}(f_1 + g_{11}u_1) + \frac{\partial f_1}{\partial x_2}f_2 + \frac{\partial f_1}{\partial x_4}f_4 \\
&\dot{\phi}_2^2(\boldsymbol{x}) = \phi_2^2(\boldsymbol{x}) \\
&\dot{\phi}_2^2(\boldsymbol{x}) = \sum_i \frac{\partial f_2}{\partial x_i}\dot{x}_i = \frac{\partial f_2}{\partial x_1}(f_1 + g_{11}u_1) + \frac{\partial f_2}{\partial x_2}f_2 + \frac{\partial f_2}{\partial x_3}(f_3 + g_{23}u_2) + \frac{\partial f_2}{\partial x_4}f_4
\end{aligned} \tag{12.162}$$

Clearly, it gives

$$\begin{aligned}
\begin{bmatrix} \dot{\phi}_2^1(\boldsymbol{x}) \\ \dot{\phi}_2^2(\boldsymbol{x}) \end{bmatrix} &= \begin{bmatrix} \dfrac{\partial f_1}{\partial x_1}f_1 + \dfrac{\partial f_1}{\partial x_2}f_2 + \dfrac{\partial f_1}{\partial x_4}f_4 \\ \dfrac{\partial f_2}{\partial x_1}f_1 + \dfrac{\partial f_2}{\partial x_2}f_2 + \dfrac{\partial f_2}{\partial x_3}f_3 + \dfrac{\partial f_2}{\partial x_4}f_4 \end{bmatrix} + \boldsymbol{A}(\boldsymbol{\Phi}(x)) \begin{bmatrix} u_1 \\ u_2 \end{bmatrix} \\
&= \begin{bmatrix} L_f^2 h_1 \\ L_f^2 h_2 \end{bmatrix} + \boldsymbol{A}(\boldsymbol{\Phi}(x)) \begin{bmatrix} u_1 \\ u_2 \end{bmatrix} \\
&= \boldsymbol{b}(\boldsymbol{\Phi}(x)) + \boldsymbol{A}(\boldsymbol{\Phi}(x))\,\boldsymbol{u}
\end{aligned} \tag{12.163}$$

where the fact that f_1 does not depend on x_3 is used. It results in agreement with Eq. (12.140) that the control vector law is equal to

$$\boldsymbol{u} = \boldsymbol{A}^{-1}(\boldsymbol{x})(-\boldsymbol{b} + \boldsymbol{v}) \tag{12.164}$$

where $\boldsymbol{v}$ is the external input vector which can be chosen according to pole placement, for example, or according to a PI vector. Taking into account reference trajectories, the multivariable control law becomes

$$\boldsymbol{u} = \boldsymbol{A}^{-1}(\boldsymbol{\xi}_{\text{ref}}(t))\left[\boldsymbol{y}_{\text{ref}}^{(2)}(t) - \boldsymbol{b}(\boldsymbol{\xi}_{\text{ref}}(t))\right] \tag{12.165}$$

Note that this control law does not depend on $\boldsymbol{\eta}$.

Numerical application

A pole placement has been performed on each manipulated input to set the corresponding external input as a discretized PI controller under the velocity form

$$v_{k+1} = v_k + K_c\left[(y_{k+1}^{\text{ref}} - y_{k+1}) - (y_k^{\text{ref}} - y_k)\right] + \frac{K_c}{\tau_I}\frac{T_s}{2}\left[(y_{k+1}^{\text{ref}} - y_{k+1}) + (y_k^{\text{ref}} - y_k)\right] \tag{12.166}$$

where k is the sampling instant (sampling period $T_s = 5$ s), $\boldsymbol{y}^{\text{ref}}$ the reference trajectory, K_c the proportional gain, and τ_I the integral time constant. To perform the pole placement,[15] the control law (12.165) was modified by taking the vector $\boldsymbol{b}$ as

$$\boldsymbol{b} = \begin{bmatrix} L_f^2 h_1 + c_1 L_f h_1 + c_2 (y_1 - y_1^{\text{ref}}) \\ L_f^2 h_2 + c_1 L_f h_2 + c_2 (y_2 - y_2^{\text{ref}}) \end{bmatrix} \tag{12.167}$$

Approximating the discretized PI controller as a continuous one results approximately in the following input-output dynamics

$$y_i^{(r)} + c_1 y_i^{(r-1)} + c_2 y_i = v_i \tag{12.168}$$

with a relative degree $r = 2$. This is only approximate as the control law (12.165) is indeed multivariable and the previous equation would only be valid for an SISO system. Taking into account the continuous PI controller, the corresponding characteristic equation for each input-output couple is

$$s^3 + c_1 s^2 + (c_2 + K_c)s + \frac{K_c}{\tau_I} = 0 \tag{12.169}$$

The parameters were chosen to satisfy the Integral Time weighted Average Error (ITAE) criterion as $K_c = 2.12 \times 10^{-4}$, $\tau_I = 6.84$ s, $c_1 = 0.055$, $c_2 = 1.91 \times 10^{-3}$.

Recall that both manipulated inputs are constrained in the domain [0, 1]. By applying the modified control law (12.165), it appeared that the manipulated input u_1 corresponding mainly to the concentration control rapidly reached its constraint and remained at this value. Nonlinear geometric control is not designed for handling the constraints, opposite to model predictive control, so that, in actual problems, some adaptations must be considered. When looking numerically at the elements a' of matrix $\boldsymbol{A}^{-1}$, the element a_{11}' is very large, a_{12}' is zero, a_{21}' is small, and a_{22}' is medium. The multiplication of a_{11}' by b results in a large numerical value, explaining that the constraint for u_1 is hit. Physically, this is due to the strong influence of temperature on the concentration in particular because of the Arrhenius term in the kinetic constant and also the heat of reaction. This is not the case for u_2 for which the nonlinear control law (12.165) can be implemented with only the introduction of c_i's. Thus a modification of u_1 is performed by maintaining the term b_1 of Eq. (12.167), but changing the factor a_{11}' in a way close to an antiwindup to avoid saturation. This resulted in the controlled concentration of Fig. 12.8. The control of temperature where coupling intervenes posed little problem. The manipulated inputs (Fig. 12.9) are not saturated, although the input u_1 changes abruptly after $t = 2000$ s when a concentration set point variation is imposed. The nonlinear control law (12.165) perfectly decoupled the output y_1 from the influence of u_2 but the output y_2 is

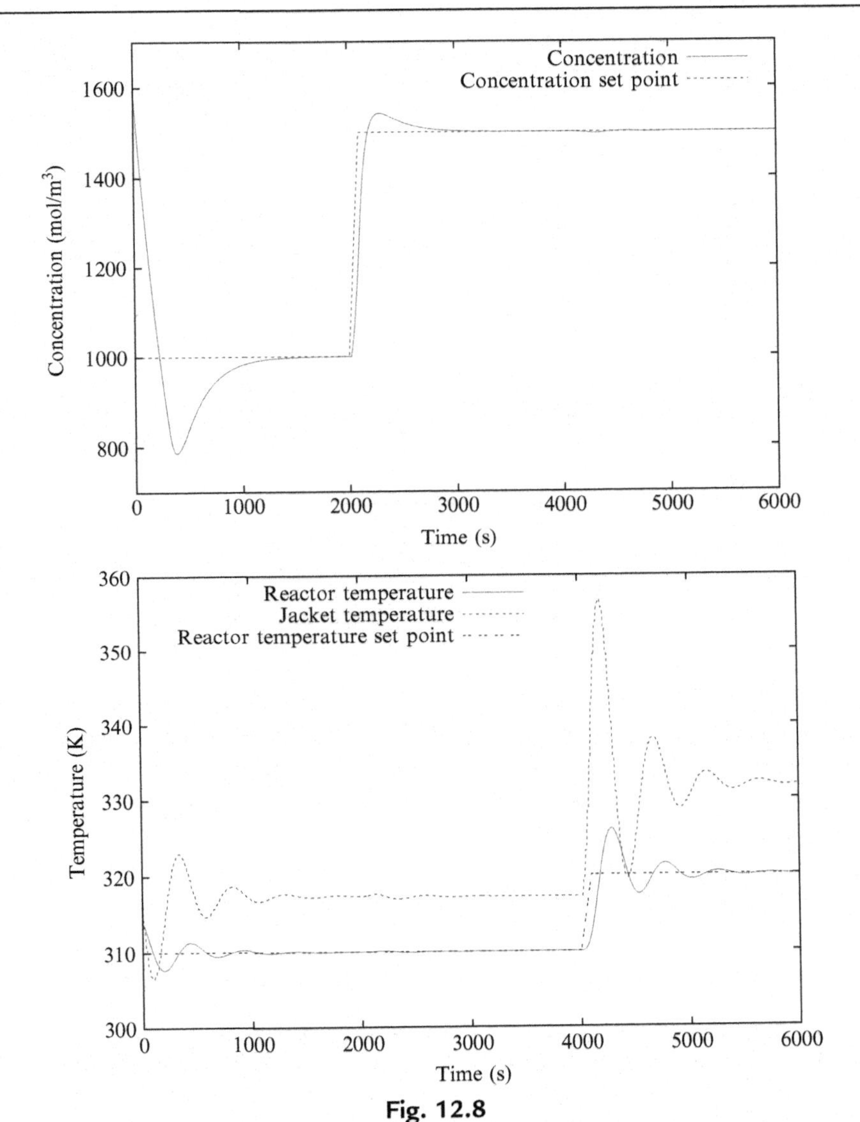

Fig. 12.8
Nonlinear geometric control of chemical reactor. *Top*: Concentration. *Bottom*: Reactor temperature.

influenced by both manipulated inputs. This coupling is hardly visible on Fig. 12.9 only because of the values of the respective elements a'_{21} and a'_{22} where a'_{22} is much larger than a'_{21}. The previous study was performed in simulation and all the states were assumed to be known. With little additional difficulty, it would be possible to estimate the states using a nonlinear observer such as the extended Kalman filter[15] or another one, assuming that the reactor temperature and the concentration are measured. Then the estimated states are used in the nonlinear control law. Furthermore, a robustness study taking into account the unavoidable uncertainty of some physical parameters would be useful.

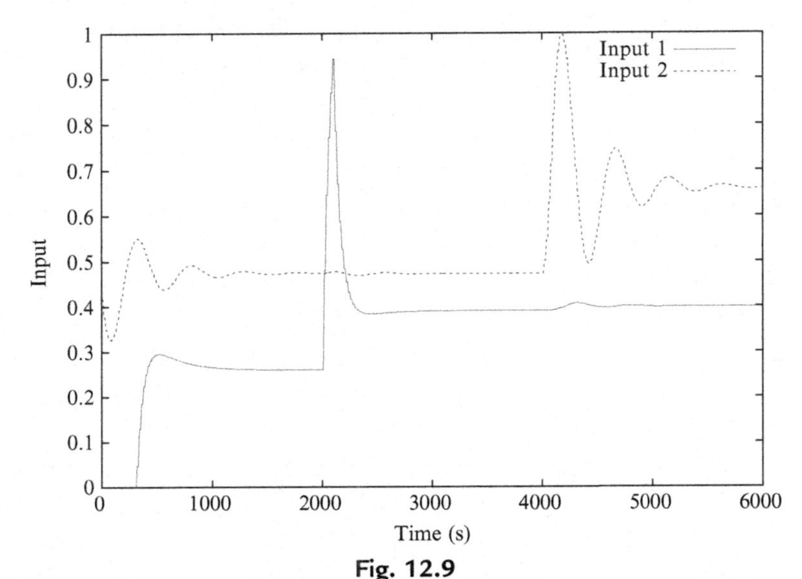

Fig. 12.9
Nonlinear geometric control of chemical reactor. Manipulated inputs.

In conclusion, the nonlinear multivariable control of the chemical reactor is possible but some adaptations were necessary because of the physical constraints of the manipulated inputs. Compared to model predictive control, the multivariable nonlinear geometric control appears more difficult to implement, but shows real efficiency. For SISO plants, nonlinear geometric control is generally easier to use.

References

1. Krstić M, Kanellakopoulos I, Kokotovic P. *Nonlinear and adaptive control design*. New York: Wiley Interscience; 1995.
2. Khalil HK. *Nonlinear systems*. Englewood Cliffs (NJ): Prentice-Hall; 1996.
3. Slotine JJE, Li W. *Applied nonlinear control*. Englewood Cliffs (NJ): Prentice-Hall; 1991.
4. Fliess M, Lévine J, Martin P, Rouchon P. Flatness and defect in nonlinear systems: introducing theory and applications. *Int J Control* 1995;**61**:1327–61.
5. Fliess M, Lévine J, Martin P, Rouchon P. Controlling nonlinear systems by flatness. In: Byrnes C, editor. *Systems and control in the twenty-first century*. Basel: Birkhäuser; 1997.
6. Allgöwer F, Zheng A. *Nonlinear model predictive control*. Basel: Birkhäuser; 2000.
7. Alamir M. *Stabilization of nonlinear systems using receding-horizon control schemes*. London: Springer; 2006.
8. Rawlings JB, Meadows ES, Muske KR. Nonlinear model predictive control: a tutorial and survey. In: *Advanced control of chemical processes*. Kyoto (Japan): IFAC; 1994. p. 203–24.
9. Isidori A. *Nonlinear control systems: an introduction*. 2nd ed. New York: Springer-Verlag; 1989.
10. Isidori A. *Nonlinear control systems*. 3rd ed. New York: Springer-Verlag; 1995.
11. Jurdjevic V. *Geometric control theory*. No. 51, *Cambridge Studies in Advanced Mathematics*, Cambridge: Cambridge University Press; 1997.
12. Fossard AJ, Normand-Cyrot D. *Systèmes non Linéaires 3. Commande*. Paris: Masson; 1993.
13. Nijmeijer H, Vander Schaft AJ. *Nonlinear dynamical systems*. New York: Springer-Verlag; 1990.
14. Vidyasagar M. *Nonlinear systems analysis*. Englewood Cliffs (NJ): Prentice-Hall; 1993.

15. Corriou JP. *Process control—theory and applications*. London: Springer; 2004.
16. Hirschorn RM. Invertibility of nonlinear control systems. *SIAM J Control Optim* 1979;**17**:289–95.
17. Kang W, Krener AJ. Normal forms of nonlinear control systems. In: *Chaos in automatic control, control engineering*. New York: Taylor and Francis; 2006.
18. Brunovsky P. A classification of linear controllable systems. *Kybernetika* 1970;**6**:173–88.
19. Hunt LR, Su R, Meyer G. Global transformations of nonlinear systems. *IEEE Trans Autom Control* 1983;**28**:24–31.
20. Su R. On the linear equivalents of nonlinear systems. *Syst Control Lett* 1982;**2**:48–52.
21. Morari M, Zafiriou E. *Robust process control*. Englewood Cliffs (NJ): Prentice-Hall; 1989.
22. Gentric C, Pla F, Latifi MA, Corriou JP. Optimization and non-linear control of a batch emulsion polymerization reactor. *Chem Eng J* 1999;**75**:31–46.
23. Kravaris C, Daoutidis P. Nonlinear state feedback control of second-order non-minimum phase nonlinear systems. *Comp Chem Eng* 1990;**14**:439–49.
24. Engell S, Klatt KU. Nonlinear control of a non-minimum-phase CSTR. In: *American Control Conference, Groningen*; 1993.
25. Kravaris C, Chung CB. Nonlinear state feedback synthesis by global input/output linearization. *AIChE J* 1987;**33**(4):592–603.
26. Kravaris C. Input-output linearization: a nonlinear analog of placing poles at process zeros. *AIChE J* 1988;**34**(11):1803–12.
27. Kravaris C, Soroush M. Synthesis of multivariable nonlinear controllers by input/output linearization. *AIChE J* 1990;**36**(2):249–64.
28. Kravaris C, Kantor JC. Geometric methods for nonlinear process control. 1. Background. *Ind Eng Chem Res* 1990;**29**:2295–310.
29. Kravaris C, Kantor JC. Geometric methods for nonlinear process control. 2. Controller synthesis. *Ind Eng Chem Res* 1990;**29**:2310–23.
30. Daoutidis P, Kravaris C. Synthesis of feedforward/state feedback controllers for nonlinear processes. *AIChE J* 1989;**35**(10):1602–16.
31. Kravaris C, Wright RA. Deadtime compensation for nonlinear processes. *AIChE J* 1989;**35**(9):1535–42.
32. Soroush M, Kravaris C. Nonlinear control of a batch polymerization reactor: an experimental study. *AIChE J* 1992;**38**(9):1429–48.
33. Kravaris C, Palanki S. Robust nonlinear state feedback under structured uncertainty. *AIChE J* 1988;**34**(7):1119–27.
34. Lee PL. *Nonlinear process control: applications of generic model control*. New York: Springer-Verlag; 1993.
35. Lee PL, Sullivan GR. Generic model control (GMC). *Comp Chem Eng* 1988;**12**:573–80.
36. Henson MA, Seborg DE. A critique of differential geometric control strategies for process control. In: *11th IFAC world congress, USSR*; 1990.
37. Bequette BW. Nonlinear control of chemical processes: a review. *Ind Eng Chem Res* 1991;**30**:1391–413.
38. Henson MA, Seborg DE. A unified differential geometric approach to nonlinear process control. In: *AIChE annual meeting*; 1989, San Francisco.
39. Henson MA, Seborg DE. Input-output linearization of general nonlinear processes. *AIChE J* 1990;**36**(11):1753–7.
40. Soroush M, Kravaris C. Multivariable nonlinear control of a continuous polymerization reactor: an experimental study. *AIChE J* 1993;**39**(12):1920–37.
41. To LC, Tadé MO, Kraetzl M, Le Page GP. Nonlinear control of a simulated industrial evaporator process. *J Process Control* 1995;**5**(3):173–82.
42. Soroush M, Kravaris C. Nonlinear control of a polymerization CSTR with singular characteristic matrix. *AIChE J* 1994;**40**(6):980–90.
43. Aström KJ. Theory and applications of adaptive control—a survey. *Automatica* 1983;**19**(5):471–86.
44. Aström KJ, Wittenmark B. *Adaptive control*. New York: Addison-Wesley; 1989.
45. Sastry S, Bodson M. *Adaptive control—stability, convergence, and robustness*. Englewood Cliffs (NJ): Prentice-Hall; 1989.
46. Tao G. *Adaptive control design and analysis*. Hoboken (NJ): Wiley-IEEE Press; 2003.

47. Bitmead RR, Gevers M, Wertz V. *Adaptive optimal control, the thinking man's GPC*. New York: Prentice-Hall; 1990.
48. Goodwin GC, Sin KS. *Adaptive filtering, prediction and control*. Englewood Cliffs (NJ): Prentice-Hall; 1984.
49. Haykin S. *Adaptive filter theory*. 3rd ed. Englewood Cliffs (NJ): Prentice-Hall; 1991.
50. Landau I. *System identification and control design*. Englewood Cliffs (NJ): Prentice-Hall; 1990.
51. Watanabe K. *Adaptive estimation and control*. London: Prentice-Hall; 1992.
52. Ioannou PA, Sun J. *Robust adaptive control*. Upper Saddle River (NJ): Prentice-Hall; 1996.
53. Gustafsson TK, Waller KV. Nonlinear and adaptive control of pH. *Ind Eng Chem Res* 1992;**31**:2681.
54. Seborg DE, Edgar TF, Shah SL. Adaptive control strategies for process control: a survey. *AIChE J* 1986;**32** (6):881–913.
55. Sung SW, Lee I, Choi JY, Lee J. Adaptive control for pH systems. *Chem Eng Sci* 1998;**53**:1941.
56. Kosanovich KA, Piovosa MJ, Rokhlenko V, Guez A. Nonlinear adaptive control with parameter estimation of a CSTR. *J Process Control* 1995;**5**(3):137–48.
57. Wang ZL, Pla F, Corriou JP. Nonlinear adaptive control of batch styrene polymerization. *Chem Eng Sci* 1995;**50** (13):2081–91.
58. Wang GB, Peng SS, Huang HP. A sliding observer for nonlinear process control. *Chem Eng Sci* 1997;**52**:787–805.

CHAPTER 13

Economic Model Predictive Control of Transport-Reaction Processes

Liangfeng Lao, Matthew Ellis, Panagiotis D. Christofides
University of California, Los Angeles, CA, United States

13.1 Introduction

The development of computationally efficient control methods for partial differential equation (PDE) systems has been a major research topic in the past 30 years (e.g., Ref. 1). The design of feedback control algorithms for PDE systems is usually achieved on the basis of finite-dimensional systems (i.e., sets of ordinary differential equations (ODEs) in time) obtained by applying a variety of spatial discretization and/or order reduction methods to the PDE system. The classification of PDE systems based on the properties of the spatial differential operator into hyperbolic, parabolic, or elliptic typically determines the finite-dimensional approximation approaches employed to derive finite-dimensional models (e.g., Refs. 1, 2). A class of processes described by PDEs within chemical process industries is transport-reaction processes. For example, tubular reactors are typically described by parabolic PDEs since both convective and diffusive transport phenomena are significant. On the other hand, plug-flow reactors are typically described by first-order hyperbolic PDEs since only convective transport is significant (i.e., diffusive transport is negligible relative to convective transport).

For parabolic PDE systems (e.g., diffusion-convective-reaction processes) whose dominant dynamics can be adequately represented by a finite number of dominant modes, Galerkin's method with spatially global basis functions is a good way among many weighted residual methods (e.g., Refs. 3, 4) to construct a reduced-order model (ROM) of the PDE system. Specifically, it can be used to derive a finite-dimensional ODE model by applying approximate inertial manifolds (AIMs) (e.g., Ref. 5) that capture the dominant dynamics of the original PDE system. The basis functions used in Galerkin's method may either be analytical or empirical eigenfunctions. After applying Galerkin's method to the PDE system, a low-order ODE is derived, and the control system can be designed by utilizing control methods for linear/nonlinear ODE systems.[1]

Coulson and Richardson's Chemical Engineering. http://dx.doi.org/10.1016/B978-0-08-101095-2.00013-8

One way to construct the empirical eigenfunctions is by applying proper orthogonal decomposition (POD) (e.g., Refs. 6–8) to PDE solution data. This data-based methodology for constructing the basis eigenfunctions has been widely adopted in the field of model-based control of parabolic PDE systems (e.g., Refs. 8–12). However, to achieve high accuracy of the ROM derived from the empirical eigenfunctions of the original PDE system, the POD method usually needs a large ensemble of solution data (snapshots) to contain as much local and global process dynamics as possible. Constructing such a large ensemble of snapshots becomes a significant challenge from a practical point of view; since currently, there is no general way to realize a representative ensemble. Based on this consideration, an adaptive proper orthogonal decomposition (APOD) methodology was proposed to recursively update the ensemble of snapshots and compute on-line the new empirical eigenfunctions in the on-line closed-loop operation of PDE systems (e.g., Refs. 13–17). While the APOD methodology of Refs. 15 and 16 demonstrated its ability to capture the dominant process dynamics by a relatively small number of snapshots which reduces the overall computational burden, these works did not address the issue of computational efficiency with respect to optimal control action calculation and input and state constraint handling. Moreover, the ROM accuracy is indeed limited by the number of the empirical eigenfunctions adopted for the ROM; in practice, when a process faces state constraints, the accuracy of the ROM based on a limited number of eigenfunctions may not be able to allow the controller to avoid a state constraint violation.

For hyperbolic PDE systems (e.g., convection-reaction processes where the convective phenomena dominate over diffusive ones), it is common that the eigenvalues of the spatial differential operator cluster along vertical or nearly vertical asymptotes in the complex plane, and thus, many modes are required to construct finite-dimensional models of desired accuracy.[1] Considering this, the Galerkin's method would be computationally expensive for this case compared to finite difference method utilizing a sufficient large number of spatial discretization points.[3] Orthogonal collocation method as another weighted residual method has also been used for the control of hyperbolic PDE systems.[18] The orthogonal collocation method is based on the collocation points chosen as roots of a series of specific orthogonal polynomial functions. Although the collocation method is easy to apply, its accuracy can be good only if the collocation points are judiciously chosen which has been a general difficulty for its wide application.[19] With respect to controller design for hyperbolic PDE systems, many researchers have proposed various control methods (e.g., Refs. 20–24); furthermore, several works (e.g., Refs. 18, 25–28) have considered the application of (tracking) model predictive control (MPC) to chemical processes modeled by hyperbolic PDE systems.

Economic model predictive control (EMPC) is a practical optimal control-based technique that has recently gained widespread popularity within the process control community and beyond because of its unique ability of effectively integrating process economics and feedback control (see Ref. 29 for an overview of recent results and references). However, most of previous EMPC systems have been designed for lumped parameter processes described by linear/nonlinear ODE systems (e.g., Refs. 30–34). In our previous work,[8,35] an EMPC system with a general economic cost function for parabolic PDE systems was proposed which operates the closed-loop system in a dynamically optimal fashion. Specifically, the EMPC scheme was developed on the basis of low-order nonlinear ODE models derived through Galerkin's method using analytical eigenfunctions[35] and empirical eigenfunctions derived by POD,[8] respectively. However, no work has been done on applying APOD techniques for model order reduction to parabolic PDE systems under EMPC and the application of EMPC to hyperbolic PDE systems. Typically, EMPC will operate a system at its constraints in order to achieve the maximum closed-loop economic performance benefit. Thus, the unifying challenge of both of these cases is to formulate EMPC schemes that can handle state constraints (i.e., prevent state constraint violation).

Motivated by the above considerations, we address two key issues in the context of applying EMPC to transport-reaction processes. In the next section, we focus on applying an EMPC scheme to nonlinear parabolic PDEs subject to state and control constraints. An EMPC algorithm is proposed which integrates APOD with a high-order finite-difference method to handle state and control constraints and computational efficiency. In the subsequent section, an EMPC system is formulated and applied to a hyperbolic PDE system. The EMPC systems formulated for parabolic and hyperbolic PDE systems are demonstrated by applying them to a tubular reactor and plug-flow reactor, respectively, and the computational efficiency and process constraint satisfaction capabilities of the EMPC systems are evaluated.

13.2 EMPC of Parabolic PDE Systems With State and Control Constraints

13.2.1 Preliminaries

13.2.1.1 Parabolic PDEs

We consider parabolic PDEs of the form:

$$\frac{\partial x}{\partial t}=A\frac{\partial x}{\partial z}+B\frac{\partial^2 x}{\partial z^2}+Wu(t)+f(x) \tag{13.1}$$

with the boundary conditions:

$$\left.\frac{\partial x}{\partial z}\right|_{z=0}=g_0x(0,t),\quad \left.\frac{\partial x}{\partial z}\right|_{z=1}=g_1x(1,t) \tag{13.2}$$

and the initial condition:

$$x(z,0)=x_0(z) \tag{13.3}$$

where $z\in[0,1]$ is the spatial coordinate, $t\in[0,\infty)$ is the time, $x'(z,t)=[x_1(z,t)\ \cdots\ x_{n_x}(z,t)]$ is the vector of the state variables (x' denotes the transpose of x), and $f(x)$ denotes a nonlinear vector function. The notation A, B, W, g_0, and g_1 is used to denote (constant) matrices of appropriate dimensions. The control input vector is denoted as $u(t) \in \mathbb{R}^{n_u}$ and is subject to the following constraints:

$$u_{\min} \leq u(t) \leq u_{\max} \tag{13.4}$$

where $u_{\min}$ and $u_{\max}$ are the lower and upper bound vectors of the manipulated input vector, $u(t)$. Moreover, the system states are also subject to the following state constraints:

$$x_{i,\min} \leq \int_0^1 r_{x_i}(z) x_i(z,t) dz \leq x_{i,\max}, \quad i = 1, \ldots, n_x \tag{13.5}$$

where $x_{i,\min}$ and $x_{i,\max}$ are the lower and upper state constraint for the ith state, respectively. The function $r_{x_i}(z) \in L_2(0,1)$ where $L_2(0, 1)$ is the space of measurable, square-integrable functions on the interval [0, 1], is the state constraint distribution function.

13.2.1.2 Galerkin's method with POD-computed basis functions

To reduce the PDE model of Eq. (13.1) into an ODE model, we take advantage of the orthogonality of the empirical eigenfunctions obtained from POD.[6,7] Specifically, using Galerkin's method,[5,36] a low-order ODE system for each of the PDEs describing the temporal evolution of the amplitudes corresponding to the first m_i eigenfunctions of the ith PDE has the following form:

$$\begin{aligned} \dot{a}_s(t) &= \mathcal{A}_s a_s(t) + \mathcal{F}_s(a_s(t)) + \mathcal{W}_s u(t) \\ x_i(z,t) &\approx \sum_{j=1}^{m_i} a_{s,ij}(t) \phi_{ij}(z), \quad i = 1, \ldots, n_x \end{aligned} \tag{13.6}$$

where $a_s'(t) = [a_{s,1}'(t) \cdots a_{s,n_x}'(t)]$ is a vector of the total eigenmodes, $a_{s,i}'(t) = [a_{s,i1}(t) \cdots a_{s,im_i}(t)]$ is a vector of the amplitudes of the first m_i eigenfunctions, $\mathcal{A}_s$ and $\mathcal{W}_s$ are constant matrices, $\mathcal{F}_s(a_s(t))$ is a nonlinear smooth vector function of the modes obtained by applying weighted residual method to Eq. (13.1), and $\{\phi_{ij}(z)\}$ are the first m_i dominant empirical eigenfunctions computed from POD for each PDE state, $x_i(z, t)$.

13.2.2 Methodological Framework for Finite-Dimensional EMPC Using APOD

13.2.2.1 Adaptive proper orthogonal decomposition

Compared with POD, APOD is a more computationally efficient algorithm, in that it only needs an ensemble of a small number of snapshots in the beginning. It can complete the recursive update of the computation of the dominant eigenfunctions, while keeping the size of the ensemble small to reduce the computational burden of updating the ensemble once a new process state measurement is available. Moreover, APOD can also adaptively adjust the number of the basis eigenfunctions under a desired energy occupation requirement to improve the computational efficiency of the control system constructed based on the ROM with the dominant eigenfunctions.[15] Since the basis eigenfunctions are updated on-line, the initial ensemble of process snapshots is not required to include too many process solution data. More details of the APOD methodology can be found in Refs. 15 and 16. The implementation steps of the APOD methodology can be summarized as follows:

1. [1.] At $t < 0$, generate an ensemble of solutions of the PDE system (e.g., Eq. (13.1) for single manipulated input value $u(t)$ from certain initial condition;
   1.1. Apply Karhunen-Loève (K-L) expansion to this ensemble to derive a set of first $m_i(t_0)$ most dominant empirical eigenfunctions for each state x_i, $i=1,\ldots,n_x$ which occupy ϵ energy of the chosen ensemble[37];
   1.2. Construct a ROM in the form of a low-dimensional nonlinear ODE system based on these empirical eigenfunctions within a Galerkin's model reduction framework from the infinite dimensional nonlinear PDE system;
2. [2.] At $t = t_k = k\Delta > 0$ where $\Delta = t_{k+1} - t_k$ is the update cycle, when the new process state measurements are available, update the ensemble by utilizing the most important snapshots approach[16] which analyzes the contribution of the current snapshots in the ensemble and replaces the snapshot that corresponds to the lowest contribution of representativeness with new state measurement to keep the size of the ensemble;
   2.1. Recompute the dominant eigenvalues corresponding to the first $m_i(t_{k-1})$ eigenfunctions by constructing small scale matrix to reduce the computational burden;
   2.2. Adopt orthogonal power iteration methodology to get the $(m_i(t_{k-1}) + 1)$th dominant eigenvalue;
   2.3. Get the new size of the basis eigenfunctions, $m_i(t_k)$ which should still occupy ϵ energy of the updated ensemble (i.e., the new size of the basis eigenfunctions, $m_i(t_k)$, may increase, decrease, or keep the same compared with $m_i(t_{k-1})$);
3. [2.] At $t = t_{k+1}$, repeat Step 2.

13.2.2.2 EMPC scheme of integrating APOD and finite-difference method to avoid state constraint violation and improve computational efficiency

Although APOD only needs an ensemble of a small number of snapshots which could greatly improve the computational efficiency of the eigenfunction update calculation, smaller size of ensemble usually results in a single or a few dominant eigenfunctions. The accuracy of the ROM based on fewer eigenfunctions computed from an ensemble of small size is usually worse than that of the ROM constructed by adopting more eigenfunctions from a large ensemble of snapshots. However, as pointed out in Ref. 38, eigenfunctions that have high frequency spatial profiles (corresponding to small eigenvalues) should be discarded because of potentially significant round-off errors. In this situation, only a single or a few eigenfunctions can be adopted from APOD keeping the dimension of the ROM low. Moreover, from the point of view of practical implementation, when the process faces some specific state constraint, we cannot judge whether the ROM underestimates or overestimates the process state values which may mislead the controller to make a wrong decision and lead to state constraint violation.

To circumvent this problem, we propose a methodology which integrates the APOD methodology in the context of EMPC to increase the computational efficiency with a high order finite-dimensional approximation of the PDE system to avoid potential state constraint violation. A flow chart illustrating the aforementioned methodology of improving the optimal control action performance from EMPC to avoid potential state constraint violation and inheriting the computational efficiency of APOD method is presented in Fig. 13.1. The detailed steps of this methodology are explained as follows:

1. [1.] At $t < 0$, generate an ensemble of solutions of the PDE system of Eq. (13.1) for single manipulated input value u starting from a certain initial condition; go to Step 2;
2. [2.] Since $t = t_{k-1} = (k-1)\Delta$, $k = 1, 2, \ldots$, use the available full spatial state measurement profile over the entire spatial domain $x(z, t_{k-1})$ at $t = t_{k-1}$ to complete the APOD procedure and get the new number of the basis eigenfunctions, $m(t_k)$ where $m'(t_k) = [m_1(t_k) \cdots m_{n_x}(t_k)]$ is a vector containing the number of basis eigenfunctions for each PDE state, for the next sampling time instant, $t = t_k$; go to Step 3;
3. [3.] If the finite-dimensional model of the process discretized by finite-difference method is adopted to get the optimal input value at $t = t_{k-1}$ to avoid state constraint violation of the ith PDE state as described by Step 5.1.1, go to Step 3.1; otherwise, go to Step 3.2;
   - 3.1. Enforce the number of the basis eigenfunctions to be increased by 1 for the ith PDE state, i.e., $m_i(t_k) = m_i(t_k) + 1$ and update the basis eigenfunctions for the ith PDE state; go to Step 4;
   - 3.2. Keep the same number of the basis eigenfunctions; go to Step 4;
4. [4.] Solve the EMPC problem using the updated basis eigenfunctions with the size of $m(t_k)$ and get the trial optimal input trajectory, u_{APOD}; go to Step 5;
5. [5.] Check whether the current state profile $x_i(z, t_k)$ of each state is located in the state constraint violation alert region, Ω_i; if $x_i(z, t_k) \in \Omega_i$, then go to Step 5.1; otherwise go to Step 5.2;
   - 5.1. Apply the trial optimal input trajectory u_{APOD} to the finite-difference model and compute the estimated state value at the next sampling time instant, $\hat{x}(z, t_{k+1})$; check whether the estimated state value $\hat{x}_i(z, t_{k+1})$ violates the state constraint; if $\hat{x}_i(z, t_{k+1}) \notin \Omega_i$, go to Step 5.1.1; otherwise $u^* = u_{APOD}$ and go to Step 5.1.2;
     - 5.1.1. Solve the EMPC problem constructed using the finite-dimensional model of the process and compute the revised optimal input trajectory, u_{FD} and use this trajectory as the final optimal input trajectory, i.e., $u^* = u_{FD}$; go to Step 5.2;
     - 5.1.2. Use the trial optimal input trajectory, u_{APOD} as the final optimal input trajectory, i.e., $u^* = u_{APOD}$; go to Step 5.2;

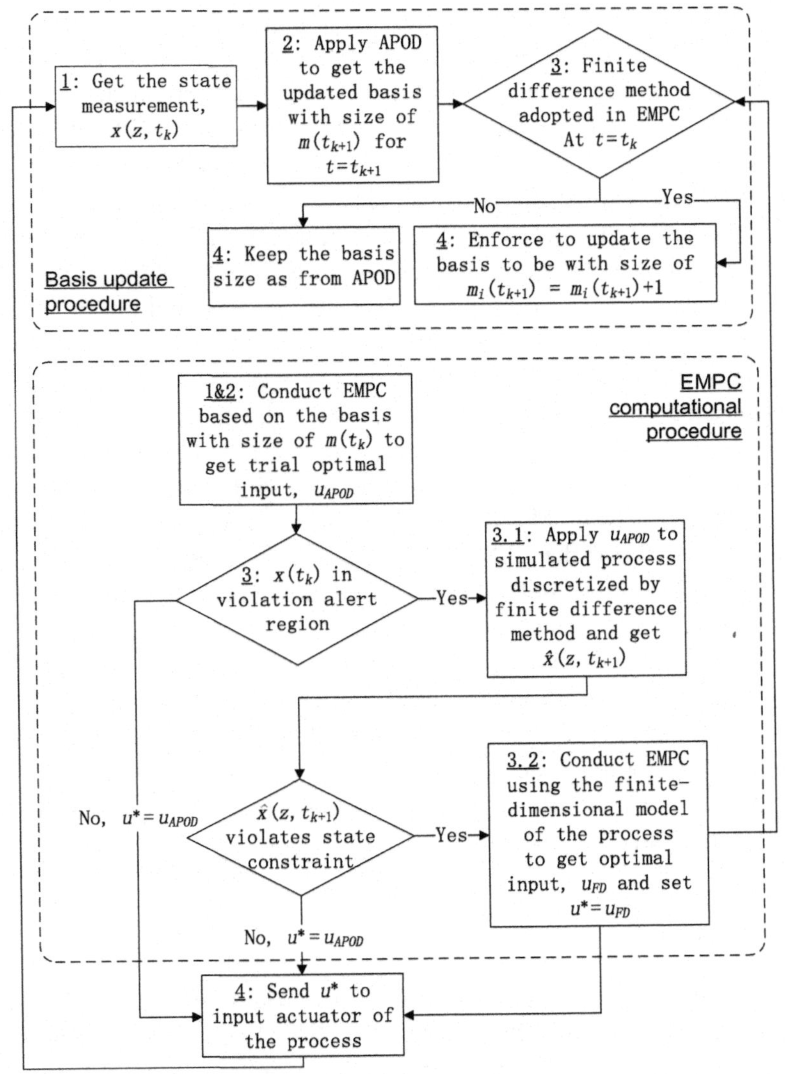

Fig. 13.1
EMPC system flow chart which integrates the APOD methodology with a finite-difference method to increase the computational efficiency and avoid potential state constraint violation.

5.2. Apply the optimal control action u^* for the time period $t = t_k$ to $t = t_{k+1}$; go to Step 2, $k \leftarrow k + 1$ and obtain the state measurement $x(z, t_{k+1})$.

Remark 1 Considering that the APOD procedure (i.e., updating the empirical eigenfunctions) is computationally expensive especially when the size of the ensemble is large, here at current time $t = t_k$ we use the previous state measurement value $x(z, t_{k-1})$ to update the basis eigenfunctions which can be completed during the sampling time between $t = t_{k-1}$ and $t = t_k$. With respect to the APOD update cycle length, the availability of the full state profile across the entire spatial domain is assumed at each sampling instance (i.e., $t = t_k = k\Delta$) and the update cycle of APOD is equal to the sampling time of EMPC, Δ. Based on the proposed methodology, the computational time of APOD procedure (Steps 2 and 3 noted as "Basis Update Procedure" in the Fig. 13.1) is not accounted for in the total EMPC calculation time (Steps 4 and 5 noted as "EMPC Computational Procedure" in the Fig. 13.1). We note here that in practice, the sampling time length Δ should be longer than the time needed to complete the APOD procedure otherwise the EMPC system will not be able to get the updated basis eigenfunctions at the new sampling time instant $t = t_{k+1}$. On the other hand, since the state measurements are available only at every Δ, large Δ may result in APOD missing the appearance of new process dynamics when the process goes through different regions in the state-space. Based on this consideration, the sampling should be chosen properly.

Remark 2 By setting the alert region of state constraint violation for ith state $x_i(z, t)$, Ω_i, the EMPC based on the ROM from APOD or POD method with few modes may lead to state constraint violation. However, the EMPC system based on a high-order discretization of the PDE system by finite-difference method can provide more accurate optimal manipulated input values to avoid potential state constraint violation.

Remark 3 In terms of the effectiveness of eigenfunctions, as we pointed out in Ref. 8, eigenfunctions that have high frequency spatial profiles (i.e., corresponding to small eigenvalues) should be discarded because of potentially significant round-off errors. So when implementing the proposed methodology, the eigenfunctions corresponding to eigenvalues smaller than $\lambda_{\min}$ are not included to avoid round-off errors. This consideration is implemented in Steps 2 and 3.1 of Fig. 13.1.

13.2.2.3 EMPC using adaptive POD

Utilizing the empirical eigenfunctions from APOD methodology, we can formulate a Lyapunov-based EMPC (LEMPC) for the system of Eq. (13.1) to dynamically optimize an economic cost function which is denoted as $L_e(t)$. We assume that the full state profile across the entire spatial domain is available synchronously at sampling instants denoted as $t_k = k\Delta$ with $k = 0, 1, \ldots$. To formulate a finite-dimensional EMPC problem, the first m_i modes of Eq. (13.6) are adopted to construct the ROM, and the EMPC formulation takes the following form:

$$\max_{u \in S(\Delta)} \int_{t_k}^{t_{k+N}} L_e(\tau)\ d\tau \tag{13.7a}$$

$$\text{s.t.} \quad \dot{a}_s(t) = \mathcal{A}_s a_s(t) + \mathcal{F}_s(a_s(t)) + \mathcal{W}_s u(t) \tag{13.7b}$$

$$a_{s,ij}(t_k) = \int_0^1 \phi_{s,ij}(z) x_i(z,t_k) dz \tag{13.7c}$$

$$\hat{x}_i(z,t) \approx \sum_{j=1}^{m_i} a_{s,ij}(t)\phi_{ij}(z), \; i = 1,\dots,n_x \tag{13.7d}$$

$$u_{\min} \le u(t) \le u_{\max}, \; \forall t \in [t_k, t_{k+N}) \tag{13.7e}$$

$$x_{i,\min} \le \int_0^1 r_{x_i}(z)\hat{x}_i(z,t) \le x_{i,\max} \tag{13.7f}$$

$$a_s'(t) P a_s(t) \le \overline{\rho} \tag{13.7g}$$

where Δ is the sampling period, $S(\Delta)$ is the family of piecewise constant functions with sampling period Δ, N is the prediction horizon, $\hat{x}_i(z,t)$ is the predicted evolution of state variables, with input $u(t)$ computed by the EMPC and $x_i(z, t_k)$ is the state measurement at the sampling time t_k. Since the empirical eigenfunctions derived from the POD procedure are all self-adjoint, we can use the empirical eigenfunction $\{\phi_{ij}(z)\}$ directly to calculate the estimated mode amplitude by taking advantage of the orthogonality property of the eigenfunctions.

In the optimization problem of Eq. (13.7), the objective function of Eq. (13.7a) describes the economics of the process which the EMPC maximizes over a horizon t_N. The constraint of Eq. (13.7b) is used to predict the future evolution of the system based on the first m_i dominant eigenfunctions with the initial condition given in Eq. (13.7c) (i.e., the estimate of $a_s(t_k)$ computed from the state measurement $x_i(z, t_k)$). The constraints of Eq. (13.7e), (13.7f) are the available control action and the state constraints, respectively. Finally, the constraint of Eq. (13.7g) ensures that the predicted state trajectory is restricted inside a predefined stability region which is a level set of the Lyapunov function (see Ref. 33 for a complete discussion of this issue). The optimal solution to this optimization problem is $u^*(t|t_k)$ defined for $t \in [t_k, t_{k+N})$. The EMPC applies the control action computed for the first sampling period to the system in a sample-and-hold fashion for $t \in [t_k, t_{k+1})$. The EMPC is resolved at the next sampling period, t_{k+1}, after receiving a new state measurement of each state, $x_i(z, t_{k+1})$ and updated basis functions, $\{\phi_{ij}(z)\}$.

13.2.3 Application to a Tubular Reactor Modeled by a Parabolic PDE System

13.2.3.1 Reactor description

We consider a tubular reactor, where an exothermic, irreversible second-order reaction of the form A → B takes place. A cooling jacket of constant temperature is used to remove heat from the reactor. The states of the tubular reactor are temperature and concentration of reactant species A in the reactor, and the input is the inlet concentration of the reactant species A. In order to simplify the presentation of our results below, we use dimensionless variables and obtain the

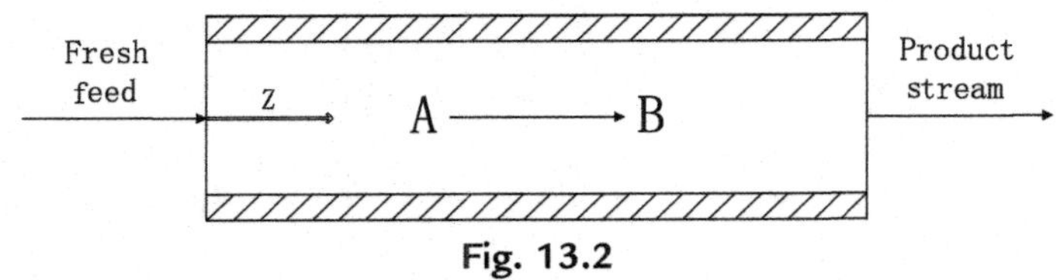

Fig. 13.2
A tubular reactor with reaction A → B.

following nonlinear parabolic PDE model for the process (details and model notation can be found in Refs. 4 and 35) (Fig. 13.2):

$$\begin{aligned}\frac{\partial x_1}{\partial t} &= -\frac{\partial x_1}{\partial z} + \frac{1}{Pe_1}\frac{\partial^2 x_1}{\partial z^2} + \beta_T(T_s - x_1) \\ &\quad + B_T B_C \exp\left(\frac{\gamma x_1}{1+x_1}\right)(1+x_2)^2 + \delta(z-0)T_i \\ \frac{\partial x_2}{\partial t} &= -\frac{\partial x_2}{\partial z} + \frac{1}{Pe_2}\frac{\partial^2 x_2}{\partial z^2} \\ &\quad - B_C \exp\left(\frac{\gamma x_1}{1+x_1}\right)(1+x_2)^2 + \delta(z-0)u\end{aligned} \tag{13.8}$$

where δ is the standard Dirac function, subject to the following boundary conditions:

$$\begin{aligned} z=0: &\frac{\partial x_1}{\partial z} = Pe_1 x_1, \ \frac{\partial x_2}{\partial z} = Pe_2 x_2 \\ z=1: &\frac{\partial x_1}{\partial z} = 0, \ \frac{\partial x_2}{\partial z} = 0 \end{aligned} \tag{13.9}$$

The following typical values are given to the process parameters: $Pe_1 = 7$, $Pe_2 = 7$, $B_T = 2.5$, $B_C = 0.1$, $\beta_T = 2$, $T_s = 0$, $T_i = 0$, and $\gamma = 10$. The following simulations were carried out using Java programming language in a Intel Core i7-2600, 3.40GHz computer with a 64-bit Windows 7 Professional operating system.

13.2.3.2 Implementation of EMPC with APOD

We formulate an EMPC system as in Refs. 35 and 8 for the tubular reactor with the ROM derived from the procedure described earlier. Ipopt[39] was used to solve the EMPC optimization problem. To numerically integrate the ODE model, explicit Euler's method was used with an integration step of 1×10^{-5} (dimensionless). Central finite difference method is adopted to discretize, in space, the two parabolic PDEs and obtain a 101st-order set of ODEs in time for each PDE state (further increase on the order of discretization led to identical open-loop and closed-loop simulation results); this discretized model was also used to describe the process dynamics. In all simulations reported later, with respect to EMPC settings, we use the prediction horizon, $N=3$ and the sampling time length, $\Delta=0.01$ (dimensionless) which can sufficiently capture the appearance of new patterns by the newly available snapshots as the process moves through different regions in the state-space.

To design the EMPC, a Lyapunov-based technique and a quadratic Lyapunov function of the following form was adopted:

$$V(a_s(t)) = a_s'(t) P a_s(t) \tag{13.10}$$

where P is an identity matrix of approximate dimension and $\overline{\rho}=3$ (see Ref. 33 for more details on LEMPC).

The cost function of Eq. (13.7) considered involves maximizing the overall reaction rate along the length of the reactor in the prediction horizon, t_k to t_{k+N} and over one operation period with $t_f=1$. The temporal economic cost along the length of the reactor then takes the form:

$$L_e(t) = \int_0^1 r(z,t)dz \tag{13.11}$$

where $r(x_1(z,t),x_2(z,t)) = B_C \exp\left(\frac{\gamma x_1(z,t)}{1+x_1(z,t)}\right)(1+x_2(z,t))^2$ is the reaction rate (dimensionless) in the tubular reactor.

The control input is subject to constraints as follows: $-1 \leq u \leq 1$. Owing to practical considerations, the amount of reactant material which can be fed to the tubular reactor over the period t_f is fixed. Specifically, $u(t)$ satisfies the following constraint over the period:

$$\frac{1}{t_f}\int_0^{t_f} u(\tau)d\tau = 0.5 \tag{13.12}$$

which will be referred to as the reactant material constraint. Details on the implementation of this constraint can by found in Refs. 35 and 8.

Furthermore, the temperature (dimensionless) along the length of the reactor is subject to the following constraint:

$$x_{1,\min} \leq \min(x_1(z,t)), \quad \max(x_1(z,t)) \leq x_{1,\max} \tag{13.13}$$

where $x_{1,\min}=-1$ and $x_{1,\max}=3$ are the lower and upper limits, respectively.

13.2.3.3 Case 1: APOD compared to POD

In order to present the effectiveness of APOD, compare the POD method and the APOD method in the context of EMPC for the tubular reactor operation. To compute the empirical eigenfunctions, we use the 101st-order discretization of each PDE of Eq. (13.8). In detail, 15 different initial conditions and arbitrary (constant) input values, $u(t)$ were applied to the process model to get the spatiotemporal solution profiles with a time length of 2 (dimensionless). Consequently, from each simulation solution profile, 200 uniformly sampled snapshots were taken and combined to generate an ensemble of 3000 solutions which is noted as Ensemble 1. The POD method was applied to the developed ensemble of solutions to compute empirical eigenfunctions that describe the dominant spatial solution patterns embedded in the ensemble where the Jacobian in the POD method is calculated through a central finite-difference method. After truncating the eigenfunctions with relatively small eigenvalues ($\lambda_{ij} < \lambda_{\text{min}} = 1 \times 10^{-5}$), we were left with the first 4 eigenvalues for each state which occupy more than 99.99% of the total energy included in the entire ensemble. These 4 eigenfunctions for each PDE state are utilized for the POD method and as the initial eigenfunctions for APOD method to construct the ROM. To demonstrate the APOD in capturing the dominant trends that appear during closed-loop process evolution as the process goes through different regions of the state-space, we use EMPC of Eq. (13.7) handling manipulated input and state constraints based on POD using Ensemble 1 and based on APOD using Ensemble 1 (as the starting ensemble) to the tubular reactor, respectively. For the POD method, we constructed 2 ROMs which use the first 3 and 4 dominant eigenfunctions of the previously constructed eigenfunctions, respectively, for the EMPC system of Eq. (13.7). The EMPC utilizing ROM with 4 eigenfunctions is denoted as EMPC based on POD 1 and the other is denoted as EMPC based on POD 2 (Fig. 13.3).

The maximum temperature (dimensionless) profiles of the tubular reactor under the EMPC systems of Eq. (13.7) based on POD 1, POD 2, and APOD using Ensemble 1 are shown in Fig. 13.4. Since the temperature directly influences the reaction rate (i.e., higher temperature leads to higher reaction rate), the optimal operating strategy is to operate the reactor at the maximum allowable temperature. From Fig. 13.4, the EMPC system based on POD 1 and

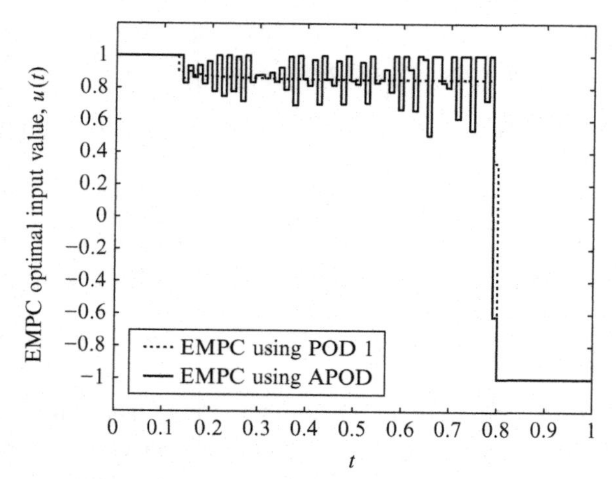

Fig. 13.3
Manipulated input profiles of the EMPC system of Eq. (13.7) based on POD 1 (*dotted line*) and APOD using Ensemble 1 (*solid line*) over one operation period.

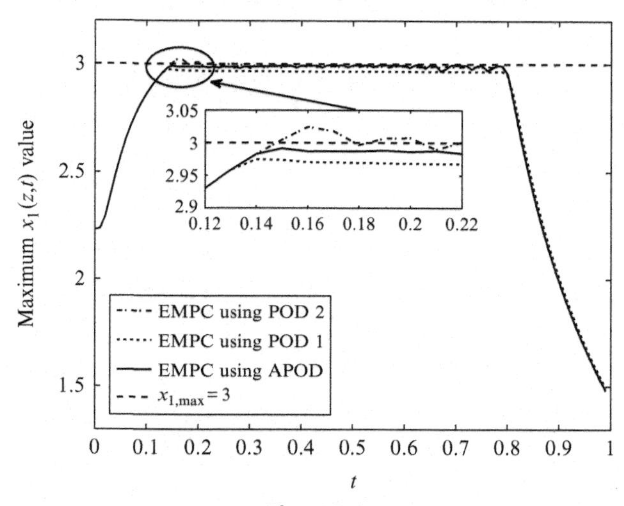

Fig. 13.4
Maximum $x_1(z, t)$ profiles of the process under the EMPC system of Eq. (13.7) based on POD 1 (*dotted line*), POD 2 (*dash-dotted line*), and APOD using Ensemble 1 (*solid line*) over one operation period.

APOD operate the tubular reactor with a maximum temperature less than the maximum allowable which is a consequence of the error associated with the ROM. However, the process under the EMPC system based on POD 2 violates the state constraint imposed on $x_1(z, t)$ due to fewer eigenfunctions used for constructing the ROM in the EMPC system of Eq. (13.7). On the other hand, since the APOD is able to more accurately compute the state profile owing to its continuously updated dominant eigenfunctions, the EMPC system formulated with the ROM using APOD eigenfunctions operates the reactor at a greater temperature than the other EMPC system as demonstrated by the magnified plot in Fig. 13.4.

The computed manipulated input profiles from the EMPC systems of Eq. (13.7) based on POD 1 and APOD using Ensemble 1, respectively, over one period are shown in Fig. 13.3. From Fig. 13.3, the EMPC system based on APOD computes a less smooth manipulated input profile than that of the EMPC system based on POD 1 due to its continuously updated dominant eigenfunctions so that new process dynamics information is included in the dominant eigenfunctions. These updated dominant eigenfunctions improved the ROM which may be different from the previous ones when compared with the dominant eigenfunctions POD 1 used which are kept the same during the whole operation period. The temporal economic cost profiles of the process under these EMPC systems are shown in Fig. 13.5. From Fig. 13.5, over one period $t_f = 1$, the total reaction rate of the process under the EMPC system based on APOD is 1.18% greater than that of the EMPC system based on POD 1.

Further, we compare the EMPC calculation time for the EMPC systems of Eq. (13.7) based on POD 1 and APOD using Ensemble 1. As displayed in Fig. 13.6, the EMPC based on APOD achieves 38.8% improvement on the average computational time compared with that of the

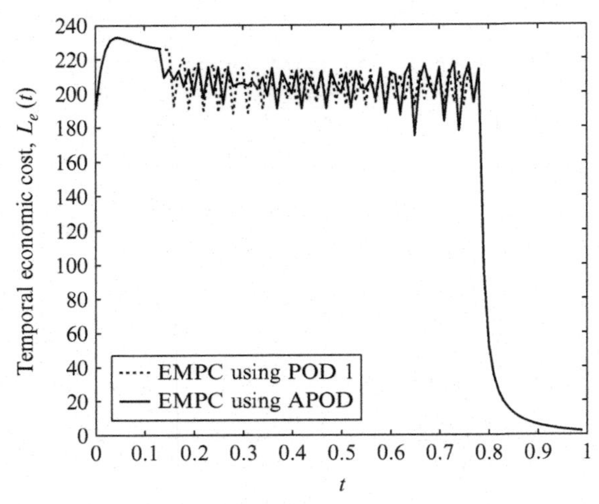

Fig. 13.5
Temporal economic cost along the length of the reactor, $L_e(t)$, under the EMPC system of Eq. (13.7) based on POD 1 (*dotted line*) and APOD using Ensemble 1 (*solid line*) over one operation period.

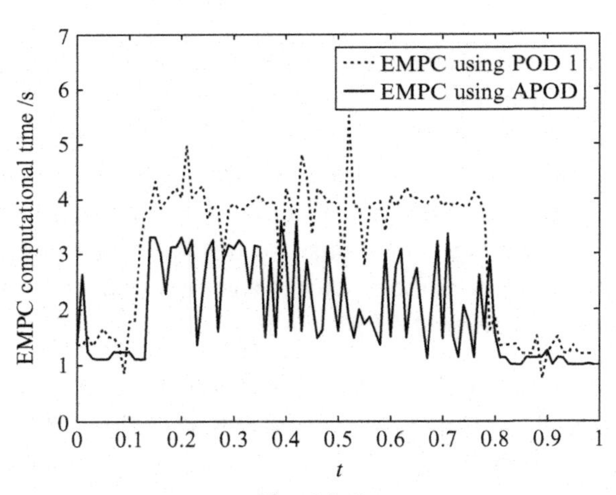

Fig. 13.6
EMPC computational time profiles for the EMPC system of Eq. (13.7) based on POD 1 (*dotted line*) and APOD using Ensemble 1 (*solid line*) over one operation period.

EMPC based on POD 1. As shown in Fig. 13.7, the APOD can adaptively adjust the required minimum number of eigenfunctions to satisfy the energy occupation requirement for each state while the number of the eigenfunctions utilized by the POD 1 is fixed at $m_1 = m_2 = 4$. ROM based on this number of eigenfunctions increases the computational burden to the EMPC optimization problem. Moreover, based on our simulation results, the recursive APOD procedure requires 45.2 s on the average for the case of EMPC based on APOD using Ensemble

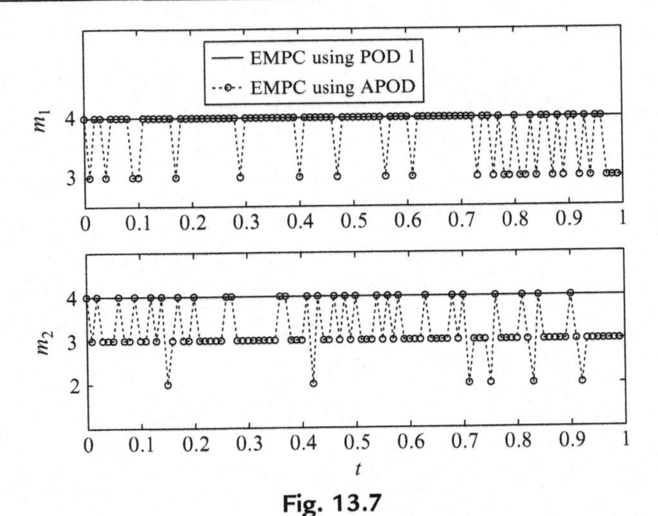

Fig. 13.7
Numbers of dominant eigenfunctions based on POD 1 (*solid line*) and APOD using Ensemble 2 (*dotted line with circles*) over one operation period.

1 which is computationally expensive. We note here that the APOD is completed before the EMPC problem is solved at $t = t_k$ $(k = 0,\ 1,\ \ldots)$ which follows the methodology we proposed in Fig. 13.1. Therefore, for the EMPC calculation time in Fig. 13.6, the time of completing the APOD procedure is not included.

Remark 4 The number of snapshots affects not only the computational burden but also the state constraint satisfaction. From simulation results which are not presented here due to space limitations, large ensemble size usually results in more dominant eigenfunctions and at the same time it increases the EMPC and recursive APOD computational time; but more eigenfunctions can improve the ROM accuracy and help the EMPC system to avoid state constraint violation. As shown by the maximum $x_1(z, t)$ value profile of POD 2, the ROM based on fewer eigenfunctions may underestimate the state evolution and mislead the EMPC system to provide a higher control input value which results in state constraint violation. Therefore, the choice of the number of snapshots (and the number of the dominant eigenfunctions) is a tradeoff between the computational efficiency and the reduced order model accuracy as we will further discuss later.

13.2.3.4 Case 2: APOD with an ensemble of small size

As we pointed in Case 1, although the EMPC system based on APOD using Ensemble 1 achieves high state approximation accuracy of ROM and the process economic performance owing to the fact that APOD continuously updates the dominant eigenfunctions, its APOD procedure and EMPC calculation is more computationally expensive when compared with that of the EMPC system based on a set of 101 ODEs for each PDE state. The computational efficiency difference is mainly caused by the number of the eigenfunctions adopted for constructing the ROM of PDE system. Based on this consideration, in this case, we reduce the size of the ensemble by adopting an ensemble of 125 snapshots denoted as Ensemble 2 and apply Ensemble 2 to the APOD procedure for EMPC system of Eq. (13.7). The required energy occupation is still the same ($\eta = 99.99\%$). Moreover, from the practical point of applying the APOD to the process, the APOD procedure is completed by using the full state profile at $t = t_{k-1}$ for the dominant eigenfunctions at $t = t_k$ as we show in Fig. 13.1 which means the APOD can be completed during the sampling time. In detail, as long as we update the APOD during the sampling time, using the state value at the previous sampling time, $x(t_{k-1})$, we can complete the APOD update and this computational time has no effect on the EMPC computational efficiency.

The closed-loop state profiles of the reactor over one period $t_f = 1.0$ under system of Eq. (13.7) based on APOD using Ensemble 2 are displayed in Figs. 13.12 and 13.13. The computed manipulated input profiles from the EMPC systems of Eq. (13.11) based on APOD using Ensemble 1 and APOD using Ensemble 2, respectively, over one period are compared in Fig. 13.8. From Fig. 13.8, the EMPC system based on APOD using Ensemble 2 computes a less smooth manipulated input profile than that of the EMPC system based on APOD using Ensemble 1 due to the fact that fewer snapshots are used to get the dominant eigenfunctions. The temporal economic cost profiles of the process under the EMPC system based on APOD using Ensemble 1 and APOD using Ensemble 2 are shown in Fig. 13.9. From Fig. 13.9, over one period $t_f = 1$, the total reaction rate of the process under the EMPC system based on APOD using Ensemble 2 is only 0.83% smaller than that of EMPC system based on APOD using Ensemble 1 and 1.74% smaller than that of the EMPC system based on the set of 101 ODEs for each PDE state.

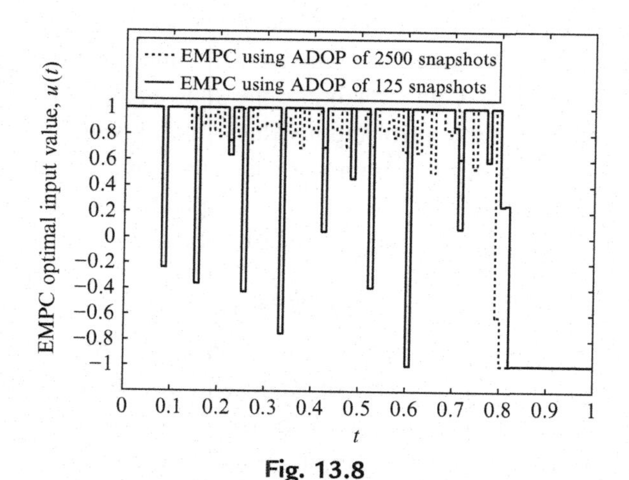

Fig. 13.8
Manipulated input profiles of the EMPC systems of Eq. (13.7) based on APOD using Ensemble 1 (*dotted line*) and on APOD (*solid line*) using Ensemble 2 over one operation period.

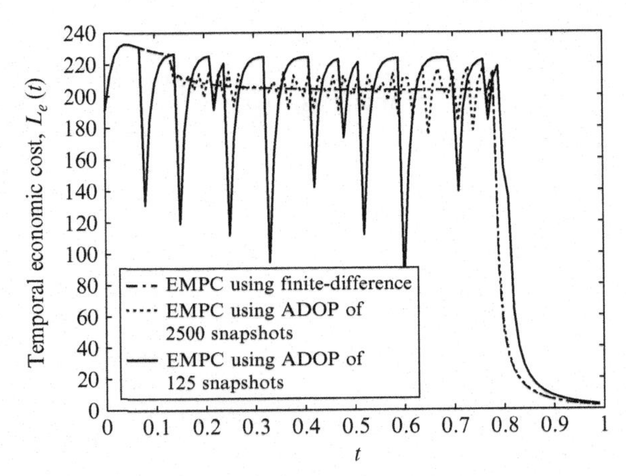

Fig. 13.9

Temporal economic cost along the length of the reactor, $L_e(t)$, under the EMPC systems of Eq. (13.7) based on APOD using Ensemble 1 (*dotted line*), on APOD using Ensemble 2 (*solid line*), and a set of 101 ODEs for each PDE state (*dashed line*) over one operation period.

We have compared the EMPC calculation time for the above EMPC system based on APOD using Ensemble 1, APOD using Ensemble 2, and a model of a set of 101 ODEs for each PDE state in Fig. 13.10. As displayed in Fig. 13.10, the EMPC calculation time for the EMPC system based on APOD using Ensemble 2 is less than compared with that of the EMPC system based on APOD using Ensemble 1. The computational time of EMPC system based on Ensemble 2 is 12.5% less than that of the EMPC system based on model of a set of 101 ODEs for each PDE state. This computational efficiency improvement of EMPC

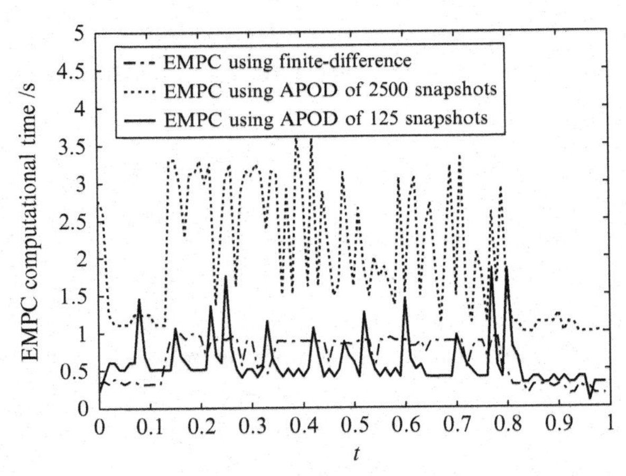

Fig. 13.10

EMPC calculation time profiles of the process under the EMPC systems of Eq. (13.7) based on APOD using Ensemble 1 (*dotted line*), on APOD using Ensemble 2 (*solid line*), and a set of 101 ODEs for each PDE state (*dash-dotted line*) over one operation period.

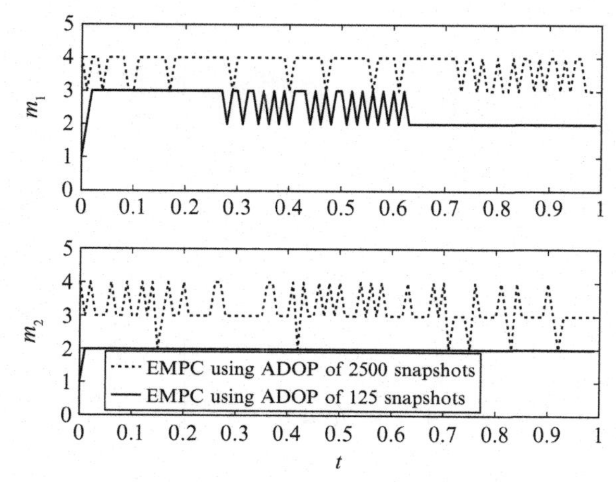

Fig. 13.11
Number of dominant eigenfunctions based on APOD using Ensemble 1 (*dotted line*) and on APOD using Ensemble 2 (*solid line*) over one operation period.

system based on APOD using Ensemble 2 comes from the fact that fewer number of dominant eigenfunctions are adopted for constructing the ROM as compared in Fig. 13.11. We set a 99.99% energy occupation requirement for the APOD using Ensemble 1 and the corresponding number of eigenfunctions for each state is kept at $m_i = 4$, $i = 1, 2$ over one operation period; while for the EMPC based on the APOD using Ensemble 2, the corresponding number of eigenfunctions for each state adaptively changes as different process dynamics are collected and integrated into the dominant eigenfunctions. Moreover, the APOD using Ensemble 2 which has a much smaller size of ensemble also decreases the computational time of the APOD update procedure to 0.24 s (the results are shown in Figs. 13.12 and 13.13).

Since Ensemble 2 with a size of 125 snapshots only reflects part of process dynamics, it may not contain enough process dynamic behavior to guarantee the accuracy of the ROM of the PDE system. Especially, when there exists a specific state constraint, the ROM may not be a good approximation of the original PDE system to help the EMPC avoid the state constraint violation due to its poor or incomplete state representation. In other words, the ROM either from POD or APOD may overestimate or underestimate the state value of Eq. (13.7b) in the EMPC optimization problem of Eq. (13.7). For the state constraint in this case, when the ROM underestimates the state value of $\hat{x}_1(t)$, it may mislead the EMPC to compute and implement a higher optimal input value to the actual process which may result in the state constraint violation on $x_1(t)$ due to the second-order exothermic reaction rate. Here, we constructed another ensemble of 125 snapshots from different process solutions which is noted as Ensemble 3. We constructed EMPC systems using both of these 2 ensembles and applied them to the process. The maximum temperature (dimensionless) profiles of the

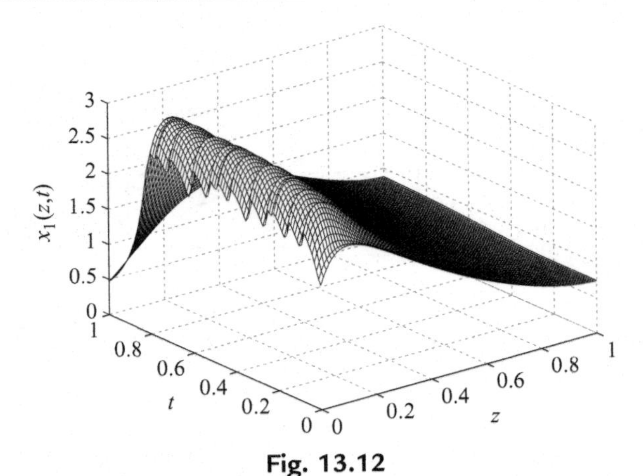

Fig. 13.12
Closed-loop profile of x_1 of the process under the EMPC system of Eq. (13.7) based on APOD using Ensemble 2 over one operation period.

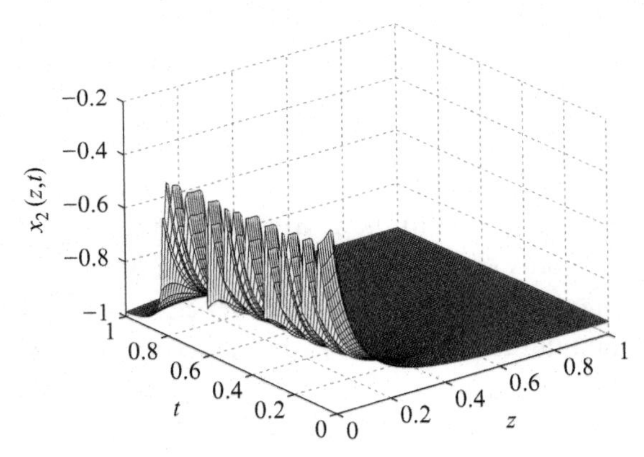

Fig. 13.13
Closed-loop profile of x_2 of the process under the EMPC system of Eq. (13.7) based on APOD using Ensemble 2 over one operation period.

tubular reactor under the EMPC systems of Eq. (13.7) based on APOD using Ensemble 2 and APOD using Ensemble 3 are shown in Fig. 13.14. From Fig. 13.14, the EMPC system based on APOD using Ensemble 2 operates the process around the maximum allowable temperature but at some points, it is close to the state constraint. While, from the magnified plot of Fig. 13.14, the EMPC system based on APOD using Ensemble 3 violates the state constraint around $t = 0.51$.

Remark 5 The number of snapshots affects not only the computational burden but also the state constraint satisfaction. A large size of ensemble usually results in more dominant eigenfunctions and at the same time, it increases the EMPC and APOD computational time; but more eigenfunctions can improve the ROM accuracy and help the EMPC system to avoid the state constraint violation. Therefore, the choice of the number of snapshots

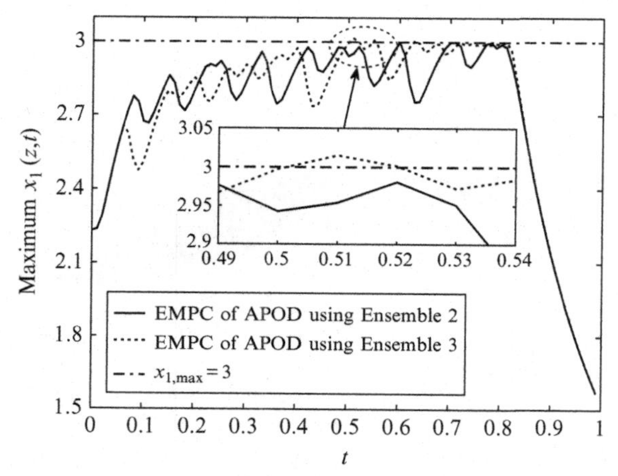

Fig. 13.14
Maximum x_1 profiles of the process under the EMPC systems of Eq. (13.7) based on APOD using Ensemble 2 (*solid line*) and on APOD using Ensemble 3 (*dotted line*) over one operation period.

(i.e., the number of the dominant eigenfunctions) is a tradeoff between the computational efficiency and the reduced order model accuracy.

1. For POD method, the ensemble must have enough snapshots which contain as much global process dynamics as possible to help the EMPC system predict the state value more accurately. Since POD is only conducted once, it has no effect on the EMPC computational burden which only depends on how many modes/energy occupation is required.
2. For APOD method, the number of snapshots depends on the model accuracy although the ROM can be updated during the closed-loop operation. More snapshots will increase the APOD computational burden. But it will help the system avoid the state constraint violation. As long as the APOD update time is less than the sampling time size, we can use as many snapshots as possible, but large number of snapshots usually decreases the computational efficiency of the EMPC system.

13.2.3.5 Case 3: Proposed flow chart of integrating APOD with finite difference method

Based on the results and analysis in Case 1, we adopt the flow chart of Fig. 13.1 integrating APOD method and finite-difference method to avoid state constraint violation and improve computational efficiency. A 101st-order ODE model for each PDE state as the result of applying central finite difference method to each PDE state is integrated into the EMPC scheme. An ensemble of 150 snapshots which is noted as Ensemble 2 is initially adopted for the EMPC system based on APOD method. We still request that the dominant eigenfunctions occupy $\epsilon = 99.99\%$ of the total energy of the ensemble. The EMPC system of Eq. (13.7) based on the finite-difference method resulting in a 101st-order ODE model for each PDE state is taken as the comparison objective for the proposed EMPC formulation. Same prediction horizon, sampling time, and integration step are adopted as the previous case. We assume the state violation alert region, Ω, is defined as:

$$\Omega := \{x \in \mathbb{R} \mid |\max(x_1(z, t_k)) - x_{1,\max}| \leq 0.05\} \quad (13.14)$$

The computed manipulated input profiles over one period $t_f = 1.0$ from the EMPC system of Fig. 13.1 and the EMPC system of Eq. (13.7) based on the finite-difference method over one period are shown in Fig. 13.15. From Fig. 13.15, the EMPC system of Eq. (13.7) based on the finite-difference method computes a smoother manipulated input profile than that of the EMPC of Fig. 13.1. The temporal economic cost profiles of the process under the EMPC of Fig. 13.1 and the EMPC system of Eq. (13.7) based on the finite-difference method are shown in Fig. 13.16. From Fig. 13.16, over one period $t_f = 1$, the total reaction rate of the process under the EMPC of Fig. 13.1 is only 0.33% smaller than that of the EMPC system of Eq. (13.7) based on the finite difference model.

With respect to the performance of the EMPC optimal input value, we compared the maximum temperature (dimensionless) profiles of the tubular reactor under the EMPC systems as shown in Fig. 13.17. From Fig. 13.17, we see that the EMPC system of Fig. 13.1 can operate the process at the maximum allowable temperature and meanwhile avoid the state constraint violation issues by adopting the integrated EMPC system based on the finite-difference method when the process state value enters into the alert region of Eq. (13.14).

We finally compare the calculation time of the EMPC system of Fig. 13.1 and the EMPC system of Eq. (13.7) based on the finite-difference method in Fig. 13.18. As displayed in Fig. 13.18, the EMPC of Fig. 13.1 achieves 8.71% improvement compared with the EMPC

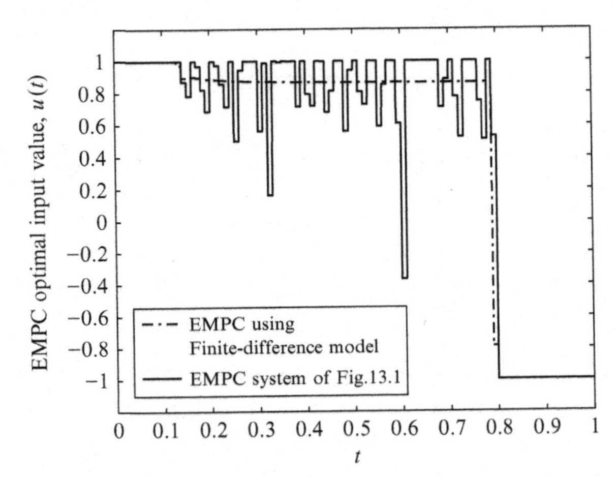

Fig. 13.15
Manipulated input profiles of the EMPC system of Fig. 13.1 (*solid line*) and the EMPC system of Eq. (13.7) based on the finite-difference method (*dash-dotted line*) over one operation period.

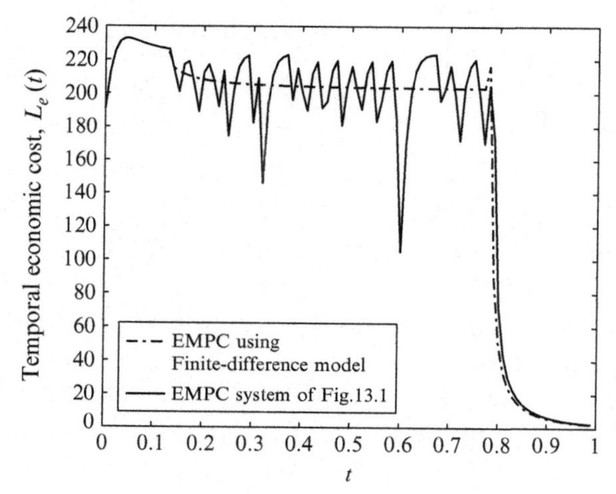

Fig. 13.16

Temporal economic cost along the length of the reactor, $L_e(t)$, of the EMPC system of Fig. 13.1 (*solid line*) and the EMPC system of Eq. (13.7) based on the finite-difference method (*dash-dotted line*) over one operation period.

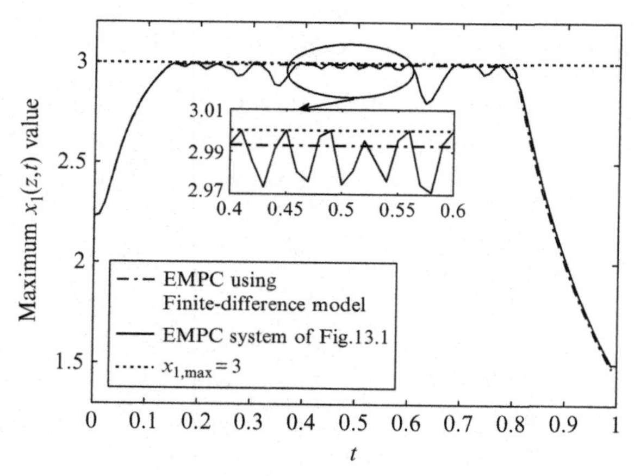

Fig. 13.17

Maximum $x_1(z, t)$ profiles under the EMPC system of Fig. 13.1 (*solid line*) and the EMPC system of Eq. (13.7) based on the finite-difference method (*dash-dotted line*) over one operation period.

system of Eq. (13.7) based on the finite-difference method. In terms of Fig. 13.18, we point out that when the state value enters in the violation alert region, both the EMPC based on APOD method (to get a trial optimal input value) and the EMPC system of Eq. (13.7) based on the finite-difference method (to get an accurate optimal input value to help the EMPC scheme to avoid the state constraint violation if the previous optimal input value leads to constraint violation) are conducted which results in a longer computational time.

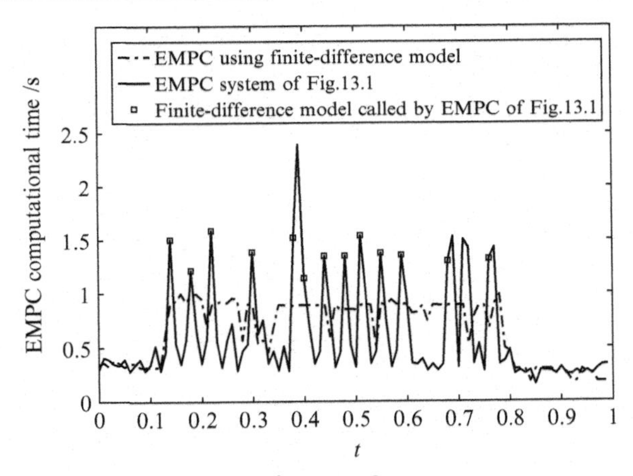

Fig. 13.18
EMPC computational time profiles for the EMPC system of Fig. 13.1 (*solid line*) and the EMPC system of Eq. (13.7) based on the finite-difference method (*dash-dotted line*) over one operation period.

Based on the above results and analysis, the proposed EMPC scheme of Fig. 13.1 successfully improves the whole computational efficiency and at the same time guarantees avoiding the state constraint violation.

13.2.3.6 Case 4: EMPC with manipulated input fluctuation control

As the results in Case 3 demonstrated, the EMPC system of Fig. 13.1 integrating APOD method and finite-difference method resulted in large fluctuations of the optimal input profile. To deal with this issue, we conduct the EMPC system of Fig. 13.1 with different prediction horizon values, N, to see how prediction horizon affects the smoothness of the manipulated input profile. Ensemble 4 is still used initially for the EMPC system based on APOD method and all the other settings are exactly the same as those in Case 3. The EMPC system of Eq. (13.7) based on the finite-difference method resulting in a set of 101 ODEs for each PDE state is taken as the comparison for the proposed EMPC formulation. We define the input fluctuation index, $||du||_2^2$, as the difference between the input profile from EMPC system of Fig. 13.1 and the input profile from the EMPC system of Eq. (13.7) based on the finite-difference method with the same prediction horizon which is formulated as follows (for periodic operation with $t_f = 1$ and $\Delta = 0.01$, the EMPC is evaluated for $M = 100$):

$$||du||_2^2 = \sum_{k=1}^{M=100} [u^*(t_k) - u_{FD}(t_k)]^2 \tag{13.15}$$

We applied prediction horizon $N = 2, \ldots, 10$ to both of the EMPC formulations. The corresponding input fluctuation index value of $||du||_2^2$ and the average economic cost of $L_e(t)$ during this one operation period is demonstrated in Fig. 13.19. From Fig. 13.19, we find that in most situations, longer prediction horizon can reduce the input profile

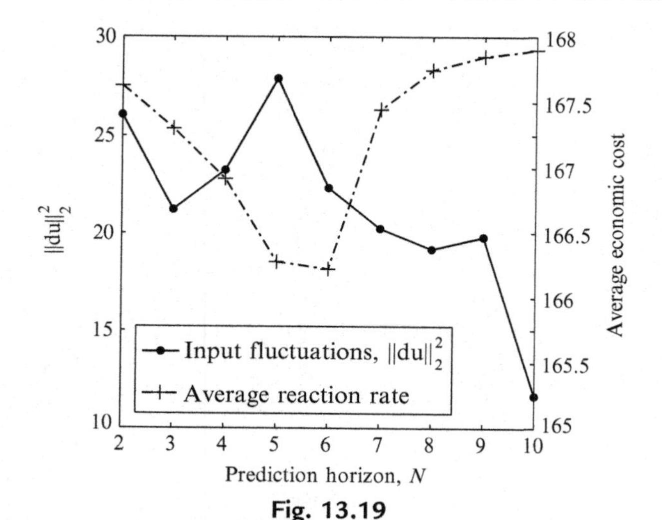

Fig. 13.19

Manipulated input value difference between the EMPC system of Fig. 13.1 (*solid line*) and the EMPC system of Eq. (13.7) based on the finite-difference method (*dash-dotted line*) and average economic cost value of the EMPC system of Fig. 13.1 with different prediction horizon lengths over one operation period.

fluctuations especially when $N \geq 7$. For example, Fig. 13.20 displays the manipulated input profiles of the EMPC system of Fig. 13.1 and the EMPC system of Eq. (13.7) based on the finite-difference method with the same prediction horizon, $N = 10$. Compared with the input profiles in Fig. 13.15, the smoothness of the input profile of the EMPC system of Fig. 13.1 in Fig. 13.20 improves significantly (no negative input values). As shown by the profile of the average economic cost in Fig. 13.19, the process economic performance can also

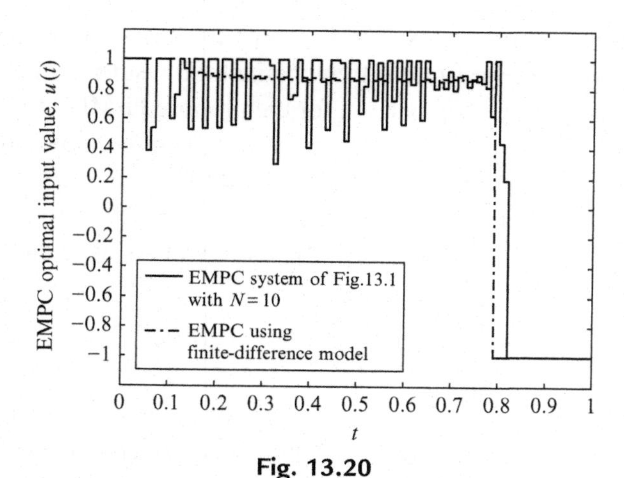

Fig. 13.20

Manipulated input profiles of the EMPC system of Fig. 13.1 (*solid line*) and the EMPC system of Eq. (13.7) based on the finite-difference method (*dash-dotted line*) over one operation period with prediction horizon, $N = 10$.

be improved with the increase of the smoothness of the input trajectories, i.e., larger input fluctuation index usually results in lower economic performance.

We also added an input fluctuation penalty cost, $P_u(t)$, in the economic cost objective function of Eq. (13.7) as another approach to reduce the input fluctuation. The revised economic cost function in the EMPC formulation of Eq. (13.7) and the input fluctuation penalty cost, $P_u(t)$, are demonstrated by the following formulations:

$$\int_{t_k}^{t_{k+N}} L_e(\tau)\ d\tau - P_u(t_k) \quad \text{with} \quad P_u(t_k) = A_{P_u}[u^*(t_k) - u^*(t_{k-1})]^2 \tag{13.16}$$

where $P_u(t_k)$ penalizes the input fluctuation between the optimal input value $u^*(t_k)$ at the current sampling time instant, t_k, and the calculated input value $u^*(t_{k-1})$ at the previous sampling time instant, $t = t_{k-1}$ with a weighting constant, $A_{P_u} = 10$.

To realize the EMPC with the input fluctuation penalty, Ensemble 4 is still used initially for the EMPC system based on APOD method and all the other settings are the exactly same as those in Case 3. The computed manipulated input profiles over one period $t_f = 1.0$ from the EMPC system of Fig. 13.1 with the new economic cost function of Eq. (13.16) and the EMPC system of Eq. (13.7) based on the finite-difference method over one period are shown in Fig. 13.21. From Fig. 13.21, the smoothness of the input profile of the EMPC system of Fig. 13.1 improves significantly when compared with the input profile in Case 3. Increasing the value of the weighting constant, A_{P_u}, can further reduce the input fluctuation as the dominance of the input fluctuation penalty cost, $P_u(t)$, increases in the new economic cost function of Eq. (13.16). The temporal economic cost profiles of the process under the EMPC of Fig. 13.1 and the EMPC system of Eq. (13.7) based on the finite-difference method are

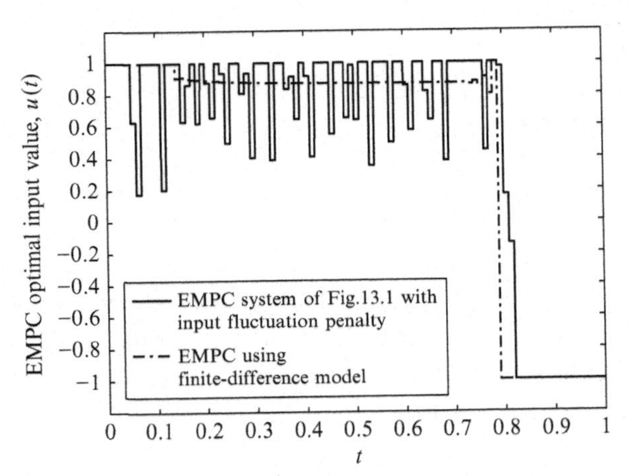

Fig. 13.21
Manipulated input profiles of the EMPC system of Fig. 13.1 considering input fluctuation penalty cost, $P_u(t)$, (*solid line*) and the EMPC system of Eq. (13.7) based on the finite-difference method (*dash-dotted line*) over one operation period.

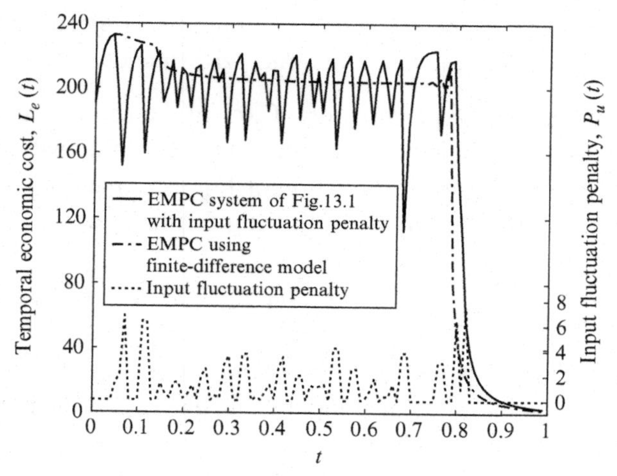

Fig. 13.22
Temporal economic cost of the EMPC system of Fig. 13.1 (*solid line*) and the EMPC system of Eq. (13.7) based on the finite-difference method (*dash-dotted line*) and input fluctuation penalty cost, $P_u(t)$ (*dotted line*), over one operation period.

shown in Fig. 13.22. From Fig. 13.22, over one period $t_f = 1$, the total reaction rate of the process under the EMPC of Fig. 13.1 is 1.37% smaller than that of the EMPC system of Eq. (13.7) based on the finite-difference model. The input fluctuation penalty cost, $P_u(t)$, is also shown in Fig. 13.22, which increases the degradation on the economic cost of the EMPC system of Fig. 13.1 since the EMPC system sacrificed economic performance to reduce the input fluctuations.

13.3 EMPC of Hyperbolic PDE Systems With State and Control Constraints

13.3.1 Reactor Description

In this section, we consider EMPC of hyperbolic PDE systems and focus on a nonisothermal plug flow reactor (PFR) where an irreversible and exothermic second-order reaction of the form $A \rightarrow B$ takes place. The process model in dimensionless variable form consists of two quasilinear hyperbolic PDEs. The process description details and model notation can be found in Ref. 35 (i.e., the PFR model is similar to the tubular reactor model presented in Ref. 35 except for the fact that the diffusion term used in the tubular reactor model of Ref. 35 is neglected in the PFR model below):

$$\begin{aligned}\frac{\partial x_1}{\partial t} &= -\frac{\partial x_1}{\partial z} + B_T B_C \exp\left(\frac{\gamma x_1}{1+x_1}\right)(1+x_2)^2 + \beta_T (T_j - x_1)\\ \frac{\partial x_2}{\partial t} &= -\frac{\partial x_2}{\partial z} - B_C \exp\left(\frac{\gamma x_1}{1+x_1}\right)(1+x_2)^2\end{aligned} \tag{13.17}$$

subject to the following boundary conditions:

$$x_1(0,t) = T_i, \quad x_2(0,t) = u(t) \tag{13.18}$$

where $x_1(z, t)$ denotes a dimensionless temperature, $x_2(z, t)$ denotes a dimensionless reactant concentration, and z and t are the dimensionless spatial coordinate and time variables, respectively. The following typical values are given to the process parameters: $B_T = 2.5$, $B_C = 0.1$, $\beta_T = 2$, $T_i = 0.5$, $T_j = -0.5$, and $\gamma = 10$. Upwind finite-difference scheme is adopted (which can guarantee the stability of the numerical spatial discretization of the process model[40]) to discretize, in space, the two hyperbolic PDEs and obtain a 101st-order system of ODEs in time for each PDE (i.e., a total of 202 ODEs in time). Further increase of the number of discretization points led to identical results. Based on this discretized model, open-loop simulation results for a constant input of $u(t) \equiv 0.0$ are shown in Figs. 13.23 and 13.24. The state profile is initialized at the steady-state profile corresponding to the steady-state input $u_s = 0.5$.

The process model of Eq. (13.17) can be formulated as follows:

$$\begin{aligned} \frac{\partial x(z,t)}{\partial t} &= A\frac{\partial x(z,t)}{\partial z} + f(x(z,t)) \\ y_j(t) = \mathcal{C}x(z,t) &= \int_0^1 c_j(z)x(z,t)\,dz, \quad j = 1,\ldots,p \end{aligned} \tag{13.19}$$

subject to the boundary conditions:

$$x(0,t) = [T_i \;\; u(t)]' \tag{13.20}$$

and the initial condition:

$$x(z,0) = x_0(z) \tag{13.21}$$

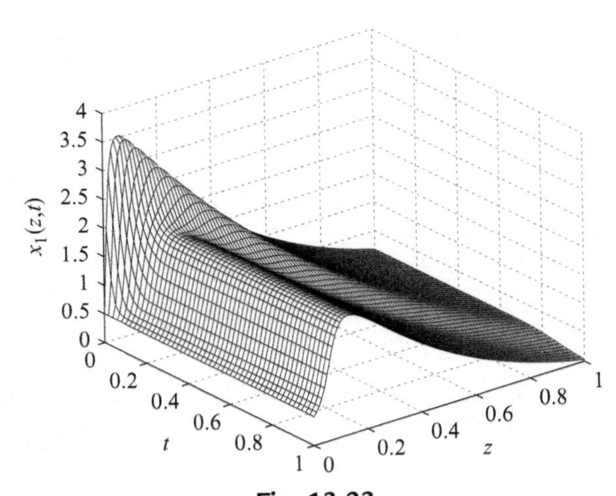

Fig. 13.23
Open-loop profile of x_1 of the process model of Eq. (13.17).

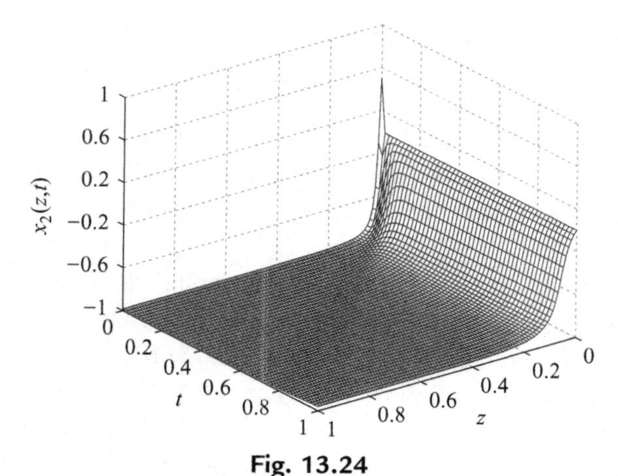

Fig. 13.24
Open-loop profile of x_2 of the process model of Eq. (13.17).

where $x(z, t) = [x_1(z, t)\, x_2(z, t)]'$ denotes the vector of the process state variables, the notation x' is the transpose of x, $z \in [0, 1]$ is the spatial coordinate, $t \in [0, \infty)$ is the time, $f(x(z, t))$ denotes a nonlinear vector function, $y_j(t)$ is the jth measured output ($y(t) = [y_1(t) \cdots y_p(t)]'$ is the measured output vector), $c_j(z)$ are known smooth functions of z ($j = 1, \ldots, p$) whose functional form depends on the type of the measurement sensors, $\mathcal{C}$ is the measured output operator, A is a constant matrix of appropriate dimensions, and $u(t)$ denotes the manipulated input.

13.3.2 EMPC System Constraints and Objective

For the PFR of Eq. (13.17), the input $u(t)$ is chosen as the reactant concentration (dimensionless) of the inlet stream at $z = 0$ (Eq. (13.18). We assume the manipulated input is subject to constraints as follows:

$$u_{\min} \leq u(t) \leq u_{\max} \tag{13.22}$$

where $u_{\min} = -1$ and $u_{\max} = 1$ are the lower and upper bound of the manipulated input, $u(t)$. We assume that there is also a limitation on the amount of reactant material (dimensionless) available over each operating period of length t_p. Specifically, the control input of $u(t)$ should satisfy:

$$\frac{1}{t_p}\int_0^{t_p} u(\tau)\, d\tau = 0.5 \tag{13.23}$$

which will be referred to as the reactant material integral constraint. To ensure that the constraint of Eq. (13.23) is satisfied over the operating period of length t_p when the prediction horizon of the EMPC does not cover the entire operating period, it is implemented according to the strategy described in Ref. 29. In the EMPC formulations below, the constraint will be denoted as $u \in g(t_k)$ to simply the notation.

In terms of the state constraint, we consider that the reactor temperature (dimensionless) is subject to:

$$x_{1,\min} \leq x_1(z,t) \leq x_{1,\max} \tag{13.24}$$

for all $z \in [0, 1]$ and $t \geq 0$ where $x_{1,\min} = -1$ and $x_{1,\max} = 4.5$ are the lower and upper limits, respectively.

To solve the EMPC problem, the open-source interior point solver Ipopt[39] was used. Explicit Euler's method was used for temporal integration of the 202nd spatial discretization of the PDE model of Eq. (13.17) with an integration step of 1×10^{-3} to numerically integrate ODE model in EMPC. The dimensionless sampling period of EMPC is $\Delta = 0.01$.

To develop EMPC formulations in the subsequent sections, we use a Lyapunov function of form

$$V(x) = \int_0^1 [x_1^2(z,t) + x_2^2(z,t)]\, dz < \overline{\rho} \tag{13.25}$$

where $x_1(z, t)$ and $x_2(z, t)$ are the spatial state values respectively and $\overline{\rho} = 15$ (see Ref. 33 for results on LEMPC designs).

The cost function that we consider is to maximize the overall reaction rate along the length of the reactor and over one operation period of length $t_p = 1$. The economic cost then takes the form:

$$L_e(t) = \int_0^1 r(z,t)\, dz \tag{13.26}$$

where $r(z,t) = B_C \exp\left(\frac{\gamma x_1(z,t)}{1 + x_1(z,t)}\right)(1 + x_2(z,t))^2$ is the reaction rate (dimensionless) in the PFR.

13.3.3 State Feedback EMPC of Hyperbolic PDE Systems

13.3.3.1 State feedback EMPC formulation

We consider the application of a state feedback EMPC to the system of Eq. (13.17) to optimize an economic cost function and handle input and state constraints. Specifically, a LEMPC system is designed using the results in Ref. 33 with the economic cost function of Eq. (13.26), the input constraint of Eq. (13.22), the state constraint of Eq. (13.24), and the reactant material integral constraint of Eq. (13.23) for the PFR. The state feedback EMPC receives the (full) state profile of $x(z, t_k)$ of the system of PDEs of Eq. (13.17) synchronously at sampling instants denoted as $t_k = k\Delta$ with $k = 0, 1, \ldots$. The state feedback EMPC control action is computed by solving the following finite-dimensional optimization problem in a receding horizon fashion:

$$\max_{u \in S(\Delta)} \int_{t_k}^{t_{k+N}} L_e(\tau)\, d\tau \tag{13.27a}$$

$$\text{s.t.} \quad \frac{\partial \tilde{x}(z,t)}{\partial t} = A \frac{\partial \tilde{x}(z,t)}{\partial z} + f(\tilde{x}(z,t)) \tag{13.27b}$$

$$\tilde{x}(z,t_k) = x(z,t_k), \forall z \in [0,1] \tag{13.27c}$$

$$\tilde{x}_1(0,t) = T_i, \ \forall t \in [t_k, t_{k+N}) \tag{13.27d}$$

$$\tilde{x}_2(0,t) = u(t), \ \forall t \in [t_k, t_{k+N}) \tag{13.27e}$$

$$u_{\min} \le u(t) \le u_{\max} \tag{13.27f}$$

$$u(t) \in g(t_k) \tag{13.27g}$$

$$x_{1,\min} \le x_1(z,t) \le x_{1,\max}, \ \forall \ z \in [0,1], \text{ and } \forall t \in [t_k, t_{k+N}) \tag{13.27h}$$

$$V(\tilde{x}) \le \bar{\rho} \tag{13.27i}$$

where $L_e(\tau)$ is the economic cost, Δ is the sampling period, $S(\Delta)$ is the family of piecewise constant functions with sampling period Δ, N is the prediction horizon, and $\tilde{x}(z,t)$ is the predicted state function evolution with input $u(t)$ computed by the state feedback EMPC.

In the optimization problem of Eq. (13.27), the cost function of Eq. (13.27a) accounts for the economics of the system of PDEs (i.e., maximization of the production of product species B). The constraint of Eq. (13.27b) is used to predict the future evolution of the system of PDEs with the initial condition given in Eq. (13.27c) and the boundary conditions given in Eq. (13.27d),(13.27e). The constraints of Eq. (13.27f)–(13.27h) are the available control action, the integral input constraint, and the state constraints, respectively. The constraint of Eq. (13.27i) ensures that the predicted state trajectory is restricted inside a predefined stability region which is a level set of the Lyapunov function (see Ref. 33 for a complete discussion of this issue). The optimal solution to this optimization problem is $u^*(t|t_k)$ defined for $t \in [t_k, t_{k+N})$. The state feedback EMPC applies the control action computed for the first sampling period to the system in a sample-and-hold fashion for $t \in [t_k, t_{k+1})$. The state feedback EMPC is resolved at the next sampling period, t_{k+1}, after receiving a measurement of the state profile, $x(z, t_{k+1})$ for $z \in [0, 1]$.

13.3.3.2 Application of state feedback EMPC system to the tubular reactor

We operate the reactor under the state feedback EMPC system of Eq. (13.27). In detail, we choose the prediction horizon, $N = 3$. The PFR is initialized with a steady-state profile (i.e., the steady-state profile corresponding to the steady-state input $u_s = 0.5$).

The closed-loop state profiles of the reactor over one period $t_p = 1.0$ under the state feedback EMPC of Eq. (13.27) is displayed in Figs. 13.25 and 13.26. The computed manipulated input profile from the state feedback EMPC system over one period is shown in Fig. 13.27.

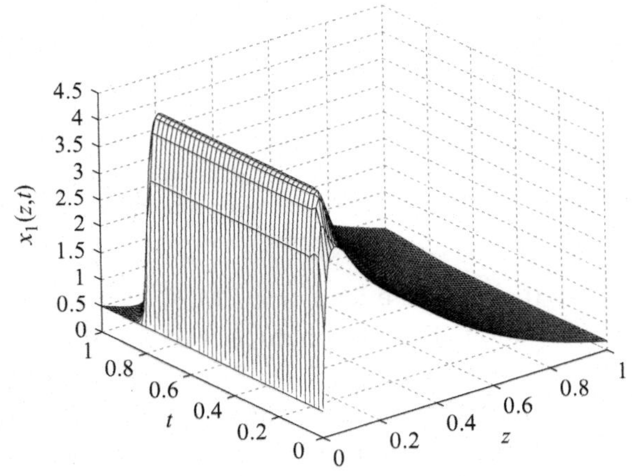

Fig. 13.25
Closed-loop profile of x_1 of the process under the state feedback EMPC system of Eq. (13.27) over one operation period.

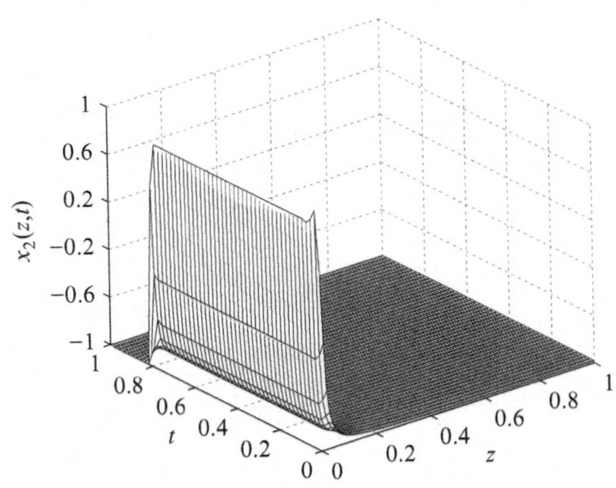

Fig. 13.26
Closed-loop profile of x_2 of the process under the state feedback EMPC system of Eq. (13.27) over one operation period.

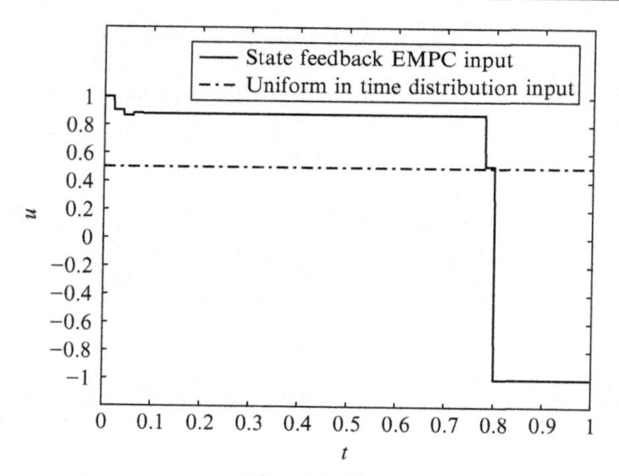

Fig. 13.27
Manipulated input profiles of the state feedback EMPC system of Eq. (13.27) and uniform in time distribution of the reactant material over one operation period.

From Fig. 13.27, we observe that the state feedback EMPC system varies the optimal input value $u(t)$ in a time-varying fashion. The maximum x_1 value profile of the PFR under the state feedback EMPC system is shown in Fig. 13.28. Since the temperature directly influences the reaction rate (i.e., the reaction is second-order and exothermic), the optimal operating strategy is to first force the reactor to operate at the maximum allowable value and then, decrease the input to its minimum value to satisfy the integral input constraint. During this one operation period, $t_p = 1$, the economic cost profile of the reactor under the state feedback

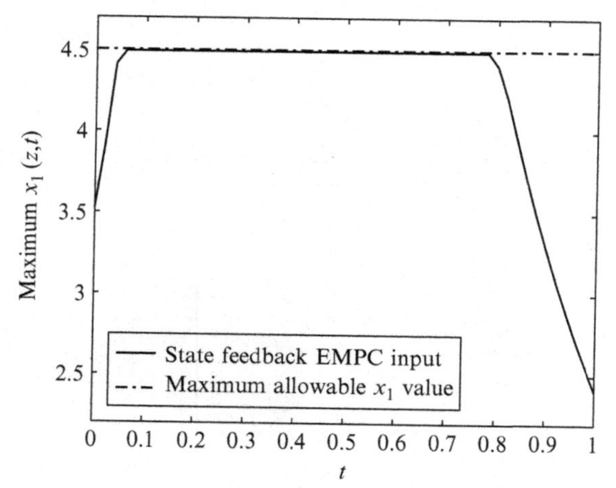

Fig. 13.28
Maximum x_1 profiles of the process under the state feedback EMPC system of Eq. (13.27) over one operation period.

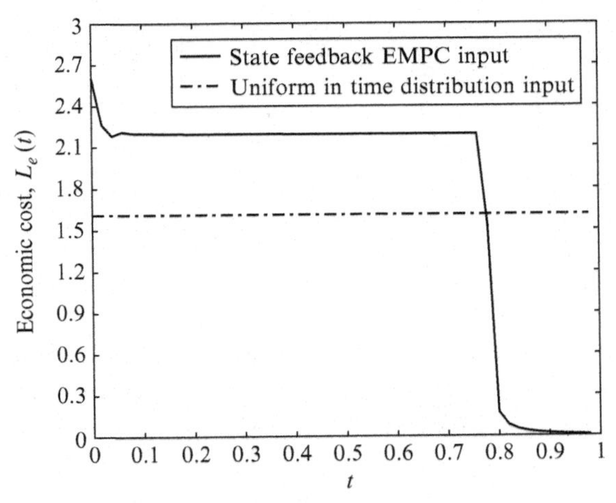

Fig. 13.29
Economic cost, $L_e(t)$, along the length of the reactor under the state feedback EMPC system of Eq. (13.27) and under uniform in time distribution of the reactant material over one operation period.

EMPC system is demonstrated in Fig. 13.29. The average economic cost over one operating period $t_p = 1$, which is denoted as $\overline{L}_e(t)$, is given by

$$\overline{L}_e(t) = \int_0^1 L_e(t)\, dt. \tag{13.28}$$

Comparing the process under the state feedback EMPC system to the process under a constant input, the average economic cost is 15.14% greater than that of the reactor under uniform in time distribution of the reactant material. The state feedback EMPC system achieves a significant advantage in maximizing the process economic performance over one operation period through operating the process in a time-varying fashion.

13.3.4 Output Feedback EMPC of Hyperbolic PDE Systems

13.3.4.1 State estimation using output feedback methodology

In this section, we consider a state estimation technique that makes use of a finite number, p, of measured outputs $y_j(t)$ ($j = 1, \ldots, p$) to estimate the state vector of the system, $x(z, t)$ in space and time.

This state estimation scheme is designed based on the distributed state observer design approach presented in Ref. 41. The nonlinear state observer is designed to guarantee local exponential convergence of the state estimates to the actual state values. In particular, based on Eq. (13.19), the following state observer will be utilized:

$$\frac{\partial \hat{x}(z,t)}{\partial t} = A\frac{\partial \hat{x}(z,t)}{\partial z} + f(\hat{x}(z,t)) + \mathcal{K}(y(t) - \mathcal{C}\hat{x}(z,t)) \tag{13.29}$$

subject to the boundary conditions:

$$\hat{x}(0,t) = [T_i \;\; u(t)]' \tag{13.30}$$

and the initial condition:

$$\hat{x}(z,0) = \overline{x}_0(z) \tag{13.31}$$

where $\hat{x}$ is the observer state vector, $\overline{x}_0(z)$ is the initial condition of the observer state which is sufficiently smooth with respect to z, $y(t)$ is the measured output vector which is assumed to be continuously available and $\mathcal{K}$ is a linear operator representing the observer gain which is designed on the basis of the linearization of the system of Eq. (13.29) so that the eigenvalues of the operator $\mathcal{L}_o = \mathcal{L} - \mathcal{K}\mathcal{C}$ lie in the left-half plane. By satisfying this condition, $\mathcal{L}_o$ can generate an exponentially stable semigroup for $\hat{x}(t)$. Specifically, the operator $\mathcal{L}$ is defined on the basis of the linearized form of the Eq. (13.19) at some specific steady-state profile, denoted as $x_s(z)$, i.e.,:

$$\mathcal{L}x = A\frac{\partial x}{\partial z} + B(z)x \tag{13.32}$$

where

$$B(z) := \left(\frac{\partial f(x)}{\partial x}\right)_{x=x_s(z)} \tag{13.33}$$

If the operator $\mathcal{K}$ is chosen by satisfying the aforementioned condition that the eigenvalues of the operator $\mathcal{L}_o = \mathcal{L} - \mathcal{K}\mathcal{C}$ lie in the left-half of the complex plane, the term $\mathcal{K}(y(t) - \mathcal{C}\hat{x})$ can enforce a fast decay of the discrepancy between the estimated state values and the actual state values of the system of PDEs. In practice, the design of the operator $\mathcal{K}$ depends on whether the output measurements are corrupted by noise or not. Pole placement is often adopted when the output measurements are free of measurement noise, while Kalman filtering theory is often adopted when the output measurements are noisy. For the present work, we assume there is no measurement noise on the output measurements. Therefore, pole placement is utilized to design an appropriate state observer gain for output feedback correction to diminish the state estimation error as fast as possible.

13.3.4.2 Implementation of state estimation

We now combine the state observer with the state feedback EMPC to derive an output feedback EMPC formulation. For the case of PFR of Eq. (13.17), we assume that only p evenly spaced measurement points are available of the state x_1 (temperature); we use these p measurements to estimate the state values of both $x_1(z, t)$ and $x_2(z, t)$ in the whole space and time domain by designing a state observer of the form of Eq. (13.29). Here, each spatial measurement point, z_j, is at $z_j = (j-1)/(p-1)$. Since point-wise measurements are assumed, the measurement distribution functions are:

$$c_j(z) = \delta(z - z_j) \tag{13.34}$$

for $j = 1, \ldots, p$ where δ is the standard Dirac function. Consequently, each output measurement point value, $y_j(t_k)$, is equal to the PDE state value, x_1 at the corresponding spatial point, z_j, i.e.,:

$$y_j(t) = x_1(z_j, t) \tag{13.35}$$

and the output measurement vector is $y'(t) = [y_1(t) \cdots y_p(t)]$. The operator $\mathcal{K} \in \mathbb{R}^{2\times p}$ is chosen to be:

$$\begin{bmatrix} 1/p & \ldots & 1/p \\ 1/p & \ldots & 1/p \end{bmatrix} \tag{13.36}$$

so that the eigenvalues of the operator $\mathcal{L}_0$ will lie in the left-half of the complex plane.

13.3.4.3 Application of output feedback EMPC system to the tubular reactor

We consider the application of output feedback EMPC to the system of Eq. (13.19) of the form:

$$\max_{u \in S(\Delta)} \int_{t_k}^{t_{k+N}} L_e(\tau)\, d\tau \tag{13.37a}$$

$$\text{s.t.} \quad \frac{\partial \tilde{x}(z,t)}{\partial t} = \frac{A \partial \tilde{x}(z,t)}{\partial z} + f(\tilde{x}(z,t)) \tag{13.37b}$$

$$\tilde{x}(z, t_k) = \hat{x}(z, t_k) \tag{13.37c}$$

$$\tilde{x}_1(0,t) = T_i, \ \forall t \in [t_k, t_{k+N}) \tag{13.37d}$$

$$\tilde{x}_2(0,t) = u(t), \ \forall t \in [t_k, t_{k+N}) \tag{13.37e}$$

$$u_{\min} \le u(t) \le u_{\max} \tag{13.37f}$$

$$u(t) \in g(t_k) \tag{13.37g}$$

$$x_{1,\min} \le x_1(z,t) \le x_{1,\max}, \ \forall \ z \in [0,1], \text{ and } \forall t \in [t_k, t_{k+N}) \tag{13.37h}$$

$$V(\tilde{x}) \le \overline{\rho} \tag{13.37i}$$

where $L_e(\tau)$ is the economic cost of Eq. (13.26), Δ is the sampling period, $S(\Delta)$ is the family of piecewise constant functions with sampling period Δ, N is the prediction horizon, $\tilde{x}(z,t)$

is the predicted state function evolution with input $u(t)$ computed by the output feedback EMPC and $\hat{x}(z,t_k)$ is the state estimate of $x(z, t_k)$ obtained from state observer of Eq. (13.29) at $t = t_k$.

13.3.4.4 Case 1: Output feedback EMPC system

For this case, we operate the reactor by using the output feedback EMPC system of Eq. (13.37). In detail, we choose the prediction horizon, $N = 3$ and assume there are $p = 11$ evenly spaced measurement points along the length of the reactor. The PFR is initialized with the steady-state profile corresponding to the steady-state input $u_s = 0.5$.

The closed-loop state profiles of the PFR over one period $t_p = 1.0$ under the output feedback EMPC of Eq. (13.37) based on 11 measurements of x_1 are shown in Figs. 13.30 and 13.31. To check the effectiveness of the state observer, the profile of the state estimation error defined as

$$dx(t) = \sum_{i=1}^{2}\left(\sum_{j=0}^{101}(x_i(z_j,t) - \hat{x}_i(z_j,t))^2\right)^{1/2} \tag{13.38}$$

where z_j, $j = 1, 2, \ldots, 101$ is the discretized spatial coordinate and $dx(t)$ is the metric that defines the estimation error. The estimation error, $dx(t)$ for this case study is shown in Fig. 13.32. From Fig. 13.32, the state estimation error becomes small over time and converges to 0 at the end of the operating period. The optimal manipulated input profile from the output feedback EMPC system based on 11 measurements of x_1 over one period is displayed in Fig. 13.33. The input profile of the output feedback EMPC displays chattering. The chattering is from inaccuracies of the state estimate provided to the EMPC from the

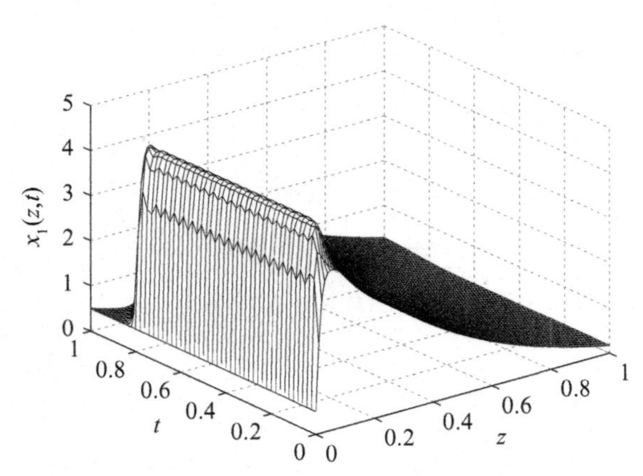

Fig. 13.30
Closed-loop profile of x_1 of the process under the output feedback EMPC system of Eq. (13.37) with $N = 3$ based on 11 measurements of x_1 over one operation period.

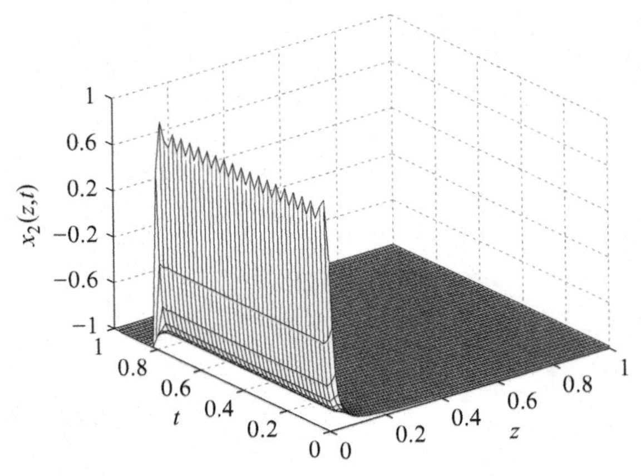

Fig. 13.31
Closed-loop profile of x_2 of the process under the output feedback EMPC system of Eq. (13.37) with $N = 3$ based on 11 measurements of x_1 over one operation period.

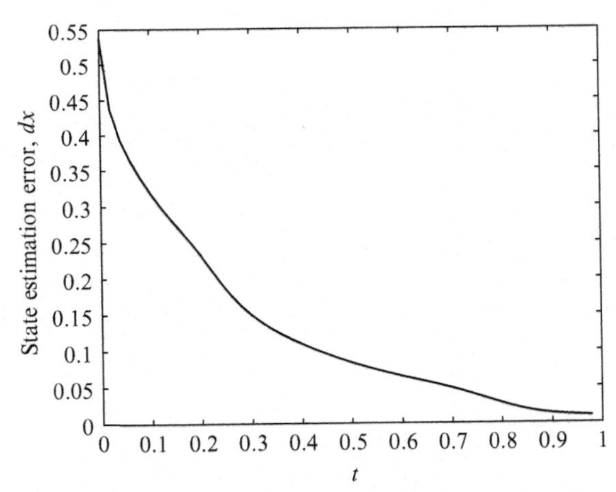

Fig. 13.32
State estimation error profile of the process under the output feedback EMPC system of Eq. (13.37) with $N = 3$ based on 11 measurements of x_1 over one operation period.

state observer. Specifically, it is desirable from an economics perspective to operate the PFR at the maximum temperature. However, the state observer cannot exactly estimate the true state. When the estimated temperature is at or above the state constraint, the EMPC decreases the amount of reactant material fed to the PFR to decrease the temperature (and satisfy the state constraint). If the estimated state is below the state constraint, more reactant material is fed to the PFR to increase the temperature and bring the temperature to the maximum allowable temperature. The maximum x_1 value profile of the PFR under the

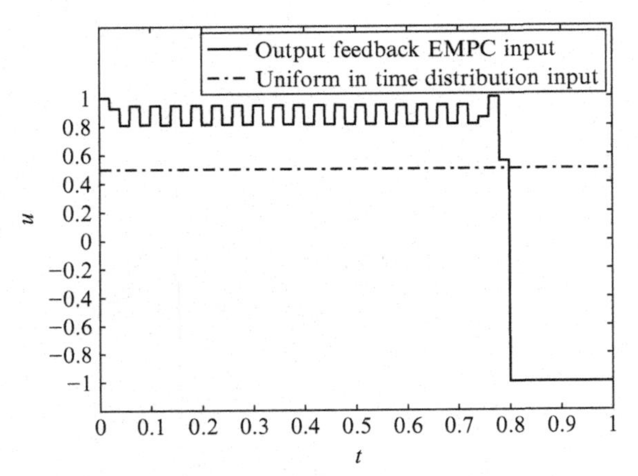

Fig. 13.33
Manipulated input profiles of the output feedback EMPC system of Eq. (13.37) with $N = 3$ based on 11 measurements of x_1 and uniform in time distribution of the reactant material over one operation period.

output feedback EMPC system is demonstrated in Fig. 13.34. From Fig. 13.34, the output feedback EMPC system is able to operate the reactor at the maximum allowable value over a period of time to maximize the economic cost while not violating the state constraint.

The economic cost profile of the reactor under the output feedback EMPC system is displayed in Fig. 13.35. During this one operation period, the process under the output feedback EMPC system achieves 14.77% improvement on the average economic cost over

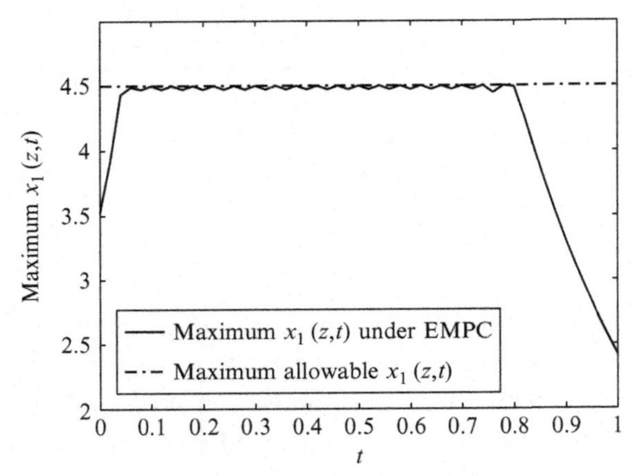

Fig. 13.34
Maximum x_1 profiles of the process under the output feedback EMPC system of Eq. (13.37) with $N = 3$ based on 11 measurements of x_1 over one operation period.

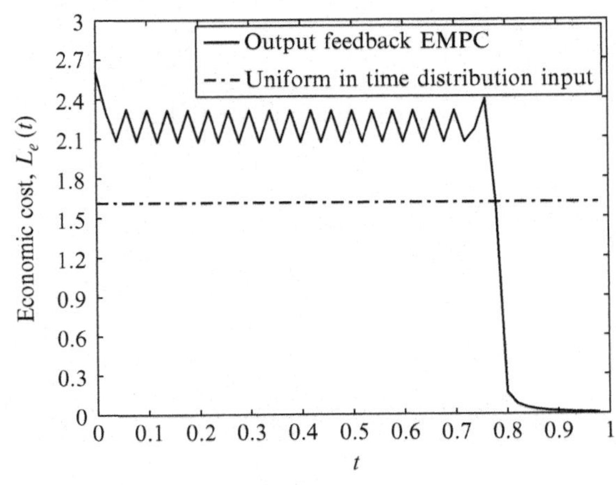

Fig. 13.35
Temporal economic cost, $E(t)$, along the length of the reactor under the output feedback EMPC system of Eq. (13.37) with $N = 3$ based on 11 measurements of x_1 and under uniform in time distribution of the reactant material over one operation period.

one operation period than that of the reactor under uniform in time distribution of the reactant material. The output feedback EMPC system demonstrates a significant advantage in maximizing the process economic performance through operating the process in a time-varying fashion over one operation period.

13.3.4.5 Case 2: Prediction horizon effect on EMPC application

We operate the process under different prediction horizons ($N < 10$ were considered because $N \geq 10$ is considered impractical based on computation time requirements to the EMPC at each sampling instance). The number of output measurement points on x_1 used was $p = 11$. The closed-loop manipulated input profiles of the output feedback EMPC system of Eq. (13.37) using the above different prediction horizons are displayed in Fig. 13.36. From Fig. 13.36, we can see that the use of a longer prediction horizon reduces the chattering of the manipulated input profiles. The corresponding average economic cost (i.e., $\overline{L}_e(t)$) profiles along the length of the reactor are displayed in Fig. 13.37. Fig. 13.37 demonstrates that the use of a prediction horizon greater than three sampling periods does not show a significant advantage of improving the process economic performance for the output feedback EMPC system of Eq. (13.37). However, as the prediction horizon increases, the computation time required to solve the EMPC at each sampling time increases. The ratio of the average computation time for a given horizon to the average computation time to solve the EMPC with $N = 3$ is shown in Fig. 13.38. The EMPC with $N = 3$ was chosen as the comparison standard because a horizon of $N > 3$ did not result in a significant improvement in the economic performance compared to $N = 3$. Comparing the EMPC with $N = 3$ and that with $N = 7$,

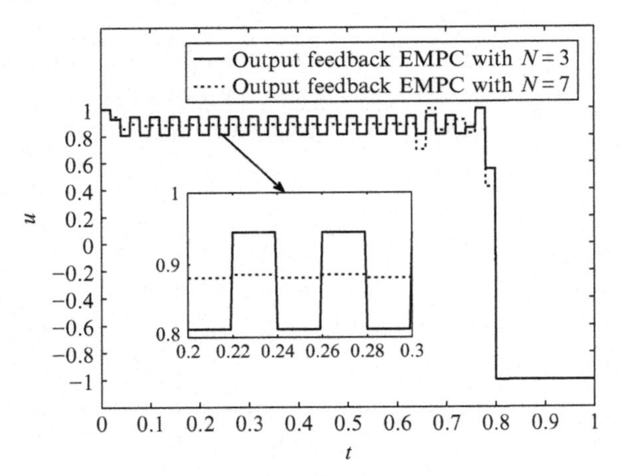

Fig. 13.36
Manipulated input profiles of the output feedback EMPC system of Eq. (13.37) based on 11 measurements of x_1 using different prediction horizons, N.

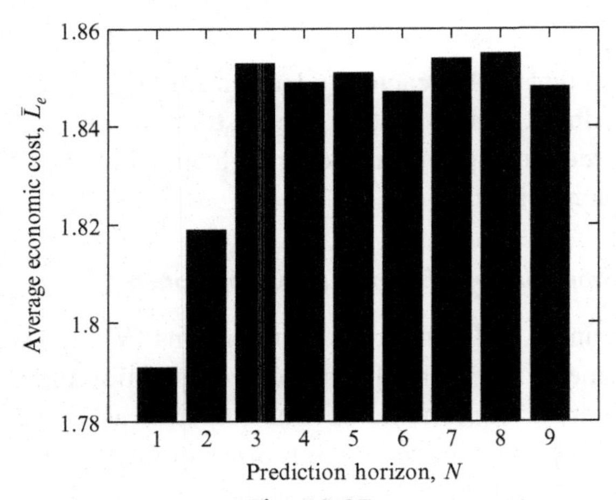

Fig. 13.37
Average economic cost, $L_e(t)$, along the length of the reactor under the output feedback EMPC system of Eq. (13.37) based on 11 measurements of x_1 using different prediction horizons, N.

for instance, the EMPC with $N = 7$ computation time is approximately six times greater than that of $N = 3$, while the increase in the economic performance is less than 0.0005 (i.e., less than 0.03%).

13.3.4.6 Case 3: Different number of output measurement points

For this case, we apply the output feedback EMPC systems of Eq. (13.37) using different numbers of available measurement points, i.e., $p = 6, 11, 21$. Operation over an operation period of $t_p = 1.0$ was considered. Different number of output measurement points bring in

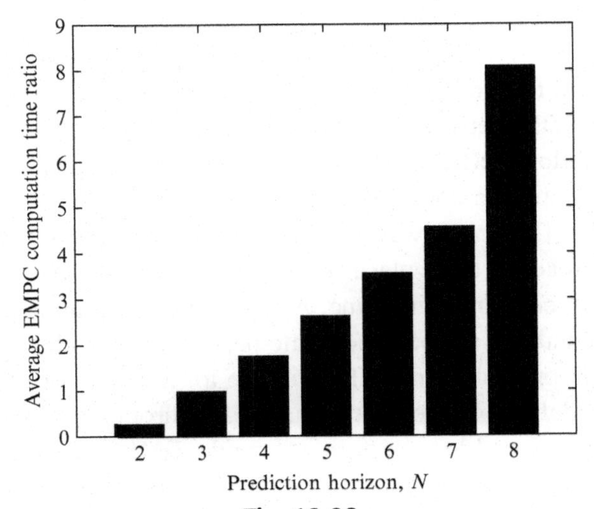

Fig. 13.38
The ratio of the average EMPC calculation time with the specified prediction horizon to the average EMPC calculation time with a prediction horizon of $N = 3$.

different state estimation error profiles for the same output feedback control formulation of Eq. (13.29) as shown in Fig. 13.39. From the magnified plot in Fig. 13.39, we can see that more available measurement points can slightly increase the state estimation error convergence rate.

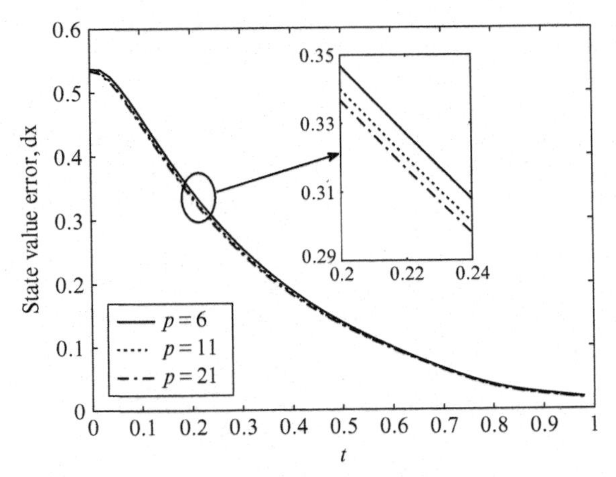

Fig. 13.39
State estimation error profiles of the process under the output feedback EMPC system of Eq. (13.37) with $N = 3$ using different numbers of measurement points, p, of x_1.

13.4 Conclusion

The first part of this work focused on developing an EMPC design for a parabolic PDE system which integrated the APOD method and a high-order finite-difference method to deal with control system computational efficiency and state constraint satisfaction. EMPC systems adopting POD, APOD, a high-order spatial discretization by central finite-difference method and the proposed EMPC flow chart were applied to a nonisothermal tubular reactor where a second-order chemical reaction takes place. These EMPC systems were compared with respect to their model accuracy, computational time, APOD update requirements, state and input constraint satisfaction, and closed-loop economic performance of the tubular reactor. The second part of this work presented an EMPC scheme for a system of two coupled hyperbolic PDEs arising in the modeling of a nonisothermal PFR. Through extensive simulations, key metrics like the economic closed-loop performance under EMPC versus steady-state operation, the impact of horizon length on computational time and economic performance, and the effect of the number of spatial discretization points used in the process model of the EMPC on closed-loop economic performance, were evaluated and discussed.

References

1. Christofides PD. *Nonlinear and robust control of PDE systems: methods and applications to transport-reaction processes*. Boston: Birkhäuser; 2001.
2. Smoller J. *Shock waves and reaction-diffusion equations*. Berlin: Springer Verlag; 1983.
3. Finlayson BA. *The method of weighted residuals and variational principles*. New York: Academic Press; 1972.
4. Ray WH. *Advanced process control*. New York: McGraw-Hill; 1981.
5. Foias C, Jolly MS, Kevrekidis IG, Sell GR, Titi ES. On the computation of inertial manifolds. *Phys Lett A* 1988;**131**:433–6.
6. Sirovich L. Turbulence and the dynamics of coherent structures. I—Coherent structures. II—Symmetries and transformations. III—Dynamics and scaling. *Q Appl Math* 1987;**45**:561–90.
7. Holmes P, Lumley JL, Berkooz G. *Turbulence, coherent structures, dynamical systems and symmetry*. New York, NY: Cambridge University Press; 1996.
8. Lao L, Ellis M, Christofides PD. Economic model predictive control of parabolic PDE systems: addressing state estimation and computational efficiency. *J Process Control* 2014;**24**:448–62.
9. Banerjee S, Cole JV, Jensen KF. Nonlinear model reduction strategies for rapid thermal processing systems. *IEEE Trans Semicond Manuf* 1998;**11**:266–75.
10. Theodoropoulou A, Adomaitis RA, Zafiriou E. Model reduction for optimization of rapid thermal chemical vapor deposition systems. *IEEE Trans Semicond Manuf* 1998;**11**:85–98.
11. Atwell J, King B. Proper orthogonal decomposition for reduced basis feedback controllers for parabolic equations. *Math Comput Model* 2001;**33**:1–19.
12. Luo B, Wu H. Approximate optimal control design for nonlinear one-dimensional parabolic PDE systems using empirical eigenfunctions and neural network. *IEEE Trans Syst Man Cybern B* 2012;**42**:1538–49.
13. Ravindran SS. Adaptive reduced-order controllers for a thermal flow system using proper orthogonal decomposition. *SIAM J Sci Comput* 2002;**23**:1924–42.
14. Singer MA, Green WH. Using adaptive proper orthogonal decomposition to solve the reaction-diffusion equation. *Appl Numer Math* 2009;**59**:272–9.
15. Varshney A, Pitchaiah S, Armaou A. Feedback control of dissipative PDE systems using adaptive model reduction. *AIChE J* 2009;**55**:906–18.

16. Pourkargar DB, Armaou A. Modification to adaptive model reduction for regulation of distributed parameter systems with fast transients. *AIChE J* 2013;**59**:4595–611.
17. Lao L, Ellis M, Christofides PD. Handling state constraints and economics in feedback control of transport-reaction processes. *J Process Control* 2015;**32**:98–108.
18. Igreja JM, Lemos JM, Silva RN. Controlling distributed hyperbolic plants with adaptive nonlinear model predictive control. In: Findeisen R, Allgower F, Biegler LT, editors. *Assessment and future directions of nonlinear model predictive control. Lecture notes in control and information sciences*, vol. 358. Springer Berlin Heidelberg; 2007. p. 435–41.
19. Finlayson BA. *Nonlinear analysis in chemical engineering*. New York: McGraw-Hill Inc.; 1980
20. Hanczyc EM, Palazoglu A. Sliding mode control of nonlinear distributed parameter chemical processes. *Ind Eng Chem Res* 1995;**34**:557–66.
21. Mordukhovich BS, Raymond JP. Dirichlet boundary control of hyperbolic equations in the presence of state constraints. *Appl Math Optim* 2004;**49**:145–57.
22. Aksikas I, Mohammadi L, Forbes JF, Belhamadia Y, Dubljevic S. Optimal control of an advection-dominated catalytic fixed-bed reactor with catalyst deactivation. *J Process Control* 2013;**23**:1508–14.
23. Castillo F, Witrant E, Prieur C, Dugard L. Boundary observers for linear and quasi-linear hyperbolic systems with application to flow control. *Automatica* 2013;**49**:3180–8.
24. Moghadam AA, Aksikas I, Dubljevic S, Forbes JF. Boundary optimal LQ control of coupled hyperbolic PDEs and ODEs. *Automatica* 2013;**49**:526–33.
25. Dufour P, Couenne F, Toure Y. Model predictive control of a catalytic reverse flow reactor. *IEEE Trans Control Syst Technol* 2003;**11**:705–14.
26. Shang H, Forbes F, Guay M. Model predictive control for quasilinear hyperbolic distributed parameter systems. *Ind Eng Chem Res* 2004;**43**:2140–9.
27. Dubljevic S, Mhaskar P, El-Farra NH, Christofides PD. Predictive control of transport-reaction processes. *Comput Chem Eng* 2005;**29**:2335–45.
28. Fuxman AM, Forbes JF, Hayes RE. Characteristics-based model predictive control of a catalytic flow reversal reactor. *Can J Chem Eng* 2007;**85**:424–32.
29. Ellis M, Durand H, Christofides PD. A tutorial review of economic model predictive control methods. *J Process Control* 2014;**24**:1156–78.
30. Amrit R, Rawlings JB, Angeli D. Economic optimization using model predictive control with a terminal cost. *Annu Rev Control* 2011;**35**:178–86.
31. Angeli D, Amrit R, Rawlings JB. On average performance and stability of economic model predictive control. *IEEE Trans Autom Control* 2012;**57**:1615–26.
32. Huang R, Harinath E, Biegler LT. Lyapunov stability of economically oriented NMPC for cyclic processes. *J Process Control* 2011;**21**:501–9.
33. Heidarinejad M, Liu J, Christofides PD. Economic model predictive control of nonlinear process systems using Lyapunov techniques. *AIChE J* 2012;**58**:855–70.
34. Huang R, Biegler LT, Harinath E. Robust stability of economically oriented infinite horizon NMPC that include cyclic processes. *J Process Control* 2012;**22**:51–9.
35. Lao L, Ellis M, Christofides PD. Economic model predictive control of transport-reaction processes. *Ind Eng Chem Res* 2014;**53**:7382–96.
36. Christofides PD, Daoutidis P. Finite-dimensional control of parabolic PDE systems using approximate inertial manifolds. *J Math Anal Appl* 1997;**216**:398–420.
37. Armaou A, Christofides PD. Dynamic optimization of dissipative PDE systems using nonlinear order reduction. *Chem Eng Sci* 2002;**57**:5083–114.
38. Pitchaiah S, Armaou A. Output feedback control of dissipative PDE systems with partial sensor information based on adaptive model reduction. *AIChE J* 2013;**59**:747–60.
39. Wächter A, Biegler LT. On the implementation of an interior-point filter line-search algorithm for large-scale nonlinear programming. *Math Program* 2006;**106**:25–57.
40. Graham MD, Rawlings JB. *Modeling and analysis principles for chemical and biological engineers*. Madison, WI: Nob Hill Publishing; 2013.
41. Christofides PD, Daoutidis P. Feedback control of hyperbolic PDE systems. *AIChE J* 1996;**42**:3063–86.

Index

Note: Page numbers followed by *f* indicate figures, and *t* indicate tables.